Graduate Texts in Physics

Graduate Texts in Physics

Graduate Texts in Physics publishes core learning/teaching material for graduate- and advanced-level undergraduate courses on topics of current and emerging fields within physics, both pure and applied. These textbooks serve students at the MS- or PhD-level and their instructors as comprehensive sources of principles, definitions, derivations, experiments and applications (as relevant) for their mastery and teaching, respectively. International in scope and relevance, the textbooks correspond to course syllabi sufficiently to serve as required reading. Their didactic style, comprehensiveness and coverage of fundamental material also make them suitable as introductions or references for scientists entering, or requiring timely knowledge of, a research field.

More information about this series at http://www.springer.com/series/8431

Harald Friedrich

Theoretical Atomic Physics

Fourth Edition

Springer

Harald Friedrich
Fachbereich Physik T 30
TU München
Garching, Germany

ISSN 1868-4513 ISSN 1868-4521 (electronic)
Graduate Texts in Physics
ISBN 978-3-319-83818-2 ISBN 978-3-319-47769-5 (eBook)
DOI 10.1007/978-3-319-47769-5

Softcover reprint of the hardcover 4th edition 2017

Originally published in the series: Advanced Texts in Physics

Printed on acid-free paper

This Springer imprint is published by Springer Nature
The registered company is Springer International Publishing AG
The registered company address is: Gewerbestrasse 11, 6330 Cham, Switzerland

Preface to the Fourth Edition

The first edition of *Theoretical Atomic Physics* was written more than a quarter of a century ago, with the aim of providing graduate students and researchers in atomic physics with the "kind of advanced quantum mechanics needed for practical applications in modern atomic physics". Since then, the unbroken advancement of improved experimental techniques and computational power has broadened the range of fascinating effects that can be studied in the laboratory and modelled in theoretical analyses. It includes the study of individual atoms in electromagnetic traps, where fundamental postulates of quantum mechanics can be tested, of degenerate quantum gases of ultracold atoms (or molecules) and of complex systems with chaotic classical dynamics, where semiclassical theories have experienced a revival and found many applications of practical relevance in the atomic domain. The aim formulated for the first edition remains valid in this context. The emphasis on theory should enable the reader to appreciate the fundamental assumptions underlying standard theoretical constructs and to embark on independent research projects.

The production of and experimentation with Bose–Einstein condensates of atomic gases is now routine in many laboratories, and this has helped to make cold and ultracold atoms (and molecules) a field of rapidly growing interest. The interaction of atoms close to the threshold between weakly bound diatomic molecular states and low-energy scattering states is important in this context, and so, many concepts of near-threshold scattering theory are used by researchers in the field. The observation that many colleagues were not aware of the origin and could not appreciate the precise meaning of such concepts as, e.g. "scattering length", motivated me to write a monograph on scattering theory, with a special focus on the relevance for cold-atom physics [H. Friedrich, *Scattering Theory*, Lecture Notes in Physics **872**, Springer, Berlin, Heidelberg, 2013, 2nd. Edition 2016]. In the fourth edition of *Theoretical Atomic Physics*, I have updated and expanded the sections and subsections involving scattering theory and/or near-threshold phenomena by incorporating the corresponding contributions from the monograph. Hence, the treatment of scattering and near-threshold phenomena has become more sophisticated. Special attention is given to the quantization of weakly bound states just below the continuum threshold and to low-energy scattering and quantum reflection just above. Particular emphasis is laid on the fundamental differences between long-ranged Coulombic potentials, where the continuum threshold represents the (semi-)classical limit of the Schrödinger equation, and shorter-ranged potentials falling off faster than $1/r^2$ at large distances r, where the threshold corresponds to the anticlassical, extreme quantum limit. Modified effective range expansions are given, even for potentials with attractive inverse-cube tails, a result derived only recently by Müller [Phys. Rev. Lett. **110** (2013) 260401], see (4.113) in Sect. 4.1.8.

A new section on scattering in two spatial dimensions is included; it is relevant not only for genuinely two-dimensional systems but also for 3D systems with translational invariance in one degree of freedom, such as an atom interacting with a cylindrical nanotube. There is also a new section on tunable near-threshold Feshbach resonances, a subject that was treated poorly in the third edition. The appendix on special mathematical functions has been expanded in order to accommodate formulas occurring in the extended treatment of scattering and near-threshold phenomena.

It is a pleasure to thank many colleagues who inspired me with numerous discussions involving atomic physics, quantum mechanics and semiclassical connections, in particular Robin Côté at the University of Connecticut, Wolfgang Domcke and Manfred Kleber at the Technical University of Munich, Gerhard Rempe and Stephan Dürr at the Max Planck Institute for Quantum Optics in Garching and Jan-Michael Rost at the Max Planck Institute for Complex Systems in Dresden. Some postdocs and several former students produced results that I have used in the book, in particular Florian Arnecke, Johannes Eiglsperger, Christopher Eltschka, Martin Fink, Georg Jacoby, Alexander Jurisch, Alexander Kaiser, Petra Meerwald, Carlo Meister, Eskender Mesfin, Javier Madroñero, Michael J. Moritz, Tim-Oliver Müller, Thomas Purr, Patrick Raab, Sebastian Schröter, Frauke Schwarz and Johannes Trost. I am grateful for the technical assistance provided by Stefan Recksiegel, our IT expert at the Physik-Department in Garching. I also thank Ute Heuser and Birgit Münch and Dr. Thorsten Schneider at Springer for their efficient help and cooperation.

Schließlich möchte ich mich bei meiner Familie bedanken, die mir immer den Zugang zur alltäglichen Welt jenseits der Physik offen gehalten hat. Vor allem bei meiner Frau Elfi, die über mehr als vier Jahrzehnte meine Arbeit mit Ermutigung, Geduld und Flexibilität unterstützt hat. Dazu hat uns in den letzten zweieinhalb Jahren das Glück drei Enkelkinder beschert, Lorenz, Alexander und Johann, die mit ihrer authentischen Lebensfreude alle Herzen höher schlagen lassen.

Garching, Germany Harald Friedrich
September 2016

Preface to the Third Edition

The one and a half decades since the publication of the first edition of *Theoretical Atomic Physics* have seen a continuation of remarkable and dramatic experimental breakthroughs. With the help of ultrashort laser pulses, special states of atoms and molecules can now be prepared and their time evolution studied on time scales shorter than femtoseconds. Trapped atoms and molecules can be cooled to temperatures on the order of a few nano-Kelvin and light fields can be used to guide and manipulate atoms, for example, in optical lattices formed as standing waves by counterpropagating laser beams. After the first production of Bose–Einstein condensates of ultracold atomic gases in 1995, degenerate quantum gases of ultracold atoms and molecules are now prepared and studied routinely in many laboratories around the world. Such progress in atomic physics has been well received and appreciated in the general academic community and was rewarded with two recent Nobel Prizes for physics. The 1997 prize was given to Steven Chu, Claude Cohen-Tannoudji and William Phillips for their work on cooling atoms, and only 4 years later Eric Cornell, Wolfgang Ketterle and Carl Wieman received the 2001 prize for the realization of the Bose–Einstein condensates mentioned above.

The prominence of modern experimental atomic physics establishes further need for a deeper understanding of the underlying theory. The continuing growth in quality and quantity of available computer power has substantially increased the effectivity of large-scale numerical studies in all fields, including atomic physics. This makes it possible to obtain some standard results such as the properties of low-lying states in many-electron atoms with good accuracy using generally applicable program packages. However, largely due to the dominant influence of long-ranged Coulomb forces, atomic systems are rather special. They can reveal a wide range of interesting phenomena in very different regimes—from near-classical states of highly excited atoms, where effects of nonlinearity and chaos are important, to the extreme quantum regime of ultracold atoms, where counterintuitive nonclassical effects can be observed. The theoretical solution of typical problems in modern atomic physics requires proficiency in the practical application of quantum mechanics at an advanced level, and a good understanding of the links to classical mechanics is almost always helpful. The aim of *Theoretical Atomic Physics* remains to provide the reader with a solid foundation of this sort of advanced quantum mechanics.

In preparing the third edition, I have again tried to do justice to the rapid development of the field. I have included references to important new work whenever this seemed appropriate and easy to do. Chapter 1 now includes a section on processes involving (wave packets of) continuum states and also an expanded treatment of the semiclassical approximation. Chapter 3 begins with a section illuminating the characteristic differences in the near-threshold properties of long-ranged and shorter-ranged potentials, and the first section of Chap. 4 contains a more elaborate discussion of scattering lengths. As a further "Special Topic" in Chap. 5 there is a section describing some aspects of atom optics, including discussions of the interactions of atoms with material surfaces and with light fields. The appendix on special mathematical functions has been slightly expanded to accommodate a few results that I repeatedly found to be useful.

I am grateful to many colleagues who continue to inspire me with numerous discussions involving atomic physics, quantum mechanics and semiclassical connections, in particular Robin Côté at the University of Connecticut, Manfred Kleber at the Technical University Munich and Jan-Michael Rost at the Max Planck Institute for Complex Systems in Dresden. Several current and former graduate students produced new results that I have used in the book, in particular Christopher Eltschka, Georg Jacoby, Alexander Jurisch, Michael J. Moritz, Thomas Purr and Johannes Trost. I thank them all for the effort and enthusiasm with which they contributed to the various projects. I also thank Thomas Mehnert for helpful comments on the previous editions. A sabbatical term at the Australian National University in Canberra during the southern summer 2002/2003 established a fruitful connection to Ken Baldwin and Stephen Gibson in the Atomic and Molecular Physics Laboratories, and I am grateful to Brian Robson and Erich Weigold who made this visit possible. Finally, I wish to thank my wife Elfi who (again) endured a hard-working and preoccupied husband during the final stages of preparation of this third edition.

Garching, Germany Harald Friedrich
June 2005

Preface to the First Edition

In the first few decades of this century, atomic physics and quantum mechanics developed dramatically from early beginnings to maturity and a degree of completeness. After about 1950 fundamental research in theoretical physics focussed increasingly on nuclear physics and high energy physics, where new conceptual insights were expected to be more probable. A further field of growing importance was theoretical solid state physics, which led to or accompanied many revolutionary technological developments. In this environment the role of atomic physics as an independent discipline of theoretical physics became somewhat subdued. In the last two decades, however, high precision experimental techniques such as high resolution laser spectroscopy have opened up new and interesting fields in atomic physics. Experiments can now be performed on individual atoms and ions in electromagnetic traps, and the dependence of their properties on their environment can be studied. Effects and phenomena which used to be regarded as small perturbations or experimentally irrelevant exceptional cases have moved into the centre of attention. At the same time it has become clear that interesting and intricate effects can occur even in seemingly simple systems with only few degrees of freedom.

The successful description and interpretation of such effects usually requires the solution of a non-trivial Schrödinger equation, and perturbative methods are often inadequate. Most lectures and textbooks which go beyond an introductory "Quantum Mechanics I" are devoted to many-body theories and field theories at a high level of abstraction. Not enough attention is given to a more practical kind of advanced quantum mechanics as required by modern atomic physics. In order to meet this demand I have taught several courses on *Theoretical Atomic Physics* at the Munich Universities since 1984. The present book grew out of these lectures. It is an updated version of the textbook *Theoretische Atomphysik*, which appeared in German in September 1990, and contains the kind of advanced quantum mechanics needed for practical applications in modern atomic physics. The level of abstraction is deliberately kept low—almost all considerations start with the Schrödinger equation in coordinate representation. The book is intended as a textbook for students who have had a first introductory contact with quantum mechanics. I have, however, aimed at a self-contained presentation which should—at least in principle—be understandable without previous knowledge.

The book contains five chapters, the first two of which present mostly conventional material as can be found in more detail in available textbooks on quantum mechanics and atomic physics. The first chapter contains a concise review of quantum mechanics and the second chapter a deliberately brief summary of traditional atomic theory. I have taken pains to treat bound states and continuum states on the same footing. This enables the inclusion of a comparatively straightforward introduction to quantum defect theory (Chap. 3), which has become a powerful and widely used tool for analyzing atomic spectra and which, up to now, has not been treated at such a basic level in a student textbook. The scope of the reaction theory presented in Chap. 4 is that of "Simple Reactions" induced by the collision of a single electron with an atom or ion. This avoids many complications otherwise occurring in the definitions of coordinates, channels and potentials. On the other hand, important concepts such as cross sections, scattering matrix, transition operator, reactance matrix, polarization effects, Born approximation and break-up channels can already be discussed in this simple framework.

The last chapter contains a selection of special topics which are currently subject to intense and sometimes controversial discussion. The interest in multiphoton processes has grown strongly with the availability of high-power lasers and underlines the importance of non-perturbative methods in quantum mechanics. The possibility of using very short laser pulses to study spatially and temporally localized excitations of individual atoms has revived interest in the relation between classical mechanics and quantum mechanics. The final section discusses "Chaos", which is currently one of the most popular and rapidly growing subfields in almost all fields of physics. While most specific investigations of chaos are numerical experiments on model systems, there are a few prominent examples in atomic physics of simple but real systems, which can be and have been observed in the laboratory and which have all the properties currently causing excitement in connection with chaos.

It is a pleasure to thank the many colleagues and friends who unselfishly helped me in the course of writing this book. Special thanks are due to Karl Blum, Wolfgang Domcke, Berthold-Georg Englert, Christian Jungen, Manfred Kleber, Achim Weiguny and Dieter Wintgen, who read through individual chapters and/or sections and suggested several improvements of the original manuscript. Valuable suggestions and hints were also provided by John S. Briggs, Hubert Klar and Peter Zoller. Gerd Handke and Markus Draeger conscientiously checked more than a thousand formulae and helped to avoid disaster. The original drawings were produced with the competent help of Mrs. I. Kuchenbecker and a plot program specially tailored for the purpose by Markus Draeger. Special thanks are also due to Dr. H.-U. Daniel from Springer-Verlag. His experience and competence contributed significantly to the success of the project. Finally I would like to thank my wife Elfi, who not only read through the German and the English manuscript word by word but also supported my work with patience and encouragement during the last three years.

Garching, Germany
June 1991

Harald Friedrich

Contents

Chapter 1
Review of Quantum Mechanics

Atomic phenomena are described mainly on the basis of non-relativistic quantum mechanics. Relativistic effects can generally be accounted for in a satisfactory way with perturbative methods. In the 1990s it became increasingly apparent, that a better understanding of the classical dynamics of an atomic system can lead to a deeper appreciation of various features in its observable quantum mechanical properties, see e.g. [RW94, CK97, FE97, BB97, SS98, BR09], Sect. 5.3. This does not, however, invalidate the generally accepted point of view, that quantum mechanics is the relevant theory for atomic physics.

This chapter gives a brief review of quantum mechanics as it is needed for use in later chapters. Although the reader is expected to have some experience in the subject already, the presentation starts at the beginning and is self-contained so that it should, at least in principle, be understandable without previous knowledge of quantum mechanics. A more thorough introduction can be found in numerous textbooks, e.g. [Sch68, Bay69, Gas74, Mes70, Sch02].

1.1 Wave Functions and Equations of Motion

1.1.1 States and Wave Functions

Non-relativistic quantum mechanics describes the state of a physical system at a given time t with a complex-valued *wave function* $\psi(X;t)$. The wave function ψ depends on the parameter t and a complete set of variables summarized as X. As an example let us think of a system of N electrons, which plays a central role in atomic physics. Then X can stand for the N spatial coordinates $\boldsymbol{r}_1, \ldots \boldsymbol{r}_N$ and the N spin coordinates $m_{s_1}, \ldots m_{s_N}$ of the electrons. The spatial coordinates $\boldsymbol{r}_i$ are ordinary (real) vectors in three-dimensional space; the spin-coordinates m_{s_i} can each assume only two values, $m_{s_i} = \pm 1/2$.

H. Friedrich, *Theoretical Atomic Physics*, Graduate Texts in Physics,
DOI 10.1007/978-3-319-47769-5_1

The set of wave functions describing a given system is closed with respect to linear superposition. This means that all multiples and sums of possible wave functions are again possible wave functions. Mathematically, the possible wave functions of a system form a vector space. The *scalar product* of two wave functions $\psi(X;t), \phi(X;t')$ in this vector space is defined as

$$\langle\psi(t)|\phi(t')\rangle = \int \psi^*(X;t)\phi(X;t')\,\mathrm{d}X\,. \tag{1.1}$$

The integral in (1.1) stands for integration over the continuous variables and summation over the discrete variables. In the above-mentioned example of an N-electron system we have

$$\int \mathrm{d}X = \int \mathrm{d}^3r_1 \cdots \int \mathrm{d}^3r_N \sum_{m_{s_1}=-1/2}^{1/2} \cdots \sum_{m_{s_N}=-1/2}^{1/2}\,.$$

The scalar product (1.1) is linear,

$$\langle\psi|\phi_1 + c\phi_2\rangle = \langle\psi|\phi_1\rangle + c\langle\psi|\phi_2\rangle, \tag{1.2}$$

and it is replaced by its complex conjugate if we interchange the wave functions,

$$\langle\phi|\psi\rangle = \langle\psi|\phi\rangle^*. \tag{1.3}$$

Two wave functions ψ and ϕ are *orthogonal* if the scalar product $\langle\psi|\phi\rangle$ vanishes. The scalar product $\langle\psi|\psi\rangle$ is a non-negative real number, and its square root is the *norm* of the wave function ψ. Square integrable wave functions, i.e. wave functions $\psi(X;t)$ with the property

$$\langle\psi|\psi\rangle = \int |\psi(X;t)|^2\,\mathrm{d}X < \infty\,, \tag{1.4}$$

are *normalizable*. This means that they become wave functions of norm unity,

$$\langle\psi|\psi\rangle = \int |\psi(X;t)|^2\,\mathrm{d}X = 1\,, \tag{1.5}$$

when multiplied by an appropriate constant. The non-negative function $|\psi(X;t)|^2$ is a *probability density*. If, at time t, a physical state is described by the wave function $\psi(X;t)$ (which is normalized to unity, $\langle\psi|\psi\rangle = 1$), then the integral

$$\int_{\delta V} |\psi(X;t)|^2\,\mathrm{d}X$$

over a part δV of the full space of values of the variable X gives the probability that a measurement of the variable X (at time t) will yield values within δV. The concept of probability densities can also be applied to wave functions which are not normalizable, as long as we only study relative probabilities.

The square integrable functions (1.4) form a subspace of the space of all wave functions. This subspace has the properties of a *Hilbert space*. In particular it is complete, meaning that the limit of each convergent sequence of wave functions in the Hilbert space is again a wave function in the Hilbert space. It also has a denumerable *basis*, i.e. there exists a sequence $\phi_1(X), \phi_2(X), \ldots$, of linearly independent square integrable functions such that any square integrable function $\psi(X)$ can be written as a linear combination

$$\psi(X) = \sum_{n=1}^{\infty} c_n \phi_n(X) \tag{1.6}$$

with uniquely determined coefficients c_n. The basis is *orthonormal* if its wave functions obey the orthonormality relation

$$\langle \phi_i | \phi_j \rangle = \delta_{i,j} \,. \tag{1.7}$$

In this case the coefficients c_n in (1.6) can be obtained by forming the scalar product with ϕ_i:

$$c_i = \langle \phi_i | \psi \rangle \,. \tag{1.8}$$

The notation can be simplified if we leave out the variables X, which often aren't specified anyhow, and write the wave functions as abstract state vectors $|\psi\rangle$. The complex conjugate wave functions ϕ^*, with which the ψ 's are multiplied to form scalar products, are written as $\langle\phi|$. From the word "*bracket*" we call the state vector $|\psi\rangle$ forming the right-hand part of a scalar product $\langle\phi|\psi\rangle$ a *ket*, and we call the left-hand part $\langle\phi|$ a *bra*. Equation (1.6) now has the simplified form

$$|\psi\rangle = \sum_{n=1}^{\infty} c_n |\phi_n\rangle \,, \tag{1.9}$$

or, with (1.8),

$$|\psi\rangle = \sum_{n=1}^{\infty} |\phi_n\rangle\langle\phi_n|\psi\rangle \,. \tag{1.10}$$

The bra-ket notation is very useful, because many statements and formulae such as (1.9), (1.10) are independent of the particular choice of variables.

1.1.2 Linear Operators and Observables

An operator $\hat{O}$ turns a possible wave function $|\psi\rangle$ into another possible wave function $\hat{O}|\psi\rangle$. A *linear operator* has the property

$$\hat{O}(|\psi_1\rangle + c|\psi_2\rangle) = \hat{O}|\psi_1\rangle + c\,\hat{O}|\psi_2\rangle\,. \tag{1.11}$$

For each linear operator $\hat{O}$ there is a *Hermitian conjugate operator* $\hat{O}^\dagger$. It is defined by the condition that the scalar product of any bra $\langle\phi|$ with the ket $\hat{O}^\dagger|\psi\rangle$ be the complex conjugate of the scalar product of the bra $\langle\psi|$ with the ket $\hat{O}|\phi\rangle$:

$$\langle\phi|\hat{O}^\dagger|\psi\rangle = \langle\psi|\hat{O}|\phi\rangle^*\,. \tag{1.12}$$

Equation (1.12) is the bra-ket notation for the equation

$$\int \phi^*(X)\{\hat{O}^\dagger\psi(X)\}\,\mathrm{d}X = \left(\int \psi^*(X)\{\hat{O}\phi(X)\}\,\mathrm{d}X\right)^*\,. \tag{1.13}$$

In quantum mechanics an especially important class of operators consists of the *Hermitian operators*. Hermitian operators are linear operators $\hat{O}$ with the property

$$\hat{O}^\dagger = \hat{O}\,. \tag{1.14}$$

Eigenstates of a linear operator $\hat{O}$ are non-vanishing wave functions $|\psi_\omega\rangle$ for which the action of the operator $\hat{O}$ merely amounts to multiplication with a number ω:

$$\hat{O}|\psi_\omega\rangle = \omega|\psi_\omega\rangle\,. \tag{1.15}$$

The number ω is called *eigenvalue* of $\hat{O}$. The *spectrum* of the operator $\hat{O}$ consists of all its eigenvalues. For a Hermitian operator

$$\langle\psi_\omega|\hat{O}|\psi_\omega\rangle = \langle\psi_\omega|\hat{O}^\dagger|\psi_\omega\rangle^* = \langle\psi_\omega|\hat{O}|\psi_\omega\rangle^* \tag{1.16}$$

and

$$\omega = \frac{\langle\psi_\omega|\hat{O}|\psi_\omega\rangle}{\langle\psi_\omega|\psi_\omega\rangle}\,, \tag{1.17}$$

so its eigenvalues are always real. Eigenstates of a Hermitian operator with different eigenvalues

$$\hat{O}|\psi_1\rangle = \omega_1|\psi_1\rangle\,, \quad \hat{O}|\psi_2\rangle = \omega_2|\psi_2\rangle \tag{1.18}$$

are always orthogonal, because the product $(\omega_1 - \omega_2)\langle\psi_2|\psi_1\rangle$ has to vanish due to

$$\langle\psi_2|\hat{O}|\psi_1\rangle = \omega_1\langle\psi_2|\psi_1\rangle = \omega_2\langle\psi_2|\psi_1\rangle\,. \tag{1.19}$$

If the eigenvalue ω is *degenerate*, this means if there is more than one linearly independent eigenstate with this eigenvalue, then we can construct orthogonal linear combinations of these eigenstates which of course stay eigenstates with eigenvalue ω.

As an example of a Hermitian operator we look at the *projection operator* $\hat{P}_\phi$. Its action on an arbitrary state vector $|\psi\rangle$ is to project out the component proportional to the state $|\phi\rangle$ (which we assume to be normalized to unity),

$$\hat{P}_\phi|\psi\rangle = \langle\phi|\psi\rangle|\phi\rangle = |\phi\rangle\,\langle\phi|\psi\rangle \tag{1.20}$$

(compare (1.6), (1.9)). In compact bra-ket notation we have

$$\hat{P}_\phi = |\phi\rangle\,\langle\phi|\,. \tag{1.21}$$

The state $|\phi\rangle$ itself is an eigenstate of $\hat{P}_\phi$ with eigenvalue unity. All states orthogonal to $|\phi\rangle$ are eigenstates of $\hat{P}_\phi$ with eigenvalue zero, which is thus highly degenerate. If we sum up the projections onto all orthogonal components of a state $|\psi\rangle$, then we must recover the state $|\psi\rangle$—see (1.10). If the states $|\phi_n\rangle$ form an (orthonormal) basis of the whole Hilbert space, then (1.10) must hold for all states $|\psi\rangle$. This can be expressed in a compact way in the *completeness relation*,

$$\sum_n |\phi_n\rangle\,\langle\phi_n| = \mathbf{1}\,. \tag{1.22}$$

The bold **1** is the *unit operator* whose action on any wave function is to leave it unchanged.

The *observables* of a physical system are described by Hermitian operators. The (real) eigenvalues are the possible results of measurement of the observable. If the state of a system is described by an eigenstate of a Hermitian operator, this means that measuring the observable will definitely yield the corresponding eigenvalue.

Any wave function must be decomposable into eigenstates of a given observable. This means that the eigenstates of an observable form a complete set. If all eigenstates of an observable are square integrable, then they form a basis of the Hilbert space of square integrable wave functions. Since eigenstates with different eigenvalues are orthogonal and degenerate eigenstates can be orthogonalized, it is then always possible to find an orthonormal basis of eigenstates:

$$\hat{O}|\psi_i\rangle = \omega_i|\psi_i\rangle\,, \quad \langle\psi_i|\psi_j\rangle = \delta_{i,j}\,. \tag{1.23}$$

An arbitrary wave function $|\psi\rangle$ in Hilbert space can be expanded in eigenstates of $\hat{O}$:

$$|\psi\rangle = \sum_n c_n |\psi_n\rangle \,. \tag{1.24}$$

If the wave function $|\psi\rangle$ is normalized to unity,

$$\langle\psi|\psi\rangle = \sum_n |c_n|^2 = 1 \,, \tag{1.25}$$

then the absolute squares

$$|c_n|^2 = |\langle\psi_n|\psi\rangle|^2 \tag{1.26}$$

of the expansion coefficients represent the probabilities for finding the system described by $|\psi\rangle$ in the respective eigenstates $|\psi_n\rangle$ and for a measurement of the observable $\hat{O}$ yielding the respective eigenvalues ω_n. The *expectation value* $\langle\hat{O}\rangle$ of the observable $\hat{O}$ in the state $|\psi\rangle$ (assumed normalized to unity) is the mean of all possible eigenvalues ω_n weighted with the probabilities (1.26):

$$\langle\hat{O}\rangle = \sum_n |c_n|^2 \omega_n = \langle\psi|\hat{O}|\psi\rangle \,. \tag{1.27}$$

The numbers $\langle\psi_i|\hat{O}|\psi_j\rangle$ defined with reference to a given basis $|\psi_i\rangle$ form the *matrix of the operator* $\hat{O}$ *in the basis* $\{|\psi_i\rangle\}$. The matrix of a Hermitian operator is Hermitian. The matrix of an operator in a basis of its own eigenstates is diagonal (provided degenerate eigenstates are orthogonalized).

Observables can also have eigenstates which are not normalizable, and whose eigenvalues are in general continuous. In this case we must replace or complement the discrete subscripts i, n in (1.23)–(1.27) by continuous subscripts, and the sums by integrals.

If a wave function $|\psi\rangle$ is simultaneously an eigenstate of two observables $\hat{A}$ and $\hat{B}$ with eigenvalues α and β respectively, then obviously

$$\hat{A}\hat{B}|\psi\rangle = \alpha\beta|\psi\rangle = \beta\alpha|\psi\rangle = \hat{B}\hat{A}|\psi\rangle \,. \tag{1.28}$$

A necessary and sufficient condition for $\hat{A}$ and $\hat{B}$ to have a common complete set of eigenstates is that $\hat{A}$ and $\hat{B}$ *commute*:

$$\hat{A}\hat{B} = \hat{B}\hat{A} \quad \text{or} \quad [\hat{A}, \hat{B}] = 0 \,. \tag{1.29}$$

$[\hat{A}, \hat{B}] = \hat{A}\hat{B} - \hat{B}\hat{A}$ is the *commutator* of $\hat{A}$ and $\hat{B}$. If $\hat{A}$ and $\hat{B}$ do not commute, then they are not simultaneously measurable, which means there is no complete set of wave functions which can simultaneously be classified by eigenvalues of $\hat{A}$ and $\hat{B}$.

In order to describe a physical system completely, we need a *complete set of commuting observables*. In this context "complete set" means that there is no further independent observable that commutes with all members of the set. The eigenvalues of the observables of a complete set form a complete set of variables for the wave functions. The choice of observables and variables is not unique; it defines the *representation* in which we study the evolution and the properties of the physical system.

For a spinless point particle in three-dimensional space, the three components $\hat{x}, \hat{y}, \hat{z}$ of the *displacement operator* $\hat{\boldsymbol{r}}$ form a complete set of observables. Application of the displacement operators merely amounts to multiplying with the respective position coordinates, e.g.

$$\hat{y}\,\psi(x, y, z; t) = y\,\psi(x, y, z; t)\,. \tag{1.30}$$

The corresponding momenta are described by the vector operator

$$\hat{\boldsymbol{p}} = \frac{\hbar}{\mathrm{i}}\nabla\,, \tag{1.31}$$

i.e.

$$\hat{p}_x = \frac{\hbar}{\mathrm{i}}\frac{\partial}{\partial x}\,, \quad \text{etc.} \tag{1.32}$$

Here we have introduced *Planck's constant* $\hbar$, which has the dimensions of an action and has the value $1.054571800(13) \times 10^{-34}$ Js $= 6.582119514(40) \times 10^{-16}$ eV s [MN16].

Position and momentum operators for the same degree of freedom do not commute:

$$[\hat{p}_x, \hat{x}] = \frac{\hbar}{\mathrm{i}}\,. \tag{1.33}$$

This means that position and momentum in the same direction are not simultaneously measurable, as is expressed quantitatively in Heisenberg's *uncertainty relation*:

$$\Delta p_x \Delta x \geq \frac{1}{2}\hbar\,. \tag{1.34}$$

The uncertainties Δp_x and Δx in a given state $|\psi\rangle$ are defined as the fluctuations of the observables around their respective expectation values $\langle\hat{x}\rangle = \langle\psi|\hat{x}|\psi\rangle$, $\langle\hat{p}_x\rangle = \langle\psi|\hat{p}_x|\psi\rangle$:

$$\Delta x = \sqrt{\langle\hat{x}^2\rangle - \langle\hat{x}\rangle^2}\,, \quad \Delta p_x = \sqrt{\langle\hat{p}_x^2\rangle - \langle\hat{p}_x\rangle^2}\,. \tag{1.35}$$

Position and momentum operators for different degrees of freedom commute, so we can write (1.33) more generally as

$$[\hat{p}_i, \hat{x}_j] = \frac{\hbar}{\mathrm{i}}\,\delta_{i,j}\,. \tag{1.36}$$

Here the subscripts i and j can stand for different coordinates of one point particle or for different particles in a many-body system.

Throughout this book relations and equations are almost always formulated in coordinate representation where the spatial coordinates are variables of the wave functions. Because of (1.30) we omit the hat $\hat{}$, which generally characterizes operators, from the position variables. The position variables are only written with a hat on a few isolated occasions, where the operator nature of the variable is intentionally emphasized.

1.1.3 *The Hamiltonian and Equations of Motion*

The Hermitian operator describing the energy of a system is the *Hamiltonian*. For a system of N spinless point particles of mass m_i, the Hamiltonian usually consists of the kinetic energy

$$\hat{T} = \sum_{i=1}^{N} \frac{\hat{\boldsymbol{p}}_i^2}{2m_i}$$

and a potential energy $\hat{V}$:

$$\hat{H} = \hat{T} + \hat{V}\,. \tag{1.37}$$

The potential energy is in general a function of the N displacement vectors, $\hat{V} = \hat{V}(\hat{\boldsymbol{r}}_1, \ldots \hat{\boldsymbol{r}}_N)$. In coordinate representation $\hat{V}$ is usually given by a real function $V(\boldsymbol{r}_1, \ldots \boldsymbol{r}_N)$ of the position variables. Applying the operator $\hat{V}$ to a wave function then simply amounts to multiplying the wave function with the function $V(\boldsymbol{r}_1, \ldots \boldsymbol{r}_N)$.

The Hamiltonian of a physical system determines its evolution in time. In the *Schrödinger picture* the evolution of a state $|\psi(t)\rangle$ is described by the *Schrödinger equation*:

$$\hat{H}|\psi(t)\rangle = \mathrm{i}\hbar\frac{\mathrm{d}|\psi\rangle}{\mathrm{d}t}\,, \tag{1.38}$$

which in coordinate representation corresponds to a partial differential equation:

$$\hat{H}\psi(X;t) = \mathrm{i}\hbar\frac{\partial\psi}{\partial t} . \qquad (1.39)$$

The evolution of a state $|\psi(t)\rangle$ can formally be described with the help of the *time evolution operator*:

$$|\psi(t)\rangle = \hat{U}(t,t_0)|\psi(t_0)\rangle . \qquad (1.40)$$

If the Hamiltonian is not explicitly time dependent, then the time evolution operator is

$$\hat{U}(t,t_0) = \exp\left[-\frac{i}{\hbar}\hat{H}(t-t_0)\right] . \qquad (1.41)$$

For a time-dependent Hamiltonian, (1.41) must be replaced by

$$\hat{U}(t,t_0) = \left[\exp\left\{-\frac{i}{\hbar}\int_{t_0}^{t}\hat{H}(t')\,\mathrm{d}t'\right\}\right]_+ , \qquad (1.42)$$

where the symbol $[\cdots]_+$ indicates time ordering of products of operators: $\left[\hat{O}(t_1)\cdots\hat{O}(t_n)\right]_+ = \hat{O}(t_{l_1})\cdots\hat{O}(t_{l_n})$ when $t_{l_1} \geq t_{l_2}\cdots \geq t_{l_n}$. The time evolution operator is *unitary*. That means

$$\hat{U}^\dagger\hat{U} = \hat{U}\hat{U}^\dagger = 1 . \qquad (1.43)$$

In the *Heisenberg picture* we regard the state vector

$$|\psi_\mathrm{H}\rangle = \hat{U}^\dagger(t,t_0)|\psi(t)\rangle = |\psi(t_0)\rangle \qquad (1.44)$$

as a time-independent quantity, and the Schrödinger equation (1.38) leads to an equation of motion for the Heisenberg representation,

$$\hat{O}_\mathrm{H}(t) = \hat{U}^\dagger(t,t_0)\,\hat{O}\,\hat{U}(t,t_0) , \qquad (1.45)$$

of the respective operators $\hat{O}$, namely:

$$\mathrm{i}\hbar\frac{\mathrm{d}\hat{O}_\mathrm{H}}{\mathrm{d}t} = \left[\hat{O}_\mathrm{H},\hat{H}_\mathrm{H}\right] + \mathrm{i}\hbar\frac{\partial\hat{O}_\mathrm{H}}{\partial t} . \qquad (1.46)$$

The expectation value of an operator does not depend on whether we work in the Schrödinger picture or in the Heisenberg picture:

$$\langle\hat{O}\rangle = \langle\psi(t)|\hat{O}|\psi(t)\rangle = \langle\psi_\mathrm{H}|\hat{O}_\mathrm{H}(t)|\psi_\mathrm{H}\rangle . \qquad (1.47)$$

The evolution of $\langle\hat{O}\rangle$ follows from (1.38) or (1.46):

$$\mathrm{i}\hbar\frac{\mathrm{d}\langle\hat{O}\rangle}{\mathrm{d}t} = \langle[\hat{O},\hat{H}]\rangle + \mathrm{i}\hbar\left\langle\frac{\partial\hat{O}}{\partial t}\right\rangle . \tag{1.48}$$

For a time-independent Hamiltonian $\hat{H}$ the wave function

$$|\psi(t)\rangle = \exp\left(-\frac{\mathrm{i}}{\hbar}Et\right)|\psi_E\rangle \tag{1.49}$$

is a solution of the Schrödinger equation (1.38) if and only if $|\psi_E\rangle$ is an eigenstate of $\hat{H}$ with eigenvalue E,

$$\hat{H}|\psi_E\rangle = E|\psi_E\rangle . \tag{1.50}$$

Equation (1.50) is the *time-independent* or *stationary Schrödinger equation*. Since any linear combination of solutions of the time-dependent Schrödinger equation (1.38) is again a solution we can use the eigenstates $|\psi_{E_n}\rangle$ of $\hat{H}$ to construct a general solution of (1.38):

$$|\psi(t)\rangle = \sum_n c_n \exp\left(-\frac{\mathrm{i}}{\hbar}E_n t\right)|\psi_{E_n}\rangle . \tag{1.51}$$

As long as the potential energy is sufficiently attractive, the Hamiltonian $\hat{H}$ has only discrete eigenvalues and normalizable eigenstates at low energies. They describe *bound* states of the system. In this energy region the time-independent Schrödinger equation (1.50) is an equation for the eigenvalues E_n and the corresponding eigenfunctions $|\psi_{E_n}\rangle$. The lowest eigenvalue is the *ground state energy* and the corresponding eigenstate the *ground state* of the system. If the potential energy $V(\boldsymbol{r}_1,\ldots,\boldsymbol{r}_N)$ converges to a constant in the asymptotic region (where at least one $|\boldsymbol{r}_i| \to \infty$), then the time-independent Schrödinger equation can be solved for all energies above this constant and the corresponding eigenstates are in general not normalizable. Such *continuum wave functions* describe *unbound* states of the system (scattering states, reactions) and their concrete meaning depends on their asymptotic properties, i.e. on the asymptotic boundary conditions.

1.2 Symmetries

1.2.1 Constants of Motion and Symmetries

If the Hamiltonian $\hat{H}$ does not depend explicitly on time, then the expectation value of $\hat{H}$ is a constant in time, as is the expectation value of any (time-independent) operator which commutes with $\hat{H}$. This follows immediately from (1.48). The energy and the observables commuting with $\hat{H}$ are the *constants of motion.* Solutions of the time-independent Schrödinger equation can be labelled by the energy and the eigenvalues of the other constants of motion. The eigenvalues of the constants of motion are often called *good quantum numbers*.

An important example is the orbital angular momentum of a point particle of mass μ:

$$\hat{\boldsymbol{L}} = \hat{\boldsymbol{r}} \times \hat{\boldsymbol{p}} \,, \tag{1.52}$$

i.e. $\hat{L}_x = \hat{y}\hat{p}_z - \hat{z}\hat{p}_y$, etc. If the potential energy $V(\boldsymbol{r})$ depends only on the length $r = |\boldsymbol{r}|$ and not on the direction of the vector $\boldsymbol{r}$,

$$\hat{H} = \frac{\hat{\boldsymbol{p}}^2}{2\mu} + V(r) \,, \tag{1.53}$$

then all components of $\hat{\boldsymbol{L}}$ commute with $\hat{H}$,

$$\left[\hat{H}, \hat{L}_x\right] = \left[\hat{H}, \hat{L}_y\right] = \left[\hat{H}, \hat{L}_z\right] = 0 \,, \tag{1.54}$$

as does the square $\hat{\boldsymbol{L}}^2 = \hat{L}_x^2 + \hat{L}_y^2 + \hat{L}_z^2$,

$$\left[\hat{H}, \hat{\boldsymbol{L}}^2\right] = 0 \,. \tag{1.55}$$

However, the components of $\hat{\boldsymbol{L}}$ themselves do not commute, rather

$$\left[\hat{L}_x, \hat{L}_y\right] = \mathrm{i}\hbar \hat{L}_z \,, \quad \left[\hat{L}_y, \hat{L}_z\right] = \mathrm{i}\hbar \hat{L}_x \,, \quad \left[\hat{L}_z, \hat{L}_x\right] = \mathrm{i}\hbar \hat{L}_y \,. \tag{1.56}$$

$\hat{\boldsymbol{L}}^2$ and all components of $\hat{\boldsymbol{L}}$ are constants of motion, but $\hat{\boldsymbol{L}}^2$ and one component alone already form a complete set of observables for the orbital angular motion of the particle. In *spherical coordinates*,

$$x = r \sin\theta \, \cos\phi \,, \quad y = r \sin\theta \, \sin\phi \,, \quad z = r \cos\theta \,, \tag{1.57}$$

the eigenstates of the angular momentum operators $\hat{\boldsymbol{L}}^2$ and $\hat{L}_z$ are the *spherical harmonics* $Y_{l,m}(\theta,\phi)$, which are labelled by the *angular momentum quantum number* l and the *azimuthal quantum number* m:

$$\hat{\boldsymbol{L}}^2 Y_{l,m} = l(l+1)\hbar^2 Y_{l,m}\,,\quad l = 0,\,1,\,2,\dots\,;$$
$$\hat{L}_z Y_{l,m} = m\hbar Y_{l,m}\,,\quad m = -l,\,-l+1,\dots,l-1,\,l\,. \tag{1.58}$$

A precise definition and some important properties of the functions $Y_{l,m}(\theta,\phi)$ are given in Appendix A.1. Here we just mention the orthonormality relation

$$\int Y^*_{l,m}(\Omega)Y_{l',m'}(\Omega)\,\mathrm{d}\Omega$$
$$= \int_0^\pi \sin\theta\,\mathrm{d}\theta \int_0^{2\pi} \mathrm{d}\phi\; Y^*_{l,m}(\theta,\phi)Y_{l',m'}(\theta,\phi)$$
$$= \delta_{l,l'}\delta_{m,m'}\,. \tag{1.59}$$

The spherical harmonics up to $l = 3$ are given explicitly in Table 1.1.

Let $\hat{K}$ be a constant of motion. The *unitary operator generated by* $\hat{K}$,

$$\hat{U}_K(k) = \exp(-\mathrm{i}k\hat{K})\,, \tag{1.60}$$

defines a transformation of the wave functions,

$$|\psi_k\rangle = \hat{U}_K(k)\,|\psi\rangle\,, \tag{1.61}$$

and of the operators,

$$\hat{O}_k = \hat{U}_K(k)\,\hat{O}\,\hat{U}_K^\dagger(k)\,. \tag{1.62}$$

Table 1.1 Spherical harmonics $Y_{l,m}(\theta,\phi)$ for $l \le 3$

l	0	1	1	2
m	0	0	± 1	0
$Y_{l,m}$	$\frac{1}{\sqrt{4\pi}}$	$\sqrt{\frac{3}{4\pi}}\cos\theta$	$\mp\sqrt{\frac{3}{8\pi}}\sin\theta\,\mathrm{e}^{\pm\mathrm{i}\phi}$	$\sqrt{\frac{5}{16\pi}}(3\cos^2\theta - 1)$
l		2	2	3
m		± 1	± 2	0
$Y_{l,m}$		$\mp\sqrt{\frac{15}{8\pi}}\sin\theta\cos\theta\mathrm{e}^{\pm\mathrm{i}\phi}$	$\sqrt{\frac{15}{32\pi}}\sin^2\theta\,\mathrm{e}^{\pm 2\mathrm{i}\phi}$	$\sqrt{\frac{7}{16\pi}}(5\cos^3\theta - 3\cos\theta)$
l		3	3	3
m		± 1	± 2	± 3
$Y_{l,m}$		$\mp\sqrt{\frac{21}{64\pi}}\sin\theta(5\cos^2\theta - 1)\,\mathrm{e}^{\pm\mathrm{i}\phi}$	$\sqrt{\frac{105}{32\pi}}\sin^2\theta\cos\theta\,\mathrm{e}^{\pm 2\mathrm{i}\phi}$	$\mp\sqrt{\frac{35}{64\pi}}\sin^3\theta\,\mathrm{e}^{\pm 3\mathrm{i}\phi}$

This transformation conserves expectation values and matrix elements:

$$\langle\psi_k|\hat{O}_k|\phi_k\rangle = \langle\psi|\hat{O}|\phi\rangle\,. \tag{1.63}$$

Since $\hat{K}$ commutes with $\hat{H}$, and hence any function of $\hat{K}$ commutes with $\hat{H}$, we have:

$$\hat{H}_k = \hat{U}_K(k)\,\hat{H}\,\hat{U}_K^\dagger(k) = \hat{H}\,, \tag{1.64}$$

that means, the Hamiltonian is *invariant under the symmetry transformation defined by* $\hat{U}_K(k)$. Conversely, if we assume the invariance (1.64) for all (real) values of the parameter k, then for infinitesimal k we have

$$(1 - ik\hat{K} + \cdots)\hat{H}(1 + ik\hat{K} + \cdots) = \hat{H} + ik[\hat{H},\hat{K}] + O(k^2) = \hat{H}\,, \tag{1.65}$$

which only works if $\hat{K}$ commutes with $\hat{H}$. Thus the Hamiltonian is invariant under the symmetry transformations (1.60) if and only if it commutes with their *generator* $\hat{K}$.

As an example let's look again at the orbital angular momentum $\hat{\boldsymbol{L}}$ of a point particle, in particular at its z-component which has the following form in spherical coordinates:

$$\hat{L}_z = \frac{\hbar}{\mathrm{i}}\frac{\partial}{\partial\phi}\,. \tag{1.66}$$

The symmetry transformations generated by $\hat{L}_z$ are rotations around the z- axis through all possible angles α:

$$\hat{\mathcal{R}}_z(\alpha) = \exp\left(-\frac{\mathrm{i}}{\hbar}\alpha\hat{L}_z\right)\,. \tag{1.67}$$

The invariance of the Hamiltonian under rotations manifests itself in the commutation of the Hamiltonian with the components of orbital angular momentum.

Mathematically, symmetry transformations which are generated by one or more generators form a *group*. This means that two symmetry transformations operating in succession form a symmetry transformation of the same kind, and to every symmetry transformation $\mathcal{R}$ there belongs an inverse symmetry transformation $\mathcal{R}^{-1}$ which undoes the original transformation: $\mathcal{R}^{-1}\mathcal{R} = \mathbf{1}$. The transformations of a *symmetry group* can be labelled by one or more continuous parameters, as in the example of rotations, or by discrete parameters, as is the case for reflections. An important example of a reflection is the reflection at the origin in coordinate space:

$$\hat{\Pi}\psi(x,y,z) = \psi(-x,-y,-z)\,. \tag{1.68}$$

Since $\hat{\Pi}^2 = \mathbf{1}$, there are only two possible eigenvalues for $\hat{\Pi}$: $+1$ and -1. The corresponding eigenstates are called states of *positive parity* and states of *negative parity* respectively. If the potential energy $V(x, y, z)$ of a point particle does not depend on the sign of the coordinates, then *parity* is a good quantum number.

Identifying constants of motion and good quantum numbers is an important step towards solving the Schrödinger equation. If $\hat{O}$ is a constant of motion we can look for eigenstates of $\hat{H}$ in subspaces consisting of eigenstates of $\hat{O}$ with given eigenvalue ω. In most cases this is much simpler than trying to solve the Schrödinger equation directly in the space of all possible wave functions, as the following example shows.

1.2.2 The Radial Schrödinger Equation

The time-independent Schrödinger equation for a point particle in a radially symmetric potential $V(r)$ is, in coordinate representation,

$$\left(-\frac{\hbar^2}{2\mu}\Delta + V(r)\right)\psi(\boldsymbol{r}) = E\psi(\boldsymbol{r})\,. \tag{1.69}$$

The Laplacian operator $\Delta = \partial^2/\partial x^2 + \partial^2/\partial y^2 + \partial^2/\partial z^2 = -\hat{\boldsymbol{p}}^2/\hbar^2$ can be expressed in spherical coordinates with the help of the orbital angular momentum $\hat{\boldsymbol{L}}$:

$$\Delta = \frac{\partial^2}{\partial r^2} + \frac{2}{r}\frac{\partial}{\partial r} - \frac{\hat{\boldsymbol{L}}^2}{r^2\hbar^2}\,. \tag{1.70}$$

Since $\hat{\boldsymbol{L}}^2$ and $\hat{L}_z$ are constants of motion, we can label the solutions of the Schrödinger equation (1.69) by the good quantum numbers l and m:

$$\psi(\boldsymbol{r}) = f_l(r)Y_{l,m}(\theta, \phi)\,. \tag{1.71}$$

Parity is also a good quantum number for the wave function (1.71), because the radial coordinate r is unaffected by the reflection $\boldsymbol{r} \to -\boldsymbol{r}$ and (see (A.6) in Appendix A.1)

$$\hat{\Pi}\, Y_{l,m}(\theta, \phi) = (-1)^l Y_{l,m}(\theta, \phi)\,. \tag{1.72}$$

Inserting (1.71) into (1.69) leads to an equation for the *radial wave function* $f_l(r)$:

$$\left[-\frac{\hbar^2}{2\mu}\left(\frac{\mathrm{d}^2}{\mathrm{d}r^2} + \frac{2}{r}\frac{\mathrm{d}}{\mathrm{d}r}\right) + \frac{l(l+1)\hbar^2}{2\mu r^2} + V(r)\right] f_l(r) = E f_l(r)\,; \tag{1.73}$$

it does not depend on the azimuthal quantum number m.

The *radial Schrödinger equation* (1.73) is an ordinary differential equation of second order for the radial wave function f_l and is thus a substantial simplification compared to the partial differential equation (1.69). A further not so substantial but very useful simplification is achieved, if we formulate an equation not for $f_l(r)$, but for $\phi_l = rf_l$, i.e. for the radial wave function $\phi_l(r)$ defined by

$$\psi(\boldsymbol{r}) = \frac{\phi_l(r)}{r}\, Y_{l,m}(\theta,\phi)\,. \tag{1.74}$$

The radial Schrödinger equation now reads

$$\left(-\frac{\hbar^2}{2\mu}\frac{\mathrm{d}^2}{\mathrm{d}r^2} + \frac{l(l+1)\hbar^2}{2\mu r^2} + V(r)\right)\phi_l(r) = E\phi_l(r)\,, \tag{1.75}$$

and this looks just like the Schrödinger equation for a point particle moving in one spatial dimension in an effective potential consisting of $V(r)$ plus the *centrifugal potential* $l(l+1)\hbar^2/(2\mu r^2)$:

$$V_{\mathrm{eff}}(r) = V(r) + \frac{l(l+1)\hbar^2}{2\mu r^2}\,. \tag{1.76}$$

Note however, that the radial Schrödinger equations (1.73) and (1.75) are only defined for non-negative values of the radial coordinate r. The boundary condition which the radial wave function $\phi_l(r)$ must fulfill at $r = 0$ can be derived by inserting an ansatz $\phi_i(r) \propto r^\alpha$ into (1.75). As long as the potential $V(r)$ is less singular than r^{-2}, the leading term on the left-hand side is proportional to $r^{\alpha-2}$ and vanishes only if $\alpha = l + 1$ or $\alpha = -l$. The latter possibility is to be discarded, because an infinite value of $\phi_l(r \to 0)$ would lead to an infinite contribution to the norm of the wave function near the origin; a finite value, as would occur for $l = 0$, leads to a delta function singularity originating from $\Delta(1/r)$ on the left-hand side of the Schrödinger equation (1.69), and this cannot be compensated by any of the other terms in the equation. The boundary condition for the radial wave function at the origin $r = 0$ is thus

$$\phi_l(0) = 0 \quad \text{for all} \quad l\,, \tag{1.77}$$

and its behaviour near the origin is given by

$$\phi_l(r) \propto r^{l+1} \quad \text{for} \quad r \to 0 \tag{1.78}$$

(as long as the potential $V(r)$ is less singular than r^{-2}).

The radial Schrödinger equation (1.75) is a one-dimensional Schrödinger equation for a particle which moves in the effective potential (1.76) for $r \geq 0$ and hits an infinite repulsive wall at $r = 0$. In a one-dimensional symmetric potential $V(|x|)$ the odd solutions, i.e. those of negative parity, automatically fulfill the condition $\phi(0) = 0$. Since the effective potential (1.76) for $l = 0$ has the same form as the potential in the one-dimensional Schrödinger equation, there is a one-to-one correspondence between the solutions of the radial equation for $l = 0$ and the negative parity solutions of the one-dimensional equation with the same potential.

Using the orthonormality (1.59) of the spherical harmonics we see that the scalar product of two wave functions $\psi_{l,m}$ and $\psi'_{l',m'}$ of type (1.74) is given by

$$\langle \psi_{l,m}|\psi'_{l',m'}\rangle = \int \psi^*_{l,m}(\mathbf{r})\psi'_{l',m'}(\mathbf{r})\mathrm{d}^3r$$
$$= \delta_{l,l'}\delta_{m,m'}\int_0^\infty \phi^*_l(r)\phi'_l(r)\mathrm{d}r\,. \tag{1.79}$$

If the potential $V(r)$ is real, the phase of the wave function (1.74) can always be chosen such that the radial wave function ϕ_l is real.

1.2.3 Example: The Radially Symmetric Harmonic Oscillator

The potential for this case is

$$V(r) = \frac{\mu}{2}\omega^2 r^2\,. \tag{1.80}$$

For angular momentum quantum numbers $l > 0$ the effective potential V_{eff} also contains the centrifugal potential. The potential tends to infinity for $r \to \infty$ and there are only bound solutions to the Schrödinger equation. For each angular momentum quantum number l there is a sequence of energy eigenvalues,

$$E_{n,l} = \left(2n + l + \frac{3}{2}\right)\hbar\omega\,, \quad n = 0,\, 1,\, 2, \ldots\,, \tag{1.81}$$

and the corresponding radial wave functions $\phi_{n,l}(r)$ (which are normalized to unity) are

$$\phi_{n,l} = 2(\sqrt{\pi}\beta)^{-\frac{1}{2}}\left[\frac{2^{n+l}n!}{(2n+2l+1)!!}\right]^{\frac{1}{2}}\left(\frac{r}{\beta}\right)^{l+1} L_n^{l+\frac{1}{2}}\left(\frac{r^2}{\beta^2}\right)$$
$$\times \exp\left(-\frac{r^2}{2\beta^2}\right)\,. \tag{1.82}$$

The polynomials $L_n^\alpha(x)$ are the *generalized Laguerre polynomials* and are polynomials of order n in x. (The ordinary Laguerre polynomials correspond to $\alpha = 0$.) For the definition and some important properties of the Laguerre polynomials see Appendix A.2. The quantity β in (1.82) is the *oscillator width* given by

$$\beta = \sqrt{\frac{\hbar}{\mu\omega}} \quad \text{or} \quad \frac{\hbar^2}{\mu\beta^2} = \hbar\omega\,. \tag{1.83}$$

Table 1.2 Radial eigenfunctions (1.82) for the harmonic oscillator, ($x = r/\beta$)

l	$n = 0$	$n = 1$	$n = 2$
0	$2x\,e^{-x^2/2}$	$\sqrt{\frac{8}{3}}x\left(\frac{3}{2} - x^2\right)e^{-x^2/2}$	$\sqrt{\frac{8}{15}}x\left(\frac{15}{4} - 5x^2 + x^4\right)e^{-x^2/2}$
1	$\sqrt{\frac{8}{3}}x^2e^{-x^2/2}$	$\frac{4}{\sqrt{15}}x^2\left(\frac{5}{2} - x^2\right)e^{-x^2/2}$	$\frac{4}{\sqrt{105}}x^2\left(\frac{35}{4} - 7x^2 + x^4\right)e^{-x^2/2}$
2	$\frac{4}{\sqrt{15}}x^3e^{-x^2/2}$	$\sqrt{\frac{32}{105}}x^3\left(\frac{7}{2} - x^2\right)e^{-x^2/2}$	$\sqrt{\frac{32}{945}}x^3\left(\frac{63}{4} - 9x^2 + x^4\right)e^{-x^2/2}$
3	$\sqrt{\frac{32}{105}}x^4e^{-x^2/2}$	$\frac{8}{\sqrt{945}}x^4\left(\frac{9}{2} - x^2\right)e^{-x^2/2}$	$\frac{8}{\sqrt{10395}}x^4\left(\frac{99}{4} - 11x^2 + x^4\right)e^{-x^2/2}$

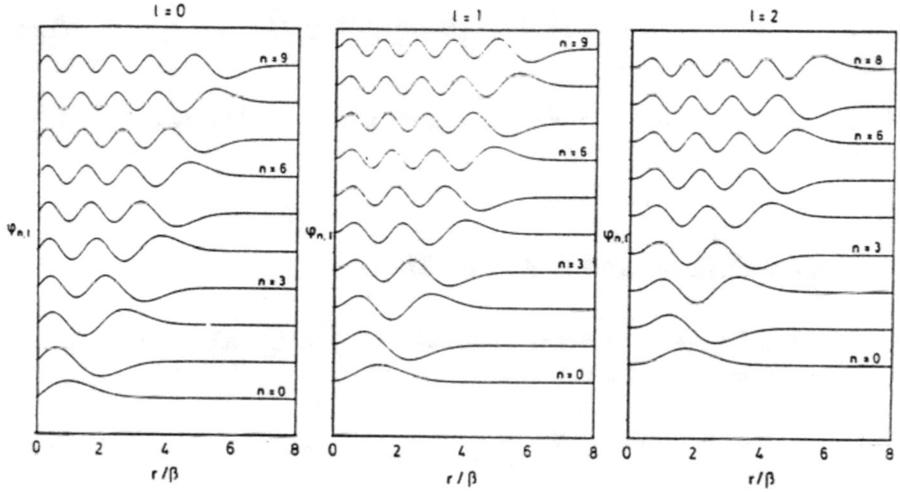

Fig. 1.1 Radial eigenfunctions $\phi_{n,l}(r)$ of the spherical harmonic oscillator (1.82) for angular momentum quantum numbers $l = 0, 1, 2$ and principal quantum numbers (1.84) up to $N = 19$

For $l = 0$ (1.81) gives us the spectrum $(2n + 3/2)\hbar\omega$, $n = 0, 1, \ldots$ of the one-dimensional oscillator states of negative parity. The radial wave functions (1.82) are summarized in Table 1.2 and illustrated in Fig. 1.1 for low values of the quantum numbers n and l.

The radial wave functions $\phi_{n,l}$ are complemented via (1.74) to give eigenfunctions of the three-dimensional Schrödinger equation for a (spinless) point particle in the potential (1.80). For every radial quantum number n and angular momentum quantum number l there are $2l + 1$ eigenfunctions corresponding to the various values of the azimuthal quantum number $m = -l, -l + 1, \ldots, l - 1, l$. These eigenfunctions all have the same energy eigenvalue $E_{n,l}$, because the radial Schrödinger equation does not depend on m. This is true in any radially symmetric potential. A peculiarity of the harmonic oscillator spectrum is its *additional degeneracy*: the energy depends not on the quantum numbers n and l independently, but only on the combination

$$N = 2n + l\,, \tag{1.84}$$

which is hence called the *principal quantum number* (of the radially symmetric harmonic oscillator). The energy eigenvalues are grouped into equidistant oscillator shells of energy $E_N = (N + 3/2)\hbar\omega$, $N = 0, 1, 2, \ldots$. The degree of degeneracy of the Nth oscillator shell is given by summation over all l values compatible with this principal quantum number; for even values of N this means all even l less or equal to N, for odd N all odd l less or equal to N. Regardless of whether N is even or odd, the number of independent eigenstates with energy eigenvalue $E_N = (N + 3/2)\hbar\omega$ is given by

$$\sum_l (2l + 1) = (N + 1)(N + 2)/2\,. \tag{1.85}$$

Due to (1.72) each oscillator shell is characterized by a definite parity, namely $(-1)^N$.

1.3 Bound States and Unbound States

Let's look at the radial Schrödinger equation (1.75) for a particle of mass μ in an effective potential $V_{\text{eff}}(r)$ which vanishes for $r \to \infty$:

$$\left(-\frac{\hbar^2}{2\mu}\frac{\mathrm{d}^2}{\mathrm{d}r^2} + V_{\text{eff}}(r)\right)\phi(r) = E\phi(r)\,. \tag{1.86}$$

The behaviour of the solutions of (1.86) depends in an essential way on whether the energy E is smaller or larger than zero.

1.3.1 Bound States

For a start let's assume that V_{eff} is short ranged, meaning that V_{eff} vanishes beyond a definite radius r_0:

$$V_{\text{eff}}(r) = 0 \quad \text{for} \;\; r \geq r_0\,. \tag{1.87}$$

This is of course only reasonable if $l = 0$, because the centrifugal potential falls off as $1/r^2$ at large r (see (1.76)).

If $E < 0$, the equation (1.86) in the outer region is simply

$$\frac{\mathrm{d}^2\phi}{\mathrm{d}r^2} = \kappa^2\phi, \quad r \geq r_0\,, \tag{1.88}$$

where κ is a (positive) constant depending on the energy $E = -|E|$:

$$\kappa = \sqrt{2\mu|E|/\hbar^2}\,. \tag{1.89}$$

Two linearly independent solutions of the ordinary second-order differential equation (1.88) are

$$\phi_+(r) = \mathrm{e}^{+\kappa r}\,, \qquad \phi_-(r) = \mathrm{e}^{-\kappa r}\,. \tag{1.90}$$

In the inner region $r \leq r_0$ the solution of (1.86) depends on the potential $V_{\mathrm{eff}}(r)$. The general solution contains two integration constants, one of which is determined by the boundary condition (1.77) at the origin, $\phi(0) = 0$; the other constant is undetermined, because any multiple of a solution $\phi(r)$ of (1.86) is again a solution. The boundary condition (1.77) determines the solution of (1.86) in the inner region uniquely, except for multiplication by an arbitrary constant.

In order to get a solution of (1.86) for all $r \geq 0$, we must connect the solution $\phi_{r\leq r_0}$ in the inner region to a linear combination of the solutions (1.90) in the outer region $r \geq 0$. We must however discard any contribution from $\phi_+(r)$, because the probability for finding the particle would otherwise grow exponentially for $r \to \infty$. The conditions that the wave function be continuous and have continuous derivative lead to the following matching conditions at the matching radius r_0:

$$\phi_{r\leq r_0}(r_0) = C\,\mathrm{e}^{-\kappa r_0}\,, \quad \phi'_{r\leq r_0}(r_0) = -\kappa\, C\,\mathrm{e}^{-\kappa r_0}\,. \tag{1.91}$$

Dividing the second of these equations by the first leads to a matching condition free of the proportionality constant C:

$$\frac{\phi'_{r\leq r_0}(r_0)}{\phi_{r\leq r_0}(r_0)} = -\kappa = -\sqrt{2\mu|E|/\hbar^2}\,. \tag{1.92}$$

For arbitrary energies $E < 0$ the matching condition (1.92) is in general not fulfilled, as is illustrated in Fig. 1.2 for a sharp-step potential. If the potential V_{eff} is sufficiently attractive, there is a discrete sequence E_1, E_2, $E_3, \ldots$ of energies for which (1.92) is fulfilled. The corresponding wave functions are square integrable and are the bound states in the potential $V_{\mathrm{eff}}(r)$.

The discussion above remains valid if the effective potential in the outer region does not vanish, but corresponds instead to a centrifugal potential with finite angular momentum quantum number $l > 0$:

$$V_{\mathrm{eff}}(r) = \frac{l(l+1)\hbar^2}{2\mu r^2}\,, \quad r \geq r_0\,. \tag{1.93}$$

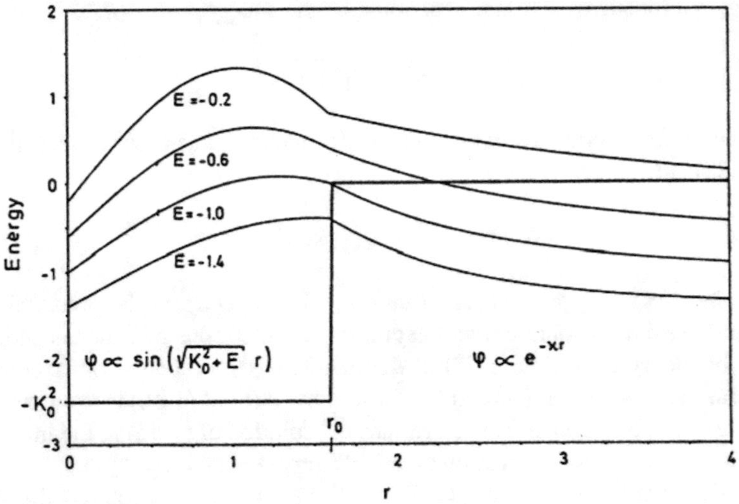

Fig. 1.2 Matching of inner and outer solutions ϕ for negative energies $E = -\kappa^2$ in an attractive sharp-step potential ($V(r) = -K_0^2$ for $r < r_0$, $V \equiv 0$ for $r > r_0$, $\hbar^2/(2\mu) = 1$). With the paremeters used in this figure, $K_0^2 = 2.5$, $r_0 = 1.6$, there is an energy between $E = -0.6$ and $E = -1.0$ at which (1.92) is fulfilled (See also Sect. 1.3.3)

Instead of the simple exponential functions (1.90), the solutions in the outer region are now *modified Bessel functions* (see Appendix A.4):

$$\phi_+(r) = \sqrt{\kappa r}\, I_{l+\frac{1}{2}}(\kappa r)\,, \qquad \phi_-(r) = \sqrt{\kappa r}\, K_{l+\frac{1}{2}}(\kappa r)\,. \tag{1.94}$$

Asymptotically $\phi_+(r)$ is again an exponentially growing solution,

$$\phi_+(r) \propto \mathrm{e}^{+\kappa r}\left(1 + O\left(\frac{1}{\kappa r}\right)\right)\,, \tag{1.95}$$

which must be discarded on physical grounds, while $\phi_-(r)$ decreases exponentially in the asymptotic region. An exact expression for $\phi_-(r)$, which is valid not only asymptotically, is

$$\phi_-(r) = \sqrt{\frac{\pi}{2}}\,\mathrm{e}^{-\kappa r}\sum_{\lambda=0}^{l}\frac{(l+\lambda)!}{\lambda!(l-\lambda)!}(2\kappa r)^{-\lambda}\,. \tag{1.96}$$

The matching condition at $r = r_0$ is now

$$\frac{\phi'_{r\le r_0}(r_0)}{\phi_{r\le r_0}(r_0)} = \frac{\phi'_-(r_0)}{\phi_-(r_0)} = -\frac{l}{r_0} - \kappa\,\frac{K_{l-\frac{1}{2}}(\kappa r_0)}{K_{l+\frac{1}{2}}(\kappa r_0)}\,, \tag{1.97}$$

where we have rewritten the derivative of $K_{l+\frac{1}{2}}$ according to (A.64) in Appendix A.4.

We can venture one step further and allow the effective potential in the outer region to contain a long-ranged Coulomb contribution proportional to $1/r$:

$$V_{\rm eff}(r) = \frac{l(l+1)\hbar^2}{2\mu r^2} - \frac{C}{r}, \quad r \geq r_0 . \tag{1.98}$$

The solutions of (1.86) in the outer region are now *Whittaker functions* (see Appendix A.5). At $r = r_0$ we now match to the wave function

$$\phi_-(r) = W_{\gamma,l+\frac{1}{2}}(2\kappa r), \tag{1.99}$$

which decreases exponentially for $r \to \infty$. The parameter

$$\gamma = \frac{\mu C}{\hbar^2 \kappa} \tag{1.100}$$

describes the relative strength of the $1/r$ term in the potential. The dependence of γ on energy E or on κ is determined by a length parameter a,

$$\gamma = \frac{1}{\kappa a} . \tag{1.101}$$

The length a, which gives a scale for the spatial extension of the bound states in the Coulomb-type potential, is called the *Bohr radius*:

$$a = \frac{\hbar^2}{\mu C} . \tag{1.102}$$

For large values of r the leading term of (1.99) is

$$\phi_-(r) = {\rm e}^{-\kappa r}(2\kappa r)^\gamma \left(1 + O\left(\frac{1}{\kappa r}\right)\right) . \tag{1.103}$$

1.3.2 Unbound States

Circumstances are quite different at positive energies $E > 0$. For a short-ranged potential (1.87) the radial Schrödinger equation in the outer region $r \geq r_0$ reads

$$\frac{{\rm d}^2\phi}{{\rm d}r^2} + k^2\phi = 0 , \tag{1.104}$$

with the *wave number*

$$k = \sqrt{2\mu E/\hbar^2}\,. \tag{1.105}$$

Two linearly independent solutions of (1.104) are

$$\phi_s(r) = \sin kr\,, \quad \phi_c(r) = \cos kr\,. \tag{1.106}$$

In the absence of the short-ranged potential, ϕ_s solves the radial Schrödinger equation for all r and fulfills the boundary condition $\phi(0) = 0$; it is called the *regular solution,* because the corresponding wave function $\psi(\boldsymbol{r})$ (c.f. (1.74)) is regular at the origin. In the presence of the short-ranged potential there is a different inner solution $\phi_{r\le r_0}(r)$ which fulfills the boundary condition $\phi(0) = 0$. This solution is unique, except for multiplication by an arbitrary constant. Matching it continuously and with continuous derivative to a linear combination of outer solutions (1.106) leads to the matching equations

$$\phi_{r\le r_0}(r_0) = A\phi_s(r_0) + B\phi_c(r_0)\,, \tag{1.107}$$

$$\phi'_{r\le r_0}(r_0) = A\phi'_s(r_0) + B\phi'_c(r_0)\,. \tag{1.108}$$

In contrast to the negative energy case, we now have no physical reasons for discarding one of the two basis functions (1.106). Thus we have two constants A and B which we can always choose such that (1.107) and (1.108) are simultaneously fulfilled. For any energy $E > 0$ there is a solution to the Schrödinger equation. Asymptotically the eigenfunctions are bounded, but they don't vanish; they describe unbound states in the potential $V_{\rm eff}(r)$.

The physical solution of the radial Schrödinger equation in the outer region thus has the form

$$\phi(r) = A\phi_s(r) + B\phi_c(r)\,, \qquad r \ge r_0\,, \tag{1.109}$$

with the constants A and B to be determined from the matching equations (1.107), (1.108). Solutions of the Schrödinger equation are in general complex. However, if the potential $V_{\rm eff}$ in (1.86) is real, we can always find real solutions ϕ and hence assume that the constants A and B are real. It is helpful to rewrite (1.109) as

$$\phi(r) = \sqrt{A^2+B^2}\,[\cos\delta\,\phi_s(r) + \sin\delta\,\phi_c(r)]\,, \qquad r \ge r_0\,, \tag{1.110}$$

where δ is the angle defined by

$$\sin\delta = \frac{B}{\sqrt{A^2+B^2}}\,, \qquad \cos\delta = \frac{A}{\sqrt{A^2+B^2}}\,. \tag{1.111}$$

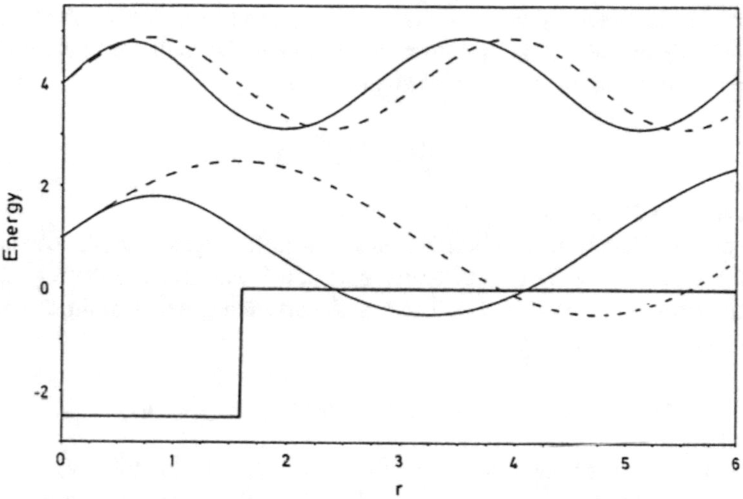

Fig. 1.3 Asymptotic phase shifts in the radial wave function, obtained by matching the inner wave function to the outer wave function at the matching radius r_0. The *dashed lines* are the regular solutions ϕ_s of the free wave equation (1.104) at two different (positive) energies; the *solid lines* are the regular physical solutions in the presence of the attractive sharp-step potential of Fig. 1.2 ($V(r) = -K_0^2 = -2.5$ for $r < r_0 = 1.6$, $V \equiv 0$ for $r > r_0$, $\hbar^2/(2\mu) = 1$) (See also Sect. 1.3.3)

Inserting (1.106) gives

$$\phi(r) = \sqrt{A^2 + B^2}\,\sin(kr + \delta)\,, \qquad r \geq r_0\,. \tag{1.112}$$

At each energy $E > 0$ the two constants A and B derived via the matching equations (1.107), (1.108) thus determine the amplitude and the phase of the physical wave function in the outer region. The amplitude is in principle an arbitrary constant, which can be fixed by a normalization condition (see Sect. 1.3.5). The phase δ, on the other hand, is a very important quantity. At each energy E it tells us how much the outer waves of the physical solution are shifted from the waves of the regular solution $\phi_s(r)$ of the "free wave equation"—see Fig. 1.3. From (1.111) we get an equation for the phase shift which no longer contains the amplitude:

$$\tan\delta = \frac{B}{A}\,. \tag{1.113}$$

Note that matching conditions determine the phase shift δ only up to an additive constant which is any integer multiple of π.

The asymptotic phase shift is a very important quantity, because it carries the information about the physical effect of the potential in the inner region into the asymptotic region. Such phase shifts determine observable cross sections in scattering and reaction experiments (see Chap. 4). The contribution $\sigma_{[l]}$ of a given angular momentum component of a scattering wave function to the elastic scattering cross section is generally proportional to the square of the sine of the asymptotic phase shift δ_l in the regular solution of the radial Schrödinger equation for the respective angular momentum quantum number l,

$$\sigma_{[l]}(E) \propto \sin^2 \delta_l \,, \tag{1.114}$$

see equation (4.39) in Sect. 4.1.3.

The above discussion of unbound states in a short-ranged potential can easily be generalized to the case that the effective potential $V_{\rm eff}(r)$ in the outer region $r \geq r_0$ is the centrifugal potential (1.93). The two linearly independent solutions of (1.86) in the outer region are now

$$\phi_{\rm s}(r) = kr\, j_l(kr)\,, \qquad \phi_{\rm c}(r) = kr\, n_l(kr)\,, \qquad r \geq r_0\,, \tag{1.115}$$

where $\phi_{\rm s}$ is again the regular solution of the free equation, in which $V_{\rm eff}(r)$ consists of the centrifugal potential alone for all r. j_l and n_l are the *spherical Bessel* and *Neumann functions* which are defined in Appendix A.4.[1] Their asymptotic behaviour is such that the wave functions and $\phi_{\rm s}$ and $\phi_{\rm c}$ asymptotically correspond to a sine and a cosine:

$$\begin{aligned} \phi_{\rm s}(r) &= \sin\left(kr - \frac{l\pi}{2}\right)\left[1 + O\left(\frac{1}{r}\right)\right] , \\ \phi_{\rm c}(r) &= \cos\left(kr - \frac{l\pi}{2}\right)\left[1 + O\left(\frac{1}{r}\right)\right] . \end{aligned} \tag{1.116}$$

All considerations following (1.104), including equations (1.107) to (1.111) and (1.113), remain valid at least asymptotically. The physical solution of the radial Schrödinger equation has the asymptotic form

$$\phi(r) \propto \sin\left(kr - \frac{l\pi}{2} + \delta_l\right) , \tag{1.117}$$

and δ_l is its asymptotic phase shift against the "free wave" $kr\, j_l(kr)$.

If we let the effective potential in the outer region include a Coulomb potential as in (1.98), then the appropriate linearly independent solutions of (1.86) in the outer region are

$$\phi_{\rm s}(r) = F_l(\eta, kr)\,, \qquad \phi_{\rm c}(r) = G_l(\eta, kr)\,, \qquad r \geq r_0\,. \tag{1.118}$$

Here F_l is the *regular Coulomb function* which solves the free equation, in which $V_{\rm eff}$ has the form (1.98) for all r. G_l is the *irregular Coulomb function*, which also solves

[1] It is also common usage to express $\phi_{\rm c}$ in terms of the *spherical Bessel functions of the second kind*, $y_l(z) = -n_l(z)$: $\phi_{\rm c}(kr) = -kr\, y_l(kr)$, cf. (A.51) in Appendix A.4.

the free equation, but which does not vanish at $r = 0$, and which asymptotically is shifted in phase by $\pi/2$ relative to F_l, see (1.120) below and (A.74), (A.75) in Appendix A.5. The Coulomb functions depend not only on kr, but also on the *Sommerfeld parameter* η (also called *Coulomb parameter*), which determines the relative strength of the Coulomb term in the Hamiltonian (see also (1.100)):

$$\eta = -\frac{\mu C}{\hbar^2 k} = -\frac{1}{ka}, \tag{1.119}$$

where a is again the Bohr radius (1.102).

Asymptotically, the regular and irregular Coulomb functions can be written as a sine and a cosine respectively, but the argument is a bit more complicated than in (1.106) and (1.116):

$$F_l(\eta, kr) \to \sin\left(kr - \eta \ln 2kr - \frac{l\pi}{2} + \sigma_l\right), \qquad \text{for} \quad r \to \infty,$$
$$G_l(\eta, kr) \to \cos\left(kr - \eta \ln 2kr - \frac{l\pi}{2} + \sigma_l\right), \qquad \text{for} \quad r \to \infty. \tag{1.120}$$

The l-dependent real constants σ_l are the *Coulomb phases*, which can be expressed with the help of the complex gamma function (see Appendix A.3):

$$\sigma_l = \arg[\Gamma(l + 1 + \mathrm{i}\eta)]\,. \tag{1.121}$$

In addition, the argument of the sine and the cosine in (1.120) contains an r-dependent term η ln $2kr$, due to which the wave length of a Coulomb wave approaches its asymptotic limit $2\pi/k$ only very slowly. This is of course a manifestation of the long-ranged nature of the Coulomb potential.

Nevertheless, the discussion following (1.104) above remains valid, even in the presence of a Coulomb potential. The physical solution of the Schrödinger equation has the asymptotic form

$$\phi(r) \propto \sin\left(kr - \eta \ln 2kr - \frac{l\pi}{2} + \sigma_l + \delta_l\right), \tag{1.122}$$

and δ_l describes its asymptotic phase shift against the "free Coulomb wave" $F_l(\eta, kr)$.

At each energy $E > 0$ the asymptotic phase shift δ_l tells us how a short-ranged deviation of the potential V_{eff} from a reference potential affects the wave function at large r. Asymptotically the physical wave function is a superposition of two solutions of the "free radial Schrödinger equation" containing the reference potential alone, namely of the regular solution ϕ_{s} and the irregular solution ϕ_{c}, asymptotically phase-shifted by $\pi/2$ relative to ϕ_{s}. The tangent of δ_l is the relative weight of the irregular component. This statement does not depend on the reference potential,

Table 1.3 Regular solutions ϕ_s and irregular solutions ϕ_c of the radial Schrödinger equation (1.86) for positive energies $E = \hbar^2 k^2/(2\mu)$. The Coulomb parameter (Sommerfeld parameter) is $\eta = -(\mu/\hbar^2)(C/k)$

$V_{\mathrm{eff}}(r)$	$\phi_s(r)$	$\phi_c(r)$
0	$\sin kr$	$\cos kr$
$\frac{l(l+1)\hbar^2}{2\mu r^2}$	$kr\, j_l(kr)$	$kr\, n_l(kr)$
Asymptotically	$\sin\left(kr - \frac{l\pi}{2}\right)$	$\cos\left(kr - \frac{l\pi}{2}\right)$
$\frac{l(l+1)\hbar^2}{2\mu r^2} - \frac{C}{r}$	$F_l(\eta, kr)$	$G_l(\eta, kr)$
Asymptotically	$\sin\left(kr - \eta \ln 2kr - \frac{l\pi}{2} + \sigma_l\right)$	$\cos\left(kr - \eta \ln 2kr - \frac{l\pi}{2} + \sigma_l\right)$

provided it vanishes asymptotically. The three cases discussed in this section are summarized in Table 1.3.

1.3.3 Examples

1.3.3.1 Sharp-Step Potential

In this case we have

$$V(r) = \begin{cases} -V_0 & \text{for } r < r_0\,, \\ 0 & \text{for } r \geq r_0\,. \end{cases} \tag{1.123}$$

If the effective potential V_{eff} consists only of $V(r)$ with no centrifugal potential and no Coulomb contribution, then for negative energies $-V_0 < E < 0$ the solution $\phi_{r \leq r_0}$ of the Schrödinger equation in the inner region is

$$\phi_{r \leq r_0}(r) = \sin Kr\,. \tag{1.124}$$

The wave number K in the inner region depends on the energy $E = -\hbar^2\kappa^2/(2\mu)$ and the potential parameter $K_0 = \sqrt{2\mu V_0/\hbar^2}$ (see Fig. 1.2):

$$K = \sqrt{K_0^2 - \kappa^2}\,. \tag{1.125}$$

The matching condition (1.92) now reads

$$K \cot K r_0 = -\kappa = -\sqrt{K_0^2 - K^2}\,, \tag{1.126}$$

and can be fulfilled at most for a finite number of wave numbers K_i or energies E_i (see Problem 1.1).

For finite angular momentum quantum number $l > 0$ the effective potential V_{eff} contains the centrifugal potential, and the regular solution in the inner region is

$$\phi_{r\leq r_0}(r) = Kr\, j_l(Kr)\,. \tag{1.127}$$

The matching condition (1.97) at $r = r_0$ now reads

$$K\frac{j_{l-1}(Kr_0)}{j_l(Kr_0)} = -\kappa\,\frac{K_{l-\frac{1}{2}}(\kappa r_0)}{K_{l+\frac{1}{2}}(\kappa r_0)}\,, \tag{1.128}$$

where we have rewritten the derivative of the spherical Bessel function according to (A.52) in Appendix A.4.

For positive energies $E = \hbar^2k^2/(2\mu)$, the regular solution in the inner region again has the form (1.124) in the absence of a centrifugal term, but the wave number in the inner region is now

$$K = \sqrt{K_0^2 + k^2} \tag{1.129}$$

(see Fig. 1.3). At $r = r_0$ the matching conditions (1.107), (1.108) can be rewritten to

$$\frac{1}{K}\tan Kr_0 = \frac{1}{k}\tan(kr_0 + \delta_0)\,, \tag{1.130}$$

from which we derive

$$\delta_0 = -kr_0 + \arctan\left(\frac{k}{K}\tan Kr_0\right)\,. \tag{1.131}$$

In the presence of a centrifugal potential, $l > 0$, we get a simple result for the case of an infinite repulsive sharp step of radius r_0, because the physical wave function must then vanish at $r = r_0$,

$$\phi_l(r_0) = A\,kr_0\, j_l(kr_0) + B\,kr_0\, n_l(kr_0) = 0\,, \tag{1.132}$$

in other words,

$$\tan\delta_l = \frac{B}{A} = -\frac{j_l(kr_0)}{n_l(kr_0)}\,. \tag{1.133}$$

1.3.4 Attractive Coulomb Potential

In this case we have

$$V(r) = -\frac{C}{r}, \tag{1.134}$$

and the constant C is e.g. for a hydrogen atom the square of the elementary electric charge, $C = e^2$.

The bound states are characterized by a *Coulomb principal quantum number*, $n = 1, 2, 3\ldots$, and the corresponding energy eigenvalues are

$$E_n = -\frac{\mathcal{R}}{n^2}. \tag{1.135}$$

$\mathcal{R}$ is the *Rydberg energy*:

$$\mathcal{R} = \frac{\mu C^2}{2\hbar^2} = \frac{1}{2}\frac{\hbar^2}{\mu a^2}, \tag{1.136}$$

where a again stands for the Bohr radius (1.102). Similar to the radially symmetric harmonic oscillator (see Sect. 1.2.3) the energy eigenvalues (1.135) in a Coulomb potential have an additional degeneracy: for a pure Coulomb potential they do not depend on the angular momentum quantum number l; values of l are however restricted to be smaller than n. Thus the angular momentum quantum numbers contributing to the n-th *Coulomb shell* of eigenvalues are

$$l = 0, 1, \ldots n-1. \tag{1.137}$$

Except for $n = 1$, the Coulomb shells have no definite parity, because they contain both even and odd angular momenta. The degeneracy of the n-th Coulomb shell is given by

$$\sum_{l=0}^{n-1}(2l+1) = n^2. \tag{1.138}$$

The radial eigenfunctions $\phi_{n,l}(r)$ are

$$\phi_{n,l}(r) = \frac{1}{n}\left[\frac{(n-l-1)!}{a\,(n+l)!}\right]^{\frac{1}{2}}\left(\frac{2r}{na}\right)^{l+1} L_{n-l-1}^{2l+1}\left(\frac{2r}{na}\right) \mathrm{e}^{-r/(na)}. \tag{1.139}$$

L_ν^α again stands for a generalized Laguerre polynomial (see Appendix A.2). In (1.139) the degree of the Laguerre polynomial, which corresponds to a radial quantum number, is $n - l - 1$. This means that the radial eigenfunction $\phi_{n,l}$ has

Table 1.4 Radial eigenfunctions (1.139) in a Coulomb potential, $x_n = 2r/(na)$

l	$n = l+1$	$n = l+2$	$n = l+3$
0	$\frac{x_1}{\sqrt{a}}\,\mathrm{e}^{-\frac{1}{2}x_1}$	$\frac{x_2}{2\sqrt{2a}}(2-x_2)\,\mathrm{e}^{-\frac{1}{2}x_2}$	$\frac{x_3}{6\sqrt{3a}}(6-6x_3+x_3^2)\,\mathrm{e}^{-\frac{1}{2}x_3}$
1	$\frac{x_2^2}{2\sqrt{6a}}\,\mathrm{e}^{-\frac{1}{2}x_2}$	$\frac{x_3^2}{6\sqrt{6a}}(4-x_3)\,\mathrm{e}^{-\frac{1}{2}x_3}$	$\frac{x_4^2}{16\sqrt{15a}}(20-10x_4+x_4^2)\,\mathrm{e}^{-\frac{1}{2}x_4}$
2	$\frac{x_3^3}{6\sqrt{30a}}\,\mathrm{e}^{-\frac{1}{2}x_3}$	$\frac{x_4^3}{48\sqrt{5a}}(6-x_4)\,\mathrm{e}^{-\frac{1}{2}x_4}$	$\frac{x_5^3}{60\sqrt{70a}}(42-14x_5+x_5^2)\,\mathrm{e}^{-\frac{1}{2}x_5}$
3	$\frac{x_4^4}{48\sqrt{35a}}\,\mathrm{e}^{-\frac{1}{2}x_4}$	$\frac{x_5^4}{120\sqrt{70a}}(8-x_5)\,\mathrm{e}^{-\frac{1}{2}x_5}$	$\frac{x_6^4}{864\sqrt{35a}}(72-18x_6+x_6^2)\,\mathrm{e}^{-\frac{1}{2}x_6}$

exactly $n-l-1$ nodes (zeros) in the region $r > 0$. The radial eigenfunctions (1.139) are tabulated in Table 1.4 and illustrated in Fig. 1.4 for angular momentum quantum numbers $l = 0,\ 1,\ 2$ and for the lowest values of n.

It is important to note that the argument $2r/(na)$ appearing in the Coulomb eigenfunctions (1.139) depends on the principal quantum number n. The reference length na increases with n. One consequence hereof is, that the wave lengths of the inner oscillations do not decrease steadily with increasing n as in the case of the harmonic oscillator (see Fig. 1.1). The wave lengths of the inner oscillations of the Coulomb functions depend strongly on the radius r, but they hardly depend on the principal quantum number n. This is easily understood:
As the principal quantum number n increases, the energy eigenvalue (1.135) approaches zero. For energies close to zero, the right-hand side $E\phi(r)$ of the radial Schrödinger equation (1.75) is only important at large values of r, where the potential energy $V(r)$ also contributes little. In the inner region, the small energy differences corresponding to the different principal quantum numbers play only a minor role. As a consequence, the radial wave functions $\phi_{n,l}$ for a given angular momentum quantum number l and large principal quantum numbers n are almost identical except for a normalization constant. This can be clearly seen in Fig. 1.5, in which the radial wave functions have been *renormalized* such that their norm becomes inversely proportional to their separation in energy at large quantum numbers:

$$\phi^{\mathrm{E}}_{n,l}(r) = \sqrt{\frac{n^3}{2\mathcal{R}}}\,\phi_{n,l}(r)\,, \tag{1.140}$$

where $\mathcal{R}$ is the Rydberg energy (1.136). In this normalization the heights of the inner maxima are independent of n for large n, and the wave functions for a given l converge to a well defined limiting wave function $\phi_l^{(E=0)}$ with infinitely many nodes in the limit $n \to \infty$. This limiting wave function is a solution of the radial Schrödinger equation (1.75) at energy $E = 0$ and has the explicit form

$$\phi_l^{(E=0)}(r) = \frac{\sqrt{r}}{a\sqrt{\mathcal{R}}}\,J_{2l+1}\left(\sqrt{\frac{8r}{a}}\right)\,. \tag{1.141}$$

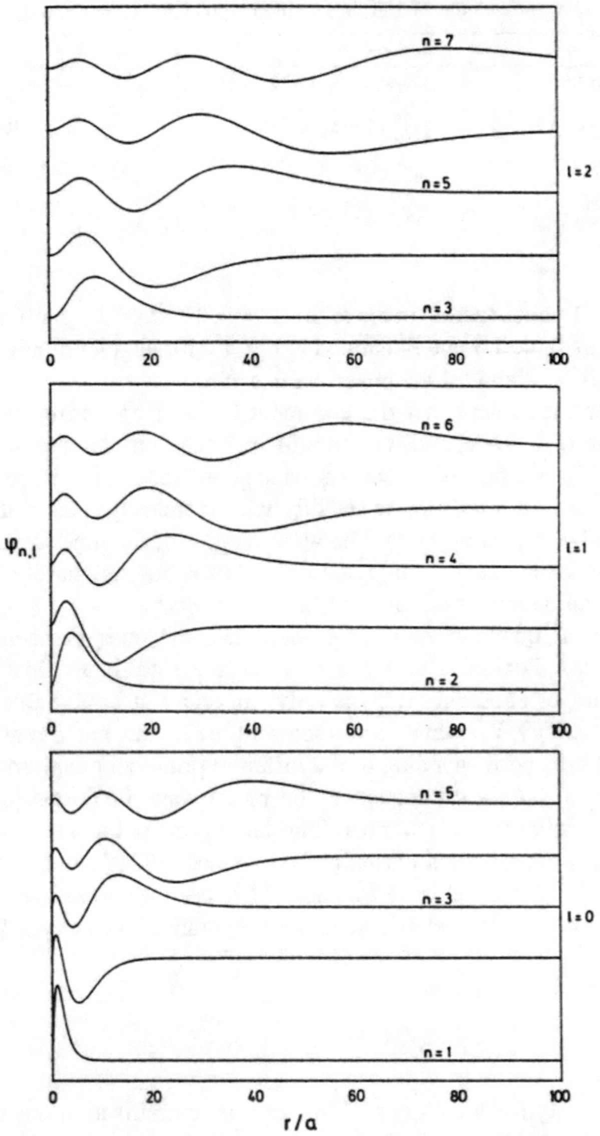

Fig. 1.4 Radial eigenfunctions $\phi_{n,l}(r)$ in a Coulomb potential (1.139) for angular momentum quantum numbers up to $l = 2$ and the lowest five values of n

$J_\nu(x)$ is the *ordinary Bessel function* (see Appendix A.4). For small arguments x we have

$$J_\nu(x) = \frac{1}{\nu!}\left(\frac{x}{2}\right)^\nu\,, \quad x \to 0\,, \tag{1.142}$$

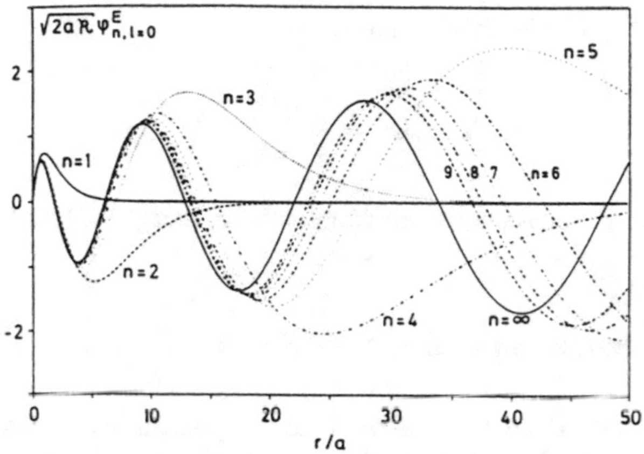

Fig. 1.5 Renormalized radial Coulomb eigenfunctions (1.140) for $l = 0$. The *solid line* labelled $n = \infty$ is the limiting wave function (1.141)

whilst asymptotically

$$J_\nu(x) = \left(\frac{\pi}{2}x\right)^{-\frac{1}{2}} \cos\left(x - \frac{\nu}{2}\pi - \frac{1}{4}\pi\right), \quad x \to \infty\,. \tag{1.143}$$

The convergence of the Coulomb eigenfunctions as $n \to \infty$ is related to the convergence of the energy eigenvalues. The energy eigenvalues (1.135) of the bound states only make up part of the spectrum of the Hamiltonian and the corresponding bound state eigenfunctions only span a part of the Hilbert space. The bound states in a Coulomb potential do not form a complete set. This becomes obvious if we try to expand a simple square integrable wave function (normalized to unity) according to (1.24). The sum $\sum_n |c_n|^2$ converges rapidly, but in general to a value noticeably less than unity (see Problem 1.2).

The eigenfunctions in a Coulomb potential only become a complete set if we include the unbound states of the continuum $E > 0$. The unbound eigenfunctions are just the regular Coulomb functions $F_l(\eta, kr)$ introduced in Sect. 1.3.2. From (A.78) in Appendix A.5 we obtain the following formula for the behaviour of the regular Coulomb functions in an attractive Coulomb potential ($\eta < 0$) at small separations ($r \to 0$) close to the continuum threshold ($k \to 0$):

$$F_l(\eta, kr) = \frac{\sqrt{\frac{\pi}{2}ka}}{(2l+1)!}\left(\frac{2r}{a}\right)^{l+1}, \quad r \to 0,\ k \to 0\,. \tag{1.144}$$

As the energy $E = \hbar^2k^2/(2\mu)$ converges to zero from above, the radial Schrödinger equation (1.75) becomes identical to the equation we obtain for negative energies $E_n = -\mathcal{R}/n^2$ when the principal quantum number n converges to infinity. Hence the continuum wave functions $F_l(\eta, kr)$ must also converge to the solution $\phi_l^{(E=0)}$ in (1.141) at the continuum threshold,

$$\lim_{E\to 0} F_l(\eta, kr) = \sqrt{\frac{\pi\hbar^2 k}{2\mu}}\, \phi_l^{(E=0)}(r)\,. \tag{1.145}$$

The proportionality constant follows from the behaviour (1.142), (1.144) for $r \to 0$.

1.3.5 Normalization of Unbound States

The orthogonality of solutions of the time-independent Schrödinger equation at different energies holds for bound states and for unbound states. Since the unbound wave functions are not square integrable, they cannot be assigned a finite norm. A natural prescription for normalizing unbound states is to require that their scalar product be proportional to a delta function. This can be done in different ways.

For radial wave functions $\phi_k(r)$ which asymptotically correspond to a sine with factor unity,

$$\phi_k(r) \to \sin(kr + \delta_{\rm as})\,, \qquad \text{for} \quad r \to \infty\,, \tag{1.146}$$

we have

$$\int_0^\infty \phi_k(r)\phi_{k'}(r)\mathrm{d}r = \frac{\pi}{2}\,\delta(k - k')\,, \tag{1.147}$$

assuming that k and k' are both positive. The phase $\delta_{\rm as}$ in (1.146) may be a constant, it may however also contain the r-dependent Coulomb modification $\eta \ln 2kr$. If we want the scalar product between two radial wave functions to be a delta function in the wave numbers without the factor $\pi/2$ in (1.147), we must normalize them to be asymptotically proportional to $\sqrt{\frac{2}{\pi}}\,\sin(kr + \delta_{\rm as})$.

In many applications we want the wave functions to be *energy normalized*, which means

$$\langle \phi_E | \phi_{E'} \rangle = \delta(E - E')\,. \tag{1.148}$$

For $E = \hbar^2 k^2/(2\mu)$ we have

$$\delta(k - k') = \frac{\mathrm{d}E}{\mathrm{d}k}\,\delta(E - E') = \frac{\hbar^2 k}{\mu}\,\delta(E - E')\,. \tag{1.149}$$

Hence energy normalized wave functions ϕ_E can be obtained from the wave functions ϕ_k in (1.146), (1.147) by the following multiplication:

$$\phi_E(r) = \left(\frac{\pi\hbar^2 k}{2\mu}\right)^{-\frac{1}{2}} \phi_k(r) . \tag{1.150}$$

The solutions of the radial Schrödinger equation are energy normalized if they have the following asymptotic form:

$$\phi_E(r) = \sqrt{\frac{2\mu}{\pi\hbar^2 k}} \sin(kr + \delta_{\mathrm{as}}) \quad \text{for} \quad r \to \infty . \tag{1.151}$$

With (1.145) we see that the energy normalized regular Coulomb functions

$$F_l^{\mathrm{E}}(\eta, kr) = \sqrt{\frac{2\mu}{\pi\hbar^2 k}} F_l(\eta, kr) \tag{1.152}$$

converge at threshold, $E \to 0$, to the wave function (1.141), which is the limiting wave function for the renormalized bound states (1.140):

$$\lim_{n\to\infty} \phi_{n,l}^{\mathrm{E}}(r) = \phi_l^{(E=0)}(r) = \lim_{E\to 0} F_l^{\mathrm{E}}(\eta, kr) . \tag{1.153}$$

Figure 1.6 shows the renormalized bound radial eigenfunctions (1.140) and the energy normalized regular Coulomb functions (1.152) together with the limiting wave function (1.141), all for angular momentum quantum number $l = 0$.

1.4 Processes Involving Unbound States

1.4.1 Wave Packets

Stationary wave functions for unbound states are generally non-vanishing all the way to infinity in coordinate space. This is an idealization of realistic physical conditions, where the probability density should be restricted to a perhaps quite large, but nevertheless finite region, so that the total probability of a particle being anywhere can be normalized to unity,

$$\int |\psi(\boldsymbol{r}, t)|^2 \, \mathrm{d}^3 r = 1 . \tag{1.154}$$

Due to the uncertainty relation (1.34), a finite localization in coordinate space implies a non-vanishing uncertainty im momentum. For an unbound state describing, e.g., the motion of a free particle, this in turn generally implies a non-vanishing

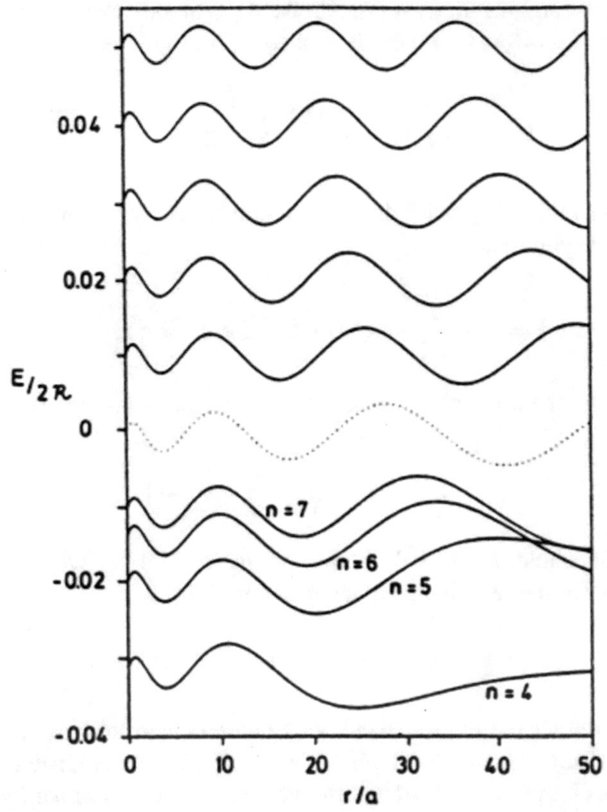

Fig. 1.6 Renormalized bound radial eigenfunctions (1.140) ($E < 0$), energy normalized regular Coulomb functions (1.152) ($E > 0$) and the limiting wave function (1.141) (*dotted line*) for $l = 0$

uncertainty in energy. The wave function $\psi(\boldsymbol{r}, t)$ is thus a superposition of many energy eigenstates—a *wave packet*—and is genuinely time dependent. The wave function for a wave packet describing a particle of mass μ moving under the influence of a (time-independent, real) potential $V(\boldsymbol{r})$ obeys the time-dependent Schrödinger equation (1.38),

$$-\frac{\hbar^2}{2\mu}\Delta\psi(\boldsymbol{r},t) + V(\boldsymbol{r})\psi(\boldsymbol{r},t) = \mathrm{i}\hbar\frac{\partial\psi(\boldsymbol{r},t)}{\partial t}\,, \tag{1.155}$$

and for the complex congugate wave function ψ^* we have

$$-\frac{\hbar^2}{2\mu}\Delta\psi^*(\boldsymbol{r},t) + V(\boldsymbol{r})\psi^*(\boldsymbol{r},t) = -\mathrm{i}\hbar\frac{\partial\psi^*(\boldsymbol{r},t)}{\partial t}\,. \tag{1.156}$$

Multiplying (1.155) by $\psi^*/(\mathrm{i}\hbar)$ and (1.156) by $\psi/(-\mathrm{i}\hbar)$ and adding the results yields,

$$\frac{\partial\rho(\boldsymbol{r},t)}{\partial t} = -\frac{\hbar}{2\mathrm{i}\mu}(\psi^*\Delta\psi - \psi\Delta\psi^*)\,, \quad \rho(\boldsymbol{r},t) = \psi^*(\boldsymbol{r},t)\psi(\boldsymbol{r},t)\,. \tag{1.157}$$

By introducing the *current density*,

$$\boldsymbol{j}(\boldsymbol{r},t) \stackrel{\text{def}}{=} \frac{\hbar}{2\mathrm{i}\mu}(\psi^*\nabla\psi - \psi\nabla\psi^*) = \Re\left\{\psi^*\left(\frac{\hat{\mathbf{p}}}{\mu}\psi\right)\right\}\,, \tag{1.158}$$

Equation (1.157) becomes a *continuity equation* connecting the time dependence of the probability density $\rho(\boldsymbol{r},t)$ with the spatial divergence $\nabla\cdot\boldsymbol{j}$ of $\boldsymbol{j}(\boldsymbol{r},t)$,

$$\nabla\cdot\boldsymbol{j}(\boldsymbol{r},t) + \frac{\partial}{\partial t}\rho(\boldsymbol{r},t) = 0\,. \tag{1.159}$$

Writing $\boldsymbol{j}$ as on the far-right-hand side of (1.158) shows up the analogy to the classical current density $\rho\,\boldsymbol{v}$ for a substance of density ρ moving with a local velocity $\boldsymbol{v}$. When integrating (1.159) over the whole of coordinate space, the contribution of the first term $\nabla\cdot\boldsymbol{j}$ vanishes, because it can be transformed to a surface integral via Gauss' Theorem, and the wave function of the localized wave packet vanishes at infinity. This implies that $\int\rho(\boldsymbol{r},t)\mathrm{d}^3r$ is time independent and thus expresses the conservation of total probability for the time-dependent wave function.

As an example consider a particle in just one spatial dimension, so the current density (1.158) and continuity equation (1.159) simplify to

$$j(x,t) = \frac{\hbar}{2\mathrm{i}\mu}\left(\psi^*\frac{\partial\psi}{\partial x} - \psi\frac{\partial\psi^*}{\partial x}\right)\,, \quad \frac{\partial j}{\partial x} + \frac{\partial\rho}{\partial t} = 0\,. \tag{1.160}$$

For a free particle with well-defined momentum $p = \hbar k$ and energy $E = \hbar^2k^2/(2\mu)$, the wave function solving the time-independent Schrödinger equation is a non-normalizable monochromatic wave,

$$\psi_k(x,t) = \phi_k(x)\,\mathrm{e}^{-(\mathrm{i}E/\hbar)t} = \frac{1}{\sqrt{2\pi}}\,\mathrm{e}^{\mathrm{i}(kx-\omega t)}\,, \quad \phi_k(x) = \frac{1}{\sqrt{2\pi}}\,\mathrm{e}^{\mathrm{i}kx}\,. \tag{1.161}$$

The wave function $\psi_k(x,t)$ propagates with the *phase velocity* $v = \omega/k$ in the direction of the positive x-axis. The parameter $\omega = E/\hbar$ defines the frequency of oscillation of the wave in time, and the wave number k defines its spatial wavelength $\lambda = 2\pi/k$, the *de Broglie wavelength*. The relation between these parameters, ω as function of k, is called the *dispersion relation*. For the present case of a free particle the dispersion relation is,

$$\omega = \frac{E}{\hbar} = \frac{\hbar k^2}{2\mu}\,. \tag{1.162}$$

With the factor $1/\sqrt{2\pi}$ in (1.161), the wave functions are normalized in the wave number k,

$$\int_{-\infty}^{\infty} \psi_k^*(x,t)\psi_{k'}(x,t)\,\mathrm{d}x = \int_{-\infty}^{\infty} \phi_k^*(x)\phi_{k'}(x)\,\mathrm{d}x = \delta(k-k')\,. \tag{1.163}$$

The probability density for the wave function (1.161) is $\rho(x) = 1/(2\pi)$ and is independent of x, as is the current density $j(x) = \hbar k/(2\pi\mu) = \upsilon\rho$, corresponding to a stationary flow of density ρ and velocity $\upsilon = p/\mu$.

A localized wave packet is described by a (normalized) wave function $\psi(x,t)$, which can be expanded in the basis of momentum eigenstates (1.161). For example, for $t = 0$,

$$\psi(x,0) = \int_{-\infty}^{\infty} \tilde{\psi}(k)\psi_k(x,0)\,\mathrm{d}k = \frac{1}{\sqrt{2\pi}}\int_{-\infty}^{\infty} \mathrm{e}^{\mathrm{i}kx}\tilde{\psi}(k)\,\mathrm{d}k\,. \tag{1.164}$$

The coefficients $\tilde{\psi}(k)$ of this expansion,

$$\tilde{\psi}(k) = \langle\psi_k(x,0)|\psi(x,0)\rangle = \frac{1}{\sqrt{2\pi}}\int_{-\infty}^{\infty} \mathrm{e}^{-\mathrm{i}kx}\psi(x,0)\,\mathrm{d}x\,, \tag{1.165}$$

constitute the *momentum representation* of the wave function $\psi(x,0)$[2]; actually $\tilde{\psi}(k)$ is just the inverse Fourier transform of $\psi(x,0)$. The time evolution of the wave packet $\psi(x,t)$ is given by the time evolution of the momentum eigenstates (1.161),

$$\psi(x,t) = \int_{-\infty}^{\infty} \tilde{\psi}(k)\psi_k(x,t)\,\mathrm{d}k = \frac{1}{\sqrt{2\pi}}\int_{-\infty}^{\infty} \mathrm{e}^{\mathrm{i}(kx-\omega t)}\tilde{\psi}(k)\,\mathrm{d}k\,. \tag{1.166}$$

For example, an initial (normalized) Gaussian wave packet

$$\psi(x,0) = (\beta\sqrt{\pi})^{-1/2}\exp\left(-\frac{(x-x_0)^2}{2\beta^2}\right)\mathrm{e}^{\mathrm{i}k_0 x} \tag{1.167}$$

---

[2] We use the term "momentum representation" when we write the wave functions as functions of momentum p or as functions of wave number $k = p/\hbar$.

is localized over a width β around the point x_0 in coordinate space and moves with a mean velocity $v_0 = \hbar k_0/\mu$. In the course of time it evolves such that

$$|\psi(x,t)|^2 = \frac{1}{b(t)\sqrt{\pi}} \exp\left(-\frac{(x-x_0-v_0 t)^2}{b(t)^2}\right), \quad b(t) = \beta\sqrt{1+\frac{\hbar^2 t^2}{\mu^2\beta^4}}, \tag{1.168}$$

i.e., the maximum of the wave packet follows the classical path $x = x_0 + v_0 t$, but the width $b(t)$ spreads with time. This *spreading* of the wave packet is a direct consequence of the fact that the different contributions to the integral in (1.166) propagate with a k-dependent phase velocity ω/k as follows from the dispersion relation (1.162). Replacing $\hbar/\mu$ by v_0/k_0 in the formula (1.168) for $b(t)$ gives: $b(t) = \beta\sqrt{1+(v_0 t)^2/(k_0\beta^2)^2}$. Spreading starts slowly, quadratically in time, and becomes appreciable when

$$v_0 t \approx \beta(k_0\beta). \tag{1.169}$$

For large times, the width of the wave packet grows linearly in time, $b(t) \overset{t\to\infty}{\sim} v_0 t/(k_0\beta)$, so that $b(t)/x(t) \overset{t\to\infty}{\sim} 1/(k_0\beta)$. The onset of spreading (1.169) and the large-time spreading rate depend on the dimensionless product $k_0\beta$. Spreading is small when $k_0\beta$ is large, meaning that the wave-number uncertainty $\Delta k = 1/(\beta\sqrt{2})$ of the initial wave packet is small compared to the mean wave number k_0—see Problem 1.4.

When a wave packet (1.166) only contains components close to a given mean wave number k_0 corresponding to a mean momentum $p_0 = \hbar k_0$, then we can expand the dispersion relation around k_0,

$$\omega(k) \approx \omega(k_0) + (k-k_0)\left.\frac{\mathrm{d}\omega}{\mathrm{d}k}\right|_{k_0}. \tag{1.170}$$

Inserting (1.170) into the expression (1.166) for the time-dependent wave packet gives

$$\psi(x,t) \approx \frac{\mathrm{e}^{-\mathrm{i}\omega_0 t}}{\sqrt{2\pi}}\int_{-\infty}^{\infty}\tilde{\psi}(k)\,\mathrm{e}^{\mathrm{i}k(x-v_g t)}\,\mathrm{d}k,$$

$$\text{with} \quad \omega_0 = \omega(k_0) - k_0\left.\frac{\mathrm{d}\omega}{\mathrm{d}k}\right|_{k_0}, \quad v_g = \left.\frac{\mathrm{d}\omega}{\mathrm{d}k}\right|_{k_0}. \tag{1.171}$$

The integral in (1.171) is a function of $x - v_g t$, so except for the oscillating phase factor $\mathrm{e}^{-\mathrm{i}\omega_0 t}$, the time evolution of the wave packet consists in propagation with the *group velocity* v_g as defined in the lower line. For the free-particle dispersion relation (1.162) we have $v_g = \hbar k_0/\mu = p_0/\mu$, as expected.

1.4.2 Transmission and Reflection

Consider a particle of mass μ moving in one spatial dimension under the influence of a (time-independent, real) potential $V(x)$. Assume that the potential has a non-trivial dependence on x in a certain "interaction region" and approaches (not necessarily equal) constant values $V_\pm$ in the asymptotic limits $x \to \pm\infty$, and that it approaches these limits faster than $1/|x|$. Then the motion of the particle approaches that of a free particle asymptotically, provided the energy is large enough, $E > V_+$, $E > V_-$. A particle incident from the left and travelling in the direction of the positive x-axis with a well-defined energy $E > V_-$ can be described by a monochromatic wave function, $\psi \overset{x\to-\infty}{\propto} \mathrm{e}^{ikx}$, with $\hbar k = \sqrt{2\mu(E - V_-)}$. The solution of the time-independent Schrödinger equation may also contain a leftward travelling contribution describing a part of the wave function reflected through the influence of the potential, $\psi \overset{x\to-\infty}{\propto} \mathrm{e}^{-ikx}$. If $E > V_+$, then the particle can also move to infinitely large distances, and the wave function may contain contributions proportional to e^{+iqx}, $\hbar q = \sqrt{2\mu(E - V_+)}$ for $x \to \infty$. If the potential approaches its asymptotic limit(s) as $1/|x|$ (e.g. for Coulombic potentials) or more slowly, then the asymptotic wave functions retain an x-dependent phase correction which does not vanish, even in the limit $|x| \to \infty$, compare (1.120) and Table 1.3 in Sect. 1.3.2.

For a potential approaching its asymptotic limits sufficiently rapidly, a particle incident from the left is described by a solution of the time-independent Schrödinger equation with the following asymptotic behaviour,

$$\psi_1(x) \overset{x\to-\infty}{=} \frac{1}{\sqrt{\upsilon_-}}\,\mathrm{e}^{ikx} + \frac{R_1}{\sqrt{\upsilon_-}}\,\mathrm{e}^{-ikx}\,, \quad \psi_1(x) \overset{x\to\infty}{=} \frac{T_1}{\sqrt{\upsilon_+}}\,\mathrm{e}^{-iqx}\,,$$
$$\upsilon_- = \frac{\hbar k}{\mu}\,, \quad k = \sqrt{2\mu(E - V_-)}\,, \quad \upsilon_+ = \frac{\hbar q}{\mu}\,, \quad q = \sqrt{2\mu(E - V_+)}\,, \tag{1.172}$$

with the *reflection amplitude* R_1 and the *transmission amplitude* T_1. The subscript "l" is to remind us that the incoming particle approaches from the left. The current density (1.160) for the contribution $\mathrm{e}^{ikx}/\sqrt{\upsilon_-}$ describing the incoming wave is $j_{\mathrm{inc}} = 1$, and for the transmitted wave at $x \to \infty$ we have $j_{\mathrm{trans}} = |T_1|^2$. For the reflected wave $j_{\mathrm{refl}} = -|R_1|^2$, where the minus sign shows that this part of the wave function describes a leftward travelling wave.

The probability P_T that the incoming particle is transmitted through the interaction region is

$$P_\mathrm{T} = \frac{j_{\mathrm{trans}}}{j_{\mathrm{inc}}} = |T_1|^2\,, \tag{1.173}$$

and the probability P_R that it is reflected is

$$P_\mathrm{R} = \frac{|j_{\mathrm{refl}}|}{j_{\mathrm{inc}}} = |R_1|^2\,. \tag{1.174}$$

When the potential $V(x)$ has a maximum which is larger than the energy E it forms a barrier, because transmission of the particle from one side of the maximum to the other side is forbidden in the framework of classical mechanics. The quantum mechanical transmission probability (1.173) need not vanish, however, because the Schrödinger equation allows non-vanishing wave functions in classically forbidden regions $V(x) > E$. Such transmission through a potential barrier—more generally, through a classically forbidden region—is called *tunnelling*. If, on the other hand, the energy E is larger than the maximum of the potential $V(x)$, then there is no turning point where the classical particle would change its direction of motion. The particle keeps its direction of motion and reflection is forbidden in the framework of classical mechanics. The quantum mechanical reflection probability (1.174) need not vanish, however, and this process of classically forbidden reflection is called *quantum reflection*, see Sect. 5.7.3 in Chap. 5.

In order to describe an incoming particle approaching from the right, we replace the asymptotic boundary conditions (1.172) by

$$\psi_{\mathrm{r}}(x) \overset{x\to-\infty}{=} \frac{T_{\mathrm{r}}}{\sqrt{v_-}}\,\mathrm{e}^{-\mathrm{i}kx}\,, \quad \psi_{\mathrm{r}}(x) \overset{x\to\infty}{=} \frac{1}{\sqrt{v_+}}\,\mathrm{e}^{-\mathrm{i}qx} + \frac{R_{\mathrm{r}}}{\sqrt{v_+}}\,\mathrm{e}^{\mathrm{i}qx}\,; \tag{1.175}$$

the subscript "r" reminds us that the incoming particle approaches from the right. Since the potential is real, the complex conjugate wave functions ψ_{l}^* and ψ_{r}^* are also solutions of the time-independent Schrödinger equation, and so is the wave function $(\psi_{\mathrm{l}}^* - R_{\mathrm{l}}^*\psi_{\mathrm{l}})/T_{\mathrm{l}}^*$—which happens to have the same asymptotic behaviour (1.175) as ψ_{r}. Comparing the amplitudes of the transmitted and reflected waves gives the *reciprocity relations*,

$$T_{\mathrm{r}} = T_{\mathrm{l}} \overset{\mathrm{def}}{=} T\,, \quad R_{\mathrm{r}} = -R_{\mathrm{l}}^*\frac{T}{T*}\,. \tag{1.176}$$

From (1.176) it immediately follows that the probabilities (1.173) for transmission and (1.174) for reflection do not depend on the side from which the incident particle approaches the interaction region.

1.4.3 Time Delays and Space Shifts

For a realistic description of transmission and reflection in the system discussed in Sect. 1.4.2, consider a wave packet which is initially ($t = 0$) localized around a large negative coordinate x_0 and approaches the interaction region with mean momentum $\hbar k_0 > 0$, e.g., the Gaussian wave packet (1.167) with $x_0 < 0$, $|x_0| \gg \beta$. The expansion coefficients $\tilde{\psi}(k)$ entering the eigenstate expansions (1.164), (1.166) are again defined according to (1.165),

$$\tilde{\psi}(k) = \langle \psi_k(x,0)|\psi(x,0)\rangle = \int_{-\infty}^{\infty} \psi_k^*(x,0)\psi(x,0)\,\mathrm{d}x\,, \tag{1.177}$$

but now the basis states $\psi_k(x,0)$ are not just the free-particle momentum eigenstates (1.161), but stationary solutions of the Schroödinger equaton including the potential $V(x)$. For a given energy $E = \hbar^2k^2/(2\mu) + V_- = \hbar^2q^2/(2\mu) + V_+$ we choose

$$\psi_k(x,0) = \sqrt{\frac{v_-}{2\pi}}\,\psi_1(x) \tag{1.178}$$

with $\psi_1(x)$ as defined in (1.172) with the appropriate wave number k. The prefactor $\sqrt{v_-/(2\pi)}$ is chosen so that the incoming-wave part of ψ_k is identical to the free-particle wave (1.161). In the following we assume that the mean momentum $\hbar k_0$ of the initial wave packet is sufficiently large and that the uncertainty Δk in the wave number is sufficiently small, so that the expansion coefficients (1.177) are only appreciable for $k > 0$, i.e. that we really only need basis functions (1.178) corresponding to a rightward travelling incoming particle. For the Gaussian wave packet (1.167) this implies $k_0 \gg \Delta k = 1/(\beta\sqrt{2})$—see Problem 1.4.

For sufficiently large values of $|x_0|$, the initial wave packet $\psi(x,0)$ is localized so far in the asymptotic region $x \to -\infty$, that only the asymptotic $x \to -\infty$ part of the basis functions (1.178) contributes to the matrix element (1.177). Furthermore, the reflected-wave part proportional to e^{-ikx} yields negligible contributions. This is because the corresponding factor e^{ikx} in $\psi_k^*(x,0)$, together with a factor e^{ik_0x} for the mean momentum of the initial wave packet [as in the Gaussian example (1.167)] produces a factor $\mathrm{e}^{\mathrm{i}(k_0+k)x}$ in the integrand on the far right-hand side of (1.177); for very large values of $|x|$ this oscillates extremely rapidly, because $k_0 + k$ is always a positive number larger than k_0, and these oscillations suppress the contributions to the integral. The incoming-wave part, on the other hand, is proportional to e^{-ikx} in $\psi_k^*(x,0)$ and together with the factor e^{ik_0x} for the mean momentum of the initial wave packet produces an exponential $\mathrm{e}^{\mathrm{i}(k_0-k)x}$ in the integrand in (1.177), and this allows for appreciable contributions to the integral when $k \approx k_0$. Consequently, the expansion coefficients (1.177) in the basis of stationary solutions with the asymptotic behaviour given by (1.172) and (1.178) are the same as those (1.165) for the free-particle basis—under the condition that the initial wave packet be localized far in the asymptotic region $x \to -\infty$ in coordinate space and in a sufficiently narrow interval around its mean momentum in momentum space. For the Gaussian example (1.167) these conditions can be formulated explicitly,

$$|x_0| \gg \beta\,,\ \frac{1}{\Delta k}\,; \qquad \frac{1}{k_0} \ll \beta\,,\ \frac{1}{\Delta k}\,. \tag{1.179}$$

For more general wave packets the conditions (1.179) still hold when we interpret β as a length of the order of the uncertainty of the initial wave packet in coordinate space.

In analogy with (1.164) the initial wave packet far to the left of the iteraction region can be written as

$$\psi(x,0) = \int_0^\infty \tilde{\psi}(k)\psi_k(x,0)\,\mathrm{d}k = \frac{1}{\sqrt{2\pi}}\int_0^\infty \tilde{\psi}(k)\,\mathrm{e}^{\mathrm{i}kx}\,\mathrm{d}k\,, \tag{1.180}$$

where $\psi_k(x,0)$ now stands for the stationary solutions defined by (1.178), (1.172). Only the incoming-wave parts of the $\psi_k(x,0)$ are relevant for the initial wave packet, and we can restrict the integration to positive k values for the reasons given above. The time evolution of this wave packet is given as in (1.166) by a factor $\mathrm{e}^{-i\omega(k)t}$ for each contribution to the integral over k,

$$\psi(x,t) = \int_0^\infty \tilde{\psi}(k)\,\psi_k(x,0)\mathrm{e}^{-\mathrm{i}\omega(k)t}\,\mathrm{d}k\,, \tag{1.181}$$

and the frequency parameter $\omega(k) = E/\hbar$ obeys

$$\omega = \frac{\hbar k^2}{2\mu} + \frac{V_-}{\hbar} = \frac{\hbar q^2}{2\mu} + \frac{V_+}{\hbar}\,. \tag{1.182}$$

Far to the right of the interaction region, $x \to \infty$, we expect contributions only from the transmitted-wave parts of the stationary basis functions (1.178), (1.172),

$$\psi_>(x,t) = \frac{1}{\sqrt{2\pi}}\int_0^\infty \sqrt{\frac{k}{q}}\,\tilde{\psi}(k)T(k)\mathrm{e}^{\mathrm{i}qx}\mathrm{e}^{-\mathrm{i}\omega(k)t}\,\mathrm{d}k\,. \tag{1.183}$$

It is helpful to decompose the transmission amplitude $T(k)$ into its modulus, which determines the transmission probability (1.173), and a phase factor,

$$T = |T|\,\mathrm{e}^{\mathrm{i}\phi_\mathrm{T}}\,. \tag{1.184}$$

If the expansion coefficients $\tilde{\psi}(k)$ are sufficiently narrowly peaked around k_0, we can replace $|T(k)|$ by $|T(k_0)|$, but because of the sensitive dependence of the integral on the phase of the integrand we include the first term of a Taylor expansion for the phase of $T(k)$,

$$\phi_\mathrm{T}(k) \approx \phi_\mathrm{T}(k_0) + (k-k_0)\left.\frac{\mathrm{d}\phi_\mathrm{T}}{\mathrm{d}k}\right|_{k_0}\,. \tag{1.185}$$

If we also expand the frequency $\omega(k)$ as in (1.170), then (1.183) becomes,

$$\psi_>(x,t) \approx T(k_0)\exp\left\{-\mathrm{i}\left[\omega(k_0)t + k_0\left(\left.\frac{\mathrm{d}\phi_\mathrm{T}}{\mathrm{d}k}\right|_{k_0} - \left.\frac{\mathrm{d}\omega}{\mathrm{d}k}\right|_{k_0} t\right)\right]\right\} \times$$
$$\frac{1}{\sqrt{2\pi}}\int_0^\infty \sqrt{\frac{k}{q}}\,\tilde{\psi}(k)\,\mathrm{e}^{\mathrm{i}qx}\exp\left[\mathrm{i}k\left(\left.\frac{\mathrm{d}\phi_\mathrm{T}}{\mathrm{d}k}\right|_{k_0} - \left.\frac{\mathrm{d}\omega}{\mathrm{d}k}\right|_{k_0} t\right)\right]\mathrm{d}k\,. \quad (1.186)$$

Let us first consider the case that the asymptotic limits V_- and V_+ of the potential are the same on both sides of the interaction region. Then $q = k$ and the lower line of (1.186) is the same as the far right-hand side of (1.180), except that $\mathrm{e}^{\mathrm{i}kx}$ is replaced by $\mathrm{e}^{\mathrm{i}k\tilde{x}(t)}$ with

$$\tilde{x}(t) = x + \left.\frac{\mathrm{d}\phi_\mathrm{T}}{\mathrm{d}k}\right|_{k_0} - \left.\frac{\mathrm{d}\omega}{\mathrm{d}k}\right|_{k_0} t\,. \quad (1.187)$$

In the upper line of (1.186), $T(k_0)$ represents the (mean) transmission amplitude for the transmitted part of the wave packet, and the exponential is an overall phase factor. The lower line represents a (normalized) wave packet with the same shape as the initial wave packet (1.180); however, it is peaked not at $x = x_0$, but at $\tilde{x} = x_0$, i.e. at

$$x = x_0 - \left.\frac{\mathrm{d}\phi_\mathrm{T}}{\mathrm{d}k}\right|_{k_0} + \left.\frac{\mathrm{d}\omega}{\mathrm{d}k}\right|_{k_0} t\,. \quad (1.188)$$

The interpretation of (1.188) is quite straightforward: the transmitted wave packet moves with the group velocity $v_\mathrm{g} = \mathrm{d}\omega/\mathrm{d}k|_{k_0} = \hbar k_0/\mu$ as follows from (1.182), but its position is shifted relative to the free particle moving with constant velocity v_g—it lags behind by a space shift x_shift,

$$x_\mathrm{shift} = \left.\frac{\mathrm{d}\phi_\mathrm{T}}{\mathrm{d}k}\right|_{k_0}\,. \quad (1.189)$$

This corresponds to a *time delay* t_delay relative to free-particle motion,

$$t_\mathrm{delay} = \frac{x_\mathrm{shift}}{v_\mathrm{g}} = \hbar\left.\frac{\mathrm{d}\phi_\mathrm{T}}{\mathrm{d}E}\right|_{E=E_0}, \quad (1.190)$$

where E_0 is the mean energy of the wave packet, $E_0 = V_- + \hbar^2(k_0)^2/(2\mu)$. If x_shift and t_delay are negative, then the transmitted wave is advanced relative to the free particle. There is nothing special about negative time delays, i.e. time gains, when they are measured relative to the motion of a free particle. An ordinary classical particle experiences a time gain when it passes through a region of negative potential energy, where it moves faster than the free particle used as reference.

The situation is a little more complicated when the asymptotic limits V_- and V_+ of the potential are different on different sides of the interaction region. The mean energy E_0 is now associated with different mean asymptotic momenta, $\hbar k_0$ to the left and $\hbar q_0$ to the right of the interaction region,

$$E_0 = \frac{\hbar^2(k_0)^2}{2\mu} + V_- = \frac{\hbar^2(q_0)^2}{2\mu} + V_+ . \tag{1.191}$$

We assume that V_+ is either smaller or not too much larger than V_-, so that E_0 is well above the transmission threshold $E = V_+$. Since the expansion coefficients $\tilde{\psi}(k)$ are appreciable only in a narrow interval of k values around k_0, corresponding to a narrow range of q values around $q(k_0) = q_0$, we can approximate $\sqrt{k/q}$ in the integral in the lower line of (1.186) as $\sqrt{k_0/q_0}$, and we approximate the oscillating exponential in the integral using $q(k) \approx q_0 + (k - k_0)k_0/q_0$,

$$\mathrm{e}^{\mathrm{i}qx} \approx \exp\left(\mathrm{i}q_0\left[1 - \left(\frac{k_0}{q_0}\right)^2\right]x\right)\exp\left(\mathrm{i}k\,\frac{k_0}{q_0}\,x\right) . \tag{1.192}$$

The first exponential on the right-hand side of (1.192) is independent of k and just adds to the overall phase of $\psi_>(x,t)$. The second exponential has the form e^{iky} for the scaled variable,

$$y = \frac{k_0}{q_0}x . \tag{1.193}$$

The lower line of (1.186) now represents a wave packet with the same shape as the initial wave packet (1.180), but only when it is considered as a function of y. It is peaked at

$$y = x_0 - \left.\frac{\mathrm{d}\phi_\mathrm{T}}{\mathrm{d}k}\right|_{k_0} + \left.\frac{\mathrm{d}\omega}{\mathrm{d}k}\right|_{k_0} t \quad \text{or} \qquad x = \frac{q_0}{k_0}\left(x_0 - \left.\frac{\mathrm{d}\phi_\mathrm{T}}{\mathrm{d}k}\right|_{k_0} + \left.\frac{\mathrm{d}\omega}{\mathrm{d}k}\right|_{k_0} t\right) . \tag{1.194}$$

The peak of the transmitted wave thus moves with the velocity

$$v_\mathrm{g}^{(\mathrm{trans})} = \frac{q_0}{k_0}\left.\frac{\mathrm{d}\omega}{\mathrm{d}k}\right|_{k_0} = \left.\frac{\mathrm{d}\omega}{\mathrm{d}q}\right|_{q_0} = \frac{\hbar q_0}{\mu} . \tag{1.195}$$

If, for example, $V_+ < V_-$, then $q_0 > k_0$ and the transmitted particle moves faster than the incoming particle. The transmitted wave packet is stretched by a factor q_0/k_0 (along the x-axis, but it remains normalized due to the factor $\sqrt{k/q} \approx \sqrt{k_0/q_0}$ in the lower line of (1.186). The norm of the whole transmitted wave packet (1.186) is $|T(k_0)|^2$. A particle moving from $x_0 < 0$ with constant velocity v_g arrives at the origin $x = 0$ at time $t_0 = -x_0/v_\mathrm{g}$. If it continues beyond $x = 0$ with the constant velocity $v_\mathrm{g}{}^{(\mathrm{trans})} = v_\mathrm{g}\,q_0/k_0$, then its position at time $t > t_0$ is given by

$x = v_g^{(\text{trans})}(t - t_0) = (x_0 + v_g t)q_0/k_0$. This is just what (1.194) predicts if we ignore the term involving the phase ϕ_T of the transmission amplitude. The term involving the phase describes the space shift

$$x_{\text{shift}} = \frac{q_0}{k_0}\left(\left.\frac{d\phi_T}{dk}\right|_{k_0}\right) = \left.\frac{d\phi_T}{dq}\right|_{q_0} \tag{1.196}$$

and the related time delay,

$$t_{\text{delay}} = \frac{x_{\text{shift}}}{v_g^{(\text{trans})}} = \hbar \left.\frac{d\phi_T}{dE}\right|_{E_0}, \tag{1.197}$$

relative to a particle moving with constant velocity $v_g = \hbar k_0$ from $x = x_0 < 0$ to $x = 0$ and continuing on with constant velocity $v_g^{(\text{trans})} = \hbar q_0$.

Far to the left of the interaction region, $x \to -\infty$, we expect contributions from the incoming- and the reflected-wave parts of the stationary basis functions (1.178), (1.172),

$$\begin{aligned}\psi_<(x,t) = &\frac{1}{\sqrt{2\pi}}\int_0^\infty \tilde{\psi}(k)e^{ikx}e^{-i\omega(k)t}\,dk \\ &+\frac{1}{\sqrt{2\pi}}\int_0^\infty \tilde{\psi}(k)R_l\,e^{-ikx}e^{-i\omega(k)t}\,dk\,.\end{aligned} \tag{1.198}$$

With the arguments used above for deriving (1.188) we conclude that the upper line of (1.198) would contribute a wave packet centred around $x = x_0 + v_g t$ and can be neglected in the regime of negative x values at large times t. For large times, only the lower line of (1.198) gives contributions to the left of the interaction region and they describe the reflected wave packet moving in the direction of the negative x axis. This reflected wave packet has the same shape as the incoming wave packet when considered as a function of $-x$ i.e. its shape is reflected in coordinate space, and it is peaked around

$$x = -x_0 + \left.\frac{d\phi_R}{dk}\right|_{k_0} - \left.\frac{d\omega}{dk}\right|_{k_0} t\,. \tag{1.199}$$

Here ϕ_R stands for the phase of the reflection amplitude,

$$R_l(k) = |R_l(k)|\,e^{i\phi_R}\,. \tag{1.200}$$

The reflected wave packet travels with the group velocity $-v_g = -\hbar k_0/\mu$. The coordinate of a free particle starting with velocity $v_g > 0$ at $x_0 < 0$ and returning to negative x values after being elastically reflected at $x = 0$ would be given by $x = -x_0 - v_g t$. The second term on the right-hand side of (1.199) thus represents a space shift relative to the reflection of a free particle at $x = 0$. The reflected wave packet lags behind by a distance

$$x_{\text{shift}} = \left.\frac{\mathrm{d}\phi_{\mathrm{R}}}{\mathrm{d}k}\right|_{k_0}, \tag{1.201}$$

corresponding to a time delay,

$$t_{\text{delay}} = \frac{x_{\text{shift}}}{v_{\mathrm{g}}} = \hbar \left.\frac{\mathrm{d}\phi_{\mathrm{R}}}{\mathrm{d}E}\right|_{E=E_0}. \tag{1.202}$$

For wave packet incident from the left, the time evolution of the reflected wave packet corresponds that of a free particle reflected not at $x = 0$ but at $x = x_{\text{shift}}/2$.

The results derived in this section are based on approximations justified by the assumption that the initial wave packet is sufficiently narrowly localized in momentum, as expressed in (1.179). These approximations are already too crude to account for the spreading of the wave packet as discussed in the Gaussian example above, (1.167)–(1.169). For general wave packets which may be strongly localized in coordinate space and widely spread in momentum, the issue of time becomes quite complicated. The basic problem is, that wave packets generally don't keep their shape in the course of time. They can reshape and/or break up into many components, and naive time definitions based, e.g., on the motion of the absolute or a relative maximum of the probability density or on its centre of mass don't lead to generally consistent results. This is an old topic which has been receiving renewed attention for several years. One school of thought is to define an operator for time as a physical observable and derive the times for quantum tunnelling and/or reflection via eigenvalues or expectation values of such operators [BK95, OR04]. An alternative and perhaps more natural approach is to accept time as a mere parameter in the time-dependent Schrödinger equation and to directly study the behaviour of its wave-packet solutions [EK87, Kle94, CN02]. A detailed discussion of time in the context of tunnelling is given in Chaps. 17 to 19 of [Raz03], and a rather comprehensive summary of the many questions associated with the general problem of time in quantum mechanics is contained in [MS02]. Notwithstanding these reservations it is worth mentioning, that in the limit of almost monochromatic wave packets discussed above, the concept of time delays (or gains) defined via the derivative of the phase of the transmission or reflection amplitude is well defined and unambiguous. A similar treatment of time delays was first discussed by Eisenbud and Wigner [Wig55] in the context of particle scattering.

1.5 Resonances and Channels

Resonances appear above the continuum threshold at energies where a bound state might have occurred, meaning that a slight modification of the Hamiltonian would have led to a bound state. In a one-dimensional potential, resonances can typically occur if almost bound states in the inner region are shielded from the outer region by a potential barrier (see Sect. 1.5.3). In systems with several degrees of freedom resonances often occur when a bound motion in one degree of freedom couples weakly to and can decay into unbound motion in another degree of freedom. These so-called *Feshbach resonances* are best described in the picture of *coupled channels*. The concept of channels is of a very fundamental importance and is introduced in a general way in the following subsection.

1.5.1 Channels

Consider a physical system whose wave functions $\psi(X, Y)$ depend on two sets X and Y of variables. Let $\hat{O}$ be an observable which only acts on functions of the variable Y, i.e. for a product wave function $\psi(X)\phi(Y)$ we have

$$\hat{O}\psi(X)\phi(Y) = \psi(X)\hat{O}\phi(Y)\,. \tag{1.203}$$

The eigenvalue problem for $\hat{O}$ is

$$\hat{O}\phi_n = \omega_n\phi_n \tag{1.204}$$

and defines a complete set of eigenfunctions $\phi_n(Y)$. $\hat{O}$ can stand for a whole set of observables; ω_n then stands for the corresponding set of eigenvalues.

If $\hat{O}$ commutes with the Hamiltonian $\hat{H}$, then the problem of solving the full Schrödinger equation can be reduced to the solution of a reduced Schrödinger equation for each eigenvalue ω_n of $\hat{O}$. Each eigenfunction $\phi_n(Y)$ of $\hat{O}$—more precisely: each eigenvalue ω_n, which is not the same in the degenerate case—defines a *channel*, and the dynamics of the reduced problem in the variable X in a given channel is not coupled to the motion in the other channels.

Coupling of channels occurs if $\hat{O}$ does not commute with $\hat{H}$. Since the functions $\phi_n(Y)$ form a complete basis in the space of all functions of Y, we can expand any wave function $\psi(X, Y)$ of the whole system in this basis:

$$\psi(X, Y) = \sum_n \psi_n(X)\phi_n(Y)\,. \tag{1.205}$$

The functions $\psi_n(X)$ are the *channel wave functions* which are to be determined by solving the Schrödinger equation. Inserting the ansatz (1.205) into the time-independent Schrödinger equation leads to

$$\sum_n \hat{H}\,\psi_n(X)\phi_n(Y) = E\sum_n \psi_n(X)\phi_n(Y)\,. \tag{1.206}$$

Multiplying from the left by $\phi_m^*(Y)$ and integrating over Y yields the *coupled-channel equations* in their most general form:

$$\hat{H}_{m,m}\psi_m(X) + \sum_{n\neq m} \hat{H}_{m,n}\psi_n(X) = E\,\psi_m(X)\,. \tag{1.207}$$

The *diagonal Hamiltonians* $\hat{H}_{m,m}$ and the *coupling operators* $\hat{H}_{m,n}$, $m \neq n$, are *reduced operators* which act only in the space of wave functions $\psi(X)$. They are defined through the eigenfunctions $\phi_n(Y)$,

$$\hat{H}_{m,n} = \langle\phi_m|\hat{H}|\phi_n\rangle_Y\,, \tag{1.208}$$

where the subscript Y on the bracket indicates integration (and/or summation) over the variable Y alone.

The coupled-channel equations (1.207) are particularly useful if the diagonal operators $\hat{H}_{m,m}$ play a dominant role, while the coupling operators $\hat{H}_{m,n}$, $m \neq n$, are "small". This happens if the operator $\hat{O}$ commutes with a dominant part of the Hamiltonian which then doesn't contribute to the coupling operators. It is also helpful if symmetry considerations restrict the number of channels coupling to each other to a finite and preferably small number, or if the expansion (1.205) can be terminated after a small number of terms on physical grounds.

For further insights let us define the situation more precisely. Assume for example, that H consists of the operators $\hat{H}_X$ and $\hat{H}_Y$, which act only on functions of X and Y respectively, together with a simple coupling potential given by the function $V(X, Y)$:

$$\hat{H} = \hat{H}_X + \hat{H}_Y + V(X, Y)\,. \tag{1.209}$$

The eigenfunctions $\phi_n(Y)$ of $\hat{H}_Y$ may be used to define channels. The diagonal Hamiltonians of the coupled-channel equations are

$$\hat{H}_{m,m} = \hat{H}_X + \langle\phi_m|\hat{H}_Y|\phi_m\rangle_Y + \langle\phi_m|V(X,Y)|\phi_m\rangle_Y\,, \tag{1.210}$$

and the coupling operators form a matrix of potentials:

$$\hat{H}_{m,n} = V_{m,n}(X) = \int \mathrm{d}Y\phi_m^*(Y)V(X,Y)\,\phi_n(Y)\,, \quad m \neq n\,. \tag{1.211}$$

The diagonal Hamiltonians (1.210) contain the operator $\hat{H}_X$, which is the same in all channels, and an additional channel-dependent potential

$$V_{m,m}(X) = \int |\phi_m(Y)|^2 V(X,Y)\,\mathrm{d}Y \tag{1.212}$$

as well as a constant energy

$$E_m = \langle \phi_m | \hat{H}_Y | \phi_m \rangle_Y , \tag{1.213}$$

corresponding to the internal energy of the Y variables in the respective channels.

To be even more precise let us assume that $\psi(X, Y)$ describes a point particle of mass μ moving in an effective radial potential $V_{\text{eff}}(r)$ and interacting with a number of other bound particles. Our ansatz for $\psi(X, Y)$ is

$$\psi = \sum_{n,l,m} \frac{\phi_{n,l,m}(r)}{r} Y_{l,m}(\theta, \phi)\, \chi_n , \tag{1.214}$$

where χ_n are the bound states of the other particles. Now X is the radial coordinate r and Y stands for the angular variables (θ, ϕ) of the point particle as well as all other degrees of freedom. The coupled-channel equations now have the form

$$\left(-\frac{\hbar^2}{2\mu} \frac{\mathrm{d}^2}{\mathrm{d}r^2} + V_{\text{eff}}(r) + V_{k,k}(r) + E_k \right) \phi_k(r) + \sum_{k' \neq k} V_{k,k'}(r) \phi_{k'}(r) = E \phi_k(r) , \tag{1.215}$$

and the channel index k covers the angular momentum quantum numbers of the point particle and all other quantum numbers of the other degrees of freedom.

If the coupling potentials vanish asymptotically ($r \to \infty$) we can distinguish between *closed* and *open* channels of the system. In closed channels the motion is bound and the channel wave functions $\phi_k(r)$ vanish asymptotically. In open channels the motion is unbound and the channel wave functions oscillate asymptotically. Assuming that the effective potential $V_{\text{eff}}(r)$ and the additional potentials $V_{k,k}(r)$ vanish asymptotically, the open channels at a given energy E of the whole system are those whose internal energy E_k is smaller than E, whilst channels with $E_k > E$ are closed. The internal energies E_k define the *channel thresholds*, above which the channel wave functions $\phi_k(r)$ in the respective channels have the properties of continuum wave functions. Bound states of the whole system and discrete energy eigenvalues occur only if all channels are closed. Thus the continuum threshold of the whole system is identical to the lowest channel threshold. For energies at which at least one channel is open, there is always a solution of the coupled channel equations. Figure 1.7 schematically illustrates a typical set of diagonal channel potentials

$$V_k(r) = V_{\text{eff}}(r) + V_{k,k}(r) + E_k , \tag{1.216}$$

as they occur in (1.215). Physical examples for systems of coupled channels are discussed in Sect. 3.3.

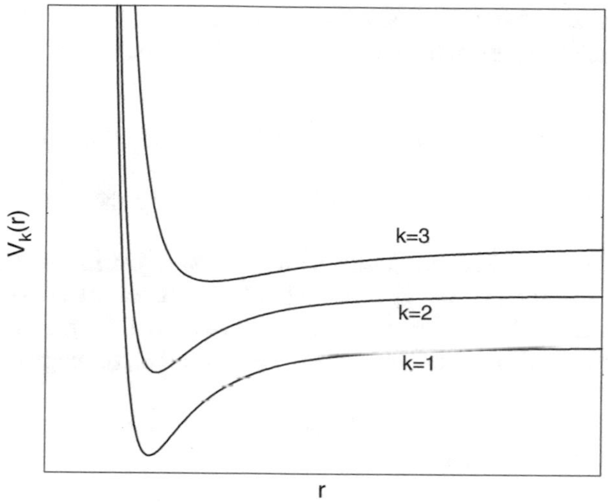

Fig. 1.7 Schematic illustration of diagonal potentials (1.216) in a system of coupled channels

1.5.2 Feshbach Resonances

For the simplest example of a Feshbach resonance consider a system of two coupled channels described by the following coupled-channel equations:

$$\left(-\frac{\hbar^2}{2\mu}\frac{\mathrm{d}^2}{\mathrm{d}r^2} + V_1(r)\right)\phi_1(r) + V_{1,2}(r)\phi_2(r) = E\phi_1(r)\,,$$
$$\left(-\frac{\hbar^2}{2\mu}\frac{\mathrm{d}^2}{\mathrm{d}r^2} + V_2(r)\right)\phi_2(r) + V_{2,1}(r)\phi_1(r) = E\phi_2(r)\,. \tag{1.217}$$

For real potentials we must require that $V_{1,2}(r) = V_{2,1}(r)$ if the two-channel Hamiltonian is to be Hermitian. Let's assume that channel 1 is open and channel 2 is closed, and that the energy scale is such that the channel threshold E_1 of the open channel lies at $E = 0$.

An almost bound state, i.e. a resonance, tends to occur near an energy at which there would be a bound state in the closed channel 2 if channel coupling were switched off. Let $\phi_0(r)$ be the wave function of such a bound state in uncoupled channel 2:

$$\left(-\frac{\hbar^2}{2\mu}\frac{\mathrm{d}^2}{\mathrm{d}r^2} + V_2(r)\right)\phi_0(r) = E_0\phi_0(r)\,. \tag{1.218}$$

The existence of such a bound state ϕ_0 has a dramatic influence on the solutions of the coupled equations (1.217) in the vicinity of the energy E_0. To see this we restrict our space of two-channel wave functions by assuming that the wave function

$\phi_2(r)$ in the closed channel 2 is simply a multiple $A\phi_0(r)$ of the bound-state wave function $\phi_0(r)$. Then the coupled equations (1.217) can be rewritten as

$$\left(E + \frac{\hbar^2}{2\mu}\frac{\mathrm{d}^2}{\mathrm{d}r^2} - V_1(r)\right)\phi_1(r) = A\,V_{1,2}(r)\phi_0(r)\,,$$
$$A(E - E_0) = \langle\phi_0|V_{2,1}|\phi_1\rangle\,. \tag{1.219}$$

Actually, exploiting (1.218) and inserting $A\phi_0$ for ϕ_2 in the lower equation (1.217) leads to $A(E - E_0)\phi_0(r) = V_{2,1}(r)\phi_1(r)$, which cannot, of course, hold in the space of arbitrary closed-channel wave functions; but in the restricted space mentioned above, all that counts is the projection onto ϕ_0, as given in the lower equation (1.219).

The upper equation (1.219) can be solved using the *Green's function* $G(r, r')$, which is defined by the relation

$$\left(E + \frac{\hbar^2}{2\mu}\frac{\mathrm{d}^2}{\mathrm{d}r^2} - V_1(r)\right)G(r, r') = \delta(r - r')\,. \tag{1.220}$$

It is immediately obvious that the wave function

$$\begin{aligned}\phi_1(r) &= \phi_{\mathrm{reg}} + A\hat{G}V_{1,2}\phi_0\\ &= \phi_{\mathrm{reg}}(r) + A\int_0^\infty G(r, r')V_{1,2}(r')\phi_0(r')\mathrm{d}r'\end{aligned} \tag{1.221}$$

is a solution of the upper equation (1.219), if $\phi_{\mathrm{reg}}(r)$ is a solution of the corresponding *homogeneous equation*:

$$\left(E + \frac{\hbar^2}{2\mu}\frac{\mathrm{d}^2}{\mathrm{d}r^2} - V_1(r)\right)\phi_{\mathrm{reg}}(r) = 0\,. \tag{1.222}$$

We take ϕ_{reg} to be the regular solution which vanishes at $r = 0$; then ϕ_1 in (1.221) also fulfills this boundary condition (see (1.228) below). If ϕ_{reg} is energy normalized, then its asymptotic form is (cf. (1.151))

$$\phi_{\mathrm{reg}}(r) = \sqrt{\frac{2\mu}{\pi\hbar^2 k}}\sin(kr + \delta_{\mathrm{bg}})\,, \quad r \to \infty\,. \tag{1.223}$$

δ_{bg} is a *background phase shift*, which originates mainly from the diagonal potential $V_1(r)$ and usually depends only weakly on the energy $E = \hbar^2k^2/(2\mu)$. If $V_1(r)$ contains a very-long-ranged Coulomb contribution, then δ_{bg} will contain the usual r-dependent Coulomb term (see Table 1.3 in Sect. 1.3.2).

By inserting the solution (1.221) for $\phi_1(r)$, the lower equation (1.219) becomes

$$A(E - E_0) = \langle\phi_0|V_{2,1}|\phi_{\mathrm{reg}}\rangle + A\langle\phi_0|V_{2,1}\hat{G}V_{1,2}|\phi_0\rangle\,. \tag{1.224}$$

Resolving for the coefficient A gives the explicit expression

$$A = \frac{\langle \phi_0 | V_{2,1} | \phi_{\rm reg} \rangle}{E - E_0 - \langle \phi_0 | V_{2,1} \hat{G} V_{1,2} | \phi_0 \rangle} . \tag{1.225}$$

The matrix element in the denominator is the double integral

$$\langle \phi_0 | V_{2,1} \hat{G} V_{1,2} | \phi_0 \rangle = \int_0^\infty {\rm d}r \int_0^\infty {\rm d}r' \, \phi_0^*(r) V_{2,1}(r) G(r, r') V_{1,2}(r') \phi_0(r') . \tag{1.226}$$

For a given diagonal potential $V_1(r)$ in the open channel 1 we can express the Green's function $G(r, r')$ through the regular solution $\phi_{\rm reg}$ of the homogeneous equation (1.222) and the corresponding irregular solution which behaves like a cosine asymptotically,

$$\phi_{\rm irr}(r) = \sqrt{\frac{2\mu}{\pi \hbar^2 k}} \cos(kr + \delta_{\rm bg}) , \quad r \to \infty . \tag{1.227}$$

The Green's function is (see Problem 1.5)

$$G(r, r') = -\pi \begin{cases} \phi_{\rm reg}(r) \phi_{\rm irr}(r') & \text{for } r \le r' , \\ \phi_{\rm reg}(r') \phi_{\rm irr}(r) & \text{for } r' \le r . \end{cases} \tag{1.228}$$

For sufficiently large values of r we can assume that the variable r' in the integral in (1.221) is always smaller than r, because $\phi_0(r')$ is a bound wave function so that the integrand vanishes for large r'. Hence we can insert the lower line of (1.228) for $G(r, r')$ and perform the integration over r'. With (1.225) this leads to the following asymptotic form of $\phi_1(r)$:

$$\begin{aligned} \phi_1(r) &= \phi_{\rm reg}(r) + \tan\delta \, \phi_{\rm irr}(r) \\ &= \frac{1}{\cos\delta} \sqrt{\frac{2\mu}{\pi \hbar^2 k}} \sin(kr + \delta_{\rm bg} + \delta) , \quad r \to \infty , \end{aligned} \tag{1.229}$$

and the angle δ ist given by

$$\tan\delta = -\pi \frac{\left| \langle \phi_0 | V_{2,1} | \phi_{\rm reg} \rangle \right|^2}{E - E_0 - \langle \phi_0 \left| V_{2,1} \hat{G} V_{1,2} \right| \phi_0 \rangle} . \tag{1.230}$$

Being solutions of a homogeneous system of differential equations, the two-channel wave functions are determined only to within multiplication by a common arbitrary constant. To obtain a continuum wave function in channel 1 which is energy normalized, we should multiply the wave function ϕ_1 of (1.229)—and simultaneously the corresponding wave function $A\phi_0$ in channel 2—by cos δ. Then the whole two-channel wave function is also energy normalized, because the normalization integrals are dominantly given by the divergent contribution of the open-channel wave function.

Coupling the bound state $\phi_0(r)$ in the closed channel 2 to the open channel 1 leads to an *additional asymptotic phase shift* δ in the open-channel wave function (1.229). This additional phase shift characterizes the resonance. Its energy dependence is determined by the position E_0 of the bound state in the uncoupled closed channel and the matrix elements

$$\langle\phi_0\left|V_{2,1}\hat{G}V_{1,2}\right|\phi_0\rangle \stackrel{\text{def}}{=} \Delta \tag{1.231}$$

and

$$2\pi\left|\langle\phi_0\left|V_{2,1}\right|\phi_{\text{reg}}\rangle\right|^2 \stackrel{\text{def}}{=} \Gamma\,. \tag{1.232}$$

The matrix elements (1.231), (1.232) are actually energy-dependent, because ϕ_{reg} and the Green's function G depend on E, but this energy dependence is insignificant compared with the energy dependence resulting from the pole structure of the formula (1.230) for $\tan\delta$. The position of the pole, i.e. the zero of the denominator, defines the *position of the resonance*, E_{R}:

$$E_{\text{R}} = E_0 + \Delta = E_0 + \langle\phi_0\left|V_{2,1}\hat{G}V_{1,2}\right|\phi_0\rangle\,. \tag{1.233}$$

It differs from the energy E_0 of the uncoupled bound state in the closed channel 2 by the *shift* Δ. Around the resonance energy E_{R} the phase δ rises more or less suddenly by π. The *width of the resonance* is Γ as defined in (1.232); at $E = E_{\text{R}} - \Gamma/2$ and $E = E_{\text{R}} + \Gamma/2$ the phase has risen by 1/4 and 3/4 of π respectively. The function

$$\delta = -\arctan\left(\frac{\Gamma/2}{E - E_{\text{R}}}\right) \tag{1.234}$$

is illustrated for constant values of the parameters E_{R} and Γ in Fig. 1.8. An isolated resonance which is described by an additional asymptotic phase shift as in (1.234) is called a *Breit-Wigner resonance*.

The derivative of the phase shift (1.234) with respect to energy is

$$\frac{\mathrm{d}\delta}{\mathrm{d}E} = \frac{\Gamma/2}{(E - E_{\text{R}})^2 + (\Gamma/2)^2} \tag{1.235}$$

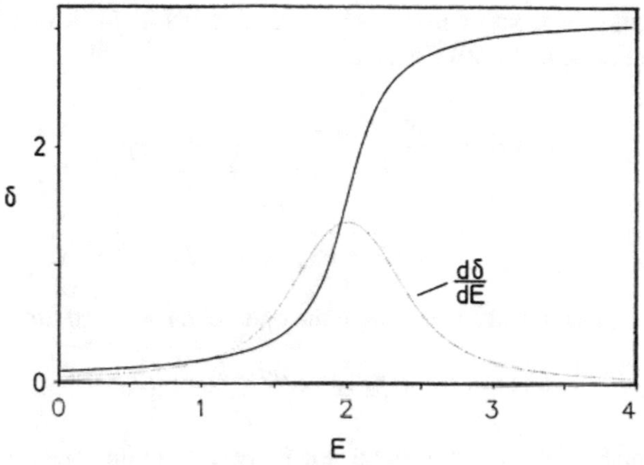

Fig. 1.8 The *solid line* shows the additional asymptotic phase shift $\delta(E)$ (without background phase shift) near an isolated Breit-Wigner resonance at $E = E_R = 2.0$ with a width $\Gamma = 0.4$ (see (1.234)). The *dotted line* is the derivative (1.235)

and has a maximum at the resonance energy E_R. According to (1.235), the width Γ is related to the maximum derivative by

$$\Gamma = 2\left(\left.\frac{d\delta}{dE}\right|_{E=E_R}\right)^{-1}. \tag{1.236}$$

In general a resonance appears as a jump in the phase shift which need not, however, have precisely the form of the Breit-Wigner resonance (1.234). In the general case, the point of maximum gradient $d\delta/dE$ serves as definition for the position E_R of the resonance, and the width can be defined via (1.236). Determining the position and width of a resonance is usually no problem as long as the resonance is so narrow that the matrix elements (1.231), (1.232) and also the background phase shift δ_{bg} can be regarded as constants over the whole width of the resonance. For a broader resonance, however, the unique definition of its position and width can become a difficult problem (see also Sect. 1.5.3).

The derivative of the phase shift with respect to energy is also a measure for the strength of the closed-channel component in the solution of the coupled-channel equations. Assuming energy normalized solutions of the coupled-channel equations (1.217) or rather (1.219), the channel wave function ϕ_2 in the closed channel 2 is

$$\phi_2(r) = A\cos\delta\,\phi_0(r)\,, \tag{1.237}$$

where the factor $\cos\delta$ stems from the energy normalization of the open-channel wave function, as explained above in the paragraph following (1.230). The strength of the closed-channel admixture is quantitatively given by the square of the amplitude $A\cos\delta$ in front of the (bound) wave function ϕ_0, which is normalized to unity. With (1.225), (1.230) we have

$$|A\cos\delta|^2 = \frac{|\langle\phi_0|V_{2,1}|\phi_{\text{reg}}\rangle|^2}{(E-E_{\text{R}})^2}\,\frac{1}{1+\tan^2\delta} = \frac{1}{\pi}\,\frac{\Gamma/2}{(E-E_{\text{R}})^2+(\Gamma/2)} = \frac{1}{\pi}\frac{\mathrm{d}\delta}{\mathrm{d}E}\,. \tag{1.238}$$

If we decompose the sine function in the open-channel wave function (1.229) as

$$\sin(kr+\delta_{\text{bg}}+\delta) \propto -\mathrm{e}^{2\mathrm{i}(\delta_{\text{bg}}+\delta)}\mathrm{e}^{\mathrm{i}kr}+\mathrm{e}^{-\mathrm{i}kr}\,, \tag{1.239}$$

then the second term on the right-hand side represents an incoming monochromatic wave and the first term an outgoing, reflected wave with the reflection amplitude –exp $[2\mathrm{i}(\delta_{\text{bg}}+\delta)]$. For wave packets narrowly localized in momentum, the energy dependence of the phase, $\phi = \pi + 2(\delta_{\text{bg}}+\delta)$, of this reflection amplitude defines the time delay of the reflected wave packet relative to a free particle reflected at $r=0$, as formulated in (1.202). Assuming an essentially energy-independent background phase shift δ_{bg} and the Breit-Wigner form (1.234), (1.235) for the energy-dependent part δ gives

$$t_{\text{delay}}(E) = \hbar\frac{\mathrm{d}\phi}{\mathrm{d}E} = 2\hbar\frac{\mathrm{d}\delta}{\mathrm{d}E} = \frac{\hbar\Gamma}{(E-E_{\text{R}})^2+(\Gamma/2)^2}\,. \tag{1.240}$$

For the formula (1.240) to be valid, the energy spread of the wave packet localized around E should be small compared to the width Γ of the resonance. The formula describes the time delay of an almost monochromatic wave packet incident with mean energy near the resonance energy E_{R}. The time delay has its maximum value when the mean energy E of the wave packet coincides with the resonance energy, $t_{\text{delay}}(E_{\text{R}}) = 4\hbar/\Gamma$, and it decreases with increasing detuning from E_{R}.

1.5.3 Potential Resonances

Another important situation which can lead to resonances occurs when a potential barrier separates the inner region of small separations r from the outer region of large r. Such potential barriers can result from the superposition of an attractive short-ranged potential and the repulsive centrifugal potential. As an example we study the potential

$$V(r) = -V_0\,\mathrm{e}^{-r^2/\beta^2} + \frac{l(l+1)\hbar^2}{2\mu r^2}\,, \tag{1.241}$$

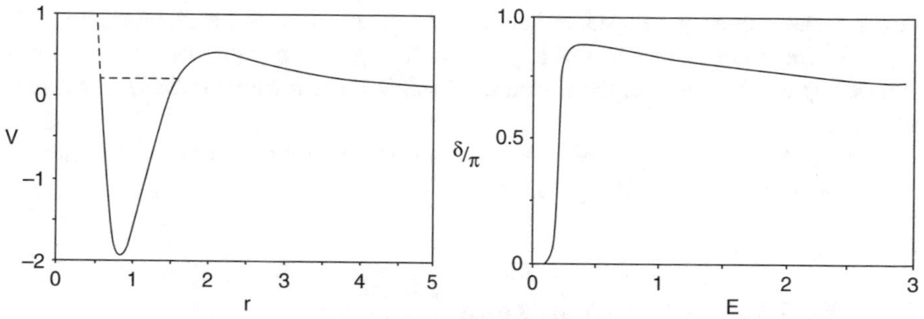

Fig. 1.9 The left half shows the potential (1.241) for angular momentum quantum number $l = 2$, $V_0 = 12.5$ und $\beta = 1.0$ ($\hbar^2/\mu = 1$). The right half shows the phase shift $\delta_{l=2}$ of the wave function (1.117) as a function of the energy E. The maximum of the gradient $d\delta/dE$ is at $E_R = 0.21$ and the width of the resonance according to (1.236) is $\Gamma \approx 0.03$

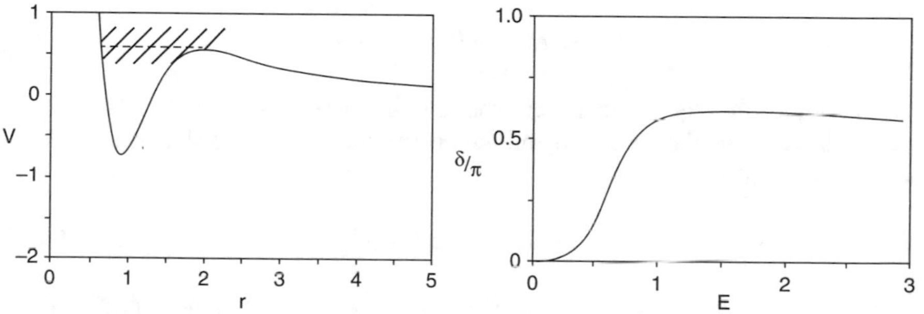

Fig. 1.10 The same as Fig. 1.9 for $V_0 = 10.0$. The maximum gradient of the phase shift is at $E_R = 0.6$ and the width of the resonance according to (1.236) is $\Gamma \approx 0.5$

which is illustrated in Figs. 1.9 and 1.10 for angular momentum quantum number $l = 2$ and two different potential strengths V_0. In Fig. 1.9 there is a resonance just above the continuum threshold and well below the maximum of the barrier. It appears as a jump of the phase shift $\delta_{l=2}$ by a little less than π. In Fig. 1.10 the potential is less attractive and the resonance lies close to the maximum of the barrier. The phase shift now jumps by appreciably less than π, but there is a point of maximum gradient and the width of the resonance can be defined via (1.236).

For a Feshbach resonance (see Sect. 1.5.2), the background phase shift due to the potential in the open channel and the additional phase shift resulting from the coupling to the bound state in the closed channel add up to give the total phase shift $\delta_{bg} + \delta$ (see (1.229)). If the energy dependence of the background phase shift and the coupling matrices is negligible and if the resonance is isolated (i.e. the width of the resonance should be smaller than the distance in energy to neighbouring resonances), then the jump of the phase shift is well described by the arctan form of the Breit-Wigner resonance. For potential resonances such as those shown in Figs. 1.9 and 1.10 it is not so straightforward to decompose the total phase shift into a weakly energy-dependent background phase shift and a resonant part. As Fig. 1.10 illustrates, the jump of a phase shift around a broad potential resonance can be appreciably smaller than what the Breit-Wigner formula (1.234) would lead to expect.

More extensive examples of potential resonances, also called *shape resonances*, are given in Sect. 4.1.9, see Figs. 4.6 and 4.7.

1.6 Methods of Approximation

1.6.1 *Time-Independent Perturbation Theory*

We are often looking for eigenvalues and eigenstates of a Hamiltonian

$$\hat{H} = \hat{H}_0 + \lambda\hat{W}\,, \quad \lambda \text{ small}\,, \tag{1.242}$$

which only differs by a "small perturbation" $\lambda\hat{W}$ from a simpler Hamiltonian $\hat{H}_0$ of which we know the spectrum and the eigenstates (which we shall assume to be normalized to unity):

$$\hat{H}_0|\psi_n^{(0)}\rangle = E_n^{(0)}|\psi_n^{(0)}\rangle\,. \tag{1.243}$$

In order to define an ordered sequence of increasingly accurate approximations of the eigenstates $|\psi_n\rangle$ of $\hat{H}$, we expand these in powers of the small parameter λ:

$$|\psi_n\rangle = |\psi_n^{(0)}\rangle + |\lambda\psi_n^{(1)}\rangle + |\lambda^2\,\psi_n^{(2)}\rangle + \dots\,. \tag{1.244}$$

Similarly for the eigenvalues E_n of $\hat{H}$:

$$E_n = E_n^{(0)} + \lambda E_n^{(1)} + \lambda^2\,E_n^{(2)} + \dots\,. \tag{1.245}$$

Inserting (1.244), (1.245) into the time-independent Schrödinger equation,

$$\begin{aligned}&(\hat{H}_0 + \lambda\hat{W})(|\psi_n^{(0)}\rangle + |\lambda\psi_n^{(1)}\rangle + \dots)\\ &= (E_n^{(0)} + \lambda E_n^{(1)} + \lambda^2\,E_n^{(2)} + \dots)(|\psi_n^{(0)}\rangle + |\lambda\psi_n^{(1)}\rangle + \dots)\,,\end{aligned} \tag{1.246}$$

and collecting powers of λ yields a hierarchy of approximations. In zeroth order we retrieve the unperturbed eigenvalue (1.243). In first order we have

$$\hat{H}_0|\lambda\psi_n^{(1)}\rangle + \lambda\hat{W}|\psi_n^{(0)}\rangle = E_n^{(0)}|\lambda\psi_n^{(1)}\rangle + \lambda E_n^{(1)}|\psi_n^{(0)}\rangle\,. \tag{1.247}$$

If we form the scalar product with the bra $\langle\psi_n^{(0)}|$, then the terms containing $|\lambda\psi_n^{(1)}\rangle$ cancel, because of

$$\langle\psi_n^{(0)}|\hat{H}_0|\lambda\psi_n^{(1)}\rangle = E_n^{(0)}\langle\psi_n^{(0)}|\lambda\psi_n^{(1)}\rangle\,, \tag{1.248}$$

and we obtain an expression for the energy shifts in first order:

$$\lambda E_n^{(1)} = \langle\psi_n^{(0)}|\lambda\hat{W}|\psi_n^{(0)}\rangle\,. \tag{1.249}$$

In order to deduce the change $|\lambda\psi_n^{(1)}\rangle$ of the wave functions in first order from (1.247) we form the scalar product with any (unperturbed) eigenstate $\langle\psi_m^{(0)}|$ of $\hat{H}_0$ as bra. Because of

$$\langle\psi_m^{(0)}|\hat{H}_0|\lambda\psi_n^{(1)}\rangle = E_m^{(0)}\langle\psi_m^{(0)}|\lambda\psi_n^{(1)}\rangle \tag{1.250}$$

this yields the following expression for the *overlap* (i.e. the scalar product) of $|\lambda\psi_n^{(1)}\rangle$ with the unperturbed states:

$$\langle\psi_m^{(0)}|\lambda\psi_n^{(1)}\rangle\left(E_n^{(0)} - E_m^{(0)}\right) = \langle\psi_m^{(0)}|\lambda\hat{W}|\psi_n^{(0)}\rangle - \lambda E_n^{(1)}\langle\psi_m^{(0)}|\psi_n^{(0)}\rangle\,. \tag{1.251}$$

For $m = n$ the left-hand side of (1.251) vanishes and we retrieve (1.249). For $m \neq n$ and provided that $E_n^{(0)}$ is *non-degenerate*, i.e. $E_m^{(0)} \neq E_n^{(0)}$ for all $m \neq n$, we obtain

$$\langle\psi_m^{(0)}|\lambda\psi_n^{(1)}\rangle = \frac{\langle\psi_m^{(0)}|\lambda\hat{W}|\psi_n^{(0)}\rangle}{E_n^{(0)} - E_m^{(0)}}\,. \tag{1.252}$$

Since the eigenstates of $\hat{H}_0$ form a complete set, (1.252) defines the expansion of $|\lambda\psi_n^{(1)}\rangle$ in the unperturbed basis (see (1.6), (1.8)). Only the coefficient of $|\psi_n^{(0)}\rangle$ is left undetermined by (1.251). It is a natural choice to set this coefficient zero, which ensures that the norm of the perturbed state $|\psi_n^{(0)} + \lambda\psi_n^{(1)}\rangle$ deviates from unity in second order at the earliest. The perturbation of the wave function in first order is thus

$$|\lambda\psi_n^{(1)}\rangle = \sum_{m\neq n}\frac{\langle\psi_m^{(0)}|\lambda\hat{W}|\psi_n^{(0)}\rangle}{E_n^{(0)} - E_m^{(0)}}|\psi_m^{(0)}\rangle\,. \tag{1.253}$$

Collecting terms of second order in λ in (1.246),

$$\begin{aligned}&\hat{H}_0|\lambda^2\psi_n^{(2)}\rangle + \lambda\hat{W}|\lambda\psi_n^{(1)}\rangle\\&= E_n^{(0)}|\lambda^2\psi_n^{(2)}\rangle + \lambda E_n^{(1)}|\lambda\psi_n^{(1)}\rangle + \lambda^2 E_n^{(2)}|\psi_n^{(0)}\rangle\,,\end{aligned} \tag{1.254}$$

and forming the scalar product with the bra $\langle\psi_n^{(0)}|$ leads to an expression for the second-order contribution to the energy shift:

$$\lambda^2 E_n^{(2)} = \langle\psi_n^{(0)}|\lambda\hat{W}|\lambda\psi_n^{(1)}\rangle = \sum_{m\neq n} \frac{|\langle\psi_n^{(0)}|\lambda\hat{W}|\psi_m^{(0)}\rangle|^2}{E_n^{(0)} - E_m^{(0)}} \,. \tag{1.255}$$

The above considerations are valid for small perturbations of non-degenerate eigenstates of the unperturbed Hamiltonian $\hat{H}_0$. In the *degenerate case* an eigenvalue $E_n^{(0)}$ has N eigenstates, $|\psi_{n,1}^{(0)}\rangle, \dots |\psi_{n,N}^{(0)}\rangle$, and each (unitary) transformation of these N states amongst each other,

$$|\psi_{n,i}^{\mathrm{d}}\rangle = \sum_{j=1}^{N} c_{i,j}|\psi_{n,j}^{(0)}\rangle \,, \tag{1.256}$$

again yields N eigenstates of $\hat{H}_0$ with the same eigenvalue $E_n^{(0)}$. A sensible choice of the coefficients $c_{i,j}$ in (1.256) is that which diagonalizes the perturbing operator $\lambda\hat{W}$ in the N-dimensional subspace spanned by the degenerate eigenstates:

$$\langle\psi_{n,i}^{\mathrm{d}}|\lambda\hat{W}|\psi_{n,j}^{\mathrm{d}}\rangle = \varepsilon_i\delta_{i,j} \,. \tag{1.257}$$

Equation (1.257) is fulfilled if the states (1.256) in the N-dimensional subspace are eigenstates of $\lambda\hat{W}$ in this subspace, i.e. if the respective "residual states" $(\lambda\hat{W} - \varepsilon_i)|\psi_{n,i}^{\mathrm{d}}\rangle$ are each orthogonal to all N states $|\psi_{n,j}^{\mathrm{d}}\rangle$ or, equivalently, to all $|\psi_{n,k}^{(0)}\rangle$, $k = 1, \dots N$. Using (1.256) this orthogonality condition can be written as a homogeneous set of simultaneous linear equations for the coefficients $c_{i,j}$:

$$\langle\psi_{n,k}^{(0)}|\lambda\hat{W} - \varepsilon_i|\psi_{n,i}^{\mathrm{d}}\rangle = \sum_{j=1}^{N} \left(\langle\psi_{n,k}^{(0)}|\lambda\hat{W}|\psi_{n,j}^{(0)}\rangle - \varepsilon_i\delta_{k,j}\right) c_{i,j} = 0 \,. \tag{1.258}$$

For each i (1.258) is a set of N equations, $k = 1, \dots N$, for the N unknowns $c_{i,1}, \dots c_{i,N}$. Non-trivial solutions exist only if the determinant of the matrix of coefficients vanishes:

$$\det\left(\langle\psi_{n,k}^{(0)}|\lambda\hat{W}|\psi_{n,j}^{(0)}\rangle - \varepsilon_i\delta_{k,j}\right) = 0 \,. \tag{1.259}$$

The *pre-diagonalized states* $|\psi_{n,i}^{\mathrm{d}}\rangle$ obtained by solving (1.258) are still only eigenstates of $\hat{H}$ to zeroth order in λ. The N roots of the *secular equation* (1.259) define the N eigenvalues $\varepsilon_1, \dots \varepsilon_N$ of $\lambda\hat{W}$ in the N-dimensional subspace spanned by the degenerate eigenstates of $\hat{H}_0$. The corresponding new energies $E_n^{(0)} + \varepsilon_i$ are the perturbed energies to first order in λ,

$$\lambda E_{n,i}^{(1)} = \varepsilon_i \,. \tag{1.260}$$

The first-order correction to the pre-diagonalized state $|\psi^{\rm d}_{n,i}\rangle$ is $|\lambda\psi^{(1)}_{n,i}\rangle$, and its projections onto the unperturbed basis states $|\psi^{(0)}_m\rangle$ with $E^{(0)}_m \neq E^{(0)}_n$ can be calculated via the same steps that led to (1.252), giving

$$\langle\psi^{(0)}_m|\lambda\psi^{(1)}_{n,i}\rangle = \frac{\langle\psi^{(0)}_m|\lambda\hat{W}|\psi^{\rm d}_{n,i}\rangle}{E^{(0)}_n - E^{(0)}_m}\,. \tag{1.261}$$

In order to obtain the projections of $|\lambda\psi^{(1)}_{n,i}\rangle$ onto the other pre-diagonalized states $|\psi^{\rm d}_{n,j}\rangle$ in the subset of degenerate unperturbed states, we insert $|\psi^{\rm d}_{n,i}\rangle$ and its first- and second-order corrections into the second-order equation (1.254) in place of $|\psi^{(0)}_n\rangle$ and its first- and second-order corrections. Forming the scalar product with the bra $\langle\psi^{\rm d}_{n,j}|$ yields $(j \neq i)$

$$\langle\psi^{\rm d}_{n,j}|\lambda\hat{W}|\lambda\psi^{(1)}_{n,i}\rangle = \lambda E^{(1)}_{n,i}\langle\psi^{\rm d}_{n,j}|\lambda\psi^{(1)}_{n,i}\rangle\,. \tag{1.262}$$

Inserting a complete set (1.22), involving the unperturbed states $|\psi^{(0)}_m\rangle$ with $E^{(0)}_m \neq E^{(0)}_m$ and the pre-diagonalized states from the degenerate subset, in between $\lambda\hat{W}$ and $|\lambda\psi^{(1)}_{n,i}\rangle$ on the left-hand side of (1.262), and remembering (1.257), (1.260) gives

$$\left(\lambda E^{(1)}_{n,i} - \lambda E^{(1)}_{n,j}\right)\langle\psi^{\rm d}_{n,j}\lambda|\psi^{(1)}_{n,i}\rangle = \sum_{E^{(0)}_m \neq E^{(0)}_n} \langle\psi^{\rm d}_{n,j}|\lambda\hat{W}|\psi^{(0)}_m\rangle\langle\psi^{(0)}_m|\lambda\psi^{(1)}_{n,i}\rangle\,. \tag{1.263}$$

With the explicit expression (1.261) for $\langle\psi^{(0)}_m|\lambda\psi^{(1)}_{n,i}\rangle$, (1.263) results in

$$\langle\psi^{\rm d}_{n,j}|\lambda\psi^{(1)}_{n,i}\rangle = \sum_{E^{(0)}_m \neq E^{(0)}_n} \frac{\langle\psi^{\rm d}_{n,j}|\lambda\hat{W}|\psi^{(0)}_m\rangle}{\lambda E^{(1)}_{n,i} - \lambda E^{(1)}_{n,j}}\,\frac{\langle\psi^{(0)}_m|\lambda\hat{W}|\psi^{\rm d}_{n,i}\rangle}{E^{(0)}_n - E^{(0)}_m}\,. \tag{1.264}$$

The first-order correction $|\lambda\psi^{(1)}_{n,i}\rangle$ to the pre-diagonalized state $|\psi^{\rm d}_{n,i}\rangle$ contains contributions from the unperturbed degenerate subset according to (1.264) and from the orthogonal subset according to (1.261) and is

$$\begin{aligned}|\lambda\psi^{(1)}_{n,i}\rangle = &\sum_{E^{(0)}_m \neq E^{(0)}_n} \frac{\langle\psi^{(0)}_m|\lambda\hat{W}|\psi^{\rm d}_{n,i}\rangle}{E^{(0)}_n - E^{(0)}_m}|\psi^{(0)}_m\rangle \\ &+\sum_{j\neq i}\sum_{E^{(0)}_m \neq E^{(0)}_n} \frac{\langle\psi^{\rm d}_{n,j}|\lambda\hat{W}|\psi^{(0)}_m\rangle}{\lambda E^{(1)}_{n,i} - \lambda E^{(1)}_{n,j}}\,\frac{\langle\psi^{(0)}_m|\lambda\hat{W}|\psi^{\rm d}_{n,i}\rangle}{E^{(0)}_n - E^{(0)}_m}|\psi^{\rm d}_{n,j}\rangle\,.\end{aligned} \tag{1.265}$$

The overlap of $|\lambda\psi_{n,i}^{(1)}\rangle$ with $|\psi_{n,i}^{\mathrm{d}}\rangle$ should vanish, so that the norm of the perturbed state deviates from unity in second order at the earliest.

The second-order correction to the energy eigenvalue is obtained by inserting $|\psi_{n,i}^{\mathrm{d}}\rangle$ and its first-order correction (1.265) into the second-order equation (1.254) and forming the scalar product with the bra $\langle\psi_{n,i}^{\mathrm{d}}|$. Because of the pre-diagonalization (1.257), the contribution of the lower line of (1.265) to the matrix element vanishes and we obtain

$$\lambda^2 E_{n,i}^{(2)} = \langle\psi_{n,i}^{\mathrm{d}}|\lambda\hat{W}|\lambda\psi_{n,i}^{(1)}\rangle = \sum_{E_m^{(0)} \neq E_n^{(0)}} \frac{|\langle\psi_{n,i}^{\mathrm{d}}|\lambda\hat{W}|\psi_m^{(0)}\rangle|^2}{E_n^{(0)} - E_m^{(0)}}\,, \tag{1.266}$$

which is essentially the same as in the non-degenerate case (1.255) with the sum taken over all unperturbed states outside the degenerate subset. The states within the degenerate subset contribute to the first-order correction (1.265) of the (pre-diagonalized) states, but not to the second-order correction (1.266) of the energies. In the sum over j in the lower line of (1.265), we assume that the unperturbed first-order energy correction $\lambda E_{n,j}^{(1)}$ is not equal to $\lambda E_{n,i}^{(1)}$. If some states of the unpeturbed degenerate subset remain degenerate after pre-diagonalization, then the first- and second-order energy shifts do not depend on the choice of basis in this still degenerate subset of first-order-corrected states.

Pre-diagonalizing a limited number of unperturbed eigenstates is a useful and valid procedure, not only in the case of exact degeneracy of the unperturbed eigenstates. In equations (1.253) and (1.255) the contributions of states with unperturbed energies $E_m^{(0)}$ close to $E_n^{(0)}$ can become very large due to the small energy denominator. Hence it can be appropriate to pre-diagonalize the states with unperturbed eigenvalues close to $E_n^{(0)}$. An unperturbed energy can be regarded as "close to $E_n^{(0)}$" if the absolute value of the energy difference $E_m^{(0)} - E_n^{(0)}$ is of the same order or smaller than the absolute value of the coupling matrix element $\langle\psi_m^{(0)}|\lambda\hat{W}|\psi_n^{(0)}\rangle$, see Problem 1.6.

In order to calculate energy shifts in second order or perturbations of the wave functions in first order, we strictly speaking need to have solved the unperturbed problem (1.243) completely, because the summations in (1.253) and (1.255) or (1.265) and (1.266) require a complete set of (unperturbed) eigenstates and eigenvalues. For unperturbed Hamiltonians with unbound eigenstates, the summations have to be replaced or complemented by integrations over the corresponding contributions of the continuum.

1.6.2 Ritz's Variational Method

The expectation value of a given Hamiltonian $\hat{H}$ in a Hilbert space of normalizable states can be regarded as a *functional* which maps each state $|\psi\rangle$ onto a real number $E[\psi]$:

$$\langle\hat{H}\rangle = \frac{\langle\psi|\hat{H}|\psi\rangle}{\langle\psi|\psi\rangle} \equiv E[\psi]\,. \tag{1.267}$$

The state $|\psi\rangle$ is an eigenstate of $\hat{H}$ if and only if $E[\psi]$ is stationary at the point $|\psi\rangle$, meaning that an infinitesimally small variation $|\psi\rangle \rightarrow |\psi + \delta\psi\rangle$ of the state leaves the energy unchanged:

$$\delta E = 0\,. \tag{1.268}$$

To see this we evaluate $\delta E = E[\psi + \delta\psi] - E[\psi]$ to first order in $|\delta\psi\rangle$,

$$\begin{aligned} \delta E &= \frac{\langle\psi|\hat{H}|\psi\rangle + \langle\delta\psi|\hat{H}|\psi\rangle + \langle\psi|\hat{H}|\delta\psi\rangle}{\langle\psi|\psi\rangle + \langle\delta\psi|\psi\rangle + \langle\psi|\delta\psi\rangle} - E \\ &= \frac{\langle\delta\psi|\hat{H} - E|\psi\rangle + \langle\psi|\hat{H} - E|\delta\psi\rangle}{\langle\psi|\psi\rangle + \langle\delta\psi|\psi\rangle + \langle\psi|\delta\psi\rangle}\,, \end{aligned} \tag{1.269}$$

and this expression vanishes if and only if

$$\langle\delta\psi|\hat{H} - E|\psi\rangle + \langle\psi|\hat{H} - E|\delta\psi\rangle = 0\,. \tag{1.270}$$

If $|\psi\rangle$ is an eigenstate of $\hat{H}$, then its eigenvalue is identical to the expectation value (1.267), and (1.270) is automatically fulfilled for all $|\delta\psi\rangle$. Conversely, if (1.270) is fulfilled for all (infinitesimal) $|\delta\psi\rangle$, then it must be fulfilled for the pair of variations $|\delta\psi\rangle$ and $\mathrm{i}|\delta\psi\rangle$; with (1.11), (1.12) we have

$$-\,\mathrm{i}\langle\delta\psi|\hat{H} - E|\psi\rangle + \mathrm{i}\langle\psi|\hat{H} - E|\delta\psi\rangle = 0\,. \tag{1.271}$$

It follows from (1.270) and (1.271) that $\langle\psi|\hat{H} - E|\delta\psi\rangle$ and $\langle\delta\psi|\hat{H} - E|\psi\rangle$ must both vanish independently. On the other hand, if $\langle\delta\psi|\hat{H} - E|\psi\rangle$ vanishes for all (infinitesimal) $|\delta\psi\rangle$ in the Hilbert space, then the state $(\hat{H} - E)|\psi\rangle$ must be orthogonal to all states in the Hilbert space und must consequently be zero. That means $|\psi\rangle$ is an eigenstate of $\hat{H}$ with eigenvalue E.

It is often much easier to calculate the energy expectation value $E[\psi]$ for a limited number of model states $|\psi\rangle$ than to solve the eigenvalue problem for the Hamiltonian $\hat{H}$. In such cases we may look for model states at which $E[\psi]$ is stationary under small variations within the space of model states and regard them as approximate eigenstates of $\hat{H}$. It is particularly sensible to search for a minimum of $E[\psi]$ in order to approximate the ground state of the system. The expectation value (1.267) can be written as a weighted mean of all exact eigenvalues of $\hat{H}$ (see (1.27)) and as such cannot be smaller than the smallest eigenvalue E_1:

$$E_1 \leq \frac{\langle\psi|\hat{H}|\psi\rangle}{\langle\psi|\psi\rangle.}, \quad \text{all } |\psi\rangle \ . \tag{1.272}$$

As a special case let's look at a set of model states forming a subspace of a Hilbert space spanned by a basis $|\psi_1\rangle, \dots |\psi_N\rangle$ (which need not be orthonormal). The general model state is then a linear combination

$$|\psi\rangle = \sum_{i=1}^{N} c_i \, |\psi_i\rangle \tag{1.273}$$

of these basis states, and the coefficients c_i are the parameters defining the model state.

The projection of the Hamiltonian $\hat{H}$ onto the subspace spanned by the $|\psi_1\rangle \dots |\psi_N\rangle$ is a reduced operator $\hat{h}$ which is defined by the matrix elements

$$h_{i,j} = \langle\psi_i|\hat{h}|\psi_j\rangle = \langle\psi_i|\hat{H}|\psi_j\rangle \, , \quad i,j = 1, \dots N \, . \tag{1.274}$$

The expectation values of $\hat{h}$ and $\hat{H}$ are the same within the model subspace:

$$\frac{\langle\psi|\hat{H}|\psi\rangle}{\langle\psi|\psi\rangle} = \frac{\langle\psi|\hat{h}|\psi\rangle}{\langle\psi|\psi\rangle} = E[\psi] \, . \tag{1.275}$$

Since the model subspace is itself a vector space of state vectors, we may apply the same reasoning as used above in full Hilbert space and conclude that the energy functional (1.275) is stationary if and only if the corresponding model state $|\psi\rangle$ is an eigenstate of the projection $\hat{h}$ of the Hamiltonian onto the model subspace. $|\psi\rangle$ is an eigenstate of $\hat{h}$ means that $(\hat{h} - E)|\psi\rangle$ vanishes, or equivalently that $(\hat{H} - E)|\psi\rangle$ is orthogonal to all basis states $|\psi_1\rangle \dots |\psi_N\rangle$ of the model subspace:

$$\langle\psi_i|\hat{H} - E|\psi\rangle = 0 \, , \quad i = 1, \dots N \, . \tag{1.276}$$

Inserting the explicit ansatz (1.273) for $|\psi\rangle$ in (1.276) we have

$$\sum_{j=1}^{N} (h_{i,j} - E n_{i,j}) \, c_j = 0 \, , \quad i = 1, \dots N \, , \tag{1.277}$$

where $h_{i,j}$ are the matrix elements of the Hamiltonian (1.274) and $n_{i,j}$ are the elements of the *overlap matrix*:

$$n_{i,j} = \langle\psi_i|\psi_j\rangle \, , \quad i,j = 1, \dots \, N \, . \tag{1.278}$$

Equation (1.277) is a homogeneous system of N simultaneous linear equations for the N unknown coefficients c_j. It contains the overlap matrix $n_{i,j}$, because we didn't assume orthonormality of the basis. The secular equation now reads

$$\det(h_{i,j} - En_{i,j}) = 0 \tag{1.279}$$

and yields N eigenvalues ε_k of $\hat{h}$ belonging to N eigenstates of the form (1.273). Each eigenstate $|\psi^{(k)}\rangle$ is characterized by an N-component vector of coefficients $c_i^{(k)}$, and as eigenstates of the Hermitian operator $\hat{h}$ they are mutually orthogonal:

$$\langle\psi^{(k)}|\psi^{(l)}\rangle = \sum_{i=1}^{N}\sum_{j=1}^{N}(c_i^{(k)})^* n_{l,j}\, c_j^{(l)} \;\propto\; \delta_{k,l}\,. \tag{1.280}$$

If they are normalized to unity we have

$$\langle\psi^{(k)}|\psi^{(l)}\rangle = \delta_{k,l}\,,$$
$$\langle\psi^{(k)}|\hat{H}|\psi^{(l)}\rangle = \varepsilon_k\delta_{k,l}, \quad k,l = 1\,,\ldots\, N\,. \tag{1.281}$$

The method of diagonalizing in a subspace is particularly useful if we are looking for approximations to describe not only the ground state of a system. Equation (1.272) sets an upper bound for the ground state energy and hence we know, the lower the value of $E[\psi]$, the closer it is to the exact ground state energy E_1. For an excited state there is in general no condition like (1.272), and it is not always a good thing to approximate it by a model state with as low an energy as possible. Bounding conditions of the form (1.272) do however hold for a set of model states, if the states don't mix among each other, i.e. if they fulfill (1.281). More precisely: Let $E_1 \le E_2 \le E_3 \cdots$ be the exact eigenvalues of $\hat{H}$ arranged in ascending order and let $\varepsilon_1 \le \varepsilon_2 \cdots \le \varepsilon_N$ be the energy expectation values of N states fulfilling the conditions (1.281). Then

$$E_i \le \varepsilon_i \quad \text{for all } i = 1,\ldots N\,. \tag{1.282}$$

This is the *Hylleraas-Undheim theorem*. With the Hylleraas-Undheim theorem it is clear that all approximate eigenvalues obtained by diagonalizing $\hat{H}$ in a subspace can only become smaller (or stay the same) when the subspace is enlarged. To see this just regard the enlarged subspace as *the* Hilbert space and apply the Hylleraas-Undheim theorem (1.282) to the eigenstates in the smaller subspace. An elegant three-line proof of the Hylleraas-Undheim theorem is contained in [New82], p. 326.

The Hylleraas-Undheim theorem can also be useful in situations more general than diagonalizing in a subspace. Assume for example, that varying $E[\psi]$ in a set of parametrized model states which don't form a closed subspace yields two (or more) stationary points, an absolute minimum at $|\psi_1\rangle$, say, and a local minimum at $|\psi_2\rangle$. In general we don't know whether $E[\psi_2]$ is larger or smaller than the exact energy of the first excited state, and furthermore, $|\psi_1\rangle$ and $|\psi_2\rangle$ need not be orthogonal. On the other hand, it is usually comparatively simple to calculate the 2×2 matrices $h_{i,j} = \langle\psi_i|\hat{H}|\psi_j\rangle$ and $n_{i,j} = \langle\psi_i|\psi_j\rangle$ and to solve the equations (1.277), (1.279). This corresponds to *post-diagonalization* of the Hamiltonian in the two-dimensional subspace spanned by $|\psi_1\rangle$ and $|\psi_2\rangle$. It yields an improved (lower) approximation ε_1 for the ground state energy and a second energy ε_2, which may lie a little above $E[\psi_2]$, but which we definitely know to be an upper bound for the exact energy of the first excited state.

Further improvements can be achieved by diagonalizing two (or more) states according to (1.277), (1.279) for different sets of values of the model parameters. Each diagonalization leads to a set $\varepsilon_1 \le \varepsilon_2 \le \ldots$ of energies and the best approximation for the ground state is the (diagonalized) wave function with the lowest value of ε_1. The best approximation for the second (the first excited) state is the wave function with the lowest value of ε_2, which may occur for a different set of values of the model parameters, etc. In this method of *variation after diagonalization* the resulting approximate eigenstates need not be orthogonal, because they emerge from different diagonalizations. The corresponding energies ε_i are however definitely upper bounds for the respective exact energies of the i-th state, because each ε_i is the i-th energy in a diagonal set of states (1.281).

1.6.3 Semiclassical Approximation

The relation between classical mechanics and quantum mechanics has interested rersearchers ever since Schrödinger formulated his wave equation in 1926. The rich structure observed in the classical dynamics of seemingly simple systems with few degrees of freedom has made the question of how such classical behaviour affects the corresponding quantum dynamics a central theme of theoretical physics in the last several years (see Sect. 5.3); the study of "simple" atoms plays an important role in this context [FE97, BB97, CK97, SS98, BR09].

The connection between classical mechanics and quantum mechanics is comparatively well understood for one-dimensional systems. One approach which relates the concept of a wave function to motion on a classical trajectory is the semiclassical approximation of Wentzel, Kramers and Brillouin, the *WKB method*.

The WKB approximation can be derived by writing the wave function $\psi(x)$ describing the one-dimensional motion of a point particle of mass μ in a (real) potential $V(x)$ as

$$\psi(x) = \exp(\mathrm{i}g(x)) \,, \tag{1.283}$$

with a complex function $g(x)$. If we write the time-independent Schrödinger equation as

$$\psi'' + \frac{p(x)^2}{\hbar^2}\psi = 0 \tag{1.284}$$

and insert (1.283), we obtain

$$(g')^2 = \frac{p^2}{\hbar^2} + \mathrm{i}\, g'' . \tag{1.285}$$

The function $p(x)$ appearing in (1.284) and (1.285) is the *local classical momentum* corresponding to a classical decomposition of the energy E into a kinetic and a potential energy:

$$E = \frac{p(x)^2}{2\mu} + V(x)\,, \quad p(x) = \sqrt{2\mu(E - V(x))}\,. \tag{1.286}$$

In the classically allowed region, $E > V(x)$, the kinetic energy is positive and we assume the convention that $p(x)$ is the positive square root of p^2. The local classical momentum is also a useful concept in the classically forbidden region, $E < V(x)$; here the kinetic energy is negative and $p(x)$ is purely imaginary.

From (1.285) we have

$$g'(x) = \pm\frac{p}{\hbar}\sqrt{1 + \mathrm{i}\frac{\hbar^2}{p^2}g''} = \pm\frac{p}{\hbar} \pm \mathrm{i}\frac{\hbar}{2p}g'' + O\left[\left(\frac{\hbar}{p}\right)^3 (g'')^2\right] . \tag{1.287}$$

Regarding $\hbar$ as a small quantity gives, to leading order, $g' = \pm p/\hbar$. Including the next term on the right-hand side of (1.287) via $g'' = \pm p'/\hbar$ yields

$$g'(x) = \pm\frac{p}{\hbar} + \mathrm{i}\frac{p'}{2p} \Rightarrow g(x) = \pm\frac{1}{\hbar}\int^x p(x')\,\mathrm{d}x' + \frac{\mathrm{i}}{2}\ln p(x) + \text{const.} \tag{1.288}$$

Inserting this expression for $g(x)$ into (1.283) defines the WKB approximation,

$$\psi_{\mathrm{WKB}}(x) \propto \frac{1}{\sqrt{p(x)}} \exp\left\{\pm\frac{\mathrm{i}}{\hbar}\int^x p(x')\mathrm{d}x'\right\} . \tag{1.289}$$

In the classically allowed region $p(x)$ is real and so is the *action integral*

$$S = \int^x p(x')\,\mathrm{d}x'\,, \tag{1.290}$$

and $\psi_{\mathrm{WKB}}(x)$ is an oscillating function characterized by the *local* de Broglie wave length

$$\lambda(x) = \frac{2\pi\hbar}{p(x)}\,. \tag{1.291}$$

The WKB wave function (1.289) depends on the lower limit for the action integral in the exponent only in the form of an overall constant. The factor exp $\left[\frac{i}{\hbar}\int^x p(x')\mathrm{d}x'\right]$ represents a rightward travelling wave with current density (1.160) equal to the classical velocity p/μ, and exp $\left[-\frac{i}{\hbar}\int^x p(x')\mathrm{d}x'\right]$ represents a leftward travelling wave with current density $-p/\mu$. The amplitude proportional to $p^{-1/2}$ in (1.289) ensures that the probability density $|\psi_{\mathrm{WKB}}|^2$ is inversely proportional to the particle's velocity, so that the current density of the WKB wave is independent of x, as required by the continuity equation for a stationary state. In the classically forbidden region where $p(x)$ is purely imaginary, the exponential in the WKB expression (1.289) is a monotonically increasing or decreasing function of x.

Semiclassical approximations are based on the assumption that Planck's constant is small, meaning that relevant observables with the same physical dimension—e.g. the action integral (1.290)—should have values which are large compared to $\hbar$. The fulfillment or violation of this condition is quite transparent in a system with any number of degrees of freedom when the potential is *homogeneous*. A homogeneous potential of degree d has the property

$$V_d(\sigma x) = \sigma^d V_d(x)\;, \tag{1.292}$$

where x may stand for any number of coordinates. For the harmonic oscillator (1.80) we have $d = 2$, whereas $d = -1$ for the Coulomb potential (1.134). Classical motion in homogeneous potentials has the property of mechanical similarity [LL58], i.e. if $x(t)$ is a valid solution of the equations of motion at energy E, then $\sigma x(\sigma^{1-d/2}t)$ is a solution at energy $E' = \sigma^d E$, see Sect. 5.3.4. This rescaling of energy with a factor $\epsilon \equiv \sigma^d$ and of the coordinates according to $s = \sigma x$ has the following effect on the classical action (1.290):

$$\begin{aligned} S(\epsilon E) &= \sqrt{2\mu}\int \mathrm{d}s\sqrt{\epsilon E - V_d(s)} = \epsilon^{\frac{1}{2}}\sqrt{2\mu}\int \mathrm{d}s\sqrt{E-\epsilon^{-1}V_d(s)} \\ &= \epsilon^{\frac{1}{2}}\sqrt{2\mu}\int \mathrm{d}s\sqrt{E - V_d(x)} \\ &= \epsilon^{\frac{1}{2}+\frac{1}{d}}\sqrt{2\mu}\int \mathrm{d}x\sqrt{E-V_d(x)} = \epsilon^{\frac{1}{2}+\frac{1}{d}} S(E)\,. \end{aligned} \tag{1.293}$$

This means that an increase in the absolute value $|E|$ of the energy, $\epsilon > 1$, results in an increase of the action S if and only if

$$\frac{1}{2}+\frac{1}{d} > 0\,,\quad \text{i.e.}\quad d > 0 \quad \text{or} \quad d < -2\,. \tag{1.294}$$

The *semiclassical limit* $\hbar/S \to 0$ is reached in the limit of large energies for all homogeneous potentials of positive degree, such as all sorts of oscillators, $V \propto |x|^d$, $d > 0$, and also for homogeneous potentials of negative degree, as long as $d < -2$. The *anticlassical* or *extreme quantum limit*, on the other hand, is defined by

$\hbar/|S| \to \infty$ and corresponds to $E \to 0$ for these systems. In contrast, for negative degrees of homogeneity in the range $-2 < d < 0$, the opposite and perhaps counterintuitive situation occurs: the limit of vanishing energy $E \to 0$ defines the semiclassical limit, whereas $|E| \to \infty$ is the anticlassical, the extreme quantum limit. All attractive or repulsive Coulomb-type potentials, for which $d = -1$, fall into this category. A discussion of the semiclassical and anticlassical limits for more general potentials containing several homogeneous terms is given in Sect. 5.3.4.

The WKB wave function (1.289) may be a good approximation to an exact solution of the Schrödinger equation, at least locally, even when the conditions of the semiclassical limit are not fulfilled for the Schrödinger equation as a whole. To see this, construct the second derivative of (1.289) and observe that $\psi_{\rm WKB}$ is a solution to the following equation:

$$\psi''_{\rm WKB} + \frac{p^2}{\hbar^2}\psi_{\rm WKB} + \left(\frac{p''}{2p} - \frac{3}{4}\frac{(p')^2}{p^2}\right)\psi_{\rm WKB} = 0\,. \tag{1.295}$$

The last term on the left-hand side of (1.295) corresponds to the contribution of an additional potential $V_{\rm add.}$ given by

$$\frac{2\mu}{\hbar^2}V_{\rm add.}(x) = \frac{3}{4}\frac{(p')^2}{p^2} - \frac{p''}{2p}\,. \tag{1.296}$$

Without this term, (1.295) is identical to the Schrödinger equation (1.284). The condition for validity of the WKB approximation is thus, that the additional term (1.296) be small compared to the function $p^2/\hbar^2$ of the potential term in the Schrödinger equation,

$$|Q(x)| \ll 1\,, \tag{1.297}$$

where

$$Q(x) = \hbar^2\left(\frac{3}{4}\frac{(p')^2}{p^4} - \frac{p''}{2p^3}\right) = \frac{1}{16\pi^2}\left[2\lambda\frac{{\rm d}^2\lambda}{{\rm d}x^2} - \left(\frac{{\rm d}\lambda}{{\rm d}x}\right)^2\right]\;; \tag{1.298}$$

here $\lambda(x)$ is the local de Broglie wave length (1.291),

The condition (1.297) for the validity of the semiclassical WKB approximation is inherently local, as expressed in the function (1.298). Where $|Q(x)|$ is small, semiclassical approximations are expected to be accurate. On the other hand, regions in coordinate space where $Q(x)$ is significantly non-vanishing are expected to show manifestly nonclassical, quantum mechanical effects. This justifies calling $Q(x)$ the *quantality function*.

An obvious problem for the WKB wave function (1.289) occurs at a classical turning point x_t, where $E = V(x_t)$ and $p(x_t)$ vanishes; $Q(x)$ diverges and $\psi_{\rm WKB}(x)$ becomes singular at x_t. If the turning point is isolated, the classically forbidden region extends indefinitely, there is no tunnelling and the wave function decays to zero on the classically forbidden side of the turning point, whereas a wave approaching the turning point on the classically allowed side is totally reflected. Under favourable conditions, the WKB approximation may be accurate away from x_t on one or both sides of the turning point. On the classically allowed side

$$\begin{aligned}\psi_{\rm WKB}^{(\phi)}(x) \propto \\ &\frac{1}{\sqrt{p(x)}}\left[\exp\left(-\frac{\rm i}{\hbar}\left|\int_{x_t}^{x} p(x')\,{\rm d}x'\right|\right) + {\rm e}^{-{\rm i}\phi}\exp\left(\frac{\rm i}{\hbar}\left|\int_{x_t}^{x} p(x')\,{\rm d}x'\right|\right)\right] \\ \propto\ &\frac{1}{\sqrt{p(x)}}\cos\left(\frac{1}{\hbar}\left|\int_{x_t}^{x} p(x')\,{\rm d}x'\right| - \frac{\phi}{2}\right),\end{aligned} \tag{1.299}$$

and on the forbidden side $\psi_{\rm WKB}(x) \propto \exp\left[-\frac{1}{\hbar}\left|\int_{x_t}^{x} p(x')\,{\rm d}x'\right|\right] / \sqrt{|p(x)|}$. Here we have chosen the classical turning point x_t, which is a natural point of reference, as the lower limit for the action integrals. The second-last line in (1.299) shows that ϕ is the phase loss in the WKB wave due to reflection at the classical turning point x_t—the *reflection phase* [FT96, FT04].

The WKB wave function (and the exact wave function) can be chosen to be real when the potential is real. The decaying WKB wave function on the classically forbidden side is uniquely defined to within an overall constant, but the ratio of the amplitudes on both sides and the phase ϕ in the oscillating wave (1.299) on the allowed side are not fixed *a priori*, they are determined by matching the WKB waves on both sides of the turning point according to the *connection formula*,

$$\frac{N}{\sqrt{|p(x)|}}\,{\rm e}^{-\frac{1}{\hbar}\left|\int_{x_t}^{x} p(x')\,{\rm d}x'\right|} \to \frac{2}{\sqrt{|p(x)|}}\cos\left(\frac{1}{\hbar}\left|\int_{x_t}^{x} p(x')\,{\rm d}x'\right| - \frac{\phi}{2}\right). \tag{1.300}$$

This form of the connection formula, i.e. with the absolute values of the action integrals in the arguments of the exponential and cosine functions, does not depend on whether the classically allowed side is to the left or to the right of the turning point x_t.

The derivation and interpretation of the connection formula (1.300) is discussed at great length in many texts on semiclassical theory [FF65, BM72, FF96, FT04]. If the WKB approximation becomes sufficiently accurate away from the turning points, then an unambiguous determination of ϕ and N can be achieved by matching the WKB wave functions to the exact solution of the Schrödinger equation. If the potential is approximately linear in a region which surrounds the classical turning point and is large enough to accommodate many de Broglie wave lengths on the allowed side and many times the penetration depth on the forbidden side, then the exact wave function is an Airy function (Appendix A.4) and the amplitude N and reflection phase ϕ in (1.300) are given by,

$$N = 1\,, \quad \phi = \frac{\pi}{2}\,, \tag{1.301}$$

see e.g. [BM72]. This is the general result of the semiclassical or short-wave limit and is the basis of conventional WKB applications involving a classical turning point. The standard choice (1.301) is, in general, far too restrictive and not related to whether or not the WKB approximation is accurate away from the turning point. Allowing more accurate values for N and ϕ in (1.300) greatly widens the range of applicability of WKB wave functions. For example for a particle reflected by an infinite steep wall the reflection coefficient is -1 and the reflection phase is π rather than $\pi/2$. This result is typical of the *long wave limit*, where the wave length on the classically allowed side of the turning point is large compared with the penetration depth of the wave function on the classically forbidden side. In more general situations it is often appropriate to use other values of the reflection phase. Regardless of whether or not the WKB approximation ever becomes accurate on the classically forbidden side, inserting the correct reflection phase ϕ on the right-hand side of the connection formula (1.300) is the key to obtaining a WKB wave function which is an accurate approximation to the exact solution of the Schrödinger equation on the classically allowed side of the classical turning point [FT96, FT04]. (See Problem 1.7.)

A particularly important case is that of a potential proportional to the inverse square of the coordinate,

$$V(x) = \frac{\hbar^2}{2\mu}\frac{\gamma}{x^2}\,, \quad \gamma \geq 0\,, \quad x > 0\,, \tag{1.302}$$

which is just the centrifugal potential for angular momentum quantum number l when x is the radial coordinate and $\gamma = l(l+1)$ [cf. (1.76)]. For homogeneous potentials of degree $d = -2$, classical action integrals are invariant under the scaling (1.293), so changing the energy does not bring us closer to or further from the semiclassical limit. As for all homogeneous potentials of negative degree, however, large absolute values of the potential strength correspond to the semiclassical and small values to the anticlassical limit of the Schrödinger equation, see (5.155)–(5.157) in Sect. 5.3.4.

The Schrödinger equation with the potential (1.302) alone can be solved exactly, and the solution is $\psi(x) \propto \sqrt{kx}J_\nu(kx)$, $k = \sqrt{2\mu E/\hbar^2}$, where J_ν is the ordinary Bessel function of index $\nu = \sqrt{\gamma + 1/4}$. The asymptotic behaviour of ψ follows from (A.37) in Appendix A.4,

$$\psi(x) \sim \cos\left(kx - \nu\frac{\pi}{2} - \frac{\pi}{4}\right)\,. \tag{1.303}$$

The classical turning point x_t is given by $kx_t = \sqrt{\gamma}$, and the action integral in the WKB wave function can be calculated analytically. The asymptotic form of the WKB wave function on the classically allowed side of the turning point, cf. (1.300), is

$$\psi_{\mathrm{WKB}}(x) \sim \cos\left(kx - \sqrt{\gamma}\frac{\pi}{2} - \frac{\phi}{2}\right), \tag{1.304}$$

where ϕ is the reflection phase.

When the conventional choice (1.301) is used for the reflection phase, $\phi = \pi/2$, the asymptotic phase of the WKB wave function (1.304) disagrees with the asymptotic phase of the exact wave function (1.303). This discrepancy can be repaired by the so-called *Langer modification*, in which the potential for the WKB calculation is manipulated by the replacement

$$\gamma \to \gamma + \frac{1}{4} \quad \text{corresponding to} \quad l(l+1) \to \left(l + \frac{1}{2}\right)^2 . \tag{1.305}$$

An alternative procedure for reconciling the phases in (1.303) and (1.304) is to leave the potential intact and to insert as reflection phase

$$\phi = \frac{\pi}{2} + \pi\left(\sqrt{\gamma + \frac{1}{4}} - \sqrt{\gamma}\right) . \tag{1.306}$$

The reflection phase (1.306) for the centrifugal potential approaches the value $\pi/2$ in the semiclassical limit $\gamma \to \infty$ and the value π in the anticlassical limit $\gamma \to 0$. This is in fact the right value for s-waves ($l = \gamma = 0$), where the node required in the wave function at $x = 0$ has the same effect as reflection by an infinite steep wall.

Although the Langer modification helps to improve the results of the WKB approximation when the reflection phase is kept fixed at $\pi/2$, leaving the potential intact and inserting the correct reflection phase (1.306) leads to wave functions which approach the exact solution of the Schrödinger equation much more rapidly in the classically allowed region [FT96].

Now consider a particle bound with total energy E in a potential $V(x)$ as illustrated in Fig. 1.11. The exact wave function is a solution of the Schrödinger equation (1.284); in the classically allowed region between the two classical turning points a and b the "kinetic energy" proportional to p^2 is positive, and the sign of the second derivative ψ'' of the wave function is opposite to the sign of ψ, i.e. the wave function oscillates and is always curved towards the x-axis. In the classically forbidden regions p^2 is negative, ψ'' and ψ have the same sign, so the wave function is curved away from the x-axis; if the entire regions to the right of b and to the left of a are classically forbidden, the wave function decays to zero in the classically

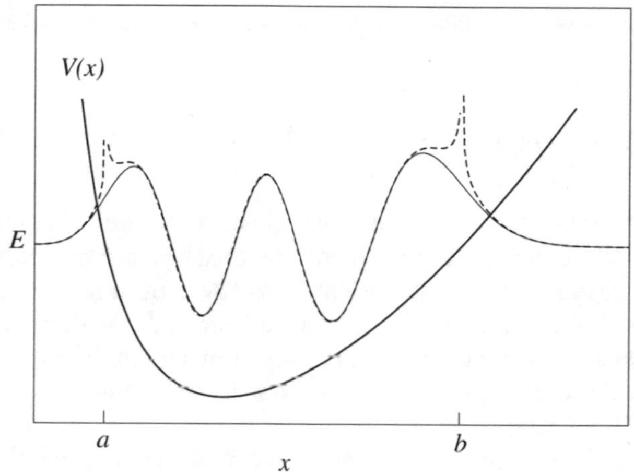

Fig. 1.11 Exact and WKB wave functions for the bound motion of a particle in a real potential $V(x)$. The *thin solid line* shows the exact solution of the Schrödinger equation (1.284); the *dashed line* shows the WKB wave function (1.289), which is singular due to the factor $p(x)^{-1/2}$ at the classical turning points a and b where $p = 0$

forbidden regions. During one whole period of oscillation the WKB wave function gains the phase

$$\frac{1}{\hbar}\oint p(x')\mathrm{d}x' \stackrel{\text{def}}{=} \frac{1}{\hbar}S(E)\,, \tag{1.307}$$

and it loses the phases ϕ_a and ϕ_b due to reflection at the classical turning points a and b. The *integrated action* $S(E)$ in (1.307) is just the area enclosed by the classical trajectory in the two-dimensional phase space spanned by the coordinate x and the momentum p. A *quantization rule* for stationary bound states can be obtained by requiring the net phase gain during one period of oscillation, viz. $\frac{1}{\hbar}S(E) - \phi_a - \phi_b$, to be an integer multiple of 2π in order that the wave function be a unique function of the coordinate. This leads to,

$$\frac{1}{2}S(E) = \int_a^b p(x)\mathrm{d}x = \pi\hbar\left(n + \frac{\mu_\phi}{4}\right), \quad n = 0, 1, 2, \dots\,. \tag{1.308}$$

In (1.308) μ_ϕ is the *Maslov index,* which is equal to the total phase loss measured in units of $\pi/2$,

$$\mu_\phi = \frac{\phi_a + \phi_b}{\pi/2}\,. \tag{1.309}$$

In conventional semiclassical theory, the reflection phases ϕ_a and ϕ_b are taken to be $\pi/2$ according to (1.301), so $\mu_\phi = 2$ and we obtain the most widely used form of the *Bohr-Sommerfeld quantization rule* corresponding to conventional WKB quantization,

$$\frac{1}{2}S(E) = \int_a^b p(x)\mathrm{d}x = \pi\hbar\left(n + \frac{1}{2}\right), \quad n = 0, 1, 2, \ldots . \tag{1.310}$$

As discussed above, this is only justified when the potential is sufficiently well approximated by a linear function near the classical turning points, which is generally the case near the semiclassical limit. Away from the semiclassical limit the reflection phases can be noninteger multiples of $\pi/2$; with the corresponding noninteger Maslov index (1.309), (1.308) represents a *modified quantization rule* which can yield accurate results beyond the restrictive assumptions of the semiclassical limit [FT96, FT04].

In order to demonstrate the power of the more general modified quantization rule, we consider a free particle of mass μ trapped within a sphere of radius R, the "spherical billiard". For given angular momentum quantum number $l = 0, 1, 2, \ldots$ the radial wave function $\phi_l(r)$ obeys the free ($V = 0$) radial Schrödinger equation (1.75), so the effective potential is just the centrifugal potential, i.e. (1.302) with $\gamma = l(l+1)$. The radial wave functions at energy $E = \hbar^2k^2/(2\mu)$ are proportional to $kr\,j_l(kr)$ as in (1.115), and the eigenvalues are given by those wave numbers $k_{n,l}$ for which the radial wave function vanishes at the confining distance $r = R$, i.e. where the spherical Bessel function $j_l(k_{n,l}R)$ vanishes,

$$E_{n,l} = \frac{\hbar^2k_{n,l}^2}{2\mu} = \frac{\hbar^2}{2\mu R^2}(x_{n,l})^2\,, \quad j_l(x_{n,l}) = 0\,, \quad n = 0,\ 1, 2, \ldots . \tag{1.311}$$

Here $x_{n,l}$ stands for the positive zeros of the spherical Bessel funtion $j_l(x)$. For $l = 0$ we have $j_l(x) = \sin x/x$ and $x_{n,l} = (n+1)\pi$. For $l > 0$ the zeros are increasingly affected by the centrifugal potential (1.302). When applying the quantization rule (1.308), the reflection phase at the outer turning point $r = R$ has to be taken as π for the hard-wall reflection. In conventional WKB quantization, the centrifugal potential (1.302) is replaced by the Langer-modified potential (1.305) and the reflection phase at the inner turning point is taken to be $\pi/2$ according to (1.301), so the Maslov index (1.309) is $\mu_\phi = 3$. In the modified quantization rule, the potential is left intact, but the condition (1.301) is relaxed and the reflection phase at the inner turning point is as given by (1.306), so the Maslov index is,

$$\mu_\phi = 3 + 2\left(l + \frac{1}{2} - \sqrt{l(l+1)}\right). \tag{1.312}$$

This gives $\mu_\phi = 4$ for $l = 0$, $\mu_\phi = 3.17157$ for $l = 1$ and $\mu_\phi = 3.10102$ for $l = 2$.

The energy eigenvalues for the spherical billiard are given in units of $E_0 = \hbar^2/(2\mu R^2)$ in Table 1.5 for $l = 0$ to 2 and $n = 0$ to 4. Next to the exact results, $E_{n,l}/E_0 = (x_{n,l})^2$, the table shows the results obtained with conventional WKB quantization (superscript "WKB") and, for $l = 1$ and $l = 2$, with the modified

Table 1.5 Energies $\epsilon_{n,l} = E_{n,l}/E_0$ in units of $E_0 = \hbar^2/(2\mu R^2)$ for the spherical billiard. The superscript "exact" labels the exact quantum mechanical eigenvalues (1.311), $\epsilon_{n,l} = (x_{n,l})^2$. The superscript "WKB" labels the eigenvalues obtained with conventional WKB quantization involving the Langer modification (1.305) of the potential and a Maslov index $\mu_\phi = 3$. The superscript "mqr" labels the eigenvalues obtained with the modified quantization rule based on the true centrifugal potential and the Maslov index $\mu_\phi = 3.17157$ for $l = 1$ and $\mu_\phi = 3.10102$ for $l = 2$ according to (1.312).

n	$\epsilon_{n,0}^{\text{exact}}$	$\epsilon_{n,0}^{\text{WKB}}$	$\epsilon_{n,1}^{\text{exact}}$	$\epsilon_{n,1}^{\text{WKB}}$	$\epsilon_{n,1}^{\text{mqr}}$	$\epsilon_{n,2}^{\text{exact}}$	$\epsilon_{n,2}^{\text{WKB}}$	$\epsilon_{n,2}^{\text{mqr}}$
0	π^2	9.6174	20.1907	19.8697	20.1390	33.2175	32.8153	33.1018
1	$(2\pi)^2$	39.2279	59.6795	59.4064	59.6625	82.7192	82.4160	82.6791
2	$(3\pi)^2$	88.5762	118.8999	118.6384	118.8914	151.8549	151.5770	151.8340
3	$(4\pi)^2$	157.6635	197.8578	197.6009	197.8527	240.7029	240.4357	240.6900
4	$(5\pi)^2$	246.4900	296.5544	296.2998	296.5510	349.2801	349.0183	349.2713

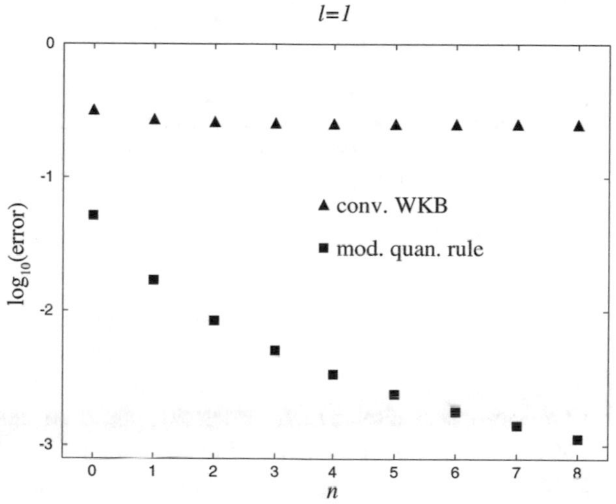

Fig. 1.12 Errors $|E_{n,l}^{\text{exact}} - E_{n,l}|$ (in units of $E_0 = \hbar^2/(2\mu R^2)$) of the energy eigenvalues for the spherical billiard for angular momentum quantum number $l = 1$. The *triangles* show the errors of the eigenvalues obtained with conventional WKB quantization involving the Langer modification (1.305) and a Maslov index $\mu_\phi = 3$. The *squares* show the errors obtained with the modified quantization rule based on the true centrifugal potential and the Maslov index $\mu_\phi = 3.17157$ according to (1.312)

quantization rule using the Maslov index (1.312) (superscript "mqr"). The energies predicted by conventional WKB quantization including the Langer modification of the potential are consistently too low by an almost n- and l-independent term near 0.25 times E_0. The results obtained with the modified quantization rule, meaning there is no Langer modification and the Maslov index is given by (1.312), are obviously exact for $l = 0$. For $l = 1$ and $l = 2$ they are much closer to the exact results than the predictions of conventional WKB quantization, and the error decreases rapidly with n as illustrated in Figs. 1.12 and 1.13.

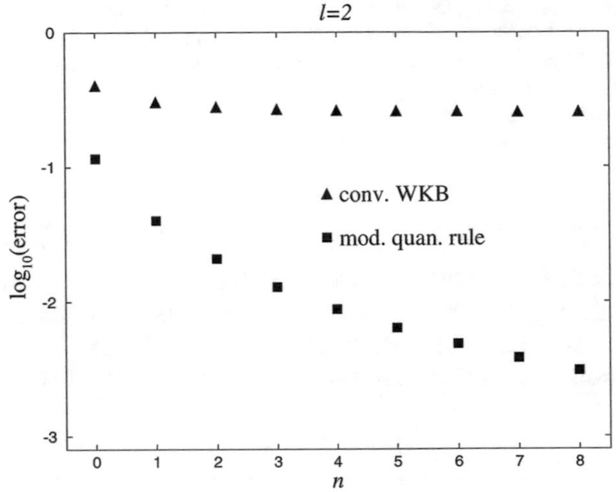

Fig. 1.13 Errors $|E_{n,l}^{\rm exact} - E_{n,l}|$ (in units of $E_0 = \hbar^2/(2\mu R^2)$) of the energy eigenvalues for the spherical billiard for angular momentum quantum number $l = 2$. The *triangles* show the errors of the eigenvalues obtained with conventional WKB quantization involving the Langer modification (1.305) and a Maslov index $\mu_\phi = 3$. The *squares* show the errors obtained with the modified quantization rule based on the true centrifugal potential and the Maslov index $\mu_\phi = 3.10102$ according to (1.312)

The results in Table 1.5 and Figs. 1.12 and 1.13 demonstrate how the accuracy of the conventional WKB approximation can be dramatically improved by relaxing the restrictions of the standard interpretation (1.301) of the connection formula (1.300) and allowing a more appropriate choice for the reflection phase. An extensive review on how such modifications of conventional WKB theory can yield accurate and even asymptotically exact results far from the semiclassical limit is given in [FT04].

1.6.4 Inverse Power-Law Potentials

Many physically interesting problems are described by a one-dimensional Schrödinger equation with a potential $V(r)$ which over a large range of distances r follows a simple inverse power law,

$$V_\alpha^{(\pm)}(r) = \pm\frac{C_\alpha}{r^\alpha}\,; \quad r > 0\,, \quad \alpha > 0\,, \quad C_\alpha > 0\,. \tag{1.313}$$

For (attractive or repulsive) Coulomb potentials $\alpha = 1$, for the centrifugal potential (1.302) $\alpha = 2$. A further example for inverse-square potentials, attractive or repulsive, is the interaction of an electric charge with an electric dipole. Examples for $\alpha = 3$ and 4 are the van der Waals interactions of polarizable atoms and a conducting or dielectric surface, neglecting or including relativistic retardation effects, see Sect. 5.7.1, and the corresponding interactions of atoms (or molecules) with each other are examples for $\alpha = 6$ and 7. For $\alpha \geq 2$, attractive potentials cannot have the form (1.313) all the way down to $r = 0$, because the energy spectrum would then be unbounded from below, so the actual potential must change to a less strongly attractive or even repulsive form at short distances. However, the regime of short distances where these deviations are appreciable may be quite small, so it is worthwhile to study not only the repulsive potentials $V_\alpha^{(+)}$ but also the attractive potentials $V_\alpha^{(-)}$ in some detail, even for $\alpha \geq 2$.

With the potential energy given by (1.313), the local classical momentum at threshold, $E = 0$, is given by $p(r) = r^{-\alpha/2}\sqrt{\pm 2\mu C_\alpha}$ and the quantality function (1.298) is easy to calculate,

$$Q(r) \stackrel{E=0}{=} \pm \frac{\hbar^2}{32\mu C_\alpha}\alpha(\alpha-4)r^{\alpha-2}\,. \tag{1.314}$$

For large distances r the quantality function diverges when $\alpha > 2$, and it vanishes when $\alpha < 2$. For inverse power-law potentials, the classical scaling discussed in Sect. 1.6.3, (1.293), leads to large distances near threshold, i.e. $\epsilon \to 0$ corresponds to $\sigma \to \infty$ when $d < 0$, so this behaviour is consistent with the observation that $E = 0$ corresponds to the anticlassical limit for $\alpha > 2$ and to the semiclassical limit for $\alpha < 2$. Inverse-square potentials, $\alpha = 2$, represent the boundary between the long-ranged potentials $0 < \alpha < 2$ and the shorter-ranged potentials $\alpha > 2$, and $Q(r)$ is constant at threshold in this case. The case $\alpha = 4$ is special, because $Q(r)$ vanishes identically at threshold. For a potential proportional to $1/r^4$, WKB wave functions are exact solutions of the Schrödinger equation at energy zero. This example of a $1/r^4$ potential at threshold shows, that the criterion (1.297) for the validity of the WKB approximation is more reliable than the commonly quoted criterion $|\mathrm{d}\lambda/\mathrm{d}r| \ll 1$.

For $r \to 0$, $Q(r)$ diverges for $\alpha < 2$ and vanishes for $\alpha > 2$. Even though the threshold represents the anticlassical, extreme quantum limit of the Schrödinger equation for $\alpha > 2$, there nevertheless is a semiclassical regime of small r values where WKB wave functions are accurate solutions of the Schrödinger equation, because the condition (1.297) is well fulfilled. The small-r behaviour of $Q(r)$ as given by (1.314) also holds for all finite energies $E \neq 0$, because the potential (1.313) diverges for $r \to 0$ and dominates over the finite energy E for sufficiently small values of r.

For the repulsive potential $V_\alpha^{(+)}(r)$ and positive energy $E = \hbar^2k^2/(2\mu)$ there is a classical turning point r_t at which the quantality function diverges,

$$r_\mathrm{t} = \left(\frac{C_\alpha}{E}\right)^{1/\alpha} \stackrel{\alpha\neq 2}{=} \beta_\alpha (k\beta_\alpha)^{-2/\alpha}\,. \tag{1.315}$$

Here we have introduced, for $\alpha \neq 2$, the potential strength parameter β_α which has the physical dimension of a length,

$$C_\alpha = \frac{\hbar^2}{2\mu}(\beta_\alpha)^{\alpha-2}, \quad \beta_\alpha = \left(\frac{2\mu\, C_\alpha}{\hbar^2}\right)^{1/(\alpha-2)} . \tag{1.316}$$

The length β_α is a *quantum length* defining a characteristic scale for the quantum mechanical properties of the potential(s) (1.313); it has no correspondence in classical mechanics.

For $\alpha > 2$, the WKB approximation gives the correct, i.e. the asymptotically exact, behaviour of the regular solution of the Schrödinger equation in the classically forbidden region near the origin,

$$\begin{aligned}\psi(r) &\overset{r\to 0}{\propto} \frac{1}{\sqrt{p(r)}} \exp\left[-\frac{1}{\hbar}\int_r^{r_0} |p(r')|\mathrm{d}r'\right] \\ &\overset{r\to 0}{\propto} r^{\alpha/4} \exp\left[-\frac{2}{\alpha-2}\left(\frac{\beta_\alpha}{r}\right)^{(\alpha-2)/2}\right],\end{aligned} \tag{1.317}$$

where r_0 is some fixed point of reference smaller than r_t. The lower line in (1.317) follows from the r dependence of the local classical momentum in the limit $r \to 0$, where the energy E is neglible compared to the potential, $|p(r)| \overset{r\to 0}{=} \sqrt{2\mu C_\alpha}\, r^{-\alpha/2} = \hbar(\beta_\alpha)^{(\alpha-2)/2}\, r^{-\alpha/2}$. For a repulsive inverse-square potential with strength $C_2 = \gamma\hbar^2/(2\mu)$ (cf. (1.302)), we have $|p(r)| \overset{r\to 0}{=} \hbar\sqrt{\gamma}/r$ for sufficiently small r and the WKB approximation as defined in the upper line of (1.317) yields

$$\psi_{\mathrm{WKB}}(r) \overset{r\to 0}{\propto} r^{1/2+\sqrt{\gamma}} . \tag{1.318}$$

Note that for the centrifugal potential (1.76), $\gamma = l(l+1)$, the result (1.318) does not agree with the correct quantum mechanical behaviour (1.78), unless we invoke the Langer modification (1.305), $\sqrt{\gamma} \to l + 1/2$. The same holds for the centrifugal potential plus a further potential less singular than $1/r^2$ for $r \to 0$, e.g. the Coulomb potential, for which the regular quantum mechanical wave function is proportional to r^{l+1} for small r, see (A.76) in Appendix A.5. For $l = 0$, the Langer modification amounts to adding a fictitious centrifugal potential corresponding to $l = 1/2$.

The attractive potential $V_\alpha^{(-)}(r)$ may constitute the tail of a realistic potential well as illustrated in Fig 1.14. For negative energies $E = -\hbar^2\kappa^2/(2\mu) < 0$ the outer classical turning point r_t in the inverse power-law tail of the potential is given, similar to (1.315), by

$$r_t = \left(\frac{-C_\alpha}{E}\right)^{1/\alpha} \overset{\alpha\neq 2}{=} \beta_\alpha(\kappa\beta_\alpha)^{-2/\alpha} ; \tag{1.319}$$

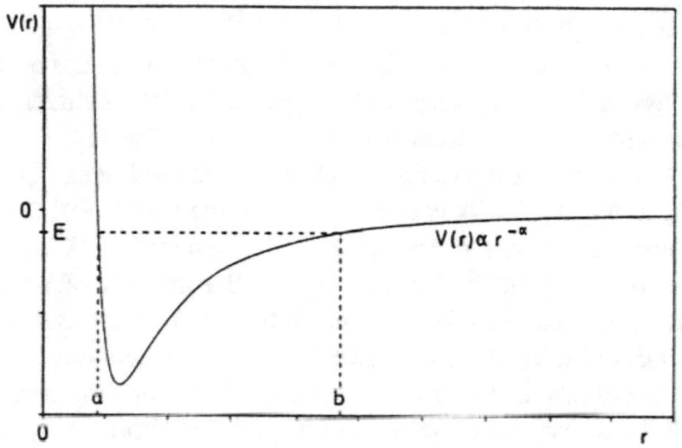

Fig. 1.14 Schematic illustration of the potential $V(r)$ with an attractive inverse power-law tail, (1.313)

r_t diverges to infinity for $E \to 0$. The inner classical turning point r_i, on the other hand, is either zero or determined by a short-ranged repulsive contribution to the potential; r_i depends only weakly on the energy E and converges to a well defined value for $E \to 0$.

The integrated action in the quantization rule (1.308) is given by

$$\frac{1}{2}S(E) = \int_{r_i}^{r_t(E)} p(r)\,\mathrm{d}r$$
$$= \int_{r_i}^{r_0} p(r)\,\mathrm{d}r + \int_{r_0}^{r_t(E)} \sqrt{2\mu\left(\frac{C_\alpha}{r^\alpha} - |E|\right)}\,\mathrm{d}r\,; \qquad (1.320)$$

here we have introduced an energy-independent distance r_0 between r_i and r_t and assume that short-ranged deviations from the inverse-power form of the potential are neglible beyond r_0. The integral $\int_{r_i}^{r_0} p(r)\mathrm{d}r$ converges to a constant for $E \to 0$. The second integral on the right-hand side of the lower line of (1.320) remains finite in the limit $E \to 0$, if the exponent α is larger than two. In this case the action (1.320) remains bounded from above as we approach the threshold $E = 0$, and the quantization rule (1.308) predicts at most a finite number of bound states. If $\alpha < 2$ however, the second integral on the right-hand side of (1.320) diverges in the limit $E \to 0$ and the integrated action S grows beyond all bounds; in this case the quantization rule (1.308) predicts infinitely many bound states. These statements are independent of the shape of the potential at small distances r and are not sensitive to bounded variations in the choice of the Maslov index, so they are quite generally valid and do not depend on the applicability of semiclassical approximations.

Thus the number of bound states in a potential $V(r)$ depends decisively on the asymptotic behaviour of the potential. *Shorter-ranged potentials*, namely those which vanish more rapidly than $1/r^2$, can support at most a finite number of bound states. *Very-long-ranged potentials*, namely those behaving asymptotically as $V(r) \overset{r\to\infty}{\to} -C_\alpha/r^\alpha$ with $0 < \alpha < 2$, always support an infinite number of bound states. This class of attractive very-long-ranged potentials includes, of course, the attractive Coulomb potential discussed in Sect. 1.3.3. The fact that the threshold $E = 0$ represents the semiclassical limit of the Schrödinger equation for such very-long-ranged potential tails is consistent with the notion that the limit of large quantum numbers, $n \to \infty$, corresponds to the semiclassical limit. For shorter-ranged potentials falling off faster than $-1/r^2$, the number of bound states may be large if the potential well is deep enough, but it is always finite, the limit $n \to \infty$ does not exist, which is consistent with the observation that the threshold represents the anticlassical, the extreme quantum limit for shorter-ranged potential tails. A detailed comparison of very-long-ranged and shorter-ranged potentials is given further on in Sect. 3.1.

Potentials asymptotically proportional to $1/r^2$ represent a special case. A potential behaving asymptotically as

$$V(r) \overset{r\to\infty}{\to} \frac{\hbar^2}{2\mu}\frac{\gamma}{r^2}, \quad \gamma < 0, \tag{1.321}$$

supports an infinite number of bound states if and only if $\gamma < -1/4$ (see [MF53] p. 1665 and Sect. 3.1.5). Note that the integrated action (1.320) is infinite in the limit $E \to 0$ for an attractive $1/r^2$-potential. The condition for supporting infinitely many bound states in a $1/r^2$ potential coincides with the requirement that the potential still be attractive after being subjected to the Langer modification (1.305). (See Problem 1.9.)

The qualitatively different properties of long-ranged and short-ranged potentials are also manifest at positive energies, $E = \hbar^2k^2/(2\mu) > 0$. Figure 1.15 shows the quantality function (1.298) for attractive inverse power-law potentials (1.313) with $\alpha = 1$, 2 and 3 at a given positive energy with scales chosen such that $E = |V_\alpha^{(-)}(r = 1)| = 1$ in all cases. For large distances, the energy term in the Schrödinger equation eventually dominates over the potential term, so $p(r) \overset{r\to\infty}{\sim}$ const $\neq 0$ and the quantality function goes to zero. For $r \to 0$, the quantality function is determined by the potential term and depends strongly on whether $\alpha < 2$, $\alpha = 2$ or $\alpha > 2$. As expected from (1.314), $Q(r)$ diverges in the limit $r \to 0$ for $\alpha = 1$ and approaches a constant finite value for $\alpha = 2$. For $\alpha = 3$, on the other hand, as for any $\alpha > 2$, $Q(r) \to 0$ for $r \to 0$.

For attractive inverse power-law potentials (1.313) with $\alpha > 2$ there is a semiclassical regime of small r values where WKB wave functions are accurate approximations to the exact solutions of the Schrödinger equation, even though the potential is a rapidly varying function of r. For sufficiently deep potentials of the type shown in Fig 1.14, this inner semiclassical regime may reach well beyond the domain of short distances where the potential neccessarily deviates from the inverse-power form. The quantum mechanical regime of the potential tail, where

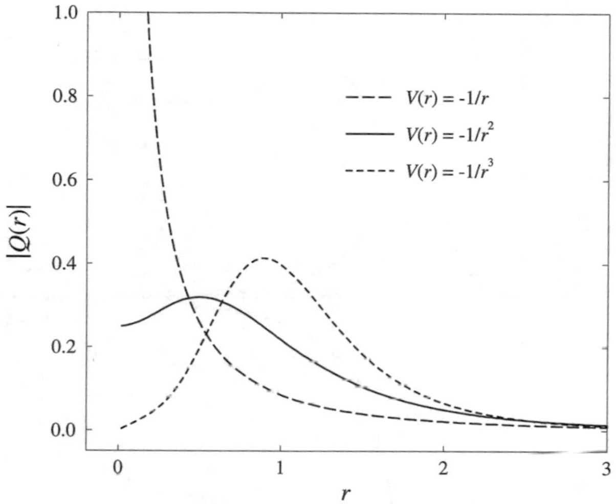

Fig. 1.15 Quantality function (1.298) for attractive inverse power-law potentials (1.313) with $\alpha = 1, 2$ and 3. Parameters were chosen as $C_\alpha = 1$, $E = 1$ and $\hbar^2/(2\mu) = 1$

the quantality function is significantly non-vanishing, is then localized between the semiclassical regimes at small and large r values.

The extent and location of the quantal region of an attractive potential tail is well understood for inverse power-law potentials (1.313). Note that, as for all homogeneous potentials, the properties of the Schrödinger equation do not depend on energy and potential strength independently. If we rewrite the Schrödinger equation at energy $E = \hbar^2k^2/(2\mu)$ in terms of the dimensionless variable $x = r/\beta_\alpha$, with the quantum length β_α given by (1.316),

$$\left(\frac{d^2}{dx^2} + (k\beta_\alpha)^2 + \frac{1}{x^\alpha}\right)\psi(x) = 0\,, \tag{1.322}$$

we see that the "scaled energy" $(k\beta_\alpha)^2$ is the one essential parameter affecting the quantum mechanical properties of the system. Figure 1.16 shows the quantality function (1.298) for the attractive inverse-cube potential, $\alpha = 3$, for three values of the scaled energy, namely 0.1, 1 and 10. The lower panel of the figure shows the scaled potential, $v = V_3^{(-)}(r) \times 2\mu(\beta_3)^2/\hbar^2 = -(\beta_3/r)^3$. The quantal region of coordinate space shrinks and moves to smaller distances as the energy increases.

For attractive inverse power-law potentials with $\alpha > 2$ and positive energies $E = \hbar^2k^2/(2\mu) > 0$, it can be shown that the maximum of $|Q(r)|$ lies close to the characteristic distance r_E, where the absolute value of the potential equals the total energy,

$$|V_\alpha^{(-)}(r_E)| = E\,; \tag{1.323}$$

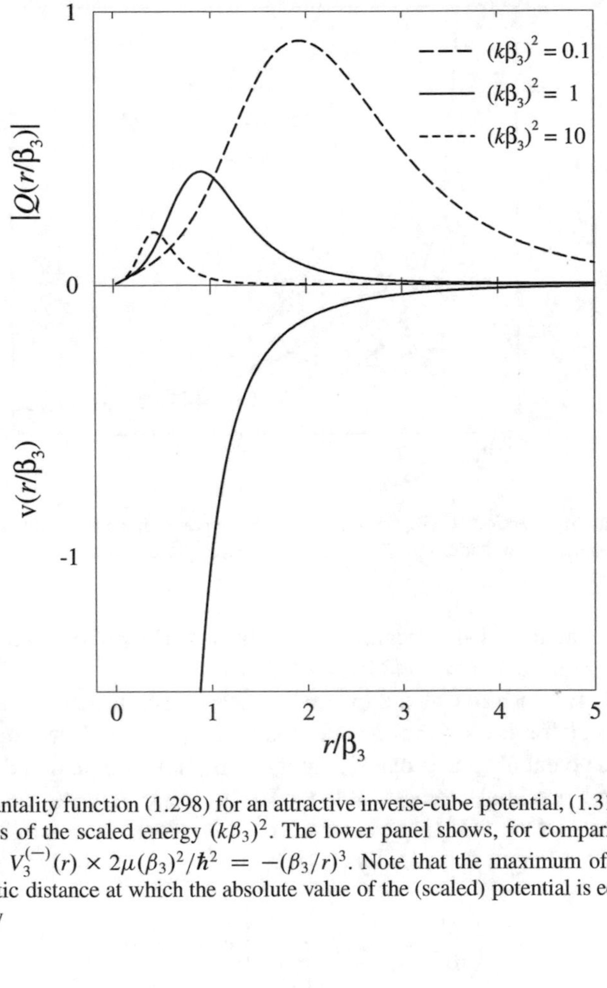

Fig. 1.16 Quantality function (1.298) for an attractive inverse-cube potential, (1.313) with $\alpha = 3$, for three values of the scaled energy $(k\beta_3)^2$. The lower panel shows, for comparison, the scaled potential $\upsilon = V_3^{(-)}(r) \times 2\mu(\beta_3)^2/\hbar^2 = -(\beta_3/r)^3$. Note that the maximum of $|Q|$ is close to the characteristic distance at which the absolute value of the (scaled) potential is equal to the total (scaled) energy

r_E is the classical turning point for the inverted potential $-V_\alpha^{(-)}(r) = V_\alpha^{(+)}(r)$ as given by (1.315),

$$r_E = \left(\frac{C_\alpha}{E}\right)^{1/\alpha} = \beta_\alpha(k\beta_\alpha)^{-2/\alpha} . \tag{1.324}$$

Indeed, the quantality function can be calculated analytically, and the maximum of $|Q(r)|$ occurs at

$$r_{\max} = c_\alpha r_E , \tag{1.325}$$

where $c_3 = 0.895$, $c_4 = 1$ and $1 < c_\alpha < 1.06$ for larger powers α, see Problem 1.10.

1.7 Angular Momentum and Spin

An angular momentum operator $\hat{\boldsymbol{J}}$ is a vector of operators $\hat{J}_x, \hat{J}_y, \hat{J}_z$ obeying the following commutation relations (see (1.56)):

$$[\hat{J}_x, \hat{J}_y] = \mathrm{i}\hbar\hat{J}_z\,, \quad [\hat{J}_y, \hat{J}_z] = \mathrm{i}\hbar\hat{J}_x\,, \quad [\hat{J}_z, \hat{J}_x] = \mathrm{i}\hbar\hat{J}_y\,. \tag{1.326}$$

These can be summarized in the suggestive if somewhat unorthodox equation

$$\hat{\boldsymbol{J}} \times \hat{\boldsymbol{J}} = \mathrm{i}\hbar\hat{\boldsymbol{J}}\,. \tag{1.327}$$

From the commutation relations (1.326) it already follows, that the eigenvalues of $\hat{\boldsymbol{J}}^2 = \hat{J}_x^2 + \hat{J}_y^2 + \hat{J}_z^2$ have the form $j(j+1)\hbar^2$ and that to each value of j there are exactly $2j+1$ different eigenvalues of $\hat{J}_z$, namely $m\hbar$ with $m = -j, -j+1, \ldots, j-1, j$. The number $2j+1$ must be a positive integer so that j itself must be integer or half-integer. For orbital angular momenta, which can be written as operators in the spatial varibles (see (1.66)), the requirement of uniqueness of the wave function in coordinate space restricts thc angular momentum quantum numbers to integers. This restriction does not hold for spin angular momenta for which there are no classical counterparts in coordinate space.

1.7.1 Addition of Angular Momenta

Let $\hat{\boldsymbol{J}}_1$ and $\hat{\boldsymbol{J}}_2$ be two commuting angular momenta ($[\hat{J}_{1x}, \hat{J}_{2x}] = [\hat{j}_{1x}, \hat{j}_{2y}] = 0$, etc.) with angular momentum quantum numbers j_1, m_1 and j_2, m_2 respectively. Since $\hat{\boldsymbol{J}}_1$ and $\hat{\boldsymbol{J}}_2$ obey the commutation relations (1.326), the sum

$$\hat{\boldsymbol{J}} = \hat{\boldsymbol{J}}_1 + \hat{\boldsymbol{J}}_2 \tag{1.328}$$

also obeys these relations and is also an angular momentum. $\hat{\boldsymbol{J}}^2$ has the eigenvalues $j(j+1)\hbar^2$ and $\hat{J}_z$ has the eigenvalues $m\hbar$.

The squares of the angular momenta commute,

$$[\hat{\boldsymbol{J}}^2, \hat{\boldsymbol{J}}_1^2] = [\hat{\boldsymbol{J}}^2, \hat{\boldsymbol{J}}_2^2] = 0\,, \tag{1.329}$$

and the components of the summed angular momentum $\hat{\boldsymbol{J}}$ commute with $\hat{\boldsymbol{J}}_1^2$ and $\hat{\boldsymbol{J}}_2^2$, e.g. for $\hat{J}_z = \hat{J}_{1z} + \hat{J}_{2z}$:

$$[\hat{J}_z, \hat{\boldsymbol{J}}_1^2] = [\hat{J}_z, \hat{\boldsymbol{J}}_2^2] = 0\,. \tag{1.330}$$

However, the components of $\hat{\boldsymbol{J}}_1$ and $\hat{\boldsymbol{J}}_2$ do not commute with the square of the summed angular momentum,

$$\hat{\boldsymbol{J}}^2 = \hat{\boldsymbol{J}}_1^2 + \hat{\boldsymbol{J}}_2^2 + 2\hat{\boldsymbol{J}}_1 \cdot \hat{\boldsymbol{J}}_2 , \tag{1.331}$$

because e.g. $\hat{J}_{1z}$ doesn't commute the terms $\hat{J}_{1x}\hat{J}_{2x}$ and $\hat{J}_{1y}\hat{J}_{2y}$ in the scalar product $\hat{\boldsymbol{J}}_1 \cdot \hat{\boldsymbol{J}}_2$.

Four mutually commuting operators are already sufficient to completely classify the angular momentum eigenstates, and these four operators can be chosen in different ways. In the *uncoupled representation* the four operators are $\hat{\boldsymbol{J}}_1^2$, $\hat{J}_{1z}$, $\hat{\boldsymbol{J}}_2^2$, $\hat{J}_{2z}$. The corresponding eigenstates $|j_1, m_1, j_2, m_2\rangle$ are also eigenstates of $\hat{J}_z = \hat{J}_{1z} + \hat{J}_{2z}$ with the eigenvalues $m\hbar$ ($m = m_1 + m_2$), but they are in general not eigenstates of $\hat{\boldsymbol{J}}^2$. In the *coupled representation* the basis states $|j, m, j_1, j_2\rangle$ are eigenstates of the four operators $\hat{\boldsymbol{J}}^2$, $\hat{J}_z$, $\hat{\boldsymbol{J}}_1^2$ and $\hat{\boldsymbol{J}}_2^2$. They are in general not eigenstates of $\hat{J}_{1z}$ and $\hat{J}_{2z}$.

For given values of j_1 and j_2, the basis states in the coupled representation can of course be expressed as linear combinations of the uncoupled basis states:

$$|j, m, j_1, j_2\rangle = \sum_{m_1, m_2} \langle j_1, m_1, j_2, m_2 | j, m\rangle \, |j_1, m_1, j_2, m_2\rangle . \tag{1.332}$$

Vice-versa we can express the uncoupled states as linear combinations of the coupled states:

$$|j_1, m_1, j_2, m_2\rangle = \sum_{j,m} \langle j, m | j_1, m_1, j_2, m_2\rangle \, |j, m, j_1, j_2\rangle . \tag{1.333}$$

The coefficients appearing in (1.332), (1.333) are the *Clebsch-Gordan coefficients* [Edm60],

$$\langle j_1, m_1, j_2, m_2 | j, m\rangle = \langle j, m | j_1, m_1, j_2, m_2\rangle^* , \tag{1.334}$$

which are real if the phases of the basis states are appropriately chosen.

Obviously the Clebsch-Gordan coefficient $\langle j_1, m_1, j_2, m_2 | j, m\rangle$ is only non-zero if

$$m_1 + m_2 = m . \tag{1.335}$$

A further selection rule is the triangle condition which determines the minimal and maximal summed angular momentum quantum number j for given values of j_1 and j_2

$$|j_1 - j_2| \le j \le j_1 + j_2 . \tag{1.336}$$

For fixed j_1 and j_2, each possible summed angular momentum quantum number j encompasses exactly $2j + 1$ eigenstates corresponding to the different eigenvalues

$m\hbar$ of $\hat{J}_z$. Since coupling cannot affect the dimension of the space spanned by the basis states, the total number of coupled states for all possible values of j (at fixed values of j_1 and j_2) is equal to the number $(2j_1 + 1) \times (2j_2 + 1)$ of states in the uncoupled basis:

$$\sum_{j=|j_1-j_2|}^{j_1+j_2} (2j + 1) = (2j_1 + 1)(2j_2 + 1) . \tag{1.337}$$

1.7.2 Spin

It is known from experimental investigations that an electron has an internal angular momentum called spin, and that the total angular momentum $\hat{\boldsymbol{J}}$ of an electron ist the sum of its orbital angular momentum $\hat{\boldsymbol{L}}$ and its spin $\hat{\boldsymbol{S}}$:

$$\hat{\boldsymbol{J}} = \hat{\boldsymbol{L}} + \hat{\boldsymbol{S}} . \tag{1.338}$$

The electron's spin has no classical counterpart and cannot be related to ordinary spatial coordinates. All physical states are eigenstates of $\hat{\boldsymbol{S}}^2$ with eigenvalue $s(s + 1)\hbar^2$, and the spin quantum number s always has the same value $s = 1/2$. Any component, e.g. $\hat{S}_z$ of $\hat{\boldsymbol{S}}$ has two eigenvalues $m_s\hbar$, namely $m_s = +1/2$ and $m_s = -1/2$.

The wave function of an electron thus depends not only on e.g. the spatial coordinate $\boldsymbol{r}$, but also on the spin variable m_s:

$$\psi = \psi(\boldsymbol{r}, m_s) . \tag{1.339}$$

Since the discrete variable m_s can only take on two values, it is convenient to write the wave function (1.339) as a pair of ordinary functions of $\boldsymbol{r}$ corresponding to the two values $m_s = 1/2$ and $m_s = -1/2$:

$$\psi = \begin{pmatrix} \psi_+(\boldsymbol{r}) \\ \psi_-(\boldsymbol{r}) \end{pmatrix} = \begin{pmatrix} \psi(\boldsymbol{r}, m_s = +\frac{1}{2}) \\ \psi(\boldsymbol{r}, m_s = -\frac{1}{2}) \end{pmatrix} . \tag{1.340}$$

These two-component entities are called *spinors* in order to distinguish them from ordinary vectors in coordinate space. If we introduce the two basis spinors

$$\chi_+ = \begin{pmatrix} 1 \\ 0 \end{pmatrix} , \quad \chi_- = \begin{pmatrix} 0 \\ 1 \end{pmatrix} , \tag{1.341}$$

we can write the general one-electron wave function (1.340) as

$$\psi = \psi_+(\boldsymbol{r})\chi_+ + \psi_-(\boldsymbol{r})\chi_- . \tag{1.342}$$

The scalar product of two spinors of the form (1.340) or (1.342) is

$$\langle\psi|\phi\rangle = \int \mathrm{d}^3r \sum_{m_s=-\frac{1}{2}}^{+\frac{1}{2}} \psi^*(\boldsymbol{r}, m_s)\phi(\boldsymbol{r}, m_s) = \langle\psi_+|\phi_+\rangle + \langle\psi_-|\phi_-\rangle . \tag{1.343}$$

States ψ normalized to unity fulfill the condition

$$\langle\psi_+|\psi_+\rangle + \langle\psi_-|\psi_-\rangle = \int \mathrm{d}^3r(|\psi_+(\boldsymbol{r})|^2 + |\psi_-(\boldsymbol{r})|^2) = 1 , \tag{1.344}$$

and $|\psi_+(\boldsymbol{r})|^2$ is e.g. the probability density for finding the electron at the position $\boldsymbol{r}$ and in the spin state χ_+.

Linear operators can not only act on the component functions ψ_+, ψ_-, they can also mix up the components in a spinor. The most general linear operators in spin space are 2×2 matrices of complex numbers. These can be expressed as linear combinations of four basis matrices; the most commonly used basis consists of the unit matrix and the three Pauli spin matrices:

$$\hat{\sigma}_x = \begin{pmatrix} 0 & 1 \\ 1 & 0 \end{pmatrix} , \quad \hat{\sigma}_y = \begin{pmatrix} 0 & -\mathrm{i} \\ \mathrm{i} & 0 \end{pmatrix} , \quad \hat{\sigma}_z = \begin{pmatrix} 1 & 0 \\ 0 & -1 \end{pmatrix} . \tag{1.345}$$

Thus the most general linear operator in the Hilbert space of one electron states has the form

$$\hat{O} = \hat{O}_0 + \hat{O}_1\hat{\sigma}_x + \hat{O}_2\hat{\sigma}_y + \hat{O}_3\hat{\sigma}_z , \tag{1.346}$$

where $\hat{O}_i$ are spin-independent operators such as $\hat{\boldsymbol{p}}$, $\hat{\boldsymbol{r}}$ and functions thereof.

The spinors χ_+ and χ_- of (1.341) are eigenstates of $\hat{\sigma}_z$ with eigenvalues $+1$ and -1 respectively. Since they are also supposed to be eigenstates of the z-component $\hat{S}_z$ of the spin with eigenvalues $+(1/2)\hbar$ and $-(1/2)\hbar$ respectively, the relation between $\hat{S}_z$ and $\hat{\sigma}_z$ must simply be:

$$\hat{S}_z = \frac{1}{2}\hbar\hat{\sigma}_z . \tag{1.347}$$

Together with the other two components,

$$\hat{S}_x = \frac{1}{2}\hbar\hat{\sigma}_x , \quad \hat{S}_y = \frac{1}{2}\hbar\hat{\sigma}_y , \tag{1.348}$$

we have the spin operator $\hat{\boldsymbol{S}}$ as

$$\hat{\boldsymbol{S}} = \frac{1}{2}\hbar\hat{\boldsymbol{\sigma}} . \tag{1.349}$$

From the commutation relations of the Pauli spin matrices,

$$\hat{\sigma}_x\hat{\sigma}_y = i\hat{\sigma}_z = -\hat{\sigma}_y\hat{\sigma}_x , \quad \text{etc.} , \tag{1.350}$$

it immediately follows, that the spin components defined by (1.347)–(1.349) obey the commutation relations (1.326) characteristic of angular momentum operators:

$$[\hat{S}_x, \hat{S}_y] = i\hbar\hat{S}_z , \quad [\hat{S}_y, \hat{S}_z] = i\hbar\hat{S}_x , \quad [\hat{S}_z, \hat{S}_x] = i\hbar\hat{S}_y . \tag{1.351}$$

Furthermore, the properties

$$\hat{\sigma}_x^2 = \hat{\sigma}_y^2 = \hat{\sigma}_z^2 = 1 \tag{1.352}$$

imply that

$$\hat{\boldsymbol{S}}^2 = \hat{S}_x^2 + \hat{S}_y^2 + \hat{S}_z^2 = \frac{3}{4}\hbar^2 , \tag{1.353}$$

which of course just means that all states are eigenstates of $\hat{\boldsymbol{S}}^2$ with eigenvalue $s(s+1)\hbar^2$ corresponding to $s = 1/2$.

The spin $\hat{\boldsymbol{S}}$ is a vector operator consisting of three components, just like the position $\hat{\boldsymbol{r}}$ and the momentum $\hat{\boldsymbol{p}}$. The components of $\hat{\boldsymbol{S}}$ are however, in contrast to position and momentum, not ordinary operators acting on functions, but 2×2 matrices which linearly transform the spinor components. The spinor components must not be confused with the components of ordinary vectors in coordinate space.

1.7.3 Spin-Orbit Coupling

In addition to the usual kinetic and potential energy terms, the Hamiltonian for an electron in a radially symmetric potential $V(r)$ contains a further term which couples the spin and spatial degrees of freedom:

$$\hat{H} = -\frac{\hbar^2}{2\mu}\Delta + V(r) + V_{LS}(r)\hat{\boldsymbol{L}}\cdot\hat{\boldsymbol{S}} . \tag{1.354}$$

The spin-orbit coupling term can be physically understood as the interaction energy of two magnetic dipoles associated with the orbital angular momentum $\hat{\boldsymbol{L}}$ and the spin $\hat{\boldsymbol{S}}$ respectively. More precisely the spin-orbit coupling appears as an additional contribution to the conventional Hamiltonian (1.53) in the non-relativistic limit of the relativistic Dirac equation (see Sect. 2.1.4, (2.45)). The coupling function V_{LS} derived in this way is

$$V_{LS}(r) = \frac{1}{2\mu^2c^2}\frac{1}{r}\frac{\mathrm{d}V}{\mathrm{d}r}\,. \tag{1.355}$$

The Hamiltonian (1.354) no longer commutes with the components of the orbital angular momentum $\hat{\boldsymbol{L}}$, but it commutes with the components of the total angular momentum $\hat{\boldsymbol{J}} = \hat{\boldsymbol{L}} + \hat{\boldsymbol{S}}$, because we can express the spin-orbit coupling operator in terms of the squares of the angular momenta

$$\hat{\boldsymbol{L}}\cdot\hat{\boldsymbol{S}} = \frac{1}{2}(\hat{\boldsymbol{J}}^2 - \hat{\boldsymbol{L}}^2 - \hat{\boldsymbol{S}}^2)\,, \tag{1.356}$$

and the components of the summed angular momentum commute with all squares (see (1.330)). Hence it is appropriate to couple the eigenstates of orbital angular momentum and spin to eigenstates of the total angular momentum $\hat{\boldsymbol{J}}$. This is done with the Clebsch-Gordan coefficients as a special case of (1.332):

$$|j,m,l,s\rangle = \sum_{m_l,m_s} \langle l,m_l,s,m_s|\,j,m\rangle\, Y_{l,ml}(\theta,\phi)\chi_{m_s}\,. \tag{1.357}$$

The quantum number s in (1.357) is of course always 1/2. Since l and m_l are always integers, j and m must always be half integers (meaning odd multiples of 1/2). Because of the triangle condition (1.336) there are exactly two possible values of j for each value of l larger than zero, namely $j = l + 1/2$ and $j = l - 1/2$. For $l = 0$ there is only one possible value of j, namely +1/2.

The coupled eigenstates $|j,m,l,s\rangle$ are called *generalized spherical harmonics* and are written as $\mathcal{Y}_{j,m,l}$. They are two-component spinors, and it is clear from (1.341) and the selection rule $m = m_l + m_s$ (see (1.335)) that the upper component corresponding to a contribution with $m_s = +1/2$ contains a spherical harmonic with $m_l = m - 1/2$, while the lower component contains a spherical harmonic with $m_l = m + 1/2$. The generalized spherical harmonics are thus essentially two-component spinors of spherical harmonics. Inserting the known Clebsch-Gordan coefficients [New82, Tin64] yields the explicit expressions

$$\mathcal{Y}_{j,m,l} = \frac{1}{\sqrt{2j}}\begin{pmatrix}\sqrt{j+m}\,Y_{l,m-\frac{1}{2}}(\theta,\phi)\\ \sqrt{j-m}\,Y_{l,m+\frac{1}{2}}(\theta,\phi)\end{pmatrix},\quad j = l + \frac{1}{2}\,,$$

$$\mathcal{Y}_{j,m,l} = \frac{1}{\sqrt{2j+2}}\begin{pmatrix}-\sqrt{j+1-m}\,Y_{l,m-\frac{1}{2}}(\theta,\phi)\\ \sqrt{j+1+m}\,Y_{l,m+\frac{1}{2}}(\theta,\phi)\end{pmatrix},\quad j = l - \frac{1}{2}\,. \tag{1.358}$$

The time-independent Schrödinger equation $\hat{H}\psi = E\psi$ with the Hamiltonian (1.354) corresponds to two coupled partial differential equations for the two components $\psi_+(\boldsymbol{r})$ and $\psi_-(\boldsymbol{r})$ of the spinor wave function (1.340). A substantial simplification can be achieved if we extend the ansatz (1.74) for separating radial and angular variables to the present case of spinor wave functions using the generalized spherical harmonics:

$$\psi(\boldsymbol{r}, m_s) = \frac{\phi_{j,l}(r)}{r}\mathcal{Y}_{j,m,l}\,. \tag{1.359}$$

In addition to the relation (1.70) (with (1.58)) we can now use the fact that the generalized spherical harmonics $\mathcal{Y}_{j,m,l}$ are eigenfunctions of the spin-orbit coupling operator (1.356),

$$\hat{\boldsymbol{L}}\cdot\hat{\boldsymbol{S}}\,\mathcal{Y}_{j,m,l} = \frac{\hbar^2}{2}[j(j+1) - l(l+1) - s(s+1)]\mathcal{Y}_{j,m,l}\,, \tag{1.360}$$

where $s(s+1) = 3/4$. For the two possible cases $j = l \pm 1/2$ we have

$$\hat{\boldsymbol{L}}\cdot\hat{\boldsymbol{S}}\,\mathcal{Y}_{j,m,l} = \frac{\hbar^2}{2}\begin{cases} l\,\mathcal{Y}_{j,m,l} & \text{for } j = l + 1/2\,, \\ -(l+1)\,\mathcal{Y}_{j,m,l} & \text{for } j = l - 1/2\,. \end{cases} \tag{1.361}$$

Thus the Schrödinger equation can be reduced to a radial Schrödinger equation

$$\left(-\frac{\hbar^2}{2\mu}\frac{\mathrm{d}^2}{\mathrm{d}r^2} + \frac{l(l+1)\hbar^2}{2\mu r^2} + V(r) + \frac{\hbar^2}{2}F(j,l)V_{LS}(r)\right)\phi_{j,l}(r) = E\phi_{j,l}(r)\,, \tag{1.362}$$

and the factor $F(j,l)$ is l or $-(l+1)$ for $j = l + 1/2$ and $j = l - 1/2$ respectively. For a given orbital angular momentum quantum number l, the spin-orbit potentials for the two possible values of j have opposite sign.

Including the spin variable in the description of an electron in a radially symmetric potential still allows us to reduce the time independent Schrödinger equation to an ordinary differential equation for the radial wave function. The radial Schrödinger equation now depends not only on the orbital angular momentum quantum number l, but also on the total angular momentum quantum number j (not, however, on m).

Problems

1.1 Consider a point particle of mass μ in a radially symmetric potential

$$V(r) = \begin{cases} -V_0 & \text{for } r \le r_0\,, \\ 0 & \text{for } r > r_0\,, \end{cases}$$

where V_0 is a positive constant considerably larger than $\hbar^2/(\mu r_0^2)$. Give an approximate (± 1) estimate for the number of bound states for $l = 0$.

1.2

a) Consider the following radial wave function which is normalized to unity:

$$\phi(r) = (\sqrt{\pi} b)^{-\frac{1}{2}} \frac{2r}{b} \mathrm{e}^{-r^2/(2b^2)} .$$

Calculate the overlaps (i.e. the scalar products) $\langle \phi | \phi_{n,l=0} \rangle$ with the radial eigenfunctions (1.82) of the harmonic oscillator with an oscillator width $\beta \neq b$.

b) Consider the following radial wave function which is normalized to unity:

$$\phi(r) = 2b^{-\frac{1}{2}} \frac{r}{b} \mathrm{e}^{-r/b} .$$

Calculate the overlaps $\langle \phi | \phi_{n,l=0} \rangle$ with the radial eigenfunctions (1.139) of the attractive Coulomb potential with a Bohr radius $a \neq b$.

c) Evaluate the first four or five terms of the sum

$$\sum_n |\langle \phi | \phi_{n,l=0} \rangle|^2$$

for the explicit values $b = \beta/2$ and $b = a/2$ respectively. Estimate the limit to which the sum converges in both cases.

d) Repeat the exercise (c) for the Coulomb potential for $b = a$ and $b = 2a$. Hint:

$$\int_0^\infty \mathrm{e}^{-sx} x^\alpha L_\nu^\alpha(x)\, \mathrm{d}x = \frac{\Gamma(\alpha + \nu + 1)(s-1)^\nu}{\nu!\, s^{\alpha+\nu+1}} ,$$

$$\int_0^\infty \mathrm{e}^{-sx} x^{\alpha+1} L_\nu^\alpha(x)\, \mathrm{d}x = -\frac{\mathrm{d}}{\mathrm{d}s} \left(\int_0^\infty \mathrm{e}^{-sx} x^\alpha L_\nu^\alpha(x)\, \mathrm{d}x \right) .$$

1.3 Use the recurrence relation (A.16) and the orthogonality relation (A.15) in Appendix A.2 to show that the expectation value of the radius r in the Coulomb eigenfunctions (1.139) (with Bohr radius a) is given by:

$$\langle \phi_{n,l} | r | \phi_{n,l} \rangle = \frac{a}{2} [3n^2 - l(l+1)] .$$

1.4 A free point particle of mass μ in one spatial dimension is described at time $t = 0$ by the wave function (1.167),

$$\psi(x, t = 0) = (\beta \sqrt{\pi})^{-1/2} \exp\left(-\frac{(x - x_0)^2}{2\beta^2} \right) \mathrm{e}^{\mathrm{i}k_0 x} .$$

Calculate the momentum representation $\tilde{\psi}(k, t = 0)$ of the initial wave function,

$$\tilde{\psi}(k, t = 0) = \frac{1}{\sqrt{2\pi}} \int_{-\infty}^{\infty} \mathrm{e}^{-ikx}\, \psi(x, t = 0)\, \mathrm{d}x\,,$$

and discuss the time evolution of $\tilde{\psi}(k, t)$ according to the time-dependent Schrödinger equation (1.38).

Calculate the time-dependent wave function $\psi(x, t)$ in coordinate space and discuss the evolution of the uncertainties Δx, Δp as defined by (1.35).

1.5 Show that the free Green's function for $l = 0$,

$$G_0(r, r') = -\frac{2\mu}{\hbar^2 k} \sin(kr_<) \cos(kr_>)\,,$$

($r_<$ is the smaller, $r_>$ the larger of the two radii r, r') fulfills the defining equation:

$$\left(E + \frac{\hbar^2}{2\mu} \frac{\mathrm{d}^2}{\mathrm{d}r^2}\right) G(r, r') = \delta(r - r')\,.$$

1.6 Consider a Hamiltonian

$$\hat{H} = \hat{H}_0 + \hat{W}$$

in a two-dimensional Hilbert space, where

$$\hat{H}_0 = \begin{pmatrix} \varepsilon_1 & 0 \\ 0 & \varepsilon_2 \end{pmatrix}, \quad \hat{W} = \begin{pmatrix} 0 & w \\ w & 0 \end{pmatrix}.$$

Calculate the eigenstates and eigenvalues of $\hat{H}$

(a) in lowest non-vanishing order perturbation theory treating $\hat{W}$ as the perturbation,
(b) by exact diagonalization of $\hat{H}$.

How do the results in both cases depend on the difference $\varepsilon_1 - \varepsilon_2$ of the unperturbed energies?

1.7

a) Use the Bohr-Sommerfeld quantization rule (1.310) to calculate the energy eigenvalues of the bound states of a one-dimensional harmonic oscillator: $V(x) = (\mu/2)\omega^2 x^2$.
b) Use the quantization rule (1.308) to calculate the energy eigenvalues of the bound states in a one-dimensional infinitely deep well,

$$V(x) = \begin{cases} 0\,, & 0 < x < L\,, \\ +\infty\,, & x < 0 \quad \text{or} \quad x > L\,. \end{cases}$$

c) Consider a particle of mass μ reflected by a one-dimensional potential step,

$$V(x) = \begin{cases} 0\,, & x < L\,, \\ +V_0\,, & x \geq L\,, \end{cases}$$

at an energy E, $0 < E < V_0$. Calculate the reflection phase ϕ and the amplitude N for the WKB wave function according to the connection formula (1.300).

Now consider a particle of mass μ bound in the finite sharp-step potential

$$V(x) = \begin{cases} 0\,, & 0 < x < L\,, \\ +V_0\,, & x \leq 0 \quad \text{or} \quad x \geq L\,. \end{cases}$$

Discuss the accuracy of the wave functions and the energy eigenvalues obtained via the quantization rule (1.308) when the appropriate reflection phases and normalization constants are used.

1.8 Consider a point particle of mass μ in a one-dimensional potential $V(x)$. Calculate the energy expectation value for the Gaussian wave function

$$\psi(x) = (\sqrt{\pi}b)^{-1/2}\,\mathrm{e}^{-x^2/(2b^2)}\,,$$

(which is normalized to unity), and think about the limit $b \to \infty$.

Show that a potential $V(x)$ with $\lim_{|x|\to\infty} V(x) = 0$, which is more attractive than repulsive, meaning

$$\int_{-\infty}^{\infty} V(x)\,\mathrm{d}x < 0\,,$$

always supports at least one bound state. Why doesn't this statement hold for a particle in three dimensions?

1.9 Consider a point particle of mass μ in a radially symmetric potential $V(r)$, which is equal to $-C/r^2$ ($C > 0$) beyond a certain radius r_0,

$$V(r) = -\frac{C}{r^2}\,, \quad r > r_0\,,$$

and which is repulsive near the origin $r = 0$. Use the WKB approximation and the Langer modification (1.305) to show that, for values of C larger than a certain l-dependent threshold $C(l)$, the energy eigenvalues for high quantum numbers n are given by

$$E_{n,l} = -c_1\mathrm{e}^{-c_2(l)n}\,.$$

Determine the constant $c_2(l)$.

1.10 Calculate the quantality function (1.298) for an attractive inverse power-law potential,

$$V_\alpha^{(-)}(r) = -\frac{\hbar^2}{2\mu}\frac{(\beta_\alpha)^{\alpha-2}}{r^\alpha}, \quad \alpha > 2,$$

at energy $E = \hbar^2 k^2/(2\mu) > 0$, and show that the maximum of $|Q(r)|$ is located at

$$r = [F(\alpha)]^{1/\alpha} r_E, \quad F_\alpha = \frac{5}{4} - \frac{9}{2\alpha+4} + \frac{9\alpha}{4\alpha+8}\sqrt{1-\frac{20}{27}\left(\frac{\alpha+2}{\alpha+1}\right)},$$

where r_E is the length defined in (1.323), (1.324). Evaluate $[F(\alpha)]^{1/\alpha}$ for integer values of α from 3 to 10.

1.11 Use (1.70) to verify the following identities:

$$[\hat{p}^2, r] = -2\hbar^2\left(\frac{\partial}{\partial r} + \frac{1}{r}\right),$$

$$[\hat{p}^2, r^2] = -2\hbar^2\left(2r\frac{\partial}{\partial r} + 3\right).$$

References

[Bay69] G. Baym, *Lectures on Quantum Mechanics* (Benjamin, New York, 1969)
[BB97] M. Brack, R.K. Bhaduri, *Semiclassical Physics* (Addison-Wesley, Reading, 1997)
[BK95] C. Bracher, M. Kleber, Ann. Phys. (Leipzig) **4**, 696 (1995)
[BM72] M.V. Berry, K.E. Mount, Semiclassical approximations in wave mechanics. Rep. Prog. Phys. **35**, 315 (1972)
[BR09] R. Blümel, W.P. Reinhardt, *Chaos in Atomic Physics* (Cambridge University Press, Cambridge, 2009)
[CK97] L.S. Cederbaum, K.C. Kulander, N.H. March (eds.), *Atoms and Molecules in Intense Fields*. Structure and Bonding, vol. 86 (Springer, Berlin, 1997)
[CN02] C.A.A. de Carvalho, H.M. Nussenzveig, Phys. Reports **364**, 83 (2002)
[Edm60] A.R. Edmonds, *Angular Momentum in Quantum Mechanics* (Princeton University Press, Princeton, 1960)
[EK87] W. Elberfeld, M. Kleber, Am. J. Phys. **56**, 154 (1988)
[FE97] H. Friedrich, B. Eckhardt (eds.), *Classical, Semiclassical and Quantum Dynamics in Atoms*. Lecture Notes in Physics, vol. 485 (Springer, Berlin, Heidelberg, New York, 1997)
[FF65] N. Fröman, P.O. Fröman, *JWKB Approximation* (North Holland, Amsterdam, 1965)
[FF96] N. Fröman, P.O. Fröman, *Phase Integral Method*. Springer Tracts in Modern Philosophy, vol. 40 (Springer, Berlin, Heidelberg, New York, 1996)
[FT96] H. Friedrich, J. Trost, Phys. Rev. Lett. **76** 4869 (1996); Phys. Rev. A **54**, 1136 (1996)
[FT04] H. Friedrich, J. Trost, Phys. Reports **397**, 359 (2004)
[Gas74] S. Gasiorowicz, *Quantum Physics* (Wiley, New York, 1974)
[Kle94] M. Kleber, Phys. Reports **236**, 331 (1994)

[LL58] L.D. Landau, E.M. Lifshitz, *Classical Mechanics* (Addison Wesley, Reading, 1958)

[Mes70] A. Messiah, *Quantum Mechanics*, vol. 1 (North Holland, Amsterdam, 1970)

[MF53] P. Morse, H. Feshbach, *Methods of Theoretical Physics*, Part II (McGraw-Hill, New York, 1953)

[MN16] P.J. Mohr, D.B. Newell, B.N. Taylor, http://physics.nist.gov/cuu/Constants/index.html

[MS02] J.G. Muga, R. Sala Mayato, I.L. Egusqiza (eds.), *Time in Quantum Mechanics*. Lecture Notes in Physics, vol. 734 (Springer, Berlin, Heidelberg, 2008)

[New82] R.G. Newton, *Scattering Theory of Waves and Particles*, 2nd edn. (Springer, Berlin, Heidelberg, New York, 1982)

[OR04] V.S. Olkhovsky, E. Recami, J. Jakiel, Phys. Reports **398**, 133 (2004)

[Raz03] M. Razavy, *Quantum Theory of Tunneling* (World Scientific, Singapore, 2003)

[RW94] H. Ruder, G. Wunner, H. Herold, F. Geyer, *Atoms in Strong Magnetic Fields* (Springer, Heidelberg, 1994)

[Sch68] L.I. Schiff, *Quantum Mechanics* (McGraw-Hill, New York, 1968)

[Sch02] F. Schwabl, *Quantum Mechanics*, 3rd edn. (Springer, Berlin, Heidelberg, New York, 2002)

[SS98] P. Schmelcher, W. Schweizer (eds.), *Atoms and Molecules in Strong External Fields*. In: Proceedings of the 172nd WE-Heraeus Seminar (Bad Honnef, 1997) (Plenum Publishing, New York, 1998)

[Tin64] M. Tinkham, *Group Theory and Quantum Mechanics* (McGraw-Hill, New York, 1964)

[Wig55] E.P. Wigner, Phys. Rev. **98**, 145 (1955)

Chapter 2
Atoms and Ions

This chapter summarizes the traditional theory of one- and many-electron systems, which has been developed and successfully applied to many atomic problems for almost a century. The presentation is deliberately brief. A more detailed introduction to atomic physics can be found in the textbook by Bransden and Joachain [BJ83]. At a much more formal level there is "Atomic Many-Body Theory" by Lindgren and Morrison [LM85]. Finally we mention "Atomic Structure" by Condon and Odabasi [CO80], where a comprehensive account of conventional atomic structure calculations can be found.

2.1 One-Electron Systems

2.1.1 The Hydrogen Atom

In non-relativistic quantum mechanics, a hydrogen atom consisting of a proton of mass m_p and an electron of mass m_e is described by the following Hamiltonian:

$$\hat{H}_\mathrm{H} = \frac{\hat{\boldsymbol{p}}_\mathrm{p}^2}{2m_\mathrm{p}} + \frac{\hat{\boldsymbol{p}}_\mathrm{e}^2}{2m_\mathrm{e}} - \frac{e^2}{|\boldsymbol{r}_\mathrm{e} - \boldsymbol{r}_\mathrm{p}|}, \tag{2.1}$$

where $\hat{\boldsymbol{p}}_\mathrm{p}$ and $\hat{\boldsymbol{p}}_\mathrm{e}$ are the momentum operators for the proton and the electron respectively, and $\boldsymbol{r}_\mathrm{p}$ and $\boldsymbol{r}_\mathrm{e}$ are the respective spatial coordinates. Introducing the centre-of-mass coordinate $\boldsymbol{R}$ and the relative distance coordinate $\boldsymbol{r}$,

$$\boldsymbol{R} = \frac{m_\mathrm{p}\boldsymbol{r}_\mathrm{p} + m_\mathrm{e}\boldsymbol{r}_\mathrm{e}}{m_\mathrm{p} + m_\mathrm{e}}, \qquad \boldsymbol{r} = \boldsymbol{r}_\mathrm{e} - \boldsymbol{r}_\mathrm{p}, \tag{2.2}$$

H. Friedrich, *Theoretical Atomic Physics*, Graduate Texts in Physics,
DOI 10.1007/978-3-319-47769-5_2

we can rewrite (2.1) as

$$\hat{H}_{\mathrm{H}} = \frac{\hat{\boldsymbol{P}}^2}{2(m_{\mathrm{p}} + m_e)} + \frac{\hat{\boldsymbol{p}}^2}{2\mu} - \frac{e^2}{r}, \tag{2.3}$$

where $\hat{\boldsymbol{P}}$ is the total momentum and $\hat{\boldsymbol{p}}$ the relative momentum in the two-body system:

$$\hat{\boldsymbol{P}} = \hat{\boldsymbol{p}}_{\mathrm{p}} + \hat{\boldsymbol{p}}_{\mathrm{e}}, \quad \frac{\hat{\boldsymbol{p}}}{\mu} = \frac{\hat{\boldsymbol{p}}_{\mathrm{e}}}{m_{\mathrm{e}}} - \frac{\hat{\boldsymbol{p}}_{\mathrm{p}}}{m_{\mathrm{p}}}. \tag{2.4}$$

In coordinate representation the momentum operators have the explicit form:

$$\hat{\boldsymbol{P}} = \frac{\hbar}{\mathrm{i}} \nabla_{\boldsymbol{R}}, \quad \hat{\boldsymbol{p}} = \frac{\hbar}{\mathrm{i}} \nabla_{\boldsymbol{r}}. \tag{2.5}$$

The mass μ appearing in (2.3) and (2.4) is the *reduced mass*

$$\mu = \frac{m_{\mathrm{e}} m_{\mathrm{p}}}{m_{\mathrm{e}} + m_{\mathrm{p}}} = \frac{m_{\mathrm{e}}}{1 + m_{\mathrm{e}}/m_{\mathrm{p}}}. \tag{2.6}$$

Since the ratio $m_{\mathrm{e}}/m_{\mathrm{p}} = 0.000544617021352(52)$ is very small (the numerical value is taken from [MN16]), the reduced mass μ is only little smaller than the rest mass m_{e} of the electron, namely by about $0.5\,^0/_{00}$.

Thus the Hamiltonian $\hat{H}_{\mathrm{H}}$ consists of a part $\hat{\boldsymbol{P}}^2/[2(m_{\mathrm{p}} + m_{\mathrm{e}})]$ describing the free motion of the centre of mass and an *internal Hamiltonian*,

$$\hat{H} = \frac{\hat{\boldsymbol{p}}^2}{2\mu} - \frac{e^2}{r}, \tag{2.7}$$

describing the motion of the electron relative to the position of the proton. Eigenfunctions $\psi_{\mathrm{cm}}(\boldsymbol{R})$ and eigenvalues E_{cm} for the centre-of-mass motion are known, $\psi_{\mathrm{cm}}(\boldsymbol{R}) \propto \exp(\mathrm{i}\boldsymbol{K}\cdot\boldsymbol{R})$, $E_{\mathrm{cm}} = \hbar^2 K^2/[2(m_{\mathrm{p}} + m_{\mathrm{e}})]$, so solving the two-body problem (2.1) or (2.3) is reduced to the problem of solving the one-body Schrödinger equation with the internal Hamiltonian (2.7).

This is just the one-body problem in an attractive Coulomb potential which was discussed in detail in Sect. 1.3.3. The energy eigenvalues are

$$\begin{aligned} E_n = -\frac{\mathcal{R}}{n^2}, \quad & n = 1, 2, 3, \ldots, \\ & l = 0, 1, \ldots, n-1, \\ & m = -l, -l+1, \ldots, l-1, l, \end{aligned} \tag{2.8}$$

where the Rydberg energy $\mathcal{R} = \mu e^4/(2\hbar^2)$ is smaller by a factor μ/m_e than the Rydberg energy $\mathcal{R}_\infty = m_e e^4/(2\hbar^2)$ corresponding to a proton of infinite mass [BN97, UH97, MN16]:

$$\begin{aligned}\mathcal{R}_\infty &= 2.179872325(27)\times 10^{-18}\,\mathrm{J} = 13.605693009(84)\,\mathrm{eV}\,,\\ \mathcal{R}_\infty/(2\pi\hbar c) &= 109737.31568508(65)\,\mathrm{cm}^{-1}\,,\\ \mathcal{R}_\infty/(2\pi\hbar) &= 3.2898419603555(19)\times 10^{15}\,\mathrm{Hz}\,.\end{aligned} \tag{2.9}$$

In coordinate space, the bound eigenfunctions of the Hamiltonian (2.7) have the form (1.74) and the radial wave functions are given by (1.139). The Bohr radius $a = \hbar^2/(\mu e^2)$ is larger by a factor m_e/μ than the Bohr radius $a_0 = \hbar^2/(m_e e^2)$ corresponding to an infinite proton mass. According to [MN16] the numerical value for a_0 is:

$$a_0 = 0.52917721067(12)\times 10^{-8}\,\mathrm{cm}\,. \tag{2.10}$$

In *atomic units* we measure energies in units of twice the Rydberg energy and lengths in units of the Bohr radius, $\boldsymbol{r} \to a\boldsymbol{r}$, $\hat{\boldsymbol{p}} \to \hat{\boldsymbol{p}}\hbar/a$, $\hat{H} \to 2\mathcal{R}\hat{H}$. The time scale in atomic units is $t_0 = \hbar/(2\mathcal{R})$. Inserting the Rydberg energy $\mathcal{R}_\infty$ corresponding to infinite proton mass we have (http://physics.nist.gov/cgi-bin/cuu/Value?aut), $t_0^\infty = \hbar/(2\mathcal{R}_\infty) = 0.2418884326509(14)\times 10^{-16}$ s.

In atomic units and coordinate representation, the (internal) Hamiltonian for the hydrogen atom is:

$$\hat{H} = -\frac{1}{2}\Delta - \frac{1}{r}\,, \tag{2.11}$$

which corresponds to $\mu = 1$, $\hbar = 1$ and $e = 1$. In atomic units, the bound spectrum of the hydrogen atom is simply $E_n = -1/(2n^2)$ and the Bohr radius is unity.

2.1.2 Hydrogenic Ions

The considerations of the preceding section apply almost without change to a system consisting of an electron and an arbitrary atomic nucleus with charge number Z. Such a system is a *hydrogenic* ion which is $(Z-1)$-fold positively charged. In the formula for the reduced mass, the mass m_p must now be replaced by the mass m_{nuc} of the nucleus which depends not only on the charge number Z, but also on the mass number A (or equivalently, on the number of neutrons $A-Z$):

$$\mu = \frac{m_e m_{nuc}}{m_e + m_{nuc}} = \frac{m_e}{1 + m_e/m_{nuc}}\,. \tag{2.12}$$

Since $m_{\text{nuc}} > m_{\text{p}}$ for all nuclei barring the proton itself, μ is now even closer to the electron mass m_{e}.

For charge numbers $Z > 1$, the essential difference between a hydrogenic ion and the hydrogen atom lies in the potential energy which is stronger by a factor Z:

$$\hat{H}_Z = \frac{\hat{p}^2}{2\mu} - \frac{Ze^2}{r} . \tag{2.13}$$

Looking at the formula (1.136) for the Rydberg energy and (1.102) for the Bohr radius we see that the formulae (2.8) for the energy eigenvalues and (1.139) for the radial wave functions still hold, provided we insert the Rydberg energy $\mathcal{R}_Z$ instead of $\mathcal{R}$,

$$\mathcal{R}_Z = \frac{Z^2 \mu e^4}{2\hbar^2} , \tag{2.14}$$

and the Bohr radius a_Z,

$$a_Z = \frac{\hbar^2}{Z\mu e^2} , \tag{2.15}$$

instead of a. In atomic units the Hamiltonian $\hat{H}_Z$ and the energy eigenvalues E_n are given by

$$\hat{H}_Z = -\frac{1}{2}\Delta - \frac{Z}{r} , \quad E_n = -\frac{Z^2}{2n^2} , \tag{2.16}$$

while the Bohr radius is $a_Z = 1/Z$.

The hydrogen atom and the hydrogenic ions He^+, Li^{++}, Be^{+++}, ..., U^{91+}, ... constitute the simplest example of an *iso-electronic sequence*: atoms and ions with the same number of electrons have very similar spectra. In sequences with more than one electron however, the energies don't follow such a simple scaling rule as (2.16), because only the electron-nucleus part of the potential energy is proportional to Z, while the electron-electron interaction is independent of Z (see Sect. 2.2 and Sect. 2.3).

2.1.3 The Dirac Equation

The time-dependent Schrödinger equation (1.39) violates the symmetry requirements of special relativity, as is already obvious from the different roles played by the spatial coordinates and the time; the Schrödinger equation contains second derivatives with respect to the spatial coordinates, but only first derivatives with respect to time. As a way out of this situation Dirac proposed a Hamiltonian containing the momentum components $\hat{p}_x = (\hbar/\mathrm{i})\partial/\partial x$ etc. linearly. For a free particle of mass m_0 Dirac's Hamiltonian is

$$\hat{H} = c\,\boldsymbol{\alpha}\cdot\hat{\boldsymbol{p}} + \beta m_0 c^2\,. \tag{2.17}$$

Here $c = 2.99792458 \times 10^8\,\mathrm{ms}^{-1}$ is the speed of light, which is included so that the coefficient β and the vector of coefficients $(\alpha_x, \alpha_y, \alpha_z) \equiv (\alpha_1, \alpha_2, \alpha_3)$ are physically dimensionless.

The square of Dirac's Hamiltonian,

$$\hat{H}^2 = c^2 \sum_{i,k=1}^{3} \frac{1}{2}(\alpha_i\alpha_k + \alpha_k\alpha_i)\hat{p}_i\hat{p}_k + m_0 c^3 \sum_{i=1}^{3} (\alpha_i\beta + \beta\alpha_i)\hat{p}_i + \beta^2 m_0^2 c^4\,, \tag{2.18}$$

can only fulfill the relativistic energy momentum relation, $E^2 = p^2c^2 + m_0^2c^4$, if the coefficients α_i, β fulfill the following *anticommutation relations*:

$$\alpha_i\alpha_k + \alpha_k\alpha_i = 2\delta_{i,k}\,, \quad \alpha_i\beta + \beta\alpha_i = 0\,, \quad \beta^2 = 1\,. \tag{2.19}$$

This means they can't simply be numbers. As square matrices they must at least be 4×4 matrices in order to fulfill (2.19). We thus replace the Schrödinger equation by an equation

$$(c\boldsymbol{\alpha}\cdot\hat{\boldsymbol{p}} + \beta m_0 c^2)\psi = \mathrm{i}\hbar\frac{\partial\psi}{\partial t} \tag{2.20}$$

for four-component quantities called *four-component spinors*:

$$\psi(\boldsymbol{r}, t) = \begin{pmatrix} \psi_1(\boldsymbol{r}, t) \\ \psi_2(\boldsymbol{r}, t) \\ \psi_3(\boldsymbol{r}, t) \\ \psi_4(\boldsymbol{r}, t) \end{pmatrix}. \tag{2.21}$$

Equation (2.20) is the *Dirac equation* representing four partial differential equations for the four components of ψ. In the so-called *standard representation* the coefficients α_i, β are expressed through the Pauli spin matrices (1.345):

$$\begin{aligned} \alpha_x &= \begin{pmatrix} 0 & \hat{\sigma}_x \\ \hat{\sigma}_x & 0 \end{pmatrix}, \quad \alpha_y = \begin{pmatrix} 0 & \hat{\sigma}_y \\ \hat{\sigma}_y & 0 \end{pmatrix}, \\ \alpha_z &= \begin{pmatrix} 0 & \hat{\sigma}_z \\ \hat{\sigma}_z & 0 \end{pmatrix}, \quad \beta = \begin{pmatrix} 1 & 0 \\ 0 & -1 \end{pmatrix}. \end{aligned} \tag{2.22}$$

Each entry in a matrix in (2.22) stands for a 2×2 matrix, e.g.

$$\alpha_x = \begin{pmatrix} 0 & 0 & 0 & 1 \\ 0 & 0 & 1 & 0 \\ 0 & 1 & 0 & 0 \\ 1 & 0 & 0 & 0 \end{pmatrix}, \quad \beta = \begin{pmatrix} 1 & 0 & 0 & 0 \\ 0 & 1 & 0 & 0 \\ 0 & 0 & -1 & 0 \\ 0 & 0 & 0 & -1 \end{pmatrix}. \tag{2.23}$$

Inserting an ansatz for a stationary solution,

$$\psi(\boldsymbol{r}, t) = \psi(\boldsymbol{r}, t = 0)\,\mathrm{e}^{-(1/\hbar)Et}, \tag{2.24}$$

turns the Dirac equation (2.20) into a time-independent Dirac equation,

$$(c\,\boldsymbol{\alpha}\cdot\hat{\boldsymbol{p}} + \beta m_0 c^2)\psi = E\psi. \tag{2.25}$$

In order to simplify notation and interpretation we write the four-component spinors ψ as pairs of two-component quantities:

$$\psi = \begin{pmatrix} \psi_A \\ \psi_B \end{pmatrix}, \quad \psi_A = \begin{pmatrix} \psi_1 \\ \psi_2 \end{pmatrix}, \quad \psi_B = \begin{pmatrix} \psi_3 \\ \psi_4 \end{pmatrix}. \tag{2.26}$$

Inserting (2.26) into (2.25) and using the representation (2.22) of the coefficients α_i, β leads to two coupled equations for the two-component spinors ψ_A and ψ_B :

$$\begin{aligned} \hat{\boldsymbol{\sigma}}\cdot\hat{\boldsymbol{p}}\,\psi_B &= \frac{1}{c}(E - m_0c^2)\psi_A\,, \\ \hat{\boldsymbol{\sigma}}\cdot\hat{\boldsymbol{p}}\,\psi_A &= \frac{1}{c}(E + m_0c^2)\psi_B\,. \end{aligned} \tag{2.27}$$

For a particle at rest, $\hat{\boldsymbol{p}}\psi_A = 0$, $\hat{\boldsymbol{p}}\psi_B = 0$, we obtain two (linearly independent) solutions of (2.27) with positive energy $E = m_0c^2$, namely $\psi_A = \begin{pmatrix} 1 \\ 0 \end{pmatrix}$ or $\begin{pmatrix} 0 \\ 1 \end{pmatrix}$ and $\psi_B = 0$, and two solutions with negative energy $E = -m_0c^2$, namely $\psi_B = \begin{pmatrix} 1 \\ 0 \end{pmatrix}$ or $\begin{pmatrix} 0 \\ 1 \end{pmatrix}$ and $\psi_A = 0$. The positive energy solutions are interpreted as the two spin states of the ordinary particle (of spin $s = 1/2$), and the negative energy solutions are related to the corresponding states of the associated *anti-particle*. (For a discussion of the concept of anti-particles see textbooks on relativistic quantum mechanics, e.g. [BD64].) In situations more general than a particle at rest, the positive energy solutions of (2.27) usually have non-vanishing lower components ψ_B, but these are small, except in the extremely relativistic case ($E \gg m_0c^2$), and are consequently called *small components* in contrast to the *large components* ψ_A.

In order to describe e.g. a hydrogen atom, we must extend the above treatment of a free particle to the case of a particle in a potential. The concept of a particle in a static potential $V(\boldsymbol{r})$ obviously contradicts the basic requirements of relativity, because it distinguishes one reference frame from all others. On the other hand, a relativistic theory does *not* allow the simple separation of a two-body problem into a centre-of-mass part and an internal relative motion part, as was possible in the non-relativistic case (Sect. 2.1.1). We can nevertheless justify the relativistic treatment of an electron in the potential of an atomic nucleus, because the nucleus is comparatively heavy and can be assumed to be at rest (in an appropriate reference frame). This picture makes sense as long as the energy of the electron is small compared with the rest energy $m_{\text{nuc}}c^2$ of the atomic nucleus.

We extend the Dirac equation (2.20) or (2.27) to a particle in a static potential $V(\boldsymbol{r})$ by simply adding $V(\boldsymbol{r})$ to the Hamiltonian. Equation (2.27) then becomes

$$
\begin{aligned}
\hat{\boldsymbol{\sigma}} \cdot \hat{\boldsymbol{p}}\, \psi_B &= \frac{1}{c}(E - V(\boldsymbol{r}) - m_0c^2)\psi_A\,, \\
\hat{\boldsymbol{\sigma}} \cdot \hat{\boldsymbol{p}}\, \psi_A &= \frac{1}{c}(E - V(\boldsymbol{r}) + m_0c^2)\psi_B\,.
\end{aligned}
\tag{2.28}
$$

If the potential is radially symmetric, $V = V(r)$, then the radial motion can be separated from the angular motion as in the non-relativistic case. To this end we use the generalized spherical harmonics $\mathcal{Y}_{j,m,l}$ introduced in Sect. 1.7.3 and make the following ansatz for the two-component spinors ψ_A and ψ_B:

$$
\psi_A = \frac{F(r)}{r}\mathcal{Y}_{j,m,l_A}\,, \quad \psi_B = \mathrm{i}\frac{G(r)}{r}\mathcal{Y}_{j,m,l_B}\,. \tag{2.29}
$$

We make use of the identity (Problem 2.1)

$$
\hat{\boldsymbol{\sigma}}\, \hat{\boldsymbol{p}} = \frac{1}{r^2}(\hat{\boldsymbol{\sigma}} \cdot \boldsymbol{r})\left(\frac{\hbar}{\mathrm{i}}\, r\frac{\partial}{\partial r} + \mathrm{i}\hat{\boldsymbol{\sigma}} \cdot \hat{\boldsymbol{L}}\right), \tag{2.30}
$$

of the properties

$$
\begin{aligned}
\frac{1}{r}(\hat{\boldsymbol{\sigma}} \cdot \boldsymbol{r})\mathcal{Y}_{j,m,l=j+1/2} &= -\mathcal{Y}_{j,m,l=j-1/2}\,, \\
\frac{1}{r}(\hat{\boldsymbol{\sigma}} \cdot \boldsymbol{r})\mathcal{Y}_{j,m,l=j-1/2} &= -\mathcal{Y}_{j,m,l=j+1/2}\,,
\end{aligned}
\tag{2.31}
$$

and of the fact that the operator $\hat{\boldsymbol{\sigma}} \cdot \hat{\boldsymbol{L}} = (2/\hbar)\hat{\boldsymbol{S}} \cdot \hat{\boldsymbol{L}}$ can be expressed through $\hat{\boldsymbol{J}}^2 - \hat{\boldsymbol{L}}^2 - \hat{\boldsymbol{S}}^2$, in other words, through $[j(j+1) - l(l+1) - 3/4]\hbar^2$ (1.360). From (2.30), (2.31) we see that each total angular momentum quantum number j allows exactly two possibilities for the orbital angular momentum quantum numbers l_A and l_B in the ansatz (2.29):

$$
\text{(i)} \quad l_A = j - \frac{1}{2}\,, \; l_B = j + \frac{1}{2}\,; \qquad \text{(ii)} \quad l_A = j + \frac{1}{2}\,, \; l_B = j - \frac{1}{2}\,. \tag{2.32}
$$

Inserting (2.29) into (2.28) and using (2.30), (2.31) leads to the *radial Dirac equation* for the radial wave functions $F(r)$ and $G(r)$:

$$\hbar c\left(\frac{\mathrm{d}F}{\mathrm{d}r}+\frac{\kappa}{r}F\right)=\left(E-V(r)+m_0c^2\right)G\,,$$
$$\hbar c\left(\frac{\mathrm{d}G}{\mathrm{d}r}-\frac{\kappa}{r}G\right)=\left(E-V(r)-m_0c^2\right)F\,. \tag{2.33}$$

The absolute value of the constant κ ist $j+1/2$; its sign depends on the orbital angular momentum numbers given by (2.32)[1]:

$$\kappa=-j-\frac{1}{2}\ \text{for (i)}\,,\quad \kappa=j+\frac{1}{2}\ \text{for (ii)}\,. \tag{2.34}$$

The radial Dirac equation (2.33) is a system of two coupled ordinary differential equations of first order. Solving the radial Dirac equation is in general no more difficult than solving the radial Schrödinger equation (1.75) or (1.362). For an attractive Coulomb potential, $V(r)=-\mathrm{Z}e^2/r$, the energy eigenvalues can be given analytically in the regime of bound particle states $0<E<m_0c^2$:

$$E_{n,j}=m_0c^2\left[1+\frac{(Z\alpha_{\mathrm{fs}})^2}{(n-\delta_j)^2}\right]^{-\frac{1}{2}}\,,$$
$$\delta_j=j+\frac{1}{2}-\sqrt{(j+1/2)^2-(Z\alpha_{\mathrm{fs}})^2}\,. \tag{2.35}$$

Here $\alpha_{\mathrm{fs}}=e^2/(\hbar c)=0.0072973525664(17)\approx 1/137$ [MN16] is the dimensionless *fine-structure constant* which characterizes the strength of the electromagnetic interaction. Note that, in atomic units corresponding to $\hbar=1$, $e=1$, the speed of light is $1/\alpha_{\mathrm{fs}}$.

The energies (2.35) depend not only on the principal quantum number $n=1,\ 2,\ 3,\ldots$, but also on the total angular momentum quantum number j, which, for given n, can assume the values $j=1/2,3/2,\ldots n-1/2$. For each j with $1/2\leq j<n-1/2$ (i.e. $j\neq n-1/2$) there are two linearly independent solutions of the radial Dirac equation characterized by the orbital angular momentum quantum numbers $l_A=j+1/2$ and $l_A=j-1/2$ in the large components. Obviously the formula (2.35) is only valid for $Z\alpha_{\mathrm{fs}}<1$. This implies $Z<137$, which is fulfilled for all known atomic nuclei.

---

[1]The constant κ is related to the factor $F(j,l)$ in front of the spin-orbit contribution in the radial Schrödinger equation (1.362) by $\kappa=-1-F(j,l_A)$.

Expanding (2.35) in powers of $Z\alpha_{\rm fs}$ yields

$$E_{n,j} = m_0c^2\left[1 - \frac{(Z\alpha_{\rm fs})^2}{2n^2} - \frac{(Z\alpha_{\rm fs})^4}{2n^3}\left(\frac{1}{j+1/2} - \frac{3}{4n}\right) + \cdots\right]. \tag{2.36}$$

The first term is simply the rest energy m_0c^2 of the particle and the second term corresponds to the non-relativistic spectrum with binding energies $\mathcal{R}/n^2$. The next term contains corrections which are smaller than the non-relativistic binding energies by at least a factor of $(Z\alpha_{\rm fs})^2/n$. This *fine structure* causes an n- and j-dependent lowering of all energy levels. For a given n the shift is largest for $j = 1/2$ and smallest for $j = n - 1/2$.

Figure 2.1 shows the fine-structure splitting of the low-lying levels of the hydrogen atom, as predicted by the Dirac equation. The standard nomenclature for hydrogenic single-particle states is as follows: Energy levels are labelled $n\,l_j$, where

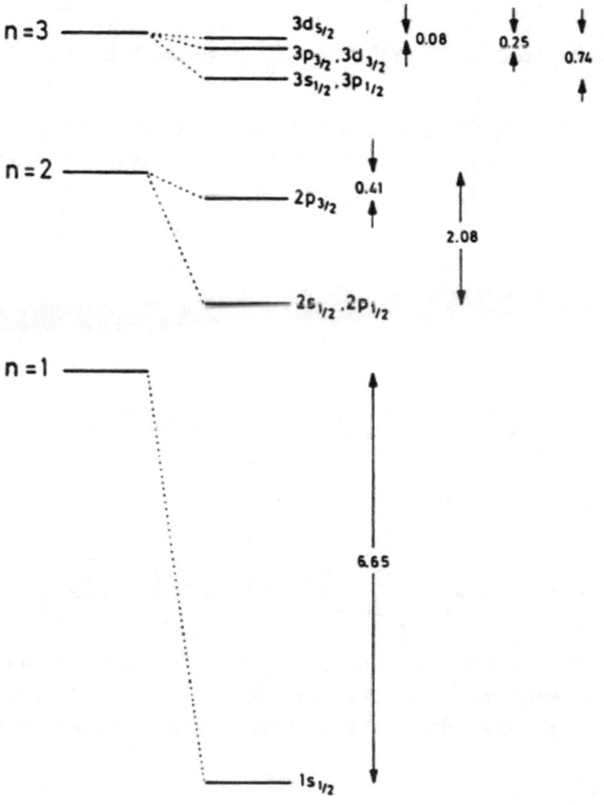

Fig. 2.1 Fine-structure splitting of the energy levels up to $n = 3$ in the hydrogen atom, as predicted by the Dirac equation (2.35). The numbers are energies in 10^{-6} atomic units; on this scale the non-relativistic binding energies are $0.5 \times 10^6/n^2$

n is the Coulomb principal quantum number and j is the total angular momentum quantum number. The orbital angular momentum quantum numbers $l_A \equiv l = 0, 1, 2, 3, \ldots$ are denoted by the letters $s, p, d, f, \ldots$ (continue alphabetically). Examples: $2s_{1/2}$ stands for $n = 2$, $l = 0$, $j = 1/2$ and $7g_{9/2}$ stands for $n = 7$, $l = 4$, $j = 9/2$.

Going beyond the Dirac equation, the electron-proton interaction can be treated with the methods of quantum electrodynamics which leads to still finer corrections to the energy levels, the *Lamb shift*. Further corrections follow from the fact that the proton is not a structureless point particle. Such higher-order corrections to the energy eigenvalues are more important for states of low angular momentum and they lift the degeneracy of the $l_A = j \pm 1/2$ states for given n and j. The Lamb shift has been measured to a high degree of precision [BH95] and amounts to about 1.24×10^{-6} atomic units for the $1s$ state in hydrogen, whereas the $2s_{1/2}$ level comes to lie 0.16×10^{-6} atomic units above the $2p_{1/2}$ level, a separation corresponding to about 10% of the fine-structure splitting of the $n = 2$ level.

2.1.4 Relativistic Corrections to the Schrödinger Equation

The Dirac equation (2.28) can be rewritten as one second-order partial differential equation for the large components ψ_A. To see this, resolve the lower equation for ψ_B,

$$\psi_B = \frac{c}{E - V(\boldsymbol{r}) + m_0c^2}\,\hat{\boldsymbol{\sigma}}\cdot\hat{\boldsymbol{p}}\,\psi_A\,, \tag{2.37}$$

and insert the result into the upper equation:

$$\hat{\boldsymbol{\sigma}}\cdot\hat{\boldsymbol{p}}\,\frac{c^2}{m_0c^2 + E - V}\,\hat{\boldsymbol{\sigma}}\cdot\hat{\boldsymbol{p}}\,\psi_A = (E - V - m_0c^2)\psi_A\,, \tag{2.38}$$

or, replacing $E - m_0c^2$ by ε:

$$\frac{1}{2m_0}\,\hat{\boldsymbol{\sigma}}\cdot\hat{\boldsymbol{p}}\left[1 + \frac{\varepsilon - V}{2m_0c^2}\right]^{-1}\hat{\boldsymbol{\sigma}}\cdot\hat{\boldsymbol{p}}\,\psi_A = (\varepsilon - V)\psi_A\,. \tag{2.39}$$

In the weakly relativistic case the energy E of the particle is not very different from its rest energy m_0c^2, so the difference $\varepsilon = E - m_0c^2$ is small compared with m_0c^2, as is the potential V. It then makes sense to expand the square bracket in (2.39), and the left-hand side becomes

$$\frac{1}{2m_0}\hat{\boldsymbol{\sigma}}\cdot\hat{\boldsymbol{p}}\left(1-\frac{\varepsilon-V}{2m_0c^2}\right)\hat{\boldsymbol{\sigma}}\cdot\hat{\boldsymbol{p}}\,\psi_A$$
$$=\left[\left(1-\frac{\varepsilon-V}{2m_0c^2}\right)\frac{(\hat{\boldsymbol{\sigma}}\cdot\hat{\boldsymbol{p}})\,(\hat{\boldsymbol{\sigma}}\cdot\hat{\boldsymbol{p}})}{2m_0}+\frac{\hbar}{\mathrm{i}}\frac{(\hat{\boldsymbol{\sigma}}\cdot\nabla V)\,(\hat{\boldsymbol{\sigma}}\cdot\hat{\boldsymbol{p}})}{4m_0^2c^2}\right]\psi_A\,. \tag{2.40}$$

Using the identity (Problem 2.1)

$$(\hat{\boldsymbol{\sigma}}\cdot\boldsymbol{A})(\hat{\boldsymbol{\sigma}}\cdot\boldsymbol{B})=\boldsymbol{A}\cdot\boldsymbol{B}+\mathrm{i}\hat{\boldsymbol{\sigma}}\cdot(\boldsymbol{A}\times\boldsymbol{B}) \tag{2.41}$$

(in particular $(\hat{\boldsymbol{\sigma}}\cdot\hat{\boldsymbol{p}})(\hat{\boldsymbol{\sigma}}\cdot\hat{\boldsymbol{p}})=\hat{\boldsymbol{p}}^2$) and assuming a radially symmetric potential, $V=V(r)$, $\nabla V=(\boldsymbol{r}/r)\,\mathrm{d}V/\mathrm{d}r$, we obtain the equation

$$\left[\left(1-\frac{\varepsilon-V}{2m_0c^2}\right)\frac{\hat{\boldsymbol{p}}^2}{2m_0}+\frac{\hbar}{\mathrm{i}}\frac{1}{4m_0^2c^2}\frac{1}{r}\frac{\mathrm{d}V}{\mathrm{d}r}(\boldsymbol{r}\cdot\hat{\boldsymbol{p}})\right.$$
$$\left.+\frac{\hbar}{4m_0^2c^2}\frac{1}{r}\frac{\mathrm{d}V}{\mathrm{d}r}\hat{\boldsymbol{\sigma}}\cdot(\boldsymbol{r}\times\hat{\boldsymbol{p}})\right]\psi_A=(\varepsilon-V)\psi_A\,. \tag{2.42}$$

In the first term on the left-hand side we approximate $\varepsilon-V$ by $\hat{\boldsymbol{p}}^2/(2m_0)$. In the last term we have $\hbar\,\hat{\boldsymbol{\sigma}}\cdot(\boldsymbol{r}\times\hat{\boldsymbol{p}})=2\hat{\boldsymbol{L}}\cdot\hat{\boldsymbol{S}}$. The middle term is not Hermitian. This is due to the fact that we are trying to account for the coupling between the large components ψ_A and the small components ψ_B in a Schrödinger-type equation for the large components alone. Darwin introduced the Hermitian average,

$$\hat{H}_\mathrm{D}=\frac{1}{8m_0^2c^2}\left[\frac{\hbar}{\mathrm{i}}\frac{1}{r}\frac{\mathrm{d}V}{\mathrm{d}r}(\boldsymbol{r}\cdot\hat{\boldsymbol{p}})-\frac{\hbar}{\mathrm{i}}(\hat{\boldsymbol{p}}\cdot\boldsymbol{r})\frac{1}{r}\frac{\mathrm{d}V}{\mathrm{d}r}\right]$$
$$=\frac{\hbar^2}{8m_0^2c^2}\left(\frac{2}{r}\frac{\mathrm{d}V}{\mathrm{d}r}+\frac{\mathrm{d}^2V}{\mathrm{d}r^2}\right)=\frac{\hbar^2}{8m_0^2c^2}\Delta V(r)\,. \tag{2.43}$$

With these manipulations we obtain a Schrödinger equation including relativistic corrections to first order in $\hat{\boldsymbol{p}}^2/(m_0c)^2$:

$$\left(\frac{\hat{\boldsymbol{p}}^2}{2m_0}-\frac{\hat{\boldsymbol{p}}^2\hat{\boldsymbol{p}}^2}{8m_0^3c^2}+V(r)+\hat{H}_{LS}+\hat{H}_\mathrm{D}\right)\psi_A=\varepsilon\psi_A\,. \tag{2.44}$$

Besides the *Darwin term* (2.43) the Hamiltonian in (2.44) contains the spin-orbit coupling

$$\hat{H}_{LS}=\frac{1}{2m_0^2c^2}\frac{1}{r}\frac{\mathrm{d}V}{\mathrm{d}r}\hat{\boldsymbol{L}}\cdot\hat{\boldsymbol{S}} \tag{2.45}$$

and a correction to the kinetic energy including the fourth power of the momentum operator. This makes (2.44) a differential equation of fourth order, which is actually no progress compared with the original Dirac equation (2.28) or (2.33). However, the effects of the relativistic corrections to the non-relativistic Schrödinger equation are small and can usually be calculated with perturbative methods.

In an attractive Coulomb potential $V(r) = -Ze^2/r$, the spin-orbit coupling and the Darwin term are explicitly:

$$\hat{H}_{LS} = \frac{Ze^2}{2m_0^2c^2}\frac{1}{r^3}\hat{\boldsymbol{L}}\cdot\hat{\boldsymbol{S}}\,, \quad \hat{H}_{\rm D} = \frac{\pi\hbar^2 Ze^2}{2m_0^2c^2}\delta(\boldsymbol{r})\,. \tag{2.46}$$

In this case the Darwin term contributes only for $l = 0$; the spin-orbit coupling always contributes only for $l > 0$. We can recover the result (2.36) using first-order perturbation theory with the perturbing operator consisting of the two terms (2.46) and the $\hat{\boldsymbol{p}}^2\hat{\boldsymbol{p}}^2$ term (Problem 2.2).

As indicated at the end of Sect. 2.1.3, further corrections can be obtained by considering that the atomic nucleus isn't a structureless point particle, but has a finite spatial size of the order of 10^{-12} cm and an internal angular momentum called the *nuclear spin*. These corrections are even smaller than the fine structure effects discussed above and appear in the spectrum as *hyperfine structure*.

2.2 Many-Electron Systems

2.2.1 The Hamiltonian

For an atom or ion consisting of N electrons and an atomic nucleus of mass $m_{\rm nuc}$ and charge number Z, the non-relativistic Hamiltonian for the whole system is

$$\hat{H}_{N,Z} = \frac{\hat{\boldsymbol{p}}_{\rm nuc}^2}{2m_{\rm nuc}} + \sum_{i=1}^{N}\left(\frac{\hat{\boldsymbol{p}}_{{\rm e}i}^2}{2m_{\rm e}} - \frac{Ze^2}{|\boldsymbol{r}_{{\rm e}i} - \boldsymbol{r}_{\rm nuc}|}\right) + \sum_{i<j}\frac{e^2}{|\boldsymbol{r}_{{\rm e}i} - \boldsymbol{r}_{{\rm e}j}|}\,; \tag{2.47}$$

$\hat{\boldsymbol{p}}_{\rm nuc}$ and $\boldsymbol{r}_{\rm nuc}$ are the momentum and the position of the nucleus, and $\hat{\boldsymbol{p}}_{{\rm e}i}$ and $\boldsymbol{r}_{{\rm e}i}$ are the momenta and position coordinates of the N electrons. We can separate the centre-of-mass motion from the internal dynamics by introducing the centre-of-mass coordinate,

$$\boldsymbol{R} = \frac{1}{M}\left(m_{\rm nuc}\boldsymbol{r}_{\rm nuc} + m_{\rm e}\sum_{i=1}^{N}\boldsymbol{r}_{{\rm e}i}\right), \quad M = m_{\rm nuc} + Nm_{\rm e}\,, \tag{2.48}$$

together with the relative distance coordinates $\boldsymbol{r}_i$, which stand for the displacement of the respective electrons from the position of the nucleus:

$$\boldsymbol{r}_i = \boldsymbol{r}_{\mathrm{e}i} - \boldsymbol{r}_{\mathrm{nuc}} \,. \tag{2.49}$$

The associated momenta are

$$\hat{\boldsymbol{P}} = \frac{\hbar}{\mathrm{i}}\nabla_{\boldsymbol{R}} \,, \quad \hat{\boldsymbol{p}}_i = \frac{\hbar}{\mathrm{i}}\nabla_{\boldsymbol{r}_i} \,. \tag{2.50}$$

Expressing the momenta $\hat{\boldsymbol{p}}_{\mathrm{nuc}}$ and $\hat{\boldsymbol{p}}_{\mathrm{e}i}$ in (2.47) in terms of the momenta (2.50),

$$\hat{\boldsymbol{p}}_{\mathrm{nuc}} = \frac{m_{\mathrm{nuc}}}{M}\hat{\boldsymbol{P}} - \sum_{i=1}^{N}\hat{\boldsymbol{p}}_i \,, \quad \hat{\boldsymbol{p}}_{\mathrm{e}i} = \frac{m_{\mathrm{e}}}{M}\hat{\boldsymbol{P}} + \hat{\boldsymbol{p}}_i \,, \tag{2.51}$$

allows us to decompose the total kinetic energy in (2.47) into a centre-of-mass part and an internal part:

$$\frac{\hat{\boldsymbol{p}}_{\mathrm{nuc}}^2}{2m_{\mathrm{nuc}}} + \sum_{i=1}^{N}\frac{\hat{\boldsymbol{p}}_{\mathrm{e}i}^2}{2m_{\mathrm{e}}} = \frac{\hat{\boldsymbol{P}}^2}{2M} + \sum_{i=1}^{N}\frac{\hat{\boldsymbol{p}}_i^2}{2\mu} + \frac{1}{m_{\mathrm{nuc}}}\sum_{i<j}\hat{\boldsymbol{p}}_i\cdot\hat{\boldsymbol{p}}_j \,. \tag{2.52}$$

Here $\mu = m_{\mathrm{e}}m_{\mathrm{nuc}}/(m_{\mathrm{e}} + m_{\mathrm{nuc}})$ again is the reduced mass of an electron relative to the atomic nucleus. The two-body potential describing the mutual electrostatic repulsion of the electrons depends on differences of two electron coordinates, and these differences do not depend on whether we use the electron coordinates $\boldsymbol{r}_{\mathrm{e}i}$ in a fixed reference frame or the displacements (2.49) from the atomic nucleus.

The Hamiltonian describing the internal structure of the atom or ion has the form

$$\hat{H} = \sum_{i=1}^{N}\frac{\hat{\boldsymbol{p}}_i^2}{2\mu} + \sum_{i=1}^{N}\hat{V}(i) + \sum_{i<j}\hat{W}(i,j) \,. \tag{2.53}$$

It differs from the Hamiltonian we would obtain for an infinitely heavy nucleus in that the kinetic energy term contains the reduced mass μ instead of the free-electron mass m_{e}. Furthermore, the last term on the right-hand side of (2.52) leads to a momentum-dependent correction $\hat{\boldsymbol{p}}_i\cdot\hat{\boldsymbol{p}}_j/m_{\mathrm{nuc}}$ to the two-body interaction. This correction is called the *mass polarization term* and originates from the fact, that the centre of mass (2.48) of the whole system is not identical to the position $\boldsymbol{r}_{\mathrm{nuc}}$ of the nucleus, from where the internal electron displacements (2.49) are measured. However, this correction is very small and can be treated perturbatively. The same is true, at least in light atoms (and ions), for the relativistic corrections such as spin-orbit coupling discussed in Sect. 2.1.4. Ignoring these corrections for the time being, we have an N-electron problem defined by the Hamiltonian (2.53) with the electrostatic attraction of the electrons by the nucleus as the one-body interaction,

$$\hat{V}(i) = -\frac{Ze^2}{r_i}\,, \tag{2.54}$$

and a two-body interaction due to the mutual electrostatic repulsion of the electrons,

$$\hat{W}(i,j) = \frac{e^2}{\left|\boldsymbol{r}_i - \boldsymbol{r}_j\right|}\,. \tag{2.55}$$

2.2.2 *Pauli Principle and Slater Determinants*

The wave functions describing the internal dynamics of an N-electron atom or ion depend on the internal spatial coordinates $\boldsymbol{r}_i$ and the spin coordinates m_{s_i}, which we shall collect in one symbol x_i. The indistinguishability of the electrons manifests itself in the fact that the Hamiltonian (2.53) does not depend on the ordering of the electron labels i. If we change a given wave function $\psi(x_1, \dots x_N)$ by permuting the electron labels,

$$\hat{P}\psi(x_1, \dots x_N) := \psi(x_{P(1)}, \dots x_{P(N)})\,, \tag{2.56}$$

then the action of the Hamiltonian on the wave function does not depend on whether it acts before or after such a permutation:

$$\hat{P}\hat{H}\psi(x_1, \dots x_N) = \hat{H}\hat{P}\psi(x_1, \dots x_N)\,. \tag{2.57}$$

Each permutation P of the numbers $1, \dots, N$ defines an operator $\hat{P}$ according to (2.56), and each such operator commutes with the Hamiltonian, because of (2.57):

$$\left[\hat{H}, \hat{P}\right] = 0\,. \tag{2.58}$$

It would seem reasonable to classify the eigenstates of $\hat{H}$ according to the eigenvalues of the permutation operators, i.e. according to their behaviour under reordering of the particle labels. In a two-body system there is only one non-trivial permutation, namely P_{21}, which replaces the pair 1, 2 by 2, 1. Obviously the corresponding operator gives the unit operator when squared, $\hat{P}_{21}\hat{P}_{21} = 1$, so its only possible eigenvalues are $+1$ and -1. In systems of more than two indistinguishable particles the situation is more complicated. Two classes of many-particle wave functions are particularly important: *totally symmetric* wave functions for which interchanging any two particle labels doesn't change the wave function at all, and totally antisymmetric wave functions for which interchanging any two particle labels multiplies the wave function by -1:

$$\begin{aligned}&\hat{P}_{ij}\psi(x_1,..,x_{i-1},x_i,..,x_{j-1},x_j,..x_N)\\=&\quad\psi(x_1,..,x_{i-1},x_j,..,x_{j-1},x_i,..x_N)\\=&-\psi(x_1,..,x_{i-1},x_i,..,x_{j-1},x_j,..x_N)\,.\end{aligned} \tag{2.59}$$

For systems of more than two indistiguishable particles, the totally symmetric or totally antisymmetric wave functions represent only a fraction of the functions one might construct mathematically, but only these two possibilities are realized in nature. Furthermore, the behaviour of the wave functions under permutations of the particle labels is an internal property of the particles and does not depend on their dynamic state or their environment. Particles with totally symmetric wave functions are called *bosons*, particles with totally antisymmetric wave functions are called *fermions*. Electrons are fermions. The statement that fermions only occur in totally antisymmetric states is called the *Pauli principle*.

Any permutation of the numbers $1,\ldots,N$ can be decomposed into a sequence of successive swaps of just two numbers. This decomposition is not unique, but the number of swaps making up a given permutation is either always even or always odd. One calls the permutation itself *even* or *odd* accordingly. The total antisymmetry of a wave function can thus be written compactly:

$$\hat{P}\psi=(-1)^P\psi\,, \tag{2.60}$$

with $(-1)^P=1$ for even permutations and $(-1)^P=-1$ for odd permutations.

From a given wave function ψ, which need not be totally antisymmetric, we can project out a totally antisymmetric part using the *antisymmetrizer*

$$\hat{\mathcal{A}}=\frac{1}{\sqrt{N!}}\sum_P(-1)^P\hat{P}\,. \tag{2.61}$$

To see that $\hat{\mathcal{A}}\psi$ is totally antisymmetric, we apply an arbitrary permutation Q:

$$\hat{Q}\hat{\mathcal{A}}\psi=\frac{1}{\sqrt{N!}}\sum_P(-1)^P\hat{Q}\hat{P}\psi\,. \tag{2.62}$$

Since the permutations mathematically form a group, the set of all permutations QP (Q fixed, P covering all permutations) again contains each permutation exactly once. Furthermore $(-1)^P=(-1)^Q(-1)^{QP}$, so that we can rewrite (2.62) using $P'=QP$:

$$\hat{Q}\hat{\mathcal{A}}\psi=(-1)^Q\frac{1}{\sqrt{N!}}\sum_{P'}(-1)^{P'}\hat{P}'\psi=(-1)^Q\hat{\mathcal{A}}\psi\,. \tag{2.63}$$

In a similar way it can be shown that

$$\hat{\mathcal{A}}\hat{\mathcal{A}} = \sqrt{N!}\,\hat{\mathcal{A}}\,, \quad \hat{\mathcal{A}}^{\dagger} = \hat{\mathcal{A}}\,. \tag{2.64}$$

This means that $\hat{\mathcal{A}}/\sqrt{N!}$ has the properties of a projection operator.

A particularly important set of totally antisymmetric wave functions consists of those constructed by antisymmetrizing simple product wave functions:

$$\Psi_0 = \prod_{i=1}^{N} \psi_i(x_i)\,. \tag{2.65}$$

Such product wave functions appear e.g. as eigenfunctions of an N-body Hamiltonian which can be written as a sum of one-body Hamiltonians (such as the Hamiltonian (2.53) if we were to leave out the two-body interaction $\hat{W}(i,j)$). Applying the antisymmetrizer to (2.65) produces the antisymmetrized product wave function

$$\hat{\mathcal{A}}\,\Psi_0 = \frac{1}{\sqrt{N!}}\sum_P (-1)^P \prod_{i=1}^{N} \psi_i(x_{P(i)}) \equiv \frac{1}{\sqrt{N!}}\det(\psi_i(x_j))\,. \tag{2.66}$$

We write det $(\psi_i(x_j))$, because the sum over the products in (2.66) can formally be written as the determinant of the $N \times N$ matrix $(\psi_i(x_j))$:

$$\det(\psi_i(x_j)) = \begin{vmatrix} \psi_1(x_1) & \psi_1(x_2) & \cdots & \psi_1(x_N) \\ \vdots & \vdots & & \vdots \\ \psi_N(x_1) & \psi_N(x_2) & \cdots & \psi_N(x_N) \end{vmatrix}\,. \tag{2.67}$$

Antisymmetrized product wave functions are called *Slater determinants*.

The determinant notation shows that an antisymmetrized product wave function vanishes identically when two or more single-particle wave functions ψ_i are the same. This leads to an alternative formulation of the Pauli principle, applicable to Slater determinants: no two fermions may occupy the same single-particle state. A more general and at the same time more precise formulation is: a Slater determinant vanishes identically if and only if the single-particle states from which it is built are linearly dependent.

Like an ordinary determinant, a Slater determinant is invariant under elementary replacements of rows:

$$\psi_i \to \psi_i' = \psi_i + \sum_{j \neq i} c_j \psi_j\,. \tag{2.68}$$

More generally: if we replace the (linearly independent) set of single-particle wave functions ψ_i by any set of linearly independent linear combinations ψ_i', then $\det(\psi_i'(x_j))$ differs from $\det(\psi_i(x_j))$ by at most a constant factor. A Slater determinant is thus characterized not so much by a particular set of single-particle states, but rather by the subspace spanned in the single-particle Hilbert space by these single-particle states.

When many-body wave functions are Slater determinants, the many-body scalar products such as (1.1) can be expressed in terms of scaler products of the single-particle wave functions involved. The overlap of two Slater determinants $\Psi = (N!)^{-1/2}\det(\psi_i(x_j))$ and $\Phi = (N!)^{-1/2}\det(\phi_i(x_j))$ is:

$$\langle\Phi|\Psi\rangle = \det\left(\langle\phi_i|\psi_j\rangle\right), \tag{2.69}$$

and the right-hand side is now an ordinary determinant of a matrix of numbers, viz. the numbers

$$A_{ij} = \langle\phi_i|\psi_j\rangle \,. \tag{2.70}$$

For a one-body operator, more precisely, for a many-body operator which can be written as a sum of single-particle operators $\hat{V}$, we have

$$\langle\Phi|\sum_{i=1}^{N}\hat{V}(i)|\Psi\rangle = \langle\Phi|\Psi\rangle \sum_{i,j=1}^{N}\langle\phi_i|\hat{V}|\psi_j\rangle\, B_{ji}\,, \tag{2.71}$$

where the matrix B is the inverse of the matrix A defined by (2.70). For an operator which can be written as a sum of two-particle operators we have

$$\langle\Phi|\sum_{i<j}\hat{W}(i,j)|\Psi\rangle$$

$$= \frac{1}{2}\langle\Phi|\Psi\rangle \sum_{i,j,k,l=1}^{N} \langle\phi_i\phi_j|\hat{W}|\psi_k\psi_l\rangle(B_{ki}B_{lj} - B_{kj}B_{li})\,. \tag{2.72}$$

The formulae (2.71), (2.72) are valid for any (not necessarily orthonormal) set of single-particle wave functions as long as $\det A \neq 0$. Simpler formulae apply when Φ and Ψ are built from the same set of orthonormal single-particle states. Then $\langle\Phi|\Psi\rangle$ is only non-vanishing if the same single-particle states are occupied in Φ and Ψ. Furthermore, $\langle\Psi|\Psi\rangle = 1$. The factor $1/\sqrt{N!}$ in (2.66) is just chosen such that a Slater determinant built from orthonormal single-particle states is normalized to unity.

For orthonormal single-particle states and $\Phi = \Psi$, the formula (2.71) is simplified to

$$\langle\Psi|\sum_{i=1}^{N}\hat{V}(i)|\Psi\rangle = \sum_{i=1}^{N}\langle\psi_i|\hat{V}|\psi_i\rangle\,. \tag{2.73}$$

There is also a non-vanishing matrix element $\langle\Phi|\sum_{i=1}^{N}\hat{V}(i)|\Psi\rangle$ when at most one of the single-particle states occupied in Ψ (ψ_h, say) is replaced in Φ by another single-particle state (ψ_p, say) which is unoccupied in Ψ. Such a Slater determinant Φ is called a *one-particle-one-hole excitation* Ψ_{ph} of Ψ. The matrix element of a one-body operator between Ψ_{ph} and Ψ is:

$$\langle\Psi_{ph}|\sum_{i=1}^{N}\hat{V}(i)|\Psi\rangle = \langle\psi_p|\hat{V}|\psi_h\rangle. \tag{2.74}$$

(Formula (2.71) cannot be applied to this case, because $\langle\Psi_{ph}|\Psi\rangle = 0$.)

For orthonormal single-particle states and $\Phi = \Psi$, the formula (2.72) for two-body operators is simplified to

$$\langle\Psi|\sum_{i<j}\hat{W}(i,j)|\Psi\rangle = \frac{1}{2}\sum_{i,j=1}^{N}\left(\langle\psi_i\psi_j|\hat{W}|\psi_i\psi_j\rangle - \langle\psi_i\psi_j|\hat{W}|\psi_j\psi_i\rangle\right). \tag{2.75}$$

The matrix element of a two-body operator between Ψ and a one-particle-one-hole excitation Ψ_{ph} is:

$$\langle\Psi_{ph}|\sum_{i<j}\hat{W}(i,j)|\Psi\rangle = \sum_{i=1}^{N}\left(\langle\psi_i\psi_p|\hat{W}|\psi_i\psi_h\rangle - \langle\psi_i\psi_p|\hat{W}|\psi_h\psi_i\rangle\right). \tag{2.76}$$

If the bra is a *two-particle-two-hole excitation* $\Psi_{p_1p_2h_1h_2}$ of Ψ, i.e. $\phi_{h1} = \psi_{p1}$, $\phi_{h_2} = \psi_{p_2}$ and $\phi_i = \psi_i$ for all other ψ_i, then there is also a non-vanishing matrix element

$$\begin{aligned}&\langle\Psi_{p_1p_2h_1h_2}|\sum_{i<j}\hat{W}(i,j)|\Psi\rangle \\ &= \langle\psi_{p_1}\psi_{p_2}|\hat{W}|\psi_{h_1}\psi_{h_2}\rangle - \langle\psi_{p_1}\psi_{p_2}|\hat{W}|\psi_{h_2}\psi_{h_1}\rangle\,. \end{aligned}\tag{2.77}$$

2.2.3 *The Shell Structure of Atoms*

If the Hamiltonian (2.53) contained only the one-body interaction and there were no two-body interactions, then it would describe independent motion of the N electrons. The Hamiltonian would be a sum of N single-particle Hamiltonians of the form (2.13) whose eigenfunctions are simply the eigenfunctions of the hydrogenic ion. Each product of N such single-particle eigenfunctions would be an eigenfunction of the N-particle Hamiltonian, and so would each Slater determinant made by antisymmetrizing such a product (because $\hat{H}$ commutes with all permutations and hence also with the antisymmetrizer (2.61)). The energy eigenvalue of such a Slater determinant would simply be the sum of the single-particle energies of the occupied states. The energetically lowest N single-particle states would make up the ground state (Pauli principle), and the excited states would be one-particle-one-hole, two-particle-two-hole, etc. excitations of the ground-state Slater determinant.

This simple picture is disturbed by the two-body interaction $\sum \hat{W}(i,j)$. It is not small and contributes significantly to the total energy of the atom or ion. However, a large part of the two-body interaction can be accounted for by a *mean single-particle potential*, often called *mean field*, which formally retains the independence of the electrons. A consistent derivation of the mean single-particle potential is given in Sect. 2.3.1. Qualitatively, the electrostatic repulsion of one given electron by all other electrons is described by an average screening potential which modifies the single-particle potential (electrostatic attraction by the nucleus) acting on that electron. Those parts of the two-body interaction which are not included in the mean single-particle potential constitute a *residual two-body interaction* and this is much less than the full two-body interaction. Take e.g. an electron in an N-electron atom or ion whose nucleus has charge number Z. At large distances from the nucleus (and the other electrons) the electron feels a screened Coulomb potential $-(Z-N+1)e^2/r$. At small separations $r < a_Z$, however, it feels the full unscreened attraction of the naked nucleus: $-Ze^2/r$. In the transition region from small to large separations the mean single-particle potential changes smoothly from the unscreened potential to the screened potential as is illustrated schematically for the case of a neutral sodium atom ($Z = N = 11$) in Fig. 2.2.

The single-particle eigenstates in such a mean single-particle potential are no longer the eigenstates of a pure Coulomb potential, but they can still be classified by the quantum numbers n, l, m. Since the mean single-particle potential is always taken to be radially symmetric, the single-particle energies for given angular momentum quantum number l are degenerate in the azimuthal quantum number m. However, eigenstates with a given principal quantum number n are no longer degenerate in l, because the potential is no longer a pure Coulomb potential.

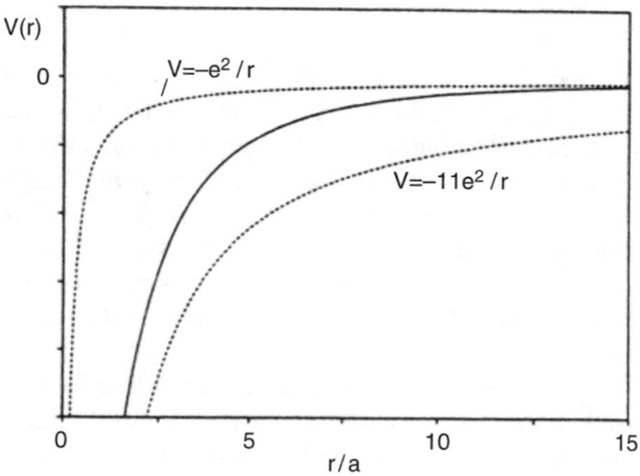

Fig. 2.2 Schematic illustration of the mean single-particle potential $V(r)$ (*solid line*) in the Na atom ($Z = N = 11$)

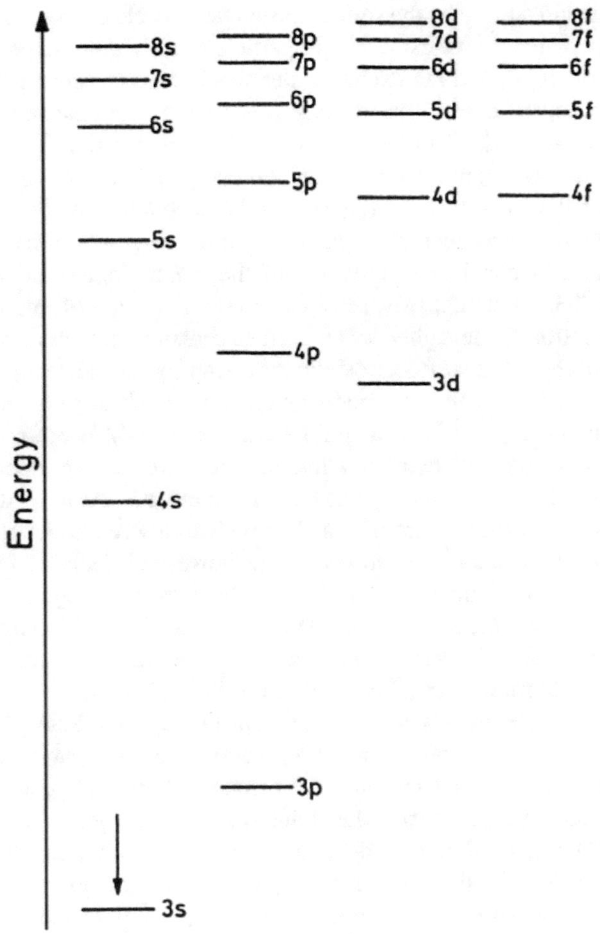

Fig. 2.3 Typical spectrum of single-particle energies in a single-particle potential as in Fig. 2.2

A glance at Fig. 2.2 shows that states with low l are most strongly influenced by the stronger attraction of the unscreened nucleus, because their wave functions have the largest amplitudes at small separations—see Fig. 1.4 and (1.78). As a result the levels with low l are shifted downwards quite strongly, relative to the levels with higher l. A typical spectrum of a single-particle Hamiltonian containing a mean single-particle potential as in Fig. 2.2 is shown in Fig. 2.3. The downward shift of the $l = 0$ levels is so large that the energy of the $4s$ state already lies below the energy of the $3d$ state. Larger gaps appear in the spectrum above the $1s$, $2p$, $3p$, ... levels.

The energy levels in Fig. 2.3 define *subshells* which accomodate a number of single-particle states according to their degeneracy, and each of these single-particle states can be occupied by at most one electron. (We reserve the term "shell" for all states belonging to one principal quantum number n.) Considering that there are two possible spin states associated with each orbital wave function $\psi(\boldsymbol{r})$, the total number of single-particle states in each nl subshell is simply $2(2l+1)$. In $s, p, d, f, \dots$ subshells there are 2, 6, 10, 14, ... etc. states.

We assume that the ground state wave functions of neutral atoms are built by successively filling the subshells of single-particle states. The electrons in the energetically lower closed (i.e. completely filled) subshells are comparatively tightly bound, and the least bound electrons are the outer electrons in the last occupied subshell. In this picture, chemically similar elements, which were grouped together in the *periodic table* long before the invention of quantum mechanics, have the same number of outer electrons, and the last partially occupied subshells within a group have the same angular momentum quantum number l. The noble gases He, Ne, Ar, Kr, Xe, Rn are built wholly of closed, i.e. fully occupied subshells and the last occupied subshell is that of a single-particle level at the lower edge of one of the larger gaps in the single-particle spectrum $1s$, $2p$, $3p$, $4p$, $5p$, $6p$.

The simple picture of the shell structure of atoms (in their ground states) following from Figs. 2.2 and 2.3 is able to explain the positions assigned to the elements in the periodic table according to their chemical properties. This is a great success of the concept of independent electrons in well defined single-particle states. Nevertheless, the exact eigenstates of the Hamiltonian (2.53) are of course much more complicated. For a quantitative description of atoms with more than one electron we need to consider correlations which go beyond the independent-single-particle picture.

2.2.4 *Classification of Atomic Levels*

In order to classify the eigenstates of the N-electron Hamiltonian it is reasonable to look for constants of motion, i.e. for good quantum numbers. Let's assume for the time being that the effects of the spin-orbit coupling are negligible. Then the total orbital angular momentum $\hat{\boldsymbol{L}}$ and the total spin $\hat{\boldsymbol{S}}$, which are made up of the single-particle orbital angular momenta $\hat{\boldsymbol{L}}_i$ and the single-particle spins $\hat{\boldsymbol{S}}_i$ of the electrons respectively,

$$\hat{\boldsymbol{L}} = \sum_{i=1}^{N} \hat{\boldsymbol{L}}_i\,, \qquad \hat{\boldsymbol{S}} = \sum_{i=1}^{N} \hat{\boldsymbol{S}}_i\,, \tag{2.78}$$

are constants of motion, i.e. their components and their squares $\hat{\boldsymbol{L}}^2$ and $\hat{\boldsymbol{S}}^2$ commute with the Hamiltonian (2.53). The eigenvalues of $\hat{\boldsymbol{L}}^2$ and $\hat{\boldsymbol{S}}^2$ are $L(L+1)\hbar^2$ and $S(S+1)\hbar^2$ respectively, and to each pair of values of L and S there are $(2L+1)\times(2S+1)$ degenerate eigenstates corresponding to the different eigenvalues of $\hat{L}_z$ and $\hat{S}_z$.

It is customary to denote the total orbital angular momentum quantum number $L = 0, 1, 2, 3, \ldots$ by the capital letters $S, P, D, F, \ldots$ (continue alphabetically), while the total spin quantum number S is noted by writing the spin multiplicity $2S + 1$ to the upper left of the letter denoting L: thus ^3P means $S = 1$ and $L = 1$, ^4D means $S = 3/2$ and $L = 2$, etc. Since all electron spins are $1/2$, the total spin quantum number S is an integer and the spin multiplicity $2S + 1$ odd for an even number N of electrons, while S is a half-integer and $2S + 1$ even if N is odd.

In the presence of a small spin-orbit term $V_{LS}(r_i)\hat{\boldsymbol{L}}_i\hat{\boldsymbol{S}}_i$ in the one-body interaction, the Hamiltonian (2.53) no longer commutes with the components of the orbital angular momenta and the spins, but it commutes with the total angular momentum of the electrons:

$$\hat{\boldsymbol{J}} = \hat{\boldsymbol{L}} + \hat{\boldsymbol{S}} . \tag{2.79}$$

We can treat the effects of the spin-orbit coupling approximately if we couple the states classified by L and S to eigenstates of $\hat{\boldsymbol{J}}^2$ and $\hat{J}_z$, similar to the one-electron case described in Sect. 1.7.3. The resulting states are now labelled by a further quantum number J for the total angular momentum, and it is written as a subscript to the letter denoting L, in analogy to the labelling of one-electron levels. Example: $^4D_{5/2}$ means $S = 3/2,\ L = 2,\ J = 5/2$. According to the triangle condition (1.336), each term ^{2S+1}L splits into $2S + 1$ (in case $S \leq L$) or $2L + 1$ (in case $L \leq S$) levels $^{2S+1}L_J$, $J = |L - S|, |L - S| + 1, \ldots, L + S - 1, L + S$, and each such level encompasses $2J + 1$ eigenstates of $\hat{J}_z$ which remain degenerate in the presence of the spin-orbit coupling. (This degeneracy is lifted if we consider the effects of the hyperfine interaction with a non-vanishing nuclear spin $\hat{\boldsymbol{I}}$, because then only the total angular momentum $\hat{\boldsymbol{I}} + \hat{\boldsymbol{J}}$ of atomic nucleus plus the orbiting electrons is a constant of motion.)

As long as the picture of independent particles is applicable, we can in addition label the atomic states by the principal and orbital angular momentum quantum numbers n, l of the occupied single-particle states. The complete set of n, l quantum numbers of the occupied single-particle states defines a *configuration*. A configuration with, say, two occupied $1s$ single-particle states, two occupied $2s$ states and three occupied $2p$ states is conventionally written as $(1s)^2(2s)^2(2p)^3$.

When constructing a many-body state out of single-particle states we must of course respect the requirements of the Pauli principle. This is still comparatively easy for atoms and ions with two electrons (or with two outer electrons), because the coupled spin states of two $s = 1/2$ particles have a well defined symmetry with respect to permutation of the two particle labels. In this special case of angular momentum coupling ($j_1 = 1/2, j_2 = 1/2$) let's abbreviate (1.332) to

$$|S, M_S\rangle = \sum_{m_{s_1}, m_{s_2}} \langle m_{s_1}, m_{s_2}|S, M_S\rangle |m_{s_1}, m_{s_2}\rangle . \tag{2.80}$$

In this notation the *triplet* of states coupled to $S = 1$ is simply

$$\begin{aligned} |1,1\rangle &= |1/2,1/2\rangle \,, \\ |1,0\rangle &= \frac{1}{\sqrt{2}}(|1/2,-1/2\rangle + |-1/2,1/2\rangle) \,, \\ |1,-1\rangle &= |-1/2,-1/2\rangle \,, \end{aligned} \tag{2.81}$$

and the $S = 0$ (*singlet*) state is

$$|0,0\rangle = \frac{1}{\sqrt{2}}(|1/2,-1/2\rangle - |-1/2,1/2\rangle) \,. \tag{2.82}$$

The three states of the triplet $S = 1$ are symmetric with respect to interchanging the two labels m_{s_1} and m_{s_2}, while the singlet state is antisymmetric. Since the whole two-particle wave function $\psi(\boldsymbol{r}_1, m_{s_1}, \boldsymbol{r}_2, m_{s_2})$ has to be antisymmetric, its behaviour with respect to interchange of the two spatial coordinates must be symmetric in the singlet state and antisymmetric in the triplet states. Thus a helium configuration in which both electrons occupy the (non-degenerate) $1s$ spatial state is only possible in the singlet spin state. Figure 2.4 shows the energy levels of the bound states of

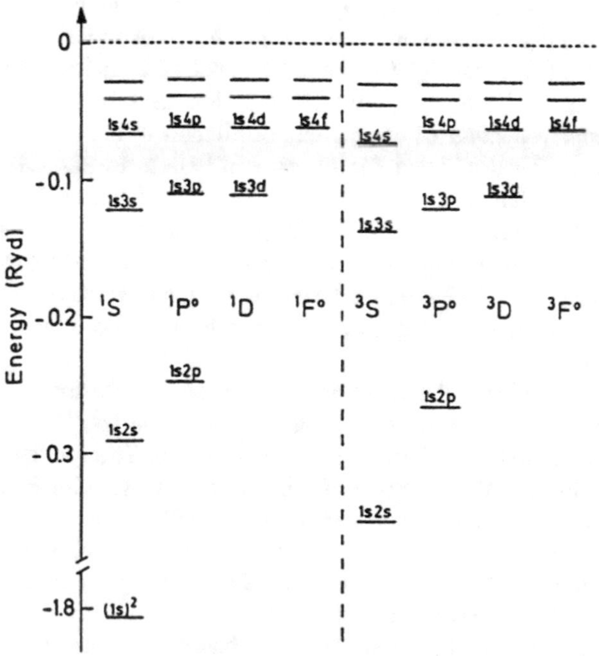

Fig. 2.4 Energies of the bound states of the helium atom. The left half of the figure shows the singlet states of para-helium and the right half of the figure shows the triplet states of ortho-helium

helium, separated according to $S = 0$ (*para-helium*) and $S = 1$ (*ortho-helium*). Provided they are allowed by the Pauli principle, the configurations of ortho-helium lie energetically lower than the corresponding configurations in para-helium. This can be understood as an effect of the residual interaction involving a short-ranged repulsion of the electrons. It is less effective in a wave function antisymmetric with respect to interchange of spatial coordinates, which has to vanish for $|\boldsymbol{r}_1 - \boldsymbol{r}_2| = 0$, than it is in a symmetric wave function, where it yields a positive contribution to the total energy. (See Problem 2.3.)

When the spins of more than one electron from a given subshell are coupled to total spin S, then the state with the largest value of S is energetically lowest, because it feels the effects of the short-ranged electron-electron repulsion least due to the symmetry properties of the spatial part of the wave function. This is *Hund's first rule*. For a given value of S the electrons can couple to different values L of the total orbital angular momentum. Amongst these states, the effect of the short-ranged repulsion is least in the states with the largest values of L. Of all states with the same value of S, the state with the maximum value of L is hence the energetically lowest. This is *Hund's second rule*.

As an example Fig. 2.5 shows the lowest-lying states of the carbon atom which has two electrons in the $2p$ subshell. The ground state triplet ^3P and the next two excited singlets ^1D and ^1S are based on the $(1s)^2(2s)^2(2p)^2$ configuration in which the lowest single-particle states are occupied. The next highest term is a quintuplet $^5S^{\mathrm{o}}$ corresponding to a $(1s)^2(2s)(2p)^3$ configuration in which the $2s$ subshell is occupied by only one electron, while the $2p$ subshell is occupied by three electrons. The small "o" at the upper right of the letter denoting L stands for *odd parity* and indicates that the whole many-body wave function has odd parity with respect to the simultaneous reflection of all spatial coordinates at the origin. This notation was already used for the P and the F states in helium in Fig. 2.4. The parity of a many-body wave function is important, because it influences the selection rules for electromagnetic transitions. A configuration characterized by the single-particle orbital angular momentum quantum numbers $l_1, \dots l_N$ has odd parity if the sum $\sum_{i=1}^{N} l_i$ is odd (see (1.72)). Note that the parity of a many-electron state is in general a good quantum number, and this is not bound to the validity of the independent particle picture.

Figure 2.5 also shows the splitting of the ground state triplet into three ^3P_J levels, $J = 0, 1, 2$ due to the spin-orbit coupling. The ^1D, ^1S and $^5S^{\mathrm{o}}$ terms do not split up, because either L or S (or both) are zero. The ground state triplet is *regular*, meaning that the energies of the levels increase with increasing values of J. Multiplets with the opposite behaviour are called *inverted*. It is empirically established, that ground state terms of atoms whose outer subshell is at most half filled form regular multiplets, while the ground state multiplets are inverted in atoms whose outer subshell is more than half filled.

The above classification of atomic states is based on the assumption that orbital and spin angular momenta are at least approximately constants of motion. This *LS coupling*, which is also called *Russell-Saunders coupling*, loses its justification when the influence of the spin-orbit coupling in the one-body interaction increases as is

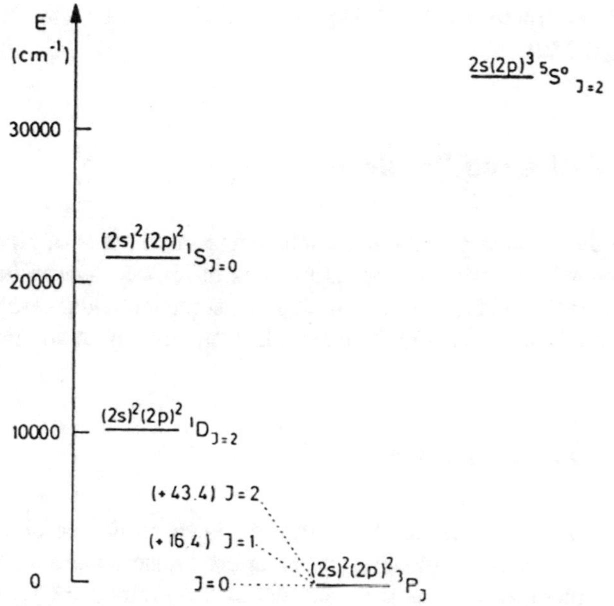

Fig. 2.5 The lowest energy levels in the carbon atom. The configuration labels should in principle carry a further $(1s)^2$ for the two occupied states in the $n = 1$ shell. Amongst the $(2s)^2(2p)^2$ states the triplet term ($S = 1$) is lowest according to Hund's first rule. The Pauli principle forbids ^3S and ^3D terms, because they would contain a spatial wave function symmetric in the particle labels in conjunction with a spin wave function which is also symmetric in the particle labels. The $L = 1$ triplet is regular, i.e. the energy increases with increasing total angular momentum quantum number J. The $L = 2$ term is the lower of the singlet states according to Hund's second rule. The first excited configuration shows up in the quintuplet term of the $2s(2p)^3$ configuration. This configuration has odd parity

the case for the heavier atoms. It may then be more appropriate to assume that the single-particle total angular momenta of the electrons

$$\hat{\boldsymbol{J}}_i = \hat{\boldsymbol{L}}_i + \hat{\boldsymbol{S}}_i\,, \tag{2.83}$$

are approximate constants of motion, and to couple these to the total angular momentum of all the electrons. For two electrons,

$$\hat{\boldsymbol{J}} = \hat{\boldsymbol{J}}_1 + \hat{\boldsymbol{J}}_2\,, \tag{2.84}$$

this can be done in a straightforward way (compare Sect. 1.7.1) and leads to the *jj coupling* scheme.

A comprehensive compilation of the known levels of atoms and ions from hydrogen to manganese can be found in [BS75, BS78, BS81, BS82]. A detailed and comprehensive discussion of the levels of atoms with one or two electrons is contained in the classic book by Bethe und Salpeter [BS77]. For a thorough discussion of the structure of low-lying states see also *Atomic Structure* by Condon and Odabasi [CO80].

2.3 The N-Electron Problem

The many-body problem poses a major challenge in all areas of physics. It is not soluble in general, but various approximations have been successful in different fields. This section contains a brief summary of some techniques which have been successful and/or widely used in the many-electron problem of atomic physics.

2.3.1 The Hartree-Fock Method

The central idea of the Hartree-Fock method is to retain the simplicity of the independent single-particle picture, and to approximate an exact solution of the N-electron problem as well as possible within this framework. This means that we describe the system by the "best" Slater determinant. In the spirit of the Ritz variational method (Sect. 1.6.2) we search for a Slater determinant Ψ for which the energy expectation value $E[\Psi]$ remains stationary under small variations $\Psi \rightarrow \Psi + \delta\Psi$ of the Slater determinant: $\delta E[\Psi] = 0$.

Let $\Psi = (1/\sqrt{N!})\,\det(\psi_i(x_j))$ be a Slater determinant of orthonormalized single-particle states ψ_i. When varying Ψ we must take care that the varied wave function is again a Slater determinant. Appropriate variations are achieved by modifying the single-particle states occupied in Ψ through small admixtures of single-particle states ψ_{p_i} which are not occupied in Ψ:

$$\psi_i \rightarrow \psi_i' = \psi_i + \lambda_i \psi_{p_i} \,. \tag{2.85}$$

Expanding the Slater determinant $\Psi' = (1/\sqrt{N!})\,\det(\psi_i'(x_j))$ around the original Slater determinant Ψ shows that the leading terms in $\delta\Psi = \Psi' - \Psi$ are those in which only one single-particle state is modified. These terms yield contributions of the form $\lambda_i \Psi_{p_i i}$, where $\Psi_{p_i i}$ is a one-particle-one-hole excitation of Ψ in which the single-particle state ψ_i is replaced by the single-particle state ψ_{p_i} (which is unoccupied in Ψ). Contributions in which more than one single-particle state are modified correspond to two-particle-two-hole, three-particle-three-hole excitations, etc. They, however, carry two, three or more factors λ_i and are hence small to higher order than the contributions of the one-particle-one-hole excitations.

The infinitesimal variations of a Slater determinant which ensure that the varied wave function is again a Slater determinant are thus one-particle-one-hole excitations. From (1.270), (1.271) it immediately follows, that the condition $\delta E[\Psi] = 0$ is equivalent to the condition that all matrix elements of $\hat{H}$ between Ψ

and one-particle-one-hole excitations vanish:

$$\delta E[\Psi] = 0 \quad \Longleftrightarrow \quad \langle \Psi_{ph}|\hat{H}|\Psi\rangle = 0 \quad \text{for all} \quad \Psi_{ph} . \tag{2.86}$$

This is *Brillouin's Theorem.*

Brillouin's Theorem leads directly to a set of equations for the "best" Slater determinant. With the Hamiltonian (2.53) as a sum of one-body and two-body operators and the formulae (2.74), (2.76) for its matrix elements with one-particle-one-hole excitations, we have

$$\langle \psi_p|\frac{\hat{\boldsymbol{p}}^2}{2\mu} + \hat{V}|\psi_h\rangle + \sum_{i=1}^{N}\Big(\langle \psi_i\psi_p|\hat{W}|\psi_i\psi_h\rangle - \langle \psi_i\psi_p|\hat{W}|\psi_h\psi_i\rangle\Big) = 0 . \tag{2.87}$$

The whole left-hand side of (2.87) can be interpreted as the matrix element of an effective one-body Hamiltonian $\hat{h}_\Psi$ between the single-particle state ψ_p, which is unoccupied in Ψ, and the single-particle state ψ_h, which is occupied in Ψ. The condition $\delta E[\Psi] = 0$ is fulfilled if the one-body operator $\hat{h}_\Psi$, which itself depends on Ψ, has no non-vanishing matrix elements between single-particle states which are occupied in Ψ and single-particle states which are unoccupied in Ψ. A sufficient (but not necessary) condition is that the one-body operator $\hat{h}_\Psi$ be diagonal in the single-particle states $\psi_1, \dots, \psi_N, \dots, \psi_p, \dots$:

$$\begin{aligned}
&\langle \psi_\alpha|\frac{\hat{\boldsymbol{p}}^2}{2\mu} + \hat{V}|\psi_\beta\rangle + \sum_{i=1}^{N}\Big(\langle \psi_i\psi_\alpha|\hat{W}|\psi_i\psi_\beta\rangle - \langle \psi_i\psi_\alpha|\hat{W}|\psi_\beta\psi_i\rangle\Big) \\
&= \langle \psi_\alpha|\hat{h}_\Psi|\psi_\beta\rangle = \varepsilon_\alpha\delta_{\alpha,\beta} .
\end{aligned} \tag{2.88}$$

Now ψ_α and ψ_β are any occupied or unoccupied single-particle states, but the sum in (2.88) runs only over the single-particle states occupied in Ψ, $\psi_1, \dots, \psi_N$. These are the *Hartree-Fock equations.*

The one-body Hamiltonian $\hat{h}_\Psi$ contains various contributions:

$$\hat{h}_\Psi = \frac{\hat{\boldsymbol{p}}^2}{2\mu} + \hat{V} + \hat{W}_{\rm d} - \hat{W}_{\rm ex} . \tag{2.89}$$

The kinetic energy $\hat{\boldsymbol{p}}^2/(2\mu)$ and the one-body potential $\hat{V}$ come from the one-body part of the N-electron Hamiltonian $\hat{H}$ and do not depend on the Slater determinant Ψ. The first terms in the bracket following the summation sign in (2.88) constitute the *direct potential* $\hat{W}_{\rm d}$, which is defined by its one-body matrix elements

$$\langle \psi_\alpha|\hat{W}_{\rm d}|\psi_\beta\rangle = \sum_{i=1}^{N}\langle \psi_i\psi_\alpha|\hat{W}|\psi_i\psi_\beta\rangle . \tag{2.90}$$

For the two-body interaction (2.55) without spin-dependent corrections, $\hat{W}_{\rm d}$ is simply a *local* potential depending on the spatial coordinate $\boldsymbol{r}$:

$$\hat{W}_{\rm d} \equiv W_{\rm d}(\boldsymbol{r}) = \sum_{i=1}^{N} \langle \psi_i | \frac{e^2}{|\boldsymbol{r}-\boldsymbol{r}'|} | \psi_i \rangle$$
$$= \int d\boldsymbol{r}' \sum_{i=1}^{N} \sum_{m_{\rm s}} |\psi_i(\boldsymbol{r}', m_{\rm s})|^2 \frac{e^2}{|\boldsymbol{r}-\boldsymbol{r}'|} . \quad (2.91)$$

The integrand in (2.91) contains the electrostatic two-body interaction $e^2/|\boldsymbol{r}-\boldsymbol{r}'|$ multiplied by the *single-particle density* ϱ at the position $\boldsymbol{r}'$:

$$\varrho(\boldsymbol{r}') \stackrel{\rm def}{=} \langle \Psi | \sum_{i=1}^{N} \delta(\boldsymbol{r}'-\boldsymbol{r}_i) | \Psi \rangle = \sum_{i=1}^{N} \sum_{m_{\rm s}} |\psi_i(\boldsymbol{r}', m_{\rm s})|^2 . \quad (2.92)$$

Thus $W_{\rm d}(\boldsymbol{r})$ is the electrostatic potential due to the N electrons of the Slater determinant Ψ.

The second terms in the bracket following the summation sign in (2.88) yield the *exchange potential* $\hat{W}_{\rm ex}$. It is also a one-body operator defined by its matrix elements,

$$\langle \psi_\alpha | \hat{W}_{\rm ex} | \psi_\beta \rangle = \sum_{i=1}^{N} \langle \psi_i \psi_\alpha | \hat{W} | \psi_\beta \psi_i \rangle , \quad (2.93)$$

but it has the much more complicated form of a *nonlocal potential*. The action of such a nonlocal potential on a single-particle wave function $\psi(\boldsymbol{r}, m_{\rm s})$ is determined by an integral kernel $W_{\rm ex}(\boldsymbol{r}, m_{\rm s}; \boldsymbol{r}', m_{\rm s}')$:

$$\hat{W}_{\rm ex}\psi(\boldsymbol{r}, m_{\rm s}) = \int d\boldsymbol{r}' \sum_{m_{\rm s}'} W_{\rm ex}(\boldsymbol{r}, m_{\rm s}; \boldsymbol{r}', m_{\rm s}') \psi(\boldsymbol{r}', m_{\rm s}') . \quad (2.94)$$

Writing out the two-body matrix elements on the right-hand side of (2.93) shows that the integral kernel in (2.94) corresponds to

$$W_{\rm ex}(\boldsymbol{r}, m_{\rm s}; \boldsymbol{r}', m_{\rm s}') = \sum_{i=1}^{N} \psi_i^*(\boldsymbol{r}', m_{\rm s}') \hat{W} \psi_i(\boldsymbol{r}, m_{\rm s}) . \quad (2.95)$$

If we neglect momentum-dependent corrections and take $\hat{W}$ simply to be the electrostatic repulsion (2.55), then

$$W_{\rm ex}(\boldsymbol{r},m_{\rm s},\boldsymbol{r}',m_{\rm s}') = \sum_{i=1}^{N}\psi_i(\boldsymbol{r},m_{\rm s})\frac{e^2}{|\boldsymbol{r}-\boldsymbol{r}'|}\psi_i^*(\boldsymbol{r}',m_{\rm s}')$$

$$= \delta_{m_{\rm s},m_{\rm s}'}\sum_{i=1}^{N}\delta_{m_{\rm s},m_{{\rm s}_i}}\psi_i(\boldsymbol{r})\frac{e^2}{|\boldsymbol{r}-\boldsymbol{r}'|}\psi_i^*(\boldsymbol{r}')\,. \tag{2.96}$$

On the right-hand side of (2.96) we assumed that the single-particle states ψ_i each correspond to a well defined spin state, $\psi_i(\boldsymbol{r},m_{\rm s}) = \psi_i(\boldsymbol{r})\chi_{m_{{\rm s}_i}}$ (compare (1.341)).

When we calculate the expectation value of $\hat{W}_{\rm d} - \hat{W}_{\rm ex}$ for a given single-particle state ψ_j occupied in Ψ, the two contributions corresponding to $i = j$ cancel, leaving

$$\langle\psi_j|\hat{W}_{\rm d} - \hat{W}_{\rm ex}|\psi_j\rangle = \sum_{i\neq j}\Big(\langle\psi_i\psi_j|\hat{W}|\psi_i\psi_j\rangle - \langle\psi_i\psi_j|\hat{W}|\psi_j\psi_i\rangle\Big)\,. \tag{2.97}$$

Thus a part of the exchange potential just cancels the unphysical *self-energies* $\langle\psi_i\psi_i|\hat{W}|\psi_i\psi_i\rangle$ in the contribution of the direct potential.

The Hartree-Fock equations (2.88) replace the N-electron problem by a one-body problem characterized by the one-body Hamiltonian $\hat{h}_\Psi$ (2.89). But $\hat{h}_\Psi$ still depends on the Slater determinant Ψ, which is to be determined by solving the Hartree-Fock equations. Thus the Hartree-Fock method involves a problem of *self-consistency*, which is usually solved iteratively. We start with a Slater determinant Ψ_0, diagonalize the one-body Hamiltonian $\hat{h}_{\Psi_0}$ defined by the single-particle states occupied in Ψ_0, obtain a new set of single-particle states and a new Slater determinant Ψ_1, diagonalize $\hat{h}_{\Psi_1}$ obtain Ψ_2, etc., until the procedure reaches convergence. A widespread simplification of this *unrestricted Hartree-Fock procedure* is the *restricted Hartree-Fock procedure*, in which we assume that the single-particle wave functions in each iteration step are eigenfunctions of the single-particle orbital angular momentum,

$$\psi_i(\boldsymbol{r},m_{\rm s}) = \frac{\phi_i^{(l)}(r)}{r}Y_{l,m}(\theta,\phi)\chi_{m_{{\rm s}_i}}\,, \tag{2.98}$$

and that all radial wave functions in a subshell are identical. The Hartree-Fock equations can then be reduced to a set of radial equations for the determination of the radial wave functions $\phi_i^{(l)}$ in each occupied subshell.

In the Hartree-Fock method, the variational method doesn't lead to diagonalization of a reduced Hamiltonian in a subspace of Hilbert space (compare Sect. 1.6.2). The reason for this is that the set of variational wave functions, i.e. of Slater determinants, is not a subspace which is closed with respect to linear superposition; a sum of Slater determinants need not itself be a Slater determinant. Consequently, two different Slater determinants which solve the Hartree-Fock equations (for the same values of the good quantum numbers of the system) need not be diagonal in $\hat{H}$. Only for the ground state (of a given symmetry) do we know that the Hartree-Fock energy $E[\Psi_{\rm HF}] = \langle\Psi_{\rm HF}|\hat{H}|\Psi_{\rm HF}\rangle$ is an upper bound for the exact energy eigenvalue.

The Hartree-Fock energy $E[\Psi_{\rm HF}]$ is not identical to the sum of single-particle energies ε_i of the occupied states, as obtained by solving the Hartree-Fock equations (2.88). This is because the summation of the single-particle energies counts the contribution of the two-body interaction between electron pairs twice. With (2.73) and (2.75) we have

$$\begin{aligned}\langle\Psi_{\rm HF}|\hat{H}|\Psi_{\rm HF}\rangle &= \sum_{i=1}^{N}\langle\psi_i|\frac{\hat{\boldsymbol{p}}^2}{2\mu}+\hat{V}|\psi_i\rangle \\ &\quad +\frac{1}{2}\sum_{i,j=1}^{N}\Big(\langle\psi_i\psi_j|\hat{W}|\psi_i\psi_j\rangle - \langle\psi_i\psi_j|\hat{W}|\psi_j\psi_i\rangle\Big) \\ &= \sum_{i=1}^{N}\varepsilon_i - \frac{1}{2}\sum_{i,j=1}^{N}\Big(\langle\psi_i\psi_j|\hat{W}|\psi_i\psi_j\rangle - \langle\psi_i\psi_j|\hat{W}|\psi_j\psi_i\rangle\Big)\,.\end{aligned} \tag{2.99}$$

In general the final Hartree-Fock wave function is not a single Slater determinant, but a sum of several Slater determinants each containing the same occupied radial single-particle states and whose spin and angular parts are coupled to good quantum numbers of the total angular momentum and perhaps also of the total orbital angular momentum and the total spin (compare Sect. 2.2.4).

For lighter atoms and ions, the effects of relativistic corrections to the non-relativistic Schrödinger equation are small and can be treated in first-order perturbation theory starting from the Hartree-Fock wave function. For heavier atoms and ions the effective fine-structure constant $Z\alpha_{\rm fs} \approx Z/137$ is no longer such a small number and, as Z becomes larger, perturbation theory becomes increasingly inadequate for describing relativistic corrections. One way of improving the description of relativistic effects is to replace the kinetic energy $\hat{\boldsymbol{p}}^2/(2\mu)$ in the one-body Hamiltonian (2.89) by Dirac's Hamiltonian (2.17) for a free particle:

$$\hat{h}^{\rm D}_{\Psi} = c\,\boldsymbol{\alpha}\cdot\hat{\boldsymbol{p}} + \beta\mu c^2 + \hat{V} + \hat{W}_{\rm d} - \hat{W}_{\rm ex}\,. \tag{2.100}$$

In this way, relativistic corrections to the one-electron problem are included consistently (cf. Sect. 2.1.4). The relativistic treatment of the two-body interaction is much more difficult, because the picture of a heavy resting mass as origin of the static potential only holds for the attraction of the electrons by the atomic nucleus (compare Sect. 2.1.3) and not for the interaction between two electrons. In practice the potentials $\hat{W}_{\rm d}$ and $\hat{W}_{\rm ex}$ are initially defined via the static interaction (2.55). *Retardation* effects due to the fact that all interactions can propagate no faster than the speed of light are subsequently treated with perturbative methods. The *Dirac-Fock method* consists in looking for self-consistent eigenfunctions of the one-body

Hamiltonian (2.100). For radially symmetric potentials this means self-consistently solving the radial Dirac equation instead of the radial Schrödinger equation.

2.3.2 Correlations and Configuration Interaction

The Hartree-Fock method (or the Dirac-Fock method) yields the best N-electron wave function compatible with the picture of N independent electrons. In order to account for *correlations*, which go beyond this picture, we have to admit variational wave functions which are more general than single Slater determinants. An obvious ansatz for a correlated N-electron wave function ψ is a sum of N_S Slater determinants Ψ_ν, which may include various different N-electron configurations (with the same values of the good quantum numbers):

$$\psi = \sum_{\nu=1}^{N_S} c_\nu \Psi_\nu \,. \qquad (2.101)$$

Effects of *configuration interaction* are included if we diagonalize the N-electron Hamiltonian in the subspace spanned by the Ψ_ν in Hilbert space. This corresponds to a variational calculation in which the mixing coefficients c_ν in (2.101) are the variational parameters (compare Sect. 1.6.2). In the *multi-configurational Hartree-Fock method* (MCHF) the energy expectation value $E[\psi]$ is minimized with respect to variations both of the coefficients c_ν in (2.101) and of the single-particle states in the Slater determinants Ψ_ν. If the sum in (2.101) includes enough terms, this procedure can in principle approximate the exact solution to any accuracy, because every totally antisymmetric N-electron wave function can be written as a sum of Slater determinants. In practice of course, the MCHF problem is most readily solved if not too many terms are included in the sum in (2.101).

Configuration interaction calculations can also be performed with the Slater determinants of the Dirac-Fock method. The corresponding generalization of the MCHF method is called *multi-configurational Dirac-Fock method* (MCDF) [IL05].

If the number of configurations included in the ansatz (2.101) is sufficiently large, then a simple diagonalization of the Hamiltonian in the subspace spanned by the Ψ_ν can yield a good approximation of the exact eigenstates, even without explicit consideration of self-consistency as in the MCHF method. If we start from a complete basis of single-particle states, then the exact eigenstates can in principle be approximated within arbitrary accuracy in this way. Such large scale diagonalizations are quite generally called "configuration interaction calculations" (CI). Usually various many-electron configurations are constructed from single-particle wave functions which are chosen so that the corresponding one-body and two-body matrix elements are not too difficult to calculate. The N-electron energies and eigenfunctions are obtained by diagonalizing the Hamiltonian matrix which now may have quite large dimensions—typically up to many thousands.

Table 2.1 Ground state energies (in atomic units) for the helium iso-electronic sequence

	E_{HF}	E_{MCHF}	E_{nr}	$E_{nr} - E_{HF}$	$E_{DF} - E_{HF}$	E_{exp}
H^-	−0.487927	−0.527510	−0.527751	−0.039824	< 0.00001	−0.52776
He	−2.861680	−2.903033	−2.903724	−0.042044	−0.00013	−2.90378
Li^+	−7.236416	−7.279019	−7.279913	−0.043497	−0.00079	−7.28041
Be^{++}	−13.611300	−13.654560	−13.655566	−0.044266	−0.00270	−13.65744
B^{3+}	−21.986235	−22.029896	−22.030972	−0.044737	−0.00692	−22.03603
C^{4+}	−32.361194	−32.405123	−32.406247	−0.045053	−0.01480	−32.41733
N^{5+}	−44.736163	−44.780287	−44.781445	−0.045282	−0.02804	−44.80351
O^{6+}	−59.111141	−59.155411	−59.156595	−0.045454	−0.04865	−59.19580
F^{7+}	−75.486124	−75.530508	−75.531712	−0.045588	−0.07898	−75.59658
Ne^{8+}	−93.861111	−93.905586	−93.906807	−0.045696	−0.12169	−94.00835

A frequent choice for the spatial part of the single-particle wave functions is based on expansions in *Slater-type orbitals*: $\phi_l(r) \propto r^m \exp(-\zeta r)$. The coefficients in such expansions as well as the coefficients ζ in the exponents are treated as variational parameters. Another basis of single-particle states, which are characterized by their similarity to the eigenfunctions (1.139) of the pure Coulomb potential, is the *Sturm-Liouville basis*. The single-particle states in this basis have the same form as in (1.139), but the number n in the argument of the Laguerre polynomial and the exponential function is replaced by a constant integer n_0 rather than varying from shell to shell. In contrast to the pure Coulomb bound states (1.139), the Sturm-Liouville states form a complete set, because of the completeness of the Laguerre polynomials. Furthermore, the single-particle states with $n = n_0$ are identical to the eigenstates of the pure Coulomb potential with this principle quantum number. On the other hand, in a Sturm-Liouville basis, single-particle states with different principle quantum numbers are no longer orthogonal.

As simplest example of a many-electron system Table 2.1 summarizes the ground state energies of the two-electron helium iso-electronic sequence from H^- to Ne^{8+} as they are obtained in various approximations, together with the experimental values E_{exp} [BS75]. The first column contains the Hartree-Fock energies[2] [Fro77, Fro87, SK88] and the second column contains the results of an MCHF calculation [SK88]. The third column contains the "exact" results E_{nr} within non-relativistic quantum mechanics, as obtained by Pekeris [Pek58] in a very clever CI calculation as early as 1958—a time when computer capacity was far less abundant than

[2]The fact that the energy of the H^- ion in the first column of Table 2.1 lies above the energy −0.5 of the H atom shows a weakness of the restricted Hartree-Fock method, which was used here and in which both electrons were restricted to having the same spatial part of the single-particle wave function. In an unrestricted Hartree-Fock calculation the Hartree-Fock energy can at least come arbitrarily close to the value −0.5. To see this construct a two-electron Slater determinant in which one occupied single-particle state is the ground state of atomic hydrogen and the other is a very distant almost plane wave with (almost) vanishing wave number.

today. The difference between the exact ground state energy and the Hartree-Fock energy (fourth column) is usually called the *correlation energy*; it is a measure of the deviation of the exact (correlated) two-body wave function from the Hartree-Fock configuration. The absolute magnitude of the correlation energy changes little within the iso-electronic sequence, because the electron-electron interaction doesn't depend on the charge number Z. On the other hand, the one-body contribution to the total binding energy increases rapidly with increasing Z, and so the relative importance of the correlations decreases with increasing charge number in the iso-electronic sequence. An estimate of the magnitude of relativistic corrections can be derived from the fifth column which lists the differences between the energies obtained in the Dirac-Fock and Hartree-Fock methods. These differences are of the same order of magnitude as the differences between the exact non-relativistic results (column 3) and the experimental data (column 6). At this level of accuracy we must however also consider the effects of *radiative corrections* which follow from a more sophisticated description of the atoms and ions in the framework of quantum electrodynamics. For precision calculations of the various corrections in the two-electron system see e.g. [KH86, Dra88, Dra01].

The art of solving the Hartree-Fock equations has been driven to a high degree of perfection [Fro77, Fro87, Fro94]. The same is true for high-dimensional CI calculations for the determination of energies and wave functions of low-lying states [Sch77, Fro94]. A thorough description of the details of such calculations for the structure of atomic many-body systems can be found in the book by Lindgren und Morrison [LM85]. (See also [CO80].)

In contrast to the substantial and comprehensive body of knowledge which has accumulated during many years of successful investigations of the electronic structure of low-lying states, our understanding of highly excited atomic states is still very incomplete. Only in the situation that just one electron is highly excited with the other electrons forming a low-lying state of the atomic (or ionic) "core", can we make far reaching and general statements concerning the structure of atomic spectra and wave functions. This case, which largely corresponds to a one-electron problem, is treated in detail in Chap. 3. The systematic understanding of the spectrum of an atom or ion already becomes a very difficult problem if two electrons are highly excited. For a detailed description of the problem of two or more highly excited electrons, see [Fan83] or Part D of the book by Fano and Rau [FR86]. The complications involved can already be appreciated by studying high doubly-excited states in the helium atom, see Sect. 5.3.5 (c) in Chap. 5.

2.3.3 The Thomas-Fermi Model

One of the simplest models of an N-electron atom or ion is the *Thomas-Fermi model,* which was developed 90 years ago [Tho27, Fer28]. The model is based on the single-particle density of a degenerate free-electron gas, in which all single-particle states up to the *Fermi energy*,

$$E_{\mathrm{F}} = \frac{\hbar^2}{2\mu} k_{\mathrm{F}}^2 , \tag{2.102}$$

are occupied and all single-particle states with higher energies are unoccupied. In $6N$-dimensional phase space, the occupied single-particle states fill a volume which is the product of the spatial volume V_{s} and the volume $\frac{4\pi}{3}(\hbar k_{\mathrm{F}})^3$ of the *Fermi sphere* in momentum space. A volume of $V_{\mathrm{s}}\frac{4\pi}{3}(\hbar k_{\mathrm{F}})^3$ is thus filled in phase space, and each cell of size $h^3 = (2\pi\hbar)^3$ can accomodate two single-particle states—one with spin up and one with spin down. The number N of occupied one-electron states is thus (see also Problem 2.4)

$$N = \frac{2}{(2\pi\hbar)^3} V_{\mathrm{s}} \frac{4\pi}{3} (\hbar k_{\mathrm{F}})^3 = k_{\mathrm{F}}^3 \frac{V_{\mathrm{s}}}{3\pi^2} . \tag{2.103}$$

This gives us a relation between the density $\varrho = N/V_{\mathrm{s}}$ and the *Fermi wave number* k_{F}:

$$k_{\mathrm{F}} = (3\pi^2 \varrho)^{1/3} . \tag{2.104}$$

In the Thomas-Fermi model we describe an atom by a radially symmetric single-particle potential $V(r)$ for the electrons, and we let the *Fermi momentum* $\hbar k_{\mathrm{F}}$ depend on the radial distance r, just like the semiclassical momentum in the WKB approximation (1.286) (see Fig. 2.6):

$$E_0 = \frac{\hbar^2}{2\mu} k_{\mathrm{F}}^2(r) + V(r) , \tag{2.105}$$

where $E_0 \leq 0$ is the total energy of the least bound electron. In this picture the kinetic energy of the least bound electron is

$$T(r) = E_0 - V(r) = \frac{\hbar^2}{2\mu} k_{\mathrm{F}}^2(r) , \tag{2.106}$$

and it depends on the spatial coordinate r, in analogy to the semiclassical approximation (1.286). The kinetic energy (2.106) of the least bound electron vanishes at the outer turning point r_0 which defines the "edge" of the atom.

We can obtain a differential equation for the single-particle potential $V(r)$, or for $T(r)$, by relating the electrostatic potential $-V/e$ to the sources of charge $-e\varrho$ (outside of the atomic nucleus at $r = 0$) via the Poisson equation:

$$\Delta V = -\Delta T = -4\pi e^2 \varrho . \tag{2.107}$$

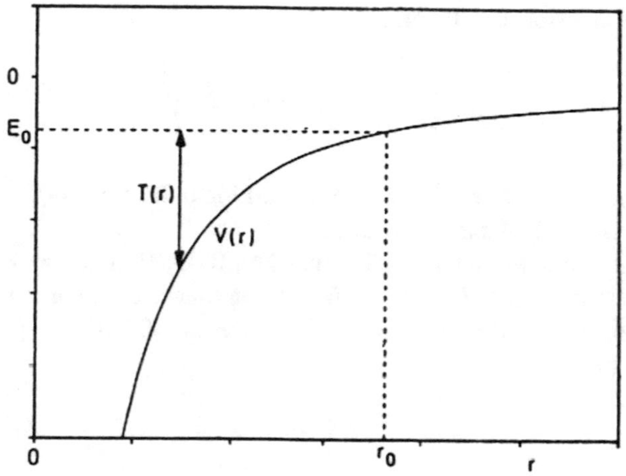

Fig. 2.6 Schematic representation of an atom or ion in the Thomas-Fermi model. All single-particle states in the single-particle potential $V(r)$ are occupied up to the energy E_0. Locally the system resembles a degenerate free-electron gas in which the states are occupied up to the Fermi energy $E_F = (\hbar^2/2\mu)k_F^2 = E_0 - V(r) = T(r)$. The local kinetic energy $T(r)$ of the least bound electron vanishes at the outer turning point r_0

We can express the density ϱ in terms of k_F via (2.104) and in terms of T via (2.106),

$$\varrho = \frac{1}{3\pi^2}\left(\frac{2\mu}{\hbar^2}T\right)^{3/2}, \tag{2.108}$$

and so we obtain the following differential equation for the function $T(r)$ (compare (1.70)):

$$\left(\frac{d^2}{dr^2} + \frac{2}{r}\frac{d}{dr}\right)T = \frac{1}{r}\frac{d^2}{dr^2}(rT) = \frac{4e^2}{3\pi}\left(\frac{2\mu}{\hbar^2}T\right)^{3/2}. \tag{2.109}$$

This equation assumes a universal form when we refer the local kinetic energy of the least bound electron $T(r)$ to the potential Coulomb energy $-Ze^2/r$ due to the atomic nucleus and introduce the dimensionless *Thomas-Fermi function*

$$\chi = \frac{rT}{Ze^2}. \tag{2.110}$$

Equation (2.109) thus becomes the *Thomas-Fermi equation*,

$$\frac{d^2\chi}{dx^2} = \frac{\chi^{3/2}}{\sqrt{x}}, \tag{2.111}$$

where x is a dimensionsless length:

$$x = \frac{r}{b}, \quad b = aZ^{-\frac{1}{3}}\left(\frac{9\pi^2}{128}\right)^{1/3}; \tag{2.112}$$

$a = \hbar^2/(\mu e^2)$ is the Bohr radius. The outer turning point $x_0 = r_0/b$ is the first zero of $\chi(x)$; χ vanishes identically beyond x_0.

The boundary condition for the Thomas-Fermi function at $x = 0$ follows from the fact that the potential $V(r)$ in (2.106) is dominated by the attractive Coulomb potential $-Ze^2/r$ near the atomic nucleus $r = 0$. From (2.110) we get the boundary condition for χ:

$$\chi(0) = 1. \tag{2.113}$$

The behaviour of $\chi(x)$ for small x is in fact [Eng88]:

$$\chi(x) \overset{x\to 0}{=} 1 + Bx + \frac{4}{3}x^{3/2} + O\left(x^{5/2}\right). \tag{2.114}$$

Since the Thomas-Fermi function χ is never zero between $x = 0$ and the outer turning point $x_0 = r_0/b$, its second derivative (2.111) never vanishes and its first derivative cannot change sign in this interval. It follows that $\chi(x)$ is a monotonically decreasing function falling from unity at $x = 0$ to zero at the outer turning point x_0. The gradient at $x = 0$ is given by the (negative) constant B in (2.114).

The outer boundary condition for the Thomas-Fermi function follows from the consideration that the integral of the single-particle density from the origin to the outer turning point must yield the total number N of electrons:

$$4\pi \int_0^{r_0} \varrho(r) r^2 \, \mathrm{d}r = N. \tag{2.115}$$

With (2.108), (2.110) and (2.112) this can be expressed in the dimensionless quantities:

$$Z \int_0^{x_0} [\chi(x)]^{3/2} \sqrt{x} \, \mathrm{d}x = N. \tag{2.116}$$

From the differential equation (2.111) we can replace $\chi^{3/2}$ by $\chi'' \sqrt{x}$ and formally integrate (2.116):

$$N = Z \int_0^{x_0} x \chi'' \, \mathrm{d}x = Z[x\chi' - \chi]_0^{x_0}. \tag{2.117}$$

With $\chi(0) = 1$ and $\chi(x_0) = 0$, (2.117) becomes

$$x_0\,\chi'(x_0) = \frac{N-Z}{Z}\,. \tag{2.118}$$

Since $\chi(x)$ is a monotonically decreasing function, the right-hand side of (2.118) cannot be positive. This means that N cannot be larger than the charge number Z of the nucleus. For $N = Z$ corresponding to a neutral atom, the outer turning point x_0 lies at infinity; the energy E_0 in (2.105), (2.106) vanishes and the single-particle potential is simply (cf. (2.106), (2.110))

$$V(r) = -\frac{Ze^2}{r}\chi_0\left(\frac{r}{b}\right). \tag{2.119}$$

All neutral atoms are described in the Thomas-Fermi model by a universal Thomas-Fermi function χ_0 which is shown as the solid line in Fig. 2.7. It is the (unique) solution of the (2.111) with the boundary conditions that $\chi(0) = 1$ and that the first zero of χ lies at infinity. The gradient at $x = 0$ in this case is $B = -1.588$ (see e.g. [Eng88], p. 65).

Solutions of (2.111) which fall off faster than χ_0 at $x = 0$ cut the x-axis at finite values of x and with finite (negative) gradient. For these solutions the right-hand side of (2.118) is a finite negative number which corresponds to a positively charged ion, $N < Z$. For example: The solution $\chi(x)$ starting with a gradient $B = -1.608$ at $x = 0$ already cuts the x-axis at $x \approx 2.9$ and the right-hand side is approximately $-1/2$. This case corresponds to an ion with half as many electrons as the associated neutral atom and is shown as the dashed line in Fig. 2.7. Solutions of (2.111) which fall off more slowly than χ_0 at $x = 0$ never reach the x-axis, not even at infinity,

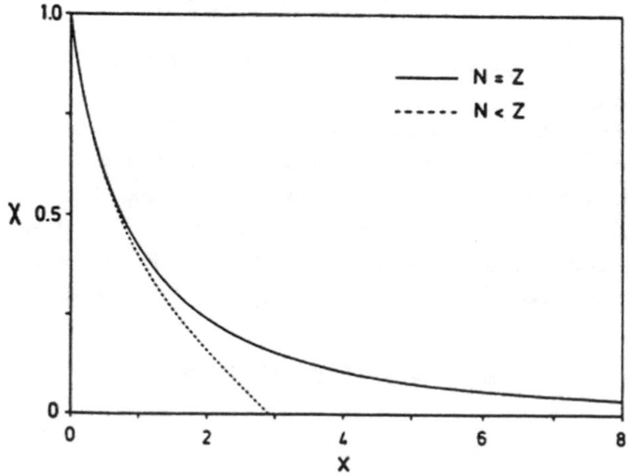

Fig. 2.7 Solutions of the Thomas-Fermi equation (2.111). The *solid line* shows the case of a neutral atom $N = Z$, the *dashed line* shows an example of a positively charged ion with $N \approx Z/2$

and are not suited for the description of isolated atoms or ions in the Thomas-Fermi model. The Thomas-Fermi model cannot describe negative ions.

Although the Thomas-Fermi model represents a drastic approximation of the N-electron problem, it is very useful for describing general trends in the properties of atoms. Equation (2.112) for example, shows that the behaviour of typical lengths as a function of charge number Z is given by proportionality to $Z^{-1/3}$. For a detailed description of Thomas-Fermi theory in particular and of semiclassical theories in atomic physics in general see [Eng88].

2.3.4 *Density Functional Methods*

The one-body contribution E_V to the potential energy of N electrons in an external local potential $V(\boldsymbol{r})$ is

$$E_V = \langle\Psi|\sum_{i=1}^{N} V(\boldsymbol{r}_i)|\Psi\rangle = \int V(\boldsymbol{r})\varrho(\boldsymbol{r})\,\mathrm{d}\boldsymbol{r}\,. \tag{2.120}$$

E_V is a unique function, i.e. a *functional,* of the single-particle density $\varrho(\boldsymbol{r})$, which is defined quite generally (and not only for Slater determinants) by the first equation (2.92). The relation (2.120) can be obtained by replacing the $V(\boldsymbol{r}_i)$ in the matrix element by $\int \delta(\boldsymbol{r}-\boldsymbol{r}_i)V(\boldsymbol{r})\mathrm{d}\boldsymbol{r}$ and then pulling the integral over the vector $\boldsymbol{r}$ out of the matrix element. If we are dealing with Slater determinants, then the direct part $\hat{W}_\mathrm{d}$ of the two-body interaction (compare (2.91)) contributes a term

$$E_\mathrm{d} = \frac{e^2}{2}\int \mathrm{d}\boldsymbol{r}_1 \int \mathrm{d}\boldsymbol{r}_2 \frac{\varrho(\boldsymbol{r}_1)\varrho(\boldsymbol{r}_2)}{|\boldsymbol{r}_1-\boldsymbol{r}_2|} \tag{2.121}$$

to the total energy (compare (2.75)), and this term is also a functional of the single-particle density ϱ.

Investigating the quite general question whether the energy of an N-electron system is a functional of the single-particle density leads to a very strong statement concerning the ground state of an N-electron system. This is the *Hohenberg-Kohn Theorem* [HK64, KS65] which states: "For a system of N electrons in an external potential $V(\boldsymbol{r})$ there is a universal functional $F[\varrho]$ of the single-particle density ϱ, which is independent of V and has the property that the expression

$$E[\varrho] = \int V(\boldsymbol{r})\varrho(\boldsymbol{r})\mathrm{d}\boldsymbol{r} + F[\varrho(\boldsymbol{r})] \tag{2.122}$$

assumes a minimum for the density corresponding to the ground state of the system, and the value at this minimum is the correct ground state energy (in this external potential)."

The first term on the right-hand side of (2.122) is the one-body contribution (2.120) to the potential energy. The universal functional F in (2.122) contains a term of the form (2.121) for the direct two-body contribution to the potential energy. Beyond this it contains a contribution $E_{\rm kin}$ of the kinetic energy as well as an "exchange and correlation" term, which collects all those contributions to the potential energy not already contained in (2.120) or (2.121). The nature of this term and of the kinetic energy contribution $E_{\rm kin}$ is in general unknown.

In the simple Thomas-Fermi model where the atom is treated locally as a degenerate electron gas (Sect. 2.3.3), it is easy to evaluate the kinetic energy as a functional of the single-particle density: The sum of the kinetic energies of all occupied single-particle states is equal to the integral of $\hbar^2k^2/(2\mu)$ over the occupied states in phase space:

$$T_{\rm F} = \frac{2}{(2\pi\hbar)^3} V_{\rm s}\, 4\pi \int_0^{k_{\rm F}} \hbar^3 k^2\, {\rm d}k \frac{\hbar^2 k^2}{2\mu} = \frac{\hbar^2}{10\pi^2\mu} V_{\rm s} k_{\rm F}^5 \,. \tag{2.123}$$

The total kinetic energy $E_{\rm kin}$ is equal to the integral of the kinetic energy density $T_{\rm F}/V_{\rm s}$ over the spatial volume of the Thomas-Fermi atom. Inserting the expression (2.104) for $k_{\rm F}$ yields $E_{\rm kin}$ as functional of ϱ (in the framework of Thomas-Fermi model):

$$(E_{\rm kin})_{\rm TF} = \frac{\hbar^2}{10\pi^2\mu} 4\pi \int_0^{r_0} \left[3\pi^2 \varrho(r)\right]^{5/3} r^2\, {\rm d}r \,. \tag{2.124}$$

Within the Thomas-Fermi model, the energy as functional of the single-particle density is thus given by a term of the form (2.120) for the potential energy of the electrons in the external potential due to the electrostatic attraction by the atomic nucleus, a term of the form (2.121) for the mutual electrostatic repulsion of the electrons and the kinetic energy term (2.124). The condition that this functional be stationary with respect to small variations of the single-particle density actually does lead to the Thomas-Fermi equation (2.111) [Eng88].

Next to the N-body Schrödinger equation, the Hohenberg-Kohn Theorem offers an alternative approach to the N-electron problem. Usually one starts with a physically or pragmatically founded ansatz for the density functional $F[\varrho(\boldsymbol{r})]$ in (2.122) and tries to minimize the energy $E[\varrho]$. In recent years, density functional theory has evolved into a sophisticated and powerful tool for accurately calculating the properties of many-electron systems in physics and chemistry [DG90]. In 1998 Walter Kohn shared the Nobel Prize in Chemistry for the development of this theory.

2.4 Electromagnetic Transitions

The Hamiltonians (2.1) and (2.7) or (2.47) and (2.53) describe the *atomic degrees of freedom* of a one- or many-electron atom (or ion) with and without inclusion of the atomic nucleus respectively. Such an atomic Hamiltonian $\hat{H}_\mathrm{A}$ possesses a spectrum of eigenvalues, and the associated eigenstates are solutions of the corresponding stationary Schrödinger equation. The eigenstates of $\hat{H}_\mathrm{A}$ are usually "seen" by observing electromagnetic radiation emitted or absorbed during a transition between two eigenstates. The fact that such transitions occur and that an atom doesn't remain in an eigenstate of $\hat{H}_\mathrm{A}$ forever, is due to the interaction between the atomic degrees of freedom and the degrees of freedom of the electromagnetic field. A Hamiltonian $\hat{H}$ able to describe electromagnetic transitions must thus account not only for the atomic degrees of freedom, but also for the degrees of freedom of the electromagnetic field. An eigenstate of the atomic Hamiltonian $\hat{H}_\mathrm{A}$ is in general not an eigenstate of the full Hamiltonian $\hat{H}$; a system which is in an eigenstate of $\hat{H}_\mathrm{A}$ at a given time will evolve as prescribed by the time evolution operator (1.40), (1.41) containing the full Hamiltonian $\hat{H}$, and may be in a different eigenstate of $\hat{H}_\mathrm{A}$ at a later time. If we look at the interaction between atom and electromagnetic field as a perturbation of the non-interacting Hamiltonian, then this perturbation causes time dependent transitions between the unperturbed eigenstates, even if the perturbation itself is time independent. Such transitions can be generally described in the framework of *time-dependent perturbation theory* which is expounded in the following section.

2.4.1 *Transitions in General, "Golden Rule"*

Consider a physical system which is described by the Hamiltonian

$$\hat{H} = \hat{H}_0 + \hat{W}\,, \tag{2.125}$$

but which is in an eigenstate ϕ_i of the Hamiltonian $\hat{H}_0$ at time $t = 0$. This Hamiltonian $\hat{H}_0$ is assumed to differ from the full Hamiltonian $\hat{H}$ by a "small perturbation" $\hat{W}$. Even if $\hat{H}_0$ isn't the exact Hamiltonian, its (orthonormalized) eigenstates ϕ_n,

$$\hat{H}_0\phi_n = E_n\phi_n\,, \tag{2.126}$$

still form a complete basis in which we can expand the exact time-dependent wave function $\psi(t)$:

$$\psi(t) = \sum_n c_n(t)\phi_n \exp\left(-\frac{\mathrm{i}}{\hbar}E_n t\right). \tag{2.127}$$

The coefficients $c_n(t)$ in this expansion are time dependent, because the time evolution of the eigenstates of $\hat{H}_0$ is, due to the perturbation $\hat{W}$, not given by the exponential functions $\exp[-(i/\hbar)E_n t]$ alone.

The initial condition that the system be in the eigenstate ϕ_i of $\hat{H}_0$ at time $t = 0$ is expressed in the following initial conditions for the coefficients $c_n(t)$:

$$c_n(t = 0) = \delta_{n,\mathrm{i}} \,. \tag{2.128}$$

At a later time t, the probability for finding the system in the eigenstate ϕ_f of $\hat{H}_0$ is:

$$w_{\mathrm{i}\to\mathrm{f}}(t) = |c_\mathrm{f}(t)|^2. \tag{2.129}$$

In order to calculate the coefficients $c_n(t)$ we insert the expansion (2.127) in the time-dependent Schrödinger equation (1.38) and obtain, using (2.125), (2.126),

$$\begin{aligned} \mathrm{i}\hbar \sum_n \phi_n \left(\frac{\mathrm{d}c_n}{\mathrm{d}t} - \frac{\mathrm{i}}{\hbar} E_n c_n \right) \exp\left(-\frac{\mathrm{i}}{\hbar} E_n t \right) \\ = \sum_n c_n \exp\left(-\frac{\mathrm{i}}{\hbar} E_n t \right) (E_n \phi_n + \hat{W}\phi_n) \,. \end{aligned} \tag{2.130}$$

If we multiply from the left with the bra $\langle \phi_m|$, (2.130) becomes a system of coupled ordinary differential equations for the coefficients $c_n(t)$:

$$\mathrm{i}\hbar \frac{\mathrm{d}c_m}{\mathrm{d}t} = \sum_n W_{mn}\, c_n \exp\left[\frac{\mathrm{i}}{\hbar}(E_m - E_n)t \right], \tag{2.131}$$

with

$$W_{mn} = \langle \phi_m | \hat{W} | \phi_n \rangle \,. \tag{2.132}$$

We can formally integrate the equations (2.131):

$$\begin{aligned} c_m(t) &= c_m(0) + \frac{1}{\mathrm{i}\hbar} \int_0^t \mathrm{d}t' \sum_n W_{mn} \exp\left[\frac{\mathrm{i}}{\hbar}(E_m - E_n)t' \right] c_n(t') \\ &= c_m(0) + \frac{1}{\mathrm{i}\hbar} \int_0^t \mathrm{d}t' \sum_n W_{mn} \exp\left[\frac{\mathrm{i}}{\hbar}(E_m - E_n)t' \right] c_n(0) \\ &\quad + \frac{1}{(\mathrm{i}\hbar)^2} \int_0^t \mathrm{d}t' \sum_n W_{mn} \exp\left[\frac{\mathrm{i}}{\hbar}(E_m - E_n)t' \right] \\ &\quad \times \int_0^{t'} \mathrm{d}t'' \sum_l W_{nl} \exp\left[\frac{\mathrm{i}}{\hbar}(E_n - E_l)t'' \right] c_l(t'') \,, \\ &\quad \text{etc.} \end{aligned} \tag{2.133}$$

To obtain the second equation (2.133) we inserted the expression given by the first equation for $c_n(t')$ in the integral (in the first equation). To obtain higher terms insert a similar expression for $c_l(t'')$ in the integral in the last row.

To first order in the matrix elements of the perturbing operator $\hat{W}$, the coefficients $c_n(t)$ are given by the second row in (2.133). Inserting the initial conditions (2.128) we obtain an expression for the *transition amplitude* $c_\mathrm{f}(t)$ to the final state ϕ_f:

$$c_\mathrm{f}(t) = \frac{1}{\mathrm{i}\hbar} \int_0^t \mathrm{d}t'\, W_\mathrm{fi} \exp\left[\frac{\mathrm{i}}{\hbar}(E_\mathrm{f} - E_\mathrm{i})t'\right] . \tag{2.134}$$

If the perturbing operator $\hat{W}$, and hence the matrix element W_fi, do not depend on time, we can integrate (2.134) directly and obtain

$$|c_\mathrm{f}(t)|^2 \equiv w_{\mathrm{i}\to\mathrm{f}}(t) = |W_\mathrm{fi}|^2 \, \frac{\sin^2[(E_\mathrm{f} - E_\mathrm{i})t/(2\hbar)]}{[(E_\mathrm{f} - E_\mathrm{i})/2]^2} . \tag{2.135}$$

For large times t, (2.135) becomes

$$w_{\mathrm{i}\to\mathrm{f}}(t) \approx |W_\mathrm{fi}|^2 \, \frac{2\pi}{\hbar} \, t \, \delta(E_\mathrm{f} - E_\mathrm{i}) . \tag{2.136}$$

This means that for large times t the *transition probability per unit time*, $P_{\mathrm{i}\to\mathrm{f}}$, becomes independent of t:

$$P_{\mathrm{i}\to\mathrm{f}} = \frac{1}{t} \, w_{\mathrm{i}\to\mathrm{f}}(t) = \frac{2\pi}{\hbar} \, |W_\mathrm{fi}|^2 \, \delta(E_\mathrm{f} - E_\mathrm{i}) . \tag{2.137}$$

It makes sense to assume that the diagonal matrix elements $\langle\phi_\mathrm{i}|\hat{W}|\phi_\mathrm{i}\rangle$ and $\langle\phi_\mathrm{f}|\hat{W}|\phi_\mathrm{f}\rangle$ vanish, because a perturbing operator diagonal in the unperturbed basis doesn't cause transitions. Then E_i and E_f are not only the eigenvalues of the unperturbed Hamiltonian $\hat{H}_0$ in the initial and final state respectively, but they are also the expectation values of the full Hamiltonian $\hat{H} = \hat{H}_0 + \hat{W}$ in the respective states. The delta function in the formula (2.137) for the transition probabilities expresses energy conservation in the long-time limit.

In many practical examples (such as the electromagnetic decay of an atomic state) the energy spectrum of the final states of the whole system (in this case of atom plus electromagnetic field) is continuous. In order to obtain the total probability per unit time for transitions from the initial state ϕ_i to all possible final states ϕ_f we must integrate over an infinitesimal energy range around E_i:

$$P_{\mathrm{i}\to\mathrm{f}} = \lim_{\varepsilon\to 0}\int_{E_\mathrm{i}-\varepsilon}^{E_\mathrm{i}+\varepsilon} \frac{2\pi}{\hbar}|\langle\phi_\mathrm{f}|\hat{W}|\phi_\mathrm{i}\rangle|^2\,\delta(E_\mathrm{f}-E_\mathrm{i})\varrho(E_\mathrm{f})\,\mathrm{d}E_\mathrm{f} \tag{2.138}$$

or rather,

$$P_{\mathrm{i}\to\mathrm{f}} = \frac{2\pi}{\hbar}|\langle\phi_\mathrm{f}|\hat{W}|\phi_\mathrm{i}\rangle|^2\,\varrho(E_\mathrm{f}=E_\mathrm{i})\,. \tag{2.139}$$

Here $\varrho(E_\mathrm{f})$ is the *density of final states.*

The formula (2.139) is Fermi's famous *Golden Rule*; it gives the probability per unit time for transitions caused by a time-independent perturbing operator in first-order perturbation theory.

The precise definition of the density $\varrho(E_\mathrm{f})$ of final states ϕ_f depends on the normalization of the final states. Consider for example a free particle in a one-dimensional box of length L. The number of bound states (normalized to unity) per unit energy is (see Problem 2.5)

$$\varrho_L(E) = \frac{L}{2\pi}\left(\frac{\hbar^2}{2\mu}E\right)^{-1/2}\,. \tag{2.140}$$

The bound states normalized to unity have the form $\sqrt{2/L}\sin kx$, where $E = \hbar^2k^2/(2\mu)$. Matrix elements like $|W_\mathrm{fi}|^2$ contain the square of the factor $\sqrt{2/L}$, so the product $|W_\mathrm{fi}|^2\varrho_L(E)$ no longer depends on the length L of the box. If we normalize the wave functions ϕ_f so that they are simply a sine with factor unity, then the density ϱ must obviously be

$$\varrho(E) = \frac{1}{\pi}\left(\frac{\hbar^2}{2\mu}E\right)^{-1/2}\,. \tag{2.141}$$

The amplitude of the wave functions and the density of states ϱ are now independent of L, and there is a smooth transition to the continuum case $L\to\infty$. If we work with (unbound) wave functions normalized in energy,

$$\big\langle\phi_\mathrm{f}(E)|\phi_\mathrm{f}(E')\big\rangle = \delta(E-E')\,, \tag{2.142}$$

their amplitude is a sine with a factor $\sqrt{2\mu/(\pi\hbar^2k)} = [2\mu/(\pi^2\hbar^2E)]^{1/4}$ (see (1.151) in Sect. 1.3.5), and the correct density of states is

$$\varrho(E) = 1\,. \tag{2.143}$$

When applying the Golden Rule (2.139) we have to take care that the density of the final states and their normalization are chosen consistently.

The Feshbach resonances discussed in Sect. 1.5.2 can also be described in the framework of time-dependent perturbation theory. If we regard the equations (1.217) without channel coupling as the (time-independent) Schrödinger equation with the unperturbed Hamiltonian $\hat{H}_0$ and the coupling potentials $V_{1,2,}$, $V_{2,1}$ as the perturbation, then the transition probability per unit time from a bound initial state $\psi_1 = 0$, $\psi_2 = \phi_0(r)$ to an unbound final state $\psi_1 = \phi_{\rm reg}(r)$, $\psi_2 = 0$ is, according to the Golden Rule,

$$P = \frac{2\pi}{\hbar} |\langle\phi_0|V_{2,1}|\phi_{\rm reg}\rangle|^2 \varrho(E)\,. \tag{2.144}$$

Since the density of final states is unity according to (2.143), the width Γ given by (1.232) is related to P by

$$P = \frac{\Gamma}{\hbar}\,, \quad \text{or} \quad \frac{1}{P} = \tau = \frac{\hbar}{\Gamma}\,. \tag{2.145}$$

P describes the time rate of change (decrease) of the occupation probability $w_{\rm i}$ of the initial state,

$$\frac{{\rm d}w_{\rm i}}{{\rm d}t} = -Pw_{\rm i}\,, \tag{2.146}$$

which corresponds to an exponential decay law:

$$w_{\rm i}(t) = w_{\rm i}(0){\rm e}^{-t/\tau}\,. \tag{2.147}$$

The time τ is the *lifetime* of the bound initial state ϕ_0 with respect to the decay into the continuum which is mediated by the coupling potential $V_{2,1}$. The second equation (2.145) states that the width Γ and the lifetime τ of a resonance fulfill a relation similar to the uncertainty relation. Note that the lifetime of the resonant state is of the same order of magnitude as the time delay suffered during scattering of an almost monochromatic wave packet whose (mean) energy lies near the resonance energy, see (1.240) in Sect. 1.5.2.

2.4.2 The Electromagnetic Field

Classically we describe the electromagnetic field with the help of the *scalar potential* $\Phi(\boldsymbol{r},t)$ and the *vector potential* $\boldsymbol{A}(\boldsymbol{r},t)$, which together define the electric field $\boldsymbol{E}(\boldsymbol{r},t)$ and the magnetic field $\boldsymbol{B}(\boldsymbol{r},t)$ (see any textbook on electrodynamics, e.g. [Jac98]):

$$\boldsymbol{E} = -\nabla\Phi - \frac{1}{c}\frac{\partial\boldsymbol{A}}{\partial t}\,, \quad \boldsymbol{B} = \nabla\times\boldsymbol{A}\,. \tag{2.148}$$

c is the speed of light (compare Sect. 2.1.3). The potentials are not unique and depend on the choice of *gauge*. The fields $\boldsymbol{E}$ and $\boldsymbol{B}$ remain unchanged when we replace the potentials Φ and $\boldsymbol{A}$ by new potentials Φ' and $\boldsymbol{A}'$ which are related to the original potentials by a *gauge transformation*:

$$\boldsymbol{A}' = \boldsymbol{A} + \nabla\Lambda\,, \quad \Phi' = \Phi - \frac{1}{c}\frac{\partial\Lambda}{\partial t}\,. \tag{2.149}$$

Λ is a scalar function of $\boldsymbol{r}$ and t. In the *Coulomb gauge,* which is also called *radiation gauge* or *transverse gauge,* we have

$$\nabla\cdot\boldsymbol{A} = 0\,, \quad \Delta\Phi = -4\pi\varrho\,, \tag{2.150}$$

where ϱ is the electric charge density. If there are no sources of charge the scalar potential vanishes in the Coulomb gauge. A physical system of electrically charged particles in an electromagnetic field is described by a Hamiltonian in which the kinetic energy is defined via the *kinetic momenta* $\hat{\boldsymbol{p}}_{\rm kin} = \hat{\boldsymbol{p}} - (q/c)\boldsymbol{A}$ while the potential energy contains the scalar potential Φ. When the electromagnetic field is included, the Hamiltonian for a system of N electrons with charge $q = -e$ and mass μ is thus

$$\hat{H} = \sum_{i=1}^{N}\left(\frac{[\hat{\boldsymbol{p}}_i + (e/c)\,\boldsymbol{A}(\boldsymbol{r}_i,t)]^2}{2\mu} - e\,\Phi(\boldsymbol{r}_i,t)\right) + \hat{V}\,. \tag{2.151}$$

Since the Hamiltonian (2.151) contains the potentials $\boldsymbol{A}$ and Φ, and not the physical fields (2.148), it depends on the particular choice of gauge, as do its absolute energy eigenvalues. Observable quantities such as energy differences and transition probabilities are however independent of the choice of gauge.

The interaction of an atom or ion with an external electromagnetic field is most easily described by treating the field classically and inserting the corresponding potentials $\boldsymbol{A}(\boldsymbol{r}_i,t)$, $\Phi(\boldsymbol{r}_i,t)$ as functions in the Hamiltonian (2.151). This procedure cannot however account for the observed phenomenon of *spontaneous emission,* in which an excited atom (or ion) emits a photon in the abscence of an external field. For a consistent description of the observed electromagnetic transitions including spontaneous emission, we must treat the electromagnetic field quantum mechanically. The full Hamiltonian then contains an interaction between atom and field which causes transitions between the eigenstates of the non-interacting Hamiltonian as described in Sect. 2.4.1, even if there is initially no field present.

To obtain a prescription for the *quantization of the electromagnetic field* we study the source-free field in a vacuum. As can be derived from Maxwell's equations, the vector potential $\boldsymbol{A}(\boldsymbol{r},t)$ fulfills the *free wave equation,*

$$\left(\frac{\partial^2}{\partial x^2} + \frac{\partial^2}{\partial y^2} + \frac{\partial^2}{\partial z^2}\right)\boldsymbol{A} = \frac{1}{c^2}\frac{\partial^2}{\partial t^2}\boldsymbol{A}\,. \tag{2.152}$$

A general solution of (2.152) can be obtained by superposing plane wave solutions, which we shall mark with a *mode label* λ. Each mode λ is characterized by a *wave vector* $\boldsymbol{k}_\lambda$ pointing in the direction of propagation of the plane wave, by an angular frequency $\omega_\lambda = c|\boldsymbol{k}_\lambda|$ and by a *polarization vector* $\boldsymbol{\pi}_\lambda$ of unit length:

$$\boldsymbol{A}_\lambda \, \mathrm{e}^{-\mathrm{i}\omega_\lambda t} = L^{-3/2} \, \boldsymbol{\pi}_\lambda \, \mathrm{e}^{\mathrm{i}(\boldsymbol{k}_\lambda \cdot \boldsymbol{r} - \omega_\lambda t)} \, . \tag{2.153}$$

Many relations are easier to formulate if we discretize the continuous distribution of wave vectors. To this end we think of the three-dimensional space as divided into large but finite cubes of side length L and require periodic boundary conditions for the plane waves. With the normalizing factor $L^{-3/2}$ on the right-hand side of (2.153), the integral of the square of the amplitude over one such cube is unity for each mode λ:

$$\int_{L^3} \mathrm{d}^3\boldsymbol{r} \, |\boldsymbol{A}_\lambda(\boldsymbol{r})|^2 = 1 \, . \tag{2.154}$$

In the Coulomb gauge (2.150) it follows from $\nabla \cdot \boldsymbol{A} = 0$ that

$$\boldsymbol{\pi}_\lambda \cdot \boldsymbol{k}_\lambda = 0 \tag{2.155}$$

in each mode λ. To each wave vector $\boldsymbol{k}_\lambda$ there are thus only two independent directions of polarization and both are perpendicular to the direction of propagation. A real polarization vector $\boldsymbol{\pi}_\lambda$ implies linearly polarized light, with the electric field vector oscillating in the direction defined by $\boldsymbol{\pi}_\lambda$. Two vectors $\boldsymbol{\pi}_{\lambda_1}$ and $\boldsymbol{\pi}_{\lambda_2}$ can serve as a basis for the possible states of polarization with the electric field vector perpendicular to the direction of propagation. Polarization vectors with complex components can be used to account for phase differences in the field components. For example, for a monochromatic wave travelling in the direction of the positive z-axis, the polarization vector

$$\boldsymbol{\pi}_{\lambda_1} \equiv \boldsymbol{\pi}_\lambda^{(\mathrm{r})} = \frac{1}{\sqrt{2}} \begin{pmatrix} 1 \\ \mathrm{i} \\ 0 \end{pmatrix} \tag{2.156}$$

describes right-handed circular polarization, while

$$\boldsymbol{\pi}_{\lambda_2} \equiv \boldsymbol{\pi}_\lambda^{(\mathrm{l})} = \frac{1}{\sqrt{2}} \begin{pmatrix} \mathrm{i} \\ 1 \\ 0 \end{pmatrix} \tag{2.157}$$

describes left-handed circular polarization. Note that the two polarization vectors (2.156) and (2.157) are related by

$$\boldsymbol{\pi}_{\lambda_2} = \boldsymbol{e}_{\boldsymbol{k}_\lambda} \times (\boldsymbol{\pi}_{\lambda_1})^* \, , \tag{2.158}$$

where e_{k_λ} is the unit vector in the direction of k_λ. Equation (2.158) represents an appropriate way of defining a second polarization vector orthogonal to a complex first one.[3]

The general (real) vector potential for a source-free electromagnetic field in a vacuum is a real superposition of the plane waves (2.153),

$$A(r,t) = \sum_\lambda (q_\lambda A_\lambda e^{-i\omega_\lambda t} + q_\lambda^* A_\lambda^* e^{+i\omega_\lambda t}) , \tag{2.159}$$

and the associated electric field E and magnetic field B are

$$E = -\frac{1}{c}\frac{\partial A}{\partial t} = \frac{i}{c}\sum_\lambda \omega_\lambda (q_\lambda A_\lambda e^{-i\omega_\lambda t} - q_\lambda^* A_\lambda^* e^{+i\omega_\lambda t}) ,$$

$$B = \nabla \times A = i\sum_\lambda k_\lambda \times (q_\lambda A_\lambda e^{-i\omega_\lambda t} - q_\lambda^* A_\lambda^* e^{+i\omega_\lambda t}) . \tag{2.160}$$

The energy $\mathcal{E}$ of the electromagnetic field is obtained by integrating the energy density $\frac{1}{8\pi}(E^2 + B^2)$ over a cube of length L:

$$\mathcal{E} = \frac{1}{8\pi}\int_{L^3} d^3r\,(E^2 + B^2) = \frac{1}{2\pi c^2}\sum_\lambda \omega_\lambda^2 q_\lambda^* q_\lambda . \tag{2.161}$$

Here we used the fact that integrals like $\int_{L^3} d^3r\ \exp(2ik_\lambda \cdot r)$ with oscillating integrands vanish because of the periodic boundary conditions.

We obtain a more familiar form of (2.161) if we replace the mode amplitudes q_λ and q_λ^* by the real variables

$$Q_\lambda = \frac{1}{\sqrt{4\pi c^2}}(q_\lambda^* + q_\lambda) , \quad P_\lambda = \frac{i\,\omega_\lambda}{\sqrt{4\pi c^2}}(q_\lambda^* - q_\lambda) , \tag{2.162}$$

namely:

$$\mathcal{E} = \sum_\lambda \frac{1}{2}(P_\lambda^2 + \omega_\lambda^2 Q_\lambda^2) . \tag{2.163}$$

This form underlines the similarity between the source-free electromagnetic field and a set of uncoupled harmonic oscillators. The correspondence of the free electromagnetic field and a set of harmonic oscillators is apparent in the energy spectrum. To each mode λ there belongs a sequence of equidistant energies $n_\lambda \hbar\omega_\lambda$, $n_\lambda = 0, 1, 2, \ldots$ representing the contribution of this mode to the total energy. In the case of the electromagnetic field, n_λ is the number of photons in the mode λ;

[3]Vectors a, b with complex components are orthogonal when $a_x^* b_x + a_y^* b_y + a_z^* b_z = 0$.

for the set of oscillators, n_λ is the quantum number determining the excitation of the oscillator in the mode λ.

To quantize the electromagnetic field we interpret the variables P_λ and Q_λ as quantum mechanical momentum and displacement operators for the oscillators in the various modes λ. So the Hamiltonian $\hat{H}_F$ for the field is

$$\hat{H}_F = \sum_\lambda \frac{1}{2}(\hat{P}_\lambda^2 + \omega_\lambda^2 \hat{Q}_\lambda^2) \,. \tag{2.164}$$

The eigenstates of this Hamiltonian are labelled by the occupation numbers $n_{\lambda_1}, n_{\lambda_2}, \ldots$ in the individual modes.

Eigenstates and eigenvalues of the Hamiltonian (2.164) can be derived elegantly if we introduce the operators

$$\begin{aligned} \hat{b}_\lambda^\dagger &= (2\hbar\omega_\lambda)^{-\frac{1}{2}}(\omega_\lambda \hat{Q}_\lambda - \mathrm{i}\hat{P}_\lambda) \equiv \sqrt{\frac{\omega_\lambda}{2\pi\hbar c^2}}\, q_\lambda^* \,, \\ \hat{b}_\lambda &= (2\hbar\omega_\lambda)^{-\frac{1}{2}}(\omega_\lambda \hat{Q}_\lambda + \mathrm{i}\hat{P}_\lambda) \equiv \sqrt{\frac{\omega_\lambda}{2\pi\hbar c^2}}\, q_\lambda \,, \end{aligned} \tag{2.165}$$

as is usually done for ordinary harmonic oscillators. (See also Sect. 5.2.2.) The commutation relations for the operators $\hat{b}_\lambda^\dagger$, $\hat{b}_\lambda$ follow from the canonical commutation relations (1.36) for the displacement and momentum operators $\hat{Q}_\lambda$, $\hat{P}_\lambda$:

$$[\hat{b}_\lambda, \hat{b}_{\lambda'}^\dagger] = \delta_{\lambda,\lambda'} \,. \tag{2.166}$$

$\hat{b}_\lambda^\dagger$ and $\hat{b}_\lambda$ are *creation* and *annihilation operators for photons* which respectively raise or lower the occupation number in the mode λ by unity (see Problem 2.6):

$$\begin{aligned} \hat{b}_\lambda^\dagger|\ldots, n_\lambda, \ldots\rangle &= \sqrt{n_\lambda + 1}\,|\ldots, n_\lambda + 1, \ldots\rangle \,, \\ \hat{b}_\lambda|\ldots, n_\lambda, \ldots\rangle &= \sqrt{n_\lambda}\,|\ldots, n_\lambda - 1, \ldots\rangle \,. \end{aligned} \tag{2.167}$$

The operator $\hat{N}_\lambda = \hat{b}_\lambda^\dagger \hat{b}_\lambda$ counts the number of quanta (photons) in the mode λ:

$$\hat{N}_\lambda\,|\ldots, n_\lambda, \ldots\rangle = n_\lambda\,|\ldots, n_\lambda, \ldots\rangle \,, \tag{2.168}$$

and the Hamiltonian for the whole electromagnetic field is

$$\hat{H}_F = \sum_\lambda \hbar\omega_\lambda\, \hat{b}_\lambda^\dagger \hat{b}_\lambda \,. \tag{2.169}$$

Going from (2.164) to (2.169) involves a *renormalization* of the Hamiltonian which consists in neglecting the constant but infinite contribution of the zero-point energies of all modes $\sum_\lambda \hbar\omega_\lambda/2$. The quantization prescription used above isn't unique anyway. In the classical formula (2.161) for the energy we could have changed the order of q_λ^* and q_λ and inserting the quantum operators (2.165) would then have given a Hamiltonian $\sum_\lambda \hbar\omega_\lambda \, \hat{b}_\lambda \hat{b}_\lambda^\dagger$ which, because of (2.166), differs from (2.169) by twice the total zero-point energy, $\sum_\lambda \hbar\omega_\lambda$.

We obtain a quantum mechanical operator corresponding to the classical vector potential $\boldsymbol{A}(\boldsymbol{r},t)$ by expanding the latter according to (2.159) and identifying the amplitudes q_λ and q_λ^* with the annihilation and creation operators of photons $\hat{b}_\lambda$ and $\hat{b}_\lambda^\dagger$ according to (2.165). The time dependence of the combinations $\hat{b}_\lambda \mathrm{e}^{-\mathrm{i}\omega_\lambda t}$ and $\hat{b}_\lambda^\dagger \, \mathrm{e}^{+\mathrm{i}\omega_\lambda t}$ appearing in this procedure is just that describing the evolution of the field operators in the Heisenberg representation (cf. (1.45) in Sect. 1.1.3). To see this recall that with $U_\mathrm{F}(t) = \exp[-(\mathrm{i}/\hbar)\hat{H}_\mathrm{F} t]$ we have

$$\hat{U}_\mathrm{F}^\dagger(t)\hat{b}_\lambda \hat{U}_\mathrm{F}(t) = \hat{b}_\lambda \, \mathrm{e}^{-\mathrm{i}\omega_\lambda t}, \quad \hat{U}_\mathrm{F}^\dagger(t)\hat{b}_\lambda^\dagger \hat{U}_\mathrm{F}(t) = \hat{b}_\lambda^\dagger \, \mathrm{e}^{+\mathrm{i}\omega_\lambda t}. \tag{2.170}$$

We have thus constructed the operator $\hat{\boldsymbol{A}}_\mathrm{H} = \hat{U}_\mathrm{F}^\dagger(t)\hat{\boldsymbol{A}}\hat{U}_\mathrm{F}(t)$ in the Heisenberg representation. To get the corresponding operator $\hat{\boldsymbol{A}}$ for the vector potential in the Schrödinger representation we just leave away the oscillating time-dependent factors $\mathrm{e}^{-\mathrm{i}\omega_\lambda t}$ and $\mathrm{e}^{+\mathrm{i}\omega_\lambda t}$:

$$\hat{\boldsymbol{A}}(\boldsymbol{r}) = \sum_\lambda \sqrt{\frac{2\pi\hbar c^2}{\omega_\lambda}} (\boldsymbol{A}_\lambda \hat{b}_\lambda + \boldsymbol{A}_\lambda^* \hat{b}_\lambda^\dagger). \tag{2.171}$$

Here the functions $\boldsymbol{A}_\lambda$ and $\boldsymbol{A}_\lambda^*$ are the spatial parts of the plane waves (2.153), normalized to a cube of length L, together with an appropriate polarization vector, e.g.:

$$\boldsymbol{A}_\lambda(\boldsymbol{r}) = L^{-3/2} \boldsymbol{\pi}_\lambda \, \mathrm{e}^{\mathrm{i}\boldsymbol{k}_\lambda \cdot \boldsymbol{r}}. \tag{2.172}$$

Later on we shall apply the Golden Rule (2.139) to electromagnetic transitions, and for this purpose it is important to know the density of photon states. The plane waves (2.153) which fit into a cube of length L (with periodic boundary conditions) have wave numbers given by $\boldsymbol{k} = (n_x, n_y, n_z)2\pi/L$ (with integer n_x, n_y and n_z). The density of possible wave vectors is thus $(2\pi/L)^{-3}$. If we ask for the number of photon states of a given polarization whose wave vector has an absolute value between k and $k + \mathrm{d}k$ and a direction in the solid angle $\mathrm{d}\Omega$, then we obtain a density $(L/2\pi)^3 k^2 \mathrm{d}\Omega$. In reference to the energy $\hbar\omega = \hbar c k$, the density ϱ_L of photon states of given polarization is given by,

$$\varrho_L \mathrm{d}\Omega = \left(\frac{L}{2\pi}\right)^3 \frac{k^2}{\hbar c} \mathrm{d}\Omega = \left(\frac{L}{2\pi}\right)^3 \frac{(\hbar\omega_\lambda)^2}{(\hbar c)^3} \mathrm{d}\Omega. \tag{2.173}$$

2.4.3 Interaction Between Atom and Field

Multiplying out the contributions in the Hamiltonian (2.151) for an N-electron atom (or ion) we obtain

$$\hat{H} = \sum_{i=1}^{N} \frac{\hat{\boldsymbol{p}}_i^2}{2\mu} + \hat{V} + \frac{e}{2\mu c} \sum_{i=1}^{N} [\hat{\boldsymbol{p}}_i \cdot \boldsymbol{A}(\boldsymbol{r}_i, t) + \boldsymbol{A}(\boldsymbol{r}_i, t) \cdot \hat{\boldsymbol{p}}_i]$$

$$+ \frac{e^2}{2\mu c^2} \sum_{i=1}^{N} \boldsymbol{A}(\boldsymbol{r}_i, t)^2 - e \sum_{i=1}^{N} \Phi(\boldsymbol{r}_i, t) . \tag{2.174}$$

For classical fields the potentials $\boldsymbol{A}(\boldsymbol{r}, t)$ and $\Phi(\boldsymbol{r}, t)$ are real-valued functions. For a fully quantum mechanical treatment of a system consisting of an atom and an electromagnetic field we need a Hamiltonian encompassing the atomic degrees of freedom and the degrees of freedom of the field. To this end we add the Hamiltonian (2.169) describing a free electromagnetic field to the expression (2.174); the interaction between atom and field is taken into account by replacing the potentials in (2.174) by the corresponding operators. For a source-free field in the radiation gauge we set $\Phi = 0$, while $\hat{\boldsymbol{A}}$ is given by the expression (2.171). The full Hamiltonian thus contains a non-interacting part $\hat{H}_0$ for the degrees of freedom of the atom plus the field (without interaction),

$$\hat{H}_0 = \hat{H}_{\mathrm{A}} + \hat{H}_{\mathrm{F}} = \sum_{i=1}^{N} \frac{\hat{\boldsymbol{p}}_i^2}{2\mu} + \hat{V} + \hat{H}_{\mathrm{F}} , \tag{2.175}$$

and an interaction term $\hat{W}$. If, in the spirit of first-order perturbation theory, we neglect the contribution quadratic in the vector potential, then

$$\hat{W} = \frac{e}{2\mu c} \sum_{i=1}^{N} [\hat{\boldsymbol{p}}_i \cdot \hat{\boldsymbol{A}}(\boldsymbol{r}_i) + \hat{\boldsymbol{A}}(\boldsymbol{r}_i) \cdot \hat{\boldsymbol{p}}_i] , \tag{2.176}$$

with $\hat{\boldsymbol{A}}(\boldsymbol{r})$ as defined in (2.171).

In most cases of interest, the *wave lengths* $2\pi/|\boldsymbol{k}_\lambda|$ of the photons emitted or absorbed by an atom are much larger than its spatial dimensions. The exponential functions entering via (2.172) in the matrix elements of the interaction operator (2.176) are thus well approximated by unity:

$$\mathrm{e}^{\mathrm{i}\boldsymbol{k}_\lambda \cdot \boldsymbol{r}_i} \approx 1 . \tag{2.177}$$

For reasons which will become clear in the next section, this approximation is called the *dipole approximation*. In the dipole approximation the interaction operator (2.176) simply becomes

$$\hat{W} = L^{-3/2}\frac{e}{\mu c}\sum_{i=1}^{N}\sum_{\lambda}\sqrt{\frac{2\pi\hbar c^2}{\omega_\lambda}}\,\hat{\boldsymbol{p}}_i\cdot(\boldsymbol{\pi}_\lambda\hat{b}_\lambda + \boldsymbol{\pi}_\lambda^*\hat{b}_\lambda^\dagger)\,. \tag{2.178}$$

2.4.4 Emission and Absorption of Photons

The Golden Rule (2.139) enables us to calculate the probabilities for the emission and absorption of photons in the dipole aproximation via matrix elements

$$W_{\rm fi} = \langle\phi_{\rm f}|\hat{W}|\phi_{\rm i}\rangle \tag{2.179}$$

of the operator (2.178). The initial state $\phi_{\rm i}$ and the final state $\phi_{\rm f}$ are eigenstates of the non-interacting Hamiltonian (2.175) and can each be written as a product of an atomic eigenstate $|\Phi_n\rangle$ of $\hat{H}_{\rm A}$ and an eigenstate of the field operator $\hat{H}_{\rm F}$ (2.169):

$$|\phi_{\rm i}\rangle = |\Phi_{\rm i}\rangle|\ldots,n_\lambda,\ldots\rangle\,,\quad |\phi_{\rm f}\rangle = |\Phi_{\rm f}\rangle|\ldots,n'_\lambda,\ldots\rangle\,. \tag{2.180}$$

The corresponding energies $E_{\rm i}$ and $E_{\rm f}$ of the initial and final state consist of respective eigenvalues $\varepsilon_{\rm i}$ or $\varepsilon_{\rm f}$ of $\hat{H}_{\rm A}$ plus the energy of the photon field. If only one mode λ has a different number of photons in the initial and final states while all other modes play a *spectator* role, then

$$\begin{aligned} E_{\rm i} &= \varepsilon_{\rm i} + n_\lambda\hbar\omega_\lambda \text{ plus energy of the spectator modes,}\\ E_{\rm f} &= \varepsilon_{\rm f} + n'_\lambda\hbar\omega_\lambda \text{ plus energy of the spectator modes.} \end{aligned} \tag{2.181}$$

The matrix element (2.179) can now be reduced to a matrix element involving only the atomic degrees of freedom:

$$\begin{aligned} W_{\rm fi} = {} & L^{-3/2}\frac{e}{\mu c}\sqrt{\frac{2\pi\hbar c^2}{\omega_\lambda}}\times\\ & \left(\langle\Phi_{\rm f}|\sum_{i=1}^{N}\boldsymbol{\pi}_\lambda\cdot\hat{\boldsymbol{p}}_i\,|\Phi_{\rm i}\rangle\,F_\lambda^{\rm abs} + \langle\Phi_{\rm f}|\sum_{i=1}^{N}\boldsymbol{\pi}_\lambda^*\cdot\hat{\boldsymbol{p}}_i\,|\Phi_{\rm i}\rangle\,F_\lambda^{\rm em}\right), \end{aligned} \tag{2.182}$$

where the factors F_λ stand for the field contribution to the transition matrix element,

$$\begin{aligned} F_\lambda^{\rm abs} &= \langle\ldots,n'_\lambda,\ldots|\hat{b}_\lambda|\ldots,n_\lambda,\ldots\rangle\,,\\ F_\lambda^{\rm em} &= \langle\ldots,n'_\lambda,\ldots|\hat{b}_\lambda^\dagger|\ldots,n_\lambda,\ldots\rangle\,, \end{aligned} \tag{2.183}$$

and can be readily evaluated via (2.167) for given values of n_λ and n'_λ. Note that $F_\lambda^{\rm abs}$ is non-zero only when $n'_\lambda = n_\lambda - 1$ while $F_\lambda^{\rm em}$ is non-zero only when $n'_\lambda = n_\lambda + 1$. The requirement of energy conservation, $E_{\rm f} = E_{\rm i}$, can also be divided into an atomic and a photonic part:

$$\varepsilon_{\rm f} - \varepsilon_{\rm i} = (n_\lambda - n'_\lambda)\,\hbar\omega_\lambda\,, \tag{2.184}$$

which merely says that the energy loss (or gain) of the atom is equal to the energy of the emitted (or absorbed) photon.

In the atomic matrix elements such as $\langle\Phi_{\rm f}|\sum_{i=1}^{N}\boldsymbol{\pi}_\lambda\cdot\hat{\boldsymbol{p}}_i|\Phi_{\rm i}\rangle$ in (2.182), the momenta $\hat{\boldsymbol{p}}_i$ can be expressed through commutators of the displacement vectors $\boldsymbol{r}_i$ with the non-interacting Hamiltonian $\hat{H}_0$. If we neglect momentum-dependent corrections such as the mass polarization term (cf. Sect. 2.2.1, Problem 2.8), only the first term on the right-hand side of (2.175) contributes to the commutator $[\hat{H}_0, \boldsymbol{r}_i]$. Then

$$\hat{\boldsymbol{p}}_i = \mu\frac{\rm i}{\hbar}[\hat{H}_0, \boldsymbol{r}_i] = \mu\frac{\rm i}{\hbar}[\hat{H}_{\rm A}, \boldsymbol{r}_i]\,, \tag{2.185}$$

and the atomic matrix element becomes a matrix element of the *electric dipole operator*

$$\hat{\boldsymbol{d}} = -e\sum_{i=1}^{N}\boldsymbol{r}_i\,, \tag{2.186}$$

e.g.,

$$-\frac{e}{\mu}\langle\Phi_{\rm f}|\sum_{i=1}^{N}\boldsymbol{\pi}_\lambda\cdot\hat{\boldsymbol{p}}_i|\Phi_{\rm i}\rangle = (\varepsilon_{\rm f} - \varepsilon_{\rm i})\frac{\rm i}{\hbar}\,\boldsymbol{\pi}_\lambda\cdot\langle\Phi_{\rm f}|\hat{\boldsymbol{d}}|\Phi_{\rm i}\rangle\,. \tag{2.187}$$

This representation of the atomic matrix element follows from the assumption (2.177) which is hence called the "dipole approximation". If we denote the vector $\langle\Phi_{\rm f}|\sum_{i=1}^{N}\boldsymbol{r}_i|\Phi_{\rm i}\rangle$ by $\boldsymbol{r}_{\rm fi}$, then

$$\langle\Phi_{\rm f}|\hat{\boldsymbol{d}}|\Phi_{\rm i}\rangle = -e\,\langle\Phi_{\rm f}|\sum_{i=1}^{N}\boldsymbol{r}_i|\Phi_{\rm i}\rangle = -\,e\boldsymbol{r}_{\rm fi}\,. \tag{2.188}$$

Inserting (2.182) and (2.187) into the Golden Rule (2.139) we now obtain with (2.188)

$$P_{\rm i\to f} = \frac{4\pi^2}{\hbar^2}L^{-3}\frac{(\varepsilon_{\rm f} - \varepsilon_{\rm i})^2}{\omega_\lambda}\,e^2\,|\boldsymbol{\pi}_\lambda\cdot\boldsymbol{r}_{\rm fi}F_\lambda^{\rm abs} + \boldsymbol{\pi}_\lambda^*\cdot\boldsymbol{r}_{\rm fi}F_\lambda^{\rm em}|^2\,\varrho_L(E_{\rm f})\,. \tag{2.189}$$

2.4.4.1 Spontaneous Emission

In order to apply the formula (2.189) to spontaneous emission we start with an initial state of the electromagnetic field containing no photons in any mode, $n_\lambda = 0$, for all λ. The transition matrix element (2.179) now differs from zero only if the final state of the field contains precisely one photon in one mode λ, $n'_\lambda = 1$, and the values of the corresponding field factors (2.183) are, according to (2.167), $F_\lambda^{\mathrm{abs}} = 0$, $F_\lambda^{\mathrm{em}} = 1$. Furthermore, the atomic energy difference $\varepsilon_\mathrm{i} - \varepsilon_\mathrm{f}$ must in this case exactly equal the energy $\hbar\omega_\lambda$ of the emitted photon. With (2.173) the probability per unit time for an atomic transition from an initial state Φ_i to a final state Φ_f accompanied by the emission of a photon of polarization $\boldsymbol{\pi}_\lambda$ into the solid angle $\mathrm{d}\Omega$ is

$$P_{\mathrm{i}\to\mathrm{f}}\,\mathrm{d}\Omega = \frac{1}{2\pi\hbar}\frac{\omega_\lambda^3 e^2}{c^3}|\boldsymbol{\pi}_\lambda^*\cdot\boldsymbol{r}_{\mathrm{fi}}|^2\,\mathrm{d}\Omega\,. \tag{2.190}$$

If, for a given wave vector $\boldsymbol{k}_\lambda$, we add the contributions (2.190) from the two possible directions of polarization perpendicular to $\boldsymbol{k}_\lambda$, then the sum of the absolute squares of the scalar product yields the absolute square of the projection of the vector $\boldsymbol{r}_{\mathrm{fi}}$ onto the plane perpendicular to $\boldsymbol{k}_\lambda$,

$$|\boldsymbol{\pi}_{\lambda_1}^*\cdot\boldsymbol{r}_{\mathrm{fi}}|^2 + |\boldsymbol{\pi}_{\lambda_2}^*\cdot\boldsymbol{r}_{\mathrm{fi}}|^2 = |\boldsymbol{r}_{\mathrm{fi}}|^2\sin^2\theta\,, \tag{2.191}$$

where θ is the angle between the wave vector $\boldsymbol{k}_\lambda$ and the real vector consisting of the magnitudes of the three components of $\boldsymbol{r}_{\mathrm{fi}}$. To derive (2.191) we assume that the two normalized polarization vectors $\boldsymbol{\pi}_{\lambda_1}$ and $\boldsymbol{\pi}_{\lambda_2}$ fulfill the relation (2.158).

Integrating over all possible directions Ω of the wave vector $\boldsymbol{k}_\lambda$ we obtain the probability per unit time $\mathbf{P}_{\mathrm{i}\to\mathrm{f}}^{\mathrm{se}}$ for the atomic transition $\Phi_\mathrm{i} \to \Phi_\mathrm{f}$ accompanied by the emission of a photon of arbitrary polarization in any direction,

$$\mathbf{P}_{\mathrm{i}\to\mathrm{f}}^{\mathrm{se}} = \int P_{\mathrm{i}\to\mathrm{f}}\,\mathrm{d}\Omega = \frac{4}{3}\frac{e^2 w_\lambda^3}{\hbar c^3}|\boldsymbol{r}_{\mathrm{fi}}|^2 \stackrel{\mathrm{def}}{=} A_{\mathrm{fi}}\,. \tag{2.192}$$

The A_{fi} are called the Einstein A coefficients for the transitions $\mathrm{i} \to \mathrm{f}$ [New02].

To obtain the total spontaneous decay rate per unit time P_i of an atomic state Φ_i we sum the decay rates (2.192) over all possible final states Φ_f:

$$P_\mathrm{i} = \sum_{\varepsilon_\mathrm{f}<\varepsilon_\mathrm{i}} \mathbf{P}_{\mathrm{i}\to\mathrm{f}}^{\mathrm{se}}\,. \tag{2.193}$$

This total decay rate corresponds to the time rate of change (decrease) of the occupation probability $w_\mathrm{i}(t)$ of the initial state Φ_i, and the reciprocal quantity

$$\tau = 1/P_\mathrm{i} \tag{2.194}$$

is, in analogy to (2.147), the lifetime of the atomic state Φ_i with respect to electromagnetic decay.

In a more complete description going beyond the framework of perturbation theory, we should not assume infinitely sharp atomic energy levels. Due to the interaction between the atom and the field only the ground state of the atom, which cannot spontaneously decay, is a truly bound state. All excited states are strictly speaking resonances in the continuum analogous to the Feshbach resonances described in Sect. 1.5.2. Thus each excited state of an atom has a *natural line width* Γ, which is related to its lifetime with respect to electromagnetic decay via the second equation (2.145).

2.4.4.2 Induced Emission

If the electromagnetic field in the initial state is not empty but contains n_λ photons in the mode λ, then a non-trivial field factor $|F_\lambda^{\mathrm{em}}|^2 = n_\lambda + 1$ has to be multiplied to the right-hand sides of (2.190) and (2.192) (cf. (2.183), (2.167)). The part proportional to n_λ describes the probability for induced emission which depends on the strength of the external field. The connection between the external field strength and the number n_λ actually to be inserted in the formulae depends on the particular physical experiment.

Let's look for example at an atom in an electromagnetic field in which all modes are occupied isotropically with an intensity distribution $I(\omega)$. Then the energy density in the frequency interval between ω and $\omega + \mathrm{d}\omega$ is equal to the number of modes with arbitrary polarization and direction of propagation $N_\omega \hbar \mathrm{d}\omega$, $N_\omega = 2 \times 4\pi \varrho_L$, multiplied by the (mean) energy density per mode, $n_\lambda \hbar\omega/L^3$. With (2.173) this means that

$$I(\omega)\mathrm{d}\omega = 8\pi \varrho_L \hbar \mathrm{d}\omega\, n_\lambda\, \hbar\omega/L^3 = \frac{\hbar}{\pi^2}\left(\frac{\omega}{c}\right)^3 n_\lambda\, \mathrm{d}\omega\,, \tag{2.195}$$

in other words,

$$n_\lambda = \pi^2 \frac{I(\omega)}{\hbar}\left(\frac{c}{\omega}\right)^3 . \tag{2.196}$$

Multiplying this factor onto the right-hand side of (2.192) gives the following formula for the probability per unit time $\mathbf{P}^{\mathrm{ie}}_{\mathrm{i}\to\mathrm{f}}$ for an atomic transition from Φ_{i} to Φ_{f} through induced emission of a photon of arbitrary polarization in any direction:

$$\mathbf{P}^{\mathrm{ie}}_{\mathrm{i}\to\mathrm{f}} = \frac{4}{3}\frac{\pi^2}{\hbar^2}\, e^2\, |\boldsymbol{r}_{\mathrm{fi}}|^2\, I(\omega)\,. \tag{2.197}$$

The factors

$$B_{\mathrm{fi}} = \frac{4}{3}\frac{\pi^2}{\hbar^2} e^2 |\boldsymbol{r}_{\mathrm{fi}}|^2 \tag{2.198}$$

are the Einstein B coefficients, which also appear in an analogous treatment of absorption [New02]. Historically, the Einstein A and B coefficients played an important role for the understanding of Planck's formula for the intensity distribution $I(\omega)$ in the particular example of black-body radiation.

2.4.4.3 Absorption

Absorption can occur only if the electromagnetic field in the initial state has a non-vanishing number n_λ of photons in at least one mode λ. After absorption of a photon from this mode the occupation number in the final state is $n'_\lambda = n_\lambda - 1$, and the field factor (2.183) is $|F_\lambda^{\rm abs}|^2 = n_\lambda$. In the case of absorption there is no additional free photon in the final state and, provided the final state of the atom lies in the discrete part of the (atomic) spectrum, we must use the discrete form (2.137) of the Golden Rule. In place of (2.189) we obtain the absorption probability per unit time as

$$P_{\rm i\to f} = 4\pi^2 L^{-3} \omega_\lambda\, e^2\, |\boldsymbol{\pi}_\lambda \cdot \boldsymbol{r}_{\rm fi}|^2\, n_\lambda\, \delta(\varepsilon_{\rm f} - \varepsilon_{\rm i} - \hbar\omega_\lambda)\,. \tag{2.199}$$

In order to describe absorption out of a uniform radiation field with an intensity distribution $I(\omega)$ we would have to integrate over the frequencies ω and over all directions, which, with the appropriate expression for n_λ, would lead to a formula analogous to (2.197).

Another experimentally important situation is the bombardment of an atom by a uniform monochromatic beam of photons (see Fig. 2.8). In this case the relevant physical quantity is the *cross section* $\sigma_{\rm abs}$ for the absorption of a photon. $\sigma_{\rm abs}$ is the absorption probability per unit time (2.199) divided by the current density of the incoming photons. This current density is simply the density n_λ/L^3 of the photons multiplied by their speed of propagation c, so we have

$$\sigma_{\rm abs}(E) = 4\pi^2 \frac{e^2}{\hbar c}\, \hbar\omega_\lambda\, |\boldsymbol{\pi}_\lambda \cdot \boldsymbol{r}_{\rm fi}|^2\, \delta(\varepsilon_{\rm f} - \varepsilon_{\rm i} - E)\,. \tag{2.200}$$

For initial and final states $\Phi_{\rm i}$ and $\Phi_{\rm f}$ normalized to unity the vector $\boldsymbol{r}_{\rm fi}$ defined by (2.188) has the dimensions of length and the cross section (2.200) has the

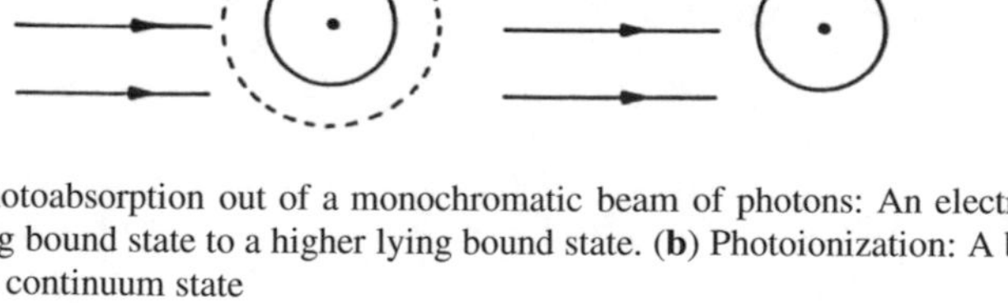

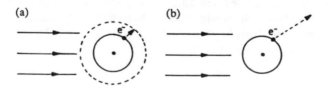

Fig. 2.8 (**a**) Photoabsorption out of a monochromatic beam of photons: An electron is elevated from a low-lying bound state to a higher lying bound state. (**b**) Photoionization: A bound electron is excited into a continuum state

dimensions of an area. Quantitatively the number of photons absorbed equals the number incident on an area of size σ_{abs} perpendicular to the direction of incidence.

2.4.4.4 Photoionization

With slight modifications, the formula (2.200) can be used to describe the ionization of an atom through absorption of a photon. In this case the wave function Φ_f of the atomic final state has the following form asymptotically (i.e. for large separations of the outgoing electron):

$$\Phi_f(x_1, \dots, x_{N-1}, x_N) = \Phi'_f(x_1, \dots, x_{N-1})\psi(x_N) . \tag{2.201}$$

Here $\psi(x_N)$ is the continuum wave function of the outgoing electron and may have the form (1.359) or (1.74) with a radial wave function of the form (1.117) or (1.122). Φ'_f is an $(N-1)$-electron wave function for the other electrons which are still bound after photoionization. Since the final states now have a continuous spectrum, we have to use the continuum version (2.139) of the Golden Rule. For energy normalized radial wave functions of the outgoing electron (cf. (1.151)) the density of final states is unity according to (2.143), and in place of (2.200) we obtain the following formula for the *photoionization cross section*:

$$\sigma_{ph}(E) = 4\pi^2 \frac{e^2}{\hbar c} \hbar\omega_\lambda \, |\boldsymbol{\pi}_\lambda \cdot \boldsymbol{r}_{fi}|^2 . \tag{2.202}$$

Due to the normalization of final states, $\langle \Phi_f(E) | \Phi_f(E') \rangle = \delta(E - E')$, the vector $\boldsymbol{r}_{fi}$ defined by (2.188) now has the dimensions of a length times the inverse square root of an energy, so that $\sigma_{ph}(E)$ again has the dimensions of an area. The constant $e^2/(\hbar c) \approx 1/137$ appearing in (2.200), (2.202) is of course the fine-structure constant which characterizes the strength of the electromagnetic interaction (see (2.35)).

In real situations the initial and/or final atomic states, Φ_i and/or Φ_f, may be members of degenerate or almost degenerate multiplets which are not resolved experimentally. This must then be taken into consideration when applying formulae like (2.200) or (2.202) for transition probabilities or cross sections. Our ignorance of the precise initial state is taken into account by *averaging over all initial states* in the multiplet. The fact that transitions to any state in a multiplet of final states contributes to the observed transion is taken into account by *summing over all final states* in the multiplet. This is performed explicitly in Sect. 3.2.3 for the particular example of one-electron atoms.

2.4.5 Selection Rules

The probability for an electromagnetic transition depends decisively on the atomic matrix element

$$\boldsymbol{r}_{\mathrm{fi}} = \langle \Phi_{\mathrm{f}} | \sum_{i=1}^{N} \boldsymbol{r}_i | \Phi_{\mathrm{i}} \rangle = \langle \Phi_{\mathrm{f}} | \hat{\boldsymbol{r}} | \Phi_{\mathrm{i}} \rangle . \tag{2.203}$$

This matrix element of the vector operator $\hat{\boldsymbol{r}} = -(1/e)\hat{\boldsymbol{d}}$ (cf. (2.186)) is conveniently evaluated via its *spherical components*

$$\hat{r}^{(\pm)} = \mp \sum_{i=1}^{N} \frac{1}{\sqrt{2}}(x_i \pm \mathrm{i}y_i) , \quad \hat{r}^{(0)} = \sum_{i=1}^{N} z_i . \tag{2.204}$$

In spherical components the scalar product of $\hat{\boldsymbol{r}}$ with another vector $\boldsymbol{a}$ is

$$\hat{\boldsymbol{r}} \cdot \boldsymbol{a} = \sum_{\nu=-1}^{+1} \left(\hat{r}^{(\nu)}\right)^* a^{(\nu)} . \tag{2.205}$$

For a one-electron atom the spherical components of $\hat{\boldsymbol{r}}$ can be expressed in terms of the radius $r = \sqrt{x^2 + y^2 + z^2}$ and the spherical harmonics $Y_{l,m}(\theta, \phi)$ defined in Sect. 1.2.1 (cf. Table 1.1):

$$\hat{r}^{(\pm)} = \sqrt{\frac{4\pi}{3}}\, r\, Y_{1,\pm 1}(\theta, \phi) , \quad \hat{r}^{(0)} = \sqrt{\frac{4\pi}{3}}\, r\, Y_{1,0}(\theta) . \tag{2.206}$$

If the atomic states Φ_{i} and Φ_{f} are simply one-electron wave functions (without spin) of the following form:

$$\Phi_{\mathrm{i}}(\boldsymbol{r}) = \frac{\phi_{l_{\mathrm{i}}}}{r} Y_{l_{\mathrm{i}},m_{\mathrm{i}}}(\theta, \phi) , \quad \Phi_{\mathrm{f}}(\boldsymbol{r}) = \frac{\phi_{l_{\mathrm{f}}}}{r} Y_{l_{\mathrm{f}},m_{\mathrm{f}}}(\theta, \phi) , \tag{2.207}$$

then we can use the formula (A.11) in Appendix A.1 for an integral over a product of three spherical harmonics to reduce the matrix elements $r_{\mathrm{fi}}^{(\nu)}$ $(\nu = +1, 0, -1)$ of the spherical components (2.206) of $\hat{\boldsymbol{r}}$ to an integral over the radial wave functions:

$$\begin{aligned} r_{\mathrm{fi}}^{(\nu)} &= \langle \Phi_{\mathrm{f}} | \hat{r}^{(\nu)} | \Phi_{\mathrm{i}} \rangle \\ &= \int_0^\infty \phi_{l_{\mathrm{f}}}^*(r)\, r\, \phi_{l_{\mathrm{i}}}(r)\, \mathrm{d}r \sqrt{\frac{4\pi}{3}} \int \mathrm{d}\Omega\, Y_{l_{\mathrm{f}},m_{\mathrm{f}}}^*(\Omega) Y_{1,\nu}(\Omega) Y_{l_{\mathrm{i}},m_{\mathrm{i}}}(\Omega) \\ &= \int_0^\infty \phi_{l_{\mathrm{f}}}^*(r)\, r\, \phi_{l_{\mathrm{i}}}(r)\, \mathrm{d}r\, F(l_{\mathrm{f}}, l_{\mathrm{i}}) \langle l_{\mathrm{f}}, m_{\mathrm{f}} | 1, \nu, l_{\mathrm{i}}, m_{\mathrm{i}} \rangle . \end{aligned} \tag{2.208}$$

Here $\langle l_\mathrm{f}, m_\mathrm{f}|1, \nu, l_\mathrm{i}, m_\mathrm{i}\rangle$ is the Clebsch-Gordan coefficient for coupling the initial angular momentum $l_\mathrm{i}, m_\mathrm{i}$ together with the angular momentum $1, \nu$ of the spherical component of the vector operator $\hat{\boldsymbol{r}}$ to the final angular momentum $l_\mathrm{f}, m_\mathrm{f}$ (see Sect. 1.7.1).

The angular momentum quantum numbers l_f, 1 und l_i must fulfill a triangle condition of the form (1.336), and this means that l_f and l_i can differ by at most unity. It furthermore follows from the parity (1.72) of the spherical harmonics that the sum of l_f, 1 und l_i must be even, since the parity of the integrand in the integral over Ω in (2.208) would otherwise be negative causing the integral itself to vanish. Together with the condition $m_\mathrm{i} + \nu = m_\mathrm{f}$ (cf. (1.335)) we obtain the following *selection rules for the one-body orbital angular momentum in dipole transitions* :

$$\Delta l = l_\mathrm{f} - l_\mathrm{i} = \pm 1\,, \qquad \Delta m = m_\mathrm{f} - m_\mathrm{i} = 0,\ \pm 1\,. \tag{2.209}$$

Transitions which do not fulfill these selection rules are *forbidden* (in the dipole approximation). The factor $F(l_\mathrm{f}, l_\mathrm{i})$ in (2.208) is explicitly

$$F(l_\mathrm{f}, l_\mathrm{i}) = \begin{cases} \sqrt{l_\mathrm{f}/(2l_\mathrm{f}+1)} & \text{for } l_\mathrm{f} = l_\mathrm{i} + 1\,, \\ -\sqrt{l_\mathrm{i}/(2l_\mathrm{f}+1)} & \text{for } l_\mathrm{f} = l_\mathrm{i} - 1\,. \end{cases} \tag{2.210}$$

If we include the spin dependence of the one-electron wave functions and assume atomic eigenstates of the form (1.359), then the formula (2.208) is replaced by an equation of the form

$$r_\mathrm{fi}^{(\nu)} = \langle \Phi_\mathrm{f}|\hat{r}^{(\nu)}|\Phi_\mathrm{i}\rangle = \langle j_\mathrm{f}\|\hat{\boldsymbol{r}}\|j_\mathrm{i}\rangle \langle j_\mathrm{f}, m_\mathrm{f}|1, \nu, j_\mathrm{i}, m_\mathrm{i}\rangle\,, \tag{2.211}$$

where the m quantum numbers now characterize the eigenvalues of the z-component of the total angular momentum $\hat{\boldsymbol{J}} = \hat{\boldsymbol{L}} + \hat{\boldsymbol{S}}$. The quantity $\langle j_\mathrm{f}\|\hat{\boldsymbol{r}}\|j_\mathrm{i}\rangle$ in (2.211) is called *reduced matrix element* of the vector operator $\hat{\boldsymbol{r}}$, and it no longer depends on the m quantum numbers of the atomic states or on the component index ν of the operator. Equation (2.211) is an illustration of the *Wigner-Eckart Theorem,* as is (2.208) above. This important theorem holds quite generally for matrix elements of the (spherical) components of a vector or tensor operator in angular momentum eigenstates. It says that the dependence of such matrix elements on the m quantum numbers and on the component index of the operator is given solely by the appropriate Clebsch-Gordan coefficients. The correct Clebsch-Gordan coefficients are those which couple the angular momentum of the initial state (here $j_\mathrm{i}, m_\mathrm{i}$) with the order and the component index of the operator (here $1, \nu$) to the angular momentum of the final state (here $j_\mathrm{f}, m_\mathrm{f}$). From the conditions (1.335), (1.336) for non-vanishing Clebsch-Gordan Coefficients we obtain the selection rules for the quantum numbers of the total angular momentum:

$$\Delta j = j_\mathrm{f} - j_\mathrm{i} = 0,\ \pm 1, \qquad \Delta m = m_\mathrm{f} - m_\mathrm{i} = 0,\ \pm 1\,. \tag{2.212}$$

The Wigner-Eckart theorem allows us to derive analogous selection rules for the angular momentum quantum numbers in many-electron atoms without knowing the precise structure of the atomic wave functions. For the total angular momentum (2.79) with the quantum numbers J, M we obviously have

$$\Delta J = J_{\rm f} - J_{\rm i} = 0,\ \pm 1\,, \quad \Delta M = M_{\rm f} - M_{\rm i} = 0\,,\ \pm 1. \tag{2.213}$$

If the atomic wave funcions are well described in LS coupling so that the total orbital angular momentum and the total spin are "good quantum numbers", then the selection rules for the orbital angular momentum quantum numbers L, M_L are

$$\Delta L = L_{\rm f} - L_{\rm i} = 0,\ \pm 1\,, \quad \Delta M_L = M_{L_{\rm f}} - M_{L_{\rm i}} = 0,\ \pm 1\,. \tag{2.214}$$

Since the interaction operator (2.178) doesn't act on the spin parts of the wave functions, the quantum numbers of the total spin cannot change in a transition,

$$\Delta S = 0\,, \quad \Delta M_S = 0\,. \tag{2.215}$$

As in a one-electron atom, the parity of the initial and final atomic states must be different for the matrix element of the dipole operator to be non-vanishing. In a many-electron atom however, the parity is not simply related to the orbital angular momentum, and hence $\Delta L = 0$ transitions aren't generally forbidden.

Above and beyond the selection rules (2.213), (2.214), all transitions in which both the initial angular momentum ($J_{\rm i}$ or $L_{\rm i}$) and the corresponding final angular momentum vanish, are forbidden. This is because the initial and final angular momenta and the order 1 of the vector operator $\hat{\boldsymbol{r}}$ must always obey a triangle condition of the form (1.336).

Transitions which are forbidden in the dipole approximation may be allowed for electromagnetic processes of higher order. If e.g. we go beyond the dipole approximation (2.177) by including the next term $\mathrm{i}\boldsymbol{k}_\lambda \cdot \boldsymbol{r}_i$ in the expansion of the exponential function we obtain the probabilities for *electric quadrupole transitions* as well as for *magnetic dipole transitions*. These are generally very small, because the absolute value of $\boldsymbol{k}_\lambda \cdot \boldsymbol{r}_i$ is very small for typical wave numbers k_λ and for displacement vectors $\boldsymbol{r}_i$ corresponding to the spatial dimensions of an atom. In order to obtain probabilities for transitions in which two or more photons are emitted or absorbed simultaneously, we have to go beyond a description based on first-order perturbation theory (see also Sect. 5.1 in Chap. 5).

2.4.6 Oscillator Strengths, Sum Rules

Dipole transitions between atomic states $\Phi_{\rm i}$ and $\Phi_{\rm f}$ can be characterized using the dimensionless *oscillator strengths*. These are the absolute squares of appropriately normalized matrix elements of the components of the vector operator $\hat{\boldsymbol{r}}$. In a cartesian basis the oscillator strength $f_{\mathrm{fi}}^{(x)}$ is, for example, defined by

$$f_{\mathrm{fi}}^{(x)} = \frac{2\mu}{\hbar^2}\,\hbar\omega\,|\langle\Phi_{\mathrm{f}}|\sum_{i=1}^{N} x_i|\Phi_{\mathrm{i}}\rangle|^2\,, \tag{2.216}$$

where $\hbar\omega = \varepsilon_{\mathrm{f}} - \varepsilon_{\mathrm{i}}$. Summed over the three cartesian components we obtain:

$$f_{\mathrm{fi}} = f_{\mathrm{fi}}^{(x)} + f_{\mathrm{fi}}^{(y)} + f_{\mathrm{fi}}^{(z)} = \frac{2\mu}{\hbar}\,\omega\,|\,\langle\Phi_{\mathrm{f}}|\hat{\boldsymbol{r}}|\Phi_{\mathrm{i}}\rangle\,|^2\,. \tag{2.217}$$

The contribution of the transition from Φ_{i} to Φ_{f} in the cross section $\sigma_{\mathrm{abs}}(E)$ for absorption of photons polarized in x-direction, $\boldsymbol{\pi}_\lambda = \hat{\boldsymbol{e}}_x$, out of a uniform beam is e.g. (cf. (2.200))

$$\sigma_{\mathrm{abs}}(E) = 4\pi^2\,\frac{e^2}{\hbar c}\,\frac{\hbar^2}{2\mu}\,f_{\mathrm{fi}}^{(x)}\,\delta(\varepsilon_{\mathrm{f}} - \varepsilon_{\mathrm{i}} - E)\,. \tag{2.218}$$

Consider a given (normalized) initial atomic state Φ_{i} and a complete set of (bound) final states Φ_n, then using the commutation relation (1.33) between position and momentum we obtain:

$$\begin{aligned}
\frac{\hbar}{\mathrm{i}}N &= \langle\Phi_{\mathrm{i}}|\sum_{i=1}^{N}(\hat{p}_{x_i}x_i - x_i\hat{p}_{x_i})\,|\Phi_{\mathrm{i}}\rangle \\
&= \sum_n\langle\Phi_{\mathrm{i}}|\sum_{i=1}^{N}\hat{p}_{x_i}|\Phi_n\rangle\langle\Phi_n|\sum_{i=1}^{N}x_i|\Phi_{\mathrm{i}}\rangle \\
&\quad - \sum_n\langle\Phi_{\mathrm{i}}|\sum_{i=1}^{N}x_i|\Phi_n\rangle\langle\Phi_n|\sum_{i=1}^{N}\hat{p}_{x_i}|\Phi_{\mathrm{i}}\rangle \\
&= \mu\,\frac{\mathrm{i}}{\hbar}\sum_n 2\,(\varepsilon_{\mathrm{i}} - \varepsilon_n)\,\langle\Phi_{\mathrm{i}}|\sum_{i=1}^{N}x_i|\Phi_n\rangle\langle\Phi_n|\sum_{i=1}^{N}x_i|\Phi_{\mathrm{i}}\rangle \\
&= \frac{2\mu}{\mathrm{i}\hbar}\sum_n \hbar\omega_n|\langle\Phi_n|\sum_{i=1}^{N}x_i|\Phi_{\mathrm{i}}\rangle|^2\,,
\end{aligned} \tag{2.219}$$

where the momentum components $\hat{p}_{x_i}$ were replaced by the commutators $\left[\hat{H}_{A}, x_i\right]$ according to (2.185) in the second last line, and we used the fact that the Φ_n are eigenfunctions of $\hat{H}_{\mathrm{A}}$ with the eigenvalues ε_n. With the definition (2.216) we obtain a *sum rule* for the oscillator strengths $f_{n\mathrm{i}}^{(x)}$:

$$\sum_n f_{n\mathrm{i}}^{(x)} = N\,. \tag{2.220}$$

Analogous sum rules obviously hold for the y- and z- components, and so we obtain the *Thomas-Reiche-Kuhn sum rule* for the oscillator strengths defined by (2.217):

$$\sum_n f_{n\mathrm{i}} = \sum_n \left(f_{n\mathrm{i}}^{(x)} + f_{n\mathrm{i}}^{(y)} + f_{n\mathrm{i}}^{(z)} \right) = 3N \,. \tag{2.221}$$

Before applying the above considerations to an atomic system we have to complement the formulae in order to take account of the fact that the complete set of final states contains continuum states. For final states Φ_E in the continuum we modify the definitions (2.216), (2.217) of the oscillator strengths,

$$\begin{aligned} \frac{\mathrm{d}f_{E\mathrm{i}}^{(x)}}{\mathrm{d}E} &= \frac{2\mu}{\hbar^2}\,\hbar\omega\, |\langle \Phi_E | \sum_{i=1}^{N} x_i | \Phi_\mathrm{i} \rangle|^2 \,, \text{ etc.}, \\ \frac{\mathrm{d}f_{E\mathrm{i}}}{\mathrm{d}E} &= \frac{\mathrm{d}f_{E\mathrm{i}}^{(x)}}{\mathrm{d}E} + \frac{\mathrm{d}f_{E\mathrm{i}}^{(y)}}{\mathrm{d}E} + \frac{\mathrm{d}f_{E\mathrm{i}}^{(z)}}{\mathrm{d}E} \,. \end{aligned} \tag{2.222}$$

If the final states Φ_E are energy normalized, then the functions $\mathrm{d}f_{E\mathrm{i}}^{(x)}/\mathrm{d}E$ and $\mathrm{d}f_{E\mathrm{i}}/\mathrm{d}E$ have the dimensions of an inverse energy. The photoionization cross section (2.202) for incoming photons polarized in x-direction is

$$\sigma_\mathrm{ph}(E) = 4\pi^2 \frac{e^2}{\hbar c} \frac{\hbar^2}{2\mu} \frac{\mathrm{d}f_{E\mathrm{i}}^{(x)}}{\mathrm{d}E} \,. \tag{2.223}$$

Inclusion of continuum states complements the sum rules (2.220), (2.221) to

$$\begin{aligned} \sum_n f_{n\mathrm{i}}^{(x)} + \int_0^\infty \frac{\mathrm{d}f_{E\mathrm{i}}^{(x)}}{\mathrm{d}E} \mathrm{d}E &= N \,, \text{ etc.}, \\ \sum_n f_{n\mathrm{i}} + \int_0^\infty \frac{\mathrm{d}f_{E\mathrm{i}}}{\mathrm{d}E} \mathrm{d}E &= 3N \,, \end{aligned} \tag{2.224}$$

where we have assumed the ionization threshold to lie at $E = 0$.

The sum rules for the oscillator strengths are a valuable help for estimating the importance of individual transitions in a particular physical system. In a numerical calculation of the transition probabilities to a finite number of final states, an estimate of the extent to which the corresponding oscillator strengths exhaust the sum rule may give valuable information on the reliability of the calculation and the importance of neglected contributions. The number N need not always be the total number of electrons. For photoabsorption by a lithium atom with one outer electron we may for example assume $N = 1$ at low energies. If the energy is large enough to excite the electrons in the low-lying 1s shell, then we must count these electrons in the formulation of the sum rule.

Problems

2.1 $\boldsymbol{A}$ and $\boldsymbol{B}$ are two vectors and $\hat{\boldsymbol{\sigma}}$ is the vector of the Pauli spin matrices (1.345). Prove the identity

$$(\hat{\boldsymbol{\sigma}} \cdot \boldsymbol{A})\,(\hat{\boldsymbol{\sigma}} \cdot \boldsymbol{B}) = \boldsymbol{A} \cdot \boldsymbol{B} + \mathrm{i}\hat{\boldsymbol{\sigma}} \cdot (\boldsymbol{A} \times \boldsymbol{B}) \ .$$

Show that the scalar product of $\hat{\boldsymbol{\sigma}}$ and the momentum operator $\hat{\boldsymbol{p}}$ can be expressed by the orbital angular momentum $\hat{\boldsymbol{L}}$ and the displacement vector $\boldsymbol{r}$ as follows:

$$\hat{\boldsymbol{\sigma}} \cdot \hat{\boldsymbol{p}} = \frac{1}{r^2}\,(\hat{\boldsymbol{\sigma}} \cdot \boldsymbol{r}) \left(\frac{\hbar}{\mathrm{i}} r \frac{\partial}{\partial r} + \mathrm{i}\hat{\boldsymbol{\sigma}} \cdot \hat{\boldsymbol{L}} \right) .$$

2.2 Use first-order perturbation theory to calculate the energy shifts due to the spin-orbit coupling $\hat{H}_{LS}$, the Darwin term $\hat{H}_{\mathrm{D}}$ and the relativistic correction $\hat{H}_{\mathrm{ke}}$ to the kinetic energy in the eigenstates of the hydrogen atom with quantum numbers up to $n = 2$.

$$\hat{H}_{LS} = \frac{Ze^2}{2m_0^2c^2}\frac{1}{r^3}\hat{\boldsymbol{L}} \cdot \hat{\boldsymbol{S}} , \quad \hat{H}_{\mathrm{D}} = \frac{\pi\hbar^2 Ze^2}{2m_0^2c^2}\delta(\boldsymbol{r}) , \quad \hat{H}_{\mathrm{ke}} = -\frac{\hat{\boldsymbol{p}}^2\hat{\boldsymbol{p}}^2}{8m_0^3c^2} .$$

2.3

a) Assume that both electrons in the helium atom or in a helium-like ion occupy the same orbital wave function

$$\psi\,(\boldsymbol{r}) = \frac{1}{\sqrt{\pi}}\beta^{-3/2}\mathrm{e}^{-r/\beta} .$$

For which value of β is the expectation value of the two-body Hamiltonian

$$\hat{H} = \sum_{i=1,2} \left(\frac{\hat{\boldsymbol{p}}_i^2}{2\mu} - \frac{Ze^2}{r_i} \right) + \frac{e^2}{|\boldsymbol{r}_1 - \boldsymbol{r}_2|}$$

a minimum? How do β and the minimal energy depend on the charge number Z?

Hint: Use (A.10) in Appendix A.1.

b) Calculate the expectation values of $\hat{H}$ in the ^1P and ^3P states of the helium atom, constructed by appropriate angular momentum coupling from the $1s\,2p$ configuration. Use hydrogenic single-particle wave functions with the parameter β as obtained in Problem 2.3 a).

2.4 Consider a "gas" of non-interacting fermions in a finite cube of side length L:

$$V = \begin{cases} 0 & \text{inside the cube} \\ +\infty & \text{outside of the cube} \end{cases}$$

a) Determine the eigenfunctions and eigenvalues of the one-body Hamiltonian

$$\hat{H} = \frac{\hat{\boldsymbol{p}}^2}{2\mu} + V \, .$$

b) Let each single-particle wave function with an energy not greater than $E_{\rm F} = \hbar^2 k_{\rm F}^2/(2\mu)$ be occupied with two fermions (spin up and spin down). How does the number N of fermions depend on the energy $E_{\rm F}$ when $E_{\rm F}$ is large?

2.5 Calculate the eigenfunctions and eigenvalues of the Hamiltonian for a particle of mass μ in a one-dimensional box of length L:

$$V(x) = \begin{cases} 0 & \text{for} \quad 0 \leq x \leq L \, , \\ +\infty & \text{for} \quad x < 0 \text{ or } x > L \, . \end{cases}$$

Show that the number of eigenstates per unit energy is given by the formula (2.140) for large E.

2.6 Let $\psi_n(x)$ be the eigenfunctions of the Hamiltonian for a one-dimensional harmonic oscillator:

$$\hat{H} = \frac{\hat{p}^2}{2} + \frac{1}{2}\omega^2 x^2 \, , \quad \hat{H}\psi_n = \left(n + \frac{1}{2}\right)\hbar\omega\,\psi_n \, .$$

Show that the operators

$$\hat{b}^\dagger = (2\hbar\omega)^{-1/2}(\omega x - \mathrm{i}\hat{p}) \, , \quad \hat{b} = (2\hbar\omega)^{-1/2}(\omega x + \mathrm{i}\hat{p})$$

act as creation and annihilation operators of oscillator quanta and, with suitable choice of phases of the eigenstates ψ_n, are given by

$$\hat{b}^\dagger\psi_n = \sqrt{n+1}\,\psi_{n+1} \, , \quad \hat{b}\psi_n = \sqrt{n}\,\psi_{n-1} \, .$$

Hint: Calculate the commutators of $\hat{b}^\dagger$ and $\hat{b}$ with $\hat{H}$.

2.7 Calculate the lifetime of the $2p$ state of the hydrogen atom with respect to electromagnetic decay.

2.8 How is the relation (2.185),

$$\hat{\boldsymbol{p}}_i = \mu\frac{\mathrm{i}}{\hbar}\left[\hat{H}_{\rm A}, \boldsymbol{r}_i\right] ,$$

affected if $\hat{H}_A$ contains not only the usual kinetic energy, but also the mass polarization term (Sect. 2.2.1)?

$$\hat{H}_A = \sum_{i=1} \frac{\hat{\boldsymbol{p}}_i^2}{2\mu} + \frac{1}{m_{\rm nuc}} \sum_{i<j} \hat{\boldsymbol{p}}_i \cdot \hat{\boldsymbol{p}}_j \quad + \quad \text{terms commuting with } \boldsymbol{r}_i \, .$$

How are formulae for transition probabilities such as (2.189) and sum rules such as (2.220) modified if the mass polarization term is taken into account?

References

[BD64] J.D. Bjorken, S.D. Drell, *Relativistic Quantum Mechanics* (McGraw-Hill, New York, 1964)

[BH95] D.J. Berkeland, E.A. Hinds, M.G. Boshier, Phys. Rev. Lett. **75**, 2470 (1995)

[BJ83] B.H. Bransden, C.J. Joachain, *Physics of Atoms and Molecules* (Longman, London, New York, 1983)

[BN97] B. de Beauvoir, F. Nez, L. Julien, B. Cagnac, F. Biraben, D. Touahri, L. Hilico, O. Acef, A. Clairon, J.J. Zondy, Phys. Rev. Lett. **78**, 440 (1997)

[BS75] S. Bashkin, J.O. Stoner, Jr., *Atomic Energy Levels and Grotrian Diagrams – vol. I. Hydrogen I – Phosphorus XV* (North Holland Publishing, Amsterdam, 1975)

[BS77] H.A. Bethe, E. Salpeter, *Quantum Mechanics of One- and Two-Electron Atoms* (Plenum Publishing, New York, 1977)

[BS78] S. Bashkin, J.O. Stoner, Jr., *Atomic Energy Levels and Grotrian Diagrams – vol. II. Sulphur I – Titanium XXII* (North Holland Publishinng, Amsterdam, 1978)

[BS81] S. Bashkin, J.O. Stoner, Jr., *Atomic Energy Levels and Grotrian Diagrams – vol. III. Vanadium I – Chromium XV* (North Holland Publishing, Amsterdam, 1981)

[BS82] S. Bashkin, J.O. Stoner, Jr., *Atomic Energy Levels and Grotrian Diagrams – vol. IV. Manganese I – XXV* (North Holland Publishing, Amsterdam, 1982)

[CO80] E.U. Condon, H. Odabasi, *Atomic Structure* (Cambridge University Press, Cambridge, 1980)

[DG90] R.M. Dreizler, E.K.U. Gross, *Density Functional Theory* (Springer, Berlin, 1990)

[Dra88] G.W.F. Drake, Nucl. Instrum. Methods B **31**, 7 (1988)

[Dra01] G.W.F. Drake, Physica Scripta **T95**, 22 (2001)

[Eng88] B.-G. Englert, *Semiclassical Theory of Atoms,* Lecture Notes in Physics vol. 300, ed. by H. Araki et al. (Springer, Berlin, Heidelberg, New York, 1988)

[Fan83] U. Fano, Rep. Prog. Phys. **46**, 97 (1983)

[Fer28] E. Fermi, Z. Physik **48**, 73 (1928)

[FR86] U. Fano, A.R.P. Rau, *Atomic Collisions and Spectra* (Academic Press, New York, 1986)

[Fro77] C. Froese Fischer, *The Hartree-Fock Method for Atoms* (Wiley, New York, 1977)

[Fro87] C. Froese Fischer, Comput. Phys. Commun. **43**, 355 (1987)

[Fro94] C. Froese Fischer, Comput. Phys. Commun. **84**, 37 (1994)

[HK64] P. Hohenberg, W. Kohn, Phys. Rev. **136**, B864 (1964)

[IL05] P. Indelicato, E. Lindroth, J.P. Desclaux, Phys. Rev. Lett. **94**, 013002 (2005)

[Jac98] J.D. Jackson, *Classical Electrodynamics*, 3rd edn. (Wiley, New York, 1998)

[KH86] A. Kono, S. Hattori, Phys. Rev. A **34**, 1727 (1986)

[KS65] W. Kohn, L.J. Sham, Phys. Rev. **140**, A1133 (1965)

[LM85] I. Lindgren, J. Morrison, *Atomic Many-Body Theory*, 2nd edn. (Springer, Berlin, Heidelberg, New York, 1985)

[MN16] P.J. Mohr, D.B. Newell, B.N. Taylor, http://physics.nist.gov/cuu/Constants/index.html
[New02] R.G. Newton, *Quantum Physics* (Springer, Berlin, Heidelberg, New York, 2002)
[Pek58] C.L. Pekeris, Phys. Rev **112**, 1649 (1958)
[Sch77] H.F. Schaefer (ed.) *Methods of Electronic Structure Theory* (Addison-Wesley, London, 1977)
[SK88] J. Styszyński, J. Karwowski, J. Phys. B **21**, 2389 (1988)
[Tho27] L.H. Thomas, Math. Proc. Camb. Phil. Soc. **23**, 542 (1927)
[UH97] T. Udem, A. Huber, B. Gross, J. Reichert, M. Prevedelli, M. Weitz, T.W. Hänsch, Phys. Rev. Lett. **79**, 2646 (1997)

Chapter 3
Atomic Spectra

A precise theoretical description of the energies and other properties of atomic states in principle requires the solution of the N-electron problem discussed in Sect. 2.2 and Sect. 2.3. This is of course not possible in general, but a lot of work based on various approximate and numerical methods has, over the years, been quite successful in explaining important properties of atomic spectra qualitatively and in simple cases quantitatively, mainly in the region of low-lying states [LM85, CO80]. On the other hand, the description of the structure of an atom or ion soon becomes very complicated when several electrons are highly excited [Fan83, FR86]. The many-electron problem in the regime of highly excited states is in fact still largely unsolved today.

The structures of atomic spectra and wave functions can be understood relatively simply and systematically if there is at most one electron in a highly excited state, while all other electrons are described by more tightly bound wave functions close to the atomic nucleus. The reason is that the interaction between the highly excited electron and the residual atom or ion is asymptotically described by a local potential. For neutral atoms and positively charged ions this local potential is the long-ranged attractive Coulomb potential, for (singly charged) negative ions it is a shorter-ranged power-law potential. Near the threshold to the continuum, the properties of the energy spectrum depend crucially on whether the potential tail behaves asymptotically as an attractive very-long-ranged potential, meaning that it vanishes slower than $-1/r^2$, or as a shorter-ranged potential falling off faster than $1/r^2$. Shorter-ranged potentials occur in the interaction of atoms (or molecules) with surfaces and with each other, and the intense research activity involving ultracold atoms and molecules has made a deeper understanding of the near-threshold properties of such shorter-ranged potentials a subject of great interest.

Section 3.1 contains a detailed discussion of spectra of shorter-ranged potentials and their near-threshold properties, while the following sections in this chapter focus

H. Friedrich, *Theoretical Atomic Physics*, Graduate Texts in Physics,
DOI 10.1007/978-3-319-47769-5_3

on potentials with Coulombic tails, as seen by a single highly excited electron in a neutral atom or positive ion.

The study of highly excited *Rydberg atoms* became a field of intense research in the late 1970's and thereafter, and this was largely due to advances in high precision experimental techniques such as laser spectroscopy. A detailed study of the general subject of Rydberg atoms can be found in the monograph by Gallagher [Gal94]. Further interest in Rydberg atoms was founded on the expectation that they may be of practical use in quantum information processing [LF01, TF04, RT05]. *Journal of Physics B* published a special issue on Rydberg physics in 2005 [CP05]. More recently, attention has also focussed on the interaction of Rydberg atoms with each other and the many-body properties of many-Rydberg-atom systems [CK07, GP08, PS09, LL13].

3.1 Long-Ranged and Shorter-Ranged Potentials

3.1.1 Very-Long-Ranged Potentials

The expressions "long-ranged" and "short-ranged" are often used with different meanings by different authors. Sometimes the term "short-ranged" is used to imply that a potential falls off exponentially or faster, whereas a potential which vanishes only as a power of the distance is termed "long-ranged" [BC02]. For a potential with an attractive tail falling off asymptotically as $-1/r^\alpha$,

$$V(r) \stackrel{r\to\infty}{\sim} V_\alpha^-(r) = -\frac{C_\alpha}{r^\alpha} = -\frac{\hbar^2}{2\mu}\frac{(\beta_\alpha)^{\alpha-2}}{r^\alpha}, \tag{3.1}$$

the structure of the quantum mechanical energy spectrum depends crucially on whether the power α is smaller or larger than two, as already discussed in Chapter 1, Sect. 1.6.4. Potentials falling off more slowly than $1/r^2$ might be called "very-long-ranged potentials" in order to distinguish them from shorter-ranged potentials with power-law tails corresponding to $\alpha > 2$. For Coulombic potential tails, which play a dominant role in atomic systems, we have $\alpha = 1$, which is the only integer in the range $0 < \alpha < 2$. Potential tails falling off asymptotically as (3.1) with non-integer powers α have little physical relevance, but studying these cases is helpful for understanding the transition from the very-long-ranged to the shorter-ranged potentials.

The generalized quantization rule as introduced in Sect. 1.6.3, Equations (1.308) and (1.309), reads

$$\frac{S(E)}{2\hbar} = \frac{1}{\hbar}\int_{r_{\rm in}(E)}^{r_{\rm out}(E)} p(r)\,{\rm d}r = n\pi + \frac{\phi_{\rm in}}{2} + \frac{\phi_{\rm out}}{2}. \tag{3.2}$$

This assumes that there is a WKB region between the inner classical turning point r_{in} and the outer classical turning point r_{out}, where WKB wave functions are accurate solutions of the Schrödinger equation. For very-long-ranged potentials this condition is always fulfilled near the threshold $E = 0$, because the threshold represents the semiclassical limit of the Schrödinger equation and the WKB approximation becomes increasingly accurate for $r \to \infty$, see (1.314) in Sect. 1.6.4.

For attractive potential tails (3.1) with $\alpha < 2$, the action integral $S(E)$ grows beyond all bounds as $E \to 0$; the potential well supports an infinite number of bound states and conventional WKB quantization, with $\phi_{out} = \pi/2$ at the outer classical turning point, becomes increasingly accurate towards threshold. For energies $E = -\hbar^2\kappa^2/(2\mu)$ close enough to threshold, the action integral can be written as

$$\frac{S(E)}{2\hbar} = C + \int_{r_0}^{r_{out}(E)} \sqrt{\frac{(\beta_\alpha)^{\alpha-2}}{r^\alpha} - \kappa^2}\, dr$$
$$\stackrel{\kappa\to 0}{\sim} C' + \frac{F(\alpha)}{(\kappa\beta_\alpha)^{(2/\alpha)-1}}, \quad F(\alpha) = \frac{\sqrt{\pi}\,\Gamma\left(\frac{1}{\alpha} - \frac{1}{2}\right)}{2\alpha\,\Gamma\left(\frac{1}{\alpha} + 1\right)}, \tag{3.3}$$

which leads to the near-threshold quantization rule,

$$n \stackrel{n\to\infty}{\sim} C'' + \frac{F(\alpha)}{\pi(\kappa\beta_\alpha)^{(2/\alpha)-1}}. \tag{3.4}$$

The point r_0 in (3.3) is to be chosen large enough for the potential to be accurately described by the leading asymptotic term proportional to $1/r^\alpha$. The constants C, C' and C'' in (3.3) and (3.4) depend on the potential at shorter distances $r < r_0$, but the energy dependent terms depend only on the potential tail beyond r_0, i.e. only on the power α and the strength parameter β_α determining the leading asymptotic behaviour of the potential tail. For a Coulombic potential tail, $\alpha = 1$, $F(1) = \pi/2$ we obtain the *Rydberg formula*,

$$E_n = -\frac{\hbar^2\kappa(n)^2}{2\mu} = -\frac{\mathcal{R}}{(n - C'')^2}, \quad \mathcal{R} = \frac{\hbar^2}{2\mu(2\beta_1)^2}, \tag{3.5}$$

with Bohr radius $2\beta_1$ and Rydberg constant $\mathcal{R}$, cf. Sect. 2.1.1.

The level density is defined as the (expected) number of energy levels per unit energy. If the quantum number n is known as a function of energy, then the level density is simply the energy derivative of the quantum number, dn/dE. Simple derivation of (3.4) with respect to $E = -\hbar^2\kappa^2/(2\mu)$ gives the near-threshold behaviour of the level density,

$$\frac{dn}{dE} \stackrel{E\to 0}{=} \frac{F(\alpha)}{\pi}\left(\frac{1}{\alpha} - \frac{1}{2}\right)\left(\frac{\hbar^2}{2\mu(\beta_\alpha)^2}\right)^{\frac{1}{\alpha}-\frac{1}{2}}\left(\frac{1}{|E|}\right)^{\frac{1}{\alpha}+\frac{1}{2}}. \tag{3.6}$$

For Coulombic tails, $\alpha = 1$, this reduces to the well known form,

$$\frac{\mathrm{d}n}{\mathrm{d}E} \stackrel{E\to 0}{=} \frac{1}{2}\frac{\sqrt{\mathcal{R}}}{|E|^{3/2}} \,. \tag{3.7}$$

3.1.2 Shorter-Ranged Potentials

We focus on the radial Schrödinger equation (1.75) for s-waves ($l = 0$),

$$\left[-\frac{\hbar^2}{2\mu}\frac{\mathrm{d}^2}{\mathrm{d}r^2} + V(r)\right]u(r) = E\,u(r)\,. \tag{3.8}$$

When the potential $V(r)$ vanishes faster than $1/r^2$ at large distances, then the action integral $S(E)$ remains bounded at threshold. The number of bound states is finite, and conventional WKB quantization deteriorates towards threshold [TE98, EF01, BA01]. The modified quantization rule (3.2), however, becomes exact in the limit $E = -\hbar^2\kappa^2/(2\mu) \to 0$, when the reflection phases are adapted according to the behaviour of the potential tail.

Consider a potential $V(r)$ which falls off faster than $1/r^2$ at large distances and is so deeply attractive at small distances that it supports a large, albeit finite number of bound states. An example with 24 bound states is shown in Fig. 3.1. Since such

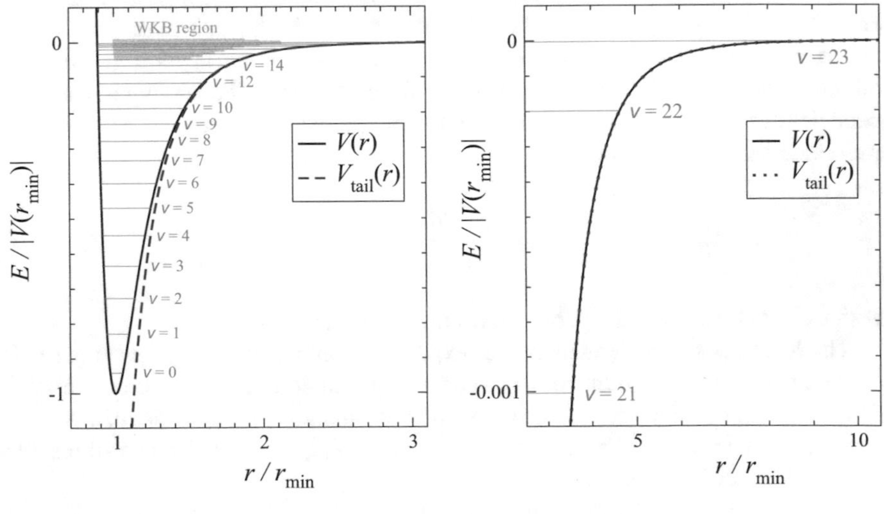

Fig. 3.1 Deep potential falling off faster than $1/r^2$ at large distances. The example is actually the Lennard-Jones potential (3.74) with strength parameter $B_{\mathrm{LJ}} = 10^4$, which supports 24 bound states, $\upsilon = 0, 1, 2, \ldots 23$. The *brown shaded area* in the left-hand panel schematically indicates where the WKB approximation is accurate at near-threshold energies

potentials typically describe the interatomic interaction in diatomic molecules, we adopt the molecular physics notation and use the letter "υ" for "vibrational" to label the bound states. The potential in Fig. 3.1 actually corresponds to a Lennard-Jones, which is discussed as example in Sect. 3.1.3. The theory below is, however, very general and does not rely on any special properties of the potential, except that it should be deep and fall off faster than $1/r^2$ at large distances.

Since the potential is deep, a total energy near threshold implies that the kinetic energy is large in a region of r-values between the inner classical turning point $r_{\rm in}(E)$ and the outer classical turning point $r_{\rm out}(E)$. This justifies the assumption, that there is a "WKB region" between $r_{\rm in}(E)$ and $r_{\rm out}(E)$, where the condition formulated as (1.297) in Sect. 1.6.3 is well fulfilled; the generalized quantization rule (3.2) in present notation reads,

$$\frac{1}{\hbar}\int_{r_{\rm in}(E_\upsilon)}^{r_{\rm out}(E_\upsilon)} p(E_\upsilon;r)\,{\rm d}r = \upsilon\pi + \frac{\phi_{\rm in}(E_\upsilon)}{2} + \frac{\phi_{\rm out}(E_\upsilon)}{2}\,, \quad \upsilon \text{ integer}\,. \tag{3.9}$$

At threshold, $E = 0$, the condition (3.9) with integer υ is fulfilled only if there a bound state exactly at threshold. For the general case, we write

$$\frac{1}{\hbar}\int_{r_{\rm in}(0)}^{\infty} p(E{=}0;r)\,{\rm d}r = \upsilon_{\rm D}\pi + \frac{\phi_{\rm in}(0)}{2} + \frac{\phi_{\rm out}(0)}{2}\,, \tag{3.10}$$

where $\upsilon_{\rm D}$ is the *threshold quantum number*, which is in general non-integer. For vibrational states of diatomic molecules or molecular ions, the bound-to-continuum threshold is the dissociation threshold, hence the subscript "D" on the threshold quantum number.

Subtracting (3.9) from (3.10) yields the *quantization rule*,

$$\upsilon_{\rm D} - \upsilon = F(E_\upsilon)\,, \tag{3.11}$$

with the *quantization function* $F(E)$ given by

$$F(E) = \frac{1}{\pi\hbar}\left[\int_{r_{\rm in}(0)}^{\infty} p(0;r)\,{\rm d}r - \int_{r_{\rm in}(E)}^{r_{\rm out}(E)} p(E;r)\,{\rm d}r\right] - \frac{\phi_{\rm in}(0)-\phi_{\rm in}(E)}{2\pi} - \frac{\phi_{\rm out}(0)-\phi_{\rm out}(E)}{2\pi}\,. \tag{3.12}$$

By definition, $F(E)$ vanishes at threshold,

$$F(E=0) = 0\,. \tag{3.13}$$

Equation (3.11) is the general form of the quantization rule for a potential which falls off faster than $1/r^2$ asymptotically and hence can support at most a finite number of bound states. The threshold quantum number $\upsilon_{\rm D}$, more precisely, its noninteger remainder $\Delta_D = \upsilon_D - \lfloor \upsilon_D \rfloor$, plays a crucial role in determining not only the precise energies of the near-threshold bound states, but also the near-threshold behaviour of s-wave scattering states, as discussed later in Sect. 4.1.

In contrast to the case of very-long-ranged potentials, the threshold solution of the radial Schrödinger equation (3.8) with a potential falling off faster than $1/r^2$ at large distances has at most a finite number of nodes. For potentials falling off faster than $1/r^3$, the asymptotic behaviour of regular solution is,

$$u(r) \overset{r\to\infty}{\propto} \text{const.} - r \propto 1 - \frac{r}{a} . \tag{3.14}$$

The constant a is the s-wave *scattering length*, which plays a prominent role in the description of near-threshold scattering, as described in detail later in Sect. 4.1. The threshold quantum number and the scattering length can both be seen as properties of the threshold ($E = 0$) solution of the radial Schrödinger equation (3.8), and they are related via (3.59) below.

Since the bound-state energies form a discrete finite set, it is always possible to find a smooth function $F(E)$ with (3.13) such that (3.11) is fulfilled at all bound-state energies E_υ. The explicit expression (3.12) is trivially valid, if we allow appropriate values of $\phi_{\text{in}}(E)$ and $\phi_{\text{out}}(E)$. If, at a given energy E, there is a WKB region between the inner and outer classical turning points where the WKB approximation is sufficiently accurate, then the reflection phases $\phi_{\text{in}}(E)$ and $\phi_{\text{out}}(E)$ can be determined precisely via the appropriate representation of wave function $u(r)$ for r-values in the WKB region,

$$u(r) \propto \frac{1}{\sqrt{p(E;r)}} \cos\left[\frac{1}{\hbar}\int_{r_{\text{in}}(E)}^{r} p(E;r')\,\mathrm{d}r' - \frac{\phi_{\text{in}}(E)}{2}\right] ,$$

$$u(r) \propto \frac{1}{\sqrt{p(E;r)}} \cos\left[\frac{1}{\hbar}\int_{r}^{r_{\text{out}}(E)} p(E;r')\,\mathrm{d}r' - \frac{\phi_{\text{out}}(E)}{2}\right] . \tag{3.15}$$

The leading near-threshold energy dependence of the quantization function (3.12) is a property of the large-distance behaviour of the potential. To be specific, we assume that the potential is accurately given at large distances by a reference potential, the "tail potential" $V_{\text{tail}}(r)$,

$$V(r) \overset{r\ \text{large}}{\sim} V_{\text{tail}}(r) . \tag{3.16}$$

As reference potential, $V_{\text{tail}}(r)$ is defined for all $r > 0$, but it only represents the true interaction for large distances. The phrase "r large" over the "$\sim$" sign in (3.16) has been chosen deliberately in order to emphasize that, in general, it is not just the leading asymptotic behaviour of $V(r)$ that is important. The radial Schrödinger equation with the reference potential $V_{\text{tail}}(r)$ alone and vanishing angular momentum reads

$$-\frac{\hbar^2}{2\mu}\frac{d^2u}{dr^2} + V_{\text{tail}}(r)\,u(r) = E\,u(r)\,. \tag{3.17}$$

Being an approximation to the full potential at large distances, the reference potential $V_{\text{tail}}(r)$ falls off faster than $1/r^2$ for $r \to \infty$. At small distances, the full interaction is not well described by the reference potential $V_{\text{tail}}(r)$, and its precise form is usually not well known anyhow. In the following we choose $V_{\text{tail}}(r)$ such that it diverges to $-\infty$ more rapidly than $-1/r^2$ for $r \to 0$. This has the advantage that the WKB representations of the solutions of (3.17), at any energy E, become increasingly accurate for decreasing r and are, in fact, exact in the limit $r \to 0$. This can be confirmed by verifying that the quantality function (1.298) vanishes for $r \to 0$ when the potential is more singular than $1/r^2$ in this limit.

The proximity to the semiclassical or anticlassical limits can be estimated via the value of a typical classically defined action in units of $\hbar$. Such a classical action is provided by the product of the momentum-like quantity $\hbar\kappa$ and the outer classical turning point $r_{\text{out}}(E)$, which is the same for the full interaction and for the reference potential $V_{\text{tail}}(r)$ at near-threshold energies and diverges to infinity at threshold,

$$r_{\text{out}}(E) \stackrel{\kappa\to 0}{\longrightarrow} \infty\,. \tag{3.18}$$

The typical action $\hbar\kappa\, r_{\text{out}}(E)$ in units of $\hbar$ is thus $\kappa\, r_{\text{out}}(E)$, a quantity that has been called the "reduced classical turning point" [TE98]. With $V_{\text{tail}}(r)r^2 \stackrel{r\to\infty}{\longrightarrow} 0$ it follows from (3.18) that

$$|V_{\text{tail}}\,(r_{\text{out}}(E))|\,r_{\text{out}}(E)^2 = \frac{\hbar^2\kappa^2}{2\mu}\,r_{\text{out}}(E)^2 \stackrel{\kappa\to 0}{\longrightarrow} 0 \quad\Longrightarrow\quad \kappa\, r_{\text{out}}(E) \stackrel{\kappa\to 0}{\longrightarrow} 0\,. \tag{3.19}$$

The threshold $E = 0$ represents the anticlassical or extreme quantum limit of the Schrödinger equation (3.17). For the singular attractive reference potential $V_{\text{tail}}(r)$, the outer classical turning point moves towards the origin for $E \to -\infty$,

$$r_{\text{out}}(E) \stackrel{\kappa\to\infty}{\longrightarrow} 0\,, \tag{3.20}$$

and with $V_{\text{tail}}(r)r^2 \stackrel{r\to 0}{\longrightarrow} -\infty$ it follows that

$$|V_{\text{tail}}\,(r_{\text{out}}(E))|\,r_{\text{out}}(E)^2 = \frac{\hbar^2\kappa^2}{2\mu}\,r_{\text{out}}(E)^2 \stackrel{\kappa\to\infty}{\longrightarrow} \infty \Longrightarrow \kappa\, r_{\text{out}}(E) \stackrel{\kappa\to\infty}{\longrightarrow} \infty\,. \tag{3.21}$$

The semiclassical limit of the Schrödinger equation (3.17) is at $\kappa \to \infty$, i.e. for large binding energies. How close the semiclassical limit is approached in a realistic potential well depends on its depth.

The quantization function (3.12) contains a contribution $F_{\text{tail}}(E)$, which is determined solely by the reference potential $V_{\text{tail}}(r)$,

$$F_{\text{tail}}(E) = \lim_{r_{\text{in}}\to 0} \frac{1}{\pi\hbar}\left[\int_{r_{\text{in}}}^{r_{\text{out}}(0)} p_{\text{tail}}(0;r)\,\mathrm{d}r - \int_{r_{\text{in}}}^{r_{\text{out}}(E)} p_{\text{tail}}(E;r)\,\mathrm{d}r\right] - \frac{\phi_{\text{out}}(0)-\phi_{\text{out}}(E)}{2\pi}\,, \tag{3.22}$$

where p_{tail} is the local classical momentum defined with $V_{\text{tail}}(r)$,

$$p_{\text{tail}}(E;r) = \sqrt{2\mu\,[E - V_{\text{tail}}(r)]}\,. \tag{3.23}$$

As the inner point of reference r_{in} tends to zero, the action integrals in (3.22) actually diverge, but their difference remains well defined in the limit. The tail part (3.22) of the quantization function contains no contribution from the inner reflection phases, because the wave functions become independent of energy for $r \to 0$ so the difference $\phi_{\text{in}}(0) - \phi_{\text{in}}(E)$ vanishes for $r_{\text{in}} \to 0$.

In addition to the tail contribution $F_{\text{tail}}(E)$, the quantization function contains a contribution $F_{\text{sr}}(E)$ arising from the deviation of the full interaction from the reference potential $V_{\text{tail}}(r)$ at small distances:

$$F(E) = F_{\text{tail}}(E) + F_{\text{sr}}(E)\,. \tag{3.24}$$

Since the full quantization function $F(E)$ vanishes at threshold according to (3.13), and since $F_{\text{tail}}(E=0)$ is obviously zero, the same must hold for $F_{\text{sr}}(E=0)$. Furthermore, $F_{\text{sr}}(E)$ is defined in the short-range region of the potential, where the bound-to-continuum threshold is not an outstanding value of the energy, so it must be a smooth function of energy near threshold. Hence we can write

$$F_{\text{sr}}(E) \overset{\kappa\to 0}{\sim} \gamma_{\text{sr}} E + O\left(E^2\right)\,, \tag{3.25}$$

where γ_{sr} is a constant with the dimension of an inverse energy.

As will be seen in the following, the leading near-threshold behaviour of $F_{\text{tail}}(E)$ is of lower order than E, so this is also the leading near-threshold behaviour of the full quantization function $F(E)$. The short-range contribution $F_{\text{sr}}(E)$ is of higher order, namely $O(E)$, and its magnitude depends on how accurately the reference potential $V_{\text{tail}}(r)$ describes the full interaction at finite distances. Its influence is small if $V_{\text{tail}}(r)$ is a good approximation of the full interaction down to distances where the WKB representation, on which the definition of $F_{\text{tail}}(E)$ is based, accurately describes the solutions of (3.17). Since the WKB approximation breaks down at the outer classical turning point $r_{\text{out}}(E)$, this implies that the reference potential be a good approximation of the full interaction down to distances somewhat smaller than $r_{\text{out}}(E)$.

If the quantization function is known accurately for a reasonable range of near-threshold energies, then a small number of energy eigenvalues in this range can be used to complement the spectrum and extrapolate to the dissociation threshold. This can, for example, make it possible to reliably predict the energy of the dissociation threshold from the relative separations of a few observed energy levels some distance away from threshold.

With the quantization function decomposed into a tail contribution and a short-range part as in (3.24), and with the ansatz (3.25) for the short-range part, the quantization rule (3.11) can be rewritten as

$$\upsilon + F_{\rm tail}(E_\upsilon) = \upsilon_{\rm D} - F_{\rm sr}(E) \overset{E\to 0}{\sim} \upsilon_{\rm D} - \gamma_{\rm sr} E_\upsilon \; . \tag{3.26}$$

As expressed on the far right of (3.26), the effects of the short-range deviation of the full interaction from the reference potential $V_{\rm tail}(r)$ are contained in two parameters, the threshold quantum number $\upsilon_{\rm D}$ and the short-range correction coefficient $\gamma_{\rm sr}$; the next term is of order E^2. According to (3.26), a plot of $\upsilon + F_{\rm tail}(E_\upsilon)$ against E_υ should approach a straight-line behaviour towards threshold; $\upsilon_{\rm D}$ and $\gamma_{\rm sr}$ can be deduced from the interception of this line with the ordinate and the gradient of the line, respectively.

The decomposition (3.24) of the full quantization function into a tail contribution and a short-range part and the representation (3.26) of the quantization rule are always valid. There is *no* semiclassical approximation involved, even though the tail contribution $F_{\rm tail}(E)$ to the quantization function is expressed in terms of WKB wave functions. For the short-range correction term to be small, however, the deviation of the full interaction from the reference potential $V_{\rm tail}(r)$ should be restricted to sufficiently small distances, at which the WKB representations of the solutions of (3.17) are accurate.

The near-threshold behaviour of $F_{\rm tail}(E)$ is crucially determined by the near-threshold energy dependence of the outer reflection phase. This can be derived under very general conditions, as described in detail in [RF08] and summarized below.

The solution of (3.17) obeying bound-state boundary conditions,

$$u^{(\kappa)}(r) \overset{r\to\infty}{\sim} {\rm e}^{-\kappa r} \; , \tag{3.27}$$

is accurately the represented for $r \to 0$ by the WKB expression

$$u^{(\kappa)}(r) \overset{r\to 0}{\sim} \frac{\mathcal{D}(\kappa)}{\sqrt{p_{\rm tail}(E;r)}} \cos\left(\frac{1}{\hbar}\int_r^{r_{\rm out}(E)} p_{\rm tail}(E;r')\,{\rm d}r' - \frac{\phi_{\rm out}(E)}{2}\right) \; . \tag{3.28}$$

Similar to the established derivation of the effective-range expansion for scattering phase shifts (see Sect. 4.1.7 in Chapter 4), we introduce two wave functions $u^{(\kappa)}(r)$ and $u^{(0)}(r)$ which solve (3.17) at the energies $E = -\hbar^2\kappa^2/(2\mu)$ and $E = 0$, respectively. We also introduce two solutions $w^{(\kappa)}$ and $w^{(0)}$, which have the same large-r boundary conditions, but are solutions of the free equation, without $V_{\rm tail}(r)$,

$$\begin{aligned} w^{(\kappa)}(r) &= {\rm e}^{-\kappa r}, \qquad w^{(0)}(r) \equiv 1, \\ u^{(\kappa)}(r) &\stackrel{r\to\infty}{\sim} w^{(\kappa)}(r)\,, \quad u^{(0)}(r) \stackrel{r\to\infty}{\sim} w^{(0)}(r)\,. \end{aligned} \tag{3.29}$$

From the radial Schrödinger equation we obtain,

$$\begin{aligned} \int_{r_l}^{r_u} \left(u^{(\kappa)}u^{(0)\prime\prime} - u^{(\kappa)\prime\prime}u^{(0)}\right) {\rm d}r &= \left[u^{(\kappa)}u^{(0)\prime} - u^{(\kappa)\prime}u^{(0)}\right]_{r_l}^{r_u} \\ &= -\kappa^2 \int_{r_l}^{r_u} u^{(\kappa)}u^{(0)}{\rm d}r \end{aligned} \tag{3.30}$$

for arbitrary lower and upper integration limits r_l and r_u. The contribution of the upper integration limit r_u to the square bracket in the middle part of (3.30) vanishes in the limit $r_u \to \infty$, because of the exponential decay of $u^{(\kappa)}(r)$ at large r. The contribution from the lower integration limit r_l follows from the WKB representation of the wave function (3.28) and its derivative,

$$\begin{aligned} u^{(\kappa)\prime}(r) = {} & \frac{\mathcal{D}(\kappa)}{\sqrt{p_{\rm tail}(E;r)}} \times \\ & \left[-\frac{1}{2}\frac{p'_{\rm tail}(E;r)}{p_{\rm tail}(E;r)} \cos\left(\frac{1}{\hbar}\int_r^{r_{\rm out}(E)} p_{\rm tail}(E;r')\,{\rm d}r' - \frac{\phi_{\rm out}(E)}{2}\right)\right. \\ & \left. + \frac{p_{\rm tail}(E;r)}{\hbar} \sin\left(\frac{1}{\hbar}\int_r^{r_{\rm out}(E)} p_{\rm tail}(E;r')\,{\rm d}r' - \frac{\phi_{\rm out}(E)}{2}\right)\right]. \end{aligned} \tag{3.31}$$

Equations (3.28) and (3.31) also apply for $u^{(0)}$ if we insert $E = 0$. Since $V_{\rm tail}(r)$ is more singular than $-1/r^2$ at the origin, $1/p_{\rm tail}(E;r)$ vanishes faster than r, and the contributions from the cosine in (3.31) to the products $u^{(\kappa)}u^{(0)\prime}$ and $u^{(\kappa)\prime}u^{(0)}$ in (3.30) vanish for $r_l \to 0$. With the abbreviations

$$S_{\rm tail}(E) = \int_{r_l}^{r_{\rm out}(E)} p_{\rm tail}(E;r)\,{\rm d}r\,, \quad I_\kappa = \frac{S_{\rm tail}(E)}{\hbar} - \frac{\phi_{\rm out}(E)}{2} \tag{3.32}$$

we obtain from (3.28) and (3.31),

$$\begin{aligned} \left[u^{(\kappa)}u^{(0)\prime} - u^{(\kappa)\prime}u^{(0)}\right]_{r_l\to 0} &= \frac{\mathcal{D}(\kappa)\mathcal{D}(0)}{\hbar} \sin(I_0 - I_\kappa)\bigg|_{r_l\to 0} \\ &= -\kappa^2 \int_0^\infty u^{(\kappa)}u^{(0)}{\rm d}r\,. \end{aligned} \tag{3.33}$$

For the free-particle solutions we obtain

$$\left[w^{(\kappa)}w^{(0)\prime} - w^{(\kappa)\prime}w^{(0)}\right]_{r_l}^{r_u} = -\kappa^2 \int_{r_l}^{r_u} w^{(\kappa)}w^{(0)}\mathrm{d}r \,. \tag{3.34}$$

Again, the contributions from r_u vanish for $r_u \to \infty$ while the contribution from r_l is

$$\left[w^{(\kappa)}w^{(0)\prime} - w^{(\kappa)\prime}w^{(0)}\right]_{r_l\to 0} = \kappa = -\kappa^2 \int_0^\infty w^{(\kappa)}w^{(0)}\mathrm{d}r \,. \tag{3.35}$$

Combining (3.33) and (3.35) gives

$$\begin{aligned}
&\frac{\mathcal{D}(\kappa)\mathcal{D}(0)}{\hbar}\sin(I_0 - I_\kappa) \\
&\quad= \frac{\mathcal{D}(\kappa)\mathcal{D}(0)}{\hbar}\sin\left(\frac{S_{\mathrm{tail}}(0) - S_{\mathrm{tail}}(E)}{\hbar} - \frac{\phi_{\mathrm{out}}(0) - \phi_{\mathrm{out}}(E)}{2}\right) \\
&\quad= \kappa + \kappa^2 \int_0^\infty \left[u^{(\kappa)}(r)u^{(0)}(r) - w^{(\kappa)}(r)w^{(0)}(r)\right]\mathrm{d}r \,.
\end{aligned} \tag{3.36}$$

Resolving for $\phi_{\mathrm{out}}(E)$ gives

$$\frac{\phi_{\mathrm{out}}(E)}{2} = \frac{\phi_{\mathrm{out}}(0)}{2} - \frac{S_{\mathrm{tail}}(0) - S_{\mathrm{tail}}(E)}{\hbar} + \arcsin\left[\frac{\kappa - \rho(E)\,\kappa^2}{\mathcal{D}(0)\mathcal{D}(\kappa)/\hbar}\right] , \tag{3.37}$$

with the length $\rho(E)$ defined by

$$\rho(E) = \int_0^\infty \left[w^{(\kappa)}(r)w^{(0)}(r) - u^{(\kappa)}(r)u^{(0)}(r)\right]\mathrm{d}r \,. \tag{3.38}$$

The action integrals $S_{\mathrm{tail}}(0)$ and $S_{\mathrm{tail}}(E)$ diverge as the lower integration limit tends to zero, but the difference $S_{\mathrm{tail}}(0) - S_{\mathrm{tail}}(E)$ tends to a well defined value in this limit.

In order to account correctly for the contributions of order κ^2 in the arcsin term in (3.37), it is necessary to know the zero-energy limit of $\rho(E)$,

$$\rho(0) = \int_0^\infty \left[\left(w^{(0)}(r)\right)^2 - \left(u^{(0)}(r)\right)^2\right]\mathrm{d}r \stackrel{\mathrm{def}}{=} \rho_{\mathrm{eff}} \,, \tag{3.39}$$

as well as the behaviour of $\mathcal{D}(\kappa)$ up to first order in κ. This can be obtained, as described in [FT04], on the basis of the two linearly independent threshold ($E = 0$) solutions $u_0^{(0)}(r)$ and $u_1^{(0)}(r)$ of the Schrödinger equation (3.17) which are defined by the following large-r boundary conditions,

$$u_0^{(0)}(r) \stackrel{r\to\infty}{\sim} 1 \,, \quad u_1^{(0)}(r) \stackrel{r\to\infty}{\sim} r \,. \tag{3.40}$$

For $r \to 0$, these wave functions can be written as WKB waves,

$$u_{0,1}^{(0)}(r) \overset{r\to 0}{\sim} \frac{D_{0,1}}{\sqrt{p_{\rm tail}(0;r)}} \cos\left(\frac{1}{\hbar}\int_r^\infty p_{\rm tail}(0;r')\,{\rm d}r' - \frac{\phi_{0,1}}{2}\right) , \tag{3.41}$$

which exactly defines the amplitudes $D_{0,1}$ and the phases $\phi_{0,1}$. The amplitude D_0 is the threshold value $\mathcal{D}(0)$ of the amplitude defined in (3.28), and ϕ_0 is the threshold value of the outer reflection phase $\phi_{\rm out}(E)$. For small but non-vanishing values of κ, the solution $u^{(\kappa)}(r)$ obeying the bound-state boundary condition (3.27) is given, up to and including the first order in κ, by

$$u^{(\kappa)}(r) \overset{\kappa r\to 0}{\sim} u_0^{(0)}(r) - \kappa\, u_1^{(0)}(r) \overset{r\to\infty}{\sim} 1 - \kappa r\,. \tag{3.42}$$

The WKB representation of the wave function (3.42), which is valid for small r and exact in the limit $r \to 0$, follows via (3.41),

$$\begin{aligned} u^{(\kappa)}(r) \overset{r\to 0}{\sim}& \frac{D_0}{\sqrt{p_{\rm tail}(0;r)}} \times \\ &\left[\cos\left(\frac{S_{\rm tail}(0)}{\hbar} - \frac{\phi_0}{2}\right) - \frac{D_1}{D_0}\kappa \cos\left(\frac{S_{\rm tail}(0)}{\hbar} - \frac{\phi_1}{2}\right)\right] \\ =& \frac{D_0}{\sqrt{p_{\rm tail}(0;r)}}\left[1 - \frac{D_1}{D_0}\kappa\cos\left(\frac{\phi_0-\phi_1}{2}\right)\right] \times \\ &\cos\left(\frac{S_{\rm tail}(0)}{\hbar} - \frac{\phi_1}{2} - \frac{D_1}{D_0}\kappa\sin\left(\frac{\phi_0-\phi_1}{2}\right)\right) + O\left(\kappa^2\right). \end{aligned} \tag{3.43}$$

Comparing amplitude and phase of the right-hand sides of (3.28) and (3.43) gives

$$\mathcal{D}(\kappa) = D_0\left[1 - \frac{D_1}{D_0}\kappa\cos\left(\frac{\phi_0 - \phi_1}{2}\right)\right] + O\left(\kappa^2\right) , \tag{3.44}$$

$$\frac{\phi_{\rm out}(E)}{2} = \frac{\phi_0}{2} - \frac{S_{\rm tail}(0) - S_{\rm tail}(E)}{\hbar} + b\kappa + O\left(\kappa^2\right) , \tag{3.45}$$

with the length b in (3.45) defined as

$$b = \frac{D_1}{D_0}\sin\left(\frac{\phi_0 - \phi_1}{2}\right). \tag{3.46}$$

Expanding the arcsin term on the right-hand side of (3.37) gives the near-threshold expansion of the outer reflection phase up to and including second order in κ as

$$\frac{\phi_{\rm out}(E)}{2} \overset{\kappa\to 0}{\sim} \frac{\phi_{\rm out}(0)}{2} - \frac{S_{\rm tail}(0) - S_{\rm tail}(E)}{\hbar} + b\,\kappa - \frac{(d\kappa)^2}{2} ; \tag{3.47}$$

the length d is defined by

$$\frac{d^2}{2} = b\,(\rho_{\rm eff} - \bar{a}) \quad \text{with} \quad \bar{a} = \frac{D_1}{D_0}\cos\left(\frac{\phi_0 - \phi_1}{2}\right) = b\,\cot\left(\frac{\phi_0 - \phi_1}{2}\right). \tag{3.48}$$

In deriving (3.47) we compared the linear terms in (3.37) and (3.45) to deduce $\hbar/\mathcal{D}(0)^2 = b$.

Away from threshold, $\kappa \to \infty$, the outer reflection approaches its semiclassical limit $\frac{\pi}{2}$. A measure for the proximity to the semiclassical limit is given by the reduced classical turning point $\kappa\, r_{\rm out}(E)$, see discussion involving (3.18) to (3.21) above, so it is reasonable to assume that the leading high-κ behaviour of the outer reflection phase is given by

$$\phi_{\rm out}(E) \overset{\kappa\to\infty}{\sim} \frac{\pi}{2} + \frac{D}{\kappa\, r_{\rm out}(E)}, \tag{3.49}$$

with some dimensionless constant D characteristic for the reference potential $V_{\rm tail}(r)$.

A remarkable feature of the near-threshold expansion (3.47) of the outer reflection phase is, that the term containing the difference of the action integrals exactly cancels the corresponding contribution to the quantization function, as represented by the big square bracket in the expression (3.22). The near-threshold behaviour of $F_{\rm tail}(E)$ is thus given by

$$F_{\rm tail}(E) \overset{\kappa\to 0}{\sim} \frac{b\kappa}{\pi} - \frac{(d\kappa)^2}{2\pi}. \tag{3.50}$$

The leading term on the right-hand side of (3.50), linear in κ, is reminiscent of Wigner's threshold law for s-waves, see Sect. 4.1.7 in Chapter 4. Since the short-range correction $F_{\rm sr}(E)$ is of order E at threshold, this term also represents the leading energy dependence of the full quantization function $F(E)$:

$$F(E) \overset{\kappa\to 0}{\sim} \frac{b\kappa}{\pi}, \tag{3.51}$$

which is universally valid for all potentials falling off faster than $1/r^2$ at large distances. The second term on the right-hand side of (3.50), quadratic in κ, is only well defined for reference potentials falling off faster than $1/r^3$, see the paragraph after (3.59) below.

For a potential $V(r)$ falling off faster than $1/r^3$ at large distances, the asymptotic behaviour of the regular solution $u(r)$ of the Schrödinger equation (3.8) exactly at threshold is given by (3.14) containing the s-wave scattering length a. When there is an s-wave bound state exactly at threshold, $u(r)$ approaches a finite value asymptotically, so the scattering length a in (3.14) diverges. The threshold quantum number $v_{\rm D}$ is an integer in this case. The derivation above enables us to formulate an explicit relation connecting the scattering length a with the threshold quantum number υ_D.

The asymptotic behaviour of the regular solution $u(r)$ of the Schrödinger equation with the full potential $V(r)$ is, according to (3.14) and (3.40),

$$u(r) \overset{r\to\infty}{\propto} 1 - \frac{r}{a} \quad \Longrightarrow \quad u(r) \overset{r\text{ large}}{\propto} u_0^{(0)}(r) - \frac{1}{a} u_1^{(0)}(r) \,. \tag{3.52}$$

The phrase "r large" refers to distances which are large enough for the full potential to be well approximated by $V_{\text{tail}}(r)$ and at the same time small enough for the WKB representations (3.41) to be accurate representations of $u_0^{(0)}(r)$ and $u_1^{(0)}(r)$. For such values of r,

$$\begin{aligned} u(r) \propto & \frac{D_1}{\sqrt{p(0;r)}} \cos\left(\frac{1}{\hbar}\int_r^\infty p(0;r')\,\mathrm{d}r' - \frac{\phi_1}{2}\right) \\ & - \frac{a\,D_0}{\sqrt{p(0;r)}} \cos\left(\frac{1}{\hbar}\int_r^\infty p(0;r')\,\mathrm{d}r' - \frac{\phi_0}{2}\right) \\ \propto & \frac{1}{\sqrt{p(0;r)}} \cos\left(\frac{1}{\hbar}\int_r^\infty p(0;r')\,\mathrm{d}r' - \frac{\phi_+}{4} - \eta\right) , \end{aligned} \tag{3.53}$$

with the angles $\phi_\pm$ and η given by

$$\phi_\pm = \phi_0 \pm \phi_1 \,, \quad \tan\eta = \frac{a + D_1/D_0}{a - D_1/D_0} \tan\left(\frac{\phi_-}{4}\right) . \tag{3.54}$$

Taking the inner classical turning point as reference gives

$$u(r) \propto \frac{1}{\sqrt{p(0;r)}} \cos\left(\frac{1}{\hbar}\int_{r_{\text{in}}(0)}^r p(0;r')\,\mathrm{d}r' - \frac{\phi_{\text{in}}(0)}{2}\right) , \tag{3.55}$$

and compatibility of (3.53) and (3.55) implies

$$\frac{1}{\hbar}\int_{r_{\text{in}}(0)}^\infty p(0;r)\,\mathrm{d}r = \frac{\phi_{\text{in}}(0)}{2} + \eta + \frac{\phi_+}{4} \ (\text{mod } \pi) \,. \tag{3.56}$$

Comparison with (3.10) gives

$$\eta = \upsilon_D \pi + \frac{\phi_-}{4} \ (\text{mod } \pi) \,. \tag{3.57}$$

Resolving the second equation (3.54) for a and inserting (3.57) for η yields

$$a = \frac{D_1}{D_0} \frac{\tan\left(\upsilon_D\pi + \frac{\phi_-}{4}\right) + \tan\left(\frac{\phi_-}{4}\right)}{\tan\left(\upsilon_D\pi + \frac{\phi_-}{4}\right) - \tan\left(\frac{\phi_-}{4}\right)}$$

$$= \frac{D_1}{D_0} \sin\left(\frac{\phi_-}{2}\right) \left[\frac{1}{\tan\left(\frac{\phi_-}{2}\right)} + \frac{1}{\tan(\upsilon_D\pi)}\right] . \tag{3.58}$$

In terms of the parameters b and $\bar{a}$ as defined in (3.46) and (3.48), this relation simplifies to

$$a = \bar{a} + \frac{b}{\tan(\upsilon_D\pi)} = \bar{a} + \frac{b}{\tan(\Delta_D\pi)} , \quad \Delta_D = \upsilon_D - \lfloor \upsilon_D \rfloor . \tag{3.59}$$

Equation (3.59) is very fundamental, giving an explicit relation between the s-wave scattering length a and the threshold quantum number υ_D. Because of the periodicity of the tangent, it is actually only the *remainder* Δ_D that counts. The remainder can assume values between zero and unity and quantifies the proximity of the most weakly bound state to threshold. A value of Δ_D very close to zero indicates a bound state very close to threshold, while a value very close to unity indicates that the potential just fails to support a further bound state.

Equation (3.59) enables a physical interpretation of the parameters entering the derivation of the expression (3.50) for the near-threshold behaviour of the tail contribution $F_{\text{tail}}(E)$ of the quantization function. In an ensemble of potentials characterized by evenly distributed values of the remainder Δ_D, the values of the scattering length will be evenly distributed around the mean value $\bar{a}$, hence $\bar{a}$ is called the *mean scattering length*, a term first introduced by Gribakin and Flambaum in [GF93]. We call the length b, which determines the leading term in the near-threshold behaviour (3.50) of the quantization function and the second term on the right-hand side of (3.59), the *threshold length*. The definition (3.39) of ρ_{eff} resembles, except for a factor two, the definition of the effective range r_{eff} in scattering theory, see (4.100) in Sect. 4.1.7; we call ρ_{eff} the *subthreshold effective range*. Note however, that the wave functions $u^{(0)}$ and $w^{(0)}$ that enter in the definition of ρ_{eff} remain bounded for $r \to \infty$, according to (3.29), so the expression (3.39) gives a well defined value for ρ_{eff} for any reference potential falling off faster than $1/r^3$ at large distances. The length d, which defines the next-to-leading term in the near-threshold behaviour (3.50) of the quantization function, is related to the mean scattering length $\bar{a}$, the threshold length b and the subthreshold effective range ρ_{eff} via the first equation (3.48). We use the term *effective length* for the parameter d.

The relation (3.59) makes it possible to derive a relation connecting the asymptotic inverse penetration depth κ_b of a bound state very near threshold and the scattering length a. With the quantization rule (3.11) we can rewrite (3.59) as

$$a = \bar{a} + \frac{b}{\tan\left[\pi F(E_b)\right]} = \bar{a} + \frac{b}{\tan\left[\pi\left(F_{\text{tail}}(E_b) + F_{\text{sr}}(E_b)\right)\right]}, \tag{3.60}$$

where $E_b = -\hbar^2\kappa_b^2/(2\mu)$ is the energy of the very weakly bound state. Replacing $F_{\text{sr}}(E_b)$ by its leading term $\gamma_{\text{sr}}E_b$ according to (3.25) and using the leading two terms of the Taylor expansion of the tangent yields [RF08]

$$a = \frac{1}{\kappa_b} + \rho_{\text{eff}} + \pi\,\frac{\hbar^2\gamma_{\text{sr}}}{2\mu\, b} + O(\kappa_b)\,. \tag{3.61}$$

It is interesting to observe, that the next-to-leading term in the expansion (3.61), i.e. the term of order κ_b^0, is *not* the mean scattering length $\bar{a}$, as one might guess from (3.59) [GF93], but the subthreshold effective range ρ_{eff}, *plus* a contribution which comes from short-range effects and is proportional to the constant γ_{sr}. In this light, one might ask what sense it makes to extend the near-threshold expansion (3.50) of $F_{\text{tail}}(E)$ up to second order in κ, when short-range effects bring in a term of the same order. The answer lies in the observation, that the length scales associated with the potential tail are generally very large, so that both ρ_{eff} and b are much larger than typical length scales associated with γ_{sr}. The dimensionless ratio $\pi\gamma_{\text{sr}}\hbar^2/(2\mu b\rho_{\text{eff}})$ of the third term on the right-hand side of (3.61) to the second term is thus usually very small, see also Example 1 below. Furthermore, a clean identification of the tail function $F_{\text{tail}}(E)$ over the whole range of energies from threshold to $-\infty$ is a prerequisite for the identification of the short-range correction $F_{\text{sr}}(E)$ due to the deviation of the full interaction from the reference potential at small distances.

At energies far from threshold, $\kappa \to \infty$, the outer reflection phase approaches its semiclassical limit according to (3.49), so the leading high-κ behaviour of $F_{\text{tail}}(E)$ is,

$$F_{\text{tail}}(E) \overset{\kappa\to\infty}{\sim} \frac{S_{\text{tail}}(0) - S_{\text{tail}}(E)}{\pi\hbar} - \left(\frac{\phi_0}{2\pi} - \frac{1}{4}\right) + \frac{D/(2\pi)}{\kappa\, r_{\text{out}}(E)}\,. \tag{3.62}$$

The zero-energy value ϕ_0 of the outer reflection phase, the lengths defining its low-κ expansion (3.47), i.e. b, $\bar{a}$, ρ_{eff} and d, and the parameter D in (3.49), (3.62) are *tail parameters*; they are properties of the reference potential $V_{\text{tail}}(r)$ alone. For a reference potential V_{tail} for which the Schrödinger equation (3.17) has analytically known solutions at threshold, $E=0$, the tail parameters can be derived analytically. The exact behaviour of $F_{\text{tail}}(E)$ in between the near-threshold regime and the high-κ, semiclassical regime is generally not known analytically, but it can be calculated numerically by a straightforward evaluation of (3.22).

The application of the theory described in this section is particularly elegant for potentials with a large-distance behaviour that is well described by a single-power tail,

$$V_{\text{tail}}(r) \equiv V_\alpha^{\text{att}}(r) = -\frac{C_\alpha}{r^\alpha} = -\frac{\hbar^2}{2\mu}\frac{(\beta_\alpha)^{\alpha-2}}{r^\alpha}\,, \quad C_\alpha > 0, \quad \alpha > 2\,. \tag{3.63}$$

The potential strength coefficient C_α in (3.63) is expressed in terms of the characteristic quantum length

$$\beta_\alpha = \left(\frac{2\mu C_\alpha}{\hbar^2}\right)^{1/(\alpha-2)}\,, \tag{3.64}$$

which does not exist in classical mechanics. The beauty of single-power reference potentials (3.63) is that the properties of the solution of the Schrödinger equation (3.17) depend only on the dimensionless product $\kappa\,\beta_\alpha$ and not on energy and potential strength independently, see (1.322) in Sect. 1.6.4. For example, the reduced classical turning point is given by

$$\kappa\, r_{\text{out}}(E) = (\kappa\beta_\alpha)^{1-2/\alpha}\,, \tag{3.65}$$

and the difference of the action integrals appearing in (3.22), (3.62) is

$$\lim_{r_{\text{in}}\to 0}\frac{1}{\pi\hbar}\left[\int_{r_{\text{in}}}^{\infty} p_{\text{tail}}(0;r)\,\mathrm{d}r - \int_{r_{\text{in}}}^{r_{\text{out}}(E)} p_{\text{tail}}(E;r)\,\mathrm{d}r\right] = \frac{(\kappa\beta_\alpha)^{1-2/\alpha}}{(\alpha-2)\sqrt{\pi}}\frac{\Gamma\left(\frac{1}{2}+\frac{1}{\alpha}\right)}{\Gamma\left(1+\frac{1}{\alpha}\right)}\,. \tag{3.66}$$

The tail parameters $\phi_{\text{out}}(0) \equiv \phi_0$, b, $\bar{a}$, ρ_{eff} and d defining the low-κ expansion (3.47) of the outer reflection phase, and the parameter D in (3.49) are explicitly given for attractive inverse-power tails (3.63) by [FT04, RF08],

$$\begin{aligned}
\phi_0 &= \left(\nu+\frac{1}{2}\right)\pi\,, \quad \frac{b}{\beta_\alpha} = \nu^{2\nu}\frac{\Gamma(1-\nu)}{\Gamma(1+\nu)}\sin(\pi\nu)\,,\\
\frac{\bar{a}}{\beta_\alpha} &= \nu^{2\nu}\frac{\Gamma(1-\nu)}{\Gamma(1+\nu)}\cos(\pi\nu)\,,\\
\frac{\rho_{\text{eff}}}{\beta_\alpha} &= \frac{\pi(2\nu)^{2\nu}\,\nu\,\Gamma\left(\frac{1}{2}+2\nu\right)}{\sin(\pi\nu)\,\Gamma\left(\frac{1}{2}+\nu\right)\,\Gamma(1+3\nu)}\,, \quad D = \frac{\sqrt{\pi}}{12}\frac{\alpha+1}{\alpha}\frac{\Gamma\left(\frac{1}{2}-\frac{1}{\alpha}\right)}{\Gamma\left(1-\frac{1}{\alpha}\right)}\,,
\end{aligned} \tag{3.67}$$

with the abbreviation $\nu = 1/(\alpha-2)$. The expression for d follows from those for b, $\bar{a}$ and ρ_{eff} via (3.48). Numerical values are given in Table 3.1.

The behaviour of the outer reflection phase $\phi_{\text{out}}(E)$ is illustrated in Fig. 3.2 for powers $\alpha = 3, \dots 7$. The abscissa is linear in $\kappa\, r_{\text{out}} = (\kappa\beta_\alpha)^{1-2/\alpha}$, so the initial

Table 3.1 Numerical values of tail parameters for single-power reference potentials (3.63), as given analytically in (3.67). The last row contains the values of the dimensionless parameter B_α governing the exponential fall-off of the modulus of the amplitude for quantum reflection according to (5.344) in Sect. 5.7.3

α	3	4	5	6	7	$\alpha \to \infty$
ϕ_0	$\frac{3}{2}\pi$	π	$\frac{5}{6}\pi$	$\frac{6}{8}\pi$	$\frac{7}{10}\pi$	$\left(\frac{1}{2}+\frac{1}{\alpha-2}\right)\pi$
b/β_α	$\frac{3}{2}$	1	0.6313422	0.4779888	0.3915136	$\frac{1}{\alpha-2}\pi$
$\bar{a}/\beta_\alpha$	–	0	0.3645056	0.4779888	0.5388722	1
$\rho_{\text{eff}}/\beta_\alpha$	–	$\frac{\pi}{3}$	0.7584176	0.6973664	0.6826794	1
d/β_α	–	$\sqrt{\frac{2\pi}{3}}$	0.7052564	0.4579521	0.3355665	$\frac{6.43}{(\alpha-2)^{3/2}}$
D	0.8095502	0.5462620	0.4554443	0.4089698	0.3806186	$\frac{1}{12}\pi$
B_α	2.24050	1.69443	1.35149	1.12025	0.95450	$\frac{2}{\alpha}\pi$

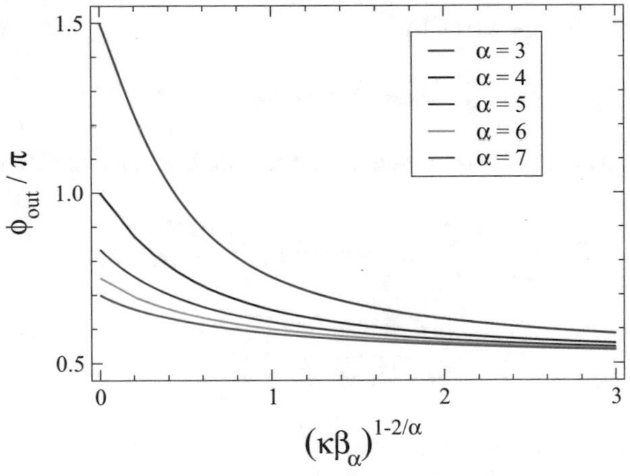

Fig. 3.2 Outer reflection phase ϕ_{out} for attractive inverse-power potentials (3.63) as function of the reduced classical turning point $\kappa\, r_{\text{out}} = (\kappa\beta_\alpha)^{1-2/\alpha}$ (Adapted from [TE98])

decrease is linear in the plot, compare (3.45) and (3.66). The threshold values ϕ_0 depend on the power α as given in the first equation (3.67).

For a given power $\alpha > 2$, one quantization function $F_{\text{tail}}(E) \equiv F_\alpha(\kappa\,\beta_\alpha)$ applies for all potential strengths. An expression for $F_\alpha(\kappa\,\beta_\alpha)$ which is accurate all the way from threshold to the semiclassical limit of large κ, can be obtained by interpolating between the near-threshold expression (3.50) and the high-κ limit (3.62). With (3.65) and (3.66), the high-κ limit of $F_\alpha(\kappa\,\beta_\alpha)$ is,

$$F_\alpha(E) \overset{\kappa\to\infty}{\sim} \frac{(\kappa\beta_\alpha)^{1-2/\alpha}}{(\alpha-2)\sqrt{\pi}}\,\frac{\Gamma\left(\frac{1}{2}+\frac{1}{\alpha}\right)}{\Gamma\left(1+\frac{1}{\alpha}\right)} - \frac{1}{2(\alpha-2)} + \frac{D/(2\pi)}{(\kappa\,\beta_\alpha)^{1-2/\alpha}}\,. \tag{3.68}$$

For $\alpha = 6$, an analytical expression involving one dimensionless fitting parameter B was derived in [RF08],

$$F_{\alpha=6}(E) = \frac{2b\kappa - (d\kappa)^2}{2\pi[1 + (\kappa B)^4]} + \frac{(\kappa B)^4}{1 + (\kappa B)^4}\left[-\frac{1}{8} + \frac{D}{2\pi(\kappa\beta_6)^{2/3}}\frac{\Gamma\left(\frac{2}{3}\right)(\kappa\beta_6)^{2/3}}{4\sqrt{\pi}\,\Gamma\left(\frac{7}{6}\right)}\right]. \quad (3.69)$$

All other parameters appearing in (3.69) are as given in (3.67) and Table 3.1 for $\alpha = 6$. With the value $B = 0.9363\,\beta_6$, the expression (3.69) reproduces the exact values, calculated by evaluating (3.22) numerically, to within an accuracy near 10^{-4} or better in the whole range from threshold to the high-κ limit [RF08].

For $\alpha = 4$, a more sophisticated treatment of the semiclassical, high-κ limit is needed to achieve a comparable accuracy on the basis of a small number of fitting parameters. An extension of the high-κ expansion (3.49) of the outer reflection phase to higher inverse powers of the reduced classical turning point $(\kappa\beta_4)^{1/2}$,

$$\phi_{\text{out}}(E) \stackrel{\kappa\beta_4\to\infty}{\sim} \frac{\pi}{2} + \sum_{j=1,3,5,7} \frac{D^{(j)}}{(\kappa\beta_4)^{j/2}}, \quad (3.70)$$

leads to the following analytical expression based on two fitting parameters, the lengths B_6 and B_7,

$$F_{\alpha=4}(E) = \frac{[2b\kappa - (d\kappa)^2]/(2\pi)}{1+(\kappa B_6)^6+(\kappa B_7)^7} + \frac{(\kappa B_6)^6+(\kappa B_7)^7}{1+(\kappa B_6)^6+(\kappa B_7)^7} \times\left[-\frac{1}{4} + \frac{\Gamma\left(\frac{3}{4}\right)}{\Gamma\left(\frac{5}{4}\right)}\frac{(\kappa\beta_4)^{1/2}}{2\sqrt{\pi}} + \frac{D^{(1)}/(2\pi)}{(\kappa\beta_4)^{1/2}} + \frac{D^{(3)}/(2\pi)}{(\kappa\beta_4)^{3/2}} + \frac{D^{(5)}/(2\pi)}{(\kappa\beta_4)^{5/2}} + \frac{D^{(7)}/(2\pi)}{(\kappa\beta_4)^{7/2}}\right]. \quad (3.71)$$

The coefficients $D^{(j)}$, which determine the expansion (3.70), are given analytically and numerically in Table 3.2. With the values $B_6 = 1.622\,576\,\beta_4$ and $B_7 = 1.338\,059\,\beta_4$ for the fitted lengths, the expression (3.71) reproduces the exact values,

Table 3.2 Coefficients $D^{(j)}$ in the high-κ expansion (3.70) of the outer reflection phase for a $-1/r^4$ reference potential

$D^{(1)}$	$D^{(3)}$	$D^{(5)}$	$D^{(7)}$
$\frac{5\sqrt{\pi}}{48}\Gamma\left(\frac{1}{4}\right)/\Gamma\left(\frac{3}{4}\right)$	$-\frac{35\sqrt{\pi}}{384}\Gamma\left(\frac{3}{4}\right)/\Gamma\left(\frac{1}{4}\right)$	$\frac{475\sqrt{\pi}}{3584}\Gamma\left(\frac{5}{4}\right)/\Gamma\left(-\frac{1}{4}\right)$	$-\frac{63\,305\sqrt{\pi}}{221\,184}\Gamma\left(\frac{7}{4}\right)/\Gamma\left(-\frac{3}{4}\right)$
0.5462620	−0.0546027	−0.0434388	0.0964461

calculated by evaluating (3.22) numerically, to within an accuracy near 10^{-4} or better in the whole range from threshold to the high-κ limit [RF09].

For $\alpha = 3$, it turned out to be more practical [MF11] to approximate $F_{\alpha=3}(E)$ by a rational function of the reduced classical turning point $(\kappa\beta_3)^{1/3}$,

$$F_{\alpha=3}(E) = \frac{\Gamma\left(\frac{5}{6}\right)}{\sqrt{\pi}\,\Gamma\left(\frac{4}{3}\right)}(\kappa\beta_3)^{1/3} + \frac{3+\sum_{i=1}^{i_{\max}} c_i(\kappa\beta_3)^{i/3}}{4+\sum_{i=1}^{i_{\max}} d_i(\kappa\beta_3)^{i/3}} - \frac{3}{4}\,. \tag{3.72}$$

With expansions up to $i_{\max} = 8$ in the numerator and the denominator of the second term on the right-hand side of (3.72), the formula is able to reproduce the exact quantization function, calculated by evaluating (3.22) numerically, to within an accuracy near $5 \cdot 10^{-8}$ or better in the whole range from threshold to the high-κ limit [MF11]. The coefficients c_i and d_i with which this is achieved are listed in Table 3.3.

The quantization functions (3.22) for the single-power tails (3.63) are shown for the cases $\alpha = 6$, 4 and 3 in Fig. 3.3 as functions of $\kappa\beta_\alpha$. The solid blue lines show exact functions, which are accurately approximated by the expressions (3.69), (3.71) and (3.72) all the way from threshold to the high-κ limit. The dashed green lines show the LeRoy-Bernstein functions $F_\alpha^{\rm LB}(E)$ [LB70, Stw70], which are obtained by ignoring the contribution from the outer reflection phase in (3.22). The LeRoy-Bernstein function is given explicitly by the first term on the right-hand side of (3.68). It is wrong at threshold, because it misses the energy-dependence (3.47) cancelling the contribution from the action integrals, and it is also wrong in the high-κ, semiclassical limit, because it misses contribution

$$-\frac{\phi_{\rm out}(0)}{2\pi} - \frac{\pi/2}{2\pi} = -\frac{1}{2(\alpha-2)}\,. \tag{3.73}$$

This leads to significant errors when extrapolating from bound-state energies to threshold, e.g. in order to determine the value of the dissociation threshold or of the scattering length from spectroscopic energies [RF08, RF09, MF11].

The dashed red lines in the three right-hand panels in Fig. 3.3 show the low-energy expansion (3.50) of $F_\alpha(E)$, including both terms, linear and quadratic in $\kappa\beta_\alpha$ for $\alpha = 6$ and $\alpha = 4$ and only the leading linear term for $\alpha = 3$. They allow us to estimate the extent of the near-threshold quantum regime. From the quantization rule (3.11) it is clear, that the value of $F(E_\upsilon)$ lies between zero and unity for the

Table 3.3 Coefficients c_i, d_i in the expression (3.72) for $F_{\alpha=3}(E)$

i	c_i	d_i	i	c_i	d_i
1	8.198 894 514 574	7.367 727 350 550	5	185.465 618 264 420	242.028 021 052 411
2	38.229 531 850 326	32.492 317 936 470	6	141.484 936 909 078	250.115 055 730 896
3	85.724 646 494 548	85.380 005 002 970	7	60.927 524 697 423	63.749 260 455 229
4	147.081 920 247 084	169.428 485 967 491	8	56.372 265 754 601	112.744 531 509 202

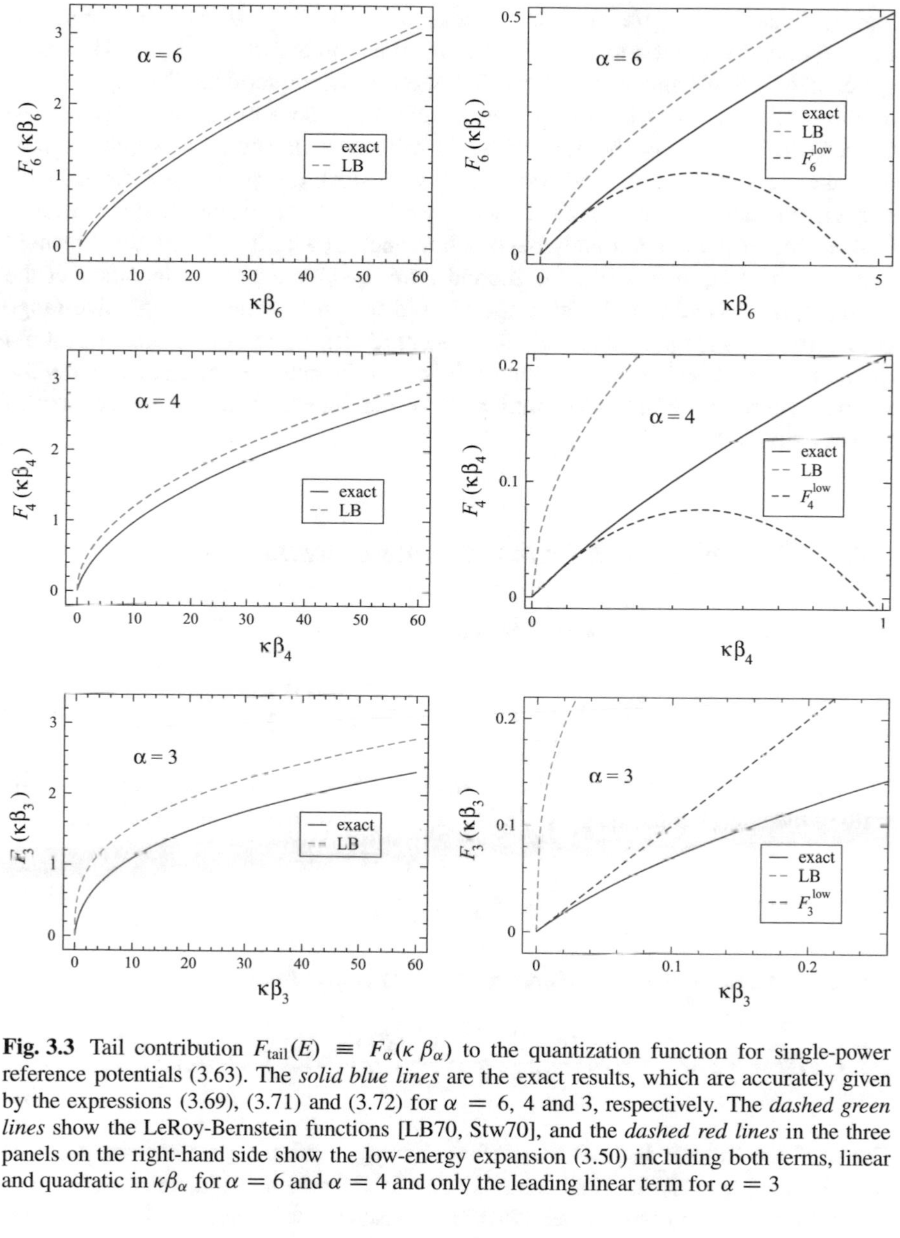

Fig. 3.3 Tail contribution $F_{\rm tail}(E) \equiv F_\alpha(\kappa\,\beta_\alpha)$ to the quantization function for single-power reference potentials (3.63). The *solid blue lines* are the exact results, which are accurately given by the expressions (3.69), (3.71) and (3.72) for $\alpha = 6$, 4 and 3, respectively. The *dashed green lines* show the LeRoy-Bernstein functions [LB70, Stw70], and the *dashed red lines* in the three panels on the right-hand side show the low-energy expansion (3.50) including both terms, linear and quadratic in $\kappa\beta_\alpha$ for $\alpha = 6$ and $\alpha = 4$ and only the leading linear term for $\alpha = 3$

highest bound state with quantum number $\upsilon_{\max} = \lfloor \upsilon_{\rm D} \rfloor$, between one and two for the second-highest bound state with quantum number $\upsilon_{\max} - 1$, etc. The range covered in the left-hand panels of Fig. 3.3 thus only accommodates the highest three bound states of a potential with the respective single-power tail. The enlargements in the right-hand part of the figure show that the near-threshold linear behaviour of the quantization function is restricted to a very small energy range indeed; in the majority of cases, it does not even contain the highest bound state, and the second-highest bound state is definitely beyond the range of validity of the near-threshold expansion (3.50), even when the second term, quadratic in κ, is included in the examples $\alpha = 6$ and $\alpha = 4$. The range of validity of near-threshold, effective-range type expansions is *tiny*. Nevertheless, an accurate description of this near-threshold quantum regime and a reliable interpolation to the large-κ semiclassical regime are paramount to a practicable application of the quantization-function concept in realistic situations.

3.1.3 Example 1. The Lennard-Jones Potential

We consider the Lennard-Jones potential,

$$V_{\rm LJ}(r) = \mathcal{E}\left[\left(\frac{r_{\min}}{r}\right)^{12} - 2\left(\frac{r_{\min}}{r}\right)^{6}\right] . \tag{3.74}$$

The quantum mechanical properties of the potential (3.74) are characterized by the ratio of the energy $\mathcal{E}$ to energy scale $\hbar^2/(2\mu\, r_{\min}^2)$ provided by the length $r_{\min}$,

$$B_{\rm LJ} = \frac{\mathcal{E}}{\hbar^2/(2\mu\, r_{\min}^2)} . \tag{3.75}$$

The natural definition of the reference potential $V_{\rm tail}(r)$ in this case is

$$V_{\rm tail}(r) \equiv V_6^{\rm att}(r) = -2\mathcal{E}\frac{(r_{\min})^6}{r^6} = -\frac{\hbar^2}{2\mu}\frac{(\beta_6)^4}{r^6} \quad \text{with } \beta_6 = r_{\min}\,(2B_{\rm LJ})^{1/4} . \tag{3.76}$$

For $B_{\rm LJ} = 10^4$ we have $\beta_6 = 10 \times 2^{1/4}\, r_{\min}$, and the potential supports 24 bound states, $\upsilon = 0, 1, \ldots 23$. This is actually the potential illustrated in Fig. 3.1. It was used by Paulsson et al. [PK83] to discuss the accuracy of higher-order WKB approximations. The energies of the highest twelve bound states are listed in Table 3.4.

According to (3.26), a plot of $\upsilon + F_6(\kappa_\upsilon \beta_6)$ against E_υ should approach a straight-line behaviour towards threshold, κ_υ being the asymptotic inverse penetration depth at the energy E_υ. This is illustrated impressively in Fig. 3.4. The solid squares represent the highest ten bound states in the left-hand part and the highest five

Table 3.4 Energies in units of $\mathcal{E}$ of the highest twelve bound states in the Lennard-Jones potential (3.74) with $B_{\mathrm{LJ}} = 10^4$ [PK83]

υ	E_υ	υ	E_υ	υ	E_υ
12	−0.115225890999	16	−0.031813309316	20	−0.003047136244
13	−0.087766914229	17	−0.020586161356	21	−0.001052747695
14	−0.064982730497	18	−0.012350373216	22	−0.000198340301
15	−0.046469911358	19	−0.006657024344	23	−0.000002696883

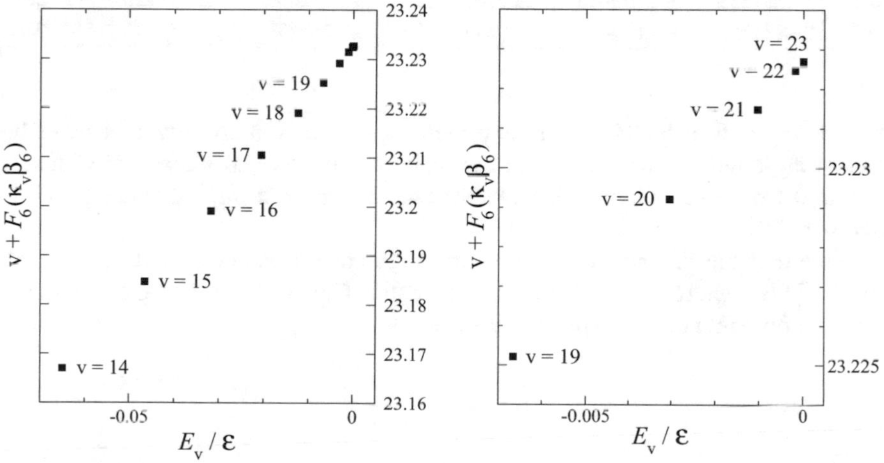

Fig. 3.4 Plot of $\upsilon + F_6(\kappa_\upsilon \beta_6)$ against energy for the highest ten bound states in the Lennard-Jones potential (3.74) with $B_{\mathrm{LJ}} = 10^4$. The energies are as listed in Table 3.4 and the quantization function $F_6(\kappa\beta_6)$ is as given by (3.69)

bound states in the right-hand part. The x-coordinate of each square is its energy eigenvalue E_υ (in units of $\mathcal{E}$), and the y-coordinate is $\upsilon + F_6(\kappa_\upsilon \beta_6)$, where $F_6(\kappa\beta_6)$ is the quantization function (3.69), and β_6 is as given in (3.76).

The fact that the linear behaviour in Fig. 3.4 reaches from threshold down to several states below threshold shows that the quantization rule based on (3.69) is reliable over this large energy range. To demonstrate this more quantitatively, Table 3.5 lists the values of the threshold quantum number υ_{D} and the short-range correction parameter γ_{sr} as obtained by fitting a straight line through two successive points, υ and $\upsilon + 1$. The values both of υ_{D} and of γ_{sr} converge rapidly and smoothly with increasing quantum number υ. The value of the threshold quantum number obtained by extrapolating from the sixth- and fifth-highest states ($\upsilon = 18$ and $\upsilon = 19$) already lies within 0.0004 of the value extrapolated via the highest two states, $\upsilon_{\mathrm{D}} = 23.23270$. This is also reflected in the similarly rapid and smooth convergence of the values of the scattering length a, as derived from the respective values of the threshold quantum number υ_{D} and the tail parameters $\bar{a}$ and b according to (3.59). In the present case of a $1/r^6$ reference potential, $\bar{a}$ and b are identical and both approximately equal to $0.478\,\beta_6$, see Table 3.1. With β_6 as given in (3.76),

Table 3.5 Values of the threshold quantum number υ_D and the short-range correction parameter γ_{sr} [in units of $\mathcal{E}^{-1}$] as obtained by fitting a straight line through two successive bound states, υ and $\upsilon + 1$, according to (3.26), see Fig. 3.4. Also listed are the values of the scattering length a [in units of r_{min}] as obtained via (3.59) with the respective values of υ_D

υ	υ_D	$\gamma_{sr}\mathcal{E}$	a/r_{min}	υ	υ_D	$\gamma_{sr}\mathcal{E}$	a/r_{min}
13	23.227230	−0.926599	12.2461	18	23.232378	−1.075980	12.0355
14	23.229053	−0.954646	12.1706	19	23.232591	−1.107941	12.0270
15	23.230401	−0.983664	12.1155	20	23.232685	−1.138876	12.0232
16	23.231354	−1.013615	12.0768	21	23.232699	−1.151726	12.0227
17	23.231988	−1.044432	12.0512	22	23.232700	−1.159540	12.0226

we have $\bar{a} = b \approx 5.684\, r_{min}$ in the present case. The well converged value of the scattering length, as obtained with the highest two states, is already predicted to within 0.1 % when extrapolating from the sixth- and fifth-highest states ($\upsilon = 18$ and $\upsilon = 19$).

Note that the magnitude of the short-range correction coefficient γ_{sr} is of the order of $1/\mathcal{E}$, where $\mathcal{E}$ is the depth of the potential. Characteristic energies associated with the potential tail are typically of the order

$$E_{\beta_6} = \frac{\hbar^2}{2\mu(\beta_6)^2} . \tag{3.77}$$

In the present example, $E_{\beta_6} \approx 0.7 \times 10^{-6}\mathcal{E}$, so the short-range correction coefficient γ_{sr} is near to six powers of ten smaller than typical inverse energies associated with the scale set by the reference potential $V_6(r)$. This justifies carrying the near-threshold expansions of the outer reflection phase (3.47) and the quantization function (3.50) to second order in κ, even though the short-range corrections come in at the same order.

The results above show, that the quantization function (3.69) for a $1/r^6$ reference potential accurately accounts for the level progression of the high-lying bound states in the deep Lennard-Jones potential (3.74), with the large value of B_{LJ} allowing the full potential to support 24 bound states. With only two parameters, υ_D and γ_{sr}, to account for all short-range effects, an accurate extrapolation to threshold, e.g. to deduce the value of the scattering length, is possible from several states below threshold. Such a clean separation of short-range effects from the influence of the potential tail is possible, when the distances at which the full interaction deviates significantly from the reference potential $V_{tail}(r)$ are small compared to characteristic length scales of $V_{tail}(r)$. In the present example, it was sufficient to take the leading single-power term of the potential as reference potential, because the deviation of $V(r)$ from $V_{tail}(r)$ is only given by the repulsive $1/r^{12}$ contribution, which is of very short range. In more realistic cases, a more sophisticated choice of reference potential may be needed to describe a range of near-threshold energies containing more than one or two bound states. This is demonstrated as Example 2 for the H_2^+ molecular ion below.

3.1.4 Example 2. The H_2^+ Molecular Ion

The H_2^+ ion, consisting of a proton and a neutral hydrogen atom, is one of the most fundamental molecular systems. Since its properties have been thoroughly examined in experiments and *ab initio* calculations, the system is ideally suited for testing and demonstrating the strengths and possible weaknesses of a theory focussing on the role of the potential tail, as done recently in [KM13].

Highly accurate energy eigenvalues of bound states of H_2^+ have been calculated by Hilico et al. [HB00]; the energies of the highest ten $L = 0$, $1s\sigma_g$ bound states are listed in Table 3.6.

The p-H potential at large distances can be decomposed into a polarisation term $V_{\rm pol}(r)$, and an exchange term $V_{\rm ex}(r)$ which is responsible for the energy splitting of the states with gerade and with ungerade parity (see [LL65], p. 81). The present example focusses on the $1s\sigma_g$ configuration, where the polarisation term is attractive,

$$V_{1s\sigma_g}(r) = V_{\rm pol}(r) - V_{\rm ex}(r) \, . \tag{3.78}$$

The expansion of $V_{\rm pol}(r)$ and $V_{\rm ex}(r)$ for large internuclear separations r was given to a large number of terms in 1968 by Damburg and Propin [DP68]. Leading terms, in atomic units, are

$$V_{\rm pol}^{\rm DP}(r) = -\frac{9}{4\,r^4} - \frac{15}{2\,r^6} - \frac{213}{4\,r^7} \, , \quad V_{\rm ex}^{\rm DP}(r) = 2r\,{\rm e}^{-r-1}\left(1 + \frac{1}{2\,r} - \frac{25}{8\,r^2}\right) . \tag{3.79}$$

Including only the leading asymptotic term of the polarisation potential to define the reference potential gives a single-power tail (3.63) with $\alpha = 4$,

$$V_{\rm tail}^{(1)}(r) = -\frac{9}{4r^4} \equiv -\frac{\hbar^2}{2\mu}\frac{(\beta_4)^2}{r^4} \, . \tag{3.80}$$

With the reduced mass $\mu = 918.32627$ a.u. this translates into a quantum length $\beta_4 = 64.2843$ a.u.

Table 3.6 Energy eigenvalues (in atomic units) relative to the dissociation threshold of the highest ten bound states in the $L = 0$, $1s\sigma_g$ series of the H_2^+ molecular ion according to Hilico et al. [HB00]

υ	E_υ	υ	E_υ	υ	E_υ	υ	E_υ
10	−0.021970529704	13	−0.009458409007	16	−0.001967933877	18	−0.000109592359
11	−0.017272525961	14	−0.006373841570	17	−0.000709200873	19	$-3.39093933\cdot10^{-6}$
12	−0.013097363811	15	−0.003867245551				

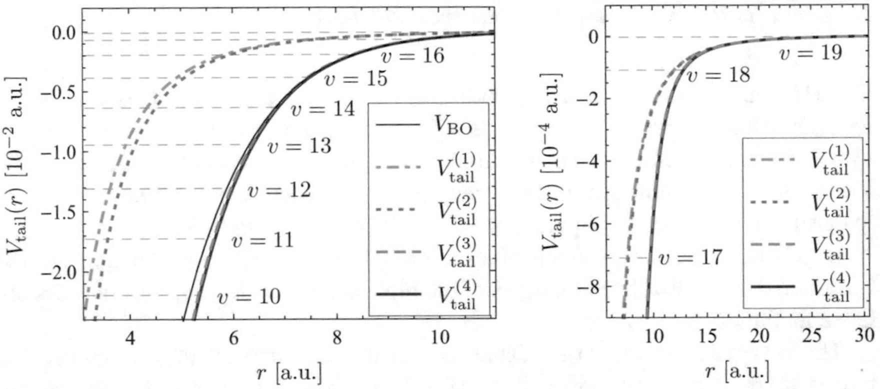

Fig. 3.5 Reference potentials $V^{(1)}_{\rm tail}(r)$ [(3.80)], $V^{(2)}_{\rm tail}(r)$ [(3.81)], $V^{(3)}_{\rm tail}(r)$ [(3.82)] and $V^{(4)}_{\rm tail}(r)$ [(3.83)] in an energy range encompassing the highest ten bound states in the $L = 0$, $1s\sigma_g$ configuration, see Table 3.6. The corresponding energy levels are shown as *horizontal dashed lines*. The potential $V_{\rm BO}(r)$ corresponds to the minimal electronic energy at internuclear separation r; this should be a good approximation to the full interaction for the range of r-values in the figure (From [KM13])

The reference potential (3.80) is shown in Fig. 3.5 (dot-dashed blue line) together with the potential $V_{\rm BO}(r)$ (solid black line), which represents the electronic ground-state energy at each internuclear separation r [Pee65] and should be a good approximation to the full interaction in the range of distances in the figure. The energies of the highest ten bound states, as listed in Table 3.6, are shown as horizontal dashed lines in the figure. The single-power reference potential (3.80) is clearly far too weak for distances less than about 12 a.u., while the outer classical turning point lies in this range at the energies E_υ of all states with $\upsilon \leq 17$. Since the dominance of $F_{\rm tail}(E)$ over short-range corrections requires the reference potential to be an accurate approximation of the full interaction down to distances somewhat smaller than the outer classical turning point, the usefulness of the single-power tail (3.80) is expected to be limited to a very narrow range of near-threshold energies, encompassing at most the highest one or two levels.

In order to understand how the choice of reference potential affects the separation of short-range and tail effects, the authors of [KM13] investigated three further versions for $V_{\rm tail}(r)$:

$$V^{(2)}_{\rm tail}(r) = -\frac{9}{4r^4} - \frac{15}{2r^6}\,, \tag{3.81}$$

$$V^{(3)}_{\rm tail}(r) = -\frac{9}{4r^4} - 2r\,{\rm e}^{-r-1}\,, \tag{3.82}$$

$$V^{(4)}_{\rm tail}(r) = -\frac{9}{4r^4} - \frac{15}{2r^6} - \frac{213}{4\,r^7} - 2r\,{\rm e}^{-r-1}\left(1 + \frac{1}{2\,r} - \frac{25}{8\,r^2}\right)\,. \tag{3.83}$$

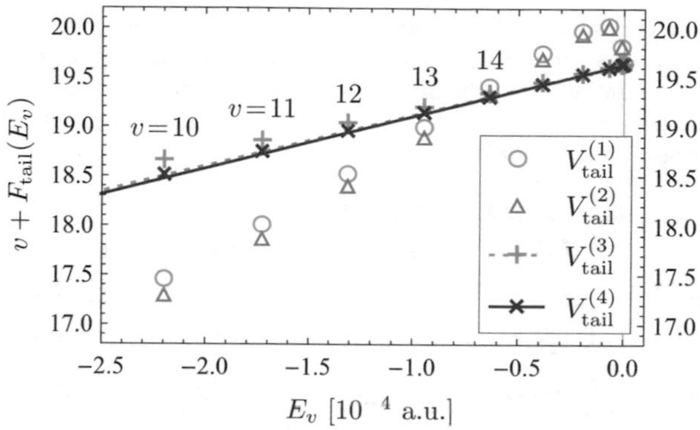

Fig. 3.6 Plots of $\upsilon+F_{\rm tail}(E_\upsilon)$ against E_υ with the quantization function $F_{\rm tail}(E)$ defined via (3.22), (3.23) on the basis of the definitions (3.80)–(3.83) of $V_{\rm tail}$. The *straight dashed green* and *solid red lines* are fitted according to (3.26) through the highest two states, $\upsilon=18$ and $\upsilon=19$, with $F_{\rm tail}(E)$ based on $V_{\rm tail}^{(3)}$ and $V_{\rm tail}^{(4)}$, respectively (Adapted from [KM13])

Table 3.7 Values $\upsilon+F_{\rm tail}(E_\upsilon)$ at the energies E_υ given in Table 3.6 for the quantization functions based on the definitions (3.80)–(3.83) of $V_{\rm tail}(r)$

υ	$V_{\rm tail}^{(1)}$	$V_{\rm tail}^{(2)}$	$V_{\rm tail}^{(3)}$	$V_{\rm tail}^{(4)}$	υ	$V_{\rm tail}^{(1)}$	$V_{\rm tail}^{(2)}$	$V_{\rm tail}^{(3)}$	$V_{\rm tail}^{(4)}$
10	17.4612	17.2870	18.6570	18.5089	15	19.7486	19.6804	19.4491	19.4310
11	18.0115	17.8571	18.8562	18.7444	16	19.9740	19.9285	19.5374	19.5304
12	18.5268	18.3929	19.0367	18.9557	17	20.0268	20.0028	19.5976	19.5968
13	18.9980	18.8853	19.1968	19.1416	18	19.8143	19.8073	19.6291	19.6287
14	19.4120	19.3213	19.3349	19.3007	19	19.6468	19.6467	19.6346	19.6343

These further reference potentials are shown as dotted orange $\left[V_{\rm tail}^{(2)}(r)\right]$, dashed green $\left[V_{\rm tail}^{(3)}(r)\right]$ and solid red $\left[V_{\rm tail}^{(4)}(r)\right]$ lines in Fig. 3.5. The addition of the next-order dispersion term $-15/(2r^6)$, which defines $V_{\rm tail}^{(2)}(r)$, is not a significant improvement over $V_{\rm tail}^{(1)}(r)$, but $V_{\rm tail}^{(3)}(r)$ and $V_{\rm tail}^{(4)}(r)$, which include a contribution from the polarisation potential, offer a far better representation of the full potential in the whole range $r > 5$ a.u.

The quality with which the reference potentials $V_{\rm tail}^{(i)}(r)$ approximate the full potential is reflected in the accuracy with which a plot of $\upsilon + F_{\rm tail}(E_\upsilon)$ against E_υ yields a straight line with a small gradient according to (3.26). The plots are shown in Fig. 3.6, and the numerical values on which they are based are listed in Table 3.7.

As already seen in Fig. 3.5, the potential tails $V_{\rm tail}^{(1)}(r)$ and $V_{\rm tail}^{(2)}(r)$ are only a fair approximation of the full potential for distances larger than about 12 a.u. The energy levels for which the outer classical turning point lies in this range are the highest state $\upsilon = 19$ and the second-highest state $\upsilon = 18$, only. Correspondingly, the behaviour of $\upsilon + F_{\rm tail}(E_\upsilon)$ for $\upsilon \leq 17$ and for $\upsilon \geq 18$ cannot, not even approximately, be reconciled to one straight line, see blue circles and red triangles

Table 3.8 For the definitions (3.80)–(3.83) of the reference potential, the table lists the values of the threshold quantum number υ_D and the short-range correction coefficient γ_{sr} as obtained by fitting a straight line through the highest two states $\upsilon = 18$ and $\upsilon = 19$ according to (3.26), together with the tail parameters $\bar{a}$, b and ϕ_0. The last column shows the value obtained for the scattering length according to (3.59)

V_{tail}	υ_D	γ_{sr} [a.u.]	$\bar{a}$ [a.u.]	b [a.u.]	ϕ_0	a [a.u.]
$V_{tail}^{(1)}$	19.6414	1577.3	0	64.28	π	−30.60
$V_{tail}^{(2)}$	19.6410	1517.4	$O(10^{-15})$	64.27	3.14396	−30.49
$V_{tail}^{(3)}$	19.6348	−51.57	−2.49	63.09	3.07548	−30.93
$V_{tail}^{(4)}$	19.6345	−52.91	−2.38	63.12	3.06881	−30.77

in Fig. 3.6. In contrast the points based on $V_{tail}^{(3)}(r)$ show a much smoother energy dependence, while for $V_{tail}^{(4)}(r)$ the behaviour of $\upsilon + F_{tail}(E_\upsilon)$ is quite close to linear down to $\upsilon = 10$.

Table 3.8 lists the values of the threshold quantum number υ_D and the short-range correction coefficient γ_{sr} as obtained by fitting a straight line through the last two states $\upsilon = 18$ and $\upsilon = 19$ according to (3.26) for the various choices of reference potential. Also listed are the tail parameters $\bar{a}$ (mean scattering length), b (threshold length) and ϕ_0 (threshold value of the outer reflection phase). The last column shows the respective values of the scattering length a that follow via (3.59). Although the choice of reference potential strongly influences the energy range over which the tail contribution to the quantization function governs the energy progression of the near-threshold bound states, the extrapolation to $E = 0$ yields a very stable value of the threshold quantum number υ_D, which turns out to be quite insensitive to the choice of $V_{tail}(r)$. This puts rather tight bounds on the value of the scattering length, which follows via (3.59) and is seen to lie in the range between −31 and −30.5 a.u. Interestingly, this range does not include the value $a = -29.3$ a.u., which was derived in [CL03] by solving the appropriate Faddeev equations for the three-body *ppe* system. Two of the authors of [HB00], who obtained the energy eigenvalues in Table 3.6, were also coauthors of [CL03]. It seems that the scattering length given there is not quite consistent with the progression of near-threshold energy levels given in [HB00]. The same applies to the value $a = -28.8$ a.u., which was obtained in [BZ08] by calculating *p*-H scattering cross sections down to very low energies.

Figure 3.7 shows the scattering length derived via (3.59), with the threshold quantum number υ_D obtained by fitting a straight line through two bound states υ and $\upsilon + 1$ according to (3.26), as function of the quantum number υ. For the reference potentials (3.80) and (3.81), the predictions are outside the range of the figure for $\upsilon \leq 17$. With the more sophisticated choices (3.82) and (3.83) of reference potential, a rapid and smooth convergence with υ is observed, similar to the case of the Lennard-Jones potential, see Table 3.5. With the reference potential $V_{tail}^{(4)}(r)$, the scattering length obtained from the fifth and fourth highest state ($\upsilon = 15$ and $\upsilon = 16$) already lies within 0.3 a.u. of the value obtained with the highest two states.

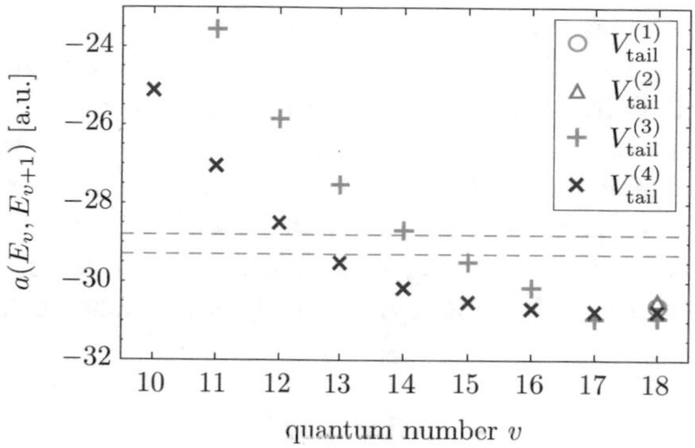

Fig. 3.7 Scattering length a according to (3.59) with v_D obtained by fitting a straight line through the points v and $v+1$ in Fig. 3.6 according to (3.26). The *blue circle* and the *red triangle* at $v=18$ are based on $V_{\rm tail}^{(1)}(r)$ and $V_{\rm tail}^{(2)}(r)$. The *upright green and diagonal red crosses* are based on $V_{\rm tail}^{(3)}(r)$ and $V_{\rm tail}^{(4)}(r)$, respectively. The *dashed horizontal lines* show the values $a=-29.3$ a.u. and $a=-28.8$ a.u. given in [CL03] and [BZ08] (Adapted from[KM13])

This example shows, how a sufficiently sophisticated choice of reference potential can substantially increase the energy range over which the progression of near-threshold energy levels can be understood as a property of $V_{\rm tail}(r)$. The "bad news" is, that any choice of $V_{\rm tail}(r)$ beyond the single-power form (3.63) destroys the universality of the quantization function. Whereas the quantization function $F_\alpha(\kappa\beta_\alpha)$ for a single-power tail caters for all values of the potential strength, expressed through the quantum length β_α, adding any further term to the definition of $V_{\rm tail}(r)$ only makes sense in an application to a specific system. For any reference potential containing two or more terms, however, the quantization function will depend on the ratios of the strengths of the various terms. These ratios are most likely to be unique to a particular system, so the quantization function derived for a given system will be applicable to this special case only.

3.1.5 The Transition From a Finite Number to Infinitely Many Bound States, Inverse-Square Tails

The near-threshold quantization rule (3.4) for an attractive potential tail vanishing as $-1/r^\alpha$ with $0 < \alpha < 2$ becomes meaningless as α approaches the value 2 (from below). For potential tails vanishing faster than $1/r^2$, the universal near-threshold quantization rule follows from (3.11) and (3.51),

$$n \overset{\kappa\to 0}{\sim} n_{\rm th} - \frac{b}{\pi}\kappa + O(\kappa^2)\,; \tag{3.84}$$

here $n_{\rm th}$ is the finite but not necessarily integer threshold quantum number, which was called $\upsilon_{\rm D}$ in Sects. 3.1.2–3.1.4. The threshold length b in (3.84) diverges to infinity for a (homogeneous) tail vanishing as $-1/r^\alpha$, when the power α approaches 2 from above implying $\nu \to \infty$, see (3.67). In order to understand the transition from potential wells with tails vanishing faster than $-1/r^2$, which support at most a finite number of bound states, to those with tails vanishing more slowly than $-1/r^2$, which support infinitely many bound states, it is necessary to look in some detail at potentials with tails asymptotically proportional to the inverse square of the distance,

$$V(r) \overset{r\to\infty}{\sim} V_\gamma(r) \overset{\rm def}{=} \frac{\hbar^2}{2\mu}\frac{\gamma}{r^2}. \tag{3.85}$$

Inverse-square potentials of the form (3.85) with positive or negative values of the strength parameter γ occur in various physically relevant situations. For a one-particle Schrödinger equation in f-dimensional coordinate space ($f \geq 2$), radial and angular motion can be separated via an ansatz,

$$\psi(r) = \frac{\psi_{\rm rad}(r)}{r^{(f-1)/2}} \times Y(\text{angles}); \tag{3.86}$$

the one-dimensional radial Schrödinger equation for $\psi_{\rm rad}(r)$ contains a centrifugal potential of the form (3.85) with

$$\gamma = \left(l_f + \frac{f-1}{2}\right)\left(l_f + \frac{f-3}{2}\right). \tag{3.87}$$

In three-dimensional space we have the well-known result $\gamma = l_3(l_3+1)$, $l_3 = 0, 1, 2, \ldots$ [see (1.75) in Sect. 1.2.2], and in two-dimensional space we have

$$\gamma = (l_2)^2 - \frac{1}{4}, \quad l_2 = 0, \pm 1, \pm 2, \ldots. \tag{3.88}$$

Attractive and repulsive inverse-square potentials occur in the interaction of an electrically charged particle with a dipole, e.g. in the interaction of an electron with a polar molecule or with a hydrogen atom in a parity-mixed excited state, see Sect. 3.1.6.

If the inverse-square tail (3.85) is sufficiently attractive, more precisely, if

$$\gamma \overset{\rm def}{=} -g < -\frac{1}{4}, \quad g > \frac{1}{4}, \tag{3.89}$$

then the potential supports an infinite number of bound states, usually called a "dipole series", and towards threshold, $E = 0$, the energy eigenvalues of the bound states depend exponentially on the quantum number n, [GD63, MF53]

$$E_n \overset{n\to\infty}{\sim} -E_0 \exp\left(-\frac{2\pi n}{\sqrt{g-\frac{1}{4}}}\right). \tag{3.90}$$

The strength g of the attractive inverse-square tail determines the asymptotic value of the ratio of successive energy eigenvalues,

$$\frac{E_n}{E_{n+1}} \overset{n\to\infty}{\sim} \exp\left(\frac{2\pi}{\sqrt{g-\frac{1}{4}}}\right), \tag{3.91}$$

but not the explicit positions of the energy levels, which are fixed by the constant E_0 in (3.90). This reflects the fact that there is no energy scale in a Schrödinger equation with a kinetic energy and an inverse-square potential; if $\psi(r)$ is a solution at energy ε, then $\psi(sr)$ is a solution at energy $s^2\varepsilon$. For a pure $-1/r^2$ potential one can obtain a discrete bound-state spectrum corresponding to the right-hand side of (3.90) by requiring orthogonality of the bound-state wave functions at different energies, but the resulting spectrum is unbounded from below, $E_n \to -\infty$ for $n \to -\infty$ [MF53].

In a realistic potential well with a sufficiently attractive inverse-square tail, (3.89), the actual positions of the bound-state energies are determined by the behaviour of the potential at small distances, where it must necessarily deviate from the pure $-1/r^2$ form. If the potential tail contains a further attractive term proportional to $-1/r^m$, $m > 2$, then as r decreases this term becomes dominant and the WKB approximation becomes increasingly accurate. Potential wells with two-term tails,

$$V(r) \overset{r\to\infty}{\sim} V_{g,m}(r) = -\frac{\hbar^2}{2\mu}\left(\frac{(\beta_m)^{m-2}}{r^m} + \frac{g}{r^2}\right), \quad m > 2, \quad g > \frac{1}{4}, \tag{3.92}$$

support an infinite dipole series of bound states and there may be a WKB region at moderate r values if the well is deep enough. In this case, near-threshold properties of the bound states can be derived [ME01] by matching the asymptotic ($r \to \infty$) solutions of the Schrödinger equation with the inverse-square term alone to zero-energy solutions of the tail (3.92), which are then expressed as WKB waves in the WKB region in a procedure similar to that used in Sect. 3.1.2. This results [ME01] in an explicit expression for the factor E_0, which asymptotically ($n \to \infty$) determines the positions of the energy levels in the dipole series (3.90),

$$E_0 = \frac{2\hbar^2(m-2)^{\frac{4}{m-2}}}{\mu(\beta_m)^2}\exp\left\{\frac{2}{\tau}\left[\theta + \chi + \frac{\pi}{2} + \arctan\left(\frac{\tan(\tilde{S}_0/(2\hbar))}{\tanh(\xi\pi/2)}\right)\right]\right\}. \tag{3.93}$$

The parameters τ, ξ, θ and χ appearing in (3.93) are,

$$\tau = \sqrt{g-\frac{1}{4}}, \quad \xi = \frac{2\tau}{m-2}, \quad \theta = \arg\Gamma(\mathrm{i}\tau), \quad \chi = \arg\Gamma(\mathrm{i}\xi), \tag{3.94}$$

and $\tilde{S}_0$ is essentially the threshold value of the action integral from the inner classical turning point $r_{\rm in}(0)$ to a point r in the WKB region,

$$\frac{\tilde{S}_0}{2\hbar} = \frac{1}{\hbar}\int_{r_{\rm in}(0)}^{r} p_0(r')\mathrm{d}r' + \frac{2}{m-2}\left(\frac{\beta_m}{r}\right)^{\frac{m-2}{2}} - \frac{\phi_{\rm in}(0)}{2} - \frac{\pi}{4}\,. \tag{3.95}$$

A condition for the applicability of the formula (3.93) is, that near the point r in the WKB region the potential must be dominated by the $-1/r^m$ term so that the inverse square contribution can be neglected; the sum of the integral and the term proportional to $1/r^{(m-2)/2}$ in (3.95) is then independent of the choice of r.

More explicit solutions are available when the whole potential consists of an attractive inverse-square tail and a repulsive $1/r^m$ core,

$$V(r) = \frac{\hbar^2}{2\mu}\left(\frac{(\beta_m)^{m-2}}{r^m} - \frac{g}{r^2}\right), \quad m > 2\,. \tag{3.96}$$

The existence of a WKB region in the well is not necessary in this case, because analytical zero-energy solutions of the Schrödinger equation are available for the whole potential. The zero-energy solution which vanishes at $r = 0$ approximates finite energy solutions to order less than $O(E)$ in the region of small and moderate r values, and it can be matched to the solution which vanishes asymptotically in the presence of the attractive inverse-square tail. This yields the following expression for the factor E_0 defining the energies of the near-threshold bound states of the dipole series (3.90) [ME01],

$$E_0 = \frac{2\hbar^2(m-2)^{\frac{4}{m-2}}}{\mu(\beta_m)^2}\exp\left(2\frac{\theta+\chi}{\tau}\right), \tag{3.97}$$

where τ, θ and χ are as already defined in (3.94). Note that E_0 is only defined to within a factor consisting of an integer power of the right-hand side of (3.91); multiplying E_0 by an integer power of $\exp(2\pi/\sqrt{g-\frac{1}{4}})$ does not affect the energies in the dipole series (3.90) except for an appropriate shift in the quantum number n labelling the bound states.

A potential with an attractive inverse-square tail (3.85) no longer supports an infinite series of bound states, when the strength parameter γ is equal to (or larger than) $-\frac{1}{4}$. This can be expected from the breakdown of formulae such as (3.90), (3.93) and (3.97) when $-\gamma = g = \frac{1}{4}$. It is also physically reasonable, considering that the inverse-square potential $V_{\gamma=-1/4}$ is the s-wave ($l_2 = 0$) centrifugal potential for a particle moving in two spatial dimensions, see (3.88). It is difficult to imagine a physical mechanism that would bind a free particle in a flat plane, so the discontinuation of dipole series of bound states at the value $-\frac{1}{4}$ of the strength parameter γ seems more than reasonable.

A potential well with a *weakly attractive* inverse-square tail, i.e. with a strength parameter in the range

$$-\frac{1}{4} \le \gamma < 0, \tag{3.98}$$

can support a (finite) number of bound states if supplemented by an additional attractive potential. If the additional potential is regular at the origin, then the action integral from the origin to the outer classical turning point diverges because of the $-1/r^2$ singularity of the potential at $r = 0$, so a naive application of the generalized quantization rule (3.2) doesn't work. This can be overcome by shifting the inner classical turning point to a small positive value and adjusting the reflection phase ϕ_{in} accordingly [FT99].

The near-threshold quantization rule for a weakly attractive inverse-square tail was studied in some detail by Moritz et al. [ME01], and analytical results were derived for tails of the form

$$V(r) \overset{r\to\infty}{\sim} V_{g,m}^{(\text{weak})}(r) = -\frac{\hbar^2}{2\mu}\left(\frac{(\beta_m)^{m-2}}{r^m} + \frac{g}{r^2}\right), \quad m > 2, \quad g < \frac{1}{4}. \tag{3.99}$$

For $g < \frac{1}{4}$ (i.e., excluding the limiting case $g = \frac{1}{4}$), the near-threshold quantization rule is [ME01],

$$\begin{aligned} n = n_{\text{th}} &- \frac{\pi(\kappa\beta_m/2)^{2\mu_\gamma}}{\sin(\pi\mu_\gamma)(m-2)^{2\nu}\mu_\gamma\nu[\Gamma(\mu_\gamma)\Gamma(\nu)]^2} \\ &+ O\left((\kappa\beta_m)^{4\mu_\gamma}\right) + O(\kappa^2), \end{aligned} \tag{3.100}$$

with

$$\mu_\gamma = \sqrt{\frac{1}{4} - g} = \sqrt{\gamma + \frac{1}{4}} \quad \text{and} \quad \nu = \frac{2\mu_\gamma}{m-2}. \tag{3.101}$$

The threshold quantum number n_{th} in (3.100) is given by

$$n_{\text{th}}\pi = \frac{1}{\hbar}\int_{r_{\text{in}}(0)}^{r} p_0(r')\,\mathrm{d}r' + \frac{2}{m-2}\left(\frac{\beta_m}{r}\right)^{\frac{m-2}{2}} - \frac{\phi_{\text{in}}(0)}{2} - \frac{\pi}{4} - \frac{\nu}{2}\pi. \tag{3.102}$$

As in the discussion of (3.93) and (3.95), the point r defining the upper limit of the action integral must lie in a region of the potential well where the WKB approximation is sufficiently accurate and the potential is dominated by the $-1/r^m$ term, so the inverse-square contribution can be neglected; the sum of the integral and the term proportional to $1/r^{(m-2)/2}$ in (3.102) is then independent of the choice of r.

When we express κ in terms of the energy $E = -\hbar^2\kappa^2/(2\mu)$, the near-threshold quantization rule (3.100) becomes

$$n = n_{\text{th}} - B(-E)^{\mu_\gamma}, \tag{3.103}$$

with

$$B = \frac{\pi\left(\mu(\beta_m)^2/\left(2\hbar^2\right)\right)^{\mu_\gamma}}{\sin\left(\pi\mu_\gamma\right)(m-2)^{2\nu}\mu_\gamma\nu\left[\Gamma\left(\mu_\gamma\right)\Gamma\left(\nu\right)\right]^2}\,. \tag{3.104}$$

The limiting case $\gamma = -\frac{1}{4}$ corresponding to $\mu_\gamma = 0$ and $\nu = 0$ requires special treatment; the near-threshold quantization rule in this case is [ME01],

$$n = n_{\text{th}} + \frac{2/(m-2)}{\ln(-E/B)} + O\left(\frac{1}{[\ln(-E/B)]^2}\right), \quad B = \frac{\hbar^2}{2\mu(\beta_m)^2}\,. \tag{3.105}$$

Again, n_{th} is given by the expression (3.102); note that ν vanishes in this case.

We now have a very comprehensive overview of near-threshold quantization in potential wells with attractive tails. Potentials falling off as $-1/r^\alpha$ with a power $0 < \alpha < 2$ support an infinite number of bound states, and the limit of infinite quantum numbers is the semiclassical limit. The near-threshold quantization rule (3.4) contains a leading term proportional to $1/(-E)^{\frac{1}{\alpha}-\frac{1}{2}}$ in the expression for the quantum number n. For $\alpha = 2$, the threshold $E = 0$ no longer represents the semiclassical limit of the Schrödinger equation, but the potential still supports an infinite number of bound states, if the attractive inverse-square tail is strong enough, (3.89); the near-threshold quantization rule now contains $\sqrt{g - \frac{1}{4}}\ \ln(-E)$ in the expression for the quantum number n, see (3.90). The attractive inverse-square tail ceases to support an infinite series of bound states at the value $g = -\gamma = \frac{1}{4}$ of the strength parameter, which corresponds to the strength of the (attractive) s-wave centrifugal potential for a particle in a plane. In the near-threshold quantization rule, the leading term in the expression for the quantum number now is a finite number n_{th} related to the total number of bound states, and the next-to-leading term contains the energy as $1/\ln(-E)$ for $\gamma = -\frac{1}{4}$ [(3.105)], or as $(-E)^{\sqrt{\gamma+\frac{1}{4}}}$ for $\gamma > -\frac{1}{4}$, see (3.103).

It is interesting to note, that the properties of potential wells with shorter-ranged tails falling off faster than $1/r^2$ fit smoothly into the picture elaborated for inverse-square tails when we take the strength of the inverse-square term to be zero. The near-threshold quantization rule (3.100) acquires the form (3.84) when $\gamma = 0$, $\mu_\gamma = \frac{1}{2}$, and the coefficient of κ becomes $b^{(m)}/\pi$ with $b^{(m)}$ given by the second equation in the upper line of (3.67) when we also insert $\nu = 1/(m-2)$.

The discussion of weakly attractive inverse-square tails, defined by the condition (3.98), can be continued without modification into the range of *weakly repulsive* inverse-square tails, defined by strength parameters in the range

$$0 < \gamma < \frac{3}{4} . \tag{3.106}$$

The parameter $\mu_\gamma = \sqrt{\gamma + \frac{1}{4}}$ determining the leading energy dependence on the right-hand sides of (3.100) and (3.103) then lies in the range

$$\frac{1}{2} < \mu_\gamma < 1, \tag{3.107}$$

and the leading energy dependence $(-E)^{\mu_\gamma}$ expressed in these equations is still dominant compared to the contributions of order $O(E)$, which come from the analytical dependence of all short-ranged features on the energy E and were neglected in the derivation of the leading near-threshold terms. We can thus complete the comprehensive overview of near-threshold quantization by extending it to repulsive potential tails. For weakly repulsive inverse-square tails (3.106), the formulae (3.100) and (3.103) remain valid. The upper boundary of this range is given by

$$\gamma = \frac{3}{4}, \quad \mu_\gamma = \sqrt{\gamma + \frac{1}{4}} = 1, \tag{3.108}$$

which corresponds to the p-wave centrifugal potential in two spatial dimensions, $l_2 = \pm 1$, see (3.88). At this limit, the near-threshold quantization rule has the form,

$$n = n_{\text{th}} - O(E), \tag{3.109}$$

and this structure prevails for more strongly repulsive inverse-square tails, $\gamma > \frac{3}{4}$, and for *repulsive* potential tails falling off more slowly than $1/r^2$. Repulsive tails falling off more rapidly than $1/r^2$ comply with the case $\gamma = 0$, i.e. of vanishing strength of the inverse-square term in the potential, and, provided there is a sufficiently attractive well at moderate r values, the quantization rule has the form (3.84) with the threshold length b and a threshold quantum number n_{th} which also depends on the shorter-ranged part of the potential.

Note that the condition (3.108) also defines the boundary between systems with a singular and a regular level density at threshold. For attractive potential tails and for repulsive potential tails falling off more rapidly than $1/r^2$ or as an inverse-square potential with $\gamma < \frac{3}{4}$, the level density $\mathrm{d}n/\mathrm{d}E$ is singular at threshold, and the leading singular term is determined by the tail of the potential. For a repulsive inverse-square tail with $\gamma \geq \frac{3}{4}$, and for a repulsive tail falling off more slowly than $1/r^2$, the level density is regular at threshold, and the leading (constant) term depends also on the shorter-ranged part of the potential.

A summary of the near-threshold quantization rules reviewed in this section is given in Table 3.9.

Table 3.9 Summary of near-threshold quantization rules for attractive and repulsive potential tails. The second column gives the leading term(s) to the quantization rule in the limit of vanishing energy, $E = -\hbar^2\kappa^2/(2\mu) \to 0$. The third column lists equations where explicit expressions for the constants appearing in the second column can be found; these can apply quite generally, as in the first row, or to special models of potential tails with the asymptotic behaviour given in the first column

$V(r)$ for $r \to \infty$	Quantization rule for $E \to 0$	Refs. for constants
$-\frac{\hbar^2}{2\mu}(\beta_\alpha)^{\alpha-2}/r^\alpha,\ 0<\alpha<2$	$n \sim \frac{1}{\pi}F(\alpha)/(\kappa\beta_\alpha)^{(2/\alpha)-1}$	$F(\alpha)$: (3.3)
$\frac{\hbar^2}{2\mu}\gamma/r^2,\ \gamma<-\frac{1}{4}$	$n \sim -\frac{1}{2\pi}\ln(-E/E_0)\sqrt{\lvert\gamma\rvert-\frac{1}{4}}$	E_0: (3.92), (3.96)
$\gamma=-\frac{1}{4}$	$n \sim n_{\text{th}} + A/\ln(-E/B)$	n_{th}: (3.101)
		A, B: (3.104)
$-\frac{1}{4}<\gamma<\frac{3}{4}$	$n \sim n_{\text{th}} - B(-E)^{\sqrt{\gamma+1/4}}$	n_{th}: (3.101)
		B: (3.103)
$\gamma \ge \frac{3}{4}$	$n \sim n_{\text{th}} - O(E)$	n_{th}: (3.101)
$\propto +1/r^\alpha,\ 0<\alpha<2$	$n \sim n_{\text{th}} - O(E)$	
		n_{th}: (3.9)
$\propto \pm 1/r^\alpha,\ \alpha>2$	$n \sim n_{\text{th}} - \frac{1}{\pi}b\kappa$	b: (3.66)
		Table 3.1

3.1.6 *Example: Truncated Dipole Series in the* H$^-$ *Ion*

The interaction between an electron and a neutral atom usually behaves asymptotically as a shorter-ranged potential falling off faster than $1/r^2$. The excited energy levels of the hydrogen atom are an exception, because the degenerate eigenstates of different parity can mix to form states with a finite dipole moment, and the interaction of such a dipole with the negatively charged electron is given by a potential proportional to $1/r^2$.

Consider a system consisting of a hydrogen atom and an additional, "outer" electron at a distance $\boldsymbol{r}$. Let us ignore the spin degrees of freedom for the time being, so the Hamiltonian is

$$\hat{H} = \frac{\hat{\boldsymbol{p}}_1^2}{2\mu} - \frac{e^2}{r_1} + \frac{\hat{\boldsymbol{p}}^2}{2\mu} - \frac{e^2}{r} + \frac{e^2}{|\boldsymbol{r}-\boldsymbol{r}_1|}, \tag{3.110}$$

and the total wave function is a function of the coordinate vector $\boldsymbol{r} \equiv (r, \Omega)$ for the outer electron and of $\boldsymbol{r}_1 \equiv (r_1, \Omega_1)$ for the "inner" electron in the atom. If we restrict the study to excited states of hydrogen in a given shell n, then the wave function for the inner electron is an eigenstate of the Hamiltonian $\hat{\boldsymbol{p}}_1^2/(2\mu) - e^2/r_1$ with eigenvalue $E_n = -1/(2n^2)$ a.u., and the total wave function can be written as,

$$\Psi(\boldsymbol{r}_1, \boldsymbol{r}) = \sum_{l_1,m_1} \frac{\phi_{n,l_1}(r_1)}{r_1} Y_{l_1,m_1}(\Omega_1) \sum_{l,m} \frac{\chi_{l_1,m_1;l,m}(r)}{r} Y_{l,m}(\Omega)\,. \tag{3.111}$$

The right-hand side of (3.111) represents an expansion in channels as described in Sect. 1.5.1, where the radial coordinate r of the outer electron describes the dynamic degree of freedom, whereas the radial coordinate r_1 of the inner electron together with all angular degrees of freedom describes the internal variables whose eigenstates label the various channels. For given values of the (conserved) total orbital angular momentum quantum numbers L, M, there is only a finite number of channels. E.g., for $n = 2$ the internal angular momentum l_1 can be either zero or one, and the angular momentum l of the outer electron can, for given L, only have the values $L-1$, L or $L+1$. For $L = 0$ there are only two possibilities, namely $l = 0$ which implies that l_1 must also be zero, and $l = 1$ which can only couple to $L = 0$ when $l_1 = 1$.

For the simple case $n = 2$ and $L = 0$, the channel expansion (3.111) thus reduces to

$$\Psi_{L=0} = \frac{\phi_{2,0}(r_1)}{r_1}\frac{\chi_0(r)}{r}\mathcal{Y}_0 + \frac{\phi_{2,1}(r_1)}{r_1}\frac{\chi_1(r)}{r}\mathcal{Y}_1, \tag{3.112}$$

where $\mathcal{Y}_0$ and $\mathcal{Y}_1$ stand for the normalized angular parts of the wave function, coupled to $L = 0$, $M = 0$ [Edm60]:

$$\begin{aligned}
\mathcal{Y}_0 &= Y_{0,0}(\Omega_1)\,Y_{0,0}(\Omega) = \frac{1}{4\pi},\\
\mathcal{Y}_1 &= \sum_{m=-1}^{1}\langle 1,m,1,-m|0,0\rangle\, Y_{1,m}(\Omega_1)\,Y_{1,-m}(\Omega)\\
&= \sum_{m=-1}^{1}\frac{(-1)^{1-m}}{\sqrt{3}}Y_{1,m}(\Omega_1)\,Y_{1,-m}(\Omega) = -\frac{1}{\sqrt{3}}\sum_{m=-1}^{1}Y_{1,m}(\Omega_1)\,Y^*_{1,m}(\Omega)\\
&= -\frac{\sqrt{3}}{4\pi}\cos(\theta),
\end{aligned} \tag{3.113}$$

where θ is the angle between $\boldsymbol{r}$ and $\boldsymbol{r}_1$. The last line in (3.113) follows from the properties of the spherical harmonics, see (A.9) in Appendix A.1.

We insert the two-channel wave function (3.112) into the two-electron Schrödinger equation with the Hamiltonian (3.110) and take matrix elements with the channel states $\mathcal{Y}_0\phi_{2,0}(r_1)/r_1$ and $\mathcal{Y}_1\phi_{2,1}(r_1)/r_1$. This involves integrating over r_1 and all angular variables and leads to the coupled-channel equations for the wave functions $\chi_0(r)$ and $\chi_1(r)$,

$$\begin{aligned}
-\frac{\hbar^2}{2\mu}\frac{d^2\chi_0}{dr^2} + \left(-\frac{e^2}{r} + V_{0,0}\right)\chi_0 + V_{0,1}\chi_1 &= (E-I)\,\chi_0,\\
-\frac{\hbar^2}{2\mu}\frac{d^2\chi_1}{dr^2} + \left(\frac{\hbar^2}{2\mu}\frac{2}{r^2} - \frac{e^2}{r} + V_{1,1}\right)\chi_1 + V_{1,0}\,\chi_0 &= (E-I)\,\chi_1\,.
\end{aligned} \tag{3.114}$$

The potentials $V_{i,j,}$, $i, j = 0,\ 1$, are defined by,

$$V_{i,j}(r) = \left\langle \frac{\phi_{2,i}(r_1)}{r_1} Y_i \left| \frac{e^2}{|\boldsymbol{r} - \boldsymbol{r}_1|} \right| \frac{\phi_{2,j}(r_1)}{r_1} Y_j \right\rangle_r , \tag{3.115}$$

where the subscript r on the matrix elements implies integration over all variables except r. The threshold I below which both channels are closed is just the energy eigenvalue $E_2 = -\frac{1}{8}$ a.u. of the isolated hydrogen atom in an $n = 2$ excited state.

At large distances r of the outer electron we can assume $|\boldsymbol{r}_1| < |\boldsymbol{r}|$ and expand the inverse separation in (3.115) using (A.9) and (A.10) in Appendix A.1,

$$\begin{aligned} \frac{e^2}{|\boldsymbol{r} - \boldsymbol{r}_1|} &\overset{r_1 < r}{=} \sum_{l=0}^{\infty} \frac{e^2 (r_1)^l}{r^{l+1}} \frac{4\pi}{2l+1} \sum_{m=-l}^{l} Y_{l,m}(\Omega)\, Y_{l,m}^*(\Omega_1) \\ &\overset{r\to\infty}{\sim} \frac{e^2}{r} + \frac{e^2 r_1}{r^2} \frac{4\pi}{3} \sum_{m=-1}^{1} Y_{1,m}(\Omega)\, Y_{1,m}^*(\Omega_1) + O\left(\frac{1}{r^3}\right) . \end{aligned} \tag{3.116}$$

For $i = j = 0$, only the leading term in the lower line of (3.116) contributes to the matrix element (3.115), due to the orthogonality of the spherical harmonics. For $i = j = 1$ the contribution of the second term in the lower line of (3.116) involves integrals over products of three spherical harmonics of the same odd order $l = 1$, and these vanish according to (A.11), (A.12) in Appendix A.1. For the diagonal potentials we thus have,

$$V_{0,0}(r) = \frac{e^2}{r}, \quad V_{1,1}(r) = \frac{e^2}{r} + O\left(\frac{1}{r^3}\right) . \tag{3.117}$$

For the non-diagonal coupling potential we have

$$\begin{aligned} V_{0,1}(r) = V_{1,0}(r) &\overset{r\to\infty}{\sim} \frac{e^2}{r^2} M_{\mathrm{rad}} \frac{1}{3\sqrt{3}} \times (-1) \times \\ &\sum_{m,m'} \int \mathrm{d}\Omega_1 \int \mathrm{d}\Omega\, Y_{1,m}(\Omega)\, Y_{1,m}^*(\Omega_1)\, Y_{1,m'}(\Omega_1)\, Y_{1,m'}^*(\Omega) , \end{aligned} \tag{3.118}$$

where

$$M_{\mathrm{rad}} = \int_0^\infty \phi_{2,0}(r_1)\, r_1\, \phi_{2,1}(r_1)\, \mathrm{d}r_1 = -3\sqrt{3}\, a \tag{3.119}$$

is the radial matrix element involving the $n = 2$ radial eigenfunctions of the hydrogen atom, as defined in (1.139) and Table 1.4 in Sect. 1.3.3; here $a = \hbar^2/(\mu e^2)$ is the Bohr radius. The sum of the integrals over all the spherical harmonics in the lower line of (3.118) amounts to three.

In the coupled-channel equations (3.114), the terms $-e^2/r$ describing the attraction of the outer electron by the atomic nucleus are exactly cancelled by the terms e^2/r in (3.117), which are the leading contributions from the repulsive interaction with the inner electron. The electron-atom potential is thus asymptotically dominated by the terms proportional to $1/r^2$, which consist of the centrifugal potential for $l = 1$, $l(l+1) = 2$ in the second equation (3.114) and the non-diagonal coupling potential (3.118),

$$V_{0,1}(r) = V_{1,0}(r) \overset{r\to\infty}{\sim} 3\frac{e^2 a}{r^2} = 6\frac{\hbar^2}{2\mu r^2} . \tag{3.120}$$

The coupled-channel equations (3.114) can be written as a matrix equation for the two-component vectors consisting of the channel wave functions χ_0 and χ_1, and at large distances this matrix equation is

$$-\frac{\hbar^2}{2\mu}\left(\mathbf{1}\frac{\mathrm{d}^2}{\mathrm{d}r^2} - \mathbf{v}\frac{1}{r^2}\right)\begin{pmatrix}\chi_0\\ \chi_1\end{pmatrix} = \begin{pmatrix}E-I & 0\\ 0 & E-I\end{pmatrix}\begin{pmatrix}\chi_0\\ \chi_1\end{pmatrix},$$
$$\text{where} \quad \mathbf{1} = \begin{pmatrix}1 & 0\\ 0 & 1\end{pmatrix} \quad \text{and} \quad \mathbf{v} = \begin{pmatrix}0 & 6\\ 6 & 2\end{pmatrix} . \tag{3.121}$$

The constant matrix $\mathbf{v}$ in (3.121) can be diagonalized by replacing χ_0 and χ_1 by appropriate linear combinations; its eigenvalues are $\gamma_\pm = 1 \pm \sqrt{37}$. Asymptotically, the coupled-channel equations (3.114) thus decouple into two independent equations for channels corresponding to parity-mixed superpositions of the $2s$ and the $2p$ states of the hydrogen atom. Through this mixing, the inverse-square terms in the non-diagonal coupling potential (3.120) contribute to the asymptotic potentials in the decoupled channels. The superposition corresponding to the eigenvalue $\gamma_- = 1 - \sqrt{37} \approx -5.08$ asymptotically obeys a Schrödinger equation with an attractive inverse-square potential (3.85), which is clearly strong enough (3.89) to support a dipole series of states as described in Sect. 3.1.5. Strictly speaking, these states are not bound states, because they lie above the threshold $E_1 = -\frac{1}{2}$ a.u. for decay into the $n = 1$ ground state of the hydrogen atom; instead of a dipole series of bound states we actually expect a dipole series of resonant states converging to the series limit, the $n = 2$ threshold at $E = I = -\frac{1}{8}$ a.u. One $L = 0$ resonance has actually been observed at an energy of about 0.024 atomic units below the $n = 2$ threshold in H^-, and further states have been derived from *ab initio* calculations by several authors, for details see [Pur99].

The low-lying states are, of course, affected by further details of the electron-atom interaction, in particular the shorter-ranged contributions of the electron-electron repulsion (3.116) as well as effects due to the Pauli principle and the spin-orbit interaction. The observed state at $E - I = -0.024$ a.u. is actually a spin singlet state, so the orbital wave function should be symmetric with respect to exchange of $\boldsymbol{r}$ and $\boldsymbol{r}_1$, see Sect. 2.2.4. Towards the threshold I, the short-ranged part of the wave functions is expected to change only marginally within the series, so the higher states differ only in their long-ranged tails which are strongly influenced by the attractive inverse-square potential. In this respect, the highly excited states in a dipole series resemble the highly excited states in an attractive Coulomb potential as discussed in Sect. 1.3.3, see Figs. 1.4–1.6.

The ratio of successive energy eigenvalues in a dipole series (relative to the threshold) approaches a constant value R near threshold, see (3.90). For $g = -\gamma_- \approx 5.08$ we have,

$$\frac{E_i - I}{E_{i+1} - I} \overset{n\to\infty}{\sim} R = \mathrm{e}^{2\pi/\sqrt{g-1/4}} \approx 17.43\,. \tag{3.122}$$

With the lowest 2^1S state of H^- roughly 0.024 atomic units below I, we would expect further states of the dipole series near $0.024/R^{i-1}$, $i = 2, 3\ldots$. As the distance to threshold, $|E_i - I|$, becomes smaller and smaller, we have to focus on smaller contributions to the Hamiltonian of the system, which were ignored so far. Indeed, the degeneracy of the $n = 2$ states of the hydrogen atom, which is an essential ingredient in decoupling the coupled-channel equations (3.114), (3.121) at large r, is only a reasonable assumption as long as we neglect the fine-structure splitting of roughly 1.7×10^{-6} a.u. between the $j = 1/2$ and $j = 3/2$ states, see Fig. 2.1. If the "internal energies" are not equal in both channels, then the matrix containing the energy relative to threshold on the right-hand side of (3.121) is still diagonal, but it no longer commutes with $\mathbf{v}$, so a superposition of the $2s$ and the $2p$ channels cannot decouple the two equations. However, including relativistic effects as prescribed by the Dirac equation in Sect. 2.1.3 still leaves a degenerate pair of parity-mixed $n = 2$ states in the hydrogen atom, namely the $2s_{1/2}$ and the $2p_{1/2}$ states. At this stage, the appropriate good quantum number is the total angular momentum J rather than the total orbital angular momentum L. For the simplest case $J = 0$ and positive parity (meaning $l_1 + l$ is even), the $2s_{1/2}$ state of the inner electron can combine with $l = 0$ and $j = 1/2$ of the outer electron, whilst the $2p_{1/2}$ state combines with $l = 1, j = 1/2$. The potential matrix determining the strength of the $1/r^2$ term in the coupled-channel equations (3.121) again has one negative eigenvalue which fulfills the condition (3.89) for supporting a dipole series, namely $\gamma = 1 - \sqrt{13} \approx -2.6$ [PF98, Pur99]. The near-threshold value of the ratio of successive energies relative to threshold increases to $R \approx 60$ for this value of γ, meaning that the energies approach the threshold even more rapidly than in (3.122). At still finer energy resolution we have to consider the Lamb shift which splits the $2s_{1/2}$ and $2p_{1/2}$ states by 0.16×10^{-6} a.u. When this small energy splitting is taken into account, decoupled equations can no longer be generated via linear combinations of the $2s_{1/2}$ and $2p_{1/2}$ channels, and there is no attractive inverse square term in the electron-atom potential.

A rough estimate of where the various corrections to the simple picture of degenerate $n = 2$ states become important in coordinate space can be obtained by looking at the position where the inverse-square potential, naively calculated assuming degeneracy, reaches values comparable to the correction in the energies of the channel thresholds. The absolute value of the potential $\hbar^2\gamma/\left(2\mu r^2\right)$, with

$\gamma = 1 - \sqrt{37} \approx -5.08$—as obtained assuming degeneracy of all states in the $n = 2$ shell—reaches the energy 1.7×10^{-6} a.u. corresponding to the fine-structure splitting between the $j = 1/2$ and $j = 3/2$ states near $r \approx 1200$ a.u. For $\gamma = 1 - \sqrt{13} \approx -2.6$—as obtained with the channel thresholds given by the Dirac equation but neglecting the Lamb shift—the potential reaches values near the Lamb shift (1.6×10^{-7} au) for $r \approx 3000$ a.u.

The low-lying states of the electron-atom system are insensitive to the long-ranged features of the potentials, because the wave functions χ_0 and χ_1 solving the coupled equations (3.114) decay rapidly for large or moderate values of $|E - I|$. The highly excited states, on the other hand, are quite similar to the lower states at small distances and depend with increasing sensitivity on the potential tail as the energy approaches threshold. The splitting of the $n = 2$ states in the hydrogen atom due to fine structure and the Lamb shift leads to a truncation of the otherwise expected infinite dipole series when the energy relative to the threshold is of the order of this splitting.

Purr et al. [PF98, Pur99] studied this truncation of dipole series using a simple model for the short-ranged part of the (diagonal and non-diagonal) electron-atom potentials together with the exact longer-ranged terms proportional to $1/r^2$ and $1/r^3$, as well as the exact binding energies of the isolated hydrogen atom, including fine structure and the Lamb shift. The short-ranged model potential was adjusted to reproduce the lowest two states below the $n = 2$ threshold, which are known from experiment and/or *ab initio* calculations neglecting relativistic and quantum electrodynamic effects. Higher states, for which these effects are essential, were then obtained by solving the coupled channel equations with the previously determined potentials. Results for the $J^\pi = 0^+$ series starting with the experimentally known ^1S state below the $n = 2$ threshold are summarized in Table 3.10. The ratio of successive energies (relative to threshold) is fairly close to the value (3.122) expected from the strength of the attractive inverse-square potential obtained by asymptotically decoupling the equations (3.114), (3.121), in particular the ratio R_2 for the $i = 2$ and $i = 3$ states. The ratio R_1 is a bit larger, which can be attributed to effects of the short-ranged part of the interaction on the lowest state of the series. The ratio R_3 is also a bit larger, which can be attributed to the influence of fine-structure splitting on the fourth state in the series. This state is separated from the lowest

Table 3.10 Energies (in atomic units) of electron-hydrogen $J^\pi = 0^+$ resonances below the $n = 2$ threshold relative to the unperturbed threshold energy $I = -\frac{1}{8}$ a.u. Due to fine-structure splitting the lowest channel threshold actually lies at $I_{2p_{1/2}} = I - 2.08 \times 10^{-6}$ a.u. The last column lists the ratios $R_i = (E_i - I_{2p_{1/2}})/(E_{i+1} - I_{2p_{1/2}})$

i	$E_i - I$	R_i
1	-2.379×10^{-2}	23.4
2	-1.018×10^{-3}	17.4
3	-6.05×10^{-5}	20.3
4	-4.95×10^{-6}	

channel threshold $I_{2p_{1/2}}$ by only 2.9×10^{-6} a.u., and a hypothetical fifth state should be closer to threshold by a factor between 17 and 60. On such a fine energy scale, the corrections lifting the degeneracy of the s and p states can no longer be ignored in the channel thresholds, the picture of an attractive inverse-square potential breaks down and the dipole series is truncated. The $i = 4$ state is actually the last state obtained below $I_{2p_{1/2}}$ by solving the coupled-channel equations (3.114).

The investigation sketched above can be applied for other values J^π in the electron-atom system. The $J^\pi = 1^-$ states below the $n = 2$ threshold were studied in [LB98, PF98, Pur99]. The coupled-channel equations now encompass seven channels and the degenerate-threshold approximation yields two decoupled channels with attractive inverse-square potentials of sufficient strength, (3.89), which can roughly be identified as $^1P^{\mathrm{o}}$ and $^3P^{\mathrm{o}}$ channels in standard LS-notation appropriate at small electron-atom separations, see Sect. 2.2.4. The ^1P states are more easily accessible that S states, because they can be reached via laser excitation from the ^1S ground state of the negative hydrogen ion. Two $^1P^{\mathrm{o}}$ states below the $n = 2$ threshold of the hydrogen atom have actually been observed using laser spectroscopy [AB97], and a coupled-channel calculation as outlined above predicts one third state of dominantly of $^1P^{\mathrm{o}}$ character and a series of four states dominantly of $^3P^{\mathrm{o}}$ character. Interestingly, fine-structure and Lamb shift corrections lead to appreciable mixing of singlet and triplet configurations for the higher states in the series [PF98].

3.2 One Electron in a Modified Coulomb Potential

3.2.1 Rydberg Series, Quantum Defects

In a neutral atom or a positive ion, a highly excited electron at large separations r from the residual ion moves in an attractive Coulomb field, i.e., in a very-long-ranged potential proportional to $1/r$. For an electron with orbital angular momentum quantum number l in a pure Coulomb potential,

$$V_{\mathrm{C}}(r) = I - \frac{Ze^2}{r} + \frac{l(l+1)\hbar^2}{2\mu r^2}, \tag{3.123}$$

the solutions of the radial Schrödinger equation have the energy eigenvalues (cf. (1.135), (2.8))

$$E_n = I - \frac{\mathcal{R}}{n^2}, \quad n = l+1,\ l+2, \ldots, \tag{3.124}$$

where $\mathcal{R}$ is the Rydberg energy; I is the continuum threshold. If the potential $V(r)$ differs from the pure Coulomb form (3.123) only through a shorter-ranged potential,

$$V(r) = V_{\mathrm{C}}(r) + V_{\mathrm{sr}}(r), \quad \lim_{r\to\infty} r^2 V_{\mathrm{sr}}(r) = 0, \tag{3.125}$$

then the energy eigenvalues can still be written in the form (3.124) [compare (3.5)] if we replace the quantum number n by an *effective quantum number*

$$n^* = n - \mu_n \,. \tag{3.126}$$

Explicitly:

$$E_n = I - \frac{\mathcal{R}}{(n^*)^2} = I - \frac{\mathcal{R}}{(n-\mu_n)2} \,. \tag{3.127}$$

The corrections μ_n are called *quantum defects* and the energies (3.127) form a *Rydberg series*.

The usefulness of the Rydberg formula (3.127) follows from the fact that the quantum defects μ_n depend only weakly on n for large n and converge to a finite value in the limit $n \to \infty$. That this is so can be understood most easily in the framework of the semiclassical approximation which was discussed in Sects 1.6.3 and 3.1.1.

For an energy $E < I$ the relevant action integral in the quantization condition (1.308) is, in the pure Coulomb case,

$$S_C(E) = 2\int_a^b \sqrt{2\mu(E - V_C(r))}\,\mathrm{d}r \,. \tag{3.128}$$

For $l = 0$ the inner classical turning point is the origin. The outer classical turning point b grows larger and larger as $E \to I$:

$$b(E) = \frac{Ze^2}{I - E}, \tag{3.129}$$

and

$$S_C(E) = 2\int_0^b \sqrt{2\mu\left(E - I + \frac{Ze^2}{r}\right)}\,\mathrm{d}r = 2\sqrt{2\mu Ze^2}\int_0^b \sqrt{\frac{1}{r} - \frac{1}{b}}\,\mathrm{d}r$$
$$= 2\pi\sqrt{\frac{b(E)\mu Ze^2}{2}}, \tag{3.130}$$

or, with (3.129),

$$E = I - \frac{\mu Z^2 e^4}{2}\left(\frac{2\pi}{S_C(E)}\right)^2 . \tag{3.131}$$

The quantization condition reads $S_C(E) = 2\pi\hbar(n + \mu_\phi/4)$ [cf. (1.308), (3.2)]. With the appropriate Maslov index μ_ϕ, which must obviously must be four in the present case, it yields the energy formula (3.124) with the correct Rydberg energy $\mathcal{R} = \mu Z^2 e^4/(2\hbar^2)$. The Maslov index four can be interpreted as sum of a contribution one, coming from the reflection at the outer classical turning point where the potential is smooth, and a contribution three, coming from the reflection at the attractive $1/r$ singularity at the origin [MK69].

For $l > 0$ the inner and outer classical turning points a and b are given by,

$$a(E) = \frac{1}{\kappa^2 a_Z} - \frac{1}{\kappa}\sqrt{\frac{1}{(\kappa a_Z)^2} - \gamma}, \quad b(E) = \frac{1}{\kappa^2 a_Z} + \frac{1}{\kappa}\sqrt{\frac{1}{(\kappa a_Z)^2} - \gamma}, \tag{3.132}$$

where we have introduced the abbreviations,

$$\kappa = \frac{1}{\hbar}\sqrt{2\mu(I-E)}\,, \quad a_Z = \frac{\mu Z e^2}{\hbar^2}, \quad \gamma = l(l+1)\,. \tag{3.133}$$

The action integral (3.128) can still be evaluated in closed form, and the quantization condition $S_C(E) = 2\pi\hbar(n+\mu_\phi/4)$ actually reproduces the exact energy eigenvalues if the Maslov index μ_ϕ is taken to be two and the centrifugal potential is subjected to the Langer modification, $l(l+1) \to (l+1/2)^2$; this trick even works for $l = 0$, where it corresponds to introducing an otherwise absent inverse-square potential $\hbar^2/(8\mu r^2)$. The fact that WKB quantization with Langer modification yields the exact bound state energies in a superposition of centrifugal and attractive Coulomb potentials [Lan37], is a coincidence which should not be given too much weight [Tro97]. For a Coulomb potential, the quantality function (1.298) becomes arbitrarily large as $r \to 0$, see (1.314) in Sect. 1.6.4. For a Coulomb potential, the WKB ansatz cannot be expected to be a good approximation for the wave function for small values of the radial coordinate.

Close to threshold $E \to I$ however, the energy dependence of the bound-state wave functions is dominated by the regime of large values of r, where the semiclassical approximation is increasingly reliable. The influence of an additional shorter-ranged potential on the spectrum near threshold can be found out by replacing the action S_C in the quantization condition by the full action

$$S(E) = 2\int_{a'}^{b} \sqrt{2\mu(E - V(r))}\,\mathrm{d}r\,. \tag{3.134}$$

This involves an additional contribution $S_{sr}(E)$ given by

$$\begin{aligned} S_{sr}(E) &= S(E) - S_C(E) \\ &= 2\int_{a'}^{b} \sqrt{2\mu(E - V(r))}\,\mathrm{d}r - 2\int_{a}^{b} \sqrt{2\mu(E - V_C(r)}\,\mathrm{d}r\,. \end{aligned} \tag{3.135}$$

The inner turning classical point is a in the absence and a' in the presence of the additional shorter-ranged potential; near threshold the outer classical turning point b is determined by the long-ranged Coulomb potential according to (3.129) in both

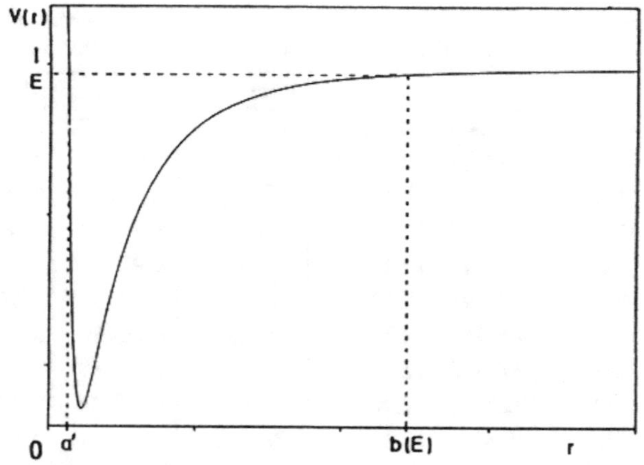

Fig. 3.8 Radial modified Coulomb potential (3.125) (including centrifugal potential) with inner classical turning point a' and outer classical turning point $b(E)$. The energy dependence of the outer classical turning point is given by (3.129) close to threshold

cases, cf. Fig. 3.8. The quantization condition now connects the integer n not to S_C but to $S_C + S_{sr}$; in place of (3.124) we now obtain the Rydberg formula (3.127) and the quantum defects are, in semiclassical approximation,

$$\mu_n^{sc} = \frac{1}{2\pi\hbar} S_{sr}(E_n) \,. \tag{3.136}$$

In the limit $E \to I$, $b \to \infty$ the diverging contributions to the two integrals in (3.135) cancel and their difference converges to a finite value.[1]

As an example for Rydberg series Table 3.11 lists the spectrum of one-electron excitations in potassium (see [Ris56]). In order to derive the quantum defects from the experimental term energies with sufficient accuracy, the corrections to the Rydberg energy which arise from the mass of the nucleus (cf. (2.12)) must be taken into account. With the nuclear masses from [WB77] and the Rydberg energy $\mathcal{R}_\infty$ from (2.9) we obtain the following result for the isotope K^{39} : $\mathcal{R} = \mathcal{R}_\infty \mu/m_e = 109735.771\,\text{cm}^{-1}$. The continuum threshold is at $I = 35009.77\,\text{cm}^{-1}$.

The quantum defects of the excited states in potassium are shown as functions of the energy relative to the continuum threshold, $E - I$, in Fig. 3.9. For each set of quantum numbers $S\,(=\frac{1}{2})$, L, J we obtain a Rydberg series of states nl in which the quantum defects depend only weakly on the principal quantum number n or energy E. The energy dependence in each series can be reproduced very accurately by a straight line. The quantum defects decrease rapidly with increasing angular

[1]These considerations still hold if the "shorter-ranged potential" falls off a little more slowly than required by (3.125), e.g. if it contains inverse-square contributions, $\lim_{r\to\infty} r^2 V_{sr}(r) = \text{const.} \neq 0$.

Table 3.11 Excitation energies E (in cm^{-1}), effective quantum numbers n^* and associated quantum defects $\mu_n = n - n^*$ for one-electron excitations in the potassium atom (from [Ris56])

Term	E	n^*	μ_n	Term	E	n^*	μ_n
$4s\,^2S_{1/2}$	0.00	1.77043	2.22957	$4p\,^2P_{1/2}$	12985.17	2.23213	1.76787
$5s\,^2S_{1/2}$	21026.58	2.80137	2.19863	$^2P_{3/2}$	13042.88	2.23506	1.76494
$6s\,^2S_{1/2}$	27450.69	3.81013	2.18987	$5p\,^2P_{1/2}$	24701.43	3.26272	1.73728
$7s\,^2S_{1/2}$	30274.28	4.81384	2.18616	$^2P_{3/2}$	24720.17	3.26569	1.73431
$8s\,^2S_{1/2}$	31765.37	5.81577	2.18423	$6p\,^2P_{1/2}$	28999.27	4.27286	1.72714
$9s\,^2S_{1/2}$	32648.35	6.81691	2.18309	$^2P_{3/2}$	29007.71	4.27587	1.72413
$10s\,^2S_{1/2}$	33214.22	7.81763	2.18237	$7p\,^2P_{1/2}$	31069.90	5.27756	1.72244
$11s\,^2S_{1/2}$	33598.54	8.81810	2.18190	$^2P_{3/2}$	31074.40	5.28058	1.71942
$12s\,^2S_{1/2}$	33817.46	9.81847	2.18153	$8p\,^2P_{1/2}$	32227.44	6.28015	1.71985
$13s\,^2S_{1/2}$	34072.22	10.8187	2.1813	$^2P_{3/2}$	32230.11	6.28316	1.71684
				$9p\,^2P_{1/2}$	32940.21	7.28174	1.71826
				$^2P_{3/2}$	32941.94	7.28478	1.71522
$3d\,^2D_{5/2}$	21534.70	2.85370	0.14630	$10p\,^2P_{1/2}$	33410.23	8.28279	1.71721
$^2D_{3/2}$	21537.00	2.85395	0.14605	$^2P_{3/2}$	33411.39	8.28579	1.71421
$4d\,^2D_{5/2}$	27397.10	3.79669	0.20331				
$^2D_{3/2}$	27398.14	3.79695	0.20305	$4f\,^2F$	28127.85	3.99318	0.00682
$5d\,^2D_{5/2}$	30185.24	4.76921	0.23079				
$^2D_{3/2}$	30185.74	4.76946	0.23054	$5f\,^2F$	30606.73	4.99227	0.00773
$6d\,^2D_{5/2}$	31695.89	5.75448	0.24552				
$^2D_{3/2}$	31696.15	5.75470	0.24530	$6f\,^2F$	31953.17	5.99177	0.00823
$7d\,^2D_{5/2}$	32598.30	6.74580	0.25420				
$^2D_{3/2}$	32598.43	6.74598	0.25402	$7f\,^2F$	32764.80	6.99148	0.00852
$8d\,^2D_{5/2}$	33178.12	7.74021	0.25979				
$^2D_{3/2}$	33178.23	7.74045	0.25955	$8f\,^2F$	33291.40	7.99127	0.00873
$9d\,^2D_{5/2}$	33572.06	8.73652	0.26348				
$^2D_{3/2}$	33572.11	8.73667	0.26333	$9f\,^2F$	33652.32	8.99109	0.00891
$10d\,^2D_{5/2}$	33851.55	9.73371	0.26629				
$^2D_{3/2}$	33851.59	9.73388	0.26612	$10f\,^2F$	33910.42	9.99094	0.00906
$11d\,^2D_{5/2}$	34056.94	10.7317	0.2683				
$^2D_{3/2}$	34057.00	10.7320	0.2680	$11f\,^2F$	34101.36	10.9909	0.0091

momentum l, because the inner region, where the full potential deviates from the pure Coulomb potential, is screened more and more effectively by the centrifugal potential (see Problem 3.1).

Because of their weak energy dependence, it is useful to complement the quantum defects $\mu_n = \mu(E_n)$ defined at the discrete energies E_n to a continuous *quantum defect function* $\mu(E)$ which describes the influence of the shorter-ranged potential V_{sr}. In the semiclassical approximation an extension of the formula (3.136) to arbitrary energies $E < I$ immediately yields an explicit formula for the quantum

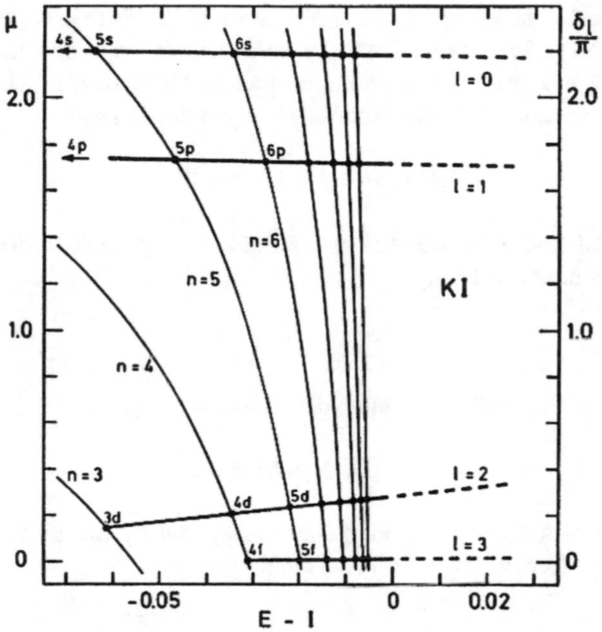

Fig. 3.9 Quantum defects (*filled circles*) of the ^2L Rydberg series in the potassium atom as functions of the energy relative to the continuum threshold (see also Table 3.11). The splitting within the individual doublets is not resolved in the figure. The almost *horizontal straight lines* are the quantum defect functions $\mu(E)$; their intersections with the set of curves (3.142) define the energies [in atomic units] of the bound states. At the continuum threshold $E = I$ the quantum defects match smoothly to the asymptotic phase shifts divided by π, which are shown as *dashed lines* in the figure. (The Roman numeral I behind the element symbol "K" indicates the neutral potassium atom. In this notation potassium ions with a single positive charge are written K II, doubly charged ions are written K III, etc.)

defect function:

$$\mu^{\text{sc}}(E) = \frac{1}{2\pi\hbar} S_{\text{sr}}(E) \,. \tag{3.137}$$

An exact definition of the quantum defect function (beyond the semiclassical approximation) can be formulated by asymptotically matching the solutions of the radial Schrödinger equation to linear combinations of Whittaker functions [Sea83]. In practice it is customary to approximate the weakly energy-dependent function $\mu(E)$ by fitting a polynomial in $E - I$ through the discrete values given by the quantum defects, $\mu(E_n) = \mu_n$.

In the bound-state region $E < I$ we can introduce the variable ν, defined by

$$\nu(E) = \sqrt{\frac{\mathcal{R}}{I - E}}, \quad E = I - \frac{\mathcal{R}}{\nu^2}, \tag{3.138}$$

as a substitute for the energy variable E. The variable ν is the *continuous effective quantum number*. In a pure Coulomb potential the condition that the energy corresponding to a given value of the continuous effective quantum number ν is one of the eigenvalues (3.124) of the Schrödinger equation reads

$$\nu(E) = n = l+1, \quad l+2, \dots . \tag{3.139}$$

For a modified Coulomb potential of the form (3.125) the condition for a bound state is, according to (3.127),

$$\nu(E) + \mu_n = n, \tag{3.140}$$

or, expressed in terms of the quantum defect function $\mu(E)$,

$$\nu(E) + \mu(E) = n . \tag{3.141}$$

Thus the energies E_n of the bound states are given by the intersections of the quantum defect function with the set of curves

$$\mu^{(n)} = n - \nu(E) = n - \sqrt{\frac{\mathcal{R}}{I - E}} \tag{3.142}$$

in the μ-E plane, as shown in Fig. 3.9.

The technology of high resolution laserspectroscopy has made the observation of very highly excited Rydberg states possible. The left-hand part of Fig. 3.10 shows an observed photoabsorption spectrum (cf. (2.200) in Sect. 2.4.4) with lines up to $n = 310$ in the $6snd\,{}^1D_2$ Rydberg series in barium. The right-hand part of the figure shows the energy differences $E_{n+l} - E_n$ as a function of the effective quantum number n^* on a logarithmic scale. The straight line shows the proportionality to $(n^*)^{-3}$ following from the Rydberg formula (3.127). Apart from resolving such small energy differences ($\approx 10^{-8}$ atomic units), it is a remarkable achievement that measurements involving such highly excited Rydberg atoms are possible at all. The spatial extension of a Rydberg atom grows quadratically with the principal quantum number n (see Problem 1.3) and exceeds 10^5 Bohr radii for $n \approx 300$, this means that the Rydberg atoms observed in Fig. 3.10 are almost one hundredth of a millimetre in size! In further measurements, states in this Rydberg series with principal quantum numbers $n > 500$ were identified [NR87].

3.2.2 Seaton's Theorem, One-Channel Quantum Defect Theory

Below the continuum threshold, the shorter-ranged deviation of the full potential from a pure Coulomb potential is described by the quantum defects or the quantum

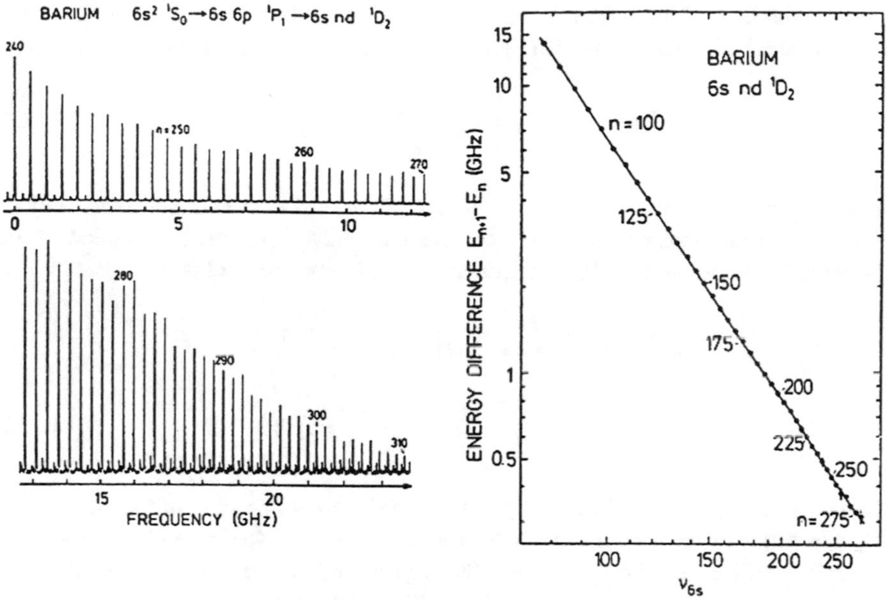

Fig. 3.10 The left-hand part shows photoabsorption cross sections with final states in the $6s\,nd\,^1D_2$ Rydberg series in barium. The right-hand part shows the energy differences of successive Rydberg states as a function of the effective quantum number n_f^* on a logarithmic scale (every fifth energy difference is plotted). The *straight line* shows the proportionality $(n_f^*)^{-3}$ following from the Rydberg formula (3.127) (From [NJ88])

defect function. Above the continuum threshold the shorter-ranged deviation from the Coulomb potential manifests itself in the asymptotic phase shifts (cf. Sect. 1.3.2, (1.122)). At the continuum threshold, the quantum defects are related to the phase shifts, because the appropriately normalized solutions of the radial Schrödinger equation in the limit $n \to \infty$ (i.e. $E \to I$ from below) and in the limit $E \to I$ (from above) converge to the same well defined solution at $E = I$, just as in the pure Coulomb case (see (1.153)). The quantitative connection between the quantum defects and the phase shifts at threshold is given by *Seaton's Theorem*:

$$\lim_{n\to\infty} \mu_n = \mu(E = I) = \frac{1}{\pi} \lim_{E\to I} \delta(E) . \tag{3.143}$$

The factor $1/\pi$ appears on the right-hand side of (3.143), because a shift of one half-wave in the asymptotic part of a wave function corresponds to a change of unity in the effective quantum number and the quantum defect below threshold, while it corresponds to a change of π in the phase shift above threshold.

The relation (3.143) can immediately be verified in the framework of the semiclassical approximation. There the radial wave function has the form (1.289)

$$\phi(r) \propto p(r)^{-1/2} \exp\left[\frac{\mathrm{i}}{\hbar}\int^{r} p(r')dr'\right], \tag{3.144}$$

and the phase of the wave is just the action integral in the exponent divided by $\hbar$. The asymptotic phase shift caused by a shorter-ranged potential $V_{\rm sr}$ added to the pure Coulomb potential is the difference of the phases with and without $V_{\rm sr}$:

$$\begin{aligned}\delta^{\rm sc}(E) = & \frac{1}{\hbar}\int_{a'}^{r} \sqrt{2\mu(E - V_{\rm C}(r') - V_{\rm sr}(r'))}\,{\rm d}r' \\ & -\frac{1}{\hbar}\int_{a}^{r} \sqrt{2\mu(E - V_{\rm C}(r'))}\,{\rm d}r' .\end{aligned} \tag{3.145}$$

(Again, the inner classical turning point is a in the absence and a' in the presence of $V_{\rm sr}$.) Because of the short range of $V_{\rm sr}$ the difference (3.145) becomes independent of r for sufficiently large r. Thus the asymptotic phase shift in semiclassical approximation is just $1/(2\hbar)$ times the additional contribution to the action due to the shorter-ranged potential (cf. (3.135)). In the limit $E \to I$ this is precisely π times the right-hand side of (3.136) in the limit $b \to \infty$ corresponding to $n \to \infty$.

The close connection between the "quasi-continuum" of the bound states just below threshold and the genuine continuum above threshold is characteristic for long-ranged Coulomb-type potentials. The wave functions consist mainly of a large number of oscillations far out in the $1/r$ potential, be it a large finite number just below threshold or an infinite number above threshold. The main influence of an additional shorter-ranged potential $V_{\rm sr}$ is to shift these outer oscillations, and this manifests itself in the phase shift above threshold and in the quantum defect function and the quantum defects below threshold.

The two mathematically similar but physically different situations just below and just above the continuum threshold can be summarized in one uniform equation of *one-channel quantum defect theory* (QDT),

$$\tan\left[\pi(\nu + \mu)\right] = 0 . \tag{3.146}$$

Here $\mu(E)$ is the function which describes the physical effects of the additional shorter-ranged potential: below threshold μ is the quantum defect function described above, and above threshold $\mu(E)$ is the asymptotic phase shift $\delta(E)$ divided by π. Below threshold $\nu(E)$ is a variable corresponding to the energy, namely the continuous effective quantum number (3.138). Above threshold, ν stands for the asymptotic phase shift divided by $-\pi$:

$$\nu(E) = \sqrt{\frac{\mathcal{R}}{I - E}} \quad \text{for}\, E < I, \quad \nu(E) = -\frac{1}{\pi}\delta(E) \quad \text{for}\ E \geq I . \tag{3.147}$$

With the identification (3.147) the QDT equation (3.146) above threshold is, for the present one-channel case, a trivial identity $\delta(E) = \delta(E)$. Below threshold (3.146) simply means that $\nu(E) + \mu(E)$ must be an integer n—this is just the condition (3.141) for the existence of a bound state.

Just as the asymptotic phase shifts are defined only to within an additive multiple of π, the quantum defects and the quantum defect function are only unique modulo unity. The particular choice of quantum defects or the quantum defect function determines where to start counting in a given Rydberg series.

3.2.3 *Photoabsorption and Photoionization*

The cross sections (2.200) for photoabsorption and (2.202) for photoionization are given, as discussed in Sect. 2.4.6, by the respective oscillator strengths $f_{\rm fi}^{(i)}$ and $df_{E\rm i}^{(i)}/dE$. The relation between the cross sections and the oscillator strengths depends on the polarization of the incoming light and on the orientation, i.e. on the azimuthal quantum numbers, of the initial and final atomic states. In order to get rid of these geometric dependences it is convenient to define *mean oscillator strengths,* which is quite easily done for one-electron atoms with wave functions of the form (1.74).

For initial and final state wave functions

$$\Phi_{n_{\rm i},l_{\rm i},m_{\rm i}}(\boldsymbol{r}) = \frac{\phi_{n_{\rm i},l_{\rm i}}(r)}{r} Y_{l_{\rm i},m_{\rm i}}(\theta,\phi),$$

$$\Phi_{n_{\rm f},l_{\rm f},m_{\rm f}}(\boldsymbol{r}) = \frac{\phi_{n_{\rm f},l_{\rm f}}(r)}{r} Y_{l_{\rm f},m_{\rm f}}(\theta,\phi), \tag{3.148}$$

we define the mean oscillator strength for transitions from the initial multiplet $n_{\rm i}, l_{\rm i}$ to the final state multiplet $n_{\rm f}, l_{\rm f}$ by averaging over the initial states and summing over the final states (cf. last paragraph in Sect. 2.4.4) as well as averaging over the three spatial directions x, y, z :

$$\begin{aligned} \bar{f}_{n_{\rm f}l_{\rm f},n_{\rm i}l_{\rm i}} &= \frac{1}{2l_{\rm i}+1} \sum_{m_{\rm i}=-l_{\rm i}}^{+l_{\rm i}} \sum_{m_{\rm f}=-l_{\rm f}}^{+l_{\rm f}} \frac{1}{3} \sum_{i=1}^{3} f^{(i)}_{n_{\rm f}l_{\rm f}m_{\rm f},n_{\rm i}l_{\rm i}m_{\rm i}} \\ &= \frac{2\mu}{3\hbar}\,\omega \sum_{m_{\rm f}=-l_{\rm f}}^{+l_{\rm f}} \frac{1}{2l_{\rm i}+1} \sum_{m_{\rm i}=-l_{\rm i}}^{+l_{\rm i}} |\langle \Phi_{n_{\rm f},l_{\rm f},m_{\rm f}} |\boldsymbol{r}|\, \Phi_{n_{\rm i},l_{\rm i},m_{\rm i}}\rangle|^2 . \end{aligned} \tag{3.149}$$

We can rewrite the absolute square in (3.149) in spherical components (2.204):

$$|\langle \Phi_{n_{\rm f},l_{\rm f},m_{\rm f}}|\boldsymbol{r}|\Phi_{n_{\rm i},l_{\rm i},m_{\rm i}}\rangle|^2 = \sum_{\nu=-1}^{+1} |\langle \Phi_{n_{\rm f},l_{\rm f},m_{\rm f}}|r^{(\nu)}|\Phi_{n_{\rm i},l_{\rm i},m_{\rm i}}\rangle|^2. \tag{3.150}$$

With the expression (2.208) for the matrix elements of the spherical components of $\boldsymbol{r}$ we have

$$\begin{aligned}
&\sum_{m_\mathrm{i}=-l_\mathrm{i}}^{l_\mathrm{i}} |\langle \Phi_{n_\mathrm{f},l_\mathrm{f},m_\mathrm{f}}|\,\boldsymbol{r}\,|\Phi_{n_\mathrm{i},l_\mathrm{i},m_\mathrm{i}}\rangle|^2 \\
&= \sum_{m_\mathrm{i}=-l_\mathrm{i}}^{l_\mathrm{i}} \left(\int_0^\infty \phi_{n_\mathrm{f},l_\mathrm{f}}(r)\, r\, \phi_{n_\mathrm{i},l_\mathrm{i}}(r)\,\mathrm{d}r\right)^2 F(l_\mathrm{f},l_\mathrm{i})^2 \\
&\quad\times \sum_{\nu=-1}^{+1} \langle l_\mathrm{f}, m_\mathrm{f}\,|1,\nu,l_\mathrm{i},m_\mathrm{i}\rangle^2 \\
&= \left(\int_0^\infty \phi_{n_\mathrm{f},l_\mathrm{f}}(r)\, r\, \phi_{n_\mathrm{i},l_\mathrm{i}}(r)\,\mathrm{d}r\right)^2 \frac{l_>}{2l_\mathrm{f}+1}\,. \qquad (3.151)
\end{aligned}$$

Here we have assumed that l_f is either $l_\mathrm{i}+1$ or $l_\mathrm{i}-1$ and used the fact that the sum of the squares of the Clebsch-Gordan coefficients over m_i and ν gives unity [Edm60]. For the factors $F(l_\mathrm{f}, l_i)$ we inserted the explicit expression (2.210); $l_>$ is the larger of the two angular momentum quantum numbers l_i and l_f. Since the expression (3.151) no longer depends on the azimuthal quantum number m_f of the final state, the factor $1/(2l_\mathrm{f}+1)$ cancels with the summation over m_f in (3.149), and the expression for the mean oscillator strengths is simplified to

$$\bar{f}_{n_\mathrm{f}l_\mathrm{f},n_\mathrm{i}l_\mathrm{i}} = \frac{2\mu}{3\hbar}\,\omega\,\frac{l_>}{2l_\mathrm{i}+1}\left(\int_0^\infty \phi_{n_\mathrm{f},l_\mathrm{f}}(r)\, r\, \phi_{n_\mathrm{i},l_\mathrm{i}}(r)\,\mathrm{d}r\right)^2 . \qquad (3.152)$$

The frequency $\omega = (\varepsilon_\mathrm{f}-\varepsilon_\mathrm{i})/\hbar$, and hence also the oscillator strengths, are positive for $\varepsilon_\mathrm{f} > \varepsilon_\mathrm{i}$ (absorption) and negative for $\varepsilon_\mathrm{f} < \varepsilon_\mathrm{i}$ (emission). From (3.152) it is easy to see that the mean oscillator strengths fulfill the relation

$$(2l_\mathrm{i}+1)\bar{f}_{n_\mathrm{f}l_\mathrm{f},n_\mathrm{i}l_\mathrm{i}} + (2l_\mathrm{f}+1)\bar{f}_{n_\mathrm{i}l_\mathrm{i},n_\mathrm{f}l_\mathrm{f}} = 0\,. \qquad (3.153)$$

For the mean oscillator strengths, where the sum over the azimuthal quantum numbers is already contained in their definition, we obtain a sum rule of the form (see also (3.161) below)

$$\sum_{n_\mathrm{f},l_\mathrm{f}} \bar{f}_{n_\mathrm{f}l_\mathrm{f},n_\mathrm{i}l_\mathrm{i}} = 1\,. \qquad (3.154)$$

The sum may further be decomposed into contributions from the two possible angular momentum quantum numbers in the final states, $l_\mathrm{f} = l_\mathrm{i} + 1$ and $l_\mathrm{f} = l_\mathrm{i} - 1$, yielding [BS57] (see also (3.162) below):

$$\sum_{n_\mathrm{f}} \bar{f}_{n_\mathrm{f} l_\mathrm{i}+1, n_\mathrm{i} l_\mathrm{i}} = \frac{1}{3} \frac{(l_\mathrm{i}+1)(2l_\mathrm{i}+3)}{2l_\mathrm{i}+1},$$

$$\sum_{n_\mathrm{f}} \bar{f}_{n_\mathrm{f} l_\mathrm{i}-1, n_\mathrm{i} l_\mathrm{i}} = -\frac{1}{3} \frac{l_\mathrm{i}(2l_\mathrm{i}-1)}{2l_\mathrm{i}+1}. \tag{3.155}$$

The matrix elements (3.150) contain no spin dependence and allow no spin changing transitions. If, however, the final state multiplets with good total angular momentum quantum number j, which are split due to the effects of spin-orbit coupling, can be resolved in the experiment, then the (mean) oscillator strength for given final state quantum numbers n_f and l_f will be distributed among the various j multiplets in proportion to their multiplicity $2j+1$. In an $n_\mathrm{i}\,{}^2S_{1/2} \to n_\mathrm{f}\,{}^2P_j$ transition, for example, the transition to the $j = 3/2$ states $(2j+1 = 4)$ is twice as strong altogether as the transition to the $j = 1/2$ states $(2j+1 = 2)$.

The cross section for the absorption of photons of arbitrary polarization by a one-electron atom of undetermined orientation is a series of sharp spikes, whose strength is given by the mean oscillator strength (3.152) (multiplied by the constant factor $2\pi^2 e^2\hbar/(\mu c)$ from (2.218)). Only comparatively small distances r contribute in the radial integral in (3.152), because the initial wave function $\phi_{n_\mathrm{i},l_\mathrm{i}}(r)$ vanishes for large r. With increasing principal quantum number n_f of the final states the amplitudes of the radial wave functions $\phi_{n_\mathrm{f},l_\mathrm{f}}(r)$ of the final states (which are normalized to unity) become smaller and smaller in the inner region, just as for the pure Coulomb functions in Fig. 1.4. Hence the oscillator strengths also become smaller and smaller with increasing principal quantum number n_f of the final states.

In order to expose the dependence of the cross sections and oscillator strengths on the principal quantum number for large principal quantum numbers, we renormalize the radial wave functions of the final states in analogy to (1.140) so that the square of their norm becomes inversely proportional to the separation $2\mathcal{R}/\left(n_\mathrm{f}^*\right)^3$ between successive energy eigenvalues:

$$\phi^\mathrm{E}_{n_\mathrm{f},l_\mathrm{f}} = \sqrt{\frac{\left(n_\mathrm{f}^*\right)^3}{2\mathcal{R}}}\, \phi_{n_\mathrm{f},l_\mathrm{f}}. \tag{3.156}$$

Here n_f^* are the effective quantum numbers $n_\mathrm{f} - \mu_{n_\mathrm{f}}$ which determine the energies of the final states according to (3.127). In the limit $n_\mathrm{f}^* \to \infty$ corresponding to $E \to I$ (from below), the wave functions (3.156) merge smoothly into the energy normalized continuum wave functions ϕ_{E,l_f} for $E \geq I$. With (3.156) we can rewrite the (mean) oscillator strengths (3.152) as

$$\bar{f}_{n_\mathrm{f} l_\mathrm{f}, n_\mathrm{i} l_\mathrm{i}} = \frac{2\mathcal{R}}{(n_\mathrm{f}^*)^3} \frac{2\mu}{3\hbar}\, \omega\, \frac{l_>}{2l_\mathrm{i}+1} \left(\int_0^\infty \phi^\mathrm{E}_{n_\mathrm{f},l_\mathrm{f}}(r)\, r \phi_{n_\mathrm{i},l_\mathrm{i}}(r)\, \mathrm{d}r \right)^2, \tag{3.157}$$

where the radial matrix element now converges to a finite number at the continuum threshold $E = I$.

The representation (3.157) of the discrete oscillator strengths facilitates their smooth matching to the photoabsorption cross sections and oscillator strengths to final states in the continuum. We define mean oscillator strengths in analogy to (3.149) as

$$\frac{\mathrm{d}\bar{f}_{El_\mathrm{f},n_\mathrm{i}l_\mathrm{i}}}{\mathrm{d}E} = \frac{1}{2l_\mathrm{i}+1}\sum_{m_\mathrm{i}=-l_\mathrm{i}}^{+l_\mathrm{i}}\sum_{m_\mathrm{f}=-l_\mathrm{f}}^{+l_\mathrm{f}}\frac{1}{3}\sum_{i=1}^{3}\frac{\mathrm{d}f^{(i)}_{El_\mathrm{f}m_\mathrm{f},n_\mathrm{i}l_\mathrm{i}m_\mathrm{i}}}{\mathrm{d}E}$$

$$= \frac{2\mu}{3\hbar}\,\omega\sum_{m_\mathrm{f}=-l_\mathrm{f}}^{+l_\mathrm{f}}\frac{1}{2l_\mathrm{i}+1}\sum_{m_\mathrm{i}=-l_\mathrm{i}}^{+l_\mathrm{i}}|\langle\Phi_{E,l_\mathrm{f},m_\mathrm{f}}|\boldsymbol{r}|\,\Phi_{n_\mathrm{i},l_\mathrm{i},m_\mathrm{i}}\rangle|^2, \tag{3.158}$$

and the same manipulations which led from (3.149) to (3.152) yield

$$\frac{\mathrm{d}\bar{f}_{n_\mathrm{i}l_\mathrm{i},El_\mathrm{f}}}{\mathrm{d}E} = \frac{2\mu}{3\hbar}\,\omega\,\frac{1_>}{2l_\mathrm{i}+1}\left(\int_0^\infty \phi_{E,l_\mathrm{f}}(r)\,r\,\phi_{n_\mathrm{i},l_\mathrm{i}}(r)\,\mathrm{d}r\right)^2. \tag{3.159}$$

The cross section for the photoionization of atoms of unknown orientation by photons of arbitrary polarization is given by the mean oscillator strength (3.159) (multiplied by the constant factor $2\pi^2e^2\hbar/(\mu c)$ from (2.223)). From (3.157) we see that the discrete oscillator strengths multiplied by the density $(n_\mathrm{f}^*)^3/(2\mathcal{R})$ of the final states merge smoothly into the continuous form (3.159) at the threshold $E = I$:

$$\lim_{n_\mathrm{f}\to\infty}\frac{(n_\mathrm{f}^*)^3}{2\mathcal{R}}\bar{f}_{n_\mathrm{f}l_\mathrm{f},n_\mathrm{i}l_\mathrm{i}} = \lim_{E\to I}\frac{\mathrm{d}\bar{f}_{El_\mathrm{f},n_\mathrm{i}l_\mathrm{i}}}{\mathrm{d}E}\,. \tag{3.160}$$

An example of the smooth transition from the discrete line spectrum below the continuum threshold to the continuous photoionization spectrum above threshold is shown in Fig. 3.11 for the case of sodium. The left-hand part of the figure shows the discrete oscillator strengths (3.157) multiplied by $(n_\mathrm{f}^*)^3/(2\mathcal{R})$ ($2\mathcal{R}$ ist unity in atomic units), and the right-hand part shows the photoionization cross sections divided by $2\pi^2e^2\hbar/(\mu c)$. The pronounced minimum at an energy around 0.05 atomic units above the threshold is attributed to a zero with an accompanying sign change in the radial matrix element in (3.159). In the discrete part of the spectrum the separation between successive lines in the near-threshold region is just $2\mathcal{R}/(n_\mathrm{f}^*)^3$, so that the areas under the dashed lines correspond to the original oscillator strengths.

The transitions to the continuum must of course be taken into account when formulating sum rules. Thus (3.154) correctly reads

$$\sum_{n_\mathrm{f},l_\mathrm{f}}\bar{f}_{n_\mathrm{f}l_\mathrm{f},n_\mathrm{i}l_\mathrm{i}} + \int_I^\infty\frac{\mathrm{d}\bar{f}_{El_\mathrm{f},n_\mathrm{i}l_\mathrm{i}}}{\mathrm{d}E}\,\mathrm{d}E = 1\,. \tag{3.161}$$

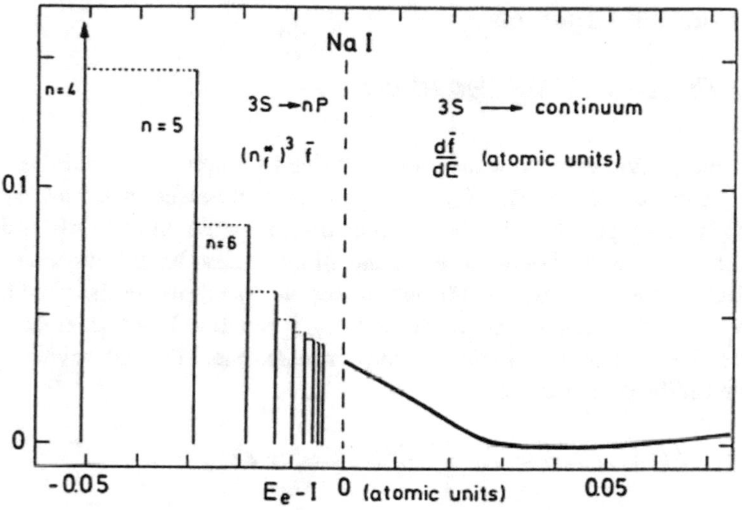

Fig. 3.11 Measured oscillator strengths for $3S \to nP$ transitions in sodium. The discrete oscillator strengths (from [KB32]) are multiplied by the respective factors $(n_f^*)^3$. Near threshold the areas under the *dashed lines* correspond to the original oscillator strengths. The continuous oscillator strengths above threshold are the photoionization cross sections from [HC67] divided by the factor $2\pi^2 e^2 \hbar/(\mu c)$

The correct form of (3.155) is

$$\sum_{n_f} \bar{f}_{n_f l_i+1, n_i l_i} + \int_I^\infty \frac{d\bar{f}_{E\, l_i+1, n_i l_i}}{dE}\, dE = \frac{1}{3}\frac{(l_i+1)(2l_i+3)}{2l_i+1},$$

$$\sum_{n_f} \bar{f}_{n_f l_i-1, n_i l_i} + \int_I^\infty \frac{d\bar{f}_{E\, l_i-1, n_i l_i}}{dE}\, dE = -\frac{1}{3}\frac{l_i(2l_i-1)}{2l_i+1}\,. \tag{3.162}$$

Finally it should be mentioned, that the derivation of the sum rules in Sect. 2.4.6 relied on a commutation relation of the form (2.185), in particular on the commuting of the dipole operator and the potential energy. In a one-electron atom this is only fulfilled if the potential energy is a local function of the displacement variable. For nonlocal one-body potentials as they occur in the Hartree-Fock method (see Sect. 2.3.1), the Thomas-Reiche-Kuhn sum rule can, strictly speaking, not be applied with $N = 1$. We find a way out of this problem by realizing that the nonlocality in the Hartree-Fock potential originates from the Pauli principle, which requires e.g. the wave function of the valence electron in an alkali atom to be orthogonal to the occupied single-particle states in the lower closed shells. The sum rule with $N = 1$ holds approximately, if we include fictitious transitions to the states forbidden by the Pauli principle. Since $\varepsilon_f < \varepsilon_i$ for such transitions and hence $\bar{f}_{n_f l_f, n_i l_i}$ is negative, the sum of the oscillator strengths for the allowed transitions becomes larger.

3.3 Coupled Channels

3.3.1 *Close-Coupling Equations*

The simple picture of one electron in a modified Coulomb potential described in Sect. 3.2 can be transferred, to a large part, to many-electron atoms, when one electron is in a highly excited loosely bound state, while all other electrons form a more or less tightly bound *core*. In the simplest case we can assume that the core electrons are not excited and only affect the spectrum via their influence on the mean single-particle potential for the "outer" electron. In a further step we can allow a finite number of excitations of the core electrons. The total wave function of the atom (or ion) then has the form

$$\Psi(x_1,\dots,x_N) = \hat{\mathcal{A}}_1 \sum_j \psi_{\text{int}}^{(j)}(m_{s_1},x_2,\dots x_N)\psi_j(\boldsymbol{r}_1), \tag{3.163}$$

where the summation index j labels the various internal states of the core whose wave functions $\psi_{\text{int}}^{(j)}$ each define a "channel" and depend on the *internal variables*. The internal variables are all variables except the spatial coordinate $\boldsymbol{r}_1$ of the outer electron; we are counting the spin of the outer electron as one of the internal variables. In each channel j, $\psi_j(\boldsymbol{r}_1)$ is the associated *channel wave function*; it is simply a one-electron wave function for the outer electron. We shall later include the angular coordinates of the outer electron among the internal coordinates, so the channel wave functions will depend only on the radial coordinate of the outer electron. To begin with however, we shall derive the equations of motion for the full orbital one-electron wave functions $\psi_j(\boldsymbol{r}_1)$.

We assume that the wave functions of the core are antisymmetric with respect to exchange amongst the particle labels 2 to N; the total wave function is made fully antisymmetric by the *residual antisymmetrizer* which takes care of the exchange of the outer electron with the electrons $2,\dots N$ of the core:

$$\hat{\mathcal{A}}_1 = \mathbf{1} - \sum_{\nu=2}^{N} \hat{P}_{1\leftrightarrow\nu} \,. \tag{3.164}$$

In order to derive equations of motion for the channel wave functions, we rewrite $\psi_j(\boldsymbol{r}_1)$ as $\int \mathrm{d}\boldsymbol{r}'\delta(\boldsymbol{r}_1-\boldsymbol{r}')\psi_j(\boldsymbol{r}')$ in (3.163) giving

$$\Psi(x_1,\dots,x_N) = \sum_j \int \mathrm{d}\boldsymbol{r}'\, \Phi_j(\boldsymbol{r}')\, \psi_j(\boldsymbol{r}') \tag{3.165}$$

with

$$\Phi_j(\boldsymbol{r}') = \hat{\mathcal{A}}_1 \left\{ \psi_{\text{int}}^{(j)}\,(m_{s_1},\, x_2,\dots x_N)\,\delta\left(\boldsymbol{r}_1-\boldsymbol{r}'\right)\right\}.$$

Equation (3.165) represents an expansion of the total wave function in the basis of states $\Phi_j(\boldsymbol{r}')$ which are labelled by the discrete index j numbering the channels and the continuous vector parameter $\boldsymbol{r}'$ corresponding to the position of the outer electron. Due to the effect of the residual antisymmetrizer (3.164), this basis is not orthogonal, i.e. the overlap matrix

$$\begin{aligned}\langle\Phi_i(\boldsymbol{r})|\Phi_j(\boldsymbol{r}')\rangle &= \langle\psi_{\rm int}^{(i)}\delta(\boldsymbol{r}_1-\boldsymbol{r})|\hat{\mathcal{A}}_1^\dagger\hat{\mathcal{A}}_1|\psi_{\rm int}^{(j)}\delta(\boldsymbol{r}_1-\boldsymbol{r}')\rangle\\ &= N\langle\psi_{\rm int}^{(i)}\delta(\boldsymbol{r}_1-\boldsymbol{r})|\hat{\mathcal{A}}_1|\psi_{\rm int}^{(j)}\delta(\boldsymbol{r}_1-\boldsymbol{r}')\rangle\end{aligned}\tag{3.166}$$

need not necessarily vanish for $i\neq j$, or $\boldsymbol{r}\neq\boldsymbol{r}'$. For the second line of (3.166) we have used the property $\hat{\mathcal{A}}_1^\dagger\hat{\mathcal{A}}_1 = N\hat{\mathcal{A}}_1$ of the residual antisymmetrizer (3.164) (which is defined without a normalization factor—cf. (2.61), (2.64)).

Diagonalizing the Hamiltonian $\hat{H}$ in the basis (3.165) leads to an equation of the type (1.277), generalized to the case of discrete and continuous basis state labels:

$$\sum_j\int d\boldsymbol{r}'\,(H_{i,j}(\boldsymbol{r},\boldsymbol{r}') - E\,A_{i,j}(\boldsymbol{r},\boldsymbol{r}'))\,\psi(\boldsymbol{r}') = 0,\tag{3.167}$$

with

$$\begin{aligned}H_{i,j}(\boldsymbol{r},\boldsymbol{r}') &= \langle\psi_{\rm int}^{(i)}\delta(\boldsymbol{r}_1-\boldsymbol{r})|\hat{H}\hat{\mathcal{A}}_1|\psi_{\rm int}^{(j)}\delta(\boldsymbol{r}_1-\boldsymbol{r}')\rangle,\\ A_{i,j}(\boldsymbol{r},\boldsymbol{r}') &= \langle\psi_{\rm int}^{(i)}\delta(\boldsymbol{r}_1-\boldsymbol{r})|\hat{\mathcal{A}}_1|\psi_{\rm int}^{(j)}\delta(\boldsymbol{r}_1-\boldsymbol{r}')\rangle\,.\end{aligned}\tag{3.168}$$

In (3.167) we have cancelled the common factor N appearing in front of the matrix elements, cf. (3.166).

Equation (3.167) represents a set of coupled-channel equations

$$\hat{H}_{i,i}\psi_i + \sum_{j\neq i}\hat{H}_{i,j}\psi_j = E\left(\hat{A}_{i,i}\psi_i + \sum_{j\neq i}\hat{A}_{i,j}\psi_j\right),\tag{3.169}$$

for the channel wave functions ψ_j, and the Hamiltonian and overlap operators $\hat{H}_{i,j}$ and $\hat{A}_{i,j}$ are integral operators defined by the integral kernels (3.168). These equations look a little more complicated than the coupled-channel equations (1.207) which were derived under quite general assumptions in Sect. 1.5.1. This is due to the non-orthogonality of the basis states which results from the fact that our present ansatz (3.163) already takes into account the indistinguishability of all electrons and collects all *equivalent channels*, which differ only by rearrangement of the electron labels, into one channel.

The overlap kernels $A_{i,j}(\boldsymbol{r},\boldsymbol{r}')$ can be decomposed into a *direct part* originating from the **1** in the residual antisymmetrizer (3.164), and an *exchange part* $K_{i,j}(\boldsymbol{r},\boldsymbol{r}')$ coming from the genuine permutations in (3.164). Because of the orthonormality of the core states the direct part of the overlap kernels is simply a Kronecker symbol in the channel indices and a delta function in the spatial coordinate, but the exchange parts are genuinely nonlocal:

$$A_{i,j}(\boldsymbol{r},\boldsymbol{r}') = \delta_{i,j}\delta(\boldsymbol{r}-\boldsymbol{r}') - K_{i,j}(\boldsymbol{r},\boldsymbol{r}'),$$
$$K_{i,j}(\boldsymbol{r},\boldsymbol{r}') = \sum_{\nu=2}^{N} \langle \psi_{\rm int}^{(i)}\delta(\boldsymbol{r}_1-\boldsymbol{r})|\hat{P}_{1\leftrightarrow\nu}|\psi_{\rm int}^{(j)}\delta(\boldsymbol{r}_1-\boldsymbol{r}')\rangle\,. \quad (3.170)$$

In each contribution to the exchange part, the spatial coordinate $\boldsymbol{r}_1$ of the outer electron is exchanged with one of the coordinates $\boldsymbol{r}_2,\dots\boldsymbol{r}_N$ of the core electrons, and the matrix element vanishes for large $|\boldsymbol{r}'|$ (or large $|\boldsymbol{r}|$) because of the exponential decay of the wave function of the bound core state $\psi_{\rm int}^{(j)}$ (or $\psi_{\rm int}^{(i)}$). For large separations the overlap operator thus becomes the unit operator.

Similar considerations apply for the Hamiltonian operators $\hat{H}_{i,j}$. They too can be decomposed into a direct part $\hat{H}_{\rm d}$ arising from the **1** in the residual anti-symmetrizer (3.164), and an exchange part $\hat{H}_{\rm ex}$ described by an integral kernel $\hat{H}_{{\rm ex}\,i,j}(\boldsymbol{r},\boldsymbol{r}')$, which is short ranged and nonlocal just as in the case of the overlap operators.

In order to expose the structure of the direct part $\hat{H}_{\rm d}$ of the one-body Hamiltonian, it is useful to decompose the N-electron Hamiltonian (2.53) as follows:

$$\hat{H} = \hat{H}_1 + \hat{H}_{2-N} + \hat{H}_W\,. \quad (3.171)$$

$\hat{H}_1$ acts only on functions of $\boldsymbol{r}_1$ and $\hat{H}_{2-N}$ acts only on functions of the remaining, the internal, variables:

$$\hat{H}_1 = \frac{\hat{\boldsymbol{p}}_1^2}{2\mu} + V(\boldsymbol{r}_1),$$
$$\hat{H}_{2-N} = \sum_{\nu=2}^{N}\frac{\hat{\boldsymbol{p}}_\nu^2}{2\mu} + \sum_{\nu=2}^{N} V(\boldsymbol{r}_\nu) + \sum_{1<\nu<\nu'} \hat{W}(\nu,\nu')\,. \quad (3.172)$$

$\boldsymbol{r}_1$ is coupled to the other degrees of freedom by the interaction term

$$\hat{H}_W = \sum_{\nu=2}^{N} \hat{W}(1,\nu)\,. \quad (3.173)$$

Since $\hat{H}_1$ does not act on the internal wave functions, the integration over the internal variables in $\langle \psi_{\rm int}^{(i)}\delta(\boldsymbol{r}-\boldsymbol{r}_1)|\hat{H}_1|\psi_{\rm int}^{(j)}\delta(\boldsymbol{r}'-\boldsymbol{r}_1)\rangle$ yields a Kronecker symbol in the channel indices and the diagonal matrix elements are simply the one-body matrix elements of the kinetic energy $\hat{\boldsymbol{p}}_1^2/(2\mu)$ plus the potential energy $V(\boldsymbol{r}_1)$ of the outer electron. From $\hat{H}_{2-N}$ we obtain diagonal contributions corresponding to the *internal energies* in the respective channels,

$$E_i = \langle \psi_{\rm int}^{(i)} | \hat{H}_{2-N} | \psi_{\rm int}^{(i)} \rangle' , \tag{3.174}$$

multiplied by $\delta(\boldsymbol{r} - \boldsymbol{r}')$. The prime on the ket bracket indicates integration and summation over the internal variables. We assume that the internal wave functions are eigenfunctions of $\hat{H}_{2-N}$, or at least that $\hat{H}_{2-N}$ is diagonal in the internal states included in the expansion (3.163); then its contributions to the direct one-body Hamiltonian which are non-diagonal in the channel labels vanish. The contribution from the interaction term $\hat{H}_W$ couples the channels. With the expression (2.55) for the electron-electron interaction this contribution consists of the local coupling potentials

$$V_{i,j}(\boldsymbol{r}) = \langle \psi_{\rm int}^{(i)} | \sum_{\nu=2}^{N} \frac{e^2}{|\boldsymbol{r} - \boldsymbol{r}_\nu|} | \psi_{\rm int}^{(j)} \rangle' . \tag{3.175}$$

The decomposition of the Hamiltonian and overlap operators into direct and exchange parts exposes the structure of (3.169) as a system of coupled Schrödinger-like equations:

$$\begin{aligned} &\left(\frac{\hat{\boldsymbol{p}}^2}{2\mu} + V(\boldsymbol{r}) \right) \psi_i(\boldsymbol{r}) + \sum_j V_{i,j}(\boldsymbol{r}) \psi_j(\boldsymbol{r}) \\ &\quad + \sum_j \int H_{\mathrm{ex}\,i,j}(\boldsymbol{r}, \boldsymbol{r}') \psi_j(\boldsymbol{r}')\, \mathrm{d}\boldsymbol{r}' \\ &= (E - E_i) \psi_i(\boldsymbol{r}) - E \sum_j \int K_{i,j}(\boldsymbol{r}, \boldsymbol{r}') \psi_j(\boldsymbol{r}')\, \mathrm{d}\boldsymbol{r}' . \end{aligned} \tag{3.176}$$

The coupled equations (3.176) are generally known under the name of *close-coupling equations*. They consist of a set of coupled integro-differential equations for the channel wave functions $\psi_i(\boldsymbol{r})$. The interactions consist of a direct local potential and a nonlocal exchange potential. The explicit energy dependence of the nonlocal contribution on the right-hand side is due to the fact that the equation of motion has the form of a generalized eigenvalue problem (3.167). This energy dependence can be removed by redefining the channel wave functions, so that they absorb the short-ranged exchange parts of the overlap kernels in (3.170), see e.g. [Fri81]. In the space of such renormalized channel wave functions, the many-channel Hamiltonian defining the coupled-channel equations is Hermitian, as long as it does not contain terms accounting for coupling to many-body wave functions not included in the expansion (3.165) [cf. Sect. 4.1.14 on Feshbach projection].

The longest-ranged contributions to the potential energy in (3.176) are the direct diagonal potential $V(\boldsymbol{r})$ describing the electrostatic attraction by the atomic nucleus,

$$V(\boldsymbol{r}) = -\frac{Ze^2}{r}, \tag{3.177}$$

and the direct interaction potentials (3.175). Using the multipole expansion (A.10) in Appendix A.1,

$$\frac{1}{|\boldsymbol{r}-\boldsymbol{r}_\nu|} = \sum_{l=0}^{\infty} \frac{[\min\{r,r_\nu\}]^l}{[\max\{r,r_\nu\}]^{l+1}} P_l(\cos\theta_\nu), \tag{3.178}$$

we can expand the potentials (3.175) in a series for large $|\boldsymbol{r}|$,

$$V_{i,j}(\boldsymbol{r}) = \sum_{l=0}^{\infty} \frac{e^2}{r^{l+1}} \langle \psi_{\text{int}}^{(i)} | \sum_{\nu=2}^{N} r_\nu^l P_l(\cos\theta_\nu) | \psi_{\text{int}}^{(j)} \rangle', \quad |\boldsymbol{r}| \to \infty . \tag{3.179}$$

Here P_l are the Legendre polynomials and θ_ν is the angle between $\boldsymbol{r}$ and $\boldsymbol{r}_\nu$. The $l = 0$ contribution in (3.179) yields a potential which is diagonal in the channel indices and describes the electrostatic repulsion by the core electrons. It ensures that the outer electron only sees the net charge $Z - (N - 1)$ of atomic nucleus plus core electrons at large distances. The higher contributions corresponding to $l > 0$ depend on the *multipole moments* and *multipole matrix elements*

$$M_{i,j}^{(l)} = \langle \psi_{\text{int}}^{(i)} | \sum_{\nu=2}^{N} r_\nu^l P_l(\cos\theta_\nu) | \psi_{\text{int}}^{(j)} \rangle' \tag{3.180}$$

of the internal states. Since the internal states are usually eigenstates of the parity operator for the $N - 1$ core electrons (cf. Sect. 2.2.4), the diagonal multipole moments $M_{i,i}^{(l)}$ vanish for odd l. For neutral atoms and positive ions (i.e. for $Z \geq N$), the structure of the close-coupling equations is thus dominated by the diagonal long-ranged Coulomb potential $-(Z - N + 1)e^2/r$ describing the attraction of the outer electron by the net charge of nucleus plus core electrons. The next contributions depend on multipole moments and multipole matrix elements of the internal core states; they fall off at least as fast as $1/r^3$ in the diagonal potentials and at least as fast as $1/r^2$ in the nondiagonal coupling potentials. Due to the exponential decay of the bound state wave functions of the internal core states, the nonlocal exchange potentials fall off exponentially at large distances.

The internal states $\psi_{\text{int}}^{(i)}$ defining the channels generally have a well defined angular momentum, the *channel spin*. It is made up of the orbital angular momenta of the core electrons $2, \ldots N$ together with the spin angular momenta of all electrons. The resulting channel spin must still be coupled with the orbital angular momentum of the outer electron to form the total angular momentum of all electrons, which is a good quantum number. When we separate the close coupling equations (3.176) into radial and angular parts, there will only be coupling between terms belonging to the same values of the total angular momentum quantum numbers J, M_J and, if the perturbation due to spin-orbit coupling is sufficiently small, to the same values L, M_L of the total orbital angular momentum and S, M_S of the total spin. The coupled equations (3.176) thus fall into various sets of coupled radial equations which, apart from the nonlocal exchange potentials, each have the general form (1.215). Each such set of coupled radial equations is characterized by the quantum numbers J, L, S and the N-electron parity, as was described for atomic states in general in Sect. 2.2.4. The transition from the coupled equations (3.176) to coupled radial equations will be discussed in more detail in connection with inelastic scattering in Sect. 4.4.2.

3.3.2 Autoionizing Resonances

The internal energy E_2 of an excited state $\psi_{\text{int}}^{(2)}$ of the core electrons lies higher than the internal energy E_1 of the ground state $\psi_{\text{int}}^{(1)}$. The channel threshold I_1 in channel 1 coincides with the continuum threshold of the whole system, and the channel threshold I_2, above which channel 2 is open, lies higher by the amount $E_2 - E_1$ corresponding to the internal excitation energy of the core.

$$I_1 = I, \quad I_2 = I + E_2 - E_1 \,. \tag{3.181}$$

At energies between I_1 and I_2 there can be states in channel 2 which would be bound if there were no coupling to the open channel 1. In the independent-particle picture such a state corresponds to a *two-electron excitation*: firstly a core electron is excited defining the internal state $\psi_{\text{int}}^{(2)}$; secondly the outer electron occupies an (excited) state in the electron-core potential (see Fig. 3.12). Due to channel coupling, the excited core electron can impart its excitation energy $E_2 - E_1$ to the outer electron, which thus attains an energy above the continuum threshold and can be ejected without absorption or emission of electromagnetic radiation. This process is called *autoionization*.

Such *autoionizing states* appear in the coupled-channel equations as Feshbach resonances which were described in Sect. 1.5.2. The bound state in the uncoupled channel 2 is described by a bound radial wave function $\phi_0(r)$, and all other

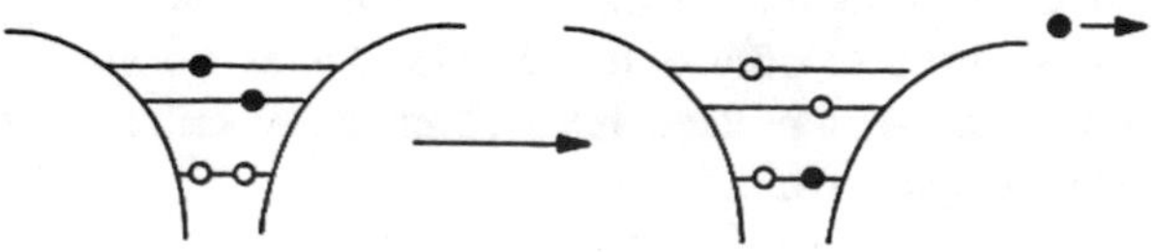

Fig. 3.12 Schematic illustration of an autoionizing resonance in the single-particle picture. Electrons are indicated by *filled circles*, unoccupied single-particle states by *empty circles*

coordinates (including the angular coordinates of the outer electron) are accounted for in the internal wave function of the excited core state $\psi_{\rm int}^{(2)}$. The radial wave function $\psi_{\rm reg}$ in the uncoupled open channel 1 has the asymptotic form $[2\mu/(\pi\hbar^2 k)]^{1/2}\sin(kr+\delta_{\rm bg})$ (cf. (1.223)), where $\delta_{\rm bg}$ is the background phase shift due to the diagonal potential. The factor $[2\mu/(\pi\hbar^2 k)]^{1/2}$ ensures normalization in energy. The effects of the channel coupling can easily be calculated if we assume that the channel wave function ϕ_2 in the closed channel 2 is always proportional to the wave function of the bound state ϕ_0. We then obtain a solution of the coupled equations in the following form:

$$\phi_1(r) = \cos\delta_{\rm res}\,\phi_{\rm reg}(r) + \sin\delta_{\rm res}\,\Delta\phi_1(r),$$
$$\phi_2(r) = \cos\delta_{\rm res}\,\frac{\langle\phi_0|V_{2,1}|\phi_{\rm reg}\rangle}{E-E_{\rm R}}\,\phi_0(r)\,. \tag{3.182}$$

The modification of the wave function in channel 1 is described by the term $\sin\delta_{\rm res}\,\Delta\phi_1$, in which $\Delta\phi_1(r)$ asymptotically merges into the irregular solution of the uncoupled equation:

$$\Delta\phi_1(r) = \sqrt{2\mu/(\pi\hbar^2 k)}\,\cos(kr+\delta_{\rm bg}),\qquad r\to\infty\,. \tag{3.183}$$

$\delta_{\rm res}$ is the additional asymptotic phase due to coupling of the bound state in channel 2 to the open channel 1. Near the energy $E_{\rm R}$ of the autoionizing resonance it rises more or less suddenly by π and is quantitatively given rather well by the formula (1.234):

$$\tan\,\delta_{\rm res} = -\frac{\Gamma/2}{E-E_{\rm R}}\,. \tag{3.184}$$

According to (1.232), the width Γ is given by

$$\Gamma = 2\pi\,|\langle\phi_{\rm reg}|V_{1,2}|\phi_0\rangle\,|^2, \tag{3.185}$$

and it determines the lifetime of the state with respect to autoionization according to (2.145). The potential $V_{1,2}$ encompasses all contributions which couple the channels, including nonlocal exchange contributions, and we assume $\langle\phi_0|V_{2,1}|\phi_{\rm reg}\rangle = \langle\phi_{\rm reg}|V_{1,2}|\phi_0\rangle^*$.

The channel wave functions (3.182) correspond exactly to the solutions of the two-channel equations in Sect. 1.5.2, together with the common factor $\cos\delta_{\rm res}$ which ensures that the wave functions in the open channel 1 are energy normalized. The associated total wave functions are then also energy normalized, because the normalization integrals are dominated by the divergent contributions from the radial wave functions in the open channel.

With (3.184) and (3.185), the radial wave function ϕ_2 in (3.182) can be rewritten as

$$\phi_2(r) = -\frac{\sin\delta_{\rm res}}{\pi\left\langle\phi_{\rm reg}|V_{1,2}|\phi_0\right\rangle}\,\phi_0(r), \tag{3.186}$$

so that the entire N-electron wave function has the form

$$\begin{aligned}\Phi_E = {} & \cos\delta_{\rm res}\,\hat{\mathcal{A}}_1\left\{\psi_{\rm int}^{(1)}\frac{\phi_{\rm reg}(r)}{r}\right\} \\ & -\frac{\sin\delta_{\rm res}}{\pi\left\langle\phi_{\rm reg}|V_{1,2}|\phi_0\right\rangle}\,\hat{\mathcal{A}}_1\left\{\psi_{\rm int}^{(2)}\frac{\phi_0(r)}{r} - \pi\left\langle\phi_{\rm reg}|V_{1,2}|\phi_0\right\rangle\psi_{\rm int}^{(1)}\frac{\Delta\phi_1(r)}{r}\right\}.\end{aligned} \tag{3.187}$$

It is appropriate to normalize the radial wave function ϕ_0 of the bound state in the (uncoupled) channel 2 such that the contribution $\hat{\mathcal{A}}_1\{\psi_{\rm int}^{(2)}\phi_0(r)/r\}$ of channel 2 to the N-electron wave function (3.187) is normalized to unity. Due to antisymmetrization, this does not necessarily mean that ϕ_0 itself is normalized to unity. With (3.184) and (3.185), the absolute square of the factor in front of the contribution from channel 2 in (3.187) can be written as

$$\frac{\sin^2\delta_{\rm res}}{\pi^2|\langle\phi_{\rm reg}|V_{1,2}|\phi_0\rangle|^2} = \frac{1}{\pi}\,\frac{\Gamma/2}{(E-E_{\rm R})^2+(\Gamma/2)^2} = \frac{1}{\pi}\,\frac{{\rm d}\delta_{\rm res}}{{\rm d}E}\,. \tag{3.188}$$

Thus the admixture of channel 2 near an autoionizing resonance is described by a Lorentzian curve with a maximum at the resonance energy $E_{\rm R}$ and a width corresponding to the width of the resonance (see Fig. 1.8 in Sect. 1.5.2).

Equation (3.187) shows that the "unperturbed wave function" ($\delta_{\rm res} = 0$) acquires an admixture due to the coupling of the channels. This admixture is not merely the *naked* bound state $\hat{\mathcal{A}}_1\{\psi_{\rm int}^{(2)}\phi_0(r)/r\}$, but is itself *dressed* by a modification of the open-channel wave function.

The existence of autoionizing resonances manifests itself not only in the radiationless decay of excited states, but also in other observable quantities such as photoabsorption cross sections. In order to calculate, for example, the oscillator strength ${\rm d}f_{E{\rm i}}^{(x)}/{\rm d}E$ for photoionization from an initial N-electron state $\Phi_{\rm i}$ we have to apply the upper formula (2.222) and insert the two-channel final state wave function Φ_E from (3.187):

$$\begin{aligned}\frac{{\rm d}f_{E{\rm i}}^{(x)}}{{\rm d}E} &= \frac{2\mu}{\hbar}\,\omega\left|d_1\cos\delta_{\rm res} - d_2\frac{\sin\delta_{\rm res}}{\pi\left\langle\phi_{\rm reg}|V_{1,2}|\phi_0\right\rangle}\right|^2 \\ &= \frac{2\mu}{\hbar}\,\omega\,|d_1|^2\cos^2\delta_{\rm res}\left|1-\frac{d_2}{d_1}\frac{\tan\delta_{\rm res}}{\pi\left\langle\phi_{\rm reg}|V_{1,2}|\phi_0\right\rangle}\right|^2.\end{aligned} \tag{3.189}$$

Here d_1 and d_2 are the N-electron matrix elements which describe the dipole transitions from the initial state Φ_i to the two components of the final state (3.187):

$$d_1 = \langle \psi_{\mathrm{int}}^{(1)} \frac{\phi_{\mathrm{reg}}(r)}{r} | \sum_{\nu=1}^{N} x_\nu \hat{\mathcal{A}}_1 | \Phi_\mathrm{i} \rangle ,$$

$$d_2 = \langle \psi_{\mathrm{int}}^{(2)} \frac{\phi_0(r)}{r} - \pi \langle \phi_{\mathrm{reg}} | V_{1,2} | \phi_0 \rangle \, \psi_{\mathrm{int}}^{(1)} \frac{\Delta\phi_1(r)}{r} | \sum_{\nu=1}^{n} x_\nu \hat{\mathcal{A}}_1 | \Phi_\mathrm{i} \rangle . \quad (3.190)$$

It is worth commenting on the physical dimensions of d_1 and d_2. Due to the energy normalization of the wave function ϕ_{reg}, each occurrence of ϕ_{reg}, rather than a normalized bound-state wave function, contributes the inverse square root of an energy to the physical dimension, as is e.g. obvious in (3.185). The same holds for $\Delta\phi_1$. Hence the dipole transition strength d_1 as defined in (3.190) has the dimension of length times an inverse square root of energy, whereas d_2 is just a length.

The formula (3.189) shows that the observable photoabsorption cross sections in the vicinity of an autoionizing resonance are formed from two interfering amplitudes. The resulting line shapes can best be discussed if we rewrite it as

$$\frac{\mathrm{d}f_{E\mathrm{i}}^{(x)}}{\mathrm{d}E} = \frac{2\mu}{\hbar}\omega |d_1|^2 \frac{|q+\varepsilon|^2}{1+\varepsilon^2} , \quad (3.191)$$

where

$$\varepsilon = -\cot\delta_{\mathrm{res}} = \frac{E - E_\mathrm{R}}{\Gamma/2} \quad (3.192)$$

is the dimensionless *reduced energy*, which measures the energy relative to the position of the resonance in units of half the width of the resonance. The parameter

$$q = \frac{d_2}{d_1 \pi \langle \phi_{\mathrm{reg}} | V_{1,2} | \phi_0 \rangle} \quad (3.193)$$

is the dimensionless *shape parameter* which depends on the relative strength of the dipole transition matrix elements (3.190) and determines the shape of the absorption line. In (3.191), $(2\mu/\hbar)\omega|d_1|^2$ is a weakly energy-dependent factor corresponding to the oscillator strength we would expect in the absence of coupling to the bound state in channel 2. The energy dependence of the cross section near the resonance is dominantly given by the modulus of the *Beutler-Fano function*

$$F(q;\varepsilon) = \frac{(q+\varepsilon)^2}{1+\varepsilon^2} , \quad (3.194)$$

which is real and non-negative when the shape parameter q is real.

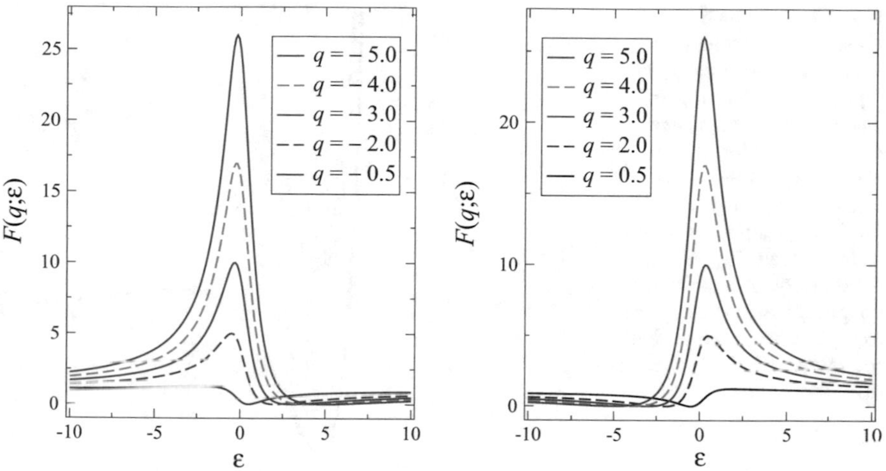

Fig. 3.13 Beutler-Fano function (3.194) for various negative (*left*) and positive (*right*) values of the shape parameter q

Beutler-Fano resonances of the form (3.191) occur not only in photoabsorption, but in all observable quantities which are determined by a transition matrix element $\langle \Phi_E|\hat{O}|\Phi_i\rangle$ as in (2.222). The matrix elements (3.190) must then be replaced by the corresponding matrix elements of the transition operator $\hat{O}$.

Different values of the shape parameter q in the Beutler-Fano function (3.194) lead to absorption lines of different shape as illustrated in Fig. 3.13. For real q, the function and hence also the oscillator strength vanish at $\varepsilon = -q$. This corresponds to completely destructive interference of the two terms in (3.189). The absorption line is steeper on the side of this zero and flatter on the other side. The sign of q determines which side is the steep one. The maximum of the Beutler-Fano function is at $\varepsilon = 1/q$; the height of the maximum is $1 + q^2$. Far from resonance, i.e. for $\varepsilon \to \pm\infty$, the Beutler-Fano function is unity and the oscillator strength merges into the oscillator strength we would expect in the absence of channel coupling. For very small values of q the Beutler-Fano function describes an almost symmetric fall off to zero around the resonance energy (*window resonance*). (See Problem 3.4.)

3.3.3 *Configuration Interaction, Interference of Resonances*

A straightforward and instructive extension of the case of a single isolated Feshbach resonance is the consideration of two bound states in different closed channels coupling to one open channel, as illustrated in Fig. 3.14. The coupled equations

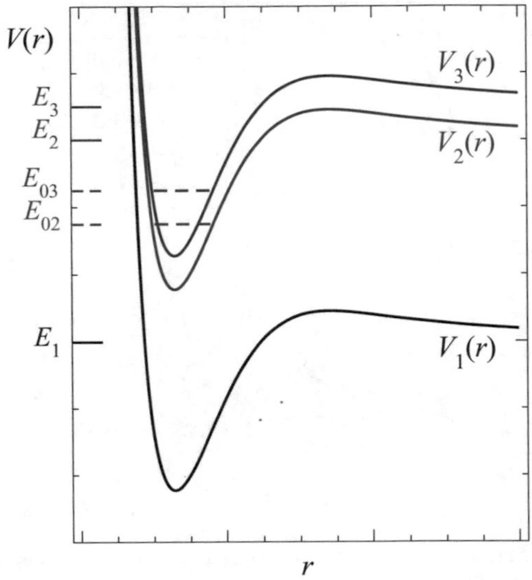

Fig. 3.14 Schematic illustration of a three-channel system. The potential $V_i(r)$ represents the potential in channel i plus the respective internal excitation energy, $V_i(r) = V_{i,i}(r) + E_i$. The uncoupled upper potentials V_2, V_3 each support a bound state at energy E_{02} and E_{03}, respectively. Due to channel coupling, these states appear as Feshbach resonances in the lower channel $i = 1$, which is open for $E > E_1$

for the three radial channel wave functions are

$$\left[-\frac{\hbar^2}{2\mu}\frac{\mathrm{d}^2}{\mathrm{d}r^2} + V_i\right]\phi_i(r) + \sum_{j\neq i} V_{i,j}\phi_j(r) = E\,\phi_i(r)\ , \tag{3.195}$$

and the energy is chosen in the interval $E_1 < E < \min\{E_2, E_3\}$ so that channel 1 is open while channels 2 and 3 are closed. The bound-state wave functions ϕ_{02} and ϕ_{03} in the closed channels 2 and 3 are, in the absence of channel coupling, solutions of the associated radial equation at the energies E_{02} and E_{03}, respectively, and they are assumed to be normalized to unity,

$$\left[-\frac{\hbar^2}{2\mu}\frac{\mathrm{d}^2}{\mathrm{d}r^2} + V_2\right]\phi_{02}(r) = E_{02}\,\phi_2(r)\ , \quad \left[-\frac{\hbar^2}{2\mu}\frac{\mathrm{d}^2}{\mathrm{d}r^2} + V_3\right]\phi_{03}(r) = E_{03}\,\phi_3(r)\ ,$$
$$\langle\phi_{02}|\phi_{02}\rangle = 1\ , \qquad \langle\phi_{03}|\phi_{03}\rangle = 1\,. \tag{3.196}$$

Again we assume, that the closed-channel wave functions are restricted to multiples of the respective uncoupled bound-state wave functions, i.e. we now look for solutions of the three-channel problem in the space of three-component wave functions,

$$\Phi \equiv \begin{pmatrix} \phi_1(r) \\ A_2\phi_{02}(r) \\ A_3\phi_{03}(r) \end{pmatrix}\,. \tag{3.197}$$

Inserting $A_2\phi_{02}$ for ϕ_2 and $A_3\phi_{03}$ for ϕ_3 in the radial equations (3.195) with $i = 2$ and $i = 3$ and projecting onto $\langle\phi_{02}|$ and $\langle\phi_{03}|$ gives, as generalization of the lower equation (1.219),

$$\begin{aligned} A_2(E - E_{02}) &= \langle\phi_{02}|V_{2,1}|\phi_1\rangle + A_3\langle\phi_{02}|V_{2,3}|\phi_{03}\rangle\ , \\ A_3(E - E_{03}) &= \langle\phi_{03}|V_{3,1}|\phi_1\rangle + A_2\langle\phi_{03}|V_{3,2}|\phi_{02}\rangle\ . \end{aligned} \tag{3.198}$$

The equation for ϕ_1, i.e. the equation (3.195) with $i = 1$, can be written as

$$\left[E + \frac{\hbar^2}{2\mu}\frac{\mathrm{d}^2}{\mathrm{d}r^2} - V_1\right]\phi_1(r) = A_2\,V_{1,2}\,\phi_{02}(r) + A_3\,V_{1,3}\,\phi_{03}(r) \tag{3.199}$$

and solved with the help of the Green's function (1.228),

$$\begin{aligned} \phi_1(r) &= \phi_{\mathrm{reg}}(r) + \int_0^\infty G(r,r')\left[A_2\,V_{1,2}(r')\,\phi_{02}(r') + A_3\,V_{1,3}(r')\,\phi_{03}(r')\right]\mathrm{d}r' \\ &\overset{r\to\infty}{\sim} \phi_{\mathrm{reg}}(r) - \pi\left[A_2\langle\phi_{\mathrm{reg}}|V_{1,2}|\phi_{02}\rangle + A_3\langle\phi_{\mathrm{reg}}|V_{1,3}|\phi_{03}\rangle\right]\phi_{\mathrm{irr}}(r), \end{aligned} \tag{3.200}$$

so the expression for the resonant contribution to the open-channel phase shift reads

$$\tan\delta_{\mathrm{res}} = -\pi\left[A_2\langle\phi_{\mathrm{reg}}|V_{1,2}|\phi_{02}\rangle + A_3\langle\phi_{\mathrm{reg}}|V_{1,3}|\phi_{03}\rangle\right]\ . \tag{3.201}$$

Inserting the upper line of (3.200) for ϕ_1 in (3.198) leads to two simultaneous equations for the coefficients A_2 and A_3,

$$\begin{aligned} A_2&\left[E - E_{02} - \langle\phi_{02}|V_{2,1}\,\hat{G}\,V_{1,2}|\phi_{02}\rangle\right] - A_3\langle\phi_{02}|V_{2,3}\,\hat{G}\,V_{1,3}|\phi_{03}\rangle \\ &= \langle\phi_{02}|V_{2,1}|\phi_{\mathrm{reg}}\rangle + A_3\langle\phi_{02}|V_{2,3}|\phi_{03}\rangle, \\ A_3&\left[E - E_{03} - \langle\phi_{03}|V_{3,1}\,\hat{G}\,V_{1,3}|\phi_{03}\rangle\right] - A_2\langle\phi_{03}|V_{3,2}\,\hat{G}\,V_{1,2}|\phi_{02}\rangle \\ &= \langle\phi_{03}|V_{3,1}|\phi_{\mathrm{reg}}\rangle + A_2\langle\phi_{03}|V_{3,2}|\phi_{02}\rangle\ . \end{aligned} \tag{3.202}$$

With the abbreviations,

$$\varepsilon_i = E_{0i} + \langle\phi_{0i}|V_{i,1}\,\hat{G}\,V_{1,i}|\phi_{0i}\rangle, \quad W_{i,1} = \langle\phi_{0i}|V_{i,1}|\phi_{\mathrm{reg}}\rangle = W_{1,i}^*, \quad i = 2, 3\ , \tag{3.203}$$

and

$$W_{2,3} = \langle\phi_{02}|V_{2,3}|\phi_{03}\rangle + \langle\phi_{02}|V_{2,1}\,\hat{G}\,V_{1,3}|\phi_{03}\rangle = W_{3,2}^*, \tag{3.204}$$

the solutions of (3.202) are

$$A_2 = \frac{(E-\varepsilon_3)W_{2,1} + W_{2,3}W_{3,1}}{(E-\varepsilon_2)(E-\varepsilon_3) - |W_{2,3}|^2}\,, \quad A_3 = \frac{(E-\varepsilon_2)W_{3,1} + W_{3,2}W_{2,1}}{(E-\varepsilon_2)(E-\varepsilon_3) - |W_{2,3}|^2}\,, \tag{3.205}$$

so (3.201) for the resonant contribution to the phase shift becomes

$$\tan\delta_{\text{res}} = \tag{3.206}$$
$$-\pi\,\frac{(E-\varepsilon_3)|W_{2,1}|^2 + W_{1,2}W_{2,3}W_{3,1} + (E-\varepsilon_2)|W_{3,1}|^2 + W_{1,3}W_{3,2}W_{2,1}}{(E-\varepsilon_2)(E-\varepsilon_3) - |W_{2,3}|^2}\,.$$

The matrix element $W_{2,3}$ defined in (3.204) describes the interaction of the closed-channel bound states, both by direct coupling through the term containing $V_{2,3}$, and by indirect coupling via the open channel 1 through the term containing $\hat{G}$. In the absence of this interaction, the resonance energies are $\varepsilon_{2/3}$, as defined in (3.203), and they are shifted from the uncoupled bound-state energies $E_{02/03}$ as in the single-resonance case (1.233). Note that the matrix element $W_{2,3}$ has the dimension of an energy, while the matrix elements $W_{1,2}$, $W_{1,3}$ have the dimension of the square root of an energy; this is due to the fact that the bound-state wave functions ϕ_{0i} are normalized to unity while the continuum wave functions ϕ_{reg} are normalized in energy.

Instead of a single pole as in (1.230), (1.234), the right-hand side of (3.206) features two poles, namely the zeros of the denominator

$$D(E) = (E-\varepsilon_2)(E-\varepsilon_3) - |W_{2,3}|^2\,, \tag{3.207}$$

which lie at

$$E_{\pm} = \frac{\varepsilon_2+\varepsilon_3}{2} \pm \sqrt{\left(\frac{\varepsilon_2-\varepsilon_3}{2}\right)^2 + |W_{2,3}|^2}\,. \tag{3.208}$$

The interaction of the resonances via the matrix element $W_{2,3}$ leads to a level repulsion of the resonance energies, an effect well known from bound two-level systems, see Problem 1.6 in Chapter 1.

The widths of the resonances are related to the residue of $\tan\delta_{\text{res}}$ at the respective poles and are explicitly given in terms of the energy derivative of δ_{res} by (1.236),

$$\Gamma_{\pm} = 2\left[\left.\frac{\mathrm{d}\delta_{\text{res}}}{\mathrm{d}E}\right|_{E=E_{\pm}}\right]^{-1}\,. \tag{3.209}$$

If we introduce the abbreviation

$$N(E) = \pi\left[(E-\varepsilon_3)|W_{2,1}|^2 + W_{1,2}W_{2,3}W_{3,1} + (E-\varepsilon_2)|W_{3,1}|^2 + W_{1,3}W_{3,2}W_{2,1}\right] \tag{3.210}$$

then (3.206) becomes

$$\tan\delta_{\text{res}} = -\frac{N(E)}{D(E)} \tag{3.211}$$

and we obtain a compact expression for the derivative at the poles, where $D(E) = 0$,

$$\left.\frac{\mathrm{d}\delta_{\text{res}}}{\mathrm{d}E}\right|_{D=0} = \left(1+\frac{N^2}{D^2}\right)^{-1}\left.\left(\frac{D'N-N'D}{D^2}\right)\right|_{D=0} = \left.\frac{D'}{N}\right|_{E_\pm}. \tag{3.212}$$

Inserting D'/N for the derivative of δ_{res} in (3.209) gives

$$\begin{aligned}\Gamma_\pm &= \pi\left(|W_{2,1}|^2+|W_{3,1}|^2\right) \\ &\pm \pi\,\frac{\frac{1}{2}(\varepsilon_2-\varepsilon_3)\left(|W_{2,1}|^2-|W_{3,1}|^2\right)+W_{1,2}W_{2,3}W_{3,1}+W_{1,3}W_{3,2}W_{2,1}}{\sqrt{\frac{1}{4}(\varepsilon_2-\varepsilon_3)^2+|W_{2,3}|^2}}\,.\end{aligned} \tag{3.213}$$

If the coupling matrix element $W_{3,2}$ vanishes, the resonance energies are $\varepsilon_{2/3}$ and the associated widths are $\Gamma_2 = 2\pi|W_{2,1}|^2$ and $\Gamma_3 = 2\pi|W_{3,1}|^2$. The closed-closed channel coupling not only leads to a level repulsion of the resonance positions, as noted after (3.208) above, it also affects the resonance widths according to (3.213). The sum of the widths of the two interfering resonances is unaffected,

$$\Gamma_+ + \Gamma_- = 2\pi\left(|W_{2,1}|^2+|W_{3,1}|^2\right) = \Gamma_2+\Gamma_3\,, \tag{3.214}$$

but the distribution of the total width over the two resonances can be strongly affected by the coupling. The extreme situation is that one resonance carries the whole width while the other resonance has exactly vanishing width and corresponds to a *bound state in the continuum*. Such a vanishing width implies an infinite energy derivative of δ_{res} according to (3.209), and it occurs when a zero of the numerator (3.210) coincides with a zero of the denominator (3.207), see (3.212). The condition for this to happen is:

$$E-\varepsilon_2 = -W_{3,2}W_{2,1}/W_{3,1} \quad \text{and} \quad E-\varepsilon_3 = -W_{2,3}W_{3,1}/W_{2,1}\,, \tag{3.215}$$

which means that the energies ε_i and matrix elements $W_{i,j}$ fulfill the relations

$$\varepsilon_2-\varepsilon_3 = W_{3,2}\frac{W_{2,1}}{W_{3,1}} - W_{2,3}\frac{W_{3,1}}{W_{2,1}}\,. \tag{3.216}$$

If the Hamiltonian governing the coupled-channel equations (3.195) in the space defined by the three-component wave functions (3.197) is not only Hermitian, but also time-reversal invariant, then its matrix representation can be based on real symmetric matrices. In this case, the right-hand side of (3.216) is real, and the

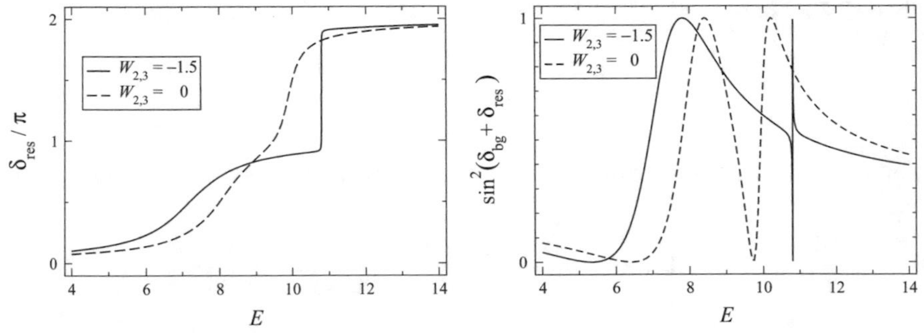

Fig. 3.15 Schematic illustration of the effect of interference of two Feshbach resonances on the phase shift (left-hand part) and on the contribution (1.114) to the scattering cross section (right-hand part). The parameter values entering (3.206) are: $\varepsilon_2 = 8.0$, $\varepsilon_3 = 10.0$, $W_{1,2} = W_{2,1} = 0.5$ and $W_{1,3} = W_{3,1} = 0.3$. The *solid lines* were obtained with a coupling matrix element $W_{2,3} = W_{3,2} = -1.5$ while the *dashed lines* show the results in the absence of coupling, $W_{2,3} = W_{3,2} = 0$. In the right-hand part, a background phase shift of $\delta_{bg} = -\pi/6$ is assumed

condition can be fulfilled if one (or more) of the parameters involved can be tuned, e.g. by varying the strength of an external field. When a bound state in the continuum is realized at a certain energy, the open-channel phase shift is indeterminate at this energy, because the open-channel wave function obeys bound-state boundary conditions.

Results for a model example are illustrated in Fig. 3.15. In the absence of closed-closed coupling ($W_{2,3} = 0$), there are two Feshbach resonances of different but comparable width, located at $E = 8$ and $E = 10$. A finite coupling matrix element $W_{2,3}$ leads to a repulsion of the resonance positions and a concentration of almost all the width in the lower resonance. In the expression $\sin^2(\delta_{bg} + \delta_{res})$, which represents the associated contribution (1.114) to the scattering cross section, the very narrow resonance at $E \approx 10.8$ is seen as a sharp cut into the Beutler-Fano profile of the broad lower resonance. Note that the strongly asymmetric distribution of the resonance widths as a consequence of the coupling does not necessarily require that the resonances be *overlapping*, i.e. that their separation be smaller than their widths. The two resonant features in the left-hand part of Fig. 3.15 are well separated, regardless of whether or not channel coupling via $W_{2,3}$ is considered.

An approximation to the formula (3.213) can be obtained with Fermi's Golden Rule (2.139), if we take the initial state to be an appropriate superposition of the bound states in the two uncoupled closed channels 2 and 3 and the final state as the (energy normalized) regular wave function in the uncoupled open channel 1 with no contributions from the closed channels; the perturbation $\hat{W}$ is defined accordingly:

$$\phi_{\mathrm{i}}^{(\pm)} = \begin{pmatrix} 0 \\ a_2^{(\pm)}\phi_{02} \\ a_3^{(\pm)}\phi_{03} \end{pmatrix}, \quad \phi_{\mathrm{f}} = \begin{pmatrix} \phi_{\mathrm{reg}} \\ 0 \\ 0 \end{pmatrix}, \quad \hat{W} = \begin{pmatrix} 0 & V_{1,2} & V_{1,3} \\ V_{2,1} & 0 & V_{2,3} \\ V_{3,1} & V_{3,2} & 0 \end{pmatrix}. \tag{3.217}$$

The initial states $\phi_{\rm i}^{(\pm)}$ are the results of diagonalizing the two-level problem defined by the bound states ϕ_{02}, ϕ_{03} and the coupling potential $V_{2,3}$. This yields energy eigenvalues $E_\pm$ as given by (3.208), except that the terms containing the open-channel propagator $\hat{G}$ are missing in the expressions for ε_i and $W_{2,3}$, compare (3.203), (3.204). The two-level diagonalization also yields the appropriate superposition amplitudes $a_{2/3}^{(\pm)}$, which are required to obey the usual orthonormality relations. Inserting the objects (3.217) into the Golden Rule (2.139) and taking ρ to be unity according to (2.143), reproduces the expression (3.213), but again, the energies and matrix elements are missing the contributions from the open-channel Green's operator $\hat{G}$. (See problem 3.3) The non-perturbative derivation of (3.206) and (3.213) above shows, that the possible existence of interference-induced exact bound states in the continuum, i.e. of resonances with exactly vanishing width, is not an artefact of the perturbative approach underlying the Golden Rule but is a real feature of systems involving two Feshbach resonances and only one open channel [FW85]. Such bound states in the continuum have recently been studied also in quantum-billiard and quantum-dot systems [SB06, SL08].

The interference of two resonances from different closed channels also affects the photoabsorption cross sections and oscillator strengths. For the three channel wave functions making up the energy normalized final state Φ_E, the corresponding generalization of (3.182) is

$$\phi_1(r) = \cos\delta_{\rm res}\phi_{\rm reg}(r) + \sin\delta_{\rm res}\Delta\phi_1(r),$$
$$\phi_2(r) = \cos\delta_{\rm res}A_2\phi_{02}(r), \quad \phi_3(r) = \cos\delta_{\rm res}A_3\phi_{03}(r)\,, \tag{3.218}$$

and the total wave function Φ_E is a corresponding generalization of (3.187). If we calculate the oscillator strengths (2.222) with this final state wave function, then we obtain, as extension of (3.189),

$$\frac{{\rm d}f_{E{\rm i}}}{{\rm d}E} = \frac{2\mu}{\hbar}\,\omega\,|d_1|^2\cos^2\delta_{\rm res}\left|1+\frac{d_2}{d_1}A_2+\frac{d_3}{d_1}A_3\right|^2. \tag{3.219}$$

Here $(2\mu/\hbar)\,\omega|d_1|^2$ is the oscillator strength we would expect in the absence of coupling of the open channel to the closed channels. The parameters d_2 and d_3 are the dipole transition matrix elements connecting the initial state to the components from the respective closed channels in the final-state wave function. These are essentially the channel wave functions ϕ_{02} and ϕ_{03}, which may be dressed with small admixtures from the open channel as in (3.190). Replacing the $\cos^2$ in (3.219) by $1/(1+\tan^2)$ and using the explicit expressions (3.205) for the amplitudes A_2, A_3, we obtain

$$\frac{{\rm d}f_{E{\rm i}}}{{\rm d}E} = \frac{2\mu}{\hbar}\,\omega\,|d_1|^2\left|D(E)+\frac{d_2}{d_1}\left[(E-\varepsilon_3)W_{2,1}+W_{2,3}W_{3,1}\right]\right.$$
$$\left.+\frac{d_3}{d_1}\left[(E-\varepsilon_2)W_{3,1}+W_{3,2}W_{2,1}\right]\right|^2\frac{1}{D(E)^2+N(E)^2}\,, \tag{3.220}$$

where $D(E)$ and $N(E)$ again stand for the denominator (3.207) and the numerator (3.210) in the expression (3.206) for the phase shift. For different values of the coupling matrix elements $W_{i,j}$, of the (shifted) energies ε_2, ε_3 of the non-interacting resonances and of the relative dipole matrix elements d_2/d_1 and d_3/d_1, the formula (3.220) for the oscillator strengths can produce very different energy dependences and line shapes. Figure 3.16 shows two examples in which qualitatively different interference effects lead to a narrow resonance. In both cases the phase shift (3.206) and the oscillator strength (3.220) were calculated using the same matrix elements $W_{2,1} = 0.5$ and $W_{3,1} = 0.3$ for the direct coupling of the closed channels to the open channel and the same relative dipole matrix elements ($d_2/d_1 = d_3/d_1 = 2.0$). In Fig. 3.16 (a) the other parameters are $\varepsilon_2 = 4.0$, $\varepsilon_3 = 6.0$, $W_{2,3} = -1.5$. In this case both resonances are clearly separated, but the lower resonance carries almost all the width while the upper resonance is very narrow, because the conditions (3.216) for a bound state in the continuum are almost fulfilled—compare also Fig. 3.15. In the oscillator strength we clearly see a broad and a narrow resonance of the Beutler-Fano type. The maximum of the narrow

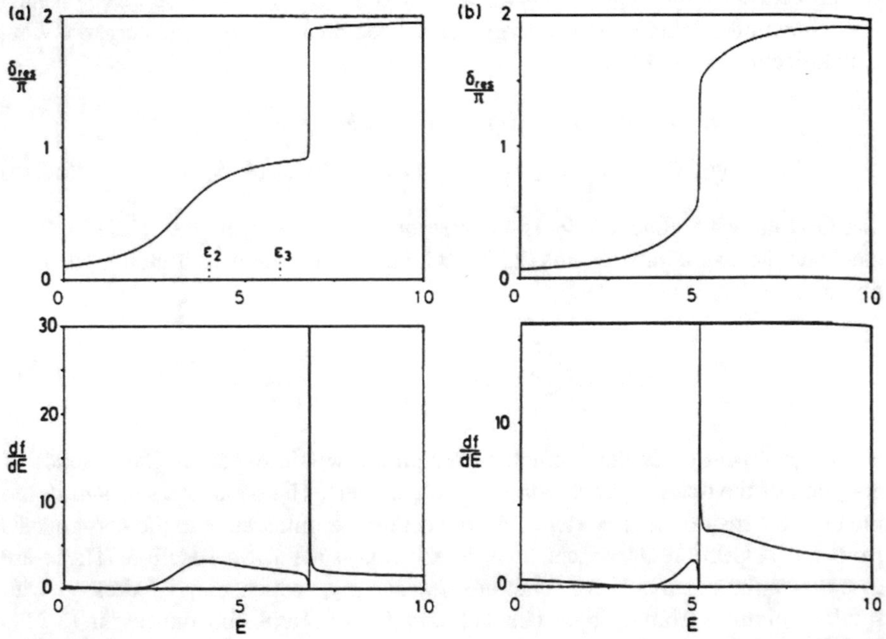

Fig. 3.16 Resonant phase shifts (3.206) and oscillator strengths (3.220) for two examples of two interfering resonances. The coupling of the two closed channels 2 and 3 to the open channel is given by the matrix elements $W_{2,1} = 0.5$, $W_{3,1} = 0.3$, and the relative dipole matrix elements are $d_2/d_1 = d_3/d_1 = 2.0$. Further parameters in case (**a**) are $\varepsilon_2 = 4.0$, $\varepsilon_3 = 6.0$, $W_{3,2} = -1.5$. In case (**b**) we have $\varepsilon_2 = 4.9$, $\varepsilon_3 = 5.1$, $W_{3,2} = 0$. The oscillator strengths in the lower parts of the figure are given in units of the oscillator strength $(2\mu/\hbar)\omega|d_1|^2$ which we would obtain in the absence of coupling to the two closed channels

resonance is very high, because the denominator $D(E)^2 + N(E)^2$ on the right-hand side of (3.220) becomes very small. The zero of the oscillator strength lies to the left of the maximum for both resonances and this corresponds to a positive shape parameter q (see Fig. 3.13).

In Fig. 3.16 (b) the matrix element $W_{3,2}$ for the direct coupling of the two closed channels was taken to vanish and the energies ε_2 and ε_3 were chosen very close together (namely at 4.9 and 5.1 respectively). This case corresponds to the superposition of two resonances which do not interact directly and whose separation is substantially smaller than their widths. Now the separation between the two poles (3.208) of $\tan\delta_{\rm res}$ is so small that it is no longer possible to identify two independent resonances. However, the phase shift is forced to rise from one half-integral multiple of π to the next half-integral multiple of π in the narrow interval between the two poles, and this also leads to a sudden jump in the phase shift (see top half of the figure). In the oscillator strengths (bottom half of the figure) we observe a very narrow (and high) Beutler-Fano resonance cutting into a broad resonance.

Apart from the examples illustrated in Figs. 3.15 and 3.16, there are many other possible line shapes corresponding to different widths, separations and shape parameters of the resonances. Observable spectra often are the product of complicated interference effects and it is by no means obvious that a maximum in a photoabsorption cross section unambiguously corresponds to a well defined autoionizing state of the atom.

So far we have assumed that only one channel is open. If e.g. two channels are open, then the asymptotic phase shifts of the continuum waves are not uniquely defined and there are two linearly independent solutions of the coupled-channel equations for each energy E. If we use the Golden Rule to calculate transition probabilities we obtain an incoherent superposition of two contributions corresponding to the two independent final states. For a detailed and comprehensive description of the theory of photoabsorption spectra see e.g. the article by Starace [Sta82].

3.3.4 Perturbed Rydberg Series

We can describe the effect of an isolated autoionizing resonance in the framework of quantum defect theory by adding a pole term to the right-hand side (3.146):

$$\tan\left[\pi(\nu+\mu)\right] = \frac{\Gamma/2}{E - E_{\rm R}} \,. \tag{3.221}$$

Again $\mu(E)$ is a weakly energy-dependent quantum defect function expressing the deviation of the potential in the open channel 1 from a pure Coulomb potential, and $\nu(E)$ has the two meanings (3.147) below and above threshold. At energies E below threshold the modified QDT equation (3.221) remains an equation for determining bound-state energies, but these energies are now given by the intersections of the set of curves (3.142) with the function

$$\tilde{\mu}(E) = \mu(E) - \frac{1}{\pi} \arctan \frac{\Gamma/2}{E - E_{\mathrm{R}}} . \tag{3.222}$$

For energies E above threshold the QDT equation (3.221) describes the resonant jump of the phase shift through π,

$$\delta = \pi\mu(E) - \arctan \frac{\Gamma/2}{E - E_{\mathrm{R}}} , \tag{3.223}$$

and the weakly energy-dependent function $\pi\mu(E)$ appears as background phase shift.

Strictly speaking we cannot simply superpose the effects of the potential and the Feshbach resonance linearly; hence the quantum defect function $\mu(E)$ in (3.221)–(3.223) may differ slightly from the quantum defect function for the open channel in the absence of the Feshbach resonance.

A typical realization of the QDT equations (3.221)–(3.223) including an autoionizing resonance at an energy E_{R} above the continuum threshold is illustrated with the blue curve in Fig. 3.17.

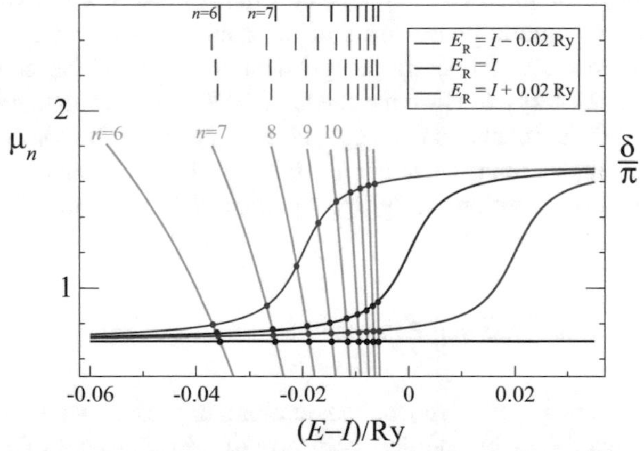

Fig. 3.17 The *solid black line* shows an almost energy independent quantum-defect function $\mu \approx 0.7$, and the *brown lines* show the family of functions $n - \nu(E)$ with $\nu(E) = \sqrt{\mathcal{R}/(I-E)}$ according to (3.142). Their intersections with $\mu(E)$ determine the energy eigenvalues E_n of the bound states and their quantum defects μ_n. The *black vertical lines* in the upper part of the figure show the unperturbed energy levels. The effects of a Feshbach resonance with $\Gamma = 0.01\,\mathcal{R}$ and $E_{\mathrm{R}} = I + 0.02\,\mathcal{R}$, $E_{\mathrm{R}} = I$ and $E_{\mathrm{R}} = I - 0.02\,\mathcal{R}$ are shown by the *blue, maroon and red lines*, which represent the modified quantum-defect function (3.222) for the respective cases. The energy levels of the correspondingly perturbed Rydberg series are shown as *vertical lines* in the same colours in the upper part of the figure. For $E_{\mathrm{R}} = I - 0.02\,\mathcal{R}$ (*red lines*), the perturbation appears as a smooth rise of roughly unity in the quantum defects corresponding to one additional bound state

Although the physical situation is quite different, the formal aspects of the considerations above change little when the energy of the two-electron excitation lies not above, but below the continuum threshold I. Since the mathematical justification of quantum defect theory holds both above and below threshold [Sea83], we can also apply the formulae (3.221), (3.222) to the case where the energy E_R, at which the bound state in channel 2 makes itself felt, lies below the threshold. Now the two-electron excitation is not an autoionizing resonance, but an additional bound state which appears as a *pseudo-resonant perturbation* of the Rydberg series of bound states. Instead of a jump by π in the phase shift we now have a more or less sudden jump by unity in the quantum defects of the bound states. This is illustrated with the red curve in Fig. 3.17. Far below the energy E_R of the perturber, the quantum defects lie on the weakly energy dependent curve $\mu(E)$. Near E_R the quantum defects become larger, so that the effective quantum numbers $n^* = n - \mu_n$ and hence also the energies (3.127) lie closer than in the unperturbed Rydberg series. Far above the energy of the perturber, the quantum defects are shifted by unity in comparison to the unperturbed states. That doesn't change their energies, but it does change their numbering: the n-th state in the unperturbed series at $E_n = I - \mathcal{R}/(n - \mu_n)^2$ is now the $(n + 1)$-st state in the perturbed series at roughly the same energy. Over an energy range corresponding approximately to the width Γ, the spectrum is compressed in order to accommodate one additional bound state. The effect of the perturber on the energy levels can also be seen in the spectra shown as vertical lines at the top of Fig. 3.17.

Finally it may happen, that the energy E_R of the resonance lies very close to the threshold so that the interval $E_R - \Gamma/2, E_R + \Gamma/2$ covers both energies below threshold and energies above threshold. In this case, which is illustrated with the maroon curve in Fig. 3.17, the bound state in the closed channel 2 manifests itself partly as a perturbation of the Rydberg series of bound states and partly as the tail of a resonance in the continuum.

A pseudo-resonant perturbation of a Rydberg series of bound states affects not only the energy eigenvalues but also other observable quantities such as photoabsorption cross sections or oscillator strengths. The effect of a perturber on the discrete oscillator strengths f_n in a Rydberg series can be described by a formula analogous to (3.191), if we replace the left-hand side by discrete oscillator strengths multiplied by the density of states (cf. (3.160)):

$$\frac{(n^*)^3}{2\mathcal{R}} f_n = \frac{(n^*)^3}{2\mathcal{R}} f_n^{(0)} \frac{|q + \varepsilon|^2}{1 + \varepsilon^2} , \tag{3.224}$$

with

$$\varepsilon = \frac{E - E_R}{\Gamma/2}, \quad q = \frac{d_2/d_1}{\pi \left\langle \phi_n^{\mathrm{E}} | V_{1,2} | \phi_0 \right\rangle} . \tag{3.225}$$

Here $f_n^{(0)}$ are the discrete oscillator strengths one would expect without the perturbation of the Rydberg series, and d_2/d_1 is a weakly energy-dependent parameter describing the relative strength of the dipole transitions to the two channels, as in (3.193). The matrix element $\langle \phi_n^{\rm E} | V_{1,2} | \phi_0 \rangle$ contains the renormalized wave functions $\phi_n^{\rm E}$ in channel 1, which merge smoothly into the energy normalized continuum wave functions at the continuum threshold; it is a weakly energy-dependent quantity describing the effective strength of the channel coupling.

3.4 Multichannel Quantum Defect Theory (MQDT)

3.4.1 Two Coupled Coulomb Channels

In this section we study a two-channel system in which the diagonal potentials both correspond to a modified Coulomb potential:

$$V_i(r) \overset{r\to\infty}{=} I_i - \frac{C}{r} \,. \tag{3.226}$$

Between the two channel thresholds I_1 and I_2 ($I_1 < I_2$) the closed channel 2 now contains not only one bound state leading to an autoionizing resonance (see Sect. 3.3.4), but an infinite number of such states which form a Rydberg series. Due to coupling to the open channel 1 this leads to a whole *Rydberg series of autoionizing resonances* at the energies

$$E_{n_2} = I_2 - \frac{\mathcal{R}}{(n_2^*)^2} = I_2 - \frac{\mathcal{R}}{[n_2 - \mu_2(n_2)]^2} \,, \tag{3.227}$$

where n_2^* and $\mu_2(n_2)$ are now effective quantum numbers and quantum defects in channel 2. The widths Γ_{n_2} of the resonances are described by a formula analogous to (3.185):

$$\Gamma_{n_2} = 2\pi \, |\langle \phi_{\rm reg} | V_{1,2} | \phi_{n_2} \rangle|^2 = 2\pi \, \frac{2\mathcal{R}}{(n_2^*)^3} |\langle \phi_{\rm reg} \, | V_{1,2} | \phi_{n_2}^{\rm E} \rangle|^2 \,. \tag{3.228}$$

Here ϕ_{n_2} are the bound radial wave functions in the closed channel 2, and

$$\phi_{n_2}^{\rm E}(r) = \sqrt{\frac{(n_2^*)^3}{2\mathcal{R}}} \, \phi_{n_2}(r) \tag{3.229}$$

are the corresponding renormalized wave functions which merge smoothly into the energy normalized continuum wave functions—now in channel 2—at the threshold I_2. Near this threshold, the matrix element on the right-hand side of (3.228) depends only weakly on energy and we see immediately, without any calculation, that the autoionization widths are inversely proportional to the third power of the effective quantum number n_2^* in channel 2 for large n_2. The autoionization widths thus decrease at the same rate as the separations between successive resonances as we approach the series limit.

The physics of a Rydberg series of autoionizing resonances as described above can be summarized in a compact and transparent way by an extension of the formula (3.221):

$$\tan\left[\pi(\nu_1+\mu_1)\right] = \frac{|R_{1,2}|^2}{\tan\left[\pi(\nu_2+\mu_2)\right]} \,. \tag{3.230}$$

In the energy range between the two channel thresholds ν_1 is just the asymptotic phase shift of the continuum wave function in the open channel 1 multiplied by $-1/\pi$ (as in (3.147)),

$$\nu_1(E) = -\frac{1}{\pi}\,\delta_1(E)\,, \quad E > I_1\,, \tag{3.231}$$

while ν_2 represents the continuous effective quantum number in the closed channel 2, which is defined via the energy separation from the channel threshold I_2:

$$\nu_2(E) = \sqrt{\frac{\mathcal{R}}{I_2 - E}}\,, \quad E < I_2\,. \tag{3.232}$$

The dimensionless quantity $R_{1,2}$ describes the strength of the coupling between the channels 1 and 2 and should depend at most weakly on energy.

In the energy region between the channel thresholds I_1 and I_2 equation (3.230) is an explicit equation for the asymptotic phase shift δ_1 of the open-channel wave function:

$$\delta_1 = \pi\mu_1(E) - \arctan\left[\frac{|R_{1,2}|^2}{\tan\left[\pi(\nu_2+\mu_2)\right]}\right] \,. \tag{3.233}$$

The term $\pi\mu_1(E)$ appears as a background phase shift due to the diagonal potential in channel 1 (more precisely: due to its deviation from a pure Coulomb potential), just like the term $\pi\mu(E)$ in (3.223). The arcus-tangent term now yields not only one single isolated jump of the phase shift through π, but a whole Rydberg series of jumps, which occur at the energies E_{n_2} where the denominator $\tan[\pi(\nu_2+\mu_2)]$ in the argument vanishes. But this condition is just the single-channel QDT equation (3.146) for the closed channel 2, and $\mu_2(E)$ now plays the role of the weakly energy-dependent quantum defect function which smoothly connects the quantum defects $\mu_2(n_2)$ in the Rydberg series of energies (3.227). Near a zero of $\tan[\pi(\nu_2+\mu_2)]$ we can expand the function in a Taylor series and, using the abbreviation

$$T_2(E) := \tan\left[\pi(\nu_2+\mu_2)\right]\,, \tag{3.234}$$

we obtain

$$T_2(E) \approx (E - E_{n_2}) \left. \frac{\mathrm{d}T_2}{\mathrm{d}E} \right|_{E=E_{n_2}} = (E - E_{n_2}) \frac{\pi}{2\mathcal{R}} (n_2^*)^3 . \tag{3.235}$$

Near the zeros of $T_2(E)$ the equation (3.233) thus simplifies to

$$\delta_1 = \pi\mu_1(E) - \arctan\left[\frac{2\mathcal{R}}{\pi(n_2^*)^3} \frac{|R_{1,2}|^2}{(E - E_{n_2})} \right] . \tag{3.236}$$

If we write $R_{1,2}$ as $-\pi$ times the coupling matrix element containing the renormalized bound-state wave functions (3.229),

$$R_{1,2} = -\pi \left\langle \phi_{\mathrm{reg}} | V_{1,2} | \phi_{n_2}^{\mathrm{E}} \right\rangle , \tag{3.237}$$

then (3.236) assumes the form (3.223) for an isolated resonance at the position E_{n_2} with the width (3.228). Transferring the picture of an isolated Feshbach resonance to a Rydberg series of autoionizing resonances thus leads to the approximate expression (3.237) for the (dimensionless) coupling parameter $R_{1,2}$. (The minus-sign, which doesn't play a role at this stage, corresponds to the most usual convention [GF84].)

Beside the behaviour of the phase shift δ_1, several other results from Sect. 3.3.2 can be adapted to the case of a Rydberg series of resonances. The explicit formulae (3.182) and (3.186) for the radial channel wave functions become

$$\begin{aligned} \phi_1(r) &\overset{r\to\infty}{=} \cos[\pi(\nu_1 + \mu_1)]\, \phi_{\mathrm{reg}}(r) - \sin[\pi(\nu_1 + \mu_1)]\, \phi_{\mathrm{irr}}(r) , \\ \phi_2(r) &= -\frac{\sin[\pi(\nu_1 + \mu_1)]}{R_{1,2}} \sqrt{\frac{(n_2^*)^3}{2\mathcal{R}}}\, \phi_{n_2}(r) \\ &= -\frac{\sin[\pi(\nu_1 + \mu_1)]}{R_{1,2}}\, \phi_{n_2}^{\mathrm{E}}(r) . \end{aligned} \tag{3.238}$$

Apart from a minus-sign, $\pi(\nu_1 + \mu_1) = -(\delta_1 - \pi\mu_1)$ is just the resonant part of the asymptotic phase shift without the weakly energy-dependent background phase shift $\pi\mu_1$, which is already accounted for in the regular and irregular solutions ϕ_{reg} and ϕ_{irr} in the (uncoupled) open channel. $\phi_{n_2}^{\mathrm{E}}$ are the renormalized bound-state wave functions (3.229) which merge smoothly into the energy normalized continuum wave functions $\phi_{\mathrm{reg}}^{(2)}$ of channel 2 at the threshold I_2.

If we use the wave functions (3.238) as final state wave functions to calculate the oscillator strengths for photoabsorption according to (2.222), then in place of (3.189) we now obtain

$$\frac{\mathrm{d}f_{E\mathrm{i}}}{\mathrm{d}E} = \frac{2\mu}{\hbar}\omega|d_1|^2\cos^2[\pi(\nu_1+\mu_1)]\left|1-\frac{d_2}{d_1}\frac{\tan[\pi(\nu_1+\mu_1)]}{R_{1,2}}\right|^2$$
$$= \frac{2\mu}{\hbar}\omega|d_1|^2\frac{|\tan[\pi(\nu_2+\mu_2)]-R_{2,1}\,d_2/d_1|^2}{\tan^2[\pi(\nu_2+\mu_2)]+|R_{1,2}|^4}\,, \tag{3.239}$$

where $(2\mu/\hbar)\omega|d_1|^2$ represents the weakly energy-dependent oscillator strength which we would expect in the absence of coupling to the closed channel 2, and the ratio d_2/d_1, which is now dimensionless, describes the relative oscillator strength for transitions from the initial state to both final state channels. In deriving the lower line of (3.239) we inserted the expression $|R_{1,2}|^2/\tan[\pi(\nu_2+\mu_2)]$ for $\tan[\pi(\nu_1+\mu_1)]$ according to (3.230). Equation (3.239) has the same form as (3.191),

$$\frac{\mathrm{d}f_{E\mathrm{i}}}{\mathrm{d}E} = \frac{2\mu}{\hbar}\omega|d_1|^2\frac{|q+\varepsilon|^2}{1+\varepsilon^2}\,, \tag{3.240}$$

provided we define the reduced energy ε as

$$\varepsilon = \frac{\tan[\pi(\nu_2+\mu_2)]}{|R_{1,2}|^2}\,. \tag{3.241}$$

The shape parameter q is now

$$q = -\frac{d_2/d_1}{R_{1,2}}\,. \tag{3.242}$$

Near a resonance energy, i.e. near a zero of $\tan[\pi(\nu_2+\mu_2)]$, the reduced energy (3.241) is a linear function of the energy E (see (3.235)), and it is indeed given by the expression on the right-hand side of (3.192) if we express $|R_{1,2}|^2$ through the width Γ (cf. (3.256) below). As the continuous effective quantum number ν_2 varies through the interval reaching from $1/2$ below to $1/2$ above a resonance position, the reduced energy (3.241) takes on values covering the entire interval $-\infty$ to $+\infty$. In the Rydberg series of resonances each individual "Beutler-Fano resonance" is thus compressed into an energy interval corresponding to an interval of unit length in the continuous effective quantum number in the closed channel 2. (See Fig. 3.18.)

The discussion above as summarized in Fig. 3.18 refers to the energy interval $I_1 < E < I_2$ in which channel 1 is open while channel 2 is closed. In order to describe the situation at energies below the threshold I_1, where both channels are closed, we must return to the interpretation of the quantity ν_1 as the continuous

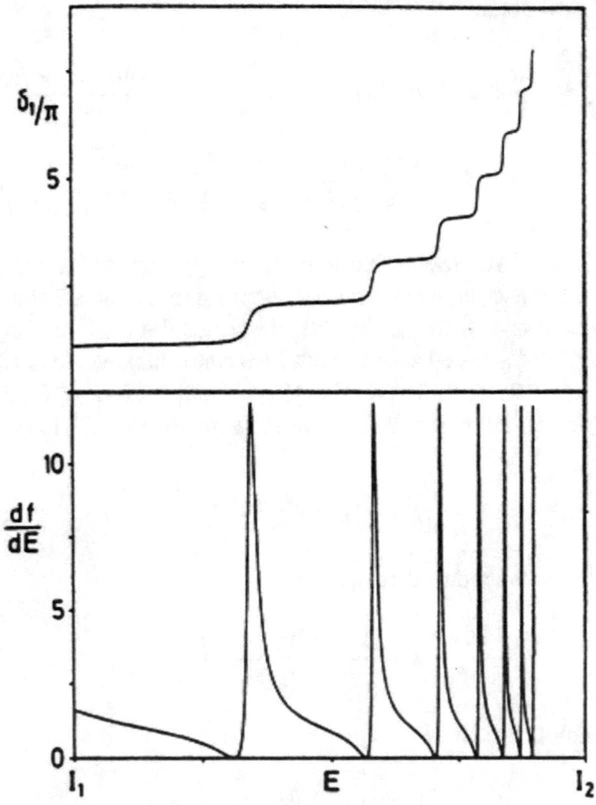

Fig. 3.18 Phase shift (3.233) and oscillator strength (3.239) in a Rydberg series of autoionizing resonances for the following values of the 2QDT parameters: $\mu_1 = 0.1$, $\mu_2 = 0.1$, $R_{1,2} = 0.3$, $d_2/d_1 = -1.0$. The separation of the two channel thresholds I_1 and I_2 is 0.1 Rydberg energies. The oscillator strengths in the lower part of the figure are given in units of the oscillator strength $(2\mu/\hbar)\omega|d_1|^2$, which we would expect in absence of coupling to the Rydberg series in the closed channel 2

effective quantum number in channel 1:

$$\nu_1(E) = \sqrt{\frac{\mathcal{R}}{I_1 - E}}\,, \quad E < I_1\,. \tag{3.243}$$

Now (3.230) is an equation for determining the energies of the bound states in the coupled two-channel system. If the (uncoupled) channel 2 supports a bound state at an energy below I_1, then the associated zero in the function $T_2(E)$ (3.234) leads to a pseudo-resonant perturbation in the Rydberg series of bound states and it manifests itself as a jump by unity in the quantum defects, as described for a single perturber in Sect. 3.3.4. As an example Fig. 3.19 shows the quantum defects of a Rydberg series in calcium consisting of $4s\,np$ states coupled to $^1P^o$. Near $n = 7$

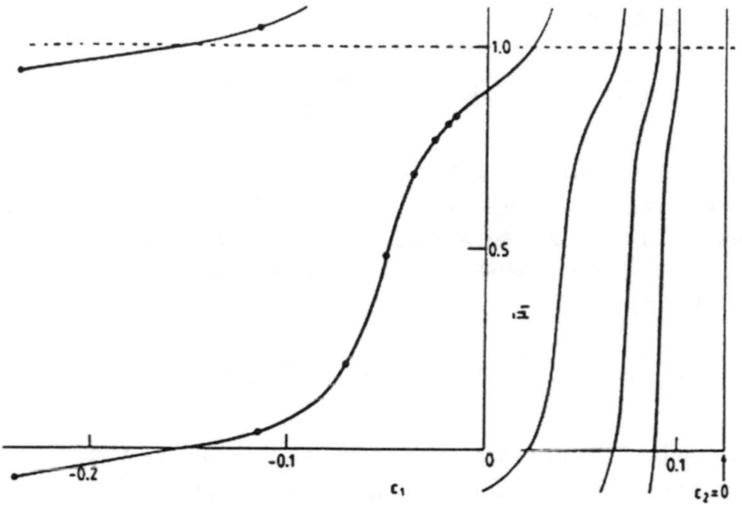

Fig. 3.19 The left-hand part of the figure shows the quantum defects (modulo unity) of the $4s\,np\,^1P^o$ Rydberg series in Ca I, which is perturbed by the lowest state in the $3d\,np\,^1P^o$ channel. The right-hand part of the figure shows $1/\pi$ times the asymptotic phase shift in the now open $4s\,np\,^1P^o$ channel (again modulo unity). The resonant jumps in the phase shift are due to higher states in the $3d$ channel and correspond to autoionizing resonances as in Fig. 3.18 (From [Sea83])

this Rydberg series is perturbed by the lowest state in the $3d\,np\,^1P^o$ channel. Above threshold the picture continues as a series of jumps of the phase shift corresponding to autoionizing resonances as in Fig. 3.18.

In the photoabsorption spectrum, the perturber below threshold appears as a modulation of the oscillator strengths (renormalized with the factor $(n^*)^3$) as described in (3.224), but we now have to insert the periodic form (3.241) for the reduced energy ε.

Above the second channel threshold I_2, both channels are open. At each energy $E > I_2 > I_1$ there are two linearly independent solutions of the coupled-channel equations and each linear combination hereof is again a solution. At a given energy the asymptotic phase shift in channel 1 is not fixed but depends on the asymptotic behaviour of the wave function in channel 2. Conversely, a definite choice of the asymptotic phase shift in channel 1 fixes the asymptotic behaviour of the wave function in channel 2.

The asymptotic behaviour of the solutions in the case of two open channels can be readily understood if we continue the explicit expressions (3.238) for the wave functions just below the channel threshold I_2 to energies above I_2. Pairs ϕ_1, ϕ_2 of wave functions containing a maximum admixture from channel 2 are characterized by $\sin[\pi(\nu_1 + \mu_1)] = \pm 1$ and $\cos[\pi(\nu_1 + \mu_1)] = 0$. When moving to energies above I_2, the renormalized bound state wave functions $\phi^{\mathrm{E}}_{n_2}$ merge into the energy normalized regular solutions $\phi^{(2)}_{\mathrm{reg}}$ in uncoupled channel 2. With the appropriate choice of sign we thus obtain a pair of channel wave functions with the following asymptotic behaviour:

$$\phi_1(r) \stackrel{r\to\infty}{=} \phi_{\rm irr}^{(1)}(r), \quad \phi_2(r) \stackrel{r\to\infty}{=} \frac{1}{R_{1,2}}\,\phi_{\rm reg}^{(2)}(r)\,. \tag{3.244}$$

(The superscript (1) has been introduced in order to distinguish the irregular (and regular) solutions in channel 1 from the corresponding solutions in channel 2.) As both channels are open we can interchange the channel labels to construct a solution of the coupled-channel equations with an asymptotic behaviour complementary to that described by (3.244):

$$\phi_1(r) \stackrel{r\to\infty}{=} \frac{1}{R_{2,1}}\,\phi_{\rm reg}^{(1)}(r), \quad \phi_2(r) \stackrel{r\to\infty}{=} \phi_{\rm irr}^{(2)}(r)\,. \tag{3.245}$$

For the coupling matrix elements $R_{1,2}$ and $R_{2,1}$ appearing in (3.244) and (3.245) we can extend the (approximate) formula (3.237) to energies $E > I_2$ and obtain the (approximate) expressions

$$R_{1,2} = -\pi\,\langle\phi_{\rm reg}^{(1)}|V_{1,2}|\phi_{\rm reg}^{(2)}\rangle\,, \quad R_{2,1} = -\pi\,\langle\phi_{\rm reg}^{(2)}|V_{2,1}|\phi_{\rm reg}^{(1)}\rangle = R_{1,2}^*\,. \tag{3.246}$$

The matrix elements are finite, because the coupling potential falls off asymptotically at least as fast as $1/r^2$.

The general solution of the coupled equations in the case of two channels is a linear combination of the two solutions with the asymptotic behaviour (3.244) and (3.245) respectively:

$$\begin{aligned} \phi_1(r) &\stackrel{r\to\infty}{=} \frac{B}{R_{2,1}}\,\phi_{\rm reg}^{(1)}(r) + A\,\phi_{\rm irr}^{(1)}(r)\,, \\ \phi_2(r) &\stackrel{r\to\infty}{=} \frac{A}{R_{1,2}}\,\phi_{\rm reg}^{(2)}(r) + B\,\phi_{\rm irr}^{(2)}(r)\,. \end{aligned} \tag{3.247}$$

For the asymptotic phase shifts $\delta_1 - \pi\mu_1 = -\pi(\nu_1+\mu_1)$, $\delta_2 - \pi\mu_2 = -\pi(\nu_2+\mu_2)$ we obtain

$$\tan\left[\pi(\nu_1+\mu_1)\right] = -\frac{A}{B}R_{2,1}\,, \quad \tan\left[\pi(\nu_2+\mu_2)\right] = -\frac{B}{A}R_{1,2}\,, \tag{3.248}$$

from which follows

$$\tan\left[\pi(\nu_1+\mu_1)\right]\,\tan\left[\pi(\nu_2+\mu_2)\right] = |R_{1,2}|^2\,. \tag{3.249}$$

Equation (3.249) again has the same form as (3.230), but it now represents a *compatibility equation* for the asymptotic phase shifts in the two open channels.

The equations (3.230), (3.233) and (3.249) can be written in a unified way as one equation of *two-channel quantum defect theory* (2QDT):

$$\begin{vmatrix} \tan\left[\pi(\nu_1+\mu_1)\right] & R_{1,2} \\ R_{2,1} & \tan\left[\pi(\nu_2+\mu_2)\right] \end{vmatrix} = 0\,. \tag{3.250}$$

Its different meanings—as an equation for determining the bound state energy eigenvalues when both channels are closed, as an explicit equation for the phase shift of the open-channel wave function when just one channel is open, or as a compatibility equation for the asymptotic phase shifts when both channels are open—follow in a straightforward way if we insert the different definitions of the quantities ν_i, namely continuous effective quantum number in channel i below the respective channel threshold I_i and $-1/\pi$ times the asymptotic phase shift above I_i:

$$\nu_i(E) = \begin{cases} \sqrt{\frac{\mathcal{R}}{I_i-E}} & \text{for} \quad E < I_i\,, \\ -\frac{1}{\pi}\,\delta_i & \text{for} \quad E > I_i\,. \end{cases} \tag{3.251}$$

The formulae in this section were derived by generalizing the considerations of Sect. 1.5.2 on isolated Feshbach resonances. The approximate expressions (3.238) for the wave functions and (3.237), (3.246) for the coupling parameter are based on this picture of isolated Feshbach resonances. A more rigorous treatment, e.g. by Seaton [Sea83] and by Giusti and Fano [GF84], shows that the 2QDT equation (3.250) is valid quite generally, even if the channel coupling parameter $|R_{1,2}|$ is large, so that resonances and perturbers aren't isolated. The only condition for the validity of the formulae of quantum defect theory is that the deviations of the diagonal potentials from the pure Coulomb potential and the non-diagonal coupling potentials fall off sufficiently fast for large r. In a rigorous derivation, the 2QDT parameters μ_1, μ_2, $R_{1,2}$ appear as weakly energy-dependent quantities whose precise definition is given by the actual solutions of the coupled-channel equations.

Finally it should be pointed out that there are various formulations of quantum defect theory in use. In this chapter we asymptotically represent the channel wave functions as superpositions of the regular and irregular solutions of the uncoupled equations including the deviations of the diagonal potentials from the pure Coulomb potential. The original formulation of Seaton was based on the regular and irregular (pure) Coulomb functions. The argument of tangent functions such as (3.234) then contains the asymptotic phase shift including the weakly energy-dependent background phase shift. The effect of deviations of the diagonal potentials from the pure Coulomb potential is contained in diagonal elements $R_{i,i}$ of the matrix appearing in the MQDT equation. For two channels the MQDT equation in Seaton's formulation reads

$$\begin{vmatrix} \tan(\pi\nu_1) + R_{1,1} & R_{1,2} \\ R_{2,1} & \tan(\pi\nu_2) + R_{2,2} \end{vmatrix} = 0\,. \tag{3.252}$$

When both channels are open, the matrix $(R_{i,j})$ is the *reactance matrix* of scattering theory, which will be defined in Sect. 4.4.2. The formulation of MQDT used in the present chapter and summarized in (3.250) can be derived from Seaton's MQDT by shifting the phases of the basis wave functions until the diagonal elements of the matrix $R_{i,j}$ vanish. The resulting matrix which has no diagonal elements is frequently referred to as the *phase shifted reactance matrix.*

A further formulation of MQDT is due to Fano [Fan70] and is based on a diagonalization of Seaton's reactance matrix. The resulting superpositions of channels are called *eigenchannels.* The eigenvalues of the reactance matrix are written as $\tan\delta$ and the angles δ are the *eigenphases* (see also Sect. 4.4.2).

3.4.2 The Lu-Fano Plot

The physical content of 2QDT can easily be illustrated graphically. Below the upper threshold I_2 channel 2 is closed and the 2QDT equation (3.250) is

$$-\nu_1 = \mu_1 - \frac{1}{\pi}\arctan\left[\frac{|R_{1,2}|^2}{\tan\left[\pi(\nu_2+\mu_2)\right]}\right] \stackrel{\text{def}}{=} \tilde{\mu}_1 \,. \tag{3.253}$$

The right-hand side is a function $\tilde{\mu}_1$ which depends on the energy E or on the continuous effective quantum number $\nu_2 = [\mathcal{R}/(I_2 - E)]^{1/2}$. It is an extension of the function (3.222) for a single perturber to the case of a whole Rydberg series of perturbers. If the 2QDT parameters μ_1, μ_2 and $R_{1,2}$ were not weakly energy dependent but constant, then $\tilde{\mu}_1$ would be exactly periodic in ν_2 with period unity. Above the lower threshold I_1 the left-hand side of (3.253) stands for $1/\pi$ times the asymptotic phase shift in the open channel 1, which is only defined modulo unity, and below I_1 it is -1 times the continuous effective quantum number in channel 1. As in the case of a single isolated perturber discussed in Sect. 3.3.4, the intersections of the function $\tilde{\mu}_1$ below I_1 with the set of curves (3.142) define the quantum defects

$$\mu_{n_1} = n_1 - \nu_1(E_{n_1}) \tag{3.254}$$

and energies E_{n_1} of the bound states. If we plot these quantum defects together with the phase $\delta_1(E)/\pi$ (both modulo unity) as functions of the continuous effective quantum number ν_2 in the upper channel 2 (also modulo unity), then—given constant 2QDT parameters—both the quantum defects and the phases lie on one period of the function $\tilde{\mu}_1$ from (3.253). This representation is called *Lu-Fano plot* [LF70]. Figure 3.20 shows three typical examples of Lu-Fano plots.

Some general properties of the Lu-Fano-plot can be formulated quantitatively if we study the derivatives of the function $\tilde{\mu}_1(\nu_2)$. With the abbreviation (3.234),

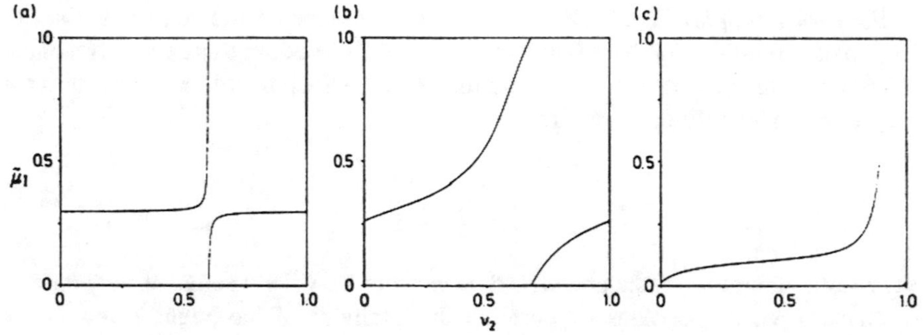

Fig. 3.20 Examples of Lu-Fano plots with the constant 2QDT parameters: **(a)** $\mu_1 = 0.3$, $\mu_2 = 0.4$, $R_{1,2} = 0.1$. **(b)** $\mu_1 = 0.3$, $\mu_2 = 0.4$, $R_{1,2} = 0.6$. The parameters in part **(c)**, $\mu_1 = 0.1$, $\mu_2 = 0.1$, $R_{1,2} = 0.3$, are the same as in Fig. 3.18. Figure 3.20 (c) is thus a reduction of the upper part of Fig. 3.18 to one period in energy (or rather ν_2) and phase

$T_2 = \tan\left[\pi(\nu_2 + \mu_2)\right]$, we have

$$\frac{\mathrm{d}(\tilde{\mu}_1)}{\mathrm{d}\nu_2} = |R_{1,2}|^2 \frac{1+T_2^2}{T_2^2 + |R_{1,2}|^4}, \quad \frac{\mathrm{d}^2(\tilde{\mu}_1)}{\mathrm{d}\nu_2^2} = 2\pi |R_{1,2}|^2 \frac{T_2(|R_{1,2}|^4 - 1)}{\left(T_2^2 + |R_{1,2}|^4\right)^2}(1+T_2^2)\,. \tag{3.255}$$

The gradient of the curve is always positive. For *weak coupling*, $|R_{1,2}| < 1$, the maximum gradient is at $T_2 = 0$ which corresponds to $\nu_2 = n_2 - \mu_2$, and we obtain resonant jumps around the zeros of T_2 as expected. The value of the maximum gradient is $1/|R_{1,2}|^2$ and defines the width of the resonance (or of the pseudo-resonant perturbation) according to the general formula (1.236):

$$\Gamma = \frac{4\mathcal{R}}{\pi \left(n_2^*\right)^3} |R_{1,2}|^2\,. \tag{3.256}$$

Here we have used the fact that

$$\frac{\mathrm{d}}{\mathrm{d}E} = \frac{\nu_2^3}{2\mathcal{R}} \frac{\mathrm{d}}{\mathrm{d}\nu_2}\,. \tag{3.257}$$

Strictly speaking a maximum of the derivative with respect to E will not lie at exactly the same position as the corresponding maximum of the derivative with respect to ν_2. This difference is generally ignored, firstly because it is very small due to the weak energy dependence of the factor $\nu_2^3/\mathcal{R}$ in (3.257), and secondly because this makes formulae such as (3.256) very much simpler. The minimal gradients of the function $\tilde{\mu}_1(\nu_2)$ lie at $T_2 = \infty$, i.e at $\nu_2 = n_2 + \frac{1}{2} - \mu_2$ (for $|R_{1,2}| < 1$) which is exactly in the middle between the resonance energies, and the value of the minimal gradient is $|R_{1,2}|^2$.

For *strong coupling,* i.e. for $|R_{1,2}| > 1$, conditions reverse: the gradient in the Lu-Fano plot is minimal for $T_2 = 0$ and maximal for $T_2 = \infty$. In this case the resonant jumps occur at $\nu_2 = n_2 + \frac{1}{2} - \mu_2$ and the associated widths of the resonances (or pseudo-resonant perturbations) are

$$\Gamma = \frac{4\mathcal{R}}{\pi \nu_2^3} \frac{1}{|R_{1,2}|^2} \,. \tag{3.258}$$

Very strong coupling of the channels thus leads to a Rydberg series of very narrow resonances whose positions lie between the positions of the bound states in the excited channel [Mie68].

The case $|R_{1,2}| \approx 1$ is somewhat special. The Lu-Fano plot is now essentially a straight line with unit gradient and it is no longer possible to uniquely define the positions of resonances.

A peculiarity of the 2QDT formula (3.253) is, that the 2QDT parameters it contains are not uniquely defined. The function $\tilde{\mu}_1(\nu_2)$ which one obtains with the parameters μ_1, μ_2, $R_{1,2}$ is not affected if we replace the parameters by

$$\mu_1' = \mu_1 + \frac{\pi}{2} \,, \quad \mu_2' = \mu_2 + \frac{\pi}{2} \,, \quad R_{1,2}' = \frac{\pm 1}{R_{1,2}} \,. \tag{3.259}$$

In real physical situations the 2QDT parameters are not constant but weakly energy dependent. Hence the function (3.253) is not exactly periodic in ν_2 and we obtain a slightly different curve in the Lu-Fano plot for each period of $\tan[\pi(\nu_2 + \mu_2)]$. This is illustrated in Fig. 3.21 for the example of the coupled $^1P^{\mathrm{o}}$ series in Ca I discussed in Sect. 3.4.1 above (cf. Fig. 3.19).

3.4.3 *More Than Two Channels*

After the detailed treatment of two-channel quantum defect theory in Sects. 3.4.1 and 3.4.2 it is now relatively easy to extend the results to the more general case of N coupled Coulomb channels. The central formula of MQDT is a generalization of the two-channel equation (3.250) and reads

$$\det\{\tan\left[\pi(\nu_i + \mu_i)\right] \delta_{i,j} + (1 - \delta_{i,j}) R_{i,j}\} = 0 \,. \tag{3.260}$$

The coupling of the various channels $i = 1, 2, \ldots N$ is described by the weakly energy-dependent Hermitian matrix $R_{i,j}$. We shall continue to use the representation corresponding to the phase shifted reactance matrix which has no diagonal elements. The diagonal effects of deviations from the pure Coulomb potential are contained in the weakly energy-dependent parameters μ_i. The quantities ν_i have one of two meanings depending on whether the respective channel i is closed or open. For energies below the channel threshold I_i, ν_i is the continuous effective quantum

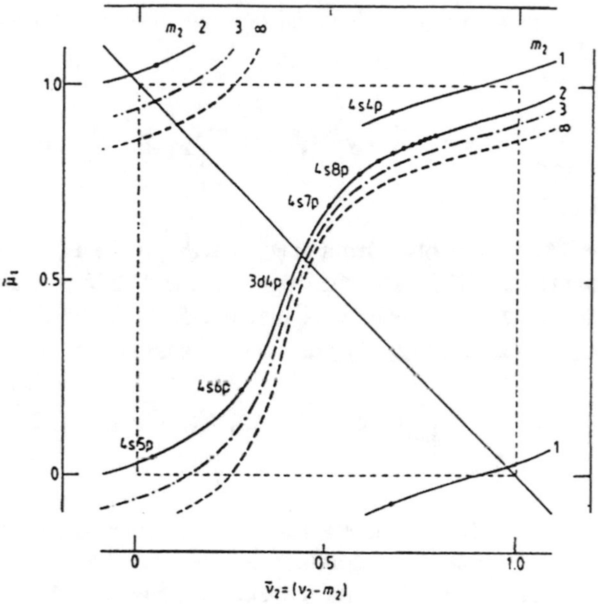

Fig. 3.21 Lu-Fano-plot for coupled $4s\,np$ and $3d\,np$ ${}^1P^{\mathrm{o}}$ channels in Ca I (From [Sea83])

number in the closed channel i; at energies above I_i the quantity $-\pi\nu_i$ is the asymptotic phase shift of the channel wave function in the open channel i, see (3.251).

Because it is so important we shall derive the MQDT equation (3.260) in another way. Let's first consider the energy range where all N channels are open. For a given energy there are then N linearly independent solutions of the coupled-channel equations, and each solution Φ has N components, namely the channel wave functions $\phi_i(r)$, $i = 1, \dots N$. We choose a basis $\Phi^{(j)}$ of solutions with channel wave functions $\phi_i^{(j)}$ whose asymptotic behaviour corresponds to a generalization of (3.244), (3.245):

$$\phi_j^{(j)}(r) \overset{r\to\infty}{=} \phi_{\mathrm{reg}}^{(j)}(r)\,, \quad \phi_i^{(j)}(r) \overset{r\to\infty}{=} R_{i,j}\,\phi_{\mathrm{irr}}^{(i)}(r)\,, \quad i \neq j\,. \tag{3.261}$$

The general solution of the coupled-channel equations is now an arbitrary superposition

$$\Phi = \sum_{j=1}^{N} Z_j \Phi^{(j)} \tag{3.262}$$

of these basis solutions. In a given channel i the channel wave function of the general solution (3.262) is

$$\phi_i(r) = \sum_{j=1}^{N} Z_j \phi_i^{(j)}(r) \stackrel{r\to\infty}{=} Z_i \phi_{\rm reg}^{(i)}(r) + \left(\sum_{j\neq i} R_{i,j} Z_j \right) \phi_{\rm irr}^{(i)}(r) . \tag{3.263}$$

The quotient of the coefficients in front of $\phi_{\rm irr}^{(i)}$ and $\phi_{\rm reg}^{(i)}$ in the asymptotic expression on the right-hand side of (3.263) is the tangent of the additional phase $\delta_i - \pi\mu_i = -\pi(\nu_i + \mu_i)$ by which the channel wave function ϕ_i is asymptotically shifted with respect to the regular solution $\phi_{\rm reg}^{(i)}$ in channel i. In other words,

$$\tan[\pi(\nu_i + \mu_i)] Z_i = -\sum_{j\neq i} R_{i,j} Z_j \iff \tan[\pi(\nu_i + \mu_i)] Z_i + \sum_{j\neq i} R_{i,j} Z_j = 0 . \tag{3.264}$$

Equation (3.264) is a homogeneous system of N linear equations for the N unknowns Z_i, and its matrix of coefficients consists of the diagonal elements $\tan[\pi(\nu_i + \mu_i)]$ and the non-diagonal elements $R_{i,j}$. Non-vanishing solutions exist when the determinant of this matrix vanishes, and this is just the content of the MQDT equation (3.260).

This derivation of the MQDT equation can be extended to lower energies at which some or all channels are closed. To this end the definitions of the regular and irregular solutions $\phi_{\rm reg}^{(i)}$ and $\phi_{\rm irr}^{(i)}$ must be continued to energies below the respective channel thresholds I_i in the closed channels. A detailed description of such a procedure has been given by Seaton for the case that $\phi_{\rm reg}^{(i)}$ and $\phi_{\rm irr}^{(i)}$ are the regular and irregular Coulomb functions [Sea83].

The MQDT equation (3.260) has different meanings in different energy ranges. For simplicity we number the channels in order of increasing channel thresholds:

$$I_1 < I_2 < \cdots < I_N . \tag{3.265}$$

For $E < I_1$ all channels are closed and (3.260) is a condition for the existence of a bound state.

For $I_1 < E < I_2$ only channel 1 is open while all other channels are closed. The MQDT equation is now an explicit equation for the asymptotic phase shift of the wave function in the open channel 1. Expanding the determinant we can rewrite (3.260) as

$$\tan[\pi(\nu_1 + \mu_1)] \det \mathbf{R}_{11} = \sum_{j=2}^{N} (-1)^j R_{1,j} \det \mathbf{R}_{1,j} , \tag{3.266}$$

or

$$\delta_1 = \pi\mu_1 - \arctan\left[\frac{\sum_{j=2}^{N}(-1)^j R_{1,j}\det\mathbf{R}_{1j}}{\det\mathbf{R}_{11}}\right] . \tag{3.267}$$

Here $\mathbf{R}_{1j}$ is the matrix which emerges from the matrix $\{\tan[\pi(\nu_i + \mu_i)]\delta_{i,j} + (1-\delta_{i,j})R_{i,j}\}$ in (3.260) if we eliminate the first row and the jth column. In particular, $\mathbf{R}_{11}$ is the matrix we would use to formulate an MQDT equation for the $N-1$ closed channels $i = 2, \dots N$ without considering coupling to the open channel 1. The zeros of $\det\mathbf{R}_{11}$ thus correspond to bound states of the mutually coupled closed channels $i = 2, \dots N$. Equation (3.267) describes $N-1$ coupled Rydberg series of autoionizing resonances due to these bound states of the coupled closed channels.

For $I_1 < \dots < I_n < E < I_{n+1} < \dots < I_N$ the lower n channels are open and the upper $N-n$ channels are closed. Now there are n linearly independent solutions of the coupled-channel equations and each solution is characterized by n asymptotic phase shifts δ_i in the open channels $i = 1, \dots n$. For $n \geq 2$ the MQDT equation (3.260) has the meaning of a compatibility equation for these asymptotic phase shifts.

The intricate and complicated structure which spectra can acquire when more than two channels couple already becomes apparent in the three-channel case [GG83, GL84, WF87]. In the energy interval $I_1 < E < I_2 < I_3$ in which channel 1 is open while channels 2 and 3 are closed, the interference of two Rydberg series of autoionizing resonances leads to quite complex spectra. In this energy range the 3QDT equation (3.260) (with $N = 3$) is an explicit equation for the phase shift δ_1 in the open channel. Using the abbreviations

$$T_2(E) = \tan[\pi(\nu_2 + \mu_2)] , \quad T_3(E) = \tan[\pi(\nu_3 + \mu_3)] , \tag{3.268}$$

the 3QDT equation reads

$$\tan(\delta_1 - \pi\mu_1) = -\frac{|R_{1,2}|^2\, T_3 + |R_{1,3}|^2\, T_2 - R_{1,2}R_{2,3}R_{3,1} - R_{1,3}R_{3,2}R_{2,1}}{T_2T_3 - |R_{2,3}|^2} . \tag{3.269}$$

The 3QDT equation (3.269) has the same form as (3.206) in Sect. 3.3.3 which describes the influence of coupling to just two bound states in different closed channels. Equation (3.206) becomes the 3QDT equation (3.269) if we replace $(E - \varepsilon_i)$, $i = 2, 3$ by $1/\pi$ times the periodic functions $T_i(E)$ defined in (3.268), and the coupling matrix elements $W_{i,j}$ by $-R_{i,j}/\pi$ as suggested by (3.246),

$$E - \varepsilon_i \to \frac{T_i}{\pi} = \frac{1}{\pi}\tan[\pi(\nu_i + \mu_i)] , \quad W_{i,j} \to -\frac{R_{i,j}}{\pi} . \tag{3.270}$$

Both $\delta_{\rm res}$ in (3.206) and $\delta_1 - \pi\mu_1$ in (3.269) refer to the additional phase shift due to the non-diagonal coupling effects.

Whereas Sect. 3.3.3 described the interference of just two autoionizing resonances, the 3QDT equation (3.269) accounts for the interference of two whole Rydberg series of resonances.

If coupling to the open channel 1 is neglected, then the closed channels 2 and 3 support a series of bound states at the energies given by

$$T_2 T_3 = |R_{2,3}|^2 , \tag{3.271}$$

which is just the two-channel QDT equation. We had assumed that $I_2 < I_3$, so this defines a Rydberg series of bound states below I_2, which is distorted by a Rydberg series of perturbers from channel 3, as described in Sect. 3.4.1. Due to their coupling to the open channel 1, these bound states appear as a perturbed Rydberg series of Feshbach resonances, and they are characterized by the poles on the right-hand side of (3.269).

Equation (3.269) has the form

$$\tan(\delta_1 - \pi\mu_1) = -\frac{N(E)}{D(E)} , \tag{3.272}$$

similar to (3.211) in Sect. 3.3.3, but numerator and denominator are now defined by

$$\begin{aligned} N(E) &= |R_{1,2}|^2\, T_3 + |R_{1,3}|^2\, T_2 - R_{1,2}R_{2,3}R_{3,1} - R_{1,3}R_{3,2}R_{2,1} , \\ D(E) &= T_2 T_3 - |R_{2,3}|^2 . \end{aligned} \tag{3.273}$$

The resonance positions are the zeros of $D(E)$, and the energy derivative of phase shift at resonance is given by

$$\left.\frac{\mathrm{d}}{\mathrm{d}E}(\delta_1 - \mu_1)\right|_{D=0} = \left.\frac{D'}{N}\right|_{D=0} , \tag{3.274}$$

as in (3.212). Neglecting a possible weak energy dependence of the QDT parameters μ_i and $R_{i,j}$ gives

$$\frac{\mathrm{d}T_2}{\mathrm{d}E} = \left(1 + T_2^2\right)\frac{\pi {\nu_2}^3}{2\mathcal{R}} , \quad \frac{\mathrm{d}T_3}{\mathrm{d}E} = \left(1 + T_3^2\right)\frac{\pi {\nu_3}^3}{2\mathcal{R}} . \tag{3.275}$$

When $D(E) = 0$ we can insert $|R_{2,3}|^2/T_3$ for T_2, so

$$\left.\frac{\mathrm{d}}{\mathrm{d}E}(\delta_1 - \pi\mu_1)\right|_{D=0} = \frac{\pi\, \nu_2^3}{2\mathcal{R}|R_{1,2}|^2}\, \frac{T_3^2 + |R_{2,3}|^4 + (1 + T_3^2)|R_{2,3}|^2(\nu_3/\nu_2)^3}{(T_3 - R_{2,3}R_{3,1}/R_{2,1})(T_3 - R_{3,2}R_{1,3}/R_{1,2})} , \tag{3.276}$$

and we obtain

$$\begin{aligned}\Gamma &= 2\left[\frac{\mathrm{d}}{\mathrm{d}E}(\delta_1 - \pi\mu_1)\Big|_{E_\mathrm{R}}\right]^{-1} \\ &= \frac{4\mathcal{R}|R_{1,2}|^2}{\pi\,\nu_2^3}\,\frac{(T_3 - R_{2,3}R_{3,1}/R_{2,1})(T_3 - R_{3,2}R_{1,3}/R_{1,2})}{T_3^2 + |R_{2,3}|^4 + (1+T_3^2)|R_{2,3}|^2(\nu_3/\nu_2)^3} \\ &= \frac{4\mathcal{R}|R_{1,2}|^2}{\pi\,\nu_2^3}\,\frac{|(T_3 - R_{2,3}R_{3,1}/R_{2,1})|^2}{T_3^2 + |R_{2,3}|^4 + (1+T_3^2)|R_{2,3}|^2(\nu_3/\nu_2)^3}\,. \end{aligned} \tag{3.277}$$

The interpretation of the expression (3.277) as the width of the Feshbach resonance associated with a given pole of $\tan(\delta_1 - \pi\mu_1)$ at E_R assumes that this width and the widths associated with neighbouring poles are not so large that the resonances overlap strongly.

We had assumed that $I_2 < I_3$, so $\nu_3(E)$ remains finite while $\nu_2(E) \to \infty$ as the energy approaches I_2 from below. When the Rydberg series of Feshbach resonances converging to the lower closed-channel threshold I_2 is perturbed by the Rydberg series of resonances from the closed channel with the higher threshold I_3, the resonance positions are perturbed as described by (3.271), and (3.277) shows how the perturbations from channel 3 affect the associated resonance widths. The factor

$$\Gamma_0 = \frac{4\mathcal{R}|R_{1,2}|^2}{\pi\,\nu_2^3} \tag{3.278}$$

describes the widths expected in an unperturbed Rydberg series of Feshbach resonances, compare (3.256) in Sect. 3.4.2, while the following quotient describes the modifications due to the perturbations from channel 3. Towards the lower closed-channel threshold I_2 the ratio $\nu_3(E)/\nu_2(E)$ tends to zero, so the last term in the denominator on the right-hand sides of (3.277) becomes negligible. The modified widths can then be written as

$$\Gamma = \frac{4\mathcal{R}|R_{1,2}|^2}{\pi\,\nu_2^3}\,\frac{|\varepsilon + p|^2}{1+\varepsilon^2}\,, \tag{3.279}$$

with

$$\varepsilon = \frac{T_3}{|R_{2,3}|^2}\,, \quad p = -\frac{R_{3,1}}{R_{3,2}R_{2,1}}\,. \tag{3.280}$$

Equation (3.279) represents the unperturbed widths (3.278) multiplied by the modulus of a Beutler-Fano function, compare (3.194) in Sect. 3.3.2. The shape parameter is p, and the role of the reduced energy is played by the quantity ε as defined in (3.280). It varies from $-\infty$ to $+\infty$ during each period of the tangent defining $T_3(E)$ according to (3.268), and it passes through zero at each energy of the Rydberg series of bound states in the uncoupled closed channel 3. If the shape parameter p is real, then in each such period corresponding to one perturber from the closed channel 3, there is a point of maximum width when $\varepsilon = 1/p$ and a point of vanishing width when $\varepsilon = -p$. If this point of vanishing width coincides with the position of one of the resonances in the perturbed series, which are the zeros of $D(E)$, cf. (3.273), then there actually is a resonance of vanishing width, i.e., a bound state in the continuum. The condition for this to occur is,

$$T_3 = \frac{R_{2,3}R_{3,1}}{R_{2,1}} , \quad T_2 = \frac{|R_{2,3}|^2}{T_3} = \frac{R_{3,2}R_{2,1}}{R_{3,1}} . \tag{3.281}$$

If (3.279) is not a good description of resonance widths in the perturbed series, either because the resonances overlap too strongly or because it is not justified to neglect the term containing $(\nu_3/\nu_2)^3$ in the denominators on the right-hand sides of (3.277), there nevertheless always is a bound state in the continuum when the conditions (3.281) are fulfilled. This is because the zeros of numerator and denominator on the right-hand side of (3.272) vanish simultaneously, compare discussion around (3.215) in Sect. 3.3.3.

Let us now estimate the maximum broadening caused according to (3.277) by perturbation of the Rydberg series of resonances. The right-hand side of (3.279) is an upper limit to the right-hand sides in (3.277), because it was obtained by neglecting the term proportional to $(\nu_3/\nu_2)^3$ in the denominator. The modulus of the Beutler-Fano function can be no larger that $1 + |p|^2$, and hence the maximum widths can be no larger than

$$\Gamma_{\rm max} = \frac{4\mathcal{R}}{\pi\nu_2^3}\left[|R_{1,2}|^2 + \left|\frac{R_{3,1}}{R_{3,2}}\right|^2\right] . \tag{3.282}$$

The zeros in the denominator on the right-hand side of (3.269) can only be interpreted as the positions of resonances if their separation is larger than the widths of the resonances. This condition is fulfilled if the maximum widths (3.282) are smaller than the separations which can be approximated by the separations $2\mathcal{R}/\nu_2^3$ in the unperturbed Rydberg series (of resonances). We thus obtain the following condition for the validity of the general formula (3.277) for the widths in a perturbed Rydberg series of autoionizing resonances:

$$|R_{1,2}|^2 + \left|\frac{R_{1,3}}{R_{2,3}}\right|^2 < \frac{\pi}{2} . \tag{3.283}$$

Remember, however, that if the conditions (3.281) for a bound state in the continuum are fulfilled exactly or approximately, then numerator and denominator on the right-hand side of (3.269) vanish at exactly or almost exactly the same energy and we obtain vanishing or very small widths irrespective of whether (3.283) is fulfilled or not.

Finally we can also give a formula for the photoabsorption cross sections or oscillator strengths in a perturbed Rydberg series of autoionizing resonances. To this end we exploit the analogy to the situation described in Sect. 3.3.3, where there are just two bound states in different closed channels and where the oscillator strengths are described by (3.220). Making the transition (3.270) we obtain

$$\frac{\mathrm{d}f_{E\mathrm{i}}}{\mathrm{d}E} = \frac{2\mu}{\hbar}\,\omega\,|d_1|^2 \times \frac{\left|D - \frac{d_2}{d_1}(T_3R_{2,1} - R_{2,3}R_{3,1}) - \frac{d_3}{d_1}(T_2R_{3,1} - R_{3,2}R_{2,1})\right|^2}{D^2 + N^2}\;; \tag{3.284}$$

N and D are again as defined in (3.273) above. This formula for oscillator strengths in a system of three coupled Coulomb channels was first derived in 1984 by Giusti and Lefebvre-Brion [GL84].

The general expression (3.284) can formally be written as a product of an unperturbed oscillator strength multiplied by the modulus of a Beutler-Fano function (cf. (3.240)):

$$\frac{\mathrm{d}f_{E\mathrm{i}}}{\mathrm{d}E} = \frac{2\mu}{\hbar}\,\omega\,|d_1|^2\,\frac{|\tilde{q} + \tilde{\varepsilon}|^2}{1 + \tilde{\varepsilon}^2}\,, \tag{3.285}$$

where the energy parameter $\tilde{\varepsilon}$ and the parameter $\tilde{q}$ are now given by

$$\tilde{\varepsilon} = D(E)/N(E) \qquad \text{and} \tag{3.286}$$

$$\tilde{q} = -\frac{\frac{d_2}{d_1}(T_3R_{2,1} - R_{2,3}R_{3,1}) + \frac{d_3}{d_1}(T_2R_{3,1} - R_{3,2}R_{2,1})}{N(E)}\,. \tag{3.287}$$

Near a zero E_R of $D(E)$, the energy parameter is again a linear function of energy,

$$\tilde{\varepsilon} = \frac{E - E_\mathrm{R}}{\Gamma/2}\,, \tag{3.288}$$

where Γ is the width given by (3.277). The formula (3.287) can be used to define shape parameters as long as it makes sense to assign a single value of $\tilde{q}$ to an individual resonance. If the widths are not too large, (3.285) again describes a series of Beutler-Fano-type resonances. In contrast to an unperturbed Rydberg series of autoionizing resonances however, the widths vary strongly within the series according to (3.277) and the shape parameters can no longer be accounted for by one energy-independent or only weakly energy-dependent number as in (3.242). If the resonances are so narrow that we can take the function $T_2(E) = \tan[\pi(\nu_2 + \mu_2)]$, which covers the whole range of values from $-\infty$ to $+\infty$ in each period, as essentially constant over the width of a resonance, then we can insert its value $T_2(E) = |R_{1,2}|^2/T_3(E)$ at each resonance energy into (3.287) and obtain a simple formula describing the variation of the shape parameter $\tilde{q}$ within a perturbed Rydberg series of autoionizing resonances:

$$\tilde{q} = \left(-\frac{d_2/d_1}{R_{1,2}}\right)\left(\frac{T_3 - R_{3,2}\,d_3/d_2}{T_3 - R_{3,2}R_{1,3}/R_{1,2}}\right). \tag{3.289}$$

The first factor on the right-hand side of (3.289) is the shape parameter q which one would expect in an unperturbed Rydberg series of autoionizing resonances according to (3.242). The second factor describes the changes due to the perturbations. If the 3QDT parameters are real, then in each period of T_3 there is a zero of $\tilde{q}$ at

$$T_3 = R_{3,2}\frac{d_3}{d_2} \tag{3.290}$$

and a pole at

$$T_3 = R_{3,2}\frac{R_{1,3}}{R_{1,2}}. \tag{3.291}$$

The pole position (3.291) is just the point of vanishing width, see (3.281). Here the height $1 + \tilde{q}^2$ of a resonance line becomes infinite in principle, but the product of the height and the width (3.277) remains finite. The sign of the shape parameter $\tilde{q}$ changes both at the zero (3.290) and at the pole (3.291). This sign change is known under the name of *q-reversal* and is conspicuous in spectra as an interchange of the steep and the flat sides of Beutler-Fano-type resonance lines (see Fig. 3.22).

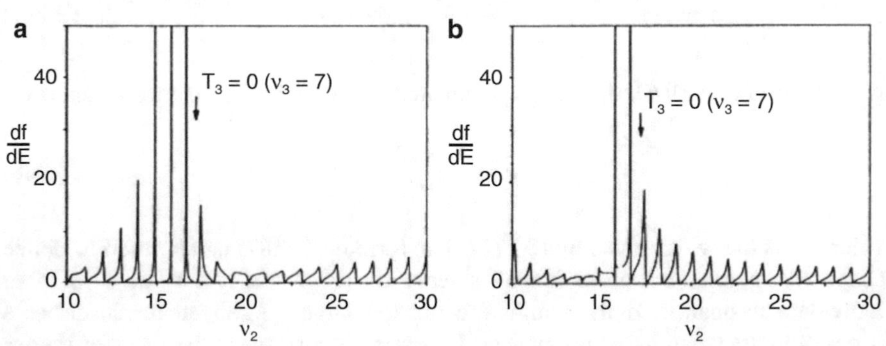

Fig. 3.22 Oscillator strengths (3.284) in a perturbed Rydberg series of autoionizing resonances. The following 3QDT parameters are common to both parts of the figure: $\mu_2 = \mu_3 = 0$, $R_{1,2} = 0.4$, $R_{1,3} = -0.2$, $R_{2,3} = 0.5$, $I_3 - I_2 = 0.017$ Rydberg energies. The centre of the perturber ($T_3 = 0$ for $\nu_3 = 7.0$) is at $\nu_2 = 17.13$. In part (**a**) the dipole transition parameters are $d_2/d_1 = d_3/d_1 = 1$, so that the point of vanishing width and the point of vanishing q-value lie on opposite sides of the centre $T_3 = 0$. In part (**b**) we have $d_2/d_1 = 0.5$, $d_3/d_1 = -1$, so that both q-reversals lie to the left of $T_3 = 0$. The oscillator strengths $\mathrm{d}f/\mathrm{d}E$ are given in units of the oscillator strength $(2\mu/\hbar)\omega|d_1|^2$ which we would expect in absence of coupling to the channels 2 and 3

The relation (3.289) was derived from (3.287) with the help of some rather crude approximations, but it does enable us to qualitatively understand some of the different structures which can appear in a perturbed Rydberg series of autoionizing resonances. Figure 3.22 shows two examples of the oscillator strength (3.284) as a function of the continuous effective quantum number ν_2 in channel 2. The series is perturbed around $\nu_2 \approx 17$ by a state with effective quantum number $\nu_3 = 7$ (in channel 3). The two q-reversals, one at the point of vanishing width and one at the zero of the shape parameter, are easy to discern in both cases.

If more than one channel is open, then a resonant state in a Rydberg series of autoionizing resonances can decay into several decay channels, and the total autoionization width is a sum of the partial widths into the individual open channels. If such a Rydberg series is perturbed by states in further closed channels, then one consequence of such perturbations is a strong energy dependence of the *branching ratios,* which are the ratios of the partial decay widths [VC88]. A detailed description of characteristic features of MQDT spectra in cases with two or more open channels was given by Cohen in [Coh98].

In real physical situations we often have to consider more than two or three Coulomb channels. Figure 3.23 shows part of a photoionization spectrum to $J = 2$ states in neutral barium in an energy interval in which both the $5d_{3/2}ns$ and the $5d_{3/2}nd$ series of autoionizing resonances are perturbed by the $5d_{5/2}14s_{1/2}$ resonance. The lower part of the figure shows the results of an MQDT calculation involving six closed and two open channels. Many significant features in the spectrum are accurately reproduced by the MQDT fit. Note however, that an application of MQDT with so many channels already involves a large number of parameters which are not easy to determine uniquely by fitting even such a rich spectrum.

Problems of non-uniqueness of the MQDT parameters typically occur when we treat them as independent empirical parameters to be determined in a fit to experimental data. In an analysis of Rydberg spectra of molecules, Jungen and Atabek implemented a frame transformation which allowed them to calculate a large number of independent elements of the reactance matrix on the basis of a few fundamental dynamical parameters. This made it possible to apply the MQDT with quite large numbers (up to 30) of channels [JA77].

An *ab initio* theory without empirical parameters requires a solution (at least an approximate solution) of the many-electron Schrödinger equation, or, in the present case, of the coupled-channel equations. In the so-called *R matrix method* [Gre83, OG85], coordinate space is divided into an inner region of radius R and an outer region. The many-electron problem is solved approximately in the inner region and, at $r = R$, the solutions are matched to the appropriate asymptotic one-electron wave functions in each channel, which consist of superpositions of regular and irregular Coulomb functions or, in closed channels, of the corresponding Whittaker functions. MQDT is still useful in connection with such *ab initio* theories, because the weakly energy-dependent MQDT parameters can be calculated (and stored) on a comparatively sparse mesh of energies and the complicated and sometimes violent energy dependences in physical observables follow from the

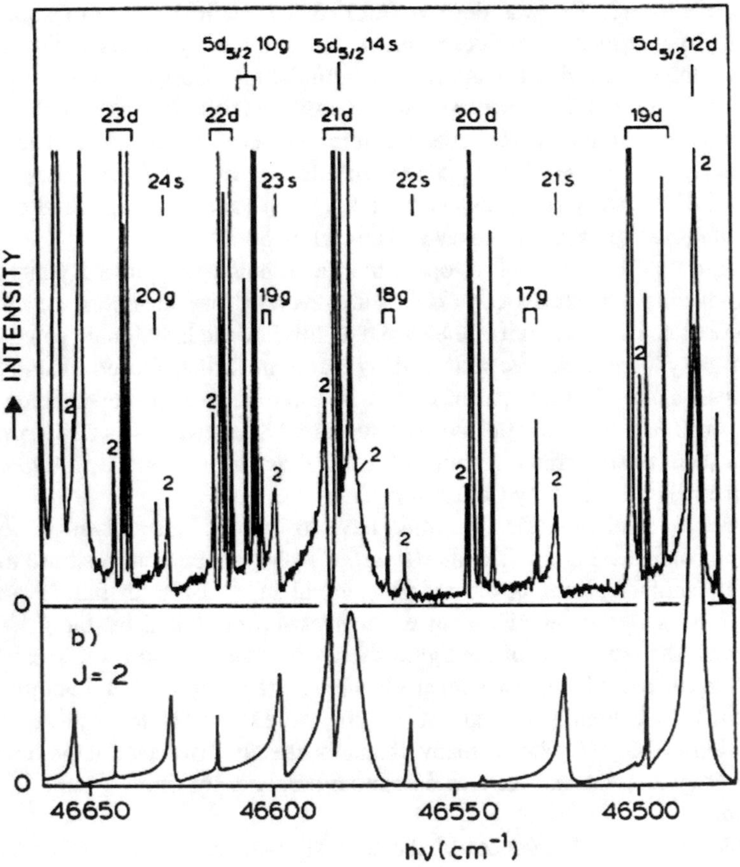

Fig. 3.23 Photoionization spectrum in barium near the $5d_{5/2}\,14s_{1/2}\,J = 2$ state which perturbs the $5d_{3/2}\,ns$ and the $5d_{3/2}\,nd$ series. The lower part of the figure shows the results of an MQDT analysis involving six closed and two open channels (From [BH89])

MQDT equations. The combination of MQDT and *R* matrix methods has been applied with considerable success, especially to the description of the spectra of alkali earth atoms [AL87, AL89, LA91, AB94, AL94, LA94, LU95, AG96].

3.5 Atoms in External Fields

Everything said up to now has to be modified more or less strongly if we consider atoms (or ions) which are not isolated, but influenced by an external electromagnetic field. For low-lying bound states of an atom the influence of external fields can often be satisfactorily accounted for with perturbative methods, but this is no longer possible for highly excited states and/or very strong fields, in which case intricate and physically interesting effects occur, even in the "simple" hydrogen atom. The study of atoms (and molecules) in strong external fields has been a topic of considerable interest for many years, and this is documented by the publication of several books on the subject [NC90, RW94, CK97, SS98].

In this section we consider a classical electromagnetic field described by the scalar potential $\Phi(\boldsymbol{r}, t)$ and the vector potential $\boldsymbol{A}(\boldsymbol{r}, t)$. The Hamiltonian for an N-electron atom or ion is then (see (2.151))

$$\hat{H} = \sum_{i=1}^{N} \left(\frac{[\hat{\boldsymbol{p}}_i + (e/c)\boldsymbol{A}(\boldsymbol{r}_i, t)]^2}{2\mu} - e\,\Phi(\boldsymbol{r}_i, t) \right) + \hat{V} . \tag{3.292}$$

An important consequence of external fields is, that the Hamiltonian (3.292) is in general no longer rotationally invariant, so that its eigenstates aren't simultaneously eigenstates of angular momentum. For spatially homogeneous fields and the appropriate choice of gauge the Hamiltonian does however remain invariant under rotations around an axis parallel to the direction of the field, so that the component of total angular momentum in the direction of the field remains a constant of motion. For an electron in a potential which is not radially symmetric but invariant under rotations around the z-axis, say, we can at least reduce the three-dimensional problem to a two-dimensional problem by transforming to *cylindrical coordinates* ϱ, z, ϕ:

$$x = \varrho \cos\phi , \quad y = \varrho \sin\phi , \quad z = z ; \quad \varrho = \sqrt{x^2 + y^2} . \tag{3.293}$$

With the ansatz

$$\psi(\boldsymbol{r}) = f_m(\varrho, z)\, \mathrm{e}^{\mathrm{i}m\phi} \tag{3.294}$$

we can reduce the stationary Schrödinger equation to an equation for the function $f_m(\varrho, z)$:

$$\left[-\frac{\hbar^2}{2\mu} \left(\frac{\partial^2}{\partial \varrho^2} + \frac{1}{\varrho}\frac{\partial}{\partial \varrho} - \frac{m^2}{\varrho^2} + \frac{\partial^2}{\partial z^2} \right) + V(\varrho, z) \right] f_m(\varrho, z) = E f_m(\varrho, z) . \tag{3.295}$$

3.5.1 *Atoms in a Static, Homogeneous Electric Field*

We describe a static homogeneous electric field $\boldsymbol{E}$, which is taken to point in the direction of the z-axis, by a time-independent scalar potential

$$\Phi(\boldsymbol{r}) = -E_z\, z \tag{3.296}$$

and a vanishing vector potential. The Hamiltonian (3.292) then has the following special form:

$$\hat{H} = \sum_{i=1}^{N} \frac{\hat{p}_i^2}{2\mu} + \hat{V} + eE_z \sum_{i=1}^{N} z_i \, . \tag{3.297}$$

The shifts in the energy eigenvalues caused by the contribution of the field in (3.297) are given in time-independent perturbation theory (see Sect. 1.6.1) to first order by (1.249),

$$\Delta E_n^{(1)} = eE_z \langle \psi_n | \sum_{i=1}^{N} z_i \, |\psi_n\rangle \, , \tag{3.298}$$

where ψ_n are the eigenstates of the unperturbed ($E_z = 0$) Hamiltonian. As mentioned in Sect. 2.2.4, these eigenstates are usually eigenstates of the N-electron parity operator, so that the expectation values (3.298) of the operator $\sum_{i=1}^{N} z_i$, which changes the parity, vanish. In second order, the energy shifts are given by (1.255) or (1.266),

$$\Delta E_n^{(2)} = (eE_z)^2 \sum_{E_m \neq E_n} \frac{\left|\langle \psi_n | \sum_{i=1}^{N} z_i |\psi_m\rangle\right|^2}{E_n - E_m} \, , \tag{3.299}$$

where E_n and E_m are the eigenvalues of the unperturbed Hamiltonian. The right-hand side of (3.299) should include the continuum, so the sum is to be replaced by an integral above the continuum threshold. The energy shifts (3.299) depend quadratically on the strength E_z of the electric field and are known under the name *quadratic Stark effect*.

The energy shifts (3.299) are closely connected with the *dipole polarizability* of the atom in an electric field. The modification of wave functions caused by an infinitesimally weak electrical field can be described in first-order perturbation theory. As long as the unperturbed state ψ_n is not degenerate with an unperturbed state of opposite parity, the modified eigenfunctions to first order are described by (1.253),

$$|\psi_n'\rangle = |\psi_n\rangle + eE_z \sum_{m \neq n} \frac{\langle \psi_m | \sum_{i=1}^{N} z_i |\psi_n\rangle}{E_n - E_m} \, |\psi_m\rangle \, . \tag{3.300}$$

The wave functions (3.300) are no longer eigenfunctions of the N-electron parity, and they have a dipole moment induced by the external field and pointing in the direction of the field (the z-direction). The z-component of the induced dipole moment is

$$d_z = -e\langle\psi_n'|\sum_{i=1}^{N} z_i|\psi_n'\rangle$$

$$= 2e^2E_z\sum_{m\neq n}\frac{|\langle\psi_m|\sum_{i=1}^{N} z_i|\psi_n\rangle|^2}{E_m - E_n} := \alpha_{\mathrm{d}}E_z\,. \tag{3.301}$$

Using the dipole polarizability α_{d} defined by (3.301) (for the state ψ_n) we can write the energy shift (3.299) of the quadratic Stark effect as

$$\Delta E_n^{(2)} = -\frac{\alpha_{\mathrm{d}}}{2}E_z^2\,. \tag{3.302}$$

In the unusual case that an eigenvalue of the unperturbed Hamiltonian is degenerate and has eigenstates of different parity, we already obtain non-vanishing energy shifts in first order, and this is called the *linear Stark effect*. The first-order energy shifts are calculated by diagonalizing the perturbing operator $eE_z\sum_{i=1}^{N} z_i$ in the subspace of the eigenstates with the degenerate (unperturbed) energy, see equation (1.258). An important example is found in the one-electron atoms, where each principal quantum number $n \geq 2$ corresponds to a degenerate energy eigenvalue with eigenfunctions of different parity $(-1)^l$. The interaction matrix elements between two degenerate eigenfunctions $\psi_1(\boldsymbol{r}) = Y_{l_1,m_1}(\Omega)\phi_{n,l_1}(r)/r$ and $\psi_2(\boldsymbol{r}) = Y_{l_2,m_2}(\Omega)\phi_{n,l_2}(r)/r$ are

$$\langle\psi_1|eE_z z|\psi_2\rangle = eE_z r_{12}^{(0)}\,, \tag{3.303}$$

where $r_{12}^{(0)}$ is the $\nu = 0$ spherical component of the vector matrix element as defined in (2.208). The matrix element (3.303) is non-vanishing only if the azimuthal quantum numbers in bra and ket are the same, $m_1 = m_2$. For $n = 2$ there is a non-vanishing matrix element between the $l = 0$ and $l = 1$ states with $m = 0$. The two further $l = 1$ states with azimuthal quantum numbers $m = +1$ and $m = -1$ are unaffected by the linear Stark effect (see Problem 3.8). Figure 3.24(a) shows the splitting of the $n = 2$ term in the hydrogen atom due to the linear Stark effect. For comparison Fig. 3.24(b) shows the energy shift (3.302) of the $n = 1$ level due to the quadratic Stark effect (see Problem 3.9).

The perturbative treatment of the Stark effect is not unproblematic. This becomes obvious when we consider that the perturbing potential $eE_z\sum_{i=1}^{N} z_i$ (positive field strength E_z assumed) tends to $-\infty$ when one of the z_i goes to $-\infty$. The perturbed Hamiltonian (3.297) is not bounded from below and has no ground state; strictly speaking it has no bound states at all and no discrete eigenvalues, but a continuous energy spectrum unbounded from above and below. In the presence of the electric field the bound states of the unperturbed Hamiltonian become resonances and the width of each such resonance is $\hbar/\tau$, where τ is the lifetime of the state with

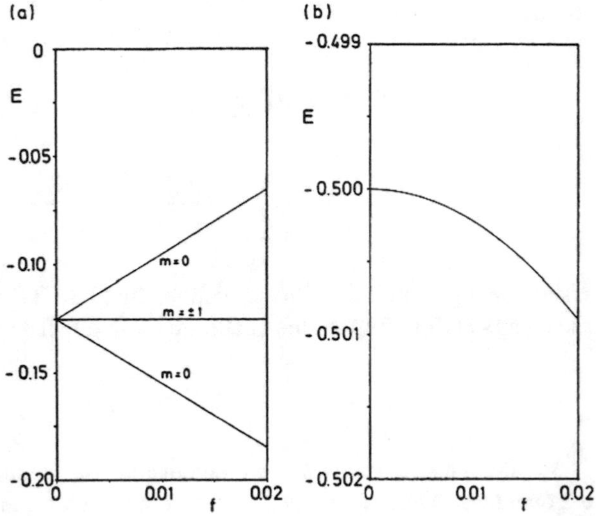

Fig. 3.24 (**a**) Splitting of the degenerate $n = 2$ level in hydrogen due to the linear Stark effect (Problem 3.8). (**b**) Energy shift of the hydrogen ground state due to the quadratic Stark effect (3.302). f is the electric field strength in units of $E_0 \approx 5.142 \times 10^9$ V/cm (3.308), and the energies are in atomic units

respect to decay via *field ionization*. For low-lying states and not too strong fields these lifetimes are so long that the states can be regarded as bound for all practical purposes, but for highly excited states and/or very strong fields the lifetimes can be short and the widths of the resonant states large. Even for an arbitrarily small but finite field strength perturbation theory loses its justification at sufficiently high excitations. The transition from vanishing to small but finite field strengths is not continuous at threshold. For vanishing strength of the external field the long-ranged Coulomb potential supports infinitely many bound states accumulating at threshold. In an arbitrarily weak but non-vanishing electric field there are no bound states.

Classically, field ionization is possible above the *Stark saddle*. For a one- electron atom,

$$V(\boldsymbol{r}) = -\frac{Ze^2}{r} + eE_z z\,, \quad E_z > 0\,, \tag{3.304}$$

the Stark saddle is located on the negative z-axis at the local maximum of $V(x{=}0, y{=}0, z)$. Here the potential energy has a minimum in the two directions perpendicular to the z-axis (see Fig. 3.25). The position z_S and energy V_S of the Stark saddle are:

$$z_\mathrm{S} = -\sqrt{\frac{Ze}{E_z}}\,, \quad V_\mathrm{S} = -2e\sqrt{ZeE_z}\,. \tag{3.305}$$

For a one-electron atom described by a pure Coulomb potential $-Ze^2/r$, it is possible in *parabolic coordinates* to decouple the Schrödinger equation into ordinary differential equations, even in the presence of an external electric field.

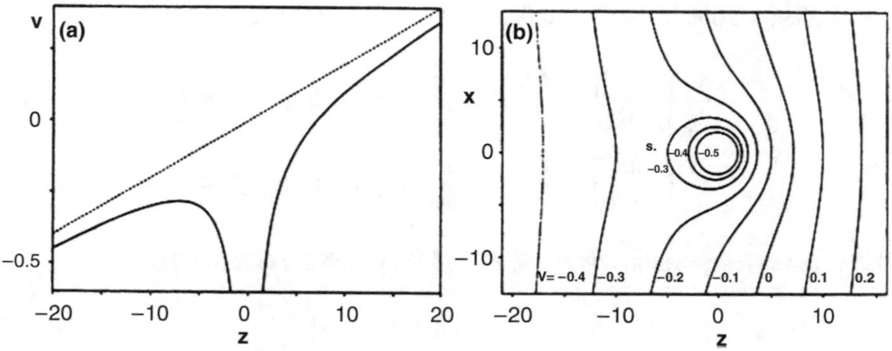

Fig. 3.25 Potential energy (3.304) for a one-electron atom in an electric field of $f = 0.02$ atomic units (see (3.308)). (**a**) Potential along the z-axis; (**b**) equipotential lines in the zx-plane. The point "S" marks the Stark saddle

As the electric field in z-direction doesn't disturb the rotational symmetry around the z-axis, it is sensible to keep the azimuthal angle ϕ as one of the coordinates. The two other coordinates ξ and η have the physical dimension of a length and are defined by

$$\xi = r + z\,, \quad \eta = r - z\,; \quad r = \frac{1}{2}(\xi + \eta)\,, \quad z = \frac{1}{2}(\xi - \eta)\,. \tag{3.306}$$

The coordinates ξ and η can assume values between zero and $+\infty$. They are called parabolic, because surfaces defined by ξ = const. and η = const. are rotational paraboloids around the z-axis.

In parabolic coordinates and atomic units the Hamiltonian for an electron under the influence of a Coulomb potential and an external electric field is

$$\hat{H} = -\frac{2}{\xi + \eta}\left[\frac{\partial}{\partial \xi}\left(\xi \frac{\partial}{\partial \xi}\right) + \frac{\partial}{\partial \eta}\left(\eta \frac{\partial}{\partial \eta}\right)\right] - \frac{1}{2\xi\eta}\frac{\partial^2}{\partial \phi^2} - \frac{2Z}{\xi + \eta} + f\,\frac{\xi - \eta}{2}\,. \tag{3.307}$$

Here f is the electric field strength in atomic units:

$$f = \frac{E_z}{E_0}, \quad E_0 = \frac{e}{a^2} = \frac{\mu^2 e^5}{\hbar^4} \approx 5.142 \times 10^9\,\mathrm{V/cm}\,. \tag{3.308}$$

If we multiply the Schrödinger equation $\hat{H}\psi = E\psi$ by $(\xi + \eta)/2$ and insert the product ansatz

$$\psi = f_1(\xi) f_2(\eta)\, \mathrm{e}^{\mathrm{i}m\phi} \tag{3.309}$$

for ψ, then we obtain two decoupled equations for $f_1(\xi)$ and $f_2(\eta)$

$$\frac{d}{d\xi}\left(\xi\frac{df_1}{d\xi}\right)+\left(\frac{E}{2}\xi-\frac{m^2}{4\xi}-\frac{f}{4}\xi^2\right)f_1+Z_1f_1=0\,,$$
$$\frac{d}{d\eta}\left(\eta\frac{df_2}{d\eta}\right)+\left(\frac{E}{2}\eta-\frac{m^2}{4\eta}+\frac{f}{4}\eta^2\right)f_2+Z_2f_2=0\,. \quad (3.310)$$

There are two separation constants, Z_1 and Z_2, which are related by

$$Z_1+Z_2=Z\,. \quad (3.311)$$

Dividing the upper equation (3.310) by 2ξ and the lower equation by 2η yields

$$\left[-\frac{1}{2}\left(\frac{d^2}{d\xi^2}+\frac{1}{\xi}\frac{d}{d\xi}-\frac{m^2}{4\xi^2}\right)-\frac{Z_1}{2\xi}+\frac{f}{8}\xi\right]f_1(\xi)=\frac{E}{4}f_1(\xi)\,, \quad (3.312)$$

$$\left[-\frac{1}{2}\left(\frac{d^2}{d\eta^2}+\frac{1}{\eta}\frac{d}{d\eta}-\frac{m^2}{4\eta^2}\right)-\frac{Z_2}{2\eta}-\frac{f}{8}\eta\right]f_2(\eta)=\frac{E}{4}f_2(\eta)\,. \quad (3.313)$$

The equations (3.312), (3.313) have the form of two cylindrical radial Schrödinger equations with azimuthal quantum number $m/2$ (cf. (3.295)). In addition to the cylindrical radial potential $(m/2)^2/(2\xi^2)$, Equation (3.312) for $f_1(\xi)$, the *uphill equation,* contains a Coulomb potential $-Z_1/2\xi$ and an increasing linear potential $(f/8)\xi$ originating from the electric field. For any (positive or negative) value of the separation constant Z_1, this *uphill potential* supports a sequence of bound solutions with discrete eigenvalues. Conversely, for each energy E there is a discrete sequence of values of Z_1 for which the uphill equation has bound solutions, which are characterized by the respective number $n_1 = 0, 1, 2, \ldots$ of nodes of $f_1(\xi)$ in the region $\xi > 0$. As Z_1 grows larger and larger (it may even be larger than the nuclear charge Z) the number n_1 increases. We obtain a minimal value of Z_1 (which may be negative if the energy E is positive) when the whole uphill potential is just deep enough to support one nodeless eigenstate. (See Fig. 3.26.)

In the field-free case $f = 0$, (3.313) has the same form as (3.312). In the negative energy regime there is a discrete sequence of energies, namely $E_n = -Z^2/(2n^2)$, $n = 1, 2, \ldots$, at which both equations (3.312) and (3.313) with appropriate values of Z_1 and Z_2 simultaneously have square integrable solutions with n_1 and n_2 nodes respectively. For a given azimuthal quantum number $m = 0, \pm1, \pm2, \ldots \pm(n-1)$ the parabolic quantum numbers $n_1 = 0, 1, 2, \ldots$ and $n_2 = 0, 1, 2, \ldots$ are related to the separation constants Z_1, Z_2 and the Coulomb principal quantum number n by [LL65]

$$n_i+\frac{|m|+1}{2}=n\frac{Z_i}{Z}\,,\quad i=1,2\,;\quad n_1+n_2+|m|+1=n\,. \quad (3.314)$$

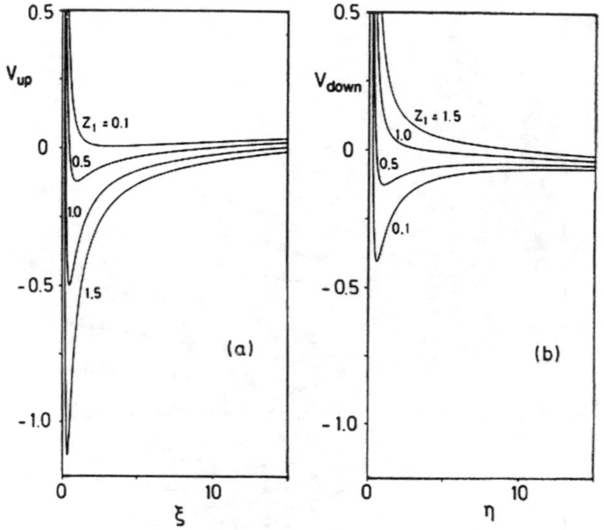

Fig. 3.26 Effective potentials in the uphill equation (3.312) (**a**) and in the downhill equation (3.313) (**b**) for $m = 1$ and four different values of the separation constant Z_1, $(Z_1 + Z_2 = 1)$. The electric field strength is $f = 0.02$ atomic units

If however, the field strength f is non-vanishing (positive), then the *downhill equation* (3.313) can be solved for a given value of the separation constant Z_2 $(= Z - Z_1)$ at any energy with the appropriate boundary conditions. The solutions $f_2(\eta)$ do not behave like regular and irregular Coulomb functions asymptotically, because the potential decreases linearly and so the kinetic energy increases linearly with η. The wave function for the asymptotic motion of an electron accelerated in such a linear potential is a superposition of Airy functions (see Appendix A.4), and it is increasingly well approximated by the semiclassical WKB expression (1.289), because the quantality function (1.298) asymptotically vanishes as the inverse cube of the coordinate. The low-lying bound states of the field-free case become narrow resonances in the presence of a finite field, and these can be identified by more or less sudden jumps through π of a phase shift describing the influence of deviations from the homogeneous linear potential [TF85], cf. Sect. 1.5. As the energy increases, so does the width of these resonances corresponding to a decreasing lifetime with respect to field ionization.

A systematic theoretical investigation of the Stark spectrum of hydrogen was published in 1980 by Luc-Koenig and Bachelier [LB80]. Figure 3.27 shows the spectrum for azimuthal quantum number $|m| = 1$. In the field-free case each principle quantum number n accommodates $n - |m|$ degenerate eigenstates, which can be labelled by the possible values of the parabolic quantum number $n_1 = 0, 1, \ldots, n - |m| - 1$. A finite field strength lifts the degeneracy in this n manifold. The energy is shifted downward most for states with small values of the (uphill) quantum number n_1, because they have the largest fraction of the wave function

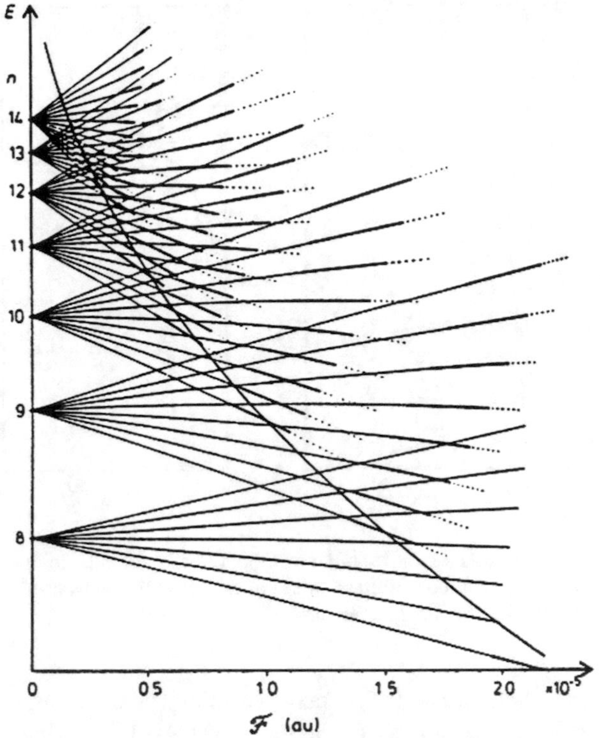

Fig. 3.27 Stark splitting and decay widths with respect to field ionization for the $|m| = 1$ states in hydrogen. The *thin lines* show states with widths less than 5×10^{-12} atomic units, the *thick lines* indicate widths between 5×10^{-12} and 5×10^{-8}, and the *dashed lines* represent resonances broader than 5×10^{-8} atomic units. The energy of the Stark saddle is shown by the thick curve running from the upper left to the lower right corner in the figure (From [LB80])

concentrated in the downhill direction. Since small values of n_1 are connected to small values of the separation constant Z_1 and correspondingly large values of $Z_2 = 1 - Z_1$, these states have the largest decay widths, because larger values of Z_2 imply a more strongly attractive Coulomb potential and a smaller potential barrier against field ionization in the downhill equation. Conversely, solutions corresponding to large values of n_1 and small values of n_2 can have very small widths and large lifetimes with respect to field ionization even above the Stark saddle. A pronounced resonance structure above the Stark saddle and even above the "field-free ionization threshold" can indeed be observed e.g. in photoionization spectra [RW86].

Stark states of hydrogen are studied by many authors with continually improving experimental and calculational techniques [GN85, RW86, Kol89, Ali92, SR13], in particular in the interesting region near the saddle and the field-free threshold. Kolosov [Kol89] calculated the positions and widths of resonant Stark states appearing as eigenstates of the Hamiltonian with complex eigenvalues. Results for resonances with maximum uphill quantum number, $n_1 = n - 1$, $n_2 = 0$, $m = 0$,

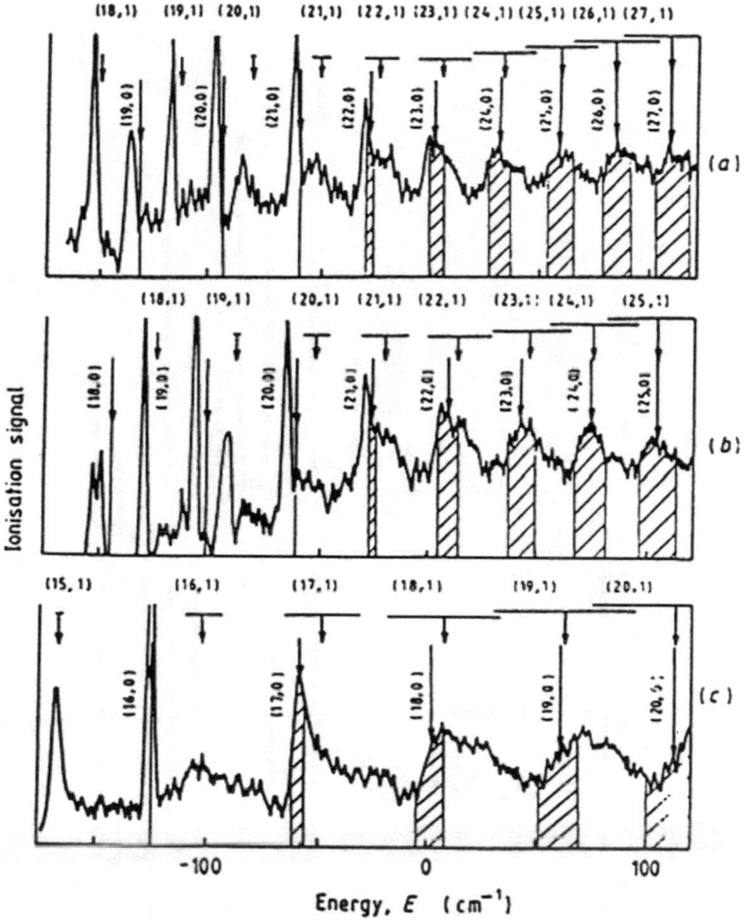

Fig. 3.28 Experimental photoionization spectra of hydrogen in a Stark field for three field strengths E_z: (**a**) 6.5 kV/cm, (**b**) 8.0 kV/cm, (**c**) 16.7 kV/cm [GN85]. The *arrows* show the calculated positions of resonant Stark states with parabolic quantum numbers (n_1, n_2) $(m = 0)$. The hatchings show the widths of the states with maximum uphill quantum number n_1 $(= n - 1)$. The widths of the states with $n_1 = n - 2$, $n_2 = 1$ are shown as *horizontal bars* (From [Kol89])

and with second largest n_1 for energies around the field-free threshold are shown in Fig. 3.28 and compared with experimental photoionization spectra from Glab et al. [GN85] at three different electric field strengths. The calculated positions of the resonances with $n_1 = n - 1$, $n_2 = 0$ and with $n_1 = n - 2$, $n_2 = 1$ are indicated by arrows, and the widths are shown as hatching or as horizontal bars. A correlation between experimental structures and calculated resonances is obvious, even at positive energies.

Alijah [Ali92] calculated the photoionization spectrum as function of energy from the wave functions obtained by direct numerical integration of the Schrödinger

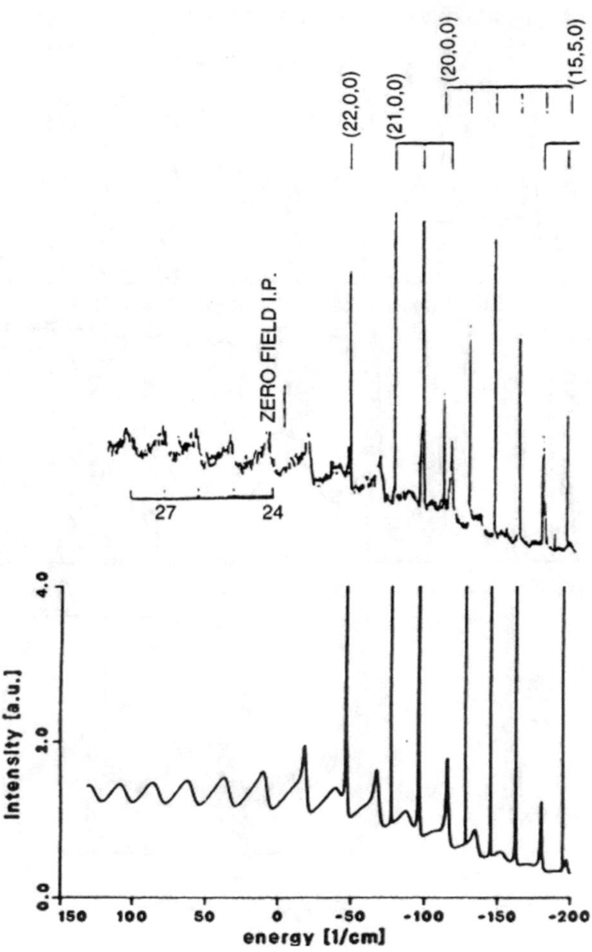

Fig. 3.29 Photoionization spectrum of hydrogen in an electric field (5.714 kV/cm) from the initial state $n_1 = 1$, $n_2 = 0$, $m = 0$ with $\Delta m = 0$. The upper part of the figure shows the experimental results of Rottke and Welge [RW86]; the *sharp lines* below the zero field ionization threshold are labelled by the quantum numbers (n_1, n_2, m), while the oscillations in the positive energy region are labelled just by n_1. The lower part of the figure shows the numerical results of Alijah (From [Ali92])

equation. His results are shown in Fig. 3.29 together with the experimental photoionization spectrum of Rottke and Welge [RW86] at a field strength of 5.714 kV/cm. The numerical calculation reproduces all the experimentally observed features.

More recently, Stodolna et al. reported photoionization microscopy experiments, in which the nodal structure of eigenstates of the Stark Hamiltonian (3.307) could be observed directly. Their experiments "provide a validation of theoretical predictions that have been made over the last three decades." [SR13]

3.5.2 Atoms in a Static, Homogeneous Magnetic Field

A static homogeneous magnetic field pointing in z direction can be described in the *symmetric gauge* by a vector potential

$$\boldsymbol{A}(\boldsymbol{r}) = -\frac{1}{2}(\boldsymbol{r} \times \boldsymbol{B}) = \frac{1}{2}\begin{pmatrix} -y \\ x \\ 0 \end{pmatrix} B_z . \tag{3.315}$$

In this gauge the Hamiltonian (3.292) keeps its axial symmetry around the z-axis and has the following special form:

$$\hat{H} = \sum_{i=1}^{N} \frac{\hat{\boldsymbol{p}}_i^2}{2\mu} + \hat{V} + \omega\hat{L}_z + \sum_{i=1}^{N} \frac{\mu\omega^2}{2}(x_i^2 + y_i^2) . \tag{3.316}$$

Here ω is one half of the cyclotron frequency which characterizes the energy eigenstates of an otherwise free electron in a magnetic field (see Problem 3.10):

$$\omega = \frac{\omega_c}{2} = \frac{eB_z}{2\mu c} . \tag{3.317}$$

In the Hamiltonian (3.316), $\hat{L}_z$ stands for the z-component of the total orbital angular momentum of the N electrons, and the contribution $\omega\hat{L}_z$ is just the energy $-\boldsymbol{\mu}_L \cdot \boldsymbol{B}$ of the magnetic moment $\boldsymbol{\mu}_L = -e/(2\mu c)\hat{\boldsymbol{L}}$ due to this orbital motion in the magnetic field $\boldsymbol{B}$. The ratio $-e/(2\mu c)$ of the magnetic moment to the orbital angular momentum is the *gyromagnetic ratio*.

If, for the time being, we neglect the term in the Hamiltonian (3.316) which is quadratic in the field strength B_z, i.e. quadratic in ω, then the external magnetic field simply leads to an additional energy $\omega\hat{L}_z$. Eigenstates of the unperturbed (field-free) Hamiltonian, in which effects of spin-orbit coupling are negligible and in which the total spin vanishes, i.e. in which the orbital angular momentum equals the total angular momentum, remain eigenstates of the Hamiltonian in the presence of the magnetic field, but the degeneracy in the quantum number M_L is lifted. Quantitatively the energies are shifted by

$$\Delta E_{M_L} = \frac{e\hbar B_z}{2\mu c} M_L . \tag{3.318}$$

This is the *normal Zeeman effect*. Note that the result (3.318) is not based on perturbation theory, but only on the neglect of spin and of the contributions quadratic in ω to the Hamiltonian (3.316). Except for the small difference between the reduced mass μ and the electron mass m_e, the constant $e\hbar/(2\mu c)$ is the *Bohr magneton* (http://physics.nist.gov/cuu/Constants/Table/allascii.txt),

$$\mu_{\rm B} = \frac{e\hbar}{2m_{\rm e}c} = 5.7883818012(26) \times 10^{-5}{\rm eV/Tesla}\,. \tag{3.319}$$

Except in states with vanishing total spin, we generally cannot neglect the contributions of the spin to the energy shifts in a magnetic field. The most important contribution comes from the magnetic moments due to the spins of the electrons. The interaction of these spin moments with a magnetic field is obtained most directly if we introduce the field (cf. Sect. 2.4.2) into the Dirac equation (2.28) via the substitution $\hat{\boldsymbol{p}}_i \rightarrow \hat{\boldsymbol{p}}_i + (e/c)\boldsymbol{A}(\boldsymbol{r}_i)$ and perform the transition to the non-relativistic Schrödinger equation (Problem 3.11). To first order we obtain the following Hamiltonian for a free electron in an external magnetic field:

$$\hat{H}_{\rm B}^{(0)} = \frac{\hat{\boldsymbol{p}}_i^2}{2\mu} + \frac{e}{2\mu c}(\hat{\boldsymbol{L}}_i + 2\hat{\boldsymbol{S}}_i)\cdot\boldsymbol{B}\,. \tag{3.320}$$

Note the factor two in front of the spin. It implies that the spin $\hbar/2$ of an electron leads to a magnetic moment just as big as that due to an orbital angular momentum of $\hbar$.

The interaction of an atom with a magnetic field is thus given to first order in the field strength by a contribution

$$\hat{W}_{\rm B} = \frac{e}{2\mu c}(\hat{\boldsymbol{L}} + 2\hat{\boldsymbol{S}})\cdot\boldsymbol{B} = \frac{eB_z}{2\mu c}(\hat{L}_z + 2\hat{S}_z) \tag{3.321}$$

in the N-Electron Hamiltonian. This corresponds to the energy of a magnetic dipole with a magnetic moment $-(\hat{\boldsymbol{L}} + 2\hat{\boldsymbol{S}})e/(2\mu c)$ in the magnetic field $\boldsymbol{B}$. The magnetic moment now is no longer simply proportional to the total angular momentum $\hat{\boldsymbol{J}} = \hat{\boldsymbol{L}} + \hat{\boldsymbol{S}}$, which means that there is no constant gyromagnetic ratio. The splitting of the energy levels in the magnetic field now depends not only on the field strength and the azimuthal quantum number as in the normal Zeeman effect (3.318); for this reason the more general case, in which the spin of the atomic electrons plays a role, is called *anomalous Zeeman effect*.

The unperturbed atomic states can be labelled by the total angular momentum quantum number J and the quantum number M_J for the z-component of the total angular momentum, and the unperturbed energies don't depend on M_J. If the atom is described in LS coupling, then the unperturbed eigenstates Ψ_{L,S,J,M_J} in a degenerate J-multiplet also have a good total orbital angular momentum quantum number L and a good total spin quantum number S (cf. Sect. 2.2.4). If the energy shifts due to the perturbing operator (3.321) are small compared with the separations of the energies of different J-multiplets, then these energy shifts are given in first-order perturbation theory by the expectation values of the perturbing operator (3.321) in the unperturbed states Ψ_{L,S,J,M_J}. As shown below, the perturbing operator is diagonal in the quantum number M_J within each J-multiplet and hence a diagonalization according to the formula (1.258) of degenerate perturbation theory is not necessary.

We can derive a quantitative formula for the matrix elements of the perturbing operator (3.321) by applying the Wigner-Eckart theorem for the components of vector operators in the angular momentum eigenstates. Thereby the dependence of the matrix elements on the (spherical) component index of the vector and on the azimuthal quantum numbers in bra and ket is the same for all vector operators and is given by appropriate Clebsch-Gordan coefficients (see Sects. 1.7.1, 2.4.5). In particular,

$$\begin{aligned}
\langle \Psi_{L,S,J,M_J}|\hat{S}_z|\Psi_{L,S,J,M_J}\rangle &= \langle LSJ\|\hat{\boldsymbol{S}}\|LSJ\rangle\langle J,M_J|1,0,J,M_J\rangle\,,\\
\langle \Psi_{L,S,J,M_J}|\hat{J}_z|\Psi_{L,S,J,M_J}\rangle &= \langle LSJ\|\hat{\boldsymbol{J}}\|LSJ\rangle\langle J,M_J|1,0,J,M_J\rangle\,,\\
\langle \Psi_{L,S,J,M_J}|\hat{\boldsymbol{J}}\cdot\hat{\boldsymbol{S}}|\Psi_{L,S,J,M_J}\rangle &= \langle LSJ\|\hat{\boldsymbol{J}}\|LSJ\rangle\langle LSJ\|\hat{\boldsymbol{S}}\|LSJ\rangle\{\mathrm{CG}\}\,,\\
\langle \Psi_{L,S,J,M_J}|\hat{\boldsymbol{J}}^2|\Psi_{L,S,J,M_J}\rangle &= \langle LSJ\|\hat{\boldsymbol{J}}\|LSJ\rangle\langle LSJ\|\hat{\boldsymbol{J}}\|LSJ\rangle\{\mathrm{CG}\}\,.
\end{aligned}\tag{3.322}$$

All matrix elements like (3.322) which are not diagonal in M_J vanish. This follows in the two lower equations, because we actually calculate matrix elements of a *scalar* product of two vector operators, and in the two upper equations, because we are dealing with the $\nu = 0$ spherical component of a vector operator in both cases. (This also implies that non-diagonal matrix elements of the perturbing operator (3.321) vanish.) Quantities such as $\langle LSJ\|\hat{\boldsymbol{S}}\|LSJ\rangle$ are the reduced matrix elements, which are characteristic of the whole J-multiplet and independent of azimuthal quantum numbers or component indices. The expression $\{\mathrm{CG}\}$ in the two lower equations stands for the same combination of Clebsch-Gordan coefficients, and its precise composition is irrelevant for the following discussion.

Dividing the first equation (3.322) by the second and the third by the fourth leads to the same number in both cases, namely the quotient of the reduced matrix elements $\langle LSJ\|\hat{\boldsymbol{S}}\|LSJ\rangle$ and $\langle LSJ\|\hat{\boldsymbol{J}}\|LSJ\rangle$. Hence the quotients of the left-hand sides must also be equal, giving

$$\begin{aligned}
&\langle \Psi_{L,S,J,M_J}|\hat{S}_z|\Psi_{L,S,J,M_J}\rangle\\
&= \frac{\langle \Psi_{L,S,J,M_J}|\hat{\boldsymbol{J}}\cdot\hat{\boldsymbol{S}}|\Psi_{L,S,J,M_J}\rangle}{\langle \Psi_{L,S,J,M_J}|\hat{\boldsymbol{J}}^2|\Psi_{L,S,J,M_J}\rangle}\langle \Psi_{L,S,J,M_J}|\hat{J}_z|\Psi_{L,S,J,M_J}\rangle\\
&= \frac{\langle \Psi_{L,S,J,M_J}|\hat{\boldsymbol{J}}\cdot\hat{\boldsymbol{S}}|\Psi_{L,S,J,M_J}\rangle}{J(J+1)\hbar^2}M_J\hbar\,.
\end{aligned}\tag{3.323}$$

We can replace the operator product $\hat{\boldsymbol{J}}\cdot\hat{\boldsymbol{S}}$ by $(\hat{\boldsymbol{J}}^2 + \hat{\boldsymbol{S}}^2 - \hat{\boldsymbol{L}}^2)/2$ in analogy to (1.356) and express the expectation value of $\hat{\boldsymbol{J}}\cdot\hat{\boldsymbol{S}}$ in terms of the eigenvalues of $\hat{\boldsymbol{J}}^2$, $\hat{\boldsymbol{S}}^2$ and $\hat{\boldsymbol{L}}^2$:

$$\langle \Psi_{L,S,J,M_J}|\hat{S}_z|\Psi_{L,S,J,M_J}\rangle = \frac{J(J+1)+S(S+1)-L(L+1)}{2J(J+1)} M_J\hbar \; . \tag{3.324}$$

With $\hat{L}_z + 2\hat{S}_z = \hat{J}_z + \hat{S}_z$ we obtain the following expression for the energy shifts of the anomalous Zeeman effect in first-order perturbation theory:

$$\begin{aligned} \Delta E_{L,S,J,M_J} &= \frac{eB_z}{2\mu c}\langle \Psi_{L,S,J,M_J}|\hat{J}_z + \hat{S}_z|\Psi_{L,S,J,M_J}\rangle \\ &= \frac{eB_z}{2\mu c}\left(1 + \frac{J(J+1)+S(S+1)-L(L+1)}{2J(J+1)}\right) M_J\hbar \\ &= \frac{e\hbar B_Z}{2\mu c}\, g\, M_J \; . \end{aligned} \tag{3.325}$$

The dependence of the gyromagnetic ratio on the J-multiplet is contained in the *Landé factor*

$$g = 1 + \frac{J(J+1)+S(S+1)-L(L+1)}{2J(J+1)} = \frac{3J(J+1)+S(S+1)-L(L+1)}{2J(J+1)} \; . \tag{3.326}$$

For $S = 0$ and $J = L$ we have $g = 1$ and recover the result of the normal Zeeman effect (3.318).

As the strength of the magnetic field increases, the interaction with the field becomes stronger than the effects of spin-orbit coupling. It is then sensible to first calculate the atomic states without spin-orbit coupling and to classify them according to the quantum numbers of the z-components of the total orbital angular momentum and the total spin: Ψ_{L,S,M_L,M_S}. The energy shifts due to the interaction with the magnetic field (3.321) are then—without any further perturbative assumptions—simply

$$\Delta E_{M_L,M_S} = \frac{e\hbar B_z}{2\mu c}(M_L + 2M_S) \; . \tag{3.327}$$

This is the *Paschen-Back-Effekt*. An example for the transition from the anomalous Zeeman effect in weak fields to the Paschen-Back effect in stronger fields is illustrated schematically in Fig. 3.30.

The perturbing operator (3.321) describes the *paramagnetic* interaction between the magnetic field and the (permanent) magnetic dipole moment of the atom. The operators $\hat{L}_z$ and $\hat{S}_z$ commute with $\hat{\boldsymbol{L}}^2$ and $\hat{\boldsymbol{S}}^2$. If the total orbital angular momentum L and the total spin S are good quantum numbers in the absence of an external magnetic field, then they remain good quantum numbers in the presence of the perturbing operator (3.321). L is no longer a good quantum number when the contribution quadratic in ω in (3.316) (the *diamagnetic* term) becomes important.

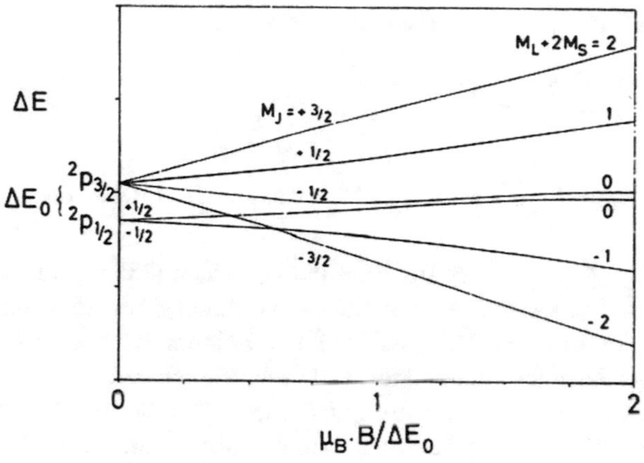

Fig. 3.30 Schematic illustration of level splitting in a magnetic field for the example of a $^2P_{1/2}$ and a $^2P_{3/2}$ multiplet, which are separated by a spin-orbit splitting ΔE_0 in the field-free case. If the product of field strength B and magneton (3.319) is smaller than ΔE_0 we obtain the level splitting of the anomalous Zeeman effect (3.325), for $\mu_B B > \Delta E_0$ we enter the region of the Paschen-Back effect (3.327)

This term is a two-dimensional harmonic oscillator potential in the two directions perpendicular to the direction of the magnetic field.

Consider the Schrödinger equation for a free electron (without spin) in an external field in the symmetric gauge. This is easy to solve in cylindrical coordinates (Problem 3.10). The eigenfunctions are

$$\psi_{N,m,k}(\varrho, \phi, z) = \Phi_{N,m}(\varrho)\, e^{im\phi}\, e^{ikz} , \qquad (3.328)$$

and the energy eigenvalues are

$$E_{N,m,k} = (2N + m + |m| + 1)\hbar\omega + \frac{\hbar^2 k^2}{2\mu} , \qquad N = 0, 1, 2, \dots ,$$
$$m = 0, \pm 1, \pm 2, \dots ,$$
$$-\infty < k < +\infty . \qquad (3.329)$$

Here $\Phi_{N,m}(\varrho)\exp(im\phi)$ are the eigenstates of the two-dimensional harmonic oscillator (*Landau states*) labelled by the cylindrical principal quantum number N and the azimuthal quantum number m for the z-component of the orbital angular momentum. The factor $\exp(ikz)$ describes the free motion of the electron parallel to the direction of the magnetic field.

We obtain a measure for the relative importance of the diamagnetic term when we compare the oscillator energy $\hbar\omega$ in (3.329) with the Rydberg energy

$\mathcal{R} = \mu e^4/(2\hbar^2)$ characterizing the atomic interactions:

$$\gamma = \frac{\hbar\omega}{\mathcal{R}} = \frac{B_z}{B_0} ,$$

$$B_0 = \frac{\mu^2 e^3 c}{\hbar^3} \approx 2.35 \times 10^9 \,\text{Gauss} = 2.35 \times 10^5 \,\text{Tesla} . \qquad (3.330)$$

For field strengths appreciably smaller than B_0, which (still) includes all fields that can be generated in a terrestrial laboratory, the diamagnetic term has no influence on low-lying atomic states. This justifies its omission in the treatment of the normal and anomalous Zeeman effects and the Paschen-Back effect. In an astrophysical context however, magnetic field strengths of the order of 10^4 to 10^8 T have been observed at the surfaces of white dwarfs and neutron stars. At such field strengths the quadratic contribution to the Hamiltonian (3.316) can by no means be neglected [WZ88]. The influence of this term is often called the *quadratic Zeeman effect.*

At field strengths of several Tesla as can be generated in the laboratory, the *magnetic field strength parameter* γ defined by (3.330) is of the order of 10^{-5} and the quadratic Zeeman effect is not important for low-lying states of atoms. It may play a role however, in the context of semiconductor physics, where electrons bound to a shallow donor are often described in a hydrogen model with an effective mass roughly one power of ten smaller than the electron mass and an effective charge roughly one power of ten smaller than the elementary charge e. In such situations effective field strength parameters near unity may be achieved at field strengths of a few Tesla [KG90].

For small field strength parameters, typically around 10^{-5} for free atoms in strong laboratory fields, the quadratic Zeeman effect does have a considerable influence on highly excited Rydberg states. Since the separation of successive terms in a Rydberg series decreases as $2\mathcal{R}/n^3$ with increasing principal quantum number n, we can already expect a significant perturbation due to the diamagnetic term near $n = 40$ or $n = 50$.

The intricacy of the quadratic Zeeman effect can already be illustrated in the simplest example of a one-electron atom, e.g. the hydrogen atom. An overview of many papers written on the H atom in a magnetic field can be found e.g. in [FW89, HR89], see also [Gay91]. A monograph devoted to this subject was published by Ruder et al. [RW94].

Ignoring spin effects the Schrödinger equation for a hydrogen atom in a uniform magnetic field is, in atomic units and cylindrical coordinates (cf. (3.316), (3.295)),

$$\left[-\frac{1}{2}\left(\frac{\partial^2}{\partial \varrho^2} + \frac{1}{\varrho}\frac{\partial}{\partial \varrho} + \frac{\partial^2}{\partial z^2} - \frac{m^2}{\varrho^2} \right) + \frac{m}{2}\gamma + \frac{1}{8}\gamma^2 \varrho^2 - \frac{1}{\sqrt{\varrho^2 + z^2}} \right] f_m(\varrho, z) = E f_m(\varrho, z) . \qquad (3.331)$$

Effects of spin-orbit coupling are mainly important for relatively weak fields, and the centre-of-mass motion, whose separation is not quite as straight-forward as in the absence of an external field, only becomes important in extremely strong fields. For values of the field strength parameter between $\gamma \approx 10^{-5}$ and $\gamma \approx 10^{+4}$, the one-electron Schrödinger equation (3.331) is a reliable description of the real physical system. The azimuthal quantum number m is a good quantum number, as is the parity π which is frequently expressed in terms of the *z-parity*, $\pi_z = (-1)^m\pi$, which describes the symmetry of the wave function with respect to a reflection at the xy-plane (perpendicular to the direction of the magnetic field). In each m^{π_z} subspace of the full Hilbert space, the Schrödinger equation remains a *non-separable* equation in two coordinates, i.e. there is no set of coordinates in which it can be reduced to ordinary differential equations as was possible for the Stark effect. If we drop the trivial normal Zeeman term $(m/2)\gamma$, the potential in (3.331) is independent of the sign of m:

$$V_m(\varrho, z) = \frac{m^2}{2\varrho^2} - \frac{1}{\sqrt{\varrho^2 + z^2}} + \frac{1}{8}\gamma^2\varrho^2 . \tag{3.332}$$

Equipotential lines of the potential (3.332) are shown in Fig. 3.31 for the case $m = 0$.

For very strong fields corresponding to field strength parameters γ near unity or larger, the energies needed to excite Landau states perpendicular to the field are larger than the typical Coulomb energies for the motion of the electron parallel to the field. In this regime it makes sense to expand the wave function $f_m(\varrho, z)$ in *Landau channels*:

$$f_m(\varrho, z) = \sum_{N=0}^{\infty} \Phi_{N,m}(\varrho)\psi_N(z) . \tag{3.333}$$

Inserting the ansatz (3.333) into the Schrödinger equation (3.331) and projecting onto the various Landau channels yields, in each m^{π_z} subspace, a set of coupled-

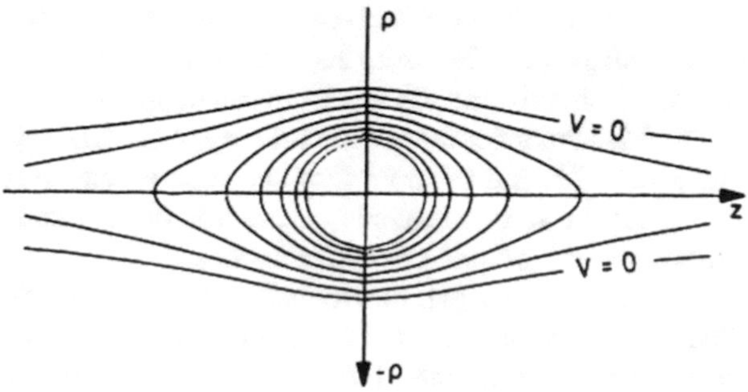

Fig. 3.31 Equipotential lines for the potential (3.332) with $m = 0$

channel equations for the channel wave functions $\psi_N(z)$, and the potentials are

$$V_{N,N'}(z) = E_{N,m}\delta_{N,N'} + \int_0^\infty \varrho\, \mathrm{d}\varrho\, \Phi_{N,m}(\varrho) \frac{-1}{\sqrt{\varrho^2 + z^2}} \Phi_{N',m}(\varrho) \,. \tag{3.334}$$

The diagonal potentials are asymptotically Coulomb potentials proportional to $1/|z|$, and the channel thresholds $E_{N,m}$ are (without the normal Zeeman term $(m/2)\gamma$)

$$E_{N,m} = [N + (|m| + 1)/2]\gamma = E_m + N\gamma \,. \tag{3.335}$$

The continuum threshold in a given m^{π_z}-subspace is at $E_m = (|m| + 1)\gamma/2$, which lies higher than the "zero-field threshold" above which the atom can ionize classically. This is because an electron escaping to $z = \pm\infty$ must at least have the zero-point energy of the lowest Landau vibration.

For very strong fields the Schrödinger equation (3.331) thus describes a system of coupled Coulomb channels, and the separation of successive channel thresholds is larger than the Coulomb binding energies in the various channels. In each m^{π_z}-subspace we obtain a Rydberg series of bound states with wave functions dominated by the lowest Landau channel $N = 0$, and a sequence of Rydberg series of autoionizing resonances corresponding to the excited Landau channels $N > 0$. Autoionization occurs, because an excited Landau state, which would be bound in the absence of channel coupling, can transfer its energy perpendicular to the field into energy parallel to the field and decay into the continuum. Autoionization doesn't require two electrons, only two (coupled) degrees of freedom! The calculation of bound-state spectra and of the energies and widths of autoionizing states is comparatively easy in the strong field regime [Fri82, FC83]. Results of numerical calculations in this region were confirmed experimentally in far-infrared magneto-optical experiments on shallow donors in the GaAs semiconductor, where a small effective mass and a small effective charge give access to effective field strength parameters near unity for laboratory field strengths of a few Tesla [KG90].

Figure 3.32 illustrates the spectrum in the regime of very strong fields for three values of the field strength parameter in the $m^{\pi_z} = 0^+$ subspace. As the field strength decreases, the separation of successive Landau thresholds becomes smaller and smaller and we get interferences between the various Landau channels. In a comparatively small range of field strengths—down to $\gamma \approx 0.01$—the coupled equations can be solved directly and the spectrum can be interpreted qualitatively in the framework of multichannel quantum defect theory. At laboratory field strengths corresponding to $\gamma \approx 10^{-5}$, the separation of successive Landau thresholds is of the order of 10^{-3} to 10^{-4} eV, so a realistic calculation in the Landau basis would involve tens of thousands of coupled Landau channels.

For weak fields $\gamma \ll 1$ and energies clearly below the zero-field threshold $E = 0$, the quadratic Zeeman effect can largely be treated with perturbative methods. In the zero-field case the degenerate states belonging to given values of the Coulomb principal quantum n and the azimuthal quantum number m can be labelled by the

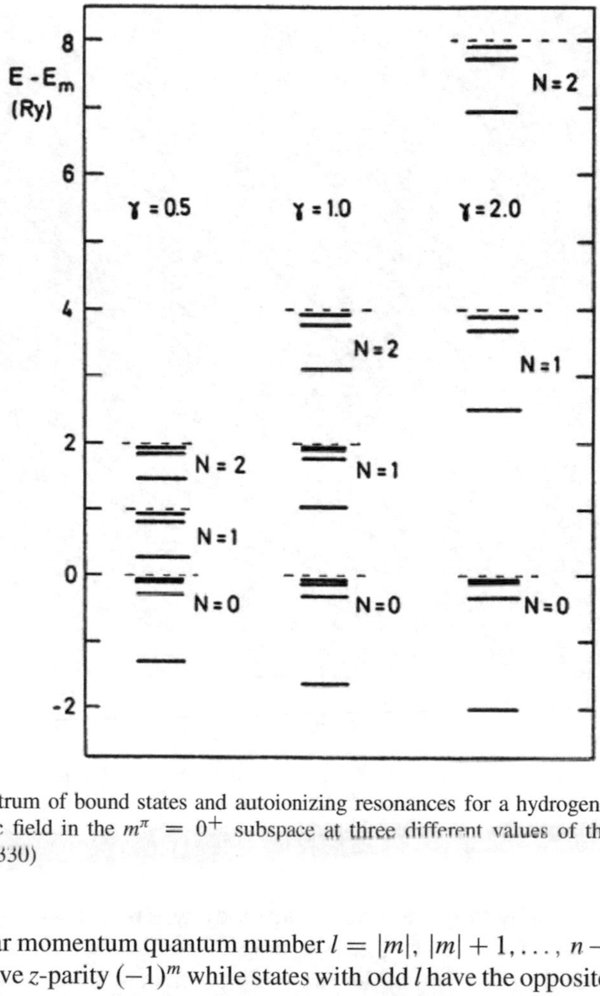

Fig. 3.32 Spectrum of bound states and autoionizing resonances for a hydrogen atom in a very strong magnetic field in the $m^{\pi} = 0^{+}$ subspace at three different values of the field strength parameter γ (3.330)

orbital angular momentum quantum number $l = |m|, |m|+1, \ldots, n-1$, and states with even l have z-parity $(-1)^m$ while states with odd l have the opposite z-parity. For finite field strengths we initially observe "l-mixing" and the degeneracy is lifted by a splitting proportional to the square of the magnetic field strength. It is customary to label the states originating from a given (n, m) manifold with an integer k, starting with $k = 0$ for the energetically highest state and ending with $k = n - |m| - 1$ for the energetically lowest state. States from successive n-manifolds in a given m^{π_z}-subspace begin to overlap as the field strength (or the principal quantum number n) increases. The interaction between different states is small at first and they can still be labelled by the two numbers n and k. With further increasing field strength or excitation energy, however, the order within the spectrum is lost more and more (see Fig. 3.33), until finally, as we approach the zero-field threshold, it becomes impossible to assign two meaningful quantum numbers to individual quantum states of this two-dimensional system. As we shall see in Sect. 5.3.5(b), this is the region where the classical dynamics becomes *chaotic*.

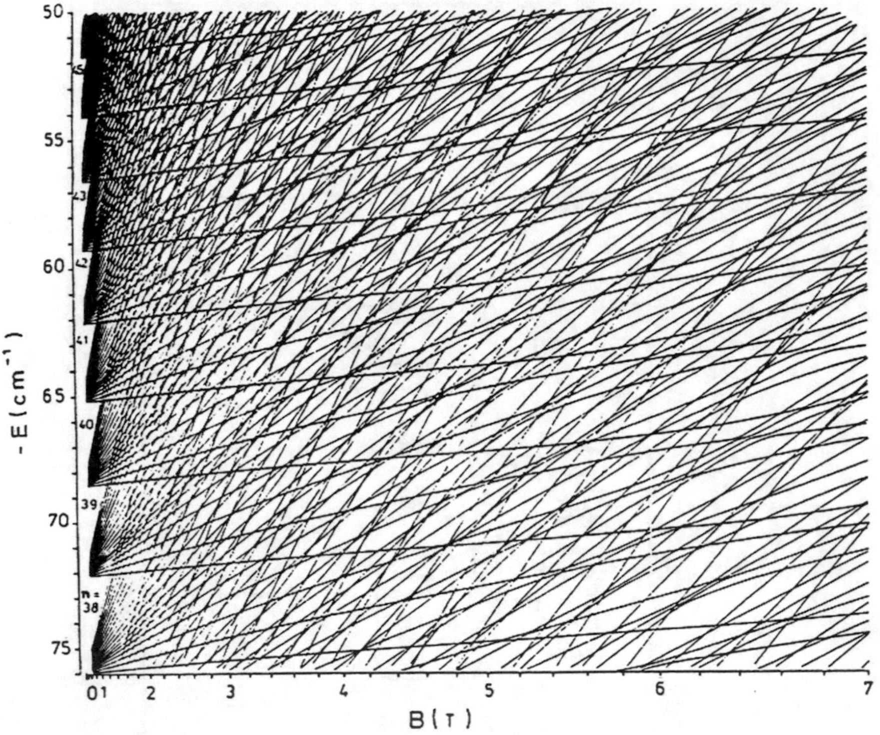

Fig. 3.33 Part of the spectrum of the hydrogen atom in a homogeneous magnetic field with field strengths up to 7 Tesla. The figure shows the bound states in the $m^{\pi_z} = 0^+$ subspace in an energy region corresponding roughly to principal quantum numbers around $n = 40$ (From [FW89])

The fact that the hydrogen atom is a two-body system has been ignored above, except for the use of the reduced mass μ in (3.330). This is, strictly speaking, not enough, because the reduction of the two-body problem to a one-body problem for the internal motion of the atom is non-trivial in the presence of an external magnetic field. The Hamiltonian for the two-body atom in a uniform magnetic field $\boldsymbol{B} = \nabla \times \boldsymbol{A}$ is,

$$\hat{H}(\boldsymbol{r}_\mathrm{e}, \boldsymbol{r}_\mathrm{p}; \hat{\boldsymbol{p}}_\mathrm{e}, \hat{\boldsymbol{p}}_\mathrm{p}) = \frac{[\hat{\boldsymbol{p}}_\mathrm{p} - \frac{e}{c}\boldsymbol{A}(\boldsymbol{r}_\mathrm{p})]^2}{2m_\mathrm{p}} + \frac{[\hat{\boldsymbol{p}}_\mathrm{e} + \frac{e}{c}\boldsymbol{A}(\boldsymbol{r}_\mathrm{e})]^2}{2m_\mathrm{e}} - \frac{e^2}{|\boldsymbol{r}_\mathrm{e} - \boldsymbol{r}_\mathrm{p}|}, \tag{3.336}$$

where $m_\mathrm{p}, \boldsymbol{r}_\mathrm{p}$ and $\hat{\boldsymbol{p}}_\mathrm{p}$ denote the mass and the displacement and momentum vectors for the proton, while $m_\mathrm{e}, \boldsymbol{r}_\mathrm{e}$ and $\hat{\boldsymbol{p}}_\mathrm{e}$ are for the electron. Neither the total canonical momentum $\hat{\boldsymbol{p}}_\mathrm{p} + \hat{\boldsymbol{p}}_\mathrm{e}$ nor the total kinetic momentum

$\hat{\boldsymbol{P}}_{\mathrm{k}} = \hat{\boldsymbol{p}}_{\mathrm{p}} - \frac{e}{c}\boldsymbol{A}(\boldsymbol{r}_{\mathrm{p}}) + \hat{\boldsymbol{p}}_{\mathrm{e}} + \frac{e}{c}\boldsymbol{A}(\boldsymbol{r}_{\mathrm{e}})$ are conserved in the presence of the external fields, but the so-called *pseudomomentum*,

$$\begin{aligned}\hat{\boldsymbol{K}} &= \hat{\boldsymbol{p}}_{\mathrm{p}} - \frac{e}{c}\boldsymbol{A}(\boldsymbol{r}_{\mathrm{p}}) + \frac{e}{c}\boldsymbol{B}\times\boldsymbol{r}_{\mathrm{p}} + \hat{\boldsymbol{p}}_{\mathrm{e}} + \frac{e}{c}\boldsymbol{A}(\boldsymbol{r}_{\mathrm{e}}) - \frac{e}{c}\boldsymbol{B}\times\boldsymbol{r}_{\mathrm{e}} \\ &= \hat{\boldsymbol{P}}_{\mathrm{k}} - \frac{e}{c}\boldsymbol{B}\times(\boldsymbol{r}_{\mathrm{e}} - \boldsymbol{r}_{\mathrm{p}}) ,\end{aligned} \tag{3.337}$$

is. Conservation of the pseudomomentum means that the total Hamiltonian can be separated into an internal part, depending only on the relative coordinate $\boldsymbol{r} = \boldsymbol{r}_{\mathrm{e}} - \boldsymbol{r}_{\mathrm{p}}$ and its canonically conjugate momentum $\hat{\boldsymbol{p}}$, and a pseudomomentum part which however depends on a combination of internal and centre-of-mass variables. This *pseudoseparation of variables* leads to the following Hamiltonian describing the internal motion of the hydrogen atom [DS94, RW94, SC97]:

$$\hat{H}_{\mathrm{int}}(\boldsymbol{r},\hat{\boldsymbol{p}}) = \frac{1}{2\mu}\left[\hat{\boldsymbol{p}} + \frac{e}{c}\frac{m_{\mathrm{p}} - m_{\mathrm{e}}}{m_{\mathrm{p}} + m_{\mathrm{e}}}\boldsymbol{A}(\boldsymbol{r})\right]^2 + \frac{[\hat{\boldsymbol{K}} + \frac{e}{c}\boldsymbol{B}\times\boldsymbol{r}]^2}{2(m_{\mathrm{e}} + m_{\mathrm{p}})} - \frac{e^2}{r} , \tag{3.338}$$

where $\mu = m_{\mathrm{e}}m_{\mathrm{p}}/(m_{\mathrm{e}} + m_{\mathrm{p}})$ is the usual reduced mass.

The Hamiltonian (3.338) contains a correction to the charge in the kinetic energy term and an additional gauge-independent potential term

$$\frac{1}{2M}\left[\hat{\boldsymbol{K}} + \frac{e}{c}\boldsymbol{B}\times\boldsymbol{r}\right]^2 = \frac{\hat{\boldsymbol{K}}^2}{2M} + \frac{e}{Mc}\hat{\boldsymbol{K}}\times\boldsymbol{B}\cdot\boldsymbol{r} + \frac{e^2}{2Mc^2}(\boldsymbol{B}\times\boldsymbol{r})^2 , \tag{3.339}$$

where the total mass $m_{\mathrm{e}} + m_{\mathrm{p}}$ of the atom has been abbreviated as M. The first term on the right-hand side of (3.339) is a constant. The last term is quadratic in $\boldsymbol{B}$ and can easily be seen in the symmetric gauge (3.315) to cancel the above-mentioned charge correction in the diamagnetic (quadratic) contribution arising from the kinetic energy. The linear term on the right-hand side of (3.339) corresponds to the effect of an external electric field,

$$\hat{\boldsymbol{E}}_{\mathrm{ms}} = \frac{1}{Mc}\hat{\boldsymbol{K}}\times\boldsymbol{B} . \tag{3.340}$$

Thus the motion of the atom as a whole in a magnetic field $\boldsymbol{B}$, more precisely: a non-vanishing component of the pseudomomentum (3.337) perpendicular to $\boldsymbol{B}$, effectively leads to an additional electric field (3.340) in the Hamiltonian describing the internal motion of the atom. This effect is called *motional Stark effect*.

The fact that the (conserved) pseudomomentum depends on both the centre of mass and the internal variables introduces a correlation between the internal motion and the motion of the centre of mass of the atom. Vanishing pseudomomentum does not mean that the centre of mass is at rest. In fact it can be shown [SC97], that the classical centre of mass meanders diffusively when the (classical) internal

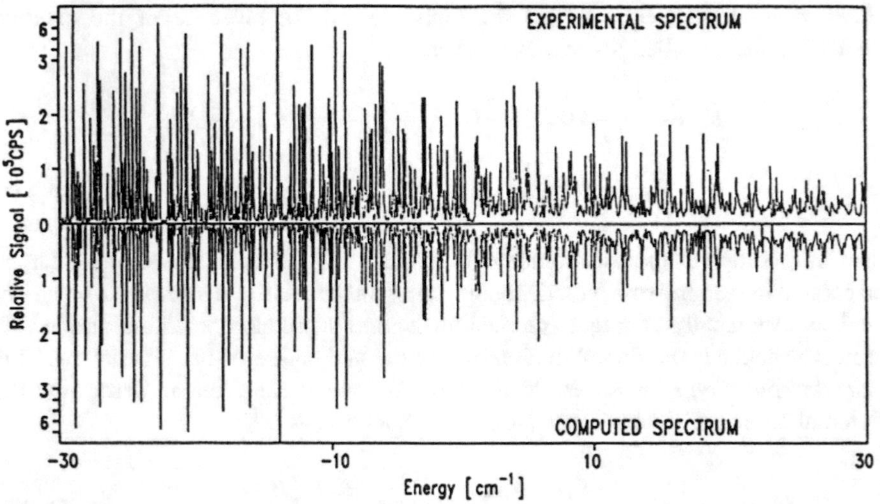

Fig. 3.34 Photoabsorption spectra for transitions from the 3s state to bound and continuum states around threshold in the $m^\pi = 0^-$ subspace in a magnetic field of 6.113 T. The upper half of the figure shows experimental results for lithium, the lower half shows the calculated spectrum for hydrogen. To facilitate the comparison, the spectra have been convoluted with a Gaussian of width 0.05 cm^{-1} (From [IW91])

motion is chaotic, which is the case for energies close to the zero-field threshold, see Sect. 5.3.5(b).

For vanishing pseudomomentum, the internal Hamiltonian (3.338) in the symmetric gauge differs from the Hamiltonian (3.316) for the one electron case $N = 1$ only in a correction of the normal Zeeman term by a factor $(m_\mathrm{p} - m_\mathrm{e})/(m_\mathrm{p} + m_\mathrm{e})$. The potential (3.332) is unaffected in this case.

The development of high resolution laser spectroscopy and advanced computer technology made detailed comparisons between measured and calculated spectra of the hydrogen atom in a uniform magnetic field possible, even in the highly irregular region close to the zero-field threshold [HW87]. Delande et al. [DB91] extended calculations to the continuum region at laboratory field strengths, which was a remarkable achievement. The bottom part of Fig. 3.34 shows their computed photoabsorption spectrum for transitions from the 3s state to bound and continuum states around threshold in the $m^\pi = 0^-$ subspace at a field strength of 6.113 T ($\gamma = 2.6 \times 10^{-5}$). The top part of the figure shows the corresponding experimental spectrum measured by Iu et al. [IW91]. The agreement is hardly short of perfect. Interestingly the experiments were performed with lithium, which is easier to handle than atomic hydrogen. Obviously the two tightly bound 1s electrons in the lithium atom have virtually no influence on the near threshold final states of the outer electron, which are very extended spatially and contain no $l = 0$ components because of their negative parity.

3.5.3 Atoms in an Oscillating Electric Field

The theory of the interaction between an atom and the electromagnetic field as discussed in Sect. 2.4 describes the resonant absorption and emission of photons between stationary eigenstates of the field-free atom. But an atom is also influenced by a (monochromatic) electromagnetic field if its frequency doesn't happen to match the energy of an allowed transition. For small intensities we obtain splitting and frequency-dependent shifts of energy levels; for sufficiently high intensities as are easily realized by modern laser technology, multiphoton processes (excitation, ionization) play an important role.

The most important contribution to the interaction of an atom with a monochromatic electromagnetic field is the influence of the oscillating electric field,

$$\boldsymbol{E}(\boldsymbol{r},t) = \boldsymbol{E}_0 \cos(\boldsymbol{k}\cdot\boldsymbol{r} - \omega t) . \tag{3.341}$$

We assume that the wave length of the field is so much larger than the dimensions of the atom that the spatial inhomogeneity of the field can be neglected, and we neglect magnetic interactions. In addition to these assumptions, which amount to the dipole approximation of Sect. 2.4.3, we take the light to be linearly polarized in z-direction:

$$\boldsymbol{E} = \boldsymbol{E}_0 \cos\omega t , \quad \boldsymbol{E}_0 = \begin{pmatrix} 0 \\ 0 \\ E_z \end{pmatrix} . \tag{3.342}$$

In the radiation gauge (2.150) such a field is given by the electromagnetic potentials

$$\boldsymbol{A} = -\frac{c}{\omega}\boldsymbol{E}_0 \sin\omega t , \quad \Phi = 0 . \tag{3.343}$$

Alternatively, in the *field gauge* we have

$$\boldsymbol{A} = 0, \quad \Phi = -\boldsymbol{E}_0\cdot\boldsymbol{r} \cos\omega t . \tag{3.344}$$

The field gauge (3.344) has the advantage that the interaction between atom and field only contributes as an additional oscillating potential energy in the Hamiltonian. In this case the Hamiltonian (3.292) has the form

$$\hat{H} = \sum_{i=1}^{N} \frac{\hat{\boldsymbol{p}}_i^2}{2\mu} + \hat{V} + eE_z \sum_{i=1}^{N} z_i \cos\omega t . \tag{3.345}$$

The periodic time dependence of the Hamiltonian (3.345) suggests looking for solutions of the time-dependent Schrödinger equation which are, to within a phase, also periodic with the same period $T = 2\pi/\omega$. If we insert the resulting ansatz

$$\psi(t) = \mathrm{e}^{-(\mathrm{i}/\hbar)\varepsilon t}\,\Phi_\varepsilon(t), \quad \Phi_\varepsilon(t+T) = \Phi_\varepsilon(t)\,, \tag{3.346}$$

into the time-dependent Schrödinger equation $\hat{H}\psi = \mathrm{i}\hbar\,\partial\psi/\partial t$, then we obtain an equation for determining the periodic function $\Phi_\varepsilon(t)$:

$$\left(\hat{H} - \mathrm{i}\hbar\frac{\partial}{\partial t}\right)\Phi_\varepsilon = \varepsilon\,\Phi_\varepsilon\,. \tag{3.347}$$

Equation (3.347) has the form of an eigenvalue equation for the operator

$$\hat{\mathcal{H}} = \hat{H} - \mathrm{i}\hbar\frac{\partial}{\partial t}\,. \tag{3.348}$$

Its eigenvalues are called *quasi-energies* and the associated solutions (3.346) are the *quasi-energy states* or *Floquet states.* They are complete in the sense that any solution of the time-dependent Schrödinger equation can be written as a superposition of Floquet states with time-independent coefficients. For each eigenstate Φ_ε of $\hat{\mathcal{H}}$ with eigenvalue ε there is a whole family of eigenstates $\Phi_\varepsilon \mathrm{e}^{\mathrm{i}k\omega t}$ with the eigenvalues $\varepsilon + k\hbar\omega$, $k = 0, \pm 1, \pm 2, \ldots$. They all belong to the same Floquet state (3.346).

The dynamics described by the Hamiltonian (3.348) become formally similar to the quantum mechanics of a time-independent Hamiltonian, if we consider the space spanned by the basis states Φ_ε as functions of the coordinates and the time in the interval $[0, T]$. The scalar product of two states ϕ_1 and ϕ_2 in this Hilbert space is defined as the time average of the ordinary scalar product over a period T and is denoted by a double bracket:

$$\langle\langle\phi_1|\phi_2\rangle\rangle := \frac{1}{T}\int_0^T \langle\phi_1(t)|\phi_2(t)\rangle\,\mathrm{d}t\,. \tag{3.349}$$

The "quasi-energy method" summarized in equations (3.346)–(3.349), and extensions thereof, have been applied to numerous problems related to the dynamics of the interaction of light with atoms. Comprehensive summaries can be found in the monograph by Delone and Krainov [DK85] and in the article by Manakov et al. [MO86].

If we want to apply perturbation theory in the spirit of Sect. 1.6.1, we start with eigenstates ψ_n of the field-free Hamiltonian $\hat{H}_0$ with eigenvalues E_n, and we take the products

$$\phi_{n,k} = \psi_n \mathrm{e}^{\mathrm{i}k\omega t} \tag{3.350}$$

as the unperturbed states. They are eigenstates of the Hamiltonian

$$\hat{\mathcal{H}}_0 = \hat{H}_0 - \mathrm{i}\hbar\frac{\partial}{\partial t} \tag{3.351}$$

with the respective eigenvalues

$$E_{n,k} = E_n + k\hbar\omega\,. \tag{3.352}$$

If we treat the oscillating potential in (3.345) as a small perturbation in the "Schrödinger equation", then we can adapt the formalism of time-independent perturbation theory as described in Sect. 1.6.1 to the present situation. In the case of non-degenerate unperturbed eigenstates, the energy shifts are given in first order by the expectation values of the perturbation, which trivially vanish, because the time average (3.349) over one period of the cosine vanishes. In second order we obtain, in analogy to (1.255),

$$\Delta E_n^{(2)} = (eE_z)^2 \sum_{E_{m,k}\neq E_n} \frac{|\langle\langle\phi_{n,0}|\sum_{i=1}^N z_i\,\cos\omega t|\phi_{m,k}\rangle\rangle|^2}{E_n - E_{m,k}}\,. \tag{3.353}$$

Time averaging over one period causes all matrix elements $\langle\langle\phi_{n,0}|\sum_{i=1}^N z_i \cos\omega t|\phi_{m,k}\rangle\rangle$ to vanish—except those for $k = +1$ and $k = -1$. In the two non-vanishing cases we obtain a factor 1/2 times the ordinary matrix element between ψ_n and ψ_m. Equation (3.353) thus becomes

$$\Delta E_n^{(2)} = \frac{(eE_z)^2}{4}\left[\sum_{E_m+\hbar\omega\neq E_n} \frac{|\langle\psi_n|\sum_{i=1}^N z_i|\psi_m\rangle|^2}{E_n - E_m - \hbar\omega} + \sum_{E_m-\hbar\omega\neq E_n} \frac{|\langle\psi_n|\sum_{i=1}^N z_i|\psi_m\rangle|^2}{E_n - E_m + \hbar\omega}\right]. \tag{3.354}$$

The energy shifts in this *ac Stark effect* thus depend on the frequency ω of the oscillating (i.e. alternating current) field. In the limit $\omega \to 0$ (3.354) reverts to the formula (3.299) for the ordinary quadratic Stark effect—except for a factor 1/2 arising from the fact that an *ac* field of amplitude E_z and intensity $E_z^2\cos^2\omega t$ corresponds, time averaged, to a *dc* field of intensity $E_z^2/2$.

Similar to (3.302) we can describe the *ac* Stark shifts via a *frequency- dependent polarizability*, which is defined in analogy to (3.301):

$$\alpha_{\rm d}(\omega) = e^2\left[\sum_{E_m+\hbar\omega\neq E_n} \frac{|\langle\psi_n|\sum_{i=1}^N z_i|\psi_m\rangle|^2}{E_m + \hbar\omega - E_n} + \sum_{E_m-\hbar\omega\neq E_n} \frac{|\langle\psi_n|\sum_{i=1}^N z_i|\psi_m\rangle|^2}{E_m - \hbar\omega - E_n}\right]. \tag{3.355}$$

For $\omega \to 0$, $\alpha_d(\omega)$ becomes the ordinary static (or *dc*) polarizability α_d. As a function of ω the frequency-dependent polarizability goes through a singularity whenever $\hbar\omega$ passes the energy of an allowed dipole transition. If the function $\alpha_d(\omega)$ is known from other sources, e.g. from a non-perturbative solution of the Schrödinger equation, then its pole structure can be used to extract the energies and other properties of the states ψ_m. An example for the calculation and analysis of frequency-dependent polarizabilities can be found in [MO88].

The derivation of the formula (3.354) was based on the choice (3.344) for the gauge of the electromagnetic field. Different gauges lead to different formulae for the energy shifts in the *ac* Stark effect. These formulae make sense despite their gauge dependence, because the physically observable quantities are not the absolute energy values but only energy differences, and they do not depend on the choice of gauge. The gauge dependence of energy shifts in the *ac* Stark effect is discussed in more detail e.g. by Mittleman [Mit82].

Beyond the observations which can be described by perturbative means, there are several experiments concerning the behaviour of matter in an external laser or microwave fields which crucially require a reliable theory for atoms (and ions) in an oscillating external field. Such a theory is necessary in order to understand e.g. multiphoton processes occurring in strong fields or the role played by "chaos" in the microwave ionization of Rydberg atoms. These special topics will be discussed in more detail in Chapter 5.

Problems

3.1 Consider an electron in a radially symmetric potential

$$V(r) = \begin{cases} -e^2/r & \text{for} \quad r > r_0, \\ -Ze^2/r & \text{for} \quad r \leq r_0, \quad Z > 1\,. \end{cases}$$

Use the semiclassical formula (3.136) to discuss how the quantum defects $\mu_{n,l}$ (n large) depend on the angular momentum quantum number l.

3.2 Use the sum rules (3.155) to show that electromagnetic dipole transitions, in which the principal quantum number n and the angular momentum quantum number l change in the same sense (i.e. both become larger or both become smaller), tend to be more probable than transitions in which n and l change in opposite sense.

Calculate the mean oscillator strengths for the $2p \to 3s$ and the $2p \to 3d$ transition in hydrogen.

3.3

a) Two bound states $\phi_{02}(r)$ and $\phi_{03}(r)$ in the closed channels 2 and 3 interact via a real channel-coupling potential $V_{2,3}(r) = V_{3,2}(r)$. Determine the eigenvalues

E_+ and E_- and the eigenstates,

$$\psi_+ = \begin{pmatrix} a_2\,\phi_{02} \\ a_3\,\phi_{03} \end{pmatrix}, \quad \psi_- = \begin{pmatrix} b_2\,\phi_{02} \\ b_3\,\phi_{03} \end{pmatrix},$$

of the Hamiltonian in the space spanned by these two states, i.e. solve the two-state problem defined by (3.198) in the absence of coupling to the open channel 1.

b) Use the Golden Rule (Sect. 2.4.1) to calculate the lifetimes and widths of the states ψ_+ and ψ_- in a) with respect to decay into the open channel 1. Compare your results with (3.213).

3.4 For two non-interacting resonances, $W_{2,3} = 0$, the formula (3.220) for the oscillator strength determining the photoabsorption cross sections simplifies to:

$$\frac{\mathrm{d}f_{E\mathrm{i}}}{\mathrm{d}E} = \frac{2\mu}{\hbar}\omega d_1^2 \frac{\left\{D + \frac{d_2}{d_1}(E-\varepsilon_3)W_{2,1} + \frac{d_3}{d_1}(E-\varepsilon_2)W_{3,1}\right\}^2}{D^2 + N^2},$$

with

$$N(E) = \pi[(E-\varepsilon_3)W_{2,1}^2 + (E-\varepsilon_2)W_{3,1}^2], \quad D(E) = (E-\varepsilon_2)\,(E-\varepsilon_3)\,.$$

(Assume real parameters d_i, $W_{i,j}$.)

Discuss the location of zeros and maxima of $\mathrm{d}f_{E\mathrm{i}}/\mathrm{d}E$ in the two special cases:

$$|W_{2,1}\frac{d_2}{d_1}| \approx |W_{3,1}\frac{d_3}{d_1}| \ll |\varepsilon_2 - \varepsilon_3|, \quad |W_{2,1}\frac{d_2}{d_1}| \approx |W_{3,1}\frac{d_3}{d_1}| \gg |\varepsilon_2 - \varepsilon_3|\,.$$

Hint: The structure of the oscillator strength function becomes clearer if written as,

$$\frac{\mathrm{d}f_{E\mathrm{i}}}{\mathrm{d}E} = \frac{2\mu}{\hbar}\omega d_1^2 \frac{\{1+\cdots\}^2}{1+(N/D)^2}\,.$$

3.5 A Rydberg series of bound states characterized by vanishing quantum defect $\mu = 0$ is perturbed by an isolated pseudo-resonant perturbation of width Γ located at $\mathrm{E}_R = I - 0.04\mathcal{R}$. Use graphical methods to determine the energies and *effective* quantum numbers of the bound states with quantum numbers $n = 3$ to $n = 10$ for the following values of the width:

$$\Gamma = 0.01\,\mathcal{R}, \quad \Gamma = 0.001\,\mathcal{R}\,, \quad \Gamma \to 0.$$

3.6

a) Extract from Fig. 3.19 numerical values for the energies of the lowest six ${}^1P^{\mathrm{o}}$ states of the calcium atom relative to the ionization threshold.

b) Give an estimate for the two-channel MQDT parameters μ_1, μ_2, (both modulo unity) and $|R_{1,2}|$ in the description of the 4(s np) and 3(d np) $^1P^{\mathrm{o}}$ series in calcium.

3.7 An isolated perturbation of constant width Γ (see (3.222)) wanders through a Rydberg series of bound states,

$$E_n = I - \frac{\mathcal{R}}{(n^*)^2},$$

i.e. its energy E_{R} is a variable parameter. Show that the minimal separation of two successive energy levels E_n and E_{n+1} relative to the unperturbed separation $2\mathcal{R}/(n^*)^3$ is given in the limit of small width Γ by

$$\left[\frac{E_{n+1} - E_n}{2\mathcal{R}/(n^*)^3}\right]_{\min} \approx \left(\frac{\Gamma(n^*)^3}{\pi\mathcal{R}}\right)^{\frac{1}{2}}.$$

3.8 The degeneracy of the four orbital wave functions with principal quantum number $n = 2$ in the hydrogen atom is lifted in the presence of an external homogeneous electric field of strength E_z. Calculate the matrix $(\langle\psi_{n=2,l,m}|eE_z z|\psi_{n=2,l',m'}\rangle)$ of the perturbing operator and determine its eigenstates and eigenvalues.

3.9 Verify that applying the commutator of the Hamiltonian $\hat{H}_0 = \hat{\boldsymbol{p}}^2/(2\mu) - e^2/r$ with the operator $\hat{b} = az(a + r/2)$ to the ground state wave function $\psi_0(\boldsymbol{r}) = \exp(-r/a)/(a\sqrt{\pi a})$ of the hydrogen atom amounts to multiplying this wave function by $(\hbar^2/\mu)z$ (see also Problem 1.11):

$$\left[\hat{H}_0, \hat{b}\right]\psi_0 = \frac{\hbar^2}{\mu} z\,\psi_0 .$$

Use the completeness relation (1.22) to calculate the static dipole polarizability

$$\alpha_{\mathrm{d}} = 2e^2 \sum_{m \neq 0} \frac{|\langle\psi_m|z|\psi_0\rangle|^2}{E_m - E_0}$$

for the hydrogen atom in its ground state.

3.10 A homogeneous magnetic field $\boldsymbol{B} = B_z\boldsymbol{e}_z$ ($\boldsymbol{e}_z$ is the unit vector in z-direction) can be described e.g. by a vector potential $\boldsymbol{A}_{\mathrm{s}}$ in the *symmetric gauge* (3.315),

$$\boldsymbol{A}_{\mathrm{s}}(\boldsymbol{r}) = -\frac{1}{2}(\boldsymbol{r} \times \boldsymbol{B}) = \frac{1}{2}\begin{pmatrix} -y \\ x \\ 0 \end{pmatrix} B_z,$$

or by a vector potential $\boldsymbol{A}_\mathrm{L}$ in the *Landau gauge*,

$$\boldsymbol{A}_\mathrm{L}(\boldsymbol{r}) = \begin{pmatrix} -y \\ 0 \\ 0 \end{pmatrix} B_z \,.$$

a) Determine the scalar function $f(\boldsymbol{r})$ which transforms one gauge into the other according to $\boldsymbol{A}_\mathrm{s} = \boldsymbol{A}_\mathrm{L} + \nabla f$.
b) Show that if the wave function ψ_L solves the stationary Schrödinger equation for a free electron in the Landau gauge,

$$\frac{1}{2\mu}\left(\hat{\boldsymbol{p}} + \frac{e}{c}\boldsymbol{A}_\mathrm{L}\right)^2 \psi_\mathrm{L} = E\psi_\mathrm{L},$$

then the *gauge-transformed wave function*

$$\psi_\mathrm{s}(\boldsymbol{r}) = \exp\left(-\frac{\mathrm{i}e}{\hbar c} f(\boldsymbol{r})\right) \psi_\mathrm{L}(\boldsymbol{r})$$

solves the corresponding equation in the symmetric gauge:

$$\frac{1}{2\mu}\left(\hat{\boldsymbol{p}} + \frac{e}{c}\boldsymbol{A}_\mathrm{s}\right)^2 \psi_\mathrm{s} = E\psi_\mathrm{s} \,.$$

c) Calculate the eigenstates and eigenvalues of the Hamiltonian for a free electron in a uniform magnetic field $\boldsymbol{B}$,

$$\hat{H} = \frac{1}{2\mu}\left(\hat{\boldsymbol{p}} + \frac{e}{c}\boldsymbol{A}(\boldsymbol{r})\right)^2,$$

in both the symmetric gauge and the Landau gauge. Discuss the spectrum and its degeneracies.

3.11 In the presence of an electromagnetic field, the Dirac equation (2.28) becomes

$$\hat{\boldsymbol{\sigma}}\cdot\left(\hat{\boldsymbol{p}} + \frac{e}{c}\boldsymbol{A}\right)\psi_B = \frac{1}{c}(E + e\Phi - m_0c^2)\psi_A,$$

$$\hat{\boldsymbol{\sigma}}\cdot\left(\hat{\boldsymbol{p}} + \frac{e}{c}\boldsymbol{A}\right)\psi_A = \frac{1}{c}(E + e\Phi + m_0c^2)\psi_B,$$

where $\boldsymbol{A}$ is the vector potential and Φ is the scalar potential.

Derive a Schrödinger equation for the large components ψ_A in the nonrelativistic limit.

Hint: Approximate the expression following from the lower equation for the small components ψ_B by replacing $c/(E + e\Phi + m_0c^2)$ by $1/(2m_0c)$.

References

[AB94] S. Assimopoulos, A. Bolovinos, A. Jimoyannis, P. Tsekeris, E. Luc-Koenig, M. Aymar, J. Phys. B **27**, 2471 (1994)

[AB97] H.H. Andersen, P. Balling, P. Kristensen, U.V. Pedersen, S.A. Aseyev, V.V. Petrunin, T. Andersen, Phys. Rev. Lett. **79**, 4770 (1997)

[AG96] M. Aymar, C.H. Greene, E. Luc-Koenig, Rev. Mod. Phys. **68**, 1015 (1996)

[AL87] M. Aymar, E. Luc-Koenig, S. Watanabe, J. Phys. B **20**, 4325 (1987)

[AL89] M. Aymar, J.M. Lecomte, J. Phys. B **22**, 223 (1989)

[AL94] M. Aymar, E. Luc-Koenig, J.-M. Lecomte, J. Phys. B **27**, 2425 (1994)

[Ali92] A. Alijah, J. Phys. B **25**, 5043 (1992)

[BA01] C. Boisseau, E. Audouard, J.P. Vigué, Phys. Rev. Lett. **86**, 2694 (2001)

[BC02] R.P.A. Bettens, M.A. Collins, J. Chem. Phys. **116**, 101 (2002)

[BH89] E.A.J.M. Bente, W. Hogervorst, Z. Phys. D **14**, 119 (1989)

[BS57] H.A. Bethe, E.E. Salpeter, *Quantum Mechanics of One- and Two-Electron Systems,* in *Encyclopedia of Physics*, vol. XXXV, ed. by S. Flügge (Springer, Berlin, Heidelberg, New York, 1957), p. 88ff

[BZ08] E. Bodo, P. Zhang, A. Dalgarno, New J. Phys. **10**, 033024 (2008)

[CK97] L.S. Cederbaum, K.C. Kulander, N.H. March (eds.), *Atoms and Molecules in Intense Fields*, Structure and Bonding, vol. 86 (Springer, Berlin, Heidelberg, New York, 1997)

[CK07] J.-H. Choi, B. Knuffman, T.C. Liebisch, A. Reinhard, G. Raithel, Adv. At. Mol. Opt. Phys. **54**, ed. by P.R. Berman et al., 131 (2007)

[CL03] J. Carbonell, R. Lasauskas, D. Delande, L. Hilico, S. Kiliç, Europhys. Lett. **64**, 316 (2003)

[CO80] E.U. Condon, H. Odabasi, *Atomic Structure* (Cambridge University Press, Cambridge, 1980)

[Coh98] S. Cohen, Europ. Phys. J. D **4**, 31 (1998)

[CP05] R. Côté, T. Pattard, M. Weidemüller (eds.), J. Phys. B **38**(2), (2005). Special issue on Rydberg physics

[DB91] D. Delande, A. Bommier, J.C. Gay, Phys. Rev. Lett. **66**, 141 (1991)

[DK85] N.B. Delone, V.P. Krainov, *Atoms in Strong Light Fields* (Springer, Berlin, Heidelberg, New York, 1985)

[DP68] R.J. Damburg, R.K. Propin, J. Phys. B **1**, 681 (1968)

[DS94] O. Dippel, P. Schmelcher, L.S. Cederbaum, Phys. Rev. A **49**, 4415 (1994)

[Edm60] A.R. Edmonds, *Angular Momentum in Quantum Mechanics* (Princeton University Press, Princeton, 1960)

[EF01] C. Eltschka, H. Friedrich, M.J. Moritz, Phys. Rev. Lett. **86**, 2693 (2001)

[Fan70] U. Fano, Phys. Rev. A **2**, 1866 (1970)

[Fan83] U. Fano, Rep. Prog. Phys. **46**, 97 (1983)

[FC83] H. Friedrich, M. Chu, Phys. Rev. A **28**, 1423 (1983)

[FR86] U. Fano, A.R.P. Rau, *Atomic Collisions and Spectra* (Academic Press, New York, 1986)

[Fri81] H. Friedrich, Phys. Reports **274**, 209 (1981)

[Fri82] H. Friedrich, Phys. Rev. A **26**, 1827 (1982)

[FT99] H. Friedrich, J. Trost, Phys. Rev. A **59**, 1683 (1999)

[FT04] H. Friedrich, J. Trost, Phys. Reports **397**, 359 (2004)

[FW85] H. Friedrich, D. Wintgen, Phys. Rev. A **32**, 3231 (1985)

[FW89] H. Friedrich, D. Wintgen, Phys. Reports **183**, 37 (1989)

[Gal94] T.F. Gallagher, *Rydberg Atoms* (Cambridge University Press, Cambridge, 1994)

[Gay91] J.-C. Gay (Guest Editor), Comm. At. Mol. Phys., vol. XXV, Special Volume on *Irregular Atomic Systems and Quantum Chaos* (1991)

[GD63] M. Gailitis, R. Damburg, Sov. Phys. JETP **17**, 1107 (1963)

[GF84] A. Giusti-Suzor, U. Fano, J. Phys. B **17**, 215 (1984)

[GF93] G.F. Gribakin, V.V. Flambaum, Phys. Rev. A **48**, 546 (1993)

[GG83] F. Gounand, T.F. Gallagher, W. Sandner, K.A. Safinya, R. Kachru, Phys. Rev. A **27**, 1925 (1983)

[GL84] A. Giusti-Suzor, H. Lefebvre-Brion, Phys. Rev. A **30**, 3057 (1984)

[GN85] W.L. Glab, K. Ng, D. Yao, M.H. Nayfeh, Phys. Rev. A **31**, 3677 (1985)

[GP08] T.F. Gallagher, P. Pillet, Adv. At. Mol. Opt. Phys. **56**, ed. by E. Arimondo et al., 161 (2008)

[Gre83] C.H. Greene, Phys. Rev. A **28**, 2209 (1983)

[HB00] L. Hilico, N. Billy, B. Grémaud, D. Delande, Eur. Phys. J. D **12**, 449 (2000)

[HC67] R.D. Hudson, V.L. Carter, J. Opt. Soc. Am. **57**, 651 (1967)

[HR89] H. Hasegawa, M. Robnik, G. Wunner, Prog. Theor. Phys. Suppl. **98**, 198 (1989)

[HW87] A. Holle, G. Wiebusch, J. Main, K.H. Welge, G. Zeller, G. Wunner, T. Ertl, H. Ruder, Z. Physik D **5**, 279 (1987)

[IW91] C.-h. Iu, G.R. Welch, M.M. Kash, D. Kleppner, D. Delande, J.C. Gay, Phys. Rev. Lett. **66**, 145 (1991)

[JA77] Ch. Jungen, O. Atabek, J. Chem. Phys. **66**, 5584 (1977)

[KB32] S.A. Korff, G. Breit, Rev. Modern Phys. **4**, 471 (1932)

[KG90] A.v. Klarenbosch, K.K. Geerinck, T.O. Klaassen, W.Th. Wenckebach, C.T. Foxon, Europhys. Lett. **13**, 237 (1990)

[KM13] A. Kaiser, T.-O. Müller, H. Friedrich, Mol. Phys. **111**, 878 (2013)

[Kol89] V.V. Kolosov, J. Phys. B **22**, 833 (1989)

[LA91] V. Lange, M. Aymar, U. Eichmann, W. Sandner, J. Phys. B **24**, 91 (1991)

[LA94] E. Luc-Koenig, M. Aymar, J.-M. Lecomte, J. Phys. B **27**, 2447 (1994)

[Lan37] R.E. Langer, Phys. Rev. **51**, 669 (1937)

[LB70] R.J. LeRoy, R.B. Bernstein, J. Chem. Phys. **52**, 3869 (1970)

[LB80] E. Luc-Koenig, A. Bachelier, J. Phys. B **13**, 1743 (1980)

[LB98] E. Lindroth, A. Bürgers, N. Brandefelt, Phys. Rev. A **57**, R685 (1998)

[LF70] K.T. Lu, U. Fano, Phys. Rev. A **2**, 81 (1970)

[LF01] M.D. Lukin, M. Fleischhauer, R. Côté, L.M. Duan, D. Jaksch, J.I. Cirac, P. Zoller, Phys. Rev. Lett. **87**, 037901 (2001)

[LL13] J. Lim, H.-g. Lee, J. Ahn, J. Korean Phys. Soc. **63**, 867 (2013)

[LL65] L.D. Landau, E.M. Lifschitz, *Lehrbuch der Theoretischen Physik III: Quantenmechanik* (Akademie Verlag, Berlin, 1965)

[LM85] I. Lindgren, J. Morrison, *Atomic Many-Body Theory*, 2nd edn. (Springer, Berlin, Heidelberg, New York, 1985)

[LU95] R. van Leeuwen, W. Ubachs, W. Hogervorst, M. Aymar, E. Luc-Koenig, Phys. Rev. A **52**, 4567 (1995)

[ME01] M.J. Moritz, C. Eltschka, H. Friedrich, Phys. Rev. A **63**, 042102 (2001); **64**, 022101 (2001)

[MF53] P. Morse, H. Feshbach, *Methods of Theoretical Physics*, Part II (McGraw-Hill, New York, 1953)

[MF11] T.-O. Müller, H. Friedrich, Phys. Rev. A **83**, 022701 (2011)

[Mie68] F.H. Mies, Phys. Rev. **175**, 1628 (1968)

[Mit82] M.H. Mittleman, *Introduction to the Theory of Laser-Atom Interactions* (Plenum Press, New York, 1982)

[MK69] A.B. Migdal, V. Krainov, *Approximation Methods in Quantum Mechanics* (W.A. Benjamin, New York, 1969), p. 131

[MO86] N.L. Manakov, V.D. Ovsiannikov, L.P. Rapoport, Phys. Reports **141**, 319 (1986)

[MO88] P.K. Mukherjee, K. Ohtsuki, K. Ohno, Theor. Chim. Acta **74**, 431 (1988)

[NC90] C.A. Nicolaides, C.W. Clark, M.H. Nayfeh (eds.), in *Atoms in Strong Fields*, Proceedings of a NATO Advanced Study Institute (Kos, 1988) (Plenum Press, New York, 1990)

[NJ88] J. Neukammer, G. Jönsson, A. König, K. Vietzke, H. Hieronymus, H. Rinneberg, Phys. Rev. A **38**, 2804 (1988)

[NR87] J. Neukammer, H. Rinneberg, K. Vietzke, A. König, H. Hieronymus, M. Kohl, H.-J. Grabka, G. Wunner, Phys. Rev. Lett. **59**, 2947 (1987)

[OG85] P.F. O'Mahony, C.H. Greene, Phys. Rev. A **31**, 250 (1985)

[Pee65] J.M. Peek, J. Chem. Phys. **43**, 3004 (1965)

[PF98] T. Purr, H. Friedrich, A.T. Stelbovics, Phys. Rev. A **57**, 308 (1998); T. Purr, H. Friedrich, Phys. Rev. A **57**, 4279 (1998)

[PK83] R. Paulsson, F. Karlsson, R.J. LeRoy, J. Chem. Phys. **79**, 4346 (1983)

[PS09] T. Pohl, H. Sadeghpour, P. Schmelcher, Phys. Reports **484**, 181 (2009)

[Pur99] T. Purr, *Abbrechende Dipolserien im* H^--Ion (Shaker Verlag, Aachen, 1999)

[RF08] P. Raab, H. Friedrich, Phys. Rev A **78**, 022707 (2008)

[RF09] P. Raab, H. Friedrich, Phys. Rev A **80**, 052705 (2009)

[Ris56] P. Risberg, Arkiv för Fysik **10**, 583 (1956)

[RT05] I.I. Ryabtsev, D.B. Tretyakov, I.I. Beterov, J. Phys. B **38**, S421 (2005)

[RW86] H. Rottke, K.H. Welge, Phys. Rev A **33**, 301 (1986)

[RW94] H. Ruder, G. Wunner, H. Herold, F. Geyer, *Atoms in Strong Magnetic Fields* (Springer, Berlin, Heidelberg, New York, 1994)

[SB06] A.F. Sadreev, E.N. Bulgakov, I. Rotter, Phys. Rev B **73**, 235342 (2006)

[SC97] P. Schmelcher, L.S. Cederbaum, in *Atoms and Molecules in Intense Fields*, ed. L.S. Cederbaum, K.C. Kulander, N.H. March, Structure and Bonding, vol. 86, 27 (1997)

[Sea83] M.J. Seaton, Rep. Prog. Phys. **46**, 167 (1983)

[SL08] B. Solis, L.N. Ladrón de Guevara, P.A. Orellana, Phys. Lett. A **372**, 4736 (2008)

[SR13] A.S. Stodolna, A Rouzée, F. Lépine, S. Cohen, F. Robicheaux, A. Gijsbertsen, J.H. Jungmann, C. Bordas, M.J.J. Vrakking, Phys. Rev. Lett. **110**, 213001 (2013)

[SS98] P. Schmelcher, W. Schweizer (eds.) *Atoms and Molecules in Strong External Fields* (Plenum, New York, 1998)

[Sta82] A.F. Starace, in *Encyclopedia of Physics,* vol. 31 (Springer, Berlin, Heidelberg, New York, 1982), p. 1

[Stw70] W. Stwalley, Chem. Phys. Lett. **6**, 241 (1970)

[TE98] J. Trost, C. Eltschka, H. Friedrich, J. Phys. B **31**, 361 (1998); Europhys. Lett. **43**, 230 (1998)

[TF85] R. Thieberger, M. Friedman, A. Rabinovitch, J. Phys. B **18**, L673 (1985)

[TF04] D. Tong, S.M. Farooqi, J. Stanojevic, S. Krishnan, Y.P. Zhang, R. Côté, E.E. Eyler, P.L. Gould, Phys. Rev. Lett. **93**, 063001 (2004)

[Tro97] J. Trost, *Nicht-ganzzahlige Maslov-Indizes* (Harri Deutsch, Frankfurt/M, 1997)

[VC88] L.D. Van Woerkom, W.E. Cooke, Phys. Rev. A **37**, 3326 (1988)

[WB77] A.H. Wapstra, K. Bos, At. Nucl. Data Tables **19**, 177 (1977)

[WF87] D. Wintgen, H. Friedrich, Phys. Rev. A **35**, 1628 (1987)

[WZ88] G. Wunner, G. Zeller, U. Woelk, W. Schweizer, R. Niemeier, F. Geyer, H. Friedrich, H. Ruder, in *Atomic Spectra and Collisions in External Fields*, ed. by K.T. Taylor, M.H. Nayfeh, C.W. Clark (Plenum Press, New York, 1988), p. 9

Chapter 4
Simple Reactions

Next to spectroscopic investigations of atoms, reactions provide one of the most important sources of information on the structure of atoms and their interactions. Reaction theory in general is a prominent and well developed field of theoretical physics [Tay72, Bur77, AJ77, New82, Bra83, Joa87, Sit91]. In this chapter we shall largely focus on the discussion of simple reactions which are induced by the collision of an electron as projectile with a target consisting of an atom or an ion. Nevertheless, many of the results are quite general and also applicable if the projectile is an ion or an atom. In the simplest case, where we can assume both projectile and target to be structureless objects, the projectile-target system is a two-body problem which can be reduced to a one-body problem for a particle with a reduced mass as described in Sect. 2.1.

4.1 Elastic Scattering

The (elastic) scattering of a particle by a potential is a time-dependent process. Under typical laboratory conditions it can, however, be adequately described using the time-independent Schrödinger equation (see e.g. [Mes70]). The precise form of the boundary conditions, which the wave function must fulfill in order to correctly describe incoming and scattered particles, depends on whether the potential is very-long-ranged or of shorter range.

4.1.1 Elastic Scattering by a Shorter-Ranged Potential

In order to describe the elastic scattering of a structureless particle of mass μ by a shorter-ranged potential $V(\boldsymbol{r})$,

$$\lim_{r\to\infty} r^2 V(\boldsymbol{r}) = 0, \tag{4.1}$$

H. Friedrich, *Theoretical Atomic Physics*, Graduate Texts in Physics,
DOI 10.1007/978-3-319-47769-5_4

at energy $E = \hbar^2k^2/(2\mu)$, we look for solutions of the time-independent Schrödinger equation,

$$\left[-\frac{\hbar^2}{2\mu}\Delta + V(\boldsymbol{r})\right]\psi(\boldsymbol{r}) = E\psi(\boldsymbol{r}), \tag{4.2}$$

which have the following asymptotic form:

$$\psi(\boldsymbol{r}) = \mathrm{e}^{\mathrm{i}kz} + f(\theta,\phi)\frac{\mathrm{e}^{\mathrm{i}kr}}{r}, \quad r \to \infty. \tag{4.3}$$

The first term on the right-hand side of (4.3) describes an incoming plane wave with particle density $\varrho = |\psi|^2 = 1$, moving with a velocity $v = \hbar k/\mu$ in direction of the positive z-axis; the current density (1.158),

$$\boldsymbol{j} = \frac{\hbar}{2\mathrm{i}\mu}(\psi^*\nabla\psi - \psi\nabla\psi^*), \tag{4.4}$$

is just $\hbar k/\mu$ times the unit vector in z-direction for such a plane wave. The second term on the right-hand side of (4.3) describes an outgoing spherical wave (see Fig. 4.1); it is modulated by a *scattering amplitude* f which depends on the polar angle θ and the azimuthal angle ϕ [see (1.57)]. This outgoing spherical wave corresponds to an outgoing current density $\boldsymbol{j}_{\mathrm{out}}$ which, according to (4.4), is given in leading order in $1/r$ by

$$\boldsymbol{j}_{\mathrm{out}} = \frac{\hbar k}{\mu}\,|f(\theta,\phi)|^2\,\frac{\boldsymbol{r}}{r^3} + O\left(\frac{1}{r^3}\right). \tag{4.5}$$

Asymptotically the particle flux scattered into the solid angle $\mathrm{d}\Omega$, i.e. through the surface $r^2\mathrm{d}\Omega = r^2 \sin\theta\,\mathrm{d}\theta\,\mathrm{d}\phi$, is simply $(\hbar k/\mu)|f(\theta,\phi)|^2\mathrm{d}\Omega$; the ratio of this flux

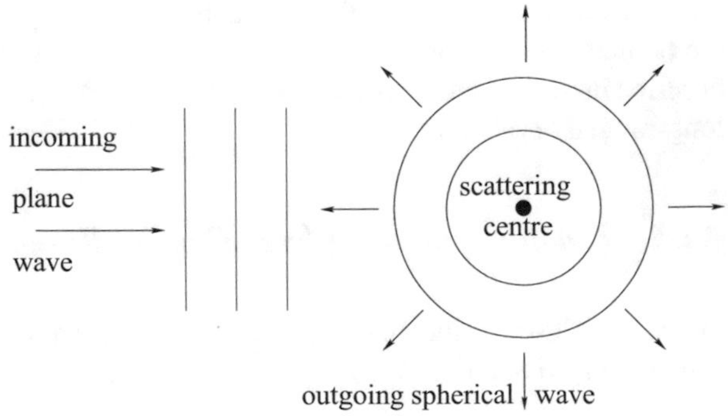

Fig. 4.1 Schematic illustration of the incoming plane wave and the outgoing spherical wave as described by a stationary solution of the Schrödinger equation obeying the boundary conditions (4.3)

to the incoming current density defines the *differential scattering cross section*,

$$\mathrm{d}\sigma = |f(\theta,\phi)|^2 \mathrm{d}\Omega, \quad \frac{\mathrm{d}\sigma}{\mathrm{d}\Omega} = |f(\theta,\phi)|^2 . \tag{4.6}$$

Integrating over all directions (θ,ϕ) yields the *integrated scattering cross section* which is also called the *total elastic scattering cross section*,

$$\sigma = \int \frac{\mathrm{d}\sigma}{\mathrm{d}\Omega}\mathrm{d}\Omega = \int_0^{2\pi} \mathrm{d}\phi \int_0^{\pi} \sin\theta \, \mathrm{d}\theta \, |f(\theta,\phi)|^2 . \tag{4.7}$$

Each solution of the stationary Schrödinger equation (4.2) fulfills the *continuity equation* in the form

$$\nabla \cdot \boldsymbol{j} = -\frac{\partial \varrho}{\partial t} = 0, \quad \text{or} \quad \oint \boldsymbol{j} \cdot \mathrm{d}\boldsymbol{s} = 0 . \tag{4.8}$$

This means that the net particle flux through a closed surface vanishes. For an asymptotically large sphere ($r \to \infty$) with surface element $\mathrm{d}\boldsymbol{s} = r^2 \mathrm{d}\Omega \, \boldsymbol{r}/r$, the integrated contribution of the incoming plane wave in (4.3) to this net flux vanishes on symmetry grounds, while the contribution I_{out} from the outgoing spherical wave is positive unless the scattering amplitude vanishes identically,

$$I_{\text{out}} = \oint \boldsymbol{j}_{\text{out}} \cdot \mathrm{d}\boldsymbol{s} = \frac{\hbar k}{\mu} \int |f(\Omega)|^2 d\Omega = \frac{\hbar k}{\mu} \sigma . \tag{4.9}$$

Since the total particle flux through the surface vanishes, the current density (4.4) must contain terms which cancel the positive contribution (4.9). The terms describing the interference between the incoming plane wave and the outgoing spherical wave do just this. An explicit calculation (Problem 4.1) shows that such interference is only important in the forward direction $\theta = 0$, and this leads to a relation between the scattering amplitude in the forward direction and the integrated scattering cross section,

$$\frac{1}{2\mathrm{i}} \left[f(\theta = 0) - f^*(\theta = 0) \right] = \Im \left[f(\theta = 0) \right] = \frac{k}{4\pi} \sigma . \tag{4.10}$$

The relation (4.10) expresses particle number conservation and is called *optical theorem*.

It is often useful to treat scattering problems using an equivalent integral equation in place of the Schrödinger equation (4.2). In order to derive the integral equation we rewrite the Schrödinger equation to make it look like an inhomogeneous differential equation,

$$\left[E + \frac{\hbar^2}{2\mu} \Delta \right] \psi(\boldsymbol{r}) = V(\boldsymbol{r}) \psi(\boldsymbol{r}) . \tag{4.11}$$

This is solved using the *free-particle Green's function*

$$\mathcal{G}\left(\boldsymbol{r},\boldsymbol{r}'\right) = -\frac{\mu}{2\pi\hbar^2}\frac{\mathrm{e}^{\mathrm{i}k|\boldsymbol{r}-\boldsymbol{r}'|}}{|\boldsymbol{r}-\boldsymbol{r}'|}, \tag{4.12}$$

which fulfills the following equation:

$$\left(E + \frac{\hbar^2}{2\mu}\Delta\right)\mathcal{G}(\boldsymbol{r},\boldsymbol{r}') = \delta(\boldsymbol{r}-\boldsymbol{r}')\,. \tag{4.13}$$

The Green's function (4.12) is an extension of the Green's function defined in Sect. 1.5.2 to three-dimensional vector arguments. It is the coordinate representation of the *Green's operator* $\hat{\mathcal{G}}$ which has the properties of an inverse operator to $E + (\hbar^2/2\mu)\Delta = E - \hat{\boldsymbol{p}}^2/(2\mu)$:

$$\hat{\mathcal{G}} = \lim_{\varepsilon\to 0}\frac{1}{E \pm \mathrm{i}\varepsilon - \hat{\boldsymbol{p}}^2/(2\mu)}\,. \tag{4.14}$$

An infinitesimally small imaginary contribution $\pm\mathrm{i}\varepsilon$ is added to the real energy E so that we can invert the operator $E - \hat{\boldsymbol{p}}^2/(2\mu)$. The plus or minus signs lead to a different asymptotic behaviour of the resulting wave function. A positive infinitesimal imaginary part of the energy corresponds to the Green's function (4.12) above and leads to a solution (4.15) below, containing an outgoing spherical wave as in (4.3); a negative imaginary part of the energy corresponds to the complex conjugate Green's function and leads to incoming spherical waves in the asymptotic region.

It is easy to verify that the wave function

$$\psi(\boldsymbol{r}) = \mathrm{e}^{\mathrm{i}kz} + \int \mathcal{G}(\boldsymbol{r},\boldsymbol{r}')V(\boldsymbol{r}')\psi(\boldsymbol{r}')\,\mathrm{d}\boldsymbol{r}' \tag{4.15}$$

solves the Schrödinger equation (4.11). Since the right-hand side of (4.11) isn't a genuinely inhomogeneous term but depends on the solution ψ, equation (4.15) isn't an explicit solution of the Schrödinger equation but a transformation into an equivalent integral equation, which is known as the *Lippmann-Schwinger equation*. Its solutions automatically fulfill the boundary conditions (4.3). For $r \gg r'$ we can approximate the free-particle Green's function (4.12) by (see Problem 4.2)

$$\mathcal{G}(\boldsymbol{r},\boldsymbol{r}') = -\frac{\mu}{2\pi\hbar^2}\frac{\mathrm{e}^{\mathrm{i}kr}}{r}\left[\mathrm{e}^{-\mathrm{i}\boldsymbol{k}_r\cdot\boldsymbol{r}'} + O\left(\frac{r'}{r}\right)\right] \tag{4.16}$$

and obtain the form (4.3) with an implicit expression for the scattering amplitude,

$$f(\theta,\phi) = -\frac{\mu}{2\pi\hbar^2}\int \mathrm{e}^{-\mathrm{i}\boldsymbol{k}_r\cdot\boldsymbol{r}'}\,V(\boldsymbol{r}')\,\psi(\boldsymbol{r}')\,\mathrm{d}\boldsymbol{r}'\,. \tag{4.17}$$

In (4.16) and (4.17) $\boldsymbol{k}_r$ is the wave vector with length k which points in direction of the radius vector $\boldsymbol{r}$ (without $'$).

We can interpret the integral in (4.17) as the matrix element of an abstract *transition operator* $\hat{T}$ between an initial state $\psi_\mathrm{i}(\boldsymbol{r}') = \exp(\mathrm{i}kz')$ and a final state $\psi_\mathrm{f}(\boldsymbol{r}') = \exp(\mathrm{i}\boldsymbol{k}_r{\cdot}\boldsymbol{r}')$,

$$T_\mathrm{fi} = \langle\psi_\mathrm{f}|\hat{T}|\psi_\mathrm{i}\rangle \stackrel{\mathrm{def}}{=} \langle\psi_\mathrm{f}|V|\psi\rangle = -\frac{2\pi\hbar^2}{\mu} f(\theta,\phi)\,. \tag{4.18}$$

Using the *T-Matrix* defined in this way, we can interpret the scattering process in the spirit of time-dependent perturbation theory (Sect. 2.4.1) as a transition from the incoming plane wave ψ_i, travelling in the direction of the z-axis, to an outgoing plane wave ψ_f, travelling outwards in the direction of the vector $\boldsymbol{r} \equiv (r,\theta,\phi)$ (Problem 4.3).

If the influence of the potential is small, it may be justified to replace the exact wave function $\psi(\boldsymbol{r}')$ in the integrand on the right-hand side of (4.15) or (4.17) by the "unperturbed" incoming plane wave $\psi_\mathrm{i}(\boldsymbol{r}') = \exp(\mathrm{i}kz')$. This assumption defines the *Born approximation*. In the Born approximation equations (4.15) and (4.17) become explicit expressions for the wave function and the scattering amplitude respectively. E.g. the scattering amplitude is, in Born approximation,

$$f^\mathrm{B} = -\frac{\mu}{2\pi\hbar^2}\int \mathrm{e}^{-\mathrm{i}\boldsymbol{q}\cdot\boldsymbol{r}'}\,V(\boldsymbol{r}')\,\mathrm{d}\boldsymbol{r}' = -\frac{\mu}{2\pi\hbar^2}\,\langle\psi_\mathrm{f}|V|\psi_\mathrm{i}\rangle\,. \tag{4.19}$$

Here $\boldsymbol{q} = k(\boldsymbol{e}_r - \boldsymbol{e}_z)$. The vector $\boldsymbol{e}_z$ is the unit vector in the direction of the positive z-axis and $\boldsymbol{e}_r$ is the unit vector in the direction of the radius vector $\boldsymbol{r}$. The formula (4.19) shows that the scattering amplitude in Born approximation is derived by a Fourier transformation from the potential. The argument $\boldsymbol{q}$ is the wave vector of the momentum transfer occurring for elastic scattering in the direction of the radius vector $\boldsymbol{r}$:

$$\hbar\boldsymbol{q} = (\hbar k)\boldsymbol{e}_r - (\hbar k)\boldsymbol{e}_z. \tag{4.20}$$

Comparing (4.18) and (4.19) shows that the Born approximation amounts to replacing the transition operator $\hat{T}$ by the potential V.

4.1.2 Partial-Waves Expansion

For a radially symmetric potential, the Schrödinger equation (4.2) is rotationally invariant, but the boundary conditions (4.3) for the scattering wave function $\psi(\boldsymbol{r})$ are not. So ψ is not an eigenfunction of angular momentum, but it can be expanded in eigenfunctions of angular momentum. Since rotational symmetry around the z-axis is conserved both by the Schrödinger equation (4.2) and the boundary conditions (4.3), the azimuthal quantum number m is conserved. Since the incoming plane wave has $m = 0$, the same can be assumed for the full wave function $\psi(\boldsymbol{r})$, which thus no longer depends on the azimuthal angle ϕ,

$$\psi(\boldsymbol{r}) = \psi(r,\theta) = \sum_{l=0}^{\infty} \frac{u_l(r)}{r} P_l(\cos\theta) \,. \tag{4.21}$$

Equation (4.21) represents an expansion of the full scattering wave $\psi(\boldsymbol{r})$ in *partial waves*, each such partial wave being labelled by its orbital angular momentum quantum number l. The contribution of each partial wave is determined by its *radial wave function* $u_l(r)$. Radial wave functions were already introduced for a single angular momentum quantum number l in Sect. 1.2.2, cf. (1.74); the present ansatz (4.21) represents a coherent superposition of contributions from all partial waves.

Inserting the expansion (4.21) into the Schrödinger equation (4.2) leads to a set of *radial Schrödinger equations* (1.75) for the radial wave functions $u_l(r)$:

$$\left[-\frac{\hbar^2}{2\mu}\frac{\mathrm{d}^2}{\mathrm{d}r^2} + \frac{l(l+1)\hbar^2}{2\mu r^2} + V(r)\right] u_l(r) = E\, u_l(r) \,. \tag{4.22}$$

The $1/r$ on the right-hand side of (4.21) ensures that the *radial Schrödinger equation* (4.22) contains only the second and not the first derivative of u_l, so it has the form of a Schrödinger equation for a particle moving in one dimension under the influence of the effective potential

$$V_{\text{eff}}(r) = V(r) + V_{\text{cent}}(r) \,, \quad V_{\text{cent}}(r) = \frac{l(l+1)\hbar^2}{2\mu r^2} \,, \tag{4.23}$$

subject to the condition that the coordinate r is non-negative, $r \geq 0$. In the space of all possible radial wave functions in the l-th partial wave, the unitary scalar product of two radial wave functions, u_l and $\tilde{u}_l$ is defined as

$$\langle u_l|\tilde{u}_l\rangle = \int_0^{\infty} u_l(r)^* \tilde{u}_l(r)\, \mathrm{d}r \,. \tag{4.24}$$

4.1.3 Scattering Phase Shifts

In the absence of the potential $V(r)$, the radial Schrödinger equation (4.22) represents the angular momentum components of the free-particle wave equation, and its solutions can be written as functions of the dimensionless product kr. Two linearly independent solutions of the radial free-particle equation are,

$$u_l^{(\mathrm{s})}(kr) = kr\, j_l(kr) \,, \quad u_l^{(\mathrm{c})}(kr) = -kr\, y_l(kr) \,, \tag{4.25}$$

where j_l and y_l stand for the *spherical Bessel functions* of the first and second kind, respectively (see Appendix A.4 and [AS70]). Their asymptotic behaviour is given by

$$u_l^{(s)}(kr) \stackrel{kr\to\infty}{=} \sin\left(kr - l\frac{\pi}{2}\right) + O\left(\frac{1}{kr}\right) ,$$
$$u_l^{(c)}(kr) \stackrel{kr\to\infty}{=} \cos\left(kr - l\frac{\pi}{2}\right) + O\left(\frac{1}{kr}\right) . \qquad (4.26)$$

For small values of kr, the radial free-particle wave functions (4.25) behave as,

$$u_l^{(s)}(kr) \stackrel{kr\to 0}{\sim} \frac{\sqrt{\pi}(kr)^{l+1}}{2^{l+1}\Gamma\left(l+\frac{3}{2}\right)}\left[1 - \frac{(kr)^2}{4l+6}\right] ,$$
$$u_l^{(c)}(kr) \stackrel{kr\to 0}{\sim} \frac{2^l\Gamma\left(l+\frac{1}{2}\right)}{\sqrt{\pi}(kr)^l}\left[1 + \frac{(kr)^2}{4l-2}\right] . \qquad (4.27)$$

The wave function $u_l^{(s)}$ is the physical, *regular* solution; $u_l^{(c)}$ is an unphysical, *irregular* solution. For $l > 0$, the irregular solution $u_l^{(c)}$ is not square integrable due to the divergence at $r \to 0$; for $l = 0$ its contribution proportional to $1/r$ in the full wave function (4.21) would lead to a delta function contribution in $\Delta\psi$, which cannot be compensated by any other term in the Schrödinger equation (4.2).

For a potential $V(r)$ less singular than $1/r^2$ at the origin, the effective potential (4.23) is dominated near $r = 0$ by the centrifugal term, so we can expect two linearly independent solutions of (4.22), u_l^{reg} and $u_l^{\text{irr}}(r)$, whose small-distance behaviour is

$$u_l^{\text{reg}}(r) \stackrel{r\to 0}{\propto} r^{l+1} , \quad u_l^{\text{irr}}(r) \stackrel{r\to 0}{\propto} r^{-l} . \qquad (4.28)$$

Here u_l^{reg} denotes the physical, regular solution; u_l^{irr} is an unphysical, irregular solution. In the following, we shall mostly be dealing with regular solutions of the radial Schrödinger equation, which vanish for $r \to 0$, and we shall dispense with the superscript "reg" unless it is explicitly needed.

At large distances, the effective potential (4.23) is again dominated by the centrifugal term, because we have assumed that $V(r)$ falls off faster than $1/r^2$. The regular solution of the radial Schrödinger equation (4.22) can, at large distances, be taken to be a superposition of the two radial free-particle wave functions (4.25) obeying (4.26),

$$u_l(r) \stackrel{r\to\infty}{\propto} A\, u_l^{(s)}(kr) + B\, u_l^{(c)}(kr) \stackrel{r\to\infty}{\propto} \sin\left(kr - l\frac{\pi}{2} + \delta_l\right) , \qquad (4.29)$$

with $\tan\delta_l = B/A$. Since the potential is real, we can assume that u_l is, except for a constant complex factor, a real function of r, so that the ratio B/A and the phase δ_l

are real. The phases δ_l, $l = 0, 1, 2, \ldots$, contain the information about the effect of the potential on the asymptotic behaviour of the wave function (4.21). They are called *scattering phase shifts*, because they determine the scattering amplitude, as shown in the following.

The partial-waves expansion of the incoming plane wave is

$$\mathrm{e}^{\mathrm{i}kz} = \sum_{l=0}^{\infty}(2l+1)\,\mathrm{i}^l\, j_l(kr)\, P_l(\cos\theta)\ , \tag{4.30}$$

where the j_l are the spherical Bessel functions of the first kind, already introduced in (4.25). At large distances, the full wave function consists of the plane wave (4.30) and an outgoing spherical wave according to (4.3). The scattering amplitude f depends only on the polar angle θ, because the whole wave function does not depend on the azimuthal angle ϕ. We expand f into partial wave contributions,

$$f(\theta) = \sum_{l=0}^{\infty} f_l\, P_l(\cos\theta)\ , \tag{4.31}$$

with constant coefficients f_l, the *partial-wave scattering amplitudes*. Expressing the sum of plane and spherical wave in the form (4.21) gives an explicit expression for the asymptotic behaviour of the radial wave functions,

$$\begin{aligned} u_l(r) &\overset{r\to\infty}{\sim} \mathrm{i}^l \left[\frac{2l+1}{k}\sin\left(kr - l\frac{\pi}{2}\right) + f_l\,\mathrm{e}^{\mathrm{i}(kr-l\pi/2)}\right] \\ &= \mathrm{i}^l \left[\left(\frac{2l+1}{k} + \mathrm{i}f_l\right)\sin\left(kr - l\frac{\pi}{2}\right) + f_l\cos\left(kr - l\frac{\pi}{2}\right)\right] . \end{aligned} \tag{4.32}$$

Comparing (4.32) and (4.29) shows that the coefficients of the sine and cosine terms in the square bracket in the lower line of (4.32) can be interpreted as the coefficients A and B in (4.29), for which $\tan\delta_l = B/A$. With the coefficients in (4.32),

$$\cot\delta_l = \frac{A}{B} \equiv \frac{2l+1}{k f_l} + \mathrm{i} \Rightarrow \cot\delta_l - \mathrm{i} = \frac{\mathrm{e}^{-\mathrm{i}\delta_l}}{\sin\delta_l} = \frac{2l+1}{k f_l}\ , \tag{4.33}$$

which leads to

$$f_l = \frac{2l+1}{k}\,\mathrm{e}^{\mathrm{i}\delta_l}\,\sin\delta_l = \frac{2l+1}{2\mathrm{i}k}\left(\mathrm{e}^{2\mathrm{i}\delta_l} - 1\right) . \tag{4.34}$$

With (4.32) the asymptotic form of the radial wave functions is,

$$u_l(r) \overset{r\to\infty}{\sim} \frac{2l+1}{k}\,\mathrm{i}^l\,\mathrm{e}^{\mathrm{i}\delta_l}\,\sin\left(kr - l\frac{\pi}{2} + \delta_l\right)\ , \tag{4.35}$$

and the asymptotic form of the full wave function (4.21) is

$$\psi(\boldsymbol{r}) \overset{r\to\infty}{\sim} \sum_{l=0}^{\infty} \frac{2l+1}{kr}\, \mathrm{i}^l\, \mathrm{e}^{\mathrm{i}\delta_l} \sin\left(kr - l\frac{\pi}{2} + \delta_l\right) P_l(\cos\theta) . \tag{4.36}$$

The explicit expression, (4.31) with (4.34), for the scattering amplitude allows us to express the differential scattering cross section in terms of the scattering phase shifts δ_l,

$$\frac{\mathrm{d}\sigma}{\mathrm{d}\Omega} = |f(\theta)|^2 = \frac{1}{k^2} \sum_{l,l'} \mathrm{e}^{\mathrm{i}(\delta_l - \delta_{l'})} (2l+1) \sin\delta_l (2l'+1) \sin\delta_{l'}\, P_l(\cos\theta) P_{l'}(\cos\theta) . \tag{4.37}$$

For the integrated scattering cross section we can exploit the orthogonality of the Legendre polynomials ((A.2) in Appendix A.1),

$$\sigma = \sum_{l=0}^{\infty} \frac{4\pi}{2l+1} |f_l|^2 = \frac{4\pi}{k^2} \sum_{l=0}^{\infty} (2l+1) \sin^2\delta_l = \frac{\pi}{k^2} \sum_{l=0}^{\infty} (2l+1) \left|\mathrm{e}^{2\mathrm{i}\delta_l} - 1\right|^2 . \tag{4.38}$$

The integrated scattering cross section is the incoherent sum of the contributions $\sigma_{[l]}$ from each partial wave,

$$\sigma = \sum_{l=0}^{\infty} \sigma_{[l]} , \quad \sigma_{[l]} = \frac{4\pi}{k^2} (2l+1) \sin^2\delta_l . \tag{4.39}$$

The maximum contribution of a given partial wave l to the integrated cross section is realized when δ_l is an odd multiple of $\frac{\pi}{2}$, so $\sin^2\delta_l = 1$,

$$\left(\sigma_{[l]}\right)_{\max} = \frac{4\pi}{k^2} (2l+1) . \tag{4.40}$$

4.1.4 *Radial Lippmann-Schwinger Equation*

The radial Schrödinger equation (4.22) can be rewritten as

$$\left[E + \frac{\hbar^2}{2\mu} \frac{\mathrm{d}^2}{\mathrm{d}r^2} - \frac{l(l+1)\hbar^2}{2\mu r^2}\right] u_l(r) = V(r)\, u_l(r) \tag{4.41}$$

and transformed into an integral equation with the help of the *radial free-particle Green's function* $\mathcal{G}_l(r,r')$, which fulfills

$$\left[E+\frac{\hbar^2}{2\mu}\frac{\mathrm{d}^2}{\mathrm{d}r^2}-\frac{l(l+1)\,\hbar^2}{2\mu\,r^2}\right]\mathcal{G}_l(r,r')=\delta(r-r') \tag{4.42}$$

and is explicitly given by

$$\mathcal{G}_l(r,r')=-\frac{2\mu}{\hbar^2 k}\,u_l^{(\mathrm{s})}\,(kr_<)\;u_l^{(\mathrm{c})}\,(kr_>)\ ; \tag{4.43}$$

here $u_l^{(\mathrm{s})}$ and $u_l^{(\mathrm{c})}$ stand for the regular and irregular free-particle radial waves as defined in (4.25), and $r_<$ stands for the smaller while $r_>$ stands for the larger of the two radial coordinates r and r'.[1] A wave function obeying the integral equation

$$u_l(r)=u_l^{(\mathrm{s})}(kr)+\int_0^\infty \mathcal{G}_l(r,r')\,V(r')\,u_l(r') \tag{4.44}$$

necessarily obeys the radial Schrödinger equation (4.41). This would also hold if the first term $u_l^{(\mathrm{s})}(kr)$ were replaced by any other solution of the "homogeneous" version $[E+\cdots]\,u_l(r)=0$ of (4.41). Similar to the situation described in (4.11), the right-hand side of (4.41) is not a genuinely inhomogeneous term, independent of the solution $u_l(r)$, and (4.44) is not an explicit solution of the radial Schrödinger equation, but an equivalent formulation as integral equation.

Equation (4.44) is the *radial Lippmann-Schwinger equation* in the l-th partial wave. Asymptotically, $r\to\infty$, we can assume $r=r_>$ and $r'=r_<$ in the radial Green's function, so the factor $u_l^{(\mathrm{c})}\,(kr_>)=u_l^{(\mathrm{c})}(kr)$ can be drawn out of the integral over r',

$$u_l(r)\overset{r\to\infty}{\sim}u_l^{(\mathrm{s})}(kr)-\left[\frac{2\mu}{\hbar^2 k}\int_0^\infty u_l^{(\mathrm{s})}(kr')\,V(r')\,u_l(r')\,\mathrm{d}r'\right]u_l^{(\mathrm{c})}(kr)\ . \tag{4.45}$$

Comparing with (4.29) shows that the coefficient of $u_l^{(\mathrm{c})}(kr)$ in (4.45) is the tangent of the scattering phase shift,

$$\tan\delta_l=-\frac{2\mu}{\hbar^2 k}\int_0^\infty u_l^{(\mathrm{s})}(kr)\,V(r)\,u_l(r)\,\mathrm{d}r\ . \tag{4.46}$$

The expression on the right-hand side of (4.46) cannot be evaluated explicitly, because it still contains the (usually unknown) exact solution u_l of the radial

[1]The prefactor on the right-hand side of (4.43) reflects the asymptotic normalization of the radial wave functions (4.26). With energy-normalized radial wave functions, the Green's function is as given in (1.228).

Schrödinger equation. It does, however, offer a possibility for approximation in the spirit of the Born approximation. Replacing $u_l(r)$ in the integrand in (4.46) by the regular free-particle radial wave $u_l^{(s)}(kr)$ gives an explicit but approximate expression for $\tan\delta_l$, in the spirit of the first-order Born approximation:

$$\tan\delta_l^{\rm Born} = -\frac{2\mu}{\hbar^2 k}\int_0^\infty \left[u_l^{(s)}(kr)\right]^2 V(r)\,{\rm d}r\,. \tag{4.47}$$

Note that the right-hand side of (4.47) is a smooth function of k that always remains finite. Hence $\delta_l^{\rm Born}$ as function of k can never cross an odd multiple of $\frac{\pi}{2}$. Equation (4.47) can only be a useful approximation when the phase shifts are restricted to a small interval around zero (or an integer multiple of π); for potentials which are bounded and short ranged, this happens both in the limit of high energies and in the limit of large angular momentum quantum numbers l.

4.1.5 S-Matrix

The asymptotic behaviour of the radial wave function (4.35) can be written as

$$\begin{aligned} u_l(r) &\stackrel{r\to 0}{\sim} \frac{2l+1}{2k}\,{\rm i}^{l+1}\left[{\rm e}^{-{\rm i}(kr-l\pi/2)} - {\rm e}^{2{\rm i}\delta_l}{\rm e}^{+{\rm i}(kr-l\pi/2)}\right] \\ &= \frac{2l+1}{2k}\,{\rm i}^{2l+1}\left[{\rm e}^{-{\rm i}kr} - (-1)^l{\rm e}^{2{\rm i}\delta_l}{\rm e}^{+{\rm i}kr}\right]. \end{aligned} \tag{4.48}$$

In both lines of (4.48), the square bracket contains an incoming radial wave proportional to ${\rm e}^{-{\rm i}kr\cdots}$ and an outgoing radial wave proportional to ${\rm e}^{+{\rm i}kr\cdots}$. The factor ${\rm e}^{2{\rm i}\delta_l}$ in the outgoing wave is the contribution of the l-th partial wave to the *scattering matrix* or S-matrix,

$$S_l = {\rm e}^{2{\rm i}\delta_l}\,. \tag{4.49}$$

For the radial potential $V(r)$, the S-matrix is diagonal, because there is no coupling between the radial Schrödinger equations (4.22) of different l.

The S-matrix is *unitary*, which, for the partial-wave contribution (4.49) means $|S_l| = 1$. This is an expression of particle conservation and is fulfilled as long as the scattering phase shifts δ_l are real. Equation (4.40) is based on the assumption, that the phase shifts are real, i.e., that the S-matrix is unitary. Its right-hand side $\left(4\pi/k^2\right)(2l+1)$ is hence called the *unitarity limit* of the contribution of the respective partial wave to the integrated scattering cross section.

For real δ_l, the scattering amplitude (4.31) with the partial-wave amplitudes (4.34) can be decomposed into real and imaginary parts as follows:

$$f(\theta) = \sum_{l=0}^{\infty}\frac{2l+1}{k}\left[\cos\delta_l\,\sin\delta_l + {\rm i}\,\sin^2\delta_l\right]P_l(\cos\theta)\,. \tag{4.50}$$

For the forward direction, $\theta = 0$, we insert $P_l(1) = 1$ and recall (4.38),

$$\Im\left[f(\theta = 0)\right] = \sum_{l=0}^{\infty} \frac{2l+1}{k} \sin^2 \delta_l = \frac{k}{4\pi} \sigma \;, \tag{4.51}$$

thus recovering the optical theorem (4.10). The unitarity of the S-matrix is an expression of particle conservation. Note that the radial Born approximation (4.47) yields real phase shifts and a unitary S-matrix, so it is compatible with particle conservation. This is in contrast to the Born approximation (4.19) for the scattering amplitude. For a radially symmetric potential V, the Born scattering amplitude (4.19) is a real function of the modulus of the momentum transfer vector (4.20) and necessarily violates the optical theorem.

4.1.6 Determination of the Scattering Phase Shifts

The boundary condition $u_l(r) \overset{r\to 0}{\propto} r^{l+1}$ (cf. (4.28)) uniquely determines the radial wave function except for a constant factor. The scattering phase shifts δ_l can be calculated by integrating the radial Schrödinger equation (4.22) with this boundary condition from small r to a finite radius $r_{\rm m}$, where the potential $V(r)$ has already fallen off sufficiently to be negligible. Matching the logarithmic derivative u_l'/u_l to the logarithmic derivative of a superposition (4.29) of the free-particle wave functions at $r = r_{\rm m}$ yields $\tan\delta_l$.

Due to the influence of the potential at short distances, the nodes (beyond $r = 0$) and antinodes of the radial wave function $u_l(r)$ are shifted relative to those of the regular free-particle wave function $u_l^{(\rm s)}$. This leads to asymptotic *spatial shifts* d_l, which are related to the phase shifts δ_l by $d_l = \delta_l/k$, as can be seen by writing u_l as

$$u_l(r) \overset{r\to\infty}{\propto} \sin\left[k\left(r + \frac{\delta_l}{k}\right) - l\frac{\pi}{2}\right] . \tag{4.52}$$

For a repulsive potential V, the radial wave function is suppressed at small distances and its nodes (beyond $r = 0$) and antinodes are pushed to larger values of r by the potential; the spatial shifts, and hence also the phase shifts, are negative. The simplest example is scattering by hard sphere of radius R. For $r > R$, the potential vanishes, and the radial wave function can be written as $A\,u_l^{(\rm s)}(kr) + B\,u_l^{(\rm c)}(kr)$, see (4.29). The wave function must vanish for $r \leq R$, so the inner boundary condition is pushed out from $r = 0$ to $r = R$. The condition $A\,u_l^{(\rm s)}(kR) + B\,u_l^{(\rm c)}(kR) = 0$ yields

$$\frac{B}{A} = -\frac{u_l^{(\rm s)}(kR)}{u_l^{(\rm c)}(kR)} = \frac{j_l(kR)}{y_l(kR)} \;, \qquad \delta_l = \arctan\left(\frac{j_l(kR)}{y_l(kR)}\right) . \tag{4.53}$$

From (4.27) and (4.26), the low- and high-energy behaviour of the hard-sphere phase shifts is

$$\delta_l \overset{kR\to 0}{\sim} -\frac{\pi}{\Gamma\left(l+\frac{3}{2}\right)\Gamma\left(l+\frac{1}{2}\right)}\left(\frac{kR}{2}\right)^{2l+1}\left[1-\left(\frac{kR}{2}\right)^2\left(\frac{1}{l-\frac{1}{2}}+\frac{1}{l+\frac{3}{2}}\right)\right],$$

$$\delta_l \overset{kR\to\infty}{\sim} -kR + l\frac{\pi}{2} \tag{4.54}$$

for $l > 0$, while $\delta_{l=0} = -kR$ for all k. Note that the high-energy behaviour in the lower line of (4.54) implies that the radial wave function (4.52) has the same asymptotic behaviour in all partial waves in the high-energy limit,

$$u_l(r) \overset{r\to\infty,\ kR\to\infty}{\propto} \sin(kr - kR)\ . \tag{4.55}$$

This is because, for any angular momentum l, the radial classical turning point always reaches the radius R of the hard sphere at a sufficiently high energy, and the influence of the centrifugal potential diminishes continuously as the energy rises further above this value. The phase shifts (4.53) for scattering by a hard sphere are shown in Fig. 4.2 for partial waves from $l = 0$ to $l = 5$.

For an attractive potential, the oscillations are of smaller wavelength in the interaction region and a given node (beyond $r = 0$) or antinode is pulled in to shorter distances by the potential; the spatial shift and the phase shift are positive. The behaviour of the phase shift depends on whether the effective potential features an attractive well that is deep enough to support one or more bound states, and the near-threshold behaviour of the phase shift depends sensitively on whether or not there is a bound state close to threshold.

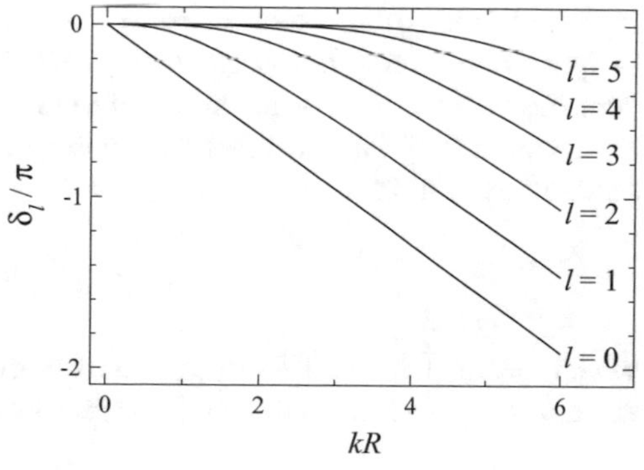

Fig. 4.2 Scattering phase shifts (4.53) for scattering by a hard sphere of radius R

4.1.7 Near-Threshold Behaviour of the Scattering Phase Shifts

The leading near-threshold behaviour of the phase shifts can be derived from the small-argument behaviour of the free-particle solutions. At distances r beyond the range of the potential, the radial wave function $u_l(r)$ is a superposition of the free-particle wave functions (4.25); towards threshold, $k \to 0$, the product kr tends to zero so we can make use of the small-argument expressions (4.27),

$$\begin{aligned} u_l(r) &\overset{kr\to 0}{\propto} u_l^{(\mathrm{s})}(kr) + \tan\delta_l\, u_l^{(\mathrm{c})}(kr) \\ &\sim \frac{\sqrt{\pi}\,k^{l+1}}{2^{l+1}\Gamma\left(l+\frac{3}{2}\right)}\left[r^{l+1} + \tan\delta_l\,\frac{2^{2l+1}\Gamma\left(l+\frac{1}{2}\right)\Gamma\left(l+\frac{3}{2}\right)}{\pi\,k^{2l+1}\,r^l}\right] . \end{aligned} \tag{4.56}$$

Directly at threshold, the radial Schrödinger equation (4.22) has a regular solution $u_l^{(0)}(r)$ which is defined up to a constant by the boundary condition $u_l^{(0)}(0) = 0$ and is function of r only. The wave function (4.56) must become proportional to this k-independent solution for $k \to 0$, so in the second term in the square bracket in the lower line of (4.56), the k-dependence of $\tan\delta_l$ must compensate the factor k^{2l+1} in the denominator, $\tan\delta \overset{k\to 0}{\propto} k^{2l+1}$. More explicitly,

$$\tan\delta_l \overset{k\to 0}{\sim} -\frac{\pi}{\Gamma\left(l+\frac{1}{2}\right)\Gamma\left(l+\frac{3}{2}\right)}\left(\frac{a_l\,k}{2}\right)^{2l+1} . \tag{4.57}$$

The characteristic length a_l appearing on the right-hand side of (4.57) is the *scattering length* in the l-th partial wave.

The proportionality to k^{2l+1} in (4.57) expresses growing suppression with increasing l due to the influence of the centrifugal barrier separating the asymptotic region of free-particle motion from the interaction region at small distances. It is typical for the l-dependence of quantum mechanical quantities involving a centrifugal barrier and is generally referred to as *Wigner's threshold law*.

Equation (4.57) implies that the leading behaviour of the partial-wave scattering amplitude (4.34) is

$$f_l \overset{k\to 0}{\propto} k^{2l} , \tag{4.58}$$

which means that small l-values dominate the scattering amplitude (4.31) and the scattering cross sections (4.37), (4.38) at low energies. For s-waves, (4.57) reads

$$\tan\delta_0 \overset{k\to 0}{\sim} -a\,k , \tag{4.59}$$

where we have dropped the subscript on the a, as is customary. The s-wave scattering length a in (4.59) is generally referred to as *the* scattering length, a concept introduced by Fermi and Marshall in 1947 [FM47]. The scattering length is a property of the threshold solution of the radial Schrödinger equation and already played an important role in the discussion of near-threshold quantization of bound states in Sect. 3.1.2

From (4.58) it follows that only the s-wave retains a nonvanishing contribution to the scattering amplitude (4.31) in the limit $k \to 0$,

$$\lim_{k\to 0} f(\theta) = f_0 P_0 \sim -a \quad \Longrightarrow \quad \lim_{k\to 0} \frac{\mathrm{d}\sigma}{\mathrm{d}\Omega} = a^2 \quad \text{and} \quad \lim_{k\to 0} \sigma = 4\pi a^2 . \tag{4.60}$$

For hard-sphere scattering, the scattering length is the radius of the sphere, and the threshold limit of the quantum mechanical integrated scattering cross section is $4\pi R^2$, which is four times the classical cross section.

The definition (4.59) of the scattering length for s-waves is universally accepted. For $l > 0$, the definitions of the scattering length vary. Some authors, e.g. [Tay72], even call the whole coefficient of k^{2l+1} in (4.57) scattering length, although this coefficient has the physical dimension of a length to the power $2l+1$. The definition (4.57) ensures that a_l is a length and that for scattering by a hard sphere of radius R we have $a_l = R$ for all l, as can be seen by comparing with (4.54).

With (4.57), the threshold solution of the radial Schrödinger equation (4.22) behaves asymptotically as,

$$u_l^{(0)}(r) \overset{r\to\infty}{\propto} r^{l+1} - \frac{a_l^{2l+1}}{r^l} , \tag{4.61}$$

so the scattering length appears as the zero of the asymptotic behaviour of the threshold solution of the radial Schrödinger equation. This was already apparent for s-waves ($l = 0$) in (3.14) in Sect. 3.1.2.

For all potentials $V(r)$ falling off faster than $1/r^2$ at large distances, the leading term proportional to r^{l+1} is a natural consequence of the repulsive centrifugal potential, which is always dominant at sufficiently large distances. The fact that the next-to-leading term is $2l+1$ powers of r lower than the leading term is nontrivial and requires a sufficiently rapid fall-off of $V(r)$ for large r. Equation (4.61) is obviously valid for all partial waves l if $V(r)$ vanishes exactly beyond some finite distance, and it also holds if $V(r)$ falls off faster than any inverse power of r. For a potential tail falling off as $1/r^\alpha$, $\alpha > 2$, its validity is limited to the partial waves

$$l < \frac{\alpha - 3}{2}, \quad \alpha > 2l + 3 , \tag{4.62}$$

as shown in the following.

Consider a potential behaving asymptotically as

$$V(r) \overset{r\to\infty}{\sim} V_\alpha(r) = \frac{C_\alpha}{r^\alpha} = \pm\frac{\hbar^2}{2\mu}\frac{(\beta_\alpha)^{\alpha-2}}{r^\alpha} , \quad \alpha > 2 . \tag{4.63}$$

Such behaviour is ubiquitous in nature. It applies, e.g., with $\alpha = 6$ for the van der Waals potential between two uncharged polarizable particles such as atoms or molecules, with $\alpha = 4$ for the interaction of a charged particle with a polarizable neutral, and with $\alpha = 3$ for the resonant dipole-dipole interaction of two identical atoms in different internal states. In quantum mechanics, the inverse-power term possesses a characteristic length β_α which does not exist in classical mechanics. It is related to the strength coefficient C_α via

$$\beta_\alpha = \left(\frac{2\mu|C_\alpha|}{\hbar^2}\right)^{1/(\alpha-2)} . \tag{4.64}$$

The length β_α has been called "van der Waals length" for attractive inverse-power tails with $\alpha = 6$ [NT07]. Since the theory does not depend on whether or not the potential tail is associated with a van der Waals interaction, it seems appropriate to choose a more general name, such as the *characteristic quantum length* associated with the inverse-power term C_α/r^α.

At large distances r, the radial Schrödinger equation at threshold ($E = 0$) for a potential fulfilling (4.63) reads

$$\left(-\frac{\mathrm{d}^2}{\mathrm{d}r^2} + \frac{l(l+1)}{r^2} \pm \frac{\beta^{\alpha-2}}{r^\alpha}\right) u_l^{(0)}(r) = 0 ; \tag{4.65}$$

we dispense with the subscript on the characteristic quantum length β, as long as only one power α is in the focus of attention. The solutions of (4.65) are known analytically,

$$u_l^{(0)}(r) = \sqrt{\frac{r}{\beta}}\,[A\,\mathcal{C}_\nu(\zeta) + B\,\mathcal{D}_\nu(\zeta)] , \tag{4.66}$$

where $\mathcal{C}_\nu$ and $\mathcal{D}_\nu$ stand for Bessel functions whose order ν and argument ζ are,

$$\nu = \frac{2l+1}{\alpha-2} , \quad \zeta = \frac{2}{\alpha-2}\left(\frac{\beta}{r}\right)^{(\alpha-2)/2} . \tag{4.67}$$

In the attractive case, for which the "$\pm$" in front of the inverse-power term in (4.65) is a "$-$", $\mathcal{C}_\nu$ and $\mathcal{D}_\nu$ are the ordinary Bessel functions J_ν and Y_ν. In the repulsive case, for which the "$\pm$" in front of the inverse-power term in (4.65) is a "+", $\mathcal{C}_\nu$ and $\mathcal{D}_\nu$ are the modified Bessel functions I_ν and K_ν, see Appendix A.4.

Large distances r correspond to small arguments ζ of the Bessel functions in (4.66). For the repulsive case, the small-argument expansions of the modified Bessel functions are [AS70],

$$I_\nu(\zeta) \overset{\zeta\to 0}{\sim} \frac{(\zeta/2)^\nu}{\Gamma(1+\nu)}\left[1 + O\left(\zeta^2\right)\right] , \quad K_\nu(\zeta) \overset{\zeta\to 0}{\sim} \frac{\Gamma(\nu)}{2(\zeta/2)^\nu}\left[1 + O\left(\zeta^2\right)\right] . \tag{4.68}$$

Since

$$\left(\frac{\zeta}{2}\right)^{\nu} = \left(\frac{1}{\alpha-2}\right)^{(2l+1)/(\alpha-2)} \left(\frac{\beta}{r}\right)^{l+1/2} \quad \text{and} \quad \zeta^2 \propto \left(\frac{\beta}{r}\right)^{\alpha-2}, \tag{4.69}$$

the asymptotic behaviour of the wave function (4.66) is,

$$u_l^{(0)}(r) \overset{r\to\infty}{\propto} A'\left(\frac{\beta}{r}\right)^{l} + B'\left(\frac{r}{\beta}\right)^{l+1}\left[1 + O\left(\left(\frac{\beta}{r}\right)^{\alpha-2}\right)\right], \tag{4.70}$$

where the first term originates from the I_ν-contribution and the second term from the K_ν-contribution in (4.66). If $2l+1 < \alpha-2$, then the term $O\left((\beta/r)^{\alpha-2}\right)$ in the square bracket leads to a contribution of higher order than l in $1/r$, the asymptotic expression (4.61) is valid and defines the scattering length in the partial wave l. On the other hand, if $2l+1 > \alpha-2$, then the term $O\left((\beta/r)^{\alpha-2}\right)$ in the square bracket leads to a contribution of lower order than l in $1/r$. In this case, (4.61) is not valid, and a scattering length in the partial wave l cannot be defined. Similar arguments apply for the attractive case, where the modified Bessel functions I_ν and K_ν are replaced by the ordinary Bessel functions J_ν and Y_ν.

For $2+1 > \alpha-2$ corresponding to $2l+3 > \alpha$, the leading near-threshold behaviour of $\tan\delta_l$ can be obtained using the expression (4.46) derived with the help of the radial Lippmann-Schwinger equation. Changing the integration variable in (4.46) from r to $\rho = kr$ yields,

$$\tan\delta_l = -\frac{2\mu}{\hbar^2 k^2}\int_0^\infty u_l^{(s)}(\rho)\, V\!\left(\frac{\rho}{k}\right) u_l(\rho)\,\mathrm{d}\rho\,. \tag{4.71}$$

Towards threshold, $k \to 0$, the regular solution of the Schrödinger equation $u_l(\rho)$ is, beyond the range of V, dominated by the centrifugal potential. The potential V is a function of the distance r and is given by the inverse-power term V_α at large distances. Its range is characterized by the quantum length β, and it shrinks to ever smaller values of ρ for $k \to 0$. Beyond this range, the regular radial wave function has the form $u_l^{(s)}(\rho)+\tan\delta_l\, u_l^{(c)}(\rho)$ and can be replaced by $u_l^{(s)}$, because $\tan\delta_l$ tends to zero for $k \to 0$. Furthermore, in the limit $k \to 0$, the integral in (4.71) is dominated by contributions from large arguments $r = \rho/k$ of the potential V, which can thus be replaced by its asymptotic form V_α defined in (4.63). Inserting V_α in (4.71) and replacing u_l by $u_l^{(s)}$ gives

$$\tan\delta_l \overset{k\to 0}{\sim} \mp(k\beta)^{\alpha-2}\int_0^\infty \frac{u_l^{(s)}(\rho)^2}{\rho^\alpha}\,\mathrm{d}\rho\,. \tag{4.72}$$

The integrand in (4.72) falls off as $1/\rho^\alpha$ at large scaled distances, because $u_l^{(s)}$ approaches $\sin\left(\rho - l\frac{\pi}{2}\right)$ for large ρ. For small ρ, $u_l^{(s)}(\rho)$ is proportional to ρ^{l+1}, so the integrand is proportional to $\rho^{2l+2-\alpha}$. The integral converges when $2l{+}2{-}\alpha > -1$, i.e. when $\alpha < 2l{+}3$.

The integral on the right-hand side of (4.72) can be evaluated analytically for the inverse-power potential V_α of (4.63) with $\alpha < 2l{+}3$,

$$\tan\delta_l \overset{k\to 0}{\sim} \mp\frac{\pi}{4}\,\frac{\Gamma(\alpha-1)\,\Gamma\left(l+\frac{3}{2}-\frac{\alpha}{2}\right)}{\left[\Gamma\left(\frac{\alpha}{2}\right)\right]^2\,\Gamma\left(l+\frac{1}{2}+\frac{\alpha}{2}\right)}\left(\frac{k\beta}{2}\right)^{\alpha-2}, \quad \alpha < 2l{+}3\,. \tag{4.73}$$

Note that the right-hand side of (4.73) is determined exclusively by the asymptotic inverse-power behaviour of the potential and does not depend on deviations from this form at smaller distances.

For $\alpha > 2l{+}3$, the asymptotic form of the threshold solution is given by (4.61). The leading near-threshold behaviour of the scattering phase shift is as given in (4.57), with a scattering length a_l depending on the whole potential, not only its asymptotic behaviour. The s-wave scattering length is well defined for potentials falling off faster than $1/r^3$, the p-wave ($l = 1$) scattering length for potentials falling off faster than $1/r^5$.

For repulsive single-power potentials,

$$V_\alpha^{(\mathrm{rep})}(r) = \frac{\hbar^2}{2\mu}\,\frac{(\beta_\alpha)^{\alpha-2}}{r^\alpha}, \quad \alpha > 2, \tag{4.74}$$

the s-wave scattering length can be derived directly from the zero-energy solution of the radial Schrödinger equation (4.22). The regular s-wave solution is

$$u^{(0)}_{l=0}(r) \propto \sqrt{\frac{r}{\beta_\alpha}}\,K_\nu\left(2\nu\left(\frac{\beta_\alpha}{r}\right)^{1/(2\nu)}\right), \quad \nu = \frac{1}{\alpha-2}, \tag{4.75}$$

where K_ν is a modified Bessel function, see Appendix A.4. The large-argument behaviour K_ν gives the small-r behaviour of $u^{(0)}_{l=0}(r)$,

$$u^{(0)}_{l=0}(r) \overset{r\to 0}{\propto} \left(\frac{r}{\beta_\alpha}\right)^{\alpha/4} \mathrm{e}^{-2\nu(\beta_\alpha/r)^{1/(2\nu)}}\,. \tag{4.76}$$

The asymptotic ($r \to \infty$) behaviour of $u^{(0)}_{l=0}(r)$ follows from the small-argument behaviour of K_ν,

$$u^{(0)}_{l=0}(r) \overset{r\to\infty}{\propto} \frac{\nu^{-\nu}}{\Gamma(1-\nu)}\,\frac{r}{\beta_\alpha} - \frac{\nu^\nu}{\Gamma(1+\nu)} + O\left(\left(\frac{\beta_\alpha}{r}\right)^{\alpha-3}\right). \tag{4.77}$$

From the zero of this threshold solution we conclude from (4.61) for $l = 0$, that the s-wave scattering length for the potential (4.74) is:

$$a = \nu^{2\nu}\,\frac{\Gamma(1-\nu)}{\Gamma(1+\nu)}\,\beta_\alpha\,, \tag{4.78}$$

Table 4.1 Scattering lengths (4.78) for repulsive inverse-power potentials (4.74) in units of β_α

α	4	5	6	7	8	$\alpha \to \infty$
a/β_α	1	0.729011	0.675978	0.666083	0.669594	1

which gives a finite result for $\alpha > 3$.

The scattering length (4.78) scales with the characteristic quantum length β_α; the prefactor depends on the power α and is in general close to unity. Numerical values of the scattering length in units of β_α are given for $\alpha = 4, \ldots 8$ in Table 4.1.

When the scattering length vanishes, the threshold solution (4.61) is asymptotically proportional to r^{l+1}, just as the regular solution of the radial Schrödinger equation for the centrifugal potential alone. An infinite scattering length, $|a_l| \to \infty$, implies that the threshold solution of the radial Schrödinger equation (4.22) decays as $1/r^l$ for large distances. For $l > 0$ this means that there is a normalizable wave function solving the radial Schrödinger equation at $E = 0$, i.e., a bound state exactly at threshold.

For s-waves, (4.61) reads

$$u_{l=0}^{(0)} \overset{r\to\infty}{\propto} r - a \propto 1 - \frac{r}{a} \,. \tag{4.79}$$

An infinite s-wave scattering length means that the threshold solution becomes constant at large distances. One speaks of a bound state at threshold in this case as well, even though the wave function is not normalizable.

The scattering length depends very sensitively on whether there is a bound state very close to threshold, or whether the potential just fails to bind a further bound state. This is easily demonstrated via the simple but instructive example of an attractive sharp-step potential,

$$V(r) = \begin{cases} -V_\text{S} & \text{for} \quad r \le L \,, \\ \;\;0 & \text{for} \quad r > L \,, \end{cases} \qquad V_\text{S} = \frac{\hbar^2 K_\text{S}^2}{2\mu} \,. \tag{4.80}$$

When $K_\text{S}L = \frac{\pi}{2}$, which corresponds to a depth V_S equal to the energy $E_0 = \left(\frac{\pi}{2}\hbar\right)^2/(2\mu L^2)$, the potential (4.80) has a threshold solution which becomes constant for $r > L$. For a slightly deeper step, $V_\text{S} = 1.4\,E_0$, the potential supports a weakly bound state at the energy $E_\text{b} \approx -0.189\,E_0$, indicated by the horizontal dotted brown line in the left half of Fig. 4.3; the associated bound-state wave function is shown as dashed brown line. As is customary in such illustrations, the zero-axis for a wave function is chosen to lie at the energy for which it solves the Schrödinger equation. The threshold solution at $E = 0$ (solid blue line) is not very different from the bound-state wave function for $r \le L$. For $r > L$ the potential vanishes, so the threshold solution assumes its asymptotic behaviour (4.79) corresponding to a linear fall-off; it cuts the r-axis at a value defining the scattering length a ($\approx 2.8\,L$ in the present case). The right half of Fig. 4.3 shows a shallower step, $V_\text{S} = 0.8\,E_0$, for

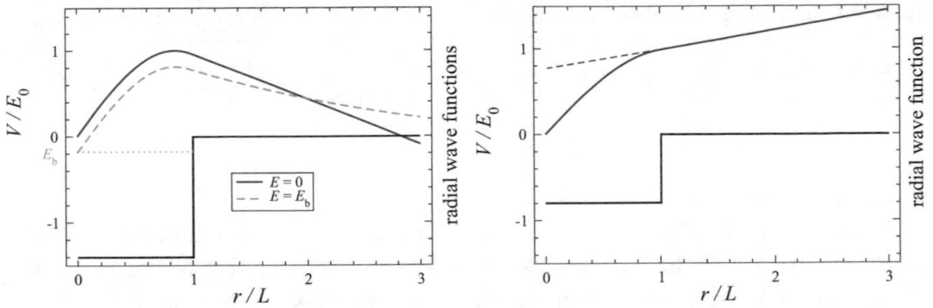

Fig. 4.3 Sharp-step potential (4.80). The energy is given in units of $E_0 = \left(\frac{\pi}{2}\hbar\right)^2/(2\mu L^2)$. For $V_S = E_0$, the s-wave radial Schrödinger equation has a zero-energy solution which becomes constant for $r \geq L$. The left half of the figure shows the case $V_S = 1.4\,E_0$, for which the potential supports a bound state at the energy $E_b \approx -0.189\,E_0$, indicated by the *horizontal dotted brown line*. The bound-state wave function is shown as *dashed brown line*, and its zero-axis lies at its energy E_b. The threshold solution is shown as *solid blue line* with zero-axis at $E = 0$; for $r > L$ it is a linear function which cuts the axis at a distance defining the scattering length a. The right half of the figure shows the case $V_S = 0.8\,E_0$, for which there is no bound state; the threshold solution (*solid blue line*) is a straight line for $r > L$, and the extrapolation of this line to smaller r-values leads to an intersection with the r-axis at a large negative value, corresponding to a large negative scattering length a

which the potential just fails to support a bound state. The threshold solution (solid blue line) now *grows* linearly for $r > L$. Extrapolation of this linear behaviour to smaller r-values eventually leads to a crossing of the r-axis at a large negative value, corresponding to a large negative scattering length.

The dependence of the scattering length on the potential depth V_S, or on the related threshold wave number $K_S = \sqrt{2\mu V_S}/\hbar$, can be easily deduced from the threshold solution $u_{l=0}^{(0)} \overset{r\leq L}{\propto} \sin(K_S r)$. Its logarithmic derivative at $r = L$ is $K_S \cot(K_S L)$ which must be equal to $1/(L - a)$ according to (4.79), so [Joa87]

$$a = L - \frac{\tan(K_S L)}{K_S} . \tag{4.81}$$

Figure 4.4 shows the behaviour of the scattering length as function of the threshold wave number K_S. It is typical for the behaviour of the scattering length of a potential as function of a parameter which can tune the number and positions of bound states in the potential. The scattering length has a pole whenever there is a bound state at threshold. Before the first pole ($K_S L < 0.5\,\pi$ in Fig. 4.4), the potential has no bound states. The number of bound states increases by one every time K_S increases through a pole.

A quantitative relation between the diverging scattering length and the vanishing eigenenergy of a near-threshold bound state can be derived quite generally as follows: Assume that there is a bound s-state at an energy $E_b = -\hbar^2\kappa_b^2/(2\mu)$ very

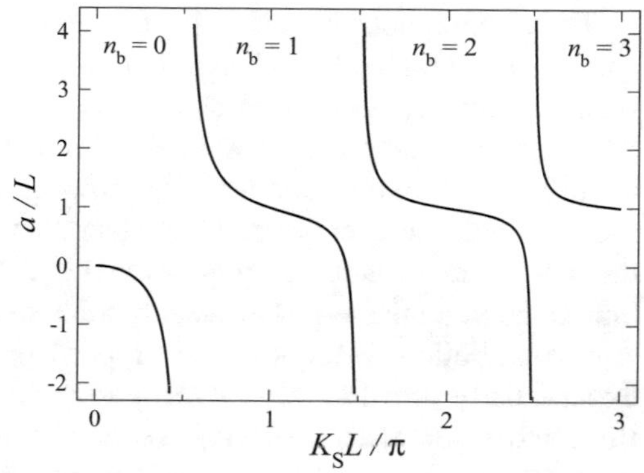

Fig. 4.4 Scattering length for the sharp-step potential as function of the threshold wave number K_S, as given by (4.81). Each pole indicates the existence a bound state at threshold; n_b is the number of bound states supported by the potential for values of K_S between successive poles

close to threshold. The radial wave function $u_{l=0}^{(\kappa_b)}$ at this energy is asymptotically proportional to $e^{-\kappa_b r}$ and behaves as

$$u_{l=0}^{(\kappa_b)}(r) \propto 1 - r\left[\kappa_b + O\left(\kappa_b^2\right)\right] \quad (\kappa_b > 0)\,, \tag{4.82}$$

beyond the range of the potential. The terms below order κ_b^2 in (4.82) are compatible with (4.79) if we assume

$$\frac{1}{a} \overset{\kappa_b \to 0}{\sim} \kappa_b + O\left(\kappa_b^2\right)\,. \tag{4.83}$$

This is plausible, since the radial Schrödinger equation at energy E_b differs from the radial Schrödinger equation at threshold by a term of order κ_b^2. Equation (4.83) implies the following relation between the scattering length a and the inverse penetration depth κ_b of a bound state very near threshold,

$$a \overset{\kappa_b \to 0}{\sim} \frac{1}{\kappa_b} + O\left(\kappa_b^0\right)\,. \tag{4.84}$$

Conversely, a large positive scattering length a implies a near-threshold bound state, whose energy is given by,

$$E_b = -\frac{\hbar^2\kappa_b^2}{2\mu} \overset{a \to \infty}{\sim} -\frac{\hbar^2}{2\mu a^2} + O\left(\frac{1}{a^3}\right)\,. \tag{4.85}$$

When the potential just fails to bind a further bound state, there may be a solution $u_{l=0}^{v}$ of the s-wave radial Schrödinger equation which is asymptotically proportional to $e^{+\kappa_v r}$ with a very small positive κ_v. By the same arguments as above, such a solution of (4.22) gives rise to a large negative scattering length, $a \overset{\kappa_v \to 0}{\sim} -1/\kappa_v + O\left(\kappa_v^0\right)$. In such a situation one speaks of a *virtual state* at the energy $E_v = -\hbar^2\kappa_v^2/(2\mu)$ [Tay72, New82].

The unambiguous identification of a virtual state poses a problem. The discrete energy of a genuine bound state is easily found via the condition that the wave function must decay to zero as $e^{-\kappa r}$ at large distances. When solving the radial Schrödinger equation, e.g. by integrating it from smaller to larger r-values, any contribution from the exponentially growing solution soon becomes dominant and indicates that the energy under consideration is not a bound-state eigenvalue. On the other hand, the solution proportional to $e^{+\kappa r}$ cannot be unambiguously defined, unless the potential vanishes exactly after some finite, preferably short, distance. Any contribution of the solution proportional to $e^{-\kappa r}$ is soon dominated by the exponentially growing term, so it is very difficult in practice to decide, whether the contribution of the decaying solution vanishes exactly or not. This problem is aggravated as κ increases, so the concept of virtual states is most useful very close to threshold.

For a potential which falls off sufficiently rapidly at large distances, the next-to-leading behaviour of the scattering phase shifts near threshold, following the leading term (4.57), can be derived from solutions of the radial Schrödinger equation at threshold [Bet49]. This is shown below for s-waves, $l = 0$. We shall drop the subscript $l = 0$, but remember that we are dealing with s-waves.

Let $u^{(0)}$ and $u^{(k)}$ be regular radial wave functions that solve the radial Schrödinger equation at threshold and for wave number $k > 0$,

$$\frac{d^2u^{(0)}}{dr^2} = \frac{2\mu}{\hbar^2} V(r)\, u^{(0)}(r) \,, \qquad \frac{d^2u^{(k)}}{dr^2} = \left(\frac{2\mu}{\hbar^2} V(r) - k^2\right) u^{(k)}(r) \,. \tag{4.86}$$

There are two alternative representations for the integral

$$I_u(r_0) = \int_0^{r_0} \left[u^{(0)}(r) \frac{d^2u^{(k)}}{dr^2} - u^{(k)}(r) \frac{d^2u^{(0)}}{dr^2} \right] dr \,. \tag{4.87}$$

One involves multiplying the first of the two equations (4.86) by $u^{(k)}$, the second by $u^{(0)}$, and integrating the difference; this leads to

$$I_u(r_0) = -k^2 \int_0^{r_0} u^{(0)}(r)\, u^{(k)}(r)\, dr \,. \tag{4.88}$$

An alternative representation of the integral (4.87) is obtained by partial integration,

$$\begin{aligned} I_u(r_0) &= \left[u^{(0)}(r) \frac{du^{(k)}}{dr} - u^{(k)}(r) \frac{du^{(0)}}{dr} \right]_0^{r_0} \\ &= u^{(0)}(r_0) \left.\frac{du^{(k)}}{dr}\right|_{r_0} - u^{(k)}(r_0) \left.\frac{du^{(0)}}{dr}\right|_{r_0} \,. \end{aligned} \tag{4.89}$$

Contributions from the lower limit of integration, $r = 0$, vanish, because the regular solutions $u(r)$ vanish for $r \to 0$.

We now repeat the procedure for two (not necessarily regular) radial wave functions, $w^{(0)}$ and $w^{(k)}$, which solve the *free-particle* radial Schrödinger equation at threshold and for wave number $k > 0$,

$$\frac{d^2 w^{(0)}}{dr^2} = 0 , \quad \frac{d^2 w^{(k)}}{dr^2} = -k^2\, w^{(k)}(r) . \tag{4.90}$$

The integral

$$I_w(r_0) = \int_0^{r_0} \left[w^{(0)}(r) \frac{d^2 w^{(k)}}{dr^2} - w^{(k)}(r) \frac{d^2 w^{(0)}}{dr^2} \right] dr \tag{4.91}$$

can, in analogy to (4.88) and (4.89), be written as

$$I_w(r_0) = -k^2 \int_0^{r_0} w^{(0)}(r)\, w^{(k)}(r)\, dr , \qquad \text{or as} \tag{4.92}$$

$$\begin{aligned} I_w(r_0) = {} & w^{(0)}(r_0) \left.\frac{dw^{(k)}}{dr}\right|_{r_0} - w^{(k)}(r_0) \left.\frac{dw^{(0)}}{dr}\right|_{r_0} \\ & - w^{(0)}(0) \left.\frac{dw^{(k)}}{dr}\right|_0 + w^{(k)}(0) \left.\frac{dw^{(0)}}{dr}\right|_0 . \end{aligned} \tag{4.93}$$

Equation (4.93) includes the contributions from the lower integration limit, $r = 0$, because the free-particle solutions $w(r)$ are *not* assumed to vanish at $r = 0$. Instead, they shall be assumed to be asymptotically equal to the regular solutions $u(r)$, which behave as (4.79) and (4.29). Explicitly and with appropriate normalization:

$$\begin{aligned} w^{(0)}(r) &= 1 - \frac{r}{a} , & w^{(k)}(r) &= -\frac{1}{ka} \sin(kr + \delta) ; \\ u^{(0)}(r) &\overset{r\to\infty}{\sim} 1 - \frac{r}{a} , & u^{(k)}(r) &\overset{r\to\infty}{\sim} -\frac{1}{ka} \sin(kr + \delta) . \end{aligned} \tag{4.94}$$

According to (4.88) and (4.92), the difference of the integrals (4.87) and (4.91) is,

$$I_u(r_0) - I_w(r_0) = k^2 \int_0^{r_0} \left[w^{(0)}(r)\, w^{(k)}(r) - u^{(0)}(r) u^{(k)}(r) \right] dr . \tag{4.95}$$

The integral converges in the limit $r_0 \to \infty$, provided that the regular solutions $u(r)$ approach their asymptotic forms $w(r)$ sufficiently fast,

$$\begin{aligned} I_u(r_0) - I_w(r_0) &\overset{r_0\to\infty}{\sim} k^2 I(k) , \\ I(k) &= \int_0^{\infty} \left[w^{(0)}(r)\, w^{(k)}(r) - u^{(0)}(r) u^{(k)}(r) \right] dr . \end{aligned} \tag{4.96}$$

Note that $I(k)$ has the physical dimension of a length. When expressing the difference $I_u(r_0) - I_w(r_0)$ via (4.89) and (4.93), the contributions from the upper integration limit r_0 vanish, so

$$I_u(r_0) - I_w(r_0) \overset{r_0\to\infty}{\sim} w^{(0)}(0)\frac{\mathrm{d}w^{(k)}}{\mathrm{d}r}\bigg|_0 - w^{(k)}(0)\frac{\mathrm{d}w^{(0)}}{\mathrm{d}r}\bigg|_0 = -\frac{\cos\delta}{a} - \frac{\sin\delta}{ka^2} . \tag{4.97}$$

Equating the right-hand side of (4.97) with $k^2 I(k)$ yields

$$ka\cot\delta = -\left[1 + k^2 a\, I(k)/\cos\delta\right]^{-1} . \tag{4.98}$$

In the limit of small wave numbers, $\cos\delta$ tends to unity and $I(k)$ assumes a certain value which is usually expressed in terms of the *effective range*, $r_{\mathrm{eff}} = 2\lim_{k\to 0} I(k)$, so the leading near-threshold behaviour of (4.98) is,

$$k\cot\delta \overset{k\to 0}{\sim} -\frac{1}{a} + \frac{1}{2} r_{\mathrm{eff}}\, k^2 + O\left(k^4\right) , \tag{4.99}$$

$$r_{\mathrm{eff}} = 2\int_0^\infty \left(\left[w^{(0)}(r)\right]^2 - \left[u^{(0)}(r)\right]^2\right) \mathrm{d}r . \tag{4.100}$$

Translating (4.99) into an expansion for the scattering phase shift itself gives,

$$\delta \overset{k\to 0}{\sim} -ka + \frac{k^3}{3}\left[a^3 - \frac{3}{2} r_{\mathrm{eff}}\, a^2\right] + O\left(k^5\right) \pmod{\pi} . \tag{4.101}$$

Equation (4.99) features the two leading terms of the *effective-range expansion*. For potentials which fall off faster than any inverse power of r at large distances, $k\cot\delta$ is known to be an analytical function of energy, i.e. of k^2. The same applies for nonvanishing angular momenta to the function $k^{2l+1}\cot\delta_l$ [CG49]. Note, however, that most realistic potentials in atomic and molecular systems do not fall off so quickly, but rather as an inverse power of r, $V(r) \overset{r\to\infty}{\propto} 1/r^\alpha$. In such cases, $k\cot\delta$ is not an analytical function of k^2, and the second term in (4.99) can only be defined in general when $\alpha > 5$; see Sect. 4.1.8 below.

The scattering length a has an immediate physical significance, because it determines the near-threshold limits of the differential and the integrated scattering cross sections according to (4.60). Only when the potential is repulsive or so weakly attractive that it is not near to supporting a bound state, can the scattering length and the effective range r_{eff} be related to a distance up to which the potential has nonnegligible values. For scattering by a hard sphere of radius R, we have $a = R$ and $r_{\mathrm{eff}} = \frac{2}{3}R$, so the k^3- term in (4.101) vanishes, as do all higher terms.

As soon as the potential is attractive enough to support one or more bound states, the proximity of a bound (or virtual) state to threshold dominantly influences the scattering length as illustrated for the attractive sharp-step potential in Fig. 4.4 above. The behaviour of the effective range is strongly correlated to the behaviour of the scattering length. When $a = 0$, for example, which happens for the sharp-step potential (4.80), (4.81) whenever $K_S L$ is an integer multiple of π, there is no bound or virtual state near threshold, the effective range diverges, but the product $a^2 r_{\text{eff}}$ in (4.101) remains finite.

4.1.8 Modified Effective-Range Expansions

For potentials falling off asymptotically as $1/r^\alpha$, the scattering length a_l can only be defined for partial waves $l < (\alpha - 3)/2$, as already discussed in Sect. 4.1.7. For s-waves ($l = 0$), *the* scattering length a can only be defined if $\alpha > 3$. The next terms in the expansions (4.99), (4.101) contain the effective range (4.100). The convergence of the integral in (4.100) depends on how rapidly $u^{(0)}(r)$ approaches its asymptotic form $w^{(0)}(r) = 1 - r/a$, and this, in turn, depends on the asymptotic fall-off of $V(r)$.

For a potential falling off as an inverse power according to (4.63), the threshold solutions of the radial Schrödinger equation are, at large distances, of the form $\sqrt{r/\beta}\, \mathcal{C}_\nu(\zeta)$; here $\mathcal{C}_\nu(\zeta)$ stands for a Bessel function whose order ν and argument ζ are given by (4.67); for $l = 0$:

$$\nu = \frac{1}{\alpha - 2}, \quad \zeta = 2\nu \left(\frac{\beta}{r}\right)^{1/(2\nu)} . \tag{4.102}$$

As already observed for arbitrary l in Sect. 4.1.7, the Bessel functions J_ν (for an attractive $1/r^\alpha$ potential) and I_ν (in the repulsive case) lead to a near-threshold wave function proportional to $(\beta/r)^l$ asymptotically, i.e. to a constant for $l = 0$, while

$$\sqrt{\frac{r}{\beta}} \begin{pmatrix} Y_\nu(\zeta) \\ K_\nu(\zeta) \end{pmatrix} \overset{r/\beta \to \infty}{\propto} \frac{r}{\beta}\left[1 + O\left(\left(\frac{\beta}{r}\right)^{\alpha-2}\right)\right] , \tag{4.103}$$

see (4.70). The asymptotic behaviour of $u^{(0)}$ is thus

$$u^{(0)}(r) \overset{r\to\infty}{\sim} 1 - \frac{r}{a} + O\left(\left(\frac{\beta}{r}\right)^{\alpha-3}\right) \Rightarrow \left[u^{(0)}(r)\right]^2 \overset{r\to\infty}{\sim} \left(1 - \frac{r}{a}\right)^2 + O\left(r^{4-\alpha}\right) . \tag{4.104}$$

Consequently, the integrand in (4.100) falls off as $r^{4-\alpha}$ asymptotically, and this must be faster than $1/r$ for the integral to converge. A finite expression for the effective range (4.100) requires $\alpha > 5$, i.e., the potential must fall off faster than $1/r^5$ asymptotically.

There is one exception to this condition. If the scattering length a is infinite, i.e., if there is a bound state exactly at threshold, then the wave function (4.66) contains only the J_ν or I_ν contribution, so $u^{(0)}(r) \overset{r\to\infty}{\sim} 1 + O\left(r^{2-\alpha}\right)$ and the integrand in (4.100) falls off as $r^{2-\alpha}$ asymptotically. In this case, $\alpha > 3$ is sufficient for the integral to converge. When there is an s-wave bound state exactly at threshold, the leading near-threshold behaviour of the s-wave scattering phase shift is, according to (4.99),

$$\cot\delta_{l=0} \overset{k\to 0}{\sim} \frac{r_{\text{eff}}k}{2}, \tag{4.105}$$

and this holds for all potentials falling off faster than $1/r^3$ asymptotically.

When the potential falls off as $1/r^\alpha$ asymptotically, with $\alpha > 3$, then the effective-range expansion for the s-wave scattering phase shift starts as in (4.99), but the expansion in powers of k^2 does not continue indefinitely. As shown above, the k^2-term (generally) requires $\alpha > 5$, and analogous considerations [LK63] show that the expansion (4.99) is only valid up to terms k^{2n} with $2n < \alpha - 3$. Higher terms include odd powers of k and can also contain non-analytic, logarithmic factors.

These observations further limit the practical use of the effective-range expansion (4.99). When both target and projectile are spherical, the highest power α of practical significance is $\alpha = 7$, which occurs in the interaction between two neutral polarizable atoms (or molecules) when the electrostatic van der Waals interaction ($\propto 1/r^6$ at large distances) is corrected for asymptotically relevant retardation effects [CP48]. In this case the leading, constant term in the expansion for $k^{2l+1}\cot\delta_l$, which defines the scattering length, exists only for $l = 0$ and $l = 1$. The expansion holds up to the second term proportional to k^2 only for $l = 0$, and the expansion is not valid up to the k^4-term, even for $l = 0$. The naïve expansion (4.99) has to be modified substantially for potentials falling off as an inverse power at large distances.

Such a *modified effective-range expansion* was formulated by O'Malley, Spruch and Rosenberg [OS61, OR62] in 1961 for the important case of a potential with an attractive tail proportional to $1/r^4$, as occurs in the interaction of a charged particle with a polarizable neutral partner,

$$V(r) \overset{r\to\infty}{\sim} V^{(-)}_{\alpha=4}(r) = -\frac{C_4}{r^3} = -\frac{\hbar^2(\beta_4)^2}{2\mu\, r^4}, \quad \beta_4 > 0. \tag{4.106}$$

Up to and including terms of order k^2, the modified effective-range expansion for s-waves reads

$$k\cot\delta_{l=0} \overset{k\to 0}{\sim} -\frac{1}{a} + \frac{\pi}{3a^2}(\beta_4)^2 k + \frac{4(\beta_4)^2}{3a} k^2 \ln\left(\frac{k\beta_4}{4}\right) \tag{4.107}$$
$$+ \left[\frac{\tilde{r}_{\text{eff}}}{2} + \frac{\pi}{3}\beta_4 + \left(\frac{20}{9} - \frac{8}{3}\psi\left(\frac{3}{2}\right)\right)\frac{(\beta_4)^2}{a} - \frac{\pi(\beta_4)^3}{3a^2} - \frac{\pi^2}{9a^3}\frac{\pi^2(\beta_4)^4}{9a^3}\right]k^2,$$

where $\psi(z) = \Gamma'(z)/\Gamma(z)$ is the digamma function, $\psi\left(\frac{3}{2}\right) = 0.0364899740\ldots$. (See Appendix A.3.) The *modified effective range* $\tilde{r}_{\text{eff}}$ is defined as in (4.100), except that $w^{(0)}(r)$ now is the, not necessarily regular, radial wave function that solves the

s-wave radial Schrödinger equation containing the attractive inverse-power potential $V^{(-)}_{\alpha=4}$, as defined in (4.106), and behaves asymptotically as $w^{(0)}(r) \overset{r\to\infty}{\sim} 1 - r/a$. Beyond the s-wave, i.e. for $l \geq 1$, the condition $2l + 3 > \alpha$ is always fulfilled for $\alpha = 4$, so the leading near-threshold behaviour of the scattering phase shifts is given by,

$$\tan\delta_l \overset{k\to 0}{\sim} \frac{\pi\ (k\beta_4)^2}{(2l+3)(2l+1)(2l-1)}\,, \quad l \geq 1\,, \tag{4.108}$$

in accordance with (4.73).

For a potential behaving asymptotically as an inverse cube of the distance,

$$V(r) \overset{r\to\infty}{\sim} V_3^{(\pm)}(r) = \pm\frac{C_3}{r^3} = \pm\frac{\hbar^2\beta_3}{2\mu\, r^3}\,, \quad C_3,\ \beta_3 > 0, \tag{4.109}$$

a finite scattering length does not exist even for $l = 0$. The leading near-threshold behaviour of the s-wave phase shift is [LK63],

$$\tan\delta_{l=0} \overset{k\to 0}{\sim} \pm k\beta_3 \ln(k\beta_3) + O(k)\,, \tag{4.110}$$

so, towards threshold, the s-wave partial-wave scattering amplitude diverges logarithmically, (see (4.34) in Sect. 4.1.3),

$$f_{l=0} \overset{k\to 0}{\sim} \pm\beta \ln(k\beta_3)\,, \tag{4.111}$$

and the differential scattering cross section also diverges logarithmically towards threshold.

The usefulness of the expressions (4.110), (4.111) is limited to a very small range in k, however, because the next term in the expansion is only of marginally higher order. For a repulsive inverse-cube potential, i.e. with a "+" sign on the right-hand side of (4.109), an extended formula was derived by Del Giudice and Galzenati [DG65] in 1965 and rederived by Gao [Gao99] in 1999. According to (12′) in [DG65],

$$\tan\delta_{l=0} = [\ln(k\beta_3) + 3\gamma_{\rm E} + \ln 2 - 3/2]\,(k\beta_3) + O\left(k^2\right)\,, \tag{4.112}$$

were $\gamma_{\rm E} = 0.577\cdots$ is Euler's constant, see Appendix A.3.

For a potential with an attractive inverse-cube tail, i.e. with a "−" sign on the right-hand side of (4.109), the near-threshold behaviour of the phase shift depends sensitively on whether or not there is a bound state close the threshold. For potentials falling off faster than $1/r^3$, a bound state near threshold manifests itself in the divergence of the scattering length, see e.g. (3.61) in Sect. 3.1.2, but the threshold quantum number $v_{\rm D}$, or rather its remainder $\Delta_{\rm D}$, can equally serve as critical parameter reflecting the proximity of a bound state to the continuum threshold, see the discussion following (3.59). For the inverse-cube potential tail, there is no finite scattering length, but the near-threshold behaviour of the scattering phase shift can be expressed using the threshold quantum number's remainder Δ_{D}. As derived only recently by Müller [Mul13]:

$$\tan\delta_0 = -\left[\ln(k\beta_3) + \frac{\pi}{\tan(\pi\Delta_{\mathrm{D}})} + 3\gamma_{\mathrm{E}} + \ln 2 - \frac{3}{2}\right](k\beta_3) \\ +\pi\left[\ln(k\beta_3) + \frac{\pi}{\tan(\pi\Delta_{\mathrm{D}})} + 3\gamma_{\mathrm{E}} + \ln 2 - \frac{19}{12}\right](k\beta_3)^2 + O(k^3)\,. \tag{4.113}$$

All nonvanishing angular momentum quantum numbers $l > 0$ fulfill the condition $2l + 3 > \alpha$ for inverse-cube tails, so $\tan\delta_l \overset{k\to 0}{\propto} k$ according to (4.73). It follows that all partial-wave scattering amplitudes f_l with $l > 0$ tend to a finite limit and all partial waves $l > 0$ give a finite contribution to the scattering cross section at threshold. The magnitude of these contributions decreases with increasing l according to (4.73),

$$\tan\delta_l \overset{k\to 0}{\sim} \mp\frac{1}{2l(l+1)}k\beta_3\,, \quad f_l \overset{k\to 0}{\sim} \mp\frac{2l+1}{2l(l+1)}\beta_3 \quad \text{for} \quad \alpha = 3,\ l > 0\,. \tag{4.114}$$

Remember that β_3 is positive in the definition (4.109); in (4.114), the "−" signs apply for repulsive and the "+" signs for attractive inverse-cube tails.

4.1.9 *Levinson's Theorem*

If the effective potential (4.23) features a sufficiently deep attractive well, then the radial wave function shows oscillations in the region of this well. As the energy approaches the threshold from above, these inner oscillations can persist all the way down to $E = 0$. Matching the radial wave function u_l to a superposition of free-particle waves at a matching radius r_{m} beyond the range of the potential only determines the phase shift δ_l to within an integer multiple of π. By comparing $u_l(r)$ to the regular free-particle wave $u_l^{(\mathrm{s})}(kr)$ in the whole range of r-values from $r = 0$ to $r = r_{\mathrm{m}}$ we can also keep track of an additional integer multiple of π corresponding to spatial shifts by as many half-waves.

This is illustrated in Fig. 4.5 which shows radial wave functions for s-waves in the model potential

$$V(r) = V_0\left[16\,\mathrm{e}^{-r^2/\beta^2} - 12\,\mathrm{e}^{-r^2/(4\beta)^2}\right], \quad V_0 = \frac{\hbar^2}{2\mu\beta^2}\,. \tag{4.115}$$

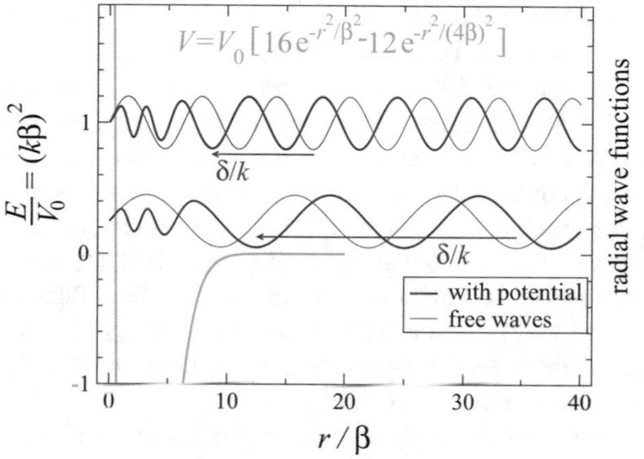

Fig. 4.5 Solutions of the radial Schrödinger equation (4.22) for $l = 0$. The *thick violet lines* show the wave functions in the potential (4.115) (*orange line*) for $E = V_0$ and $E = 0.25V_0$; the *thin blue lines* are the free waves $\propto \sin(kr)$. The zero-axes for the wave functions lie at the respective energies. The *red arrows* show the spatial shift from the third minimum of the free wave to the third minimum of the wave function in the potential

This potential consists of a repulsive Gaussian of height $16V_0$ and range β and an attractive Gaussian tail of depth $12V_0$ and range 4β. It is qualitatively similar to the Lennard-Jones potential (3.74) studied in Sect. 3.1.3, but there are two important differences: it falls off faster than any inverse power of r at large distances, and it remains bounded at small distances. The radial wave functions $u_{l=0}(r)$ (with conveniently chosen amplitudes) are shown in Fig. 4.5 as thick violet lines for the two energies $E = 0.25\,V_0$ and $E = V_0$, together with the respective free-particle wave functions $\propto \sin(kr)$ (thin blue lines). [The zero-axis for a wave function is again chosen to lie at the energy for which it solves the Schrödinger equation.] The red arrows show the spatial shift from the third minimum of the free-particle wave function to the third minimum of the radial wave function obtained with the potential. This spatial shift contains a contribution $\tilde{d}$ not larger than the range of the potential, plus an integer number $n_{\rm hw}$ of half waves, $d = \tilde{d} + n_{\rm hw}\lambda/2$. The wavelength $\lambda = 2\pi/k$ (beyond the range of the potential) diverges towards threshold, so the spatial shift $d = \tilde{d} + n_{\rm hw}\pi/k$ is dominated by the term containing $n_{\rm hw}$. For the phase shift, this implies

$$\lim_{k\to 0} \delta_l(k) = n_{\rm hw}\pi \ , \tag{4.116}$$

where $n_{\rm hw}$ is the number of additional nodes in the radial wave function, compared to the free-particle wave. Equation (4.116) is also valid for angular momenta $l > 0$. The number $n_{\rm hw}$ of additional nodes is well defined towards threshold, because those nodes of u_l (beyond $r = 0$) which are not additional nodes due to attractive behaviour of the potential V at short distances wander to infinity in the limit $k \to 0$, as do the nodes (beyond $r = 0$) of the free-particle wave $u_l^{(s)}$.

Since the potential $V(r)$ falls off faster than $1/r^2$ at large distances and is less singular than $1/r^2$ at small distances, it supports at most a finite number of bound states. The number n_b of bound states supported by the effective potential V_{eff} in the l-th partial wave is equal to the number n_{hw} of additional nodes in the radial wave function near threshold. To see this recall, that the ground-state wave function in a potential well has no nodes, and that the number of nodes increases by one for each successive excited state. A wave function solving the radial Schrödinger equation at a positive energy very near threshold has one more node in the interaction region than the highest bound state; this is a necessary condition for its orthogonality to all the bound eigenfunctions in the potential well.

If $V(r)$ is bounded, its influence becomes negligible at high energies,

$$\lim_{k\to\infty} \delta_l(k) = 0 \,. \tag{4.117}$$

In the high-energy limit, the nodes of u_l and of the free-particle wave $u_l^{(s)}$ coalesce and u_l has no additional nodes. This holds also for potentials which are not necessarily bounded, but less singular than $1/r^2$ for $r \to 0$. If we consider δ_l as a continuous function of wave number (or energy), then combining (4.117) and (4.116), with $n_{hw} = n_b$, yields

$$\lim_{k\to 0} \delta_l(k) - \lim_{k\to\infty} \delta_l(k) = n_b\pi, \tag{4.118}$$

where n_b is the number of bound states in the l-th partial wave. Equation (4.118) was first derived by Levinson in 1949 [Lev49] and is known as *Levinson's theorem.*

There is one exception to the rule (4.116), and hence also to (4.118), namely when there is an s-wave bound state exactly at threshold, with $|a| = \infty$ according to (4.79). The threshold wave function is asymptotically proportional to $\cos(kr)$ in the limit $k \to 0$, which corresponds to a phase shift of $\pi/2$ relative to the free-particle wave $\sin(kr)$, so $\delta_{l=0}(k)$ converges to an odd multiple of $\pi/2$ for $k \to 0$.

The model potential (4.115) falls off faster than any inverse power of r at large distances, so the leading near-threshold behaviour of the phase shifts is given by (4.57) for all l. Furthermore, the potential is bounded so the phase shifts obey Levinson's theorem (4.118). Figure 4.6 shows the corresponding phase shifts as functions of the scaled wave number $k\beta$ for angular momentum quantum numbers up to $l = 15$.

Some features in Fig. 4.6 can be understood by looking at the effective potentials (4.23), which are shown in Fig. 4.7. The number n_b of bound states, corresponding to the threshold value $n_b\pi$ of δ_l, decreases from five for $l = 0$ to zero for $l = 8$, where the minimum of the effective potential already lies above the threshold $E = 0$. There are no bound states above threshold, but *almost bound states* can form at certain energies below the maximum of the potential barrier formed by

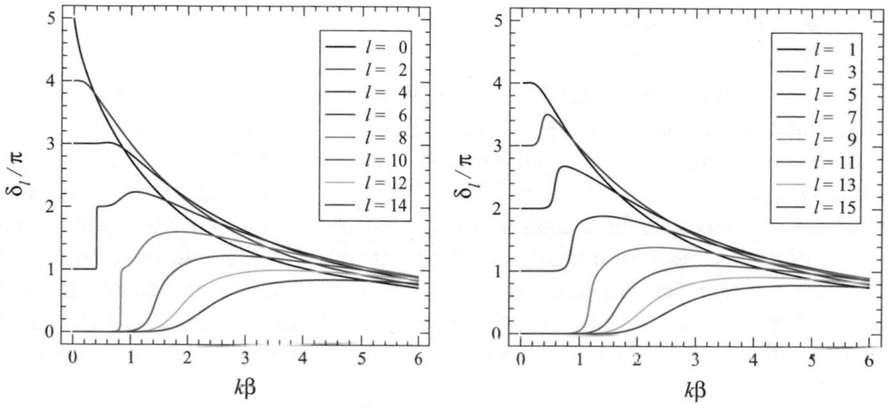

Fig. 4.6 Phase shifts for scattering by the model potential (4.115) as functions of the scaled wave number $k\beta$ for angular momentum quantum numbers up to $l = 15$. Even and odd l are shown in separate panels to avoid overcrowding in the figure

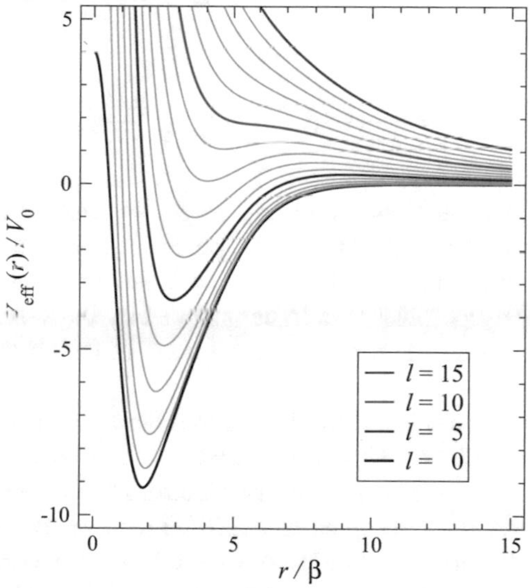

Fig. 4.7 Effective potentials (4.23) for the model potential (4.115) and angular momenta up to $l = 15$

the centrifugal potential V_{cent} and the attractive potential V. Such almost bound states above the threshold of a potential are called *potential resonances* or *shape resonances*, and were already introduced in Sect. 1.5.3; they lead to more or less sudden jumps of the phase shift by π, as seen for many of the partial waves in Fig. 4.6.

4.1.10 An Example

When calculating the interaction potential between an electron and a neutral spherical atom according to the considerations in Sect. 3.3.1, we see that there is no very-long-ranged Coulomb potential ($N = Z + 1$) and all higher ($l > 0$) direct diagonal contributions of the form (3.179) also vanish, because the internal wave function $\psi_{\rm int}$ has vanishing total angular momentum, and hence all expectation values of vector operators and higher tensors vanish according to the Wigner-Eckart theorem. At large electron-atom separations the leading contribution to the potential comes from the fact that the electric field of the electron polarizes the atom and induces a dipole moment and this leads to a $-1/r^4$ potential (4.106),

$$V(r) \overset{r\to\infty}{=} -e^2 \frac{\alpha_{\rm d}}{2r^4}, \tag{4.119}$$

where $\alpha_{\rm d}$ is the static dipole polarizability of the atom (see Problem 4.5). Comparing (4.119) and (4.106) shows that the $\alpha_{\rm d}$ is related to the potential strength parameter β_4 by

$$\alpha_{\rm d} = \frac{\hbar^2}{\mu e^2}(\beta_4)^2, \quad \beta_4 = \sqrt{\frac{\mu e^2}{\hbar^2}\alpha_{\rm d}}\,. \tag{4.120}$$

Here $\hbar^2/(\mu e^2)$ is just the Bohr radius a_0, so $\beta_4 = \sqrt{\alpha_{\rm d}/a_0}$. Note that (4.119) and (4.120) also hold for the polarization potential between any other charged particle such a positive or negative ion and a polarizable neutral atom (or molecule) in a spherical state, except that e^2 is to be replaced by the square of the charge of the charged particle; μ stands for the reduced mass of the charged and the neutral particle.

At smaller separations it is not so obvious that the electron-atom interaction can be adequately described by a simple potential. In addition to the so-called direct static potential involving the density of the electrons in the occupied states of the target atom (cf. (2.91) or, more generally (3.175) for $i = j$), the consideration of exchange effects in the elastic channel alone already leads to complicated non-local contributions. One consequence of these exchange contributions is the orthogonality of the scattering wave functions to the occupied single-particle states in the target atom as required by the Pauli principle.

Figure 4.8 shows phase shifts for the example of elastic electron scattering by neon at energies up to about $E = \hbar^2k^2/(2\mu) = 20$ eV in the partial waves $l = 0, 1$ and 2 as functions of the wave number k. The crosses are the experimental phase shifts deduced from measured elastic differential cross sections (such as the one illustrated in Fig. 4.9) by Williams [Wil79]. The solid lines are the results of solving the radial Schrödinger equation with a simple local potential consisting of the direct static terms plus a polarization potential (4.119) which merges into a constant for separations smaller than a certain value r_0. At negative energies this

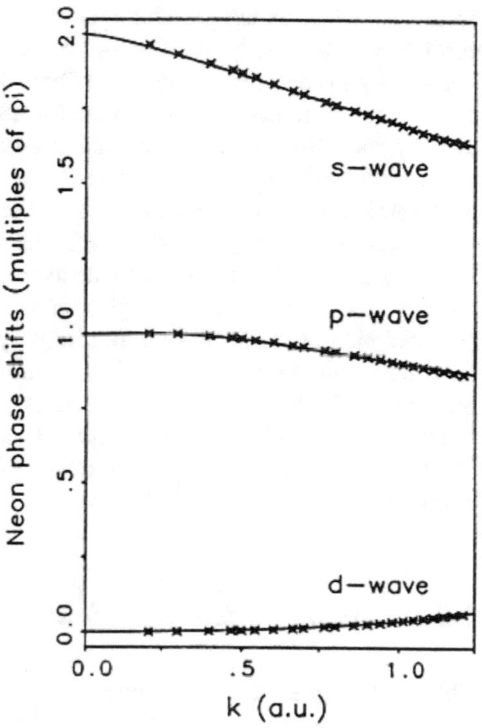

Fig. 4.8 Phase shifts for elastic scattering of electrons by neon. The *crosses* show experimental data from [Wil79]. The *solid lines* were obtained by solving the radial Schrödinger equation with a simple local potential consisting of the electrostatic terms plus a polarization potential (4.119) which merges into a constant for separations smaller than a phenomenological parameter r_0. The polarizability was taken to be the experimental value $\alpha_d = 2.66\,a_0^3$ [TP71] and the value of r_0 was $0.974\,a_0$ for $l = 0$, $1.033\,a_0$ for $l = 1$ and $1.11\,a_0$ for $l = 2$ (From [IF92])

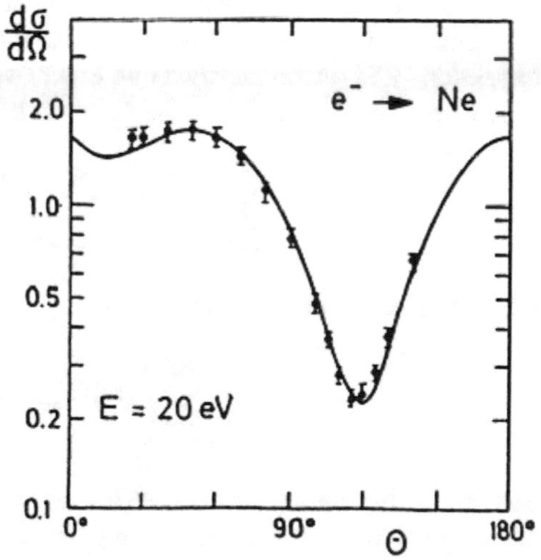

Fig. 4.9 Differential scattering cross section (in atomic units, a_0^2) for the elastic scattering of electrons by neon at $E = 20$ eV as measured by Register and Trajmar [RT84]. The *solid line* shows the cross section calculated via (4.37) with the phase shifts of Fig. 4.8 for $l \leq 2$ and the phase shifts given by (4.108) for $l > 2$ (From [IF92])

potential supports bound states quite similar to the single-particle states occupied in the target atom. The automatic orthogonality of the scattering wave functions to these bound states already accounts for a large part of the exchange effects expressing the requirements of the Pauli principle. The phase shifts δ_l are only defined to within an integral multiple of π. If we draw the function $\delta_l(E)$ [or $\delta_l(k)$] continuously from $k = 0$ to $k \to \infty$, then for a local potential the difference $\delta_l(0) - \delta_l(\infty)$ is equal to the number of bound states (in the partial wave l) multiplied by π according to Levinson's theorem (4.118). For a more sophisticated description involving non-local potentials, a generalization of Levinson's theorem [Swa55] tells us that occupied single-particle states in the target atom, which cannot be occupied by the projectile electron due to the Pauli principle, have to be included in the bound-state count when applying Levinson's theorem. The electron-neon phase shifts in Fig. 4.8 are drawn to start at threshold at 2π for $l = 0$, at π for $l = 1$ and at zero for $l = 2$, corresponding to the occupied target states $(1s, 2s, 2p)$ [Bur77]. If the simple potential picture were valid up to arbitrarily high energies, all phase shifts would tend to zero in the high energy limit in this representation. (There are no genuine bound states in the electron-neon system.)

Phase shifts for low energy elastic electron scattering by noble gas atoms can be derived with more sophisticated theories [OL83], but Figs. 4.8 and 4.9 show that simple model potentials with the correct asymptotic behaviour can work quite well.

4.1.11 Semiclassical Description of Elastic Scattering

The problem of elastic scattering by a radially symmetric potential is a convenient example for demonstrating the use of semiclassical approximations based on classical mechanics supplemented by interference effects, see e.g. [BM72]. In the semiclassical approximation of the scattering amplitude, the partial-waves expansion (4.31) is transformed into a sum over classical trajectories or rays. A basic tool for this transformation is the *Poisson summation formula*,

$$\sum_{l=0}^{\infty} g(l) = \sum_{M=-\infty}^{\infty} \int_{-1/2}^{\infty} g(l)\, \mathrm{e}^{2\pi \mathrm{i} Ml}\, \mathrm{d}l, \tag{4.121}$$

which follows from the identity,

$$\sum_{M=-\infty}^{\infty} \mathrm{e}^{2\pi \mathrm{i} Ml} \equiv \sum_{n=-\infty}^{\infty} \delta(l-n), \tag{4.122}$$

and relates the sum over discrete angular momentum quantum numbers to a sum of integrals over a continuous angular momentum. The length of the angular momentum vector is $\sqrt{l(l+1)}\hbar$, and it is well approximated by

$$L = \hbar \left(l + \frac{1}{2} \right), \tag{4.123}$$

when l is large. For large l the Legendre polynomial $P_l(\cos\theta)$ is well approximated by

$$P_l(\cos\theta) \approx \sqrt{\frac{2\hbar}{\pi L \sin\theta}} \cos\left(\frac{L\theta}{\hbar} - \frac{\pi}{4}\right), \tag{4.124}$$

except for a small range of angles θ within $\hbar/L$ of the forward direction $\theta = 0$ or the backward direction $\theta = \pi$. With this approximation the scattering amplitude (4.31) with the partial-wave amplitudes f_l given by (4.34) is

$$\begin{aligned} f(\theta) &= \sum_{l=0}^{\infty} \frac{2l+1}{2ik} (e^{2i\delta_l} - 1) P_l(\cos\theta) \\ &\approx \sum_{M=-\infty}^{\infty} e^{-i\pi M} \int_0^{\infty} \frac{L\,dL}{i\hbar^2 k} e^{\frac{i}{\hbar}2\pi ML} (e^{2i\delta_l} - 1) \sqrt{\frac{2\hbar}{\pi L \sin\theta}} \cos\left(\frac{L\theta}{\hbar} - \frac{\pi}{4}\right) \\ &= \frac{-i}{2\sqrt{\pi\hbar\mu E \sin\theta}} \sum_{M=-\infty}^{\infty} e^{-i\pi M} \left(e^{-i\pi/4} I_M^+ + e^{+i\pi/4} I_M^-\right). \end{aligned} \tag{4.125}$$

The integrals I_M^+ and I_M^- come from decomposition of the cosine of the second last line of (4.125) into two exponentials; with the abbreviation

$$\tilde{\delta}(L) = \hbar\delta_l, \tag{4.126}$$

where L is related to l via (4.123), we have

$$\begin{aligned} I_M^{\pm} &= \int_0^{\infty} \sqrt{L}\,dL \times \\ &\left\{\exp\left[\frac{i}{\hbar}\left(2\tilde{\delta}(L) \pm L\theta + 2\pi ML\right)\right] - \exp\left[\frac{i}{\hbar}(\pm L\theta + 2\pi ML)\right]\right\}. \end{aligned} \tag{4.127}$$

Equation (4.125) with (4.127) re-expresses the scattering amplitude (4.31) in terms of the angular momentum (4.123) and the phase function (4.126) which both have the dimensions of an action. If we regard (4.123) as a definition of the variable L rather than as an approximation, then the only approximation at this stage is the replacement of the Legendre polynomials according to (4.124). The phase of the first exponentials in (4.127) contains the contribution $2\tilde{\delta}(L)$ which expresses, in terms of an action, the phase shift in the radial wave function during the scattering process. The contribution $\pm L\theta + 2\pi ML$ constitutes the action of a projectile with angular momentum L which propagates through an angle $\pm\theta$, $0 < \theta < \pi$, in addition to M complete turns through 2π around the origin. The last line of (4.125) contains terms connecting all possible angular momenta L with all possible angles $\pm\theta + 2\pi M$. This expression is condensed to a sum over classical trajectories in the spirit of the semiclassical approximation by exploiting the assumption, that the actions involved are very large compared to $\hbar$. The contributions of almost all angular momenta to the integrals over L are assumed to vanish, because of cancellations due to the very rapid oscillations of the exponential factors. Non-vanishing contributions are assumed to come only from the immediate vicinity of such angular momenta for which the phase of the exponential is stationary as a function of L; this defines the *stationary phase approximation*.

The phases of the second exponentials in (4.127) depend linearly on L and have no stationary points. Points of stationary phase of the first exponentials are given by

$$2\pi M \pm \theta = -2\frac{\mathrm{d}\tilde{\delta}}{\mathrm{d}L}. \tag{4.128}$$

An explicit expression for the phase function $\tilde{\delta}$ can be obtained from the WKB approximation to the radial wave function (cf. (1.300)),

$$u_l^{\mathrm{WKB}}(r) \propto \frac{1}{\sqrt{p_l(r)}}\cos\left(\frac{1}{\hbar}\int_{r_t}^{r} p_l(r')\,\mathrm{d}r' - \frac{\phi}{2}\right), \tag{4.129}$$

defined via the radial classical momentum

$$p_l(r) = \sqrt{2\mu(E - V(r)) - \frac{l(l+1)\hbar^2}{r^2}}. \tag{4.130}$$

The phase ϕ on the right-hand side of (4.129) is the phase loss of the WKB wave due to reflection at the classical turning point r_t, which corresponds to the radius of closest approach of the projectile. Equating the asymptotic phase of the WKB wave function (4.129) and the asymptotic phase of the quantum mechanical radial wave function (4.35) yields an explicit relation between the quantum mechanical scattering phase shift δ_l and the reflection phase ϕ in the WKB wave function (cf. Sect. 1.6.3),

$$\delta_l = l\frac{\pi}{2} + \frac{\pi}{2} + \lim_{r\to\infty}\left(\frac{1}{\hbar}\int_{r_t}^{r} p_l(r')\,\mathrm{d}r' - kr\right) - \frac{\phi}{2}. \tag{4.131}$$

If, in the spirit of the semiclassical approximation, we replace $l(l+1)\hbar^2$ by L^2 (cf. (4.123)) and ϕ by $\pi/2$, then the phase function (4.126) becomes

$$\tilde{\delta}(L) = L\frac{\pi}{2} + \int_{r_t}^{\infty}(p_L(r) - p_\infty)\,\mathrm{d}r - r_t p_\infty, \tag{4.132}$$

where

$$p_L(r) = \sqrt{2\mu(E - V(r)) - L^2/r^2}, \quad p_\infty = \sqrt{2\mu E}; \tag{4.133}$$

the derivative of the phase function is

$$\frac{d\tilde{\delta}}{dL} = \frac{\pi}{2} - \int_{r_t}^{\infty} \frac{L\,dr}{r^2\sqrt{2\mu(E - V(r)) - L^2/r^2}} = \frac{1}{2}\Theta(L)\,. \tag{4.134}$$

Here we have introduced the *classical deflection function* $\Theta(L)$, which gives the total angle Θ through which a classical projectile of mass μ is scattered by the radially symmetric potential $V(r)$, as function of the (classical) angular momentum L; it is often expressed in terms of the *impact parameter* $b = L\sqrt{2\mu E}$, see e.g. [LL71], Paragraph 18, (18.1), (18.2). Thus the condition of stationary phase (4.128) is

$$\Theta(L) = \mp\theta - 2M\pi\,; \tag{4.135}$$

it selects, for each scattering angle θ, those values of angular momentum L, for which the total deflection angle Θ is equal to plus or minus θ modulo 2π. In a typical quantum mechanical scattering experiment, it is only the scattering angle θ, $0 \leq \theta \leq \pi$ between the incoming and outgoing beam of particles that can be detected. Classically we can, in addition, distinguish whether the projectile was deflected in a clockwise or anticlockwise sense and how often, if at all, it encircled the target completely. The integer M in the relation (4.135) between Θ and θ counts how many times the classical trajectory encircles the origin in the clockwise sense. The relation between θ and Θ is illustrated in Fig. 4.10.

The contribution of the vicinity of a given point L_0 of stationary phase to the integrals $I_M^{\pm}$ in (4.127) is estimated by expanding the phase of the exponential

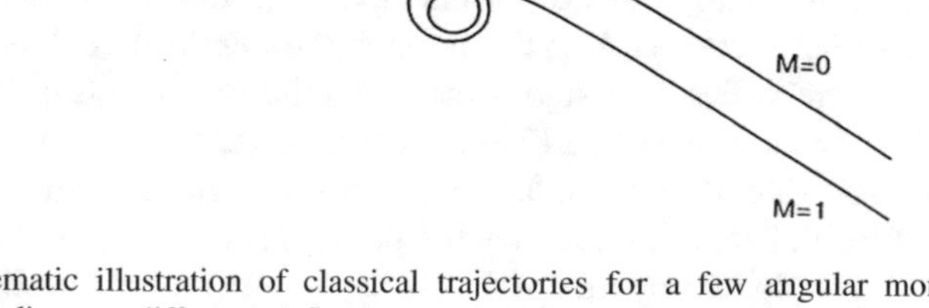

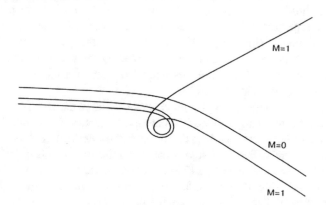

Fig. 4.10 Schematic illustration of classical trajectories for a few angular momenta (impact parameters) leading to different deflection angles Θ corresponding to the same scattering angle θ

around its stationary point,

$$2\tilde{\delta}(L) \pm L\theta + 2\pi ML \approx 2\tilde{\delta}(L_0) \pm L_0\theta + 2\pi ML_0 + \frac{\mathrm{d}^2\tilde{\delta}}{\mathrm{d}L^2}(L - L_0)^2 . \tag{4.136}$$

Extending the integral over L in the vicinity of L_0 to an integral from $-\infty$ to ∞ and ignoring the L-dependence of the factor $\sqrt{L}$ reduces the integral to a factor depending on L_0, times a simple Gaussian integral

$$\int_{\infty}^{\infty} \exp\{-a^2(L - L_0)^2\}\,\mathrm{d}L = \frac{\sqrt{\pi}}{a}, \quad a^2 = -\frac{\mathrm{i}}{\hbar}\frac{\mathrm{d}^2\tilde{\delta}}{\mathrm{d}L^2} = -\frac{\mathrm{i}}{2\hbar}\frac{\mathrm{d}\Theta}{\mathrm{d}L} . \tag{4.137}$$

Inserting this result into the integrals (4.127) in the last line of (4.125) yields the following expression for the semiclassical approximation $f_{\mathrm{sc}}(\theta)$ to the scattering amplitude:

$$f_{\mathrm{sc}}(\theta) = \frac{-\mathrm{i}}{\sqrt{2\mu E \sin\theta}} \sum_{L_i} \left(\frac{\mathrm{e}^{\mathrm{i}\alpha}\sqrt{L_i}\,\exp\left(\frac{\mathrm{i}}{\hbar}[2\tilde{\delta}(L_i) - L_i\,\Theta(L_i)]\right)}{\sqrt{|d\Theta/dL|_{L_i}}} \right) . \tag{4.138}$$

For a given scattering angle θ, the sum is to be taken over all angular momenta L_i for which the total deflection angle Θ corresponds to the (observable) scattering angle θ according to (4.135).

The expression (4.138) illustrates several features which are characteristic for semiclassical approximations to quantum mechanical amplitudes describing physical processes:

(i) The amplitude is expressed as a sum over terms each corresponding to a classical trajectory for realizing the process. Here this is a sum over (classical) angular momenta (impact parameters) leading to the given scattering angle.
(ii) Each term contains a phase essentially given by the classical action along the trajectory in units of $\hbar$. Here this phase consists of a radial and an angular contribution and is $[2\tilde{\delta} - L_i\Theta]/\hbar$.
(iii) Each term also contains a *topological phase*, which is usually a multiple of $\pi/4$ and is related to the topology of the classical trajectory. Here this phase is $\alpha = -M\pi \mp \pi/4 \pm \pi/4$, where M is the number of times the trajectory encircles the origin completely. The "$\mp$" sign comes from the coefficients $\mathrm{e}^{\mp\mathrm{i}\pi/4}$ in front of I_M^+ and I_M^- in the last line of (4.125) and corresponds to the sign in front of θ on the right-hand side of (4.135); the "$\pm$" sign stands for the sign of the gradient $\mathrm{d}\Theta/\mathrm{d}L$ of the deflection function at the point of stationary phase.
(iv) Each term is weighted by an amplitude depending on the density of classical trajectories in the vicinity of the trajectory concerned. Here this factor is $\sqrt{L_i}|\mathrm{d}\Theta/\mathrm{d}L|^{-1/2}$; it diverges at stationary points of the deflection function corresponding to an accumulation of trajectories deflected by the same angle, an effect known as *rainbow scattering*.

If there is only one classical angular momentum contributing to the scattering angle θ according to (4.135), then all the phases in (4.138) drop out of the expression for the differential cross section, giving

$$\left(\frac{d\sigma}{d\Omega}\right)_{sc} = |f_{sc}(\theta)|^2 = \frac{L}{2\mu E \sin\theta}\left|\frac{d\Theta}{dL}\right|^{-1} = \frac{b}{\sin\theta}\left|\frac{d\Theta}{db}\right|^{-1}; \tag{4.139}$$

this is exactly the same as the classical expression, see [LL71], Paragraph 18, (18.8).

If more than one classical trajectory contribute to the semiclassical approximation (4.138) of the scattering amplitude, then the corresponding approximation to the differential cross section will contain the effects of interference of the various contributions. The semiclassical cross section goes beyond the pure classical description in that it contains these quantum mechanical interference effects.

The semiclassical approximation can be useful in providing an intuitive picture of a given quantum mechanical process. An application to electron-atom scattering was given by Burgdörfer et al. in 1995. [BR95]. They studied the elastic scattering of electrons by krypton atoms using a parametrized electron-atom potential derived from Hartree-Fock calculations. Figure 4.11 shows the quantum mechanical and

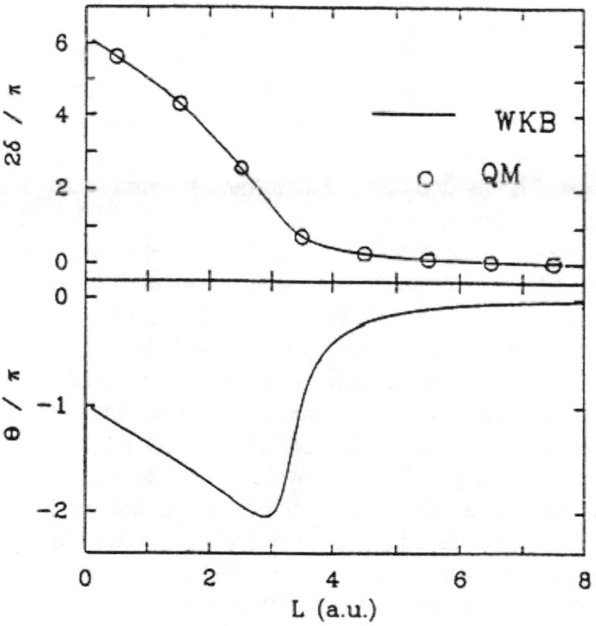

Fig. 4.11 Description of electron krypton scattering at $E = 100$ eV with a parametrized electron-atom potential. The upper panel shows quantum mechanical phase shifts (*open circles*) together with the semiclassical approximation $\tilde{\delta}(L)/\hbar$ [cf. (4.126), (4.132)] (*solid line*). The lower panel shows the classical deflection function $\Theta(L)$ in units of π. The abscissa is labelled by the angular momentum L defined by (4.123) (From [BR95])

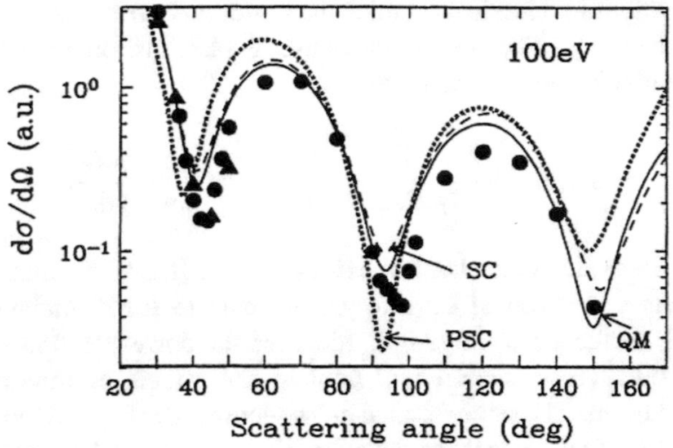

Fig. 4.12 Experimental differential cross section (*filled circles and triangles*) for electron scattering by krypton at 100 eV [WC75, JH76]. The *solid line* (QM) is the calculated quantum mechanical result (4.37), the *dashed curve* (SC) is the result obtained by calculating the scattering amplitude according to (4.125) with (4.127), and the *dotted curve* (PSC) is the result of the (primitive) semiclassical approximation based on the scattering amplitude (4.138) (From [BR95])

semiclassical (4.132) scattering phase shifts for this potential, together with the classical deflection angle as functions of the (classical) angular momentum L for an impact energy of 100 eV. The differential scattering cross section at 100 eV is shown in Fig. 4.12. The solid line is the quantum mechanical result (4.37), which agrees quite well with the experimental data. The three distinct minima indicate a dominance of the $l = 3$ partial wave in determining the shape of the cross section. The dashed curve shows the differential cross section obtained by calculating the scattering amplitude according to (4.125) with (4.127), and it reproduces the result of the direct partial wave summation very satisfactorily. This shows that the approximation of the Legendre polynomials according to (4.124) doesn't cause serious error, even though low angular momentum quantum numbers provide the dominant contributions to the cross section. Finally, the dotted line in Fig. 4.12 shows the result of the semiclassical approximation based on the scattering amplitude (4.138). Although there are noticeable deviations from the full quantum mechanical result, the semiclassical approximation does reproduce the oscillatory structure of the cross very well qualitatively. In the semiclassical description this oscillatory structure is due to the interference of amplitudes from three classical trajectories, as can be deduced from the deflection function in Fig. 4.11 with the help of (4.135). There is always one trajectory with angular momentum larger than about $3\hbar$ which is deflected by an angle Θ between zero and $-\pi$, so that $\Theta = -\theta$, $(M = 0)$ corresponds to the scattering angle θ in the interval $[0, \pi]$. There are two further trajectories, one with angular momentum close to $3\hbar$ and one with smaller angular momentum, for which the deflection angle lies between $-\pi$ and -2π, so that $\Theta = \theta - 2\pi$, $(M = 1)$ corresponds to the same scattering angle θ. The very good qualitative agreement between the quantum mechanical and the semiclassical cross sections shows that the two interpretations, one based on a purely quantum mechanical picture and attributing the oscillatory structure to the dominance of the $l = 3$ partial wave, and the other semiclassical interpretation attributing it to the interference of a small number of classical trajectories, are not mutually exclusive.

4.1.12 Elastic Scattering by a Pure Coulomb Potential

In order to describe scattering by a pure Coulomb potential,

$$V_C(\boldsymbol{r}) = -\frac{Ze^2}{r}, \tag{4.140}$$

we have to modify the description of Sect. 4.1.1 substantially because of the very-long-ranged nature of the potential. There is an analytic solution of the stationary Schrödinger equation, which is regular at the origin and fulfills asymptotic boundary conditions appropriate to the scattering problem,

$$\psi_C(\boldsymbol{r}) = \mathrm{e}^{-\pi\eta/2}\Gamma(1+\mathrm{i}\eta)\,\mathrm{e}^{\mathrm{i}kz}\,F(-\mathrm{i}\eta, 1; \mathrm{i}k(r-z))\,. \tag{4.141}$$

Here η is the Coulomb parameter (Sommerfeld parameter) as in (1.119),

$$\eta = -\frac{Z\mu e^2}{\hbar^2 k} = -\frac{1}{ka_Z}, \tag{4.142}$$

Γ is the complex gamma function defined by (A.17) in Appendix A.3, and F is the confluent hypergeometric series, (A.69) in Appendix A.5. For large values of $k(r-z)$ the wave function $\psi_C(\boldsymbol{r})$ has the form

$$\begin{aligned}\psi_C(\boldsymbol{r}) &= \mathrm{e}^{\mathrm{i}[kz+\eta\ln k(r-z)]}\left[1+\frac{\eta^2}{\mathrm{i}k(r-z)}+\dots\right]\\ &\quad + f_C(\theta)\frac{\mathrm{e}^{\mathrm{i}(kr-\eta\ln 2kr)}}{r}\left[1+\frac{(1+\mathrm{i}\eta)^2}{\mathrm{i}k(r-z)}+\dots\right]\end{aligned} \tag{4.143}$$

with

$$f_C = \frac{-\eta}{2k\sin^2(\theta/2)}\,\mathrm{e}^{-\mathrm{i}[\eta\ln(\sin^2(\theta/2))-2\sigma_0]}, \quad \sigma_0 = \arg\left[\Gamma(1+\mathrm{i}\eta)\right]. \tag{4.144}$$

To the left of the scattering centre, $z = r\cos\theta < 0$, the first term on the right-hand side of (4.143) asymptotically describes an incoming wave $\exp(\mathrm{i}k_{\mathrm{eff}}z)$, but its effective wave number $k_{\mathrm{eff}} = k + \eta[\ln k(r-z)]/z$ converges only very slowly to its asymptotic value k. For a given angle $\theta \neq 0$, the contribution j_{in} of this term to the current density according to (4.4) is, in leading order in $1/r$,

$$\boldsymbol{j}_{\rm in} = \frac{\hbar k}{\mu}\boldsymbol{e}_z + O\left(\frac{1}{r}\right). \tag{4.145}$$

The second term on the right-hand side of (4.143) describes an outgoing spherical wave with an effective wave number $k - \eta(\ln 2kr)/r$, which also converges very slowly to its asymptotic value k as $r \to \infty$. The angular modulation is asymptotically described by the function $f_{\rm C}$ of (4.144) which is called the *Coulomb scattering amplitude* or *Rutherford scattering amplitude*. The corresponding current density (again for given $\theta \neq 0$) is, in leading order in $1/r$,

$$\boldsymbol{j}_{\rm out} = \frac{\hbar k}{\mu}|f_{\rm C}(\theta)|^2\,\frac{\boldsymbol{r}}{r^3} + O\left(\frac{1}{r^3}\right). \tag{4.146}$$

The differential cross section is again defined as the asymptotic ratio of outgoing particle flux to the incoming current density and is, in analogy to (4.6),

$$\frac{\mathrm{d}\sigma_{\rm C}}{\mathrm{d}\Omega} = |f_{\rm C}(\theta)|^2 = \frac{\eta^2}{4k^2\sin^4(\theta/2)} = \frac{4}{a_Z^2\,q^4}. \tag{4.147}$$

Here q is the length of the vector $\boldsymbol{q}$ which stands for the vector difference of the outgoing and ingoing wave vectors as in (4.20), $q = 2k\sin(\theta/2)$ (see Fig. 4.13c).

The formula (4.147) is the famous Rutherford formula for elastic scattering by a pure Coulomb potential. The differential cross section doesn't depend on the sign of the potential. Furthermore, it does not depend on energy and scattering angle independently, but only on the absolute value of the momentum transfer $\hbar\boldsymbol{q}$. The Rutherford cross section (4.147) diverges strongly in the forward direction ($\theta \to 0$) so that the integrated cross section becomes infinite. This is of course due to the very long range of the Coulomb potential which even deflects particles passing the scattering centre at large distances. Figure 4.13a shows the hyperbolical classical orbits of a particle scattered by an attractive Coulomb potential, and Fig. 4.13b shows the Rutherford cross section (4.147).

The Rutherford cross section (4.147) is also obtained if the scattering amplitude $f_{\rm C}(\theta)$ is replaced by its Born approximation according to (4.19). The classical formula (4.139) also reproduces the Rutherford cross section (4.147) [LL71]. For Coulomb scattering in three dimensions we are confronted with the remarkable coincidence that quantum mechanics, classical mechanics and the Born approximation all give the same differential scattering cross section. Note however, that the scattering amplitudes are real both in the classical description and in the Born approximation, and differ from the exact quantum mechanical expression (4.144), which has a non-trivial phase. It is also interesting to note that the equality of Coulomb cross sections obtained in quantum mechanics, classical mechanics and the Born approximation is a peculiarity of three-dimensional coordinate space. It no longer holds for scattering by a potential proportional to $1/r$ e.g. in two-dimensional coordinate space, see Sect. 4.2.6.

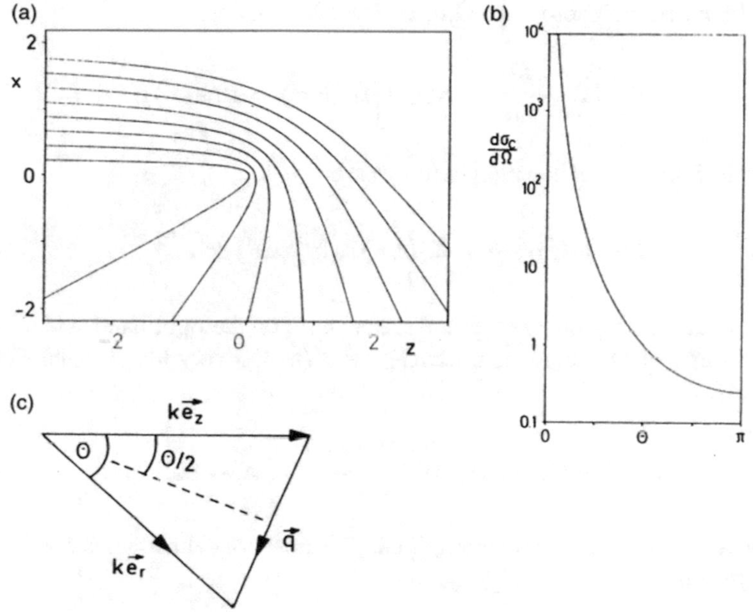

Fig. 4.13 (**a**) Hyperbolical classical orbits of a particle scattered by an attractive Coulomb potential for $k = 1/a_Z$ and $\eta = -1$. The coordinates x and z are in units of the Bohr radius a_Z. (**b**) The Rutherford cross section (4.147) in units of $(a_Z/2)^2$. Part (**c**) illustrates the identity $q = 2k\sin(\theta/2)$

4.1.13 Elastic Scattering by a Modified Coulomb Potential, DWBA

An important real situation frequently encountered in charged-particle scattering is, that the potential only corresponds to a pure Coulomb potential at large separations and that there are shorter-ranged modifications due e.g. to decreasing screening of the nucleus of the target ion by its electrons. (cf. Fig. 2.2),

$$V(\boldsymbol{r}) = V_C(r) + V_{sr}(\boldsymbol{r}), \quad \lim_{r\to\infty} r^2 V_{sr}(\boldsymbol{r}) = 0\,. \tag{4.148}$$

In order to expose the effect of the additional shorter-ranged potential we again make the Schrödinger equation look like an inhomogeneous differential equation [cf. (4.11)], but we now take the "inhomogeneous term" to be only the part due to the additional shorter-ranged potential,

$$\left[E + \frac{\hbar^2}{2\mu}\Delta - V_C(r)\right]\psi(\boldsymbol{r}) = V_{sr}(\boldsymbol{r})\psi(\boldsymbol{r})\,. \tag{4.149}$$

Let $\mathcal{G}_{\mathrm{C}}(\boldsymbol{r},\ \boldsymbol{r}')$ be the appropriate Green's function obeying

$$\left[E + \frac{\hbar^2}{2\mu}\Delta - V_{\mathrm{C}}(r)\right]\mathcal{G}_{\mathrm{C}}(\boldsymbol{r},\boldsymbol{r}') = \delta(\boldsymbol{r}-\boldsymbol{r}')\,. \tag{4.150}$$

The equivalent integral equation in this case is

$$\psi(\boldsymbol{r}) = \psi_{\mathrm{C}}(\boldsymbol{r}) + \int \mathcal{G}_{\mathrm{C}}(\boldsymbol{r},\boldsymbol{r}')V_{\mathrm{sr}}(\boldsymbol{r}')\psi(\boldsymbol{r}')\,\mathrm{d}\boldsymbol{r}'\,. \tag{4.151}$$

In the asymptotic region $r \to \infty$, the second term on the right-hand side of (4.151) has the form of an outgoing spherical wave (in the very-long-ranged Coulomb potential),

$$\psi(\boldsymbol{r}) = \psi_{\mathrm{C}}(\boldsymbol{r}) + f'(\theta,\phi)\frac{\mathrm{e}^{\mathrm{i}(kr-\eta\ln 2kr)}}{r}\,,\qquad r\to\infty\,. \tag{4.152}$$

In contrast to (4.17) the angular amplitude f' is now not defined via incoming plane waves but via *distorted waves* $\bar{\psi}_{\mathrm{C},r}$,

$$f'(\theta,\phi) = -\frac{\mu}{2\pi\hbar^2}\int \bar{\psi}^*_{\mathrm{C},r}(\boldsymbol{r}')V_{\mathrm{sr}}(\boldsymbol{r}')\psi(\boldsymbol{r}')\,\mathrm{d}\boldsymbol{r}'\,. \tag{4.153}$$

The distorted waves $\bar{\psi}_{\mathrm{C},r}(\boldsymbol{r}')$ are solutions of the Schrödinger equation with a pure Coulomb potential, but their asymptotic form [cf. (4.141), (4.143)] corresponds to an incoming modified plane wave in the direction of the radius vector $\boldsymbol{r}$ instead of in the direction of the z-axis, plus an incoming instead of an outgoing spherical wave [Bra83]. Explicitly,

$$\bar{\psi}_{\mathrm{C},\boldsymbol{r}}(\boldsymbol{r}') = \mathrm{e}^{-\pi\eta/2}\,\Gamma(1-\mathrm{i}\eta)\,\mathrm{e}^{\mathrm{i}\boldsymbol{k}_r\cdot\boldsymbol{r}'}\,F(\mathrm{i}\eta,1;-\mathrm{i}(|\boldsymbol{k}_r|r' + \boldsymbol{k}_r\cdot\boldsymbol{r}'))\,. \tag{4.154}$$

As in (4.16) and (4.17), $\boldsymbol{k}_r$ is the wave vector with length k which points in direction of the radius vector $\boldsymbol{r}$ (without $'$).

Since the first term on the right-hand side of (4.152) also contains an outgoing spherical wave [see (4.143)], modulated by the Coulomb amplitude (4.144), the total amplitude modulating the outgoing spherical wave is a sum of the Coulomb amplitude f_{C} and the additional amplitude (4.153),

$$\psi(\boldsymbol{r}) \stackrel{r\to\infty}{=} \mathrm{e}^{\mathrm{i}[kz+\eta\ln k(r-z)]} + [f_{\mathrm{C}}(\theta) + f'(\theta,\phi)]\frac{\mathrm{e}^{\mathrm{i}(kr-\eta\ln 2kr)}}{r}\,. \tag{4.155}$$

The differential cross section for elastic scattering is now

$$\frac{\mathrm{d}\sigma}{\mathrm{d}\Omega} = |f_{\mathrm{C}}(\theta) + f'(\theta,\phi)|^2\,. \tag{4.156}$$

The scattering amplitude for the elastic scattering of a particle by a potential $V_C + V_{sr}$ is thus a sum of two contributions: the first contribution is the amplitude for scattering by the "unperturbed" potential V_C, the second contribution describes the modification of the exact solution in the unperturbed potential caused by the additional perturbing potential. This decomposition is named after *Gell-Mann* and *Goldberger* and can be performed quite generally for a sum of two potentials. The cross section contains a contribution $|f_C|^2$ from the Coulomb scattering amplitude, a contribution $|f'|^2$ from the additional scattering amplitude and a further contribution $f_C f'^* + f_C^* f'$ due to interference of the two amplitudes f_C and f'.

If the influence of the additional shorter-ranged potential is small, we can replace the exact wave function ψ in the integrand on the right-hand side of (4.153) by the (distorted) incoming Coulomb wave (4.141) in the spirit of the Born approximation. This is the *distorted wave Born approximation*(DWBA) and leads to the following explicit expression for the additional scattering amplitude,

$$f^{\mathrm{DWBA}} = -\frac{\mu}{2\pi\hbar^2}\int \bar{\psi}^*_{\mathrm{C},r}(\boldsymbol{r}')V_{\mathrm{sr}}(\boldsymbol{r}')\,\psi_{\mathrm{C}}(\boldsymbol{r}')\,\mathrm{d}\boldsymbol{r}'\,. \tag{4.157}$$

If the additional shorter-ranged potential V_{sr} is radially symmetric, it makes sense to expand the wave function in partial waves. For an incoming Coulomb wave $\psi_C(\boldsymbol{r})$ travelling in z-direction we have, in analogy to (4.30),

$$\psi_{\mathrm{C}}(\boldsymbol{r}) = \sum_{l=0}^{\infty}(2l+1)\,\mathrm{i}^l\,\mathrm{e}^{\mathrm{i}\sigma_l}\,\frac{F_l(\eta,kr)}{kr}\,P_l(\cos\theta)\,. \tag{4.158}$$

Here F_l are the regular Coulomb functions, as introduced in Sect. 1.3.2, which solve the radial Schrodinger equation in a pure Coulomb potential. The Coulomb phases σ_l are defined by (1.121). The additional scattering amplitude f' doesn't depend on the azimuthal angle ϕ and can be expanded in analogy to (4.31):

$$f'(\theta) = \sum_{l=0}^{\infty} f'_l\,P_l(\cos\theta)\,. \tag{4.159}$$

The same steps that lead to (4.34) now allow us to extract from the wave function (4.152) a relation between the partial-wave amplitudes f'_l and the phase shifts δ_l due to the effect of the additional shorter-ranged potential in the respective partial waves [cf. (1.120), (1.122)]:

$$f'_l = \frac{2l+1}{2\mathrm{i}k}\,\mathrm{e}^{2\mathrm{i}\sigma_l}\left(\mathrm{e}^{2\mathrm{i}\sigma_l}-1\right) = \frac{2l+1}{k}\,\mathrm{e}^{2\mathrm{i}\sigma_l}\,\mathrm{e}^{\mathrm{i}\delta_l}\sin\delta_l\,. \tag{4.160}$$

Because of the shorter range of the potential V_{sr}, the additional phase shifts δ_l in (4.160) converge rapidly to zero (or to an integral multiple of π) as the angular momentum quantum number l increases. Hence the expansion (4.159) is rapidly convergent. On the other hand, the partial-waves expansion of the Coulomb scattering amplitude f_C converges very slowly. In order to calculate e.g. the cross section (4.156) it is thus best to use the analytically known expression (4.144) for f_C and to expand only the additional scattering amplitude f' in partial waves. The phase shifts δ_l can be extracted from the asymptotic behaviour of the radial wave functions $u_l(r)$,

$$u_l(r) \propto F_l(\eta, kr) + \tan\delta_l \, G_l(\eta, kr), \quad r \to \infty , \tag{4.161}$$

(cf. Table 1.3).

We obtain an implicit equation for the phase shifts by extending (4.46) to the case of modified Coulomb potentials,

$$\tan\delta_l = -\sqrt{\frac{2\mu\pi}{\hbar^2 k}} \int_0^\infty F_l(\eta, kr') \, V_{\rm sr}(r') \, u_l(r') \, {\rm d}r' . \tag{4.162}$$

In the spirit of the DWBA we obtain an approximate explicit expression for $\tan\delta_l$ if we replace the exact radial wave function u_l in the integrand on the right-hand side of (4.162) by the regular Coulomb function F_l (multiplied by $[2\mu/(\pi\hbar^2 k)]^{1/2}$ to ensure normalization in energy),

$$\tan\delta_l \approx -\frac{2\mu}{\hbar^2 k} \int_0^\infty [F_l(\eta, kr')]^2 \, V_{\rm sr}(r') \, {\rm d}r' . \tag{4.163}$$

If the Coulomb potential is repulsive, $\eta > 0$, then $\tan\delta_l$ vanishes at the threshold $k \to 0$ just as in the case of a shorter-ranged potential alone. For an attractive Coulomb potential, $\eta < 0$, $\tan\delta_l$ generally tends to a finite value. Remember that the phase shifts at threshold are connected to the quantum defects of the corresponding Rydberg states below threshold, as expressed in Seaton's theorem (3.143) and illustrated in Fig. 3.9 for the e^--K^+ system. A byproduct of this consideration is the insight, that the additional phase shift δ_l due to a shorter-ranged potential on top of a pure Coulomb potential cannot in general be identical to the phase shift caused by the shorter-ranged potential alone.

4.1.14 Feshbach Projection. Optical Potential

All real applications of the considerations in the preceding sections of this chapter depend crucially on the potential. For large projectile-target separations the leading terms are known, e.g. the polarization potential (4.119) for the scattering of a charged particle by a neutral atoms or the Coulomb potential (4.140) for the scattering of an electron by a target ion of charge Z. At smaller separations however, excitations of the target become important as do exchange effects, and the interaction potential may become very complicated. In this region it is not obvious that it is justified to describe e.g. an electron-atom interaction by a Schrödinger equation with a potential.

One possibility of deriving an equation of motion of the Schrödinger type is Feshbach's projection formalism. This involves projection operators $\hat{\mathcal{P}}$ and $\hat{\mathcal{Q}}$ which decompose the whole space of wave functions into a subspace of wave functions $\hat{\mathcal{P}}\Psi$, whose dynamics are to be studied further, and an orthogonal residual space, the *$\hat{\mathcal{Q}}$-space*, which is only of interest in so far as it is coupled to and influences the states in *$\hat{\mathcal{P}}$-space*:

$$\Psi = \hat{\mathcal{P}}\Psi + \hat{\mathcal{Q}}\Psi\,, \quad \hat{\mathcal{P}} + \hat{\mathcal{Q}} = 1\,, \quad \hat{\mathcal{P}}\hat{\mathcal{Q}} = 0\,. \tag{4.164}$$

Elastic scattering is usually described in a $\hat{\mathcal{P}}$-space in which the target atom is given by a fixed (generally the ground-state) wave function, while arbitrary wave functions are allowed for the projectile electron. This corresponds to a single term in the close-coupling expansion (3.163). All wave functions orthogonal to $\hat{\mathcal{P}}$-space constitute $\hat{\mathcal{Q}}$-space.

Multiplying from the left by $\hat{\mathcal{P}}$ and by $\hat{\mathcal{Q}}$ enables us to transform the stationary Schrödinger equation $\hat{H}\Psi = E\Psi$ for the wave function Ψ in (4.164) into two coupled equations for $\hat{\mathcal{P}}\Psi$ and $\hat{\mathcal{Q}}\Psi$,

$$\begin{aligned} \hat{\mathcal{P}}\hat{H}\hat{\mathcal{P}}(\hat{\mathcal{P}}\Psi) + \hat{\mathcal{P}}\hat{H}\hat{\mathcal{Q}}(\hat{\mathcal{Q}}\Psi) &= E(\hat{\mathcal{P}}\Psi), \\ \hat{\mathcal{Q}}\hat{H}\hat{\mathcal{Q}}(\hat{\mathcal{Q}}\Psi) + \hat{\mathcal{Q}}\hat{H}\hat{\mathcal{P}}(\hat{\mathcal{P}}\Psi) &= E(\hat{\mathcal{Q}}\Psi)\,. \end{aligned} \tag{4.165}$$

Here we used the property of projection operators, viz. $\hat{\mathcal{P}}\hat{\mathcal{P}} = \hat{\mathcal{P}}$ and $\hat{\mathcal{Q}}\hat{\mathcal{Q}} = \hat{\mathcal{Q}}$. If we resolve the lower equation (4.165) for $\hat{\mathcal{Q}}\Psi$,

$$\hat{\mathcal{Q}}\Psi = \frac{1}{E - \hat{\mathcal{Q}}\hat{H}\hat{\mathcal{Q}}}\hat{\mathcal{Q}}\hat{H}\hat{\mathcal{P}}(\hat{\mathcal{P}}\Psi), \tag{4.166}$$

and insert the result into the upper equation we obtain an *effective Schrödinger equation* for the component $\hat{\mathcal{P}}\Psi \equiv \psi$,

$$\hat{H}_{\text{eff}}\psi = E\psi, \quad \hat{H}_{\text{eff}} = \hat{\mathcal{P}}\hat{H}\hat{\mathcal{P}} + \hat{\mathcal{P}}\hat{H}\hat{\mathcal{Q}}\frac{1}{E - \hat{\mathcal{Q}}\hat{H}\hat{\mathcal{Q}}}\hat{\mathcal{Q}}\hat{H}\hat{\mathcal{P}}\,. \tag{4.167}$$

The first term $\hat{\mathcal{P}}\hat{H}\hat{\mathcal{P}}$ in the formula for the *effective Hamiltonian* $\hat{H}_{\text{eff}}$ contains all direct and exchange contributions of the elastic channel, but no contributions due to coupling to excited states of the target atom. These are contained in the second term $\hat{\mathcal{P}}\hat{H}\hat{\mathcal{Q}}[E - \hat{\mathcal{Q}}\hat{H}\hat{\mathcal{Q}}]^{-1}\hat{\mathcal{Q}}\hat{H}\hat{\mathcal{P}}$, which introduces an explicitly energy-dependent contribution to the *effective potential.* If the energy E lies above the continuum threshold of $\hat{\mathcal{Q}}\hat{H}\hat{\mathcal{Q}}$, it should be given an infinitesimally small imaginary part in the denominator in (4.167), similar to (4.14). This makes the effective Hamiltonian non-Hermitian. The projection of the Schrödinger equation onto a subspace of the full space of states thus leads to an explicitly energy-dependent additional potential in the effective Schroödinger equation for the projection of the total wave function onto this subspace. If the projectile electron can decay into continuum states of $\hat{\mathcal{Q}}$-space, this effective $\hat{\mathcal{P}}$-space potential is non-Hermitian. The effective potential $\hat{V}_{\text{eff}}$ in the effective Schrödinger equation in $\hat{\mathcal{P}}$-space is usually called *optical potential.*

One immediate consequence of the non-Hermitian nature of the optical potential $\hat{V}_{\text{eff}}$ is, that the continuity equation is no longer fulfilled in the form (4.8). We actually have

$$\nabla\cdot\boldsymbol{j} = \frac{\hbar}{2\mathrm{i}\mu}(\psi^*\Delta\psi - \psi\Delta\psi^*) = \frac{1}{\mathrm{i}\hbar}(\psi^*\hat{V}_{\text{eff}}\psi - \psi\hat{V}_{\text{eff}}^{\dagger}\psi^*),$$

$$\oint \boldsymbol{j}\cdot\mathrm{d}\boldsymbol{s} = \int \nabla\cdot\boldsymbol{j}\,\mathrm{d}^3r = \frac{2}{\hbar}\Im[\langle\psi|\,V_{\text{eff}}\,|\psi\rangle]\,. \tag{4.168}$$

If the boundary conditions are chosen such that the projectile electron travels outward in $\hat{\mathcal{Q}}$-space and not inward, then $\Im[\langle\psi|\hat{V}_{\text{eff}}|\psi\rangle]$ is negative, corresponding to a loss of particle flux due to *absorption* from $\hat{\mathcal{P}}$-space into $\hat{\mathcal{Q}}$-space.

If the non-Hermitian optical potential has the form of a complex radially symmetric potential $V_{\text{eff}}(r)$, then an expansion in partial waves still makes sense, but the radial wave functions u_l and the phase shifts δ_l are now complex. The imaginary part of the phase shift is generally positive for a negative imaginary part of the potential (cf. (4.47)), so the absolute value of $\exp(2\mathrm{i}\delta_l)$ is smaller than unity. The formulae (4.37) and (4.38) for the elastic scattering cross section remain valid in a (shorter-ranged) complex potential. In addition, the *total absorption cross section* σ_{abs} is defined as the loss of particle flux relative to the incoming current density $\hbar k/\mu$. The asymptotic form of the wave function (4.36) is

$$\psi(\boldsymbol{r}) \overset{r\to\infty}{=} \sum_{l=0}^{\infty} \frac{(2l+1)}{2\mathrm{i}k}\left[\mathrm{e}^{2\mathrm{i}\delta_l}\frac{\mathrm{e}^{+\mathrm{i}kr}}{r} - (-1)^l\frac{\mathrm{e}^{-\mathrm{i}kr}}{r}\right]P_l(\cos\theta), \tag{4.169}$$

and the total loss of particle flux is

$$-\oint \boldsymbol{j}\cdot d\boldsymbol{s} = \frac{\hbar}{4\mu k}\sum_{l=0}^{\infty}(2l+1)^2(1-|\mathrm{e}^{2\mathrm{i}\delta_l}|^2)\int P_l(\cos\theta)^2\,\mathrm{d}\Omega\,, \tag{4.170}$$

where we have already exploited the orthogonality of the Legendre polynomials P_l. Using $\int P_l(\cos\theta)^2\mathrm{d}\Omega = 4\pi/(2l+1)$ [cf. (A.4) in Appendix A.1, (1.59)] the loss of particle flux divided by the incoming current density, i.e. the total absorption cross section, amounts to

$$\sigma_{\text{abs}} = -\frac{\mu}{\hbar k}\oint \boldsymbol{j}\cdot d\boldsymbol{s} = \frac{\pi}{k^2}\sum_{l=0}^{\infty}(2l+1)(1-|\mathrm{e}^{2\mathrm{i}\delta_l}|^2)\,. \tag{4.171}$$

How well the Schrödinger-type equation (4.167) describes elastic scattering depends of course on how accurately the effective Hamiltonian $\hat{H}_{\text{eff}}$ or rather the optical potential $\hat{V}_{\text{eff}}$ is calculated. One of the simplest approximations consists in completely ignoring the coupling to $\hat{Q}$-space. This leads to the one-channel version of the close-coupling equations (3.176), which disregards all excitations of the target atom but includes exactly the exchange effects between projectile electron and target. The resulting potential is known as *static exchange potential.* If the target atom or ion is described by a Hartree-Fock wave function, then the static exchange potential is simply the associated Hartree-Fock potential (Sect. 2.3.1).

In order to derive the polarization potential (4.119), one has to go beyond the static exchange approximation and consider the coupling to $\hat{Q}$-space. An exact treatment of the coupling term in (4.167) would however involve an exact solution of the N-electron problem, which is of course impossible. A successful approximate access to the coupling potential is provided by replacing the whole set of eigenstates of $\hat{Q}\hat{H}\hat{Q}$ by a finite (small) number of cleverly chosen *pseudostates* [CU87, CU89]. For a calculation of optical potentials in this spirit see e.g. [BM88].

4.2 Scattering in Two Spatial Dimensions

Two-dimensional scattering problems arise naturally when the motion of projectile and target is restricted to a plane, e.g. a surface separating two bulk media. A scattering problem can also become effectively two-dimensional, if a three-dimensional configuration is translationally invariant in one direction. This is the case for a projectile scattering of a cylindrically symmetric target, e.g., an atom or molecule scattering off a cylindrical wire or nanotube. The motion of the projectile is free in the direction parallel to the cylinder axis, and we are left with a two-dimensional scattering problem in a plane perpendicular to the cylinder axis. Essential features of the two-dimensional scattering problem were illuminated by Lapidus [Lap82], Verhaar et al. [VE84] and Adhikari [Adh86] some decades ago. The recent intense activity in physics involving ultracold atoms and their interaction with nanostructures such as cylindrical nanotubes has lead to a renewed interest in this subject, in particular in the low-energy, near-threshold regime [AF08, KM09, Tic09, FE12].

We assume that the 2D scattering process occurs in the y-z plane, where the scattering angle θ varies between $-\pi$ and π, see Fig. 4.14. As in Sect. 4.1.1, the quantum mechanical description of the scattering process is based on the Schrödinger equation

$$\left[-\frac{\hbar^2}{2\mu}\Delta + V(\boldsymbol{r})\right]\psi(\boldsymbol{r}) = E\psi(\boldsymbol{r}) , \tag{4.172}$$

but $\boldsymbol{r}$ now stands for the two-component displacement vector in the y-z plane.

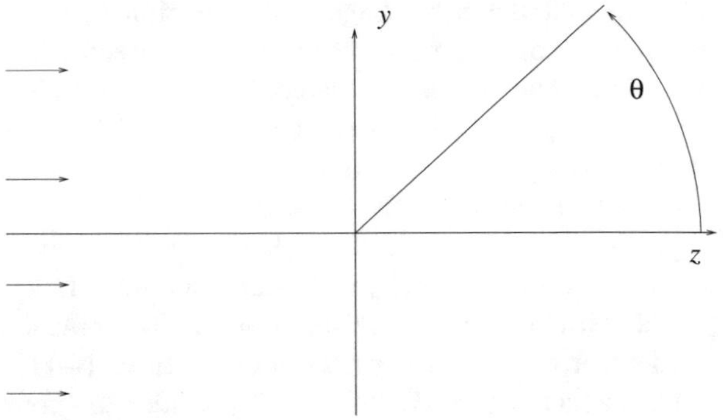

Fig. 4.14 Two-dimensional scattering in the y-z plane. The z-axis shows in the direction of incidence, and the scattering angle θ varies between $-\pi$ and π

4.2.1 Scattering Amplitude and Scattering Cross Section

We look for solutions of (4.172) with the following boundary conditions,

$$\psi(\boldsymbol{r}) \overset{r\to\infty}{\sim} \mathrm{e}^{\mathrm{i}kz} + f(\theta)\frac{\mathrm{e}^{\mathrm{i}kr}}{\sqrt{r}} . \tag{4.173}$$

Essential differences to the three-dimensional case (4.3) are, that the outgoing spherical wave becomes an outgoing circular wave whose amplitude decreases proportional to $1/\sqrt{r}$ instead of to $1/r$, and that the physical dimension of the scattering amplitude $f(\theta)$ is the square root of a length in the two-dimensional case. The current density $\boldsymbol{j}_{\mathrm{out}}(\boldsymbol{r})$ associated with the outgoing circular wave is

$$\boldsymbol{j}_{\mathrm{out}}(\boldsymbol{r}) = \frac{\hbar k}{\mu}\,|f(\theta,\phi)|^2\,\frac{\hat{\boldsymbol{e}}_r}{r} + O\left(\frac{1}{r^{3/2}}\right) , \tag{4.174}$$

while the incoming current density associated with the "plane wave" $\mathrm{e}^{\mathrm{i}kz}$ in (4.173) can again be written as $\boldsymbol{j}_{\mathrm{in}} = \hat{\boldsymbol{e}}_z\hbar k/\mu$. The surface element of a large sphere in the three-dimensional case, $r^2\mathrm{d}\Omega$, is now replaced by the arc-element of a large circle, $r\,\mathrm{d}\theta$, and the differential scattering cross section is defined by the flux scattered into this arc, $\boldsymbol{j}_{\mathrm{out}}(\boldsymbol{r})\cdot\hat{\boldsymbol{e}}_r r\,\mathrm{d}\theta$, normalized to the incoming current density $|\boldsymbol{j}_{\mathrm{in}}| = \hbar k/\mu$,

$$\mathrm{d}\lambda = |f(\theta)|^2\,\mathrm{d}\theta , \quad \frac{\mathrm{d}\lambda}{\mathrm{d}\theta} = |f(\theta)|^2 . \tag{4.175}$$

The integrated scattering cross section is

$$\lambda = \int_{-\pi}^{\pi}\frac{\mathrm{d}\lambda}{\mathrm{d}\theta}\,\mathrm{d}\theta = \int_{-\pi}^{\pi}|f(\theta)|^2\,\mathrm{d}\theta . \tag{4.176}$$

Note that the differential and the integrated scattering cross sections have the physical dimension of a length. The differential cross section can be interpreted as the length perpendicular to the direction of incidence from which the incoming particles are scattered into the differential arc defined by $d\theta$, while the integrated cross section corresponds to the length from which particles are scattered at all.

Particle conservation implies that the total flux through a circle of radius r, $\int_{-\pi}^{\pi} \boldsymbol{j}\cdot\hat{\boldsymbol{e}}_r r\, d\theta$, should vanish for large r. The contribution from the incoming wave e^{ikz} vanishes on symmetry grounds, while the contribution from the outgoing circular wave is:

$$I_{\text{out}} = \lim_{r\to\infty} \int_{-\pi}^{\pi} \boldsymbol{j}_{\text{out}}(\boldsymbol{r}) \cdot \hat{\boldsymbol{e}}_r r\, d\theta = \frac{\hbar k}{\mu} \int_{-\pi}^{\pi} |f(\theta)|^2\, d\theta = \frac{\hbar k}{\mu} \lambda\,. \tag{4.177}$$

The contribution $\boldsymbol{j}_{\text{int}}(\boldsymbol{r})$ of the interference of incoming "plane" and outgoing circular wave to the current density is,

$$\boldsymbol{j}_{\text{int}}(\boldsymbol{r}) = \frac{\hbar k}{2\mu} f(\theta) \frac{e^{ik(r-z)}}{\sqrt{r}} (\hat{\boldsymbol{e}}_r + \hat{\boldsymbol{e}}_z) + \text{cc} + \cdots\,, \tag{4.178}$$

so the interference contribution to the flux through a circle of large radius r is

$$\boldsymbol{j}_{\text{int}}(\boldsymbol{r}) \cdot \hat{\boldsymbol{e}}_r r\, d\theta = \frac{\hbar k}{2\mu} f(\theta)\, e^{ikr(1-\cos\theta)} \sqrt{r}\,(1+\cos\theta) + \text{cc}\,. \tag{4.179}$$

The integral over the right-hand side of (4.179) can be evaluated by the method of stationary phase, since the integrand contributes only around $\cos\theta = 1$ for $r \to \infty$. This gives

$$I_{\text{int}} = \int_{-\pi}^{\pi} \boldsymbol{j}_{\text{int}}(\boldsymbol{r}) \cdot \hat{\boldsymbol{e}}_r r\, d\theta = 2\frac{\hbar k}{\mu}\sqrt{\frac{\pi}{k}}\, [\Re\{f(\theta=0)\} - \Im\{f(\theta=0)\}]\,. \tag{4.180}$$

Particle conservation requires $I_{\text{out}} + I_{\text{int}} = 0$, so with (4.177) we obtain the *optical theorem for scattering in two-dimensional space*,

$$\lambda = 2\sqrt{\frac{\pi}{k}}\, [\Im\{f(\theta=0)\} - \Re\{f(\theta=0)\}]\,. \tag{4.181}$$

4.2.2 Lippmann-Schwinger Equation and Born Approximation

Adapting the treatment of Sect. 4.1.1 to the case of two spatial dimensions leads to the Lippmann-Schwinger equation

$$\psi(\boldsymbol{r}) = e^{ikz} + \int \mathcal{G}_{2D}(\boldsymbol{r},\boldsymbol{r}')V(\boldsymbol{r}')\psi(\boldsymbol{r}')\, d\boldsymbol{r}'\,, \tag{4.182}$$

which looks just like the corresponding equation (4.15) in 3D, except that the free-particle Green's function $\mathcal{G}_{2D}(\boldsymbol{r},\boldsymbol{r}')$, defined by the 2D version of (4.13), is

$$\mathcal{G}_{2D}(\boldsymbol{r},\boldsymbol{r}') = \frac{\mathrm{i}\,\mu}{2\hbar^2} H_0^{(1)}(k|\boldsymbol{r}-\boldsymbol{r}'|) \overset{k|\boldsymbol{r}-\boldsymbol{r}'|\to\infty}{\sim} \frac{\mathrm{i}\,\mu}{2\hbar^2}\,\mathrm{e}^{-\mathrm{i}\pi/4}\sqrt{\frac{2}{\pi k|\boldsymbol{r}-\boldsymbol{r}'|}}\,\mathrm{e}^{\mathrm{i}k|\boldsymbol{r}-\boldsymbol{r}'|}\,. \tag{4.183}$$

Here $H_0^{(1)}$ stands for the zero-order Hankel function of the first kind, see (A.44) and (A.45) in Appendix A.4. In the asymptotic region $|\boldsymbol{r}| \gg |\boldsymbol{r}|'$ the Green's function in (4.182) can be replaced by

$$\mathcal{G}_{2D}(\boldsymbol{r},\boldsymbol{r}') = \frac{\mu \mathrm{e}^{\mathrm{i}\pi/4}}{\hbar^2\sqrt{2\pi k}}\,\frac{\mathrm{e}^{\mathrm{i}kr}}{\sqrt{r}}\left[\mathrm{e}^{-\mathrm{i}\boldsymbol{k}_r\cdot\boldsymbol{r}'} + O\left(\frac{r'}{r}\right)\right]. \tag{4.184}$$

This is the 2D version of (4.16); $\boldsymbol{k}_r$ again stands for $k\hat{\boldsymbol{e}}_r$, but $\hat{\boldsymbol{e}}_r$ is now the radial unit vector in the y-z plane. Inserting (4.184) in (4.182) gives the asymptotic form (4.173) with

$$f(\theta) = \frac{\mu \mathrm{e}^{\mathrm{i}\pi/4}}{\hbar^2\sqrt{2\pi k}}\int \mathrm{e}^{-\mathrm{i}\boldsymbol{k}_r\cdot\boldsymbol{r}'}\,V(\boldsymbol{r}')\,\psi(\boldsymbol{r}')\,\mathrm{d}\boldsymbol{r}'\,. \tag{4.185}$$

The Born approximation is defined by replacing the exact solution $\psi(\boldsymbol{r}')$ in the integrand in (4.185) by the incoming "plane" wave $\mathrm{e}^{\mathrm{i}kz'} = \mathrm{e}^{\mathrm{i}(k\hat{\boldsymbol{e}}_z)\cdot\boldsymbol{r}'}$,

$$f^{\mathrm{Born}}(\theta) = \frac{\mu \mathrm{e}^{\mathrm{i}\pi/4}}{\hbar^2\sqrt{2\pi k}}\int \mathrm{d}\boldsymbol{r}'\mathrm{e}^{-\mathrm{i}\boldsymbol{k}_r\cdot\boldsymbol{r}'}\,V(\boldsymbol{r}')\,\mathrm{e}^{\mathrm{i}kz'} = \frac{\mu \mathrm{e}^{\mathrm{i}\pi/4}}{\hbar^2\sqrt{2\pi k}}\int \mathrm{d}\boldsymbol{r}'\mathrm{e}^{-\mathrm{i}\boldsymbol{q}\cdot\boldsymbol{r}'}\,V(\boldsymbol{r}')\,, \tag{4.186}$$

where $\hbar\boldsymbol{q}$ is the momentum transferred from the incoming wave travelling in the direction of $\hat{\boldsymbol{e}}_z$ to the outgoing wave travelling in the direction of $\hat{\boldsymbol{e}}_r$,

$$\boldsymbol{q} = k\,(\hat{\boldsymbol{e}}_r - \hat{\boldsymbol{e}}_z)\;,\quad q = |\boldsymbol{q}| = 2k\,\sin(\theta/2)\,. \tag{4.187}$$

For a radially symmetric potential $V(\boldsymbol{r}) = V(r)$, (4.186) can be simplified via an expansion of the exponential $\mathrm{e}^{-\mathrm{i}\boldsymbol{q}\cdot\boldsymbol{r}'}$ in polar variables [compare (4.197) below],

$$f^{\mathrm{Born}}(\theta) = \frac{\mu \mathrm{e}^{\mathrm{i}\pi/4}}{\hbar^2\sqrt{2\pi k}}\,2\pi\int_0^\infty V(r)\,J_0\big(2kr\,\sin(\theta/2)\big)\,r\,\mathrm{d}r\,. \tag{4.188}$$

4.2.3 Partial-Waves Expansion and Scattering Phase Shifts

For planar motion in the y-z plane, there is only one relevant component of angular momentum, namely $\hat{L} = y\hat{p}_z - z\hat{p}_y$, and in terms of the angle θ,

$$\hat{L} = \frac{\hbar}{\mathrm{i}}\frac{\partial}{\partial\theta}\,. \tag{4.189}$$

The eigenfunctions of $\hat{L}$ are $\mathrm{e}^{\mathrm{i}m\theta}$ with $m = 0, \pm 1, \pm 2, \ldots$, and the corresponding eigenvalues are $m\hbar$. Any wave function $\Psi(\boldsymbol{r}) \equiv \Psi(r, \theta)$ can be expanded in the complete basis of eigenfunctions of $\hat{L}$,

$$\Psi(\boldsymbol{r}) = \sum_{m=-\infty}^{\infty} \frac{u_m(r)}{\sqrt{r}} \, \mathrm{e}^{\mathrm{i}m\theta} \,. \tag{4.190}$$

From the polar representation of the Laplacian in 2D, we can write the kinetic energy operator in (4.172) as,

$$-\frac{\hbar^2}{2\mu}\Delta = -\frac{\hbar^2}{2\mu}\left(\frac{\partial^2}{\partial r^2} + \frac{1}{r}\frac{\partial}{\partial r}\right) + \frac{\hat{L}^2}{2\mu\, r^2} \,. \tag{4.191}$$

We assume a radially symmetric potential, $V(\boldsymbol{r}) = V(r)$. Inserting the expansion (4.190) into the Schrödinger equation (4.172) then gives, with the help of (4.191), an uncoupled set of radial equations for the radial wave functions $u_m(r)$,

$$\left[-\frac{\hbar^2}{2\mu}\frac{\mathrm{d}^2}{\mathrm{d}r^2} + \frac{\left(m^2 - \frac{1}{4}\right)\hbar^2}{2\mu\, r^2} + V(r)\right] u_m(r) = E\, u_m(r) \,. \tag{4.192}$$

The 2D radial Schrödinger equation looks similar to the 3D radial Schrödinger equation (4.22) in Sect. 4.1.2. In fact, (4.192) and (4.22) are identical, if we equate $|m| - \frac{1}{2}$ with the 3D angular momentum quantum number l:

$$l \equiv |m| - \frac{1}{2} \,. \tag{4.193}$$

Many results derived for the 3D radial waves in Sect. 4.1.2 can be carried over to the 2D radial waves simply via (4.193), but integer values of m imply half-integer values of l, so the results of Sect. 4.1.2 have to be checked to see whether they hold in these cases. This is particularly important for s-waves in 2D ($m = 0$), which correspond to $l = -\frac{1}{2}$.

For the free-particle case $V(r) \equiv 0$, two linearly independent solutions of the radial equation (4.192) are

$$u_m^{(\mathrm{s})}(kr) = \sqrt{\frac{\pi}{2}kr}\, J_{|m|}(kr) \,, \quad u_m^{(\mathrm{c})}(kr) = -\sqrt{\frac{\pi}{2}kr}\, Y_{|m|}(kr) \,, \tag{4.194}$$

where $J_{|m|}$ and $Y_{|m|}$ stand for the *ordinary Bessel functions* of the first and second kind, respectively [see Appendix A.4]. Their asymptotic behaviour is given by[2]

$$u_m^{(s)}(kr) \overset{kr\to\infty}{\sim} \sin\left[kr - \left(|m| - \frac{1}{2}\right)\frac{\pi}{2}\right] ,$$
$$u_m^{(c)}(kr) \overset{kr\to\infty}{\sim} \cos\left[kr - \left(|m| - \frac{1}{2}\right)\frac{\pi}{2}\right] . \tag{4.195}$$

The the influence of a potential $V(r)$ is manifest in the asymptotic phase shifts δ_m of the regular solutions of the radial Schrödinger equation (4.192). When $V(r)$ falls off faster than $1/r^2$ at large distances the effective potential in (4.192) is dominated by the centrifugal term at large distances, and the regular solution can be taken to be a superposition of the two radial free-particle wave functions (4.194) obeying (4.195),

$$u_m(r) \overset{r\to\infty}{\propto} A\,u_m^{(s)}(kr) + B\,u_m^{(c)}(kr) \overset{r\to\infty}{\propto} \sin\left[kr - \left(|m| - \frac{1}{2}\right)\frac{\pi}{2} + \delta_m\right] , \tag{4.196}$$

with $\tan\delta_m = B/A$.

In order to relate the scattering phase shifts to the scattering amplitude, we first expand the incoming "plane" wave of (4.173) in partial waves,

$$\mathrm{e}^{\mathrm{i}kz} = \sum_{m=-\infty}^{\infty} \mathrm{i}^m J_m(kr)\,\mathrm{e}^{\mathrm{i}m\theta} \overset{kr\to\infty}{\sim} \sum_{m=-\infty}^{\infty} \frac{1}{\sqrt{2\pi \mathrm{i}kr}}\left(\mathrm{e}^{\mathrm{i}kr} + (-1)^m \mathrm{i}\,\mathrm{e}^{-\mathrm{i}kr}\right) . \tag{4.197}$$

The appropriate partial-waves expansion for the scattering amplitude is

$$f(\theta) = \sum_{m=-\infty}^{\infty} f_m\,\mathrm{e}^{\mathrm{i}m\theta} , \tag{4.198}$$

and the constant coefficients f_m are the partial-wave scattering amplitudes. Expressing the sum of "plane" and circular wave in the form (4.190) gives an explicit expression for the asymptotic behaviour of the radial wave functions,

$$u_m(r) \overset{r\to\infty}{\sim} \frac{1}{\sqrt{2\pi \mathrm{i}k}}\left[(1 + \sqrt{2\pi \mathrm{i}k}f_m)\,\mathrm{e}^{\mathrm{i}kr} + \mathrm{i}(-1)^m\,\mathrm{e}^{-\mathrm{i}kr}\right]$$
$$= \frac{\mathrm{i}(-1)^m}{\sqrt{2\pi \mathrm{i}k}}\left[\mathrm{e}^{-\mathrm{i}kr} - \mathrm{i}(-1)^m(1 + \sqrt{2\pi \mathrm{i}k}f_m)\,\mathrm{e}^{\mathrm{i}kr}\right] . \tag{4.199}$$

[2]Due to the m-independent term $\frac{\pi}{4}$ appearing in the arguments both of $u_m^{(s)}(kr)$ and of $u_m^{(c)}(kr)$ in (4.195), there is no *a priori* preference for the assignment of an asymptotic "sine-" or "cosine-like" behaviour. The present nomenclature is chosen to make the connection to the 3D case as transparent as possible.

We can rewrite the asymptotic form of the regular solution (4.196) as

$$\begin{aligned} u_m(r) &\overset{r\to\infty}{\propto} \sin\left[kr - \left(|m| - \frac{1}{2}\right)\frac{\pi}{2} + \delta_m\right] \\ &\propto \mathrm{e}^{-\mathrm{i}\left[kr-\left(|m|-\frac{1}{2}\right)\frac{\pi}{2}+\delta_m\right]}\mathrm{e}^{+\mathrm{i}\delta_m} - \mathrm{e}^{\mathrm{i}\left[kr-\left(|m|-\frac{1}{2}\right)\frac{\pi}{2}+\delta_m\right]}\mathrm{e}^{+\mathrm{i}\delta_m} \\ &\propto \mathrm{e}^{-\mathrm{i}kr} - \mathrm{e}^{-\mathrm{i}\left(|m|-\frac{1}{2}\right)\pi}\,\mathrm{e}^{\mathrm{i}kr}\,\mathrm{e}^{2\mathrm{i}\delta_m}\,. \end{aligned} \tag{4.200}$$

Comparing the lower lines of (4.199) and (4.200) gives

$$\mathrm{e}^{2\mathrm{i}\delta_m} = 1+\sqrt{2\pi\mathrm{i}k}f_m\,, \quad f_m = \frac{1}{\sqrt{2\pi\mathrm{i}k}}\left(\mathrm{e}^{2\mathrm{i}\delta_m} - 1\right) = \sqrt{\frac{2\mathrm{i}}{\pi k}}\,\mathrm{e}^{\mathrm{i}\delta_m}\sin\delta_m\,. \tag{4.201}$$

Equation (4.201) can be used to express the scattering cross sections in terms of the scattering phase shifts,

$$\begin{aligned} \frac{\mathrm{d}\lambda}{\mathrm{d}\theta} &= |f(\theta)|^2 = \sum_{m,m'} f_m^* f_{m'}\,\mathrm{e}^{\mathrm{i}(m'-m)\theta} \\ &= \frac{2}{\pi\,k}\sum_{m,m'} \mathrm{e}^{\mathrm{i}(\delta_{m'}-\delta_m)}\sin\delta_{m'}\,\sin\delta_m \mathrm{e}^{\mathrm{i}(m'-m)\theta}\,, \end{aligned} \tag{4.202}$$

$$\lambda = \int_{-\pi}^{\pi} |f(\theta)|^2 \mathrm{d}\theta = 2\pi\sum_{m=-\infty}^{\infty} |f_m|^2 = \frac{4}{k}\sum_{m=-\infty}^{\infty}\sin^2\delta_m\,. \tag{4.203}$$

The scattering amplitude in forward direction is

$$\begin{aligned} f(\theta = 0) &= \sum_{m=-\infty}^{\infty} f_m = \sqrt{\frac{2\mathrm{i}}{\pi\,k}}\sum_{m=-\infty}^{\infty}\mathrm{e}^{\mathrm{i}\delta_m}\sin\delta_m \\ &= \sqrt{\frac{2}{\pi\,k}}\sum_{m=-\infty}^{\infty}\sin\delta_m\,\mathrm{e}^{\mathrm{i}(\delta_m+\pi/4)}\,, \end{aligned} \tag{4.204}$$

hence

$$\Im\{f(0)\} - \Re\{f(0)\} = \frac{2}{\sqrt{\pi\,k}}\sum_{m=-\infty}^{\infty}\sin^2\delta_m = \frac{1}{2}\sqrt{\frac{k}{\pi}}\,\lambda\,, \tag{4.205}$$

which again yields the optical theorem (4.181).

4.2.4 Near-Threshold Behaviour of the Scattering Phase Shifts

The leading near-threshold behaviour of the phase shifts can be derived from the small-argument behaviour of the free-particle solutions (4.194),

$$u_m^{(\mathrm{s})}(kr) \overset{kr\to 0}{\sim} \frac{\sqrt{\pi}}{\Gamma(|m|+1)} \left(\frac{kr}{2}\right)^{\frac{1}{2}+|m|}, \tag{4.206}$$

$$u_m^{(\mathrm{c})}(kr) \overset{kr\to 0}{\sim} \frac{\Gamma(|m|)}{\sqrt{\pi}} \left(\frac{kr}{2}\right)^{\frac{1}{2}-|m|} \quad \text{for} \quad m \neq 0\,. \tag{4.207}$$

The case $m = 0$ is special, because the two powers of r appearing in (4.206) and (4.206), namely $\frac{1}{2} + |m|$ and $\frac{1}{2} - |m|$ are equal in this case. We focus first on the case $m \neq 0$; the special case of s-waves in 2D is treated in Sect. 4.2.5 below.

At distances r beyond the range of the potential, the radial wave function $u_m(r)$ is a superposition of the free-particle wave functions (4.194); towards threshold, $k \to 0$, the product kr tends to zero so we can make use of the small-argument expressions (4.206), (4.207),

$$\begin{aligned} u_m(r) &\overset{kr\to 0}{\propto} u_m^{(\mathrm{s})}(kr) + \tan\delta_m\, u_m^{(\mathrm{c})}(kr) \\ &\sim \frac{\sqrt{\pi}}{\Gamma(|m|+1)} \left(\frac{k}{2}\right)^{\frac{1}{2}+|m|} \\ &\quad \times \left[r^{|m|+\frac{1}{2}} + \tan\delta_m \left(\frac{k}{2}\right)^{-2|m|} \frac{\Gamma(|m|)\Gamma(|m|+1)}{\pi\, r^{|m|-\frac{1}{2}}} \right]. \end{aligned} \tag{4.208}$$

Directly at threshold, the radial Schrödinger equation (4.192) has a regular solution $u_m^{(0)}(r)$ which is defined up to a constant by the boundary condition $u_m^{(0)}(0) = 0$ and is a function of r only. The wave function (4.208) must become proportional to this k-independent solution for $k \to 0$, so in the second term in the square bracket in the bottom line of (4.208), the k-dependence of $\tan\delta_m$ must compensate the factor $(k/2)^{-2|m|}$: $\tan\delta \overset{k\to 0}{\propto} k^{2|m|}$. More explicitly,

$$\tan\delta_m \overset{k\to 0}{\sim} \mp \frac{\pi}{\Gamma(|m|)\Gamma(|m|+1)} \left(\frac{a_m k}{2}\right)^{2|m|}. \tag{4.209}$$

The characteristic length a_m appearing on the right-hand side of (4.209) is the *scattering length* in the partial wave $m \neq 0$. Equation (4.209) is essentially identical to (4.57) in Sect. 4.1.7 if we replace $|m|$ by $l + \frac{1}{2}$, except that the power $2|m|$ in (4.209) is always even for integer m, so the possibility of having positive or negative values on the right-hand side has to be explicitly included via the $\mp$ sign.

As in the 3D case, the threshold behaviour (4.209) is only valid in all partial waves if the potential $V(r)$ in the radial Schrödinger equation (4.192) falls off faster than any power of $1/r$ at large distances. For potentials falling off as $1/r^{\alpha}$, the considerations of Sect. 4.1.7 can be carried over to the 2D case, remembering that l now stands for $|m| - \frac{1}{2}$. In particular, the condition for the validity of (4.209) now reads $2|m| < \alpha - 2$. For $2|m| > \alpha - 2$, (4.73) in Sect. 4.1.7 is applicable, provided $l + \frac{1}{2}$ is replaced by $|m|$.

4.2.5 The Case m = 0, s-Waves in Two Dimensions

The case of s-waves in two dimensions is special, because the radial Schrödinger equation (4.192) now reads

$$\left[-\frac{\hbar^2}{2\mu}\frac{\mathrm{d}^2}{\mathrm{d}r^2} - \frac{1}{4}\frac{\hbar^2}{2\mu\, r^2} + V(r)\right] u_{m=0}(r) = E\, u_{m=0}(r)\,, \tag{4.210}$$

and the centrifugal potential is *attractive*. This degree of attractivity of the inverse-square potential marks the boundary between potentials supporting at most a finite number of bound states and those supporting infinitely many bound states. If the factor $\frac{1}{4}$ in front of the inverse-square term in (4.210) were replaced by $\frac{1}{4} + \varepsilon$ with an ever so small positive ε, then the radial Schrödinger equation (4.210) would support an infinite dipole series of bound states, as described in Sect. 3.1.5.

The free-particle solutions, for $V(r) \equiv 0$ in (4.210), are

$$u^{(\mathrm{s})}_{m=0}(r) = \sqrt{\frac{\pi}{2}}kr\, J_0(kr)\,, \quad u^{(\mathrm{c})}_{m=0}(r) = -\sqrt{\frac{\pi}{2}}kr\, Y_0(kr)\,. \tag{4.211}$$

Their small-argument behaviour is

$$u^{(\mathrm{s})}_{m=0}(kr) \overset{kr\to 0}{\sim} \sqrt{\frac{\pi}{2}}\, kr\,, \tag{4.212}$$

$$u^{(\mathrm{c})}_{m=0}(kr) \overset{kr\to 0}{\sim} -\sqrt{\frac{2}{\pi}}\, kr\left[\ln\left(\frac{kr}{2}\right) + \gamma_{\mathrm{E}} + O\left((kr)^2\right)\right],$$

cf. (A.36) and (A.42) in Appendix A.4. Beyond the range of the interaction potential $V(r)$, the regular solution of the radial Schrödinger equation (4.210) is a superposition of the free-particle waves (4.211), and its asymptotic behaviour is,

$$\begin{aligned} u_{m=0}(r) &\overset{r\to\infty}{\propto} \sqrt{kr}\,[J_0(kr) - \tan\delta_{m=0} Y_0(kr)] \\ &\overset{kr\to\infty}{\propto} \sin\left(kr + \frac{\pi}{4} + \delta_{m=0}\right)\,. \end{aligned} \tag{4.213}$$

The condition, that the right-hand side of the upper line in (4.213) become, except for a simple proportionality, independent of k in the limit $k \to 0$, implies the following leading near-threshold behaviour of the s-wave scattering phase shift,

$$\cot \delta_{m=0} \overset{k \to 0}{\sim} \frac{2}{\pi} \left(\ln \left(\frac{ka}{2} \right) + \gamma_{\mathrm{E}} \right) . \tag{4.214}$$

Equation (4.214) defines the scattering length a for s-waves in two dimensions. In the limit $k \to 0$, the wave function (4.213) converges to a k-independent limit $u^{(0)}_{m=0}$,

$$u_{m=0}(r) \overset{k \to 0}{\propto} u^{(0)}_{m=0}(r) \overset{r \to \infty}{\propto} -\sqrt{r} \ln \left(\frac{r}{a} \right) . \tag{4.215}$$

The wave function on the right-hand side of (4.215) has exactly one node (beyond $r = 0$), and this node lies at $r = a$. For a potential falling off as $1/r^{\alpha}$ at large distances, a well-defined scattering length in the partial wave m exists as long as $2|m| < \alpha - 2$. For $m = 0$, this condition is fulfilled for all $\alpha > 2$. The scattering length a defined according to (4.214), (4.215) is well defined for all interaction potentials which fall off faster than $1/r^2$ at large distances.

It is worthwhile to reflect a little on the remarkable situation of s-waves in 2D. At threshold, the regular free-particle wave is proportional to $\sqrt{r}$, corresponding to r^{l+1} when $l = -\frac{1}{2}$. The "irregular" solution, which we might expect to be proportional to r^{-l}, is actually proportional to $\sqrt{r} \ln r$, which seems only marginally less regular than the regular wave. An arbitrary superposition of these two free-particle waves can be written as

$$u(r) \propto A\sqrt{r} - \sqrt{r} \ln r = -\sqrt{r} \ln \left(\frac{r}{\mathrm{e}^A} \right) , \tag{4.216}$$

which is just the form on the right-hand side of (4.215), with the scattering length given by $a = \mathrm{e}^A$. In two-dimensional scattering, the scattering length is never negative.

The leading near-threshold behaviour of the s-wave phase shift (4.214) was already given in [VE84], together with the next term of the effective-range expansion in two dimensions,

$$\cot \delta_{m=0} \overset{k \to 0}{\sim} \frac{2}{\pi} \left[\ln \left(\frac{ka}{2} \right) + \gamma_{\mathrm{E}} + \frac{(k\, r_{\mathrm{eff}})^2}{2} \right] ; \tag{4.217}$$

the effective range in 2D is defined by

$$r_{\mathrm{eff}}^2 = 2 \int_0^{\infty} \left(\left[w^{(0)}(r) \right]^2 - \left[u^{(0)}(r) \right]^2 \right) \mathrm{d}r , \tag{4.218}$$

see also [AF08]. Here $u^{(0)}(r)$ is the regular solution, at threshold, of (4.210), which behaves as the right-hand side of (4.215) asymptotically, and $w^{(0)}(r)$ is the free-particle solution which has this form for all r,

$$w^{(0)}(r) = -\sqrt{r}\,\ln\left(\frac{r}{a}\right)\,,\quad u^{0)}(r) \overset{r\to\infty}{\sim} -\sqrt{r}\,\ln\left(\frac{r}{a}\right)\,. \tag{4.219}$$

In contrast to the similar-looking definition of the effective range in 3D, see (4.100) in Sect. 4.1.7, the right-hand side of (4.218) has the physical dimension of a length squared. Note that r_{eff}^2 defined in this way can be negative. The integral on the right-hand side of (4.218) converges to a well defined limit for interaction potentials falling off faster than $1/r^4$ at large distances [AF08].

The leading near-threshold behaviour of the scattering cross sections is, naturally, dominated by the contribution from the s-wave. From (4.198), (4.201) and (4.214) we obtain

$$f(\theta) \overset{k\to 0}{\sim} f_0 \overset{k\to 0}{\sim} \frac{\sqrt{\pi\mathrm{i}/(2k)}}{\ln\left(\frac{ka}{2}\right) + \gamma_{\mathrm{E}}}\,, \tag{4.220}$$

so

$$\frac{\mathrm{d}\lambda}{\mathrm{d}\theta} \overset{k\to 0}{\sim} \frac{\pi/(2k)}{|\ln\left(\frac{ka}{2}\right) + \gamma_{\mathrm{E}}|^2} \quad\text{and}\quad \lambda \overset{k\to 0}{\sim} \frac{\pi^2/k}{|\ln\left(\frac{ka}{2}\right) + \gamma_{\mathrm{E}}|^2}\,. \tag{4.221}$$

The quantum mechanical scattering cross sections in two dimensions diverge at threshold. This divergence is essentially as $1/k$, moderated marginally by the logarithmic factor. Note that the expressions in (4.220) and (4.221), where the leading behaviour contains the logarithm in the expression $\ln(ka/2) + \gamma_{\mathrm{E}}$, are only meaningful when $ka/2$ is so small, that $\ln(ka/2) < -\gamma_{\mathrm{E}}$, i.e., for

$$ka < 2\exp(-\gamma_{\mathrm{E}})\,. \tag{4.222}$$

4.2.6 Rutherford Scattering in Two Dimensions

An instructive example showing interesting differences to the well-studied case of scattering in 3D is the case of Rutherford scattering in two dimensions, which was first treated comprehensively by Barton [Bar83]. The potential is

$$V(\boldsymbol{r}) = \frac{C}{r}\,. \tag{4.223}$$

This could be the interaction between two point particles whose motion is restricted to a two-dimensional plane embedded in three-dimensional space. It is worth remembering, however, that the Coulomb interaction in a genuinely two-dimensional space does not have this r-dependence. In terms of the scaled coordinate $\boldsymbol{\rho} = k\boldsymbol{r}$, the Schrödinger equation reads

$$\left[-\Delta_\rho + \frac{2\eta}{\rho}\right]\psi = \psi \;, \tag{4.224}$$

where η is the Sommerfeld parameter

$$\eta = \frac{\mu\, C}{\hbar^2 k} \,. \tag{4.225}$$

As in Sect. 4.1.12, we introduce the quantum mechanical length a_C, which does not exist in classical mechanics,

$$a_C = \frac{1}{|\eta| k} = \frac{\hbar^2}{\mu |C|} \;, \qquad |\eta| = \frac{1}{a_C k} \,. \tag{4.226}$$

For an attractive potential, $C < 0$ in (4.223), a_C is the usual Bohr radius.

As in the 3D case, the Schrödinger equation (4.224) has analytical solutions in 2D as well. Equations (4.141), (4.143) and (4.144) in Sect. 4.1.12 are replaced in 2D by

$$\psi_C(\boldsymbol{r}) = e^{-\frac{\pi}{2}\eta}\, \frac{\Gamma\left(\frac{1}{2} + i\eta\right)}{\Gamma\left(\frac{1}{2}\right)}\, e^{ikz}\, F\big(-i\eta, \frac{1}{2}; ik[r - z]\big) \;, \tag{4.227}$$

$$\begin{aligned}\psi_C(\boldsymbol{r}) = {} & e^{i[kz + \eta \ln(k[r-z])]}\left[1 + O\left(\frac{1}{k[r-z]}\right)\right] \\ & + f_C(\theta)\, \frac{e^{i(kr - \eta \ln 2kr)}}{\sqrt{r}}\left[1 + O\left(\frac{1}{k[r-z]}\right)\right]\end{aligned} \tag{4.228}$$

and

$$f_C(\theta) = -\frac{\eta\, e^{i\pi/4}}{\sqrt{2k\, \sin^2(\theta/2)}}\, \frac{\Gamma\left(\frac{1}{2} + i\eta\right)}{\Gamma(1 - i\eta)}\, e^{-i\eta \ln[\sin^2(\theta/2)]} \;, \tag{4.229}$$

respectively. The function F in (4.227) again denotes the confluent hypergeometric function, see Appendix A.5 . With the identities

$$\left|\Gamma\left(\frac{1}{2} + i\eta\right)\right|^2 = \frac{\pi}{\cosh(\pi\eta)} \;, \quad |\Gamma(1 - i\eta)|^2 = \frac{\pi\eta}{\sinh(\pi\eta)} \;, \tag{4.230}$$

we obtain the differential cross section for Rutherford scattering in two dimensions,

$$\frac{\mathrm{d}\lambda}{\mathrm{d}\theta} = |f_C(\theta)|^2 = \frac{\eta \tan(\pi\eta)}{2k \sin^2(\theta/2)} = \frac{|C|}{4E} \frac{\tanh(\pi|\eta|)}{\sin^2(\theta/2)} = \left(\frac{\mathrm{d}\lambda}{\mathrm{d}\theta}\right)^{\mathrm{qm}}_{\mathrm{Ruth}} . \tag{4.231}$$

In contrast to the 3D case, the quantum mechanical result (4.231) does *not* agree with the classical Rutherford cross section in two dimensions,

$$\left(\frac{\mathrm{d}\lambda}{\mathrm{d}\theta}\right)^{\mathrm{class}}_{\mathrm{Ruth}} = \frac{|C|}{4E} \frac{1}{\sin^2(\theta/2)} , \tag{4.232}$$

see, e.g., (1.55) in Sect. 1.4 of [Fri16]. On the other hand, evaluating (4.188) gives the corresponding result in Born approximation,

$$\left(\frac{\mathrm{d}\lambda}{\mathrm{d}\theta}\right)^{\mathrm{Born}}_{\mathrm{Ruth}} = \left(\frac{\mu C}{\hbar^2}\right)^2 \frac{\pi}{2\,k^3 \sin^2(\theta/2)} = \frac{|C|}{4E} \frac{\pi|\eta|}{\sin^2(\theta/2)} . \tag{4.233}$$

In terms of the quantum mechanical length a_C (the "Bohr radius") defined in (4.226),

$$\left(\frac{\mathrm{d}\lambda}{\mathrm{d}\theta}\right)^{\mathrm{qm}}_{\mathrm{Ruth}} = \frac{a_C/2}{(a_C k)^2} \frac{\tanh[\pi/(a_C k)]}{\sin^2(\theta/2)} , \tag{4.234}$$

$$\left(\frac{\mathrm{d}\lambda}{\mathrm{d}\theta}\right)^{\mathrm{class}}_{\mathrm{Ruth}} = \frac{a_C/2}{(a_C k)^2} \frac{1}{\sin^2(\theta/2)} , \quad \left(\frac{\mathrm{d}\lambda}{\mathrm{d}\theta}\right)^{\mathrm{Born}}_{\mathrm{Ruth}} = \frac{a_C/2}{(a_C k)^2} \frac{\pi/(a_C k)}{\sin^2(\theta/2)} . \tag{4.235}$$

Comparing (4.234) and (4.235) shows that the coincidence of Rutherford scattering in 3D, namely that classical mechanics, the Born approximation and the full quantum mechanical treatment all yield the same result (4.147) for the differential scattering cross section, is lifted in two spatial dimensions. The angular dependence, $\mathrm{d}\lambda/\mathrm{d}\theta \propto 1/\sin^2(\theta/2)$, is the same in all three cases, but the energy dependent prefactors of the classical cross section and of the Born approximation differ from the exact quantum mechanical result. This is illustrated in Fig. 4.15, where the respective differential cross sections, multiplied by $\sin^2(\theta/2)$, are plotted as a functions of the dimensionless product $ka_C = 1/|\eta|$.

Both the classical cross section and the Born approximation overestimate the exact quantum mechanical cross section (4.231), (4.234). As already observed by Barton [Bar83], the Born approximation becomes accurate in the high-energy limit, whereas the classical result becomes exact in the low-energy limit,

$$\left(\frac{\mathrm{d}\lambda}{\mathrm{d}\theta}\right)^{\mathrm{Born}}_{\mathrm{Ruth}} \overset{k\to\infty}{\sim} \left(\frac{\mathrm{d}\lambda}{\mathrm{d}\theta}\right)^{\mathrm{qm}}_{\mathrm{Ruth}} , \quad \left(\frac{\mathrm{d}\lambda}{\mathrm{d}\theta}\right)^{\mathrm{class}}_{\mathrm{Ruth}} \overset{k\to 0}{\sim} \left(\frac{\mathrm{d}\lambda}{\mathrm{d}\theta}\right)^{\mathrm{qm}}_{\mathrm{Ruth}} . \tag{4.236}$$

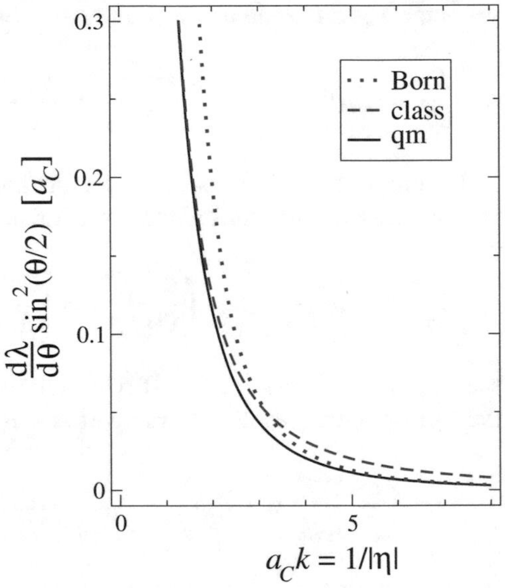

Fig. 4.15 Rutherford scattering in two spatial dimensions. The *solid black line* shows the exact quantum mechanical differential cross section (4.234) [in units of the "Bohr radius" a_C] multiplied by $\sin^2(\theta/2)$ as function of the dimensionless product $a_C k = 1/|\eta|$. The *dashed red line* and the *dotted blue line* show the corresponding classical result and the result of the Born approximation (4.235)

The example is a nice illustration of the fact that, for homogeneous potentials of degree -1, i.e., of the Coulomb type, the classical limit is at the threshold $E = 0$, and the classical treatment becomes increasingly inaccurate for large values of $|E|$. This is well accepted for bound states at negative energies, where $E \to 0$ corresponds to the limit of infinite quantum numbers, but it is not so widely appreciated for the regime of positive energies.

4.3 Spin and Polarization

In Sect. 4.1 we didn't consider the fact that electrons have spin. If the potential by which an electron is scattered is completely independent of spin, then the spin state of the electron is not affected by the scattering process and the cross sections are independent of the state of spin. In general however, electron-atom interactions at least contain a spin dependence in the form of a spin-orbit coupling—see Sect. 1.7.3. Hence the spin state is affected by scattering and the cross sections depend on the state of spin before and after the collision.

4.3.1 Consequences of Spin-Orbit Coupling

Let's assume for the time being that the spin of the incoming electron is given by the spin-up state $|\chi_+\rangle$ [cf. (1.341)]. The asymptotic form of the wave function solving the stationary Schrödinger equation (with a shorter-ranged potential) is now [instead of (4.3)]

$$\psi = \mathrm{e}^{\mathrm{i}kz}\begin{pmatrix}1\\0\end{pmatrix} + \frac{\mathrm{e}^{\mathrm{i}kr}}{r}\begin{pmatrix}f(\theta,\phi)\\g(\theta,\phi)\end{pmatrix}, \quad r\to\infty . \tag{4.237}$$

The differential cross section for elastic scattering is the outgoing particle flux, which now contains a spin-up and a spin-down contribution, divided by the incoming current density,

$$\frac{\mathrm{d}\sigma}{\mathrm{d}\Omega} = |f(\theta,\phi)|^2 + |g(\theta,\phi)|^2 . \tag{4.238}$$

Here $g(\theta,\phi)$ is the *spin-flip amplitude*, and its absolute square describes the probability that the orientation of the spin is reversed by the collision. The formula (4.238) implies that we do not separate spin-up and spin-down components for the outgoing electron, i.e. we don't measure the spin of the scattered electron.

If the target atom (or ion) itself has no spin, and if there are no further contributions in the potential breaking radial symmetry, then the spin of the projectile electron and its orbital angular momentum couple to good total angular momentum labelled by the quantum number $j = l\pm 1/2$. The stationary Schrödinger equation can be decomposed into radial Schrödinger equations (1.362), in which the potentials depend not only on the orbital angular momentum quantum number l but also on the total angular momentum quantum number j. The solutions of these radial Schrödinger equations are asymptotically characterized by phase shifts $\delta_l^{(j)}$.

We choose the quantization axis for all angular momenta to be the z-axis, which points in the direction of the incoming particle current, and we can assume the total wave function to be an eigenstate of the z-component of the total angular momentum. The corresponding quantum number must be $m = +1/2$ for consistency with the right-hand side of (4.237). We use the generalized spherical harmonics $\mathcal{Y}_{j,m,l}$ introduced in Sect. 1.7.3 to decompose the wave function (4.237) into components with good values of j, m and l. In the special case $m = +1/2$, (1.358) becomes

$$\begin{aligned} \mathcal{Y}_{l+\frac{1}{2},m,l} &= \frac{1}{\sqrt{2l+1}}\begin{pmatrix}\sqrt{l+1}\,Y_{l,0}(\theta)\\ \sqrt{l}\,Y_{l,1}(\theta,\,\phi)\end{pmatrix},\\ \mathcal{Y}_{l-\frac{1}{2},m,l} &= \frac{1}{\sqrt{2l+1}}\begin{pmatrix}-\sqrt{l}\,Y_{l,0}(\theta)\\ \sqrt{l+1}\,Y_{l,1}(\theta,\,\phi)\end{pmatrix}. \end{aligned} \tag{4.239}$$

These relations can be inverted,

$$\begin{aligned} \begin{pmatrix}Y_{l,0}(\theta)\\0\end{pmatrix} &= \sqrt{\frac{l+1}{2l+1}}\,\mathcal{Y}_{l+\frac{1}{2},m,l} - \sqrt{\frac{l}{2l+1}}\,\mathcal{Y}_{l-\frac{1}{2},m,l},\\ \begin{pmatrix}0\\Y_{l,1}(\theta,\,\phi)\end{pmatrix} &= \sqrt{\frac{l}{2l+1}}\,\mathcal{Y}_{l+\frac{1}{2},m,l} + \sqrt{\frac{l+1}{2l+1}}\,\mathcal{Y}_{l-\frac{1}{2},m,l}. \end{aligned} \tag{4.240}$$

Expanding the spatial part of the plane wave according to (4.30) and using the upper equation (4.240) yields

$$\begin{aligned} e^{ikz}\chi_+ &= \sqrt{4\pi}\sum_{l=0}^{\infty}\sqrt{2l+1}\,i^l j_l(kr)\begin{pmatrix} Y_{l,0}(\theta) \\ 0 \end{pmatrix} \\ &= \sqrt{4\pi}\sum_{l=0}^{\infty} i^l j_l(kr)\left(\sqrt{l+1}\,\mathcal{Y}_{l+\frac{1}{2},m,l} - \sqrt{l}\,\mathcal{Y}_{l-\frac{1}{2},m,l}\right) . \end{aligned} \tag{4.241}$$

If we expand the scattering amplitudes f and g in spherical harmonics in analogy to (4.31),

$$\begin{aligned} f(\theta) &= \sum_{l=0}^{\infty} f_l \sqrt{\frac{4\pi}{2l+1}}\, Y_{l,0}(\theta) , \\ g(\theta,\phi) &= \sum_{l=1}^{\infty} g_l \sqrt{l(l+1)}\sqrt{\frac{4\pi}{2l+1}}\, Y_{l,1(\theta,\phi)}, \end{aligned} \tag{4.242}$$

then we can use (4.240) and decompose the outgoing spherical wave into components with good j, m and l,

$$\begin{aligned} \begin{pmatrix} f(\theta) \\ g(\theta,\phi) \end{pmatrix} = \sum_{l=0}^{\infty}\frac{\sqrt{4\pi}}{2l+1}\Big[(f_l + l\,g_l)\,\sqrt{l+1}\,\mathcal{Y}_{l+\frac{1}{2},m,l} \\ - [f_l - (l+1)g_l]\sqrt{l}\,\mathcal{Y}_{l-\frac{1}{2},m,l}\Big] . \end{aligned} \tag{4.243}$$

If we now collect the radial parts of the incoming plane wave and the outgoing spherical wave for given values of l and j, we obtain expressions which look like the expressions in the big square brackets in (4.32), except that the coefficient f_l in (4.32) is now replaced by different linear combinations of f_l and g_l, namely $f_l + lg_l$ for $j = l + 1/2$ and $f_l - (l+1)g_l$ for $j = l - 1/2$. The same steps which led from (4.32) to (4.34) now give

$$\begin{aligned} f_l + l\,g_l &= \frac{2l+1}{2ik}\left[\exp\left(2i\delta_l^{(l+1/2)}\right) - 1\right], \\ f_l - (l+1)\,g_l &= \frac{2l+1}{2ik}\left[\exp\left(2i\delta_l^{(l-1/2)}\right) - 1\right] . \end{aligned} \tag{4.244}$$

Resolving for the partial wave amplitudes f_l and g_l yields

$$\begin{aligned} f_l &= \frac{l+1}{2ik}\left[\exp\left(2i\delta_l^{(l+1/2)}\right) - 1\right] + \frac{l}{2ik}\left[\exp\left(2i\delta_l^{(l-1/2)}\right) - 1\right], \\ g_l &= \frac{1}{2ik}\left[\exp\left(2i\delta_l^{(l+1/2)}\right) - \exp\left(2i\delta_l^{(l-1/2)}\right)\right] . \end{aligned} \tag{4.245}$$

We can use (4.243), (4.245) to deduce the direct and the spin-flip parts of the cross section (4.238) from the phase shifts, which can be obtained from the asymptotic solutions of the radial Schrödinger equations (1.362). By the way, the dependence of the spin-flip amplitude on the azimuthal angle ϕ is given simply by $\exp(i\phi)$ regardless of l, so the cross section again depends only on θ. If the effect of spin-orbit coupling is negligible, then the phase shifts are independent of j for given l; in this case g_l vanishes and f_l is again given by the expression (4.34).

For scattering by a long-ranged modified Coulomb potential we obtain formulae such as (4.243) and (4.245) for the additional scattering amplitude due to the shorter-ranged deviations from a pure Coulomb potential, which now include the effects of spin-orbit coupling. The corresponding extension of (4.160) reads

$$f_l' = \frac{l+1}{2ik}\,e^{2i\sigma_l}\left[\exp\left(2i\delta_l^{(l+1/2)}\right)-1\right]+\frac{l}{2ik}\,e^{2i\sigma_l}\left[\exp\left(2i\delta_l^{(l-1/2)}\right)-1\right],$$

$$g_l' = \frac{e^{2i\sigma_l}}{2ik}\left[\exp\left(2i\delta_l^{(l+1/2)}\right)-\exp\left(2i\delta_l^{(l-1/2)}\right)\right]. \tag{4.246}$$

We can also apply the Gell-Mann–Goldberger procedure to a decomposition of the total potential in the radial Schrödinger equation (1.362) into an unperturbed part containing everything except the spin-orbit coupling and the spin-orbit part $(\hbar^2/2)F(j,l)V_{LS}(r)$. Since the effect of the spin-orbit coupling is small, we can apply the DWBA formula (4.163) to obtain an approximate expression for the additional phase due to the spin-orbit term in the radial potential:

$$\left(\tan\delta_l^{(j)}\right)_{LS} \approx -\frac{\mu}{k}F(j,l)\int_0^\infty [u_l(r')]^2 V_{LS}(r')\,dr'. \tag{4.247}$$

Now u_l is the regular solution [asymptotically normalized to $\sin(kr+\dots)$] of the radial Schrödinger equation containing the full radial potential with the exception of the spin-orbit coupling. For a given l the j-dependence of the right-hand side of (4.247) is determined by the factor $F(j, l)$, which is simply l for $j = l+\frac{1}{2}$ and $-(l+1)$ for $j = l-\frac{1}{2}$ (see Sect. 1.7.3).

4.3.2 Application to General Pure Spin States

A *pure state* of a physical system is one which can be described by a single quantum mechanical wave function—in contrast to a *mixed state* consisting of a statistical mixture of various quantum mechanical states (see Sect. 4.3.3) below. A pure spin state of an electron is defined by a two-component spinor $\begin{pmatrix}A\\B\end{pmatrix}$. In order to describe the scattering of an electron whose incoming wave is in such a general pure spin state, we have to complement the treatment based on (4.237) above.

First we consider the case that the incoming electron is in the spin-down state $|\chi_-\rangle$. Instead of (4.237) we now have

$$\psi' = \mathrm{e}^{\mathrm{i}kz}\begin{pmatrix}0\\1\end{pmatrix} + \frac{\mathrm{e}^{\mathrm{i}kr}}{r}\begin{pmatrix}g'(\theta,\phi)\\f'(\theta,\phi)\end{pmatrix}, \quad r\to\infty\,. \tag{4.248}$$

The z-component of the total angular momentum is now $m' = -1/2$. The partial-waves expansion of the scattering amplitudes is now [instead of (4.242)]

$$f'(\theta) = \sum_{l=0}^{\infty} f_l \sqrt{\frac{4\pi}{2l+1}}\, Y_{l,0}(\theta)\,,$$
$$g'(\theta,\phi) = \sum_{l=1}^{\infty} g_l \sqrt{l(l+1)}\sqrt{\frac{4\pi}{2l+1}}\, Y_{l,-1}(\theta,\phi), \tag{4.249}$$

and leads to the same expressions (4.245) for the partial-wave amplitudes f_l and g_l (see Problem 4.6). The amplitudes f, g in (4.237) and f', g' in (4.248) thus only differ in their ϕ-dependence of the spin-flip amplitude which is given by the spherical harmonics $Y_{l,\pm1}$ and is proportional to $\mp\exp(\pm\mathrm{i}\phi)$ [see (A.4), (A.5) in Appendix A.1]. We thus have $f'(\theta) = f(\theta)$, and the spin-flip amplitudes g, g' can be expressed by a common function g_0 which depends only on the polar angle θ:

$$g(\theta,\phi) = g_0(\theta)\,\mathrm{e}^{+\mathrm{i}\phi}, \quad g'(\theta,\phi) = -g_0(\theta)\,\mathrm{e}^{-\mathrm{i}\phi}\,. \tag{4.250}$$

The wave function corresponding to an incoming electron in an arbitrary pure spin state can now be constructed as a linear combination of the two special cases (4.237) and (4.248),

$$A\psi + B\psi' \overset{r\to\infty}{=} \mathrm{e}^{\mathrm{i}kz}\begin{pmatrix}A\\B\end{pmatrix} + \frac{\mathrm{e}^{\mathrm{i}kr}}{r}\begin{pmatrix}Af(\theta) - Bg_0(\theta)\mathrm{e}^{-\mathrm{i}\phi}\\Ag_0(\theta)\mathrm{e}^{+\mathrm{i}\phi} + Bf(\theta)\end{pmatrix}. \tag{4.251}$$

The differential cross section is again defined by the ratio of the outgoing flux to the incoming current density and is now

$$\frac{\mathrm{d}\sigma}{\mathrm{d}\Omega} = \frac{\left|Af(\theta) - Bg_0(\theta)\,\mathrm{e}^{-\mathrm{i}\phi}\right|^2 + \left|Ag_0(\theta)\,\mathrm{e}^{+\mathrm{i}\phi} + Bf(\theta)\right|^2}{|A|^2 + |B|^2}$$
$$= |f(\theta)|^2 + |g_0(\theta)|^2 + 2\Im[f(\theta)g_0(\theta)^*]\frac{2\Im[AB^*\,\mathrm{e}^{\mathrm{i}\phi}]}{|A|^2 + |B|^2}\,. \tag{4.252}$$

Again this formula implies that we don't measure the spin of the scattered electron. If both A and B are different from zero, the incoming electron is no longer polarized parallel to the z-axis and the differential cross section (4.252) depends not only on the polar angle θ but also on the azimuthal angle ϕ (see Fig. 4.16). The relative

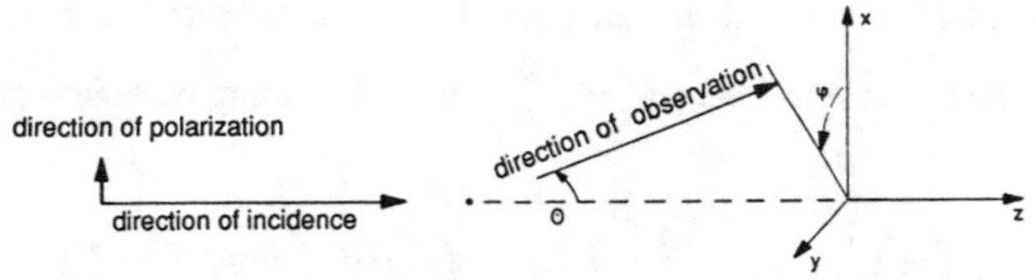

Fig. 4.16 Scattering of electrons polarized perpendicular to the direction of incidence. The cross section depends not only on the polar angle θ but also on the azimuthal angle ϕ

importance of the ϕ-dependent contribution is determined by the imaginary part of the product fg_0^* and is usually expressed with the help of the *Sherman function* $S(\theta)$,

$$S(\theta) = -2\frac{\Im[fg_0^*]}{|f|^2 + |g_0|^2} = \mathrm{i}\frac{fg_0^* - f^* g_0}{|f|^2 + |g_0|^2} . \tag{4.253}$$

It is a speciality of spin $\frac{1}{2}$ particles, that an arbitrary (pure) spin state is a spin-up (or a spin-down) state with respect to an appropriately chosen quantization axis. To see this consider an arbitrary normalized spin state $|\chi\rangle = \begin{pmatrix} A \\ B \end{pmatrix}$, $|A|^2 + |B|^2 = 1$. Using the Pauli spin matrices (1.345) we define the three-component *polarization vector*

$$\boldsymbol{P} = \langle\chi|\,\hat{\boldsymbol{\sigma}}\,|\chi\rangle . \tag{4.254}$$

In the present case its components are

$$P_x = 2\Re[A^*B], \quad P_y = 2\Im[A^*B], \quad P_z = |A|^2 - |B|^2, \tag{4.255}$$

and its length is unity. The projection of the spin operator $\hat{\boldsymbol{\sigma}}$ onto the direction of $\boldsymbol{P}$ is

$$\hat{\sigma}_P = \boldsymbol{P}\cdot\hat{\boldsymbol{\sigma}} = P_x\hat{\sigma}_x + P_y\hat{\sigma}_y + P_z\hat{\sigma}_z, \tag{4.256}$$

and it is easy to show (Problem 4.7) that the spinor $|\chi\rangle = \begin{pmatrix} A \\ B \end{pmatrix}$ is an eigenstate of $\hat{\sigma}_P$ with eigenvalue +1.[3]

[3]The deeper reason for the fact that every two-component spinor uniquely corresponds to a direction of polarization, lies in the isomorphism of the group SU(2) of special unitary transformations of two-component spinors with the group SO(3) of rotations in three-dimensional space. A similar correspondence does not apply for spinors with more than two components, i.e. for spins larger than or equal to one.

Equation (4.251) shows that scattering of the electron into the direction (θ, ϕ) transforms the initial spin state $|\chi\rangle = \begin{pmatrix} A \\ B \end{pmatrix}$ of the incoming wave into the new spin state

$$\begin{pmatrix} A' \\ B' \end{pmatrix} = \mathbf{S} \begin{pmatrix} A \\ B \end{pmatrix}, \quad \mathbf{S} = \frac{1}{\sqrt{|f|^2 + |g_0|^2}} \begin{pmatrix} f(\theta) & -g_0(\theta)e^{-i\phi} \\ g_0(\theta)e^{i\phi} & f(\theta) \end{pmatrix}. \tag{4.257}$$

The transformation is described by the transformation matrix $\mathbf{S}$, which is in general not unitary and is not to be confused with the scattering matrix to be treated later (see Sect. 4.4.2). The polarization vector $\boldsymbol{P}'$ of the scattered electron is

$$\boldsymbol{P}' = \frac{\langle\chi|\, \mathbf{S}^\dagger \hat{\sigma} \mathbf{S}\, |\chi\rangle}{\langle\chi|\, \mathbf{S}^\dagger \mathbf{S}\, |\chi\rangle}. \tag{4.258}$$

The denominator in (4.258) is needed for correct normalization, because the transformed spinor $\mathbf{S}|\chi\rangle$ is no longer normalized to unity.

4.3.3 Application to Mixed Spin States

A *mixed state* of a quantum mechanical system contains different wave functions with certain statistical probabilities. In order to describe our lack of knowledge of the precise state of a physical system we imagine a collection or *ensemble* of copies the system covering all individual states which are compatible with our limited knowledge. Such a statistical mixture of states cannot be described by a single wave function, but only by an incoherent superposition of quantities related to the individual members of the ensemble. An appropriate quantity for describing an ensemble is the *density operator*

$$\hat{\varrho} = \sum_n w_n\, |\chi_n\rangle\, \langle\chi_n|\,. \tag{4.259}$$

Here $|\chi_n\rangle$ are (orthonormalized) state vectors (wave functions) for pure quantum mechanical states, and the sum (4.259) covers all states which might be contained in the ensemble. Partial information which may make some states more probable than others is contained in the real, non-negative probabilities w_n. The sum of these probabilities is of course unity. If we have no information at all about the system, then all w_n are equal and their value is the inverse of the number of possible states, i.e. of the number of states in the ensemble.

The density operator (4.259) is a weighted mean of the projection operators $|\chi_n\rangle\langle\chi_n|$ onto the individual states $|\chi_n\rangle$. A density operator is always Hermitian and its trace is the sum of the probabilities, i.e. unity. In a particular representation the density operator becomes the *density matrix*. The *statistical expectation value*

$\langle\langle\hat{O}\rangle\rangle$ of an observable $\hat{O}$ in a mixed state is the appropriately weighted mean of the quantum mechanical expectation values in the individual states,

$$\langle\langle\hat{O}\rangle\rangle = \sum_n w_n \langle\chi_n|\,\hat{O}\,|\chi_n\rangle = \mathrm{Tr}\{\hat{O}\hat{\varrho}\}. \tag{4.260}$$

A pure state corresponds to the special case that one probability w_n is unity while all other probabilities vanish. The statistical expectation value (4.260) then is identical to the quantum mechanical expectation value in the (pure) state. The density operator $\hat{\varrho}_{\mathrm{P}}$ for a pure state is just the projection operator onto this state, in particular

$$\hat{\varrho}_{\mathrm{P}}\hat{\varrho}_{\mathrm{P}} = \hat{\varrho}_{\mathrm{P}}\,. \tag{4.261}$$

A completely unpolarized electron is one for which absolutely nothing is known about its spin state. With respect to an arbitrary axis of quantization, both spin states $|\chi_+\rangle$ and $|\chi_-\rangle$ are equally probable. The corresponding density operator is

$$\hat{\varrho} = \frac{1}{2}\,|\chi_+\rangle\,\langle\chi_+| + \frac{1}{2}\,|\chi_-\rangle\,\langle\chi_-|\,, \tag{4.262}$$

and the associated density matrix is just 1/2 times the 2×2 unit matrix. In order to describe the scattering of unpolarized electrons we have to incoherently add the cross section (4.238) from Sect. 4.3.1 and the corresponding cross section for an incoming electron in a spin-down state, both weighted with the factor 1/2. (Because of $f = f'$ and (4.250) both cross sections are equal in this case, and the sum (4.238) is unchanged.) This corresponds to averaging over the initial states compatible with the measured boundary conditions, as discussed in connection with electromagnetic transitions in Sect. 2.4.4.

A general mixed spin state is neither completely polarized like a pure state, nor completely unpolarized as in (4.262). In the spirit of (4.260) we define the polarization vector for a mixed spin state as

$$\boldsymbol{P} = \langle\langle\hat{\boldsymbol{\sigma}}\rangle\rangle = \mathrm{Tr}\{\hat{\boldsymbol{\sigma}}\,\hat{\varrho}\}\,. \tag{4.263}$$

If we take the direction of $\boldsymbol{P}$ as the axis of quantization and assume a density operator

$$\hat{\varrho} = w_+\,|\chi_+\rangle\,\langle\chi_+| + w_-\,|\chi_-\rangle\,\langle\chi_-|\,, \quad w_+ + w_- = 1, \tag{4.264}$$

then the component of $\boldsymbol{P}$ in the direction of this axis is obviously the difference of the probabilities for the spin pointing parallel and antiparallel to $\boldsymbol{P}$, namely $w_+ - w_-$. This is also the length of the polarization vector, which is smaller than unity for a mixed spin state. The length of the polarization vector serves as definition for the (degree of) polarization. The polarization can vary between zero and unity; it is unity for completely polarized electrons (pure spin state) and zero for completely unpolarized electrons.

If the incoming electron is partially polarized with respect to an axis of quantization, which need not be the z-axis, then we describe such a (mixed) spin state by a density operator like (4.264). In order to calculate the differential cross section in such a case, we must first determine the differential cross sections for the two pure states $|\chi_+\rangle$ and $|\chi_-\rangle$ with respect to the axis of quantization according to (4.252) and then incoherently superpose the results with the weights w_+ and w_-.

Scattering into the direction (θ, ϕ) transforms an incoming (pure) spin state $|\chi\rangle$ into the spin state $\mathbf{S}|\chi\rangle$ according to (4.257). Extending this result to mixed states shows that the density operator $\hat{\varrho}$ of an incoming electron is transformed into the density operator

$$\hat{\varrho}' = \frac{\mathbf{S}\hat{\varrho}\mathbf{S}^\dagger}{\mathrm{Tr}\{\mathbf{S}\hat{\varrho}\mathbf{S}^\dagger\}} \tag{4.265}$$

by the scattering process. The denominator in (4.265) ensures correct normalization, $\mathrm{Tr}\{\hat{\varrho}'\} = 1$. With (4.263) we can give a general formula for the polarization vector $\boldsymbol{P}'$ of the electron scattered into the direction (θ, ϕ),

$$\boldsymbol{P}' = \mathrm{Tr}\{\hat{\boldsymbol{\sigma}}\hat{\varrho}'\} = \frac{\mathrm{Tr}\{\hat{\boldsymbol{\sigma}}\mathbf{S}\hat{\varrho}\mathbf{S}^\dagger\}}{\mathrm{Tr}\{\mathbf{S}\hat{\varrho}\mathbf{S}^\dagger\}}\,. \tag{4.266}$$

As an application of the formula (4.266) consider the case that the incoming electron is completely unpolarized. Then $\hat{\varrho}$ is just 1/2 times the unit matrix and (4.266) simplifies to

$$\boldsymbol{P}' = \frac{\mathrm{Tr}\{\hat{\boldsymbol{\sigma}}\mathbf{S}\mathbf{S}^\dagger\}}{\mathrm{Tr}\{\mathbf{S}\mathbf{S}^\dagger\}}\,. \tag{4.267}$$

Inserting the explicit expression (4.257) for the transformation matrix $\mathbf{S}$ yields

$$\boldsymbol{P}' = S(\theta)\,\hat{\boldsymbol{e}}_\phi, \quad \text{i.e. } P_x' = -S(\theta)\sin\phi,\ P_y' = S(\theta)\cos\phi,\ P_z' = 0\,; \tag{4.268}$$

here $S(\theta)$ again stands for the Sherman function (4.253). This means that scattered electrons can have a finite polarization even if the incoming electrons are unpolarized. The direction of the polarization vector is perpendicular to the *scattering plane*, which is spanned by the direction of the incoming electron (the z-axis) and the direction of the scattered electron (θ, ϕ).

The polarization of electrons by scattering can be exploited in *double scattering experiments* in which a beam of initially unpolarized electrons is successively scattered by two targets. After scattering by the first target the electrons are (partially) polarized, and the cross section for scattering by the second target acquires a left-right asymmetry due to its dependence on the azimuthal angle ϕ. Thus polarization effects can be observed without actually having to distinguish the spin states of the electrons (see e.g. [Kes85, GK91]).

If the target atom or ion itself has a non-vanishing angular momentum, then the angular momentum coupling for the whole system becomes much more complicated. In this case we must consider various states of polarization not only of the projectile electron, but also of the target atom (or ion). In general there are several orbital angular momenta l which can couple with the spin of the projectile electron and the angular momentum of the target atom to a given total angular momentum quantum number of the system. This leads to coupled radial Schrödinger equations as they also appear in the description of inelastic scattering—see Sect. 4.4.2.

The number and quality of experiments with polarized electrons is impressive—see e.g. [Kes85, Kes91]. For a comprehensive monograph on the application of density matrix techniques see [Blu81]. The treatment of polarization effects in electron-atom scattering on the basis of the density matrix formalism is also described extensively in [Bar89]. The density matrix formalism has also been applied to situations of greater complexity, such as the scattering of electrons by optically active molecules of given orientation [Kes00].

4.4 Inelastic Scattering

4.4.1 General Formulation

In an inelastic scattering process the target atom (or ion) undergoes a change from its initial internal state to a different final internal state due to the collision with the projectile electron. In order to describe such a process, our ansatz for the wave function must contain contributions from at least two channels. A natural starting point for the description of inelastic collisions is found in the coupled channel equations (3.176), which we shall now write in the simplified form

$$\left[-\frac{\hbar^2}{2\mu}\Delta + V_{i,i}\right]\psi_i(\boldsymbol{r}) + \sum_{j\neq i} V_{i,j}\psi_j(\boldsymbol{r}) = (E - E_i)\psi_i(\boldsymbol{r})\,. \tag{4.269}$$

Again $\boldsymbol{r}$ is the spatial coordinate of the projectile electron, j (or i) labels a number of open channels defined by various internal states $\psi_{\rm int}^{(j)}$ of the target atom and E_j are the associated internal excitation energies. The potentials $V_{i,j}$ are largely given by the matrix elements of the interaction operator (3.173) between the internal states,

$$V_{i,j} = \langle\psi_{\rm int}^{(i)}|\hat{H}_{\rm W}|\psi_{\rm int}^{(j)}\rangle' + \dots\,. \tag{4.270}$$

The dots on the right-hand side stand for the (short-ranged) exchange terms in the effective potential and for possible contributions due to coupling to a not explicitly included $\hat{\mathcal{Q}}$-space. The matrix of potentials $V_{i,j}$ is an operator in the space of vectors of channel wave functions $(\psi_1,\ \psi_2, \dots)$, and it is in general explicitly energy dependent and non-Hermitian if coupling to and loss of flux into the $\hat{\mathcal{Q}}$-space is important.

As long as all interactions are shorter ranged, a solution of the coupled channel equations (4.269) describing an incoming electron in channel i fulfills the following boundary conditions:

$$\psi_j(\boldsymbol{r}) = \delta_{j,i}\mathrm{e}^{\mathrm{i}k_i z} + \frac{\mathrm{e}^{\mathrm{i}k_j r}}{r} f_{j,i}(\theta,\phi), \quad r \to \infty . \tag{4.271}$$

Here

$$k_j = \sqrt{\frac{2\mu(E-E_j)}{\hbar^2}} \tag{4.272}$$

is the asymptotic wave number of the outgoing electron in the (open) channel j. Differential cross sections are defined as in Sect. 4.1.1 by the ratio of outgoing particle flux through the surface element $r^2\mathrm{d}\Omega$ to the incoming current density. For elastic scattering $i \to i$ we again obtain the form (4.6),

$$\frac{\mathrm{d}\sigma_{i\to i}}{\mathrm{d}\Omega} = |f_{i,i}(\theta,\phi)|^2, \tag{4.273}$$

but the differential cross section for inelastic scattering $i \to j, j \neq i$ has the slightly modified form

$$\frac{\mathrm{d}\sigma_{i\to j}}{\mathrm{d}\Omega} = \frac{k_j}{k_i}|f_{j,i}(\theta,\phi)|^2 . \tag{4.274}$$

The origin of the factor k_j/k_i on the right-hand side of (4.274) is that the current density in the entrance channel is $\hbar k_i/\mu$,while the outgoing current in channel j is given by a formula similar to (4.5) but with a factor $\hbar k_j/\mu$. If we interpret the inelastic scattering amplitude (or *transition amplitude*) $f_{j,i}$ as the matrix element of a transition operator (cf. (4.18) and (4.284) below), then the expression (4.274) can be derived in the spirit of time-dependent perturbation theory (Sect. 2.4.1), and the proportionality to k_j comes in via the density of final states in the Golden Rule (2.139) (see Problem 4.3). Thus the *phase space factor* k_j/k_i accounts for the different density of states for free electrons in the exit channel j and the entrance channel i.

Integrated cross sections are defined in analogy to (4.7),

$$\sigma_{i\to j} = \int \frac{\mathrm{d}\sigma_{i\to j}}{\mathrm{d}\Omega}\,\mathrm{d}\Omega = \frac{k_j}{k_i}\int |f_{j,i}(\Omega)|^2\,\mathrm{d}\Omega . \tag{4.275}$$

The *total cross section* (with respect to channel i as the entrance channel) is a sum of the integrated elastic cross section $\sigma_{i\to i}$, the *total inelastic cross section* $\sigma_{i,\mathrm{inel}} = \sum_{j\neq i}\sigma_{i\to j}$ and the absorption cross section $\sigma_{i,\mathrm{abs}}$, which accounts for the loss of flux into open channels of $\hat{\mathcal{Q}}$-space,

$$\sigma_{i,\mathrm{tot}} = \sigma_{i\to i} + \sum_{j\neq i}\sigma_{i\to j} + \sigma_{i,\mathrm{abs}} = \sigma_{i,\mathrm{el}} + \sigma_{i,\mathrm{inel}} + \sigma_{i,\mathrm{abs}} . \tag{4.276}$$

We can also formulate a Lippmann-Schwinger equation in the many-channel case. To this end we again proceed by making the Schrödinger equation (4.269) look like an inhomogeneous differential equation,

$$\left[\mathbf{E}' + \frac{\hbar^2}{2\mu}\Delta\right]\Psi = \hat{\mathbf{V}}\Psi\,. \tag{4.277}$$

We have introduced a more compact notation using vectors and matrices: Ψ stands for the vector of channel wave functions $(\psi_1, \psi_2, \ldots)$, $\hat{\mathbf{V}}$ stands for the matrix of potentials $(V_{i,j})$, and $\mathbf{E}'$ is the diagonal matrix containing the asymptotic energies $E - E_i$ in the respective channels as the diagonal matrix elements. Since the "homogeneous equation" $\left(\hat{\mathbf{V}} \equiv 0\right)$ corresponds to a set of uncoupled free-particle Schrödinger equations, we can easily define a diagonal matrix $\mathbf{G}$ of Green's functions,

$$\mathbf{G} \equiv \begin{pmatrix} \mathcal{G}_{1,1} & 0 & 0 & \cdots \\ 0 & \mathcal{G}_{2,2} & 0 & \cdots \\ 0 & 0 & \mathcal{G}_{3,3} & \cdots \\ \cdots & \cdots & \cdots & \cdots \end{pmatrix}, \quad \mathcal{G}_{i,i} = -\frac{\mu}{2\pi\hbar^2}\frac{\mathrm{e}^{\mathrm{i}k_i|\boldsymbol{r}-\boldsymbol{r}'|}}{|\boldsymbol{r}-\boldsymbol{r}'|}, \tag{4.278}$$

which fulfills an extension of (4.13) to the many-channel case,

$$\left[\mathbf{E}' + \frac{\hbar^2}{2\mu}\Delta\right]\mathbf{G} = \mathbf{1}\,. \tag{4.279}$$

Using this multichannel Green's function we can write the general solution of (4.277) as

$$\Psi = \Psi_{\mathrm{hom}} + \hat{\mathbf{G}}\hat{\mathbf{V}}\Psi, \tag{4.280}$$

where Ψ_{hom} is a solution of the "homogeneous equation".

The wave function (4.280) fulfills boundary conditions corresponding to an incoming plane wave in channel i (and only in channel i) if we define Ψ_{hom} to have the following components:

$$\psi_i(\boldsymbol{r}) = \psi_{\mathrm{in},i}(\boldsymbol{r}) = \mathrm{e}^{\mathrm{i}k_i z}, \quad \psi_j = 0 \text{ for } j \neq i\,. \tag{4.281}$$

The components of the full wave function (4.280) are then

$$\psi_j(\boldsymbol{r}) = \delta_{j,i}\,\mathrm{e}^{\mathrm{i}k_i z} + \int \mathcal{G}_{j,j}(\boldsymbol{r},\boldsymbol{r}')\sum_n V_{j,n}\psi_n(\boldsymbol{r}')\,\mathrm{d}\boldsymbol{r}'\,. \tag{4.282}$$

Since all $\mathcal{G}_{j,j}$ have the form (4.16) asymptotically (with k_j in place of k), the channel wave functions ψ_j have the asymptotic form (4.271), and we obtain a generalization of (4.17) as an implicit equation for the scattering amplitudes,

$$f_{j,i}(\theta,\phi) = -\frac{\mu}{2\pi\hbar^2}\sum_n \int e^{-ik_j\cdot r'} V_{j,n}\psi_n(r')\,dr'. \tag{4.283}$$

Here $\boldsymbol{k}_j$ is the vector of length k_j pointing in the direction of the radius vector $\boldsymbol{r}$.

The right-hand side of (4.283) depends on the label i of the entrance channel, because the channel wave functions ψ_n to be inserted in the integrand are those which solve the Schrödinger equation (or the Lippmann-Schwinger equation) with incoming boundary conditions in the one channel i. Similar to the one-channel case, we can interpret the sum over the integrals in (4.283) as the matrix element of a transition operator $\hat{\mathbf{T}}$ between an initial state $\Psi_{\text{in},i}$, defined by just an incoming plane wave $\psi_{\text{in},i}(\boldsymbol{r}') = \exp(ik_iz')$ in channel i, and a final state $\Psi_{\text{out},j}$, defined by just a plane wave $\psi_{\text{out},j}(\boldsymbol{r}') = \exp(i\boldsymbol{k}_j\cdot\boldsymbol{r}')$ in the exit channel j:

$$T_{\text{out},j;\text{in},i} = \langle\Psi_{\text{out},j}|\hat{\mathbf{T}}|\Psi_{\text{in},i}\rangle = \langle\Psi_{\text{out},j}|\hat{\mathbf{V}}|\Psi\rangle = -\frac{2\pi\hbar^2}{\mu}f_{j,i}(\theta,\phi). \tag{4.284}$$

As in the one-channel case, the Born approximation now consists in replacing the exact channel wave functions ψ_n in (4.282), (4.283) by the "homogeneous solution" $\delta_{n,i}\exp(ik_iz')$, which is equivalent to replacing the transition operator $\hat{\mathbf{T}}$ by the potential $\hat{\mathbf{V}}$. The transition amplitudes in Born approximation are

$$\begin{aligned} f^{\text{B}}_{j,i} = -\frac{\mu}{2\pi\hbar^2}\int e^{-ik_j\cdot r'}\,V_{j,i}\,e^{ik_iz'}\,dr' &= -\frac{\mu}{2\pi\hbar^2}\langle\psi_{\text{out},j}|V_{j,i}|\psi_{\text{in},i}\rangle \\ &= -\frac{\mu}{2\pi\hbar^2}\langle\Psi_{\text{out},j}|\hat{\mathbf{V}}|\Psi_{\text{in},i}\rangle . \end{aligned} \tag{4.285}$$

If we ignore the effects of antisymmetrization etc., we can write out the matrix element in (4.285) explicitly according to (4.270). If we take the interaction $\hat{H}_{\text{W}}$ to consist of just the Coulomb attraction between the outer electron 1 and the target nucleus (charge number Z) and the Coulomb repulsion due to the other electrons $\nu = 2,\ldots N$, we obtain

$$\begin{aligned} f^{\text{B}}_{j,i} &= -\frac{\mu}{2\pi\hbar^2}\langle e^{i\boldsymbol{k}_j\cdot\boldsymbol{r}_1}\psi^{(j)}_{\text{int}}|\left(\sum_{\nu=2}^{N}\frac{e^2}{|\boldsymbol{r}_1-\boldsymbol{r}_\nu|}-\frac{Ze^2}{r_1}\right)|e^{i\boldsymbol{k}_i\cdot\boldsymbol{r}_1}\psi^{(i)}_{\text{int}}\rangle \\ &= -\frac{\mu}{2\pi\hbar^2}\int d\boldsymbol{r}_1\ldots\int d\boldsymbol{r}_N \sum_{m_{S_1},\ldots m_{S_N}} e^{i(\boldsymbol{k}_i-\boldsymbol{k}_j)\cdot\boldsymbol{r}_1} \\ &\quad\times[\psi^{(j)}_{\text{int}}(\boldsymbol{r}_2,\ldots\boldsymbol{r}_N;\ldots)]^*\left(\sum_{\nu=2}^{N}\frac{e^2}{|\boldsymbol{r}_1-\boldsymbol{r}_\nu|}-\frac{Ze^2}{r_1}\right)\psi^{(i)}_{\text{int}}(\boldsymbol{r}_2,\ldots\boldsymbol{r}_N;\ldots). \end{aligned} \tag{4.286}$$

Here $\boldsymbol{k}_i$ is the wave vector of the incoming plane wave in channel i. Because of the orthogonality of the internal states $\psi_{\rm int}^{(i)}$ in the entrance channel and $\psi_{\rm int}^{(j)}$ in the inelastic exit channel, the part of the potential which describes the attraction of electron 1 by the target nucleus, and which depends only on $\boldsymbol{r}_1$, does not contribute to the matrix element in (4.286). In order to calculate the contribution of the other part coming from the electron-electron repulsion, we exploit the fact that $1/|\boldsymbol{r}_1 - \boldsymbol{r}_\nu|$ is the Fourier-transform of $(2/\pi)^{1/2}|\boldsymbol{k}_i - \boldsymbol{k}_j|^{-2}$. The resulting identity,

$$\int \frac{\mathrm{e}^{\mathrm{i}(\boldsymbol{k}_i-\boldsymbol{k}_j)\cdot\boldsymbol{r}_1}}{|\boldsymbol{r}_1 - \boldsymbol{r}_\nu|}\,\mathrm{d}\boldsymbol{r}_1 = \frac{4\pi}{|\boldsymbol{k}_i - \boldsymbol{k}_j|^2}\,\mathrm{e}^{\mathrm{i}(\boldsymbol{k}_i-\boldsymbol{k}_j)\cdot\boldsymbol{r}_\nu}, \tag{4.287}$$

allows us to perform the integration over $\boldsymbol{r}_1$ in (4.286),

$$f_{j,i}^{\rm B} = -\frac{2\mu e^2}{\hbar^2|\boldsymbol{k}_i - \boldsymbol{k}_j|^2}\langle\psi_{\rm int}^{(j)}|\sum_{\nu=2}^{N}\mathrm{e}^{\mathrm{i}(\boldsymbol{k}_i-\boldsymbol{k}_j)\cdot\boldsymbol{r}_\nu}|\psi_{\rm int}^{(i)}\rangle'. \tag{4.288}$$

As in elastic scattering, we use the wave vector $\boldsymbol{q}$ to describe the momentum transfer, which is now the difference of two momentum vectors of different length,

$$\boldsymbol{q} = \boldsymbol{k}_j - \boldsymbol{k}_i, \tag{4.289}$$

and the formula for the inelastic scattering cross section is (in Born approximation)

$$\frac{\mathrm{d}\sigma_{i\to j}^{\rm B}}{\mathrm{d}\Omega} = \frac{k_j}{k_i}|f_{j,i}^{\rm B}|^2 = \frac{4}{q^4 a_1^2}\frac{k_j}{k_i}|\langle\psi_{\rm int}^{(j)}|\sum_{\nu=2}^{N}\mathrm{e}^{-\mathrm{i}\boldsymbol{q}\cdot\boldsymbol{r}_\nu}|\psi_{\rm int}^{(i)}\rangle'|^2. \tag{4.290}$$

The first factor $4/(q^4 a_1^2)$ on the right-hand side is a generalization of the Rutherford differential cross section (4.147) for the scattering of an electron of mass μ by a singly charged nucleus; $a_1 = \hbar^2/(\mu e^2)$ is the corresponding Bohr radius. In contrast to the elastic scattering case however, this Rutherford factor does not diverge in the forward direction, because the wave vector $\boldsymbol{q}$ (4.289) has a length of at least

$$q_{\rm min} = |k_i - k_j|. \tag{4.291}$$

The last factor on the right-hand side of (4.290) contains the information about the structure of the initial and final states of the target atom. In analogy to the oscillator strengths for electromagnetic transitions (Sect. 2.4.6) we can define *generalized oscillator strengths*,

$$F_{j,i}(\boldsymbol{q}) = \frac{2\mu}{\hbar^2}\frac{E_j - E_i}{q^2}|\langle\psi_{\rm int}^{(j)}|\sum_{\nu=2}^{N}\mathrm{e}^{-\mathrm{i}\boldsymbol{q}\cdot\boldsymbol{r}_\nu}|\psi_{\rm int}^{(i)}\rangle'|^2, \tag{4.292}$$

which depend on the length of the momentum transfer vector. These generalized oscillator strengths merge into the ordinary oscillator strengths defined by (2.216) in the hypothetic limit of vanishing momentum transfer.

The theory summarized in equations (4.286)–(4.292), which was originally formulated by Bethe, establishes a connection between the cross sections of inelastic electron scattering and photoabsorption. Conditions for the validity of the *Bethe theory* are the applicability of the Born approximation (4.285) and the negligibility of exchange contributions between the projectile electron and the electrons of the target atom. It is thus most useful for high energies of the incoming and outgoing electron. For a detailed treatise on Bethe theory see [Ino71].

When an electron is scattered by a charged ion, the diagonal potentials asymptotically have the form of a pure Coulomb potential (4.140). Let i label the entrance channel; the asymptotic boundary conditions for the channel wave functions are then [cf. (4.155), (4.271)]

$$\psi_j(\boldsymbol{r}) \stackrel{r\to\infty}{=} \delta_{j,i}\left[\mathrm{e}^{\mathrm{i}[k_i z+\eta_i \ln k_i(r-z)]} + f_{\mathrm{C}}(\theta)\frac{\mathrm{e}^{\mathrm{i}(k_i r-\eta_i \ln 2k_i r)}}{r}\right] + \frac{\mathrm{e}^{\mathrm{i}(k_j r-\eta_j \ln 2k_j r)}}{r} f'_{j,i}(\theta,\phi)\,. \tag{4.293}$$

Since the asymptotic wave number k_j depends on the channel label j via (4.272), the Coulomb parameter (Sommerfeld parameter) (4.142) also depends on the channel label, $\eta_j = -1/(k_j a_Z)$. The additional scattering amplitudes $f'_{j,i}$ in (4.293) are due to the deviations of the full potential from a pure Coulomb potential $-(Ze^2/r)\delta_{j,i}$. These deviations consist of additional contributions to the diagonal potentials ($j = i$) and all coupling potentials ($j \neq i$). They are generally shorter ranged according to the considerations of Sect. 3.3.1. As in the one channel case, the elastic scattering cross section is the absolute square of a sum consisting of the pure Coulomb scattering amplitude f_{C} and the additional scattering amplitude $f'_{i,i}$. The inelastic cross sections are given by the additional scattering amplitude $f'_{j,i}$ alone,

$$\frac{\mathrm{d}\sigma_{i\to i}}{\mathrm{d}\Omega} = \left|f_{\mathrm{C}}(\theta) + f'_{i,i}(\theta,\phi)\right|^2, \quad \frac{\mathrm{d}\sigma_{i\to j}}{\mathrm{d}\Omega} = \frac{k_j}{k_i}|f'_{j,i}(\theta,\phi)|^2, \quad j \neq i\,. \tag{4.294}$$

The additional scattering amplitudes $f'_{j,i}$ obey implicit equations of the form (4.283), except that the plane waves $\exp(-\mathrm{i}\boldsymbol{k}_j.\boldsymbol{r}')$ are now replaced by distorted (Coulomb) waves $\bar{\psi}^*_{\mathrm{C},j}$ [cf. (4.153), (4.154)]. $\bar{\psi}_{\mathrm{C},j}$ describes a Coulomb wave in channel j with an incoming modified plane wave travelling in the direction of the radius vector $\boldsymbol{r}$ plus an incoming spherical wave. With the usual assumptions of the Born approximation (for distorted waves) we obtain an explicit expression for the additional scattering amplitude in the elastic channel ($j = i$) and for the transition amplitude to the inelastic channels ($j \neq i$),

$$f^{\mathrm{DWBA}}_{j,i} = -\frac{\mu}{2\pi\hbar^2}\left\langle\bar{\psi}_{\mathrm{C},j}\left|V_{j,i}\right|\psi_{\mathrm{C}}\right\rangle. \tag{4.295}$$

Here ψ_C is the Coulomb wave (4.141) in the entrance channel i, i.e. with wave number k_i, Coulomb parameter η_i, incoming modified plane wave travelling in the direction of the positive z-axis and outgoing spherical wave.

4.4.2 *Coupled Radial Equations*

The internal states $\psi_{\rm int}^{(i)}$ of the target atom or ion are eigenstates of the total angular momentum of the $N-1$ electrons with total angular momentum quantum number J_i, and we shall assume that they are also eigenstates of the z-component of the operator with quantum number M_i. For a complete specification of all possible elastic and inelastic reactions we assume that the channel label i (or j) characterizes not only the internal state of the target atom with its angular momentum quantum numbers J_i, M_i, but also the spin state χ_+ or χ_- of the projectile electron.

When we expand the channel wave functions in partial waves we can no longer assume that the z-component of the orbital angular momentum is a good quantum number and zero (as in (4.21)), so we expand as follows:

$$\psi_i(\boldsymbol{r}) = \sum_{l=0}^{\infty} \sum_{m=-l}^{+l} \frac{u_{i,l,m}}{r} Y_{l,m}(\theta, \phi) . \tag{4.296}$$

Furthermore, the potentials no longer conserve the orbital angular momentum; their action on the angular coordinates can be expressed as follows:

$$V_{i,j} Y_{l',m'} = \sum_{l,m} Y_{l,m} V_{i,j}(l, m; l', m') . \tag{4.297}$$

The partial waves expansion (4.296) reduces matrix elements of the potentials $V_{i,j}$ to a sum of radial matrix elements of the *"radial potentials"* $V_{i,j}(l, m; l', m')$. The connection between such radial matrix elements and the matrix elements of the associated N-electron wave function is given by (4.270),

$$\begin{aligned} &\left\langle \frac{u_{i,l,m}}{r} Y_{l,m} \middle| V_{i,j} \middle| \frac{u_{j,l',m'}}{r} Y_{l',m'} \right\rangle \\ &= \left\langle \frac{u_{i,l,m}}{r} Y_{l,m}\, \psi_{\rm int}^{(i)} \middle| \hat{H}_{\rm W} \middle| \frac{u_{j,l',m'}}{r} Y_{l',m'}\, \psi_{\rm int}^{(j)} \right\rangle + \cdots \\ &= \left\langle u_{i,l,m} \middle| V_{i,j}(l, m; l', m') \middle| u_{j,l',m'} \right\rangle . \end{aligned} \tag{4.298}$$

If we insert the expansion (4.296) into the coupled channel equations (4.269) we thus obtain the coupled radial equations

$$\begin{aligned} &\left[-\frac{\hbar^2}{2\mu} \frac{\mathrm{d}^2}{\mathrm{d}r^2} + \frac{l(l+1)\hbar^2}{2\mu r^2} \right] u_{i,l,m}(r) \\ &+ \sum_{j,l',m'} V_{i,j}(l, m; l', m')\, u_{j,l',m'}(r) = (E - E_i)\, u_{i,l,m}(r) . \end{aligned} \tag{4.299}$$

How many and which combinations of channel label j and angular momentum quantum number l', m' have to be included in the sum in (4.299) for given i, l and m, depends crucially on the angular momentum quantum numbers J_i, M_i and J_j, M_j of the internal states $\psi_{\text{int}}^{(i)}$ and $\psi_{\text{int}}^{(j)}$, because they determine the action of the potentials on the spin and angular variables. Since the total angular momentum of the N-electron system is a good quantum number, the coupled channel equations (4.299) reduce to blocks belonging to different angular momentum quantum numbers of the whole system. If we start with a (truncated) expansion involving a finite number of internal states $\psi_{\text{int}}^{(j)}$, each such block contains at most a finite number of equations. A further reduction of these blocks may be possible if the N-electron Hamiltonian has further symmetries or good quantum numbers. If e.g. spin-dependent forces can be neglected, then the total orbital angular momentum and the total spin are conserved and only partial waves belonging to the same values of the corresponding quantum numbers couple.

For each block of coupled radial equations there are as many linearly independent vectors U of channel wave functions $u_{i,l,m}$ solving the equations as there are equations in the block. Asymptotically, each radial wave function of a solution is a superposition of two linearly independent solutions of the uncoupled free equation, e.g. of [cf. Table 1.3, (1.151)]

$$u_{i,l}^{\text{s}}(r) \overset{r\to\infty}{=} \sqrt{\frac{2\mu}{\pi\hbar^2 k_i}}\,\sin(k_i r - l\frac{\pi}{2}), \quad u_{i,l}^{\text{c}}(r) \overset{r\to\infty}{=} \sqrt{\frac{2\mu}{\pi\hbar^2 k_i}}\,\cos(k_i r - l\frac{\pi}{2})\,. \tag{4.300}$$

The coefficients of such superpositions can be obtained e.g. by direct numerical solution of the coupled channel equations if the potentials are known. They determine the asymptotic form of the wave function for given initial conditions and hence the observable cross sections.

A frequently used basis of vectors of solutions $U^{(i,l,m)}$ is defined by the following boundary conditions:

$$u_{j,l',m'}^{(i,l,m)}(r) \overset{r\to\infty}{=} \delta_{i,j}\delta_{l,l'}\delta_{m,m'}\,u_{i,l}^{\text{s}}(r) + R_{i,l,m;j,l',m}\,u_{j,l'}^{\text{c}}(r)\,. \tag{4.301}$$

The coefficients in front of the cosine terms define the *reactance matrix* $\mathbf{R} = (R_{i,l,m;j,l',m'})$, which is also known as the *K-matrix*.[4] In the trivial case that the coupled channel equations reduce to a single equation of the form (1.75) or (1.362), the reactance matrix is simply the tangent of the asymptotic phase shift δ due to the potential,

$$R = \tan\delta. \tag{4.302}$$

[4]Not to be confused with the *R-matrix*. This defines a particular method for solving the Schrödinger equation by first constructing bound auxiliary states in an internal region and then matching them to the appropriate scattering wave functions with the help of the *R-matrix* (see e.g. [Bra83, MW91a]).

If the potential is real, this phase shift and its tangent are also real. In the genuine many-channel case, the reactance matrix is a Hermitian matrix as long as the potential $\hat{\mathbf{V}}$ contains no non-Hermitian contributions (absorption).

We obtain an alternate basis of vectors of solutions of the coupled channel equations, $\Phi^{(i,l,m)}$, if we express the radial wave functions as superpositions not of sine and cosine functions as in (4.300), but of outgoing and incoming spherical waves,

$$\phi^{+}_{i,l}(r) = u^{c}_{i,l}(r) + \mathrm{i}u^{s}_{i,l}(r) \stackrel{r\to\infty}{=} \sqrt{\frac{2\mu}{\pi\hbar^2 k_i}}\,\mathrm{e}^{+\mathrm{i}(k_i r - l\pi/2)},$$

$$\phi^{-}_{i,l}(r) = u^{c}_{i,l}(r) - \mathrm{i}u^{s}_{i,l}(r) \stackrel{r\to\infty}{=} \sqrt{\frac{2\mu}{\pi\hbar^2 k_i}}\,\mathrm{e}^{-\mathrm{i}(k_i r - l\pi/2)}; \tag{4.303}$$

$$\phi^{(i,l,m)}_{j,l',m'}(r) \stackrel{r\to\infty}{=} \delta_{i,j}\,\delta_{l,l'}\,\delta_{m,m'}\phi^{-}_{i,l}(r) - S_{i,l,m;j,l',m'}\phi^{+}_{j,l'}(r)\,. \tag{4.304}$$

The asymptotic coefficients of the outgoing components $\phi^{+}_{j,l'}$ define the *scattering matrix* or *S-matrix*: $\mathbf{S} = (S_{i,l,m;j,l',m'})$.

Since both bases of vectors of solutions, $U^{(i,l,m)}$ and $\Phi^{(i,l,m)}$ obeying the boundary conditions (4.301) and (4.304) respectively, span the same space of solutions of the coupled channel equations, there must be a linear transformation which transforms one basis into the other. This transformation is

$$-\mathrm{i}\left(U^{(i,l,m)} + \sum_{j,l',m'} S_{i,l,m;j,l',m'}\,U^{(j,l',m')}\right) = \Phi^{(i,l,m)}\,. \tag{4.305}$$

We can see that (4.305) is correct by looking at the asymptotic behaviour of both sides of the equation in the sine-cosine basis (4.300). The coefficients of the sine terms on both sides form the same matrix $-\mathrm{i}(\mathbf{1}+\mathbf{S})$. Requiring that the coefficients of the cosine terms also be the same leads to

$$-\mathrm{i}(\mathbf{1}+\mathbf{S})\mathbf{R} = \mathbf{1}-\mathbf{S}\,. \tag{4.306}$$

This yields an explicit expression for the *S*-matrix as a function of **R**,

$$\mathbf{S} = (\mathbf{1}+\mathrm{i}\mathbf{R})(\mathbf{1}-\mathrm{i}\mathbf{R})^{-1}\,. \tag{4.307}$$

Leaving effects of absorption aside, the *S*-matrix (4.307) is unitary, because **R** is Hermitian. In the trivial case that the coupled channel equations reduce to a single equation of the form (1.75) or (1.362), the *S*-matrix is simply given by the phase shift δ due to the potential [cf. (4.302)],

$$S = \frac{1+\mathrm{i}\tan\delta}{1-\mathrm{i}\tan\delta} = \mathrm{e}^{2\mathrm{i}\delta}\,. \tag{4.308}$$

At a given energy E the Hermitian Matrix $\mathbf{R}$ can always be diagonalized. The corresponding transformation defines linear combinations of the channels labelled by i, l and m; these linear combinations are called *eigenchannels*. The eigenvalues ϱ of $\mathbf{R}$ are real and can each be written as the tangent of an angle as suggested by (4.302). The corresponding angles are called *eigenphases*. Each eigenvalue ϱ of $\mathbf{R}$ is associated with a vector of solutions of the coupled channel equations in which all radial wave functions are asymptotically proportional to a superposition of the sine and cosine functions (4.300) with the same coefficient, namely $\varrho = \tan\delta$, in front of the cosine term. If the energy dependence of an eigenphase shows a sudden rise by π, then this points to a resonant, almost bound state just as in the one-channel case. Since the S-matrix $\mathbf{S}$ is a function of $\mathbf{R}$, it is diagonal in the same basis in which $\mathbf{R}$ is diagonal and the eigenvalues of S are simply given by the eigenphases: $\exp(2\mathrm{i}\delta)$.

We can establish a relation connecting the S-matrix to observable cross sections by recalling the boundary conditions (4.271) of the channel wave functions for a typical scattering experiment. In a partial-waves expansion (4.296) of the total wave function, we obtain incoming spherical waves only from the plane wave part of the channel wave function in the entrance channel i [cf. (4.30)]. A comparison with the spherical waves in (4.303) shows that the solution of the stationary Schrödinger equation obeying the boundary conditions (4.271) is given as the following superposition of the basis vectors $\Phi^{(i,l,m=0)}$:

$$\Phi = \sum_l (-\pi\hbar)\mathrm{i}^{l-1}\sqrt{\frac{2l+1}{2\mu k_i}}\,\Phi^{(i,l,0)}\,. \tag{4.309}$$

The associated channel wave functions $\phi_j(\boldsymbol{r})$ are corresponding superpositions of the radial wave functions (4.304),

$$\begin{aligned}\phi_j(\boldsymbol{r}) &= \sum_{l',m'} Y_{l',m'}(\theta,\phi)\,\frac{1}{r}\sum_l(-\pi\hbar)\,\mathrm{i}^{l-1}\sqrt{\frac{2l+1}{2\mu k_i}}\,\phi^{(i,l,o)}_{j,l',m'}(r)\\ &\overset{r\to\infty}{=} \delta_{j,i}\,\mathrm{e}^{\mathrm{i}k_i z} + \frac{\mathrm{e}^{\mathrm{i}k_j r}}{r}\sum_{l',m'} Y_{l',m'}(\theta,\phi)\,\mathrm{i}\sum_l \mathrm{i}^{l-l'}\\ &\quad\times\sqrt{\frac{\pi(2l+1)}{k_i k_j}}\,(\delta_{j,i}\delta_{l,l'}\delta_{0,m'} - S_{i,l,0;j,l',m'})\,. \end{aligned} \tag{4.310}$$

The relation connecting the scattering amplitudes defined by (4.271) with the elements of the S-matrix is thus,

$$\begin{aligned}\mathrm{i}\sqrt{k_i k_j}\,f_{j,i}(\theta,\phi) &= \sum_{l',m'} Y_{l',m'}(\theta,\phi)\sum_l \mathrm{i}^{l-l'}\\ &\quad\times\sqrt{\pi(2l+1)}\,(S_{i,l,0;j,l',m'} - \delta_{j,i}\delta_{l,l'}\delta_{0,m'})\,. \end{aligned} \tag{4.311}$$

If the potentials conserve orbital angular momentum, the S-matrix is diagonal in l and m,

$$S_{i,l,0;j,l',m'} = S_{i,l;j,l}\delta_{l,l'}\delta_{0,m'} . \tag{4.312}$$

The expression (4.311) for the scattering amplitude then simplifies to

$$f_{j,i}(\theta,\phi) = \frac{1}{\mathrm{i}\sqrt{k_i k_j}} \sum_l \sqrt{\pi(2l+1)}(S_{i,l;j,l} - \delta_{j,i})Y_{l,0}(\theta), \tag{4.313}$$

and agrees with the result (4.31), (4.34) in the case of elastic scattering.

If the diagonal potentials contain a very-long-ranged Coulomb contribution, the preceding considerations have to be modified as in Sect. 4.1.13. The sine and cosine waves in (4.300) or the spherical waves in (4.303) have to be replaced by the appropriate distorted waves of the pure Coulomb potential. The reactance-matrix **R** now describes the influence of the short-ranged deviations from the pure Coulomb potential. This reactance matrix and its continuation to energies at which some or all channels are closed form the foundation of Seaton's formulation of multichannel quantum defect theory (see Sect. 3.4).

So far in this section we have not discussed the complications introduced by explicitly considering the spin of the electron. How to incorporate the spin of the projectile electron was discussed in Sect. 4.3 for the example of elastic scattering by a target atom with vanishing total angular momentum. In general, a target atom in an internal state $\psi_{\mathrm{int}}^{(i)}$ may have a non-vanishing angular momentum J_i associated with $2J_i + 1$ eigenstates of the z-component of the angular momentum labelled by the quantum numbers $M_i = -J_i, -J_i + 1, \ldots J_i$. An arbitrary pure or mixed spin state of the electron is described by a 2×2 density matrix, as discussed in Sect. 4.3. Correspondingly, an arbitrary pure or mixed state in the quantum numbers M_i of the target atom is described by a $(2J_i + 1) \times (2J_i + 1)$ density matrix. An arbitrary state of polarization of electron and atom (with angular momentum J_i) is thus described by a $[2(2J_i + 1)] \times [2(2J_i + 1)]$ density matrix. The theoretical description of the change induced in the polarization of electron and atom by scattering is then based on a study of the transformations which map the $[2(2J_i + 1)] \times [2(2J_i + 1)]$ density matrices in the entrance channel onto $[2(2J_j + 1)] \times [2(2J_j + 1)]$ density matrices in the respective exit channels. For a detailed discussion see [Bar89].

We shall now, for the time being, explicitly specify the quantum numbers $m_{\mathrm{s}} = \pm 1/2$ for the z-component of the electron spin and M_i for the z-component of the angular momentum of the target atom, so that the channel label i accounts only for the remaining degrees of freedom in $\psi_{\mathrm{int}}^{(i)}$. The general inelastic scattering amplitude is then $f_{j,m_{\mathrm{s}}',M_j;i,m_{\mathrm{s}},M_i}(\theta,\phi)$ for the transition from the entrance channel i to the exit channel j accompanied by a transition of the quantum numbers for the z-components from m_{s}, M_i to m_{s}', M_j. A complete experimental determination of all amplitudes for given channel labels i, j is very difficult in general, because e.g. it is not easy to prepare the target atom in a definite eigenstate of the z-component of its angular momentum. The incomplete information about the states of polarization of the electron and the target atom can be appropriately described using density matrices. The density matrix e.g. for a totally unpolarized electron and a totally unpolarized target atom (with angular momentum J_i) in the entrance channel is simply $1/[2(2J_i + 1)]$ times the unit matrix. If we also forgo measuring the components of the electron spin and the angular momentum of the atom in the exit channel, then the differential cross section for inelastic scattering from channel i to channel j is in this case an incoherent superposition of all contributions in question with a weighting factor $1/[2(2J_i + 1)]$,

$$\frac{\mathrm{d}\sigma_{i\to j}}{\mathrm{d}\Omega} = \frac{1}{2(2J_i+1)} \sum_{m_s=-\frac{1}{2}}^{+\frac{1}{2}} \sum_{M_i=-J_i}^{J_i} \sum_{m_s'=-\frac{1}{2}}^{+\frac{1}{2}} \sum_{M_j=-J_j}^{J_j} \times \frac{k_j}{k_i} |f_{j,m_s',M_j;i,m_s,M_i}(\theta,\phi)|^2 \,. \tag{4.314}$$

This corresponds to averaging over all initial states and summing over all final states compatible with the observed boundary conditions (cf. Sect. 2.4.4, last paragraph, and Sect. 4.3.3).

4.4.3 Threshold Effects

The energy dependence of the cross section (4.274) or (4.275) for inelastic scattering in the vicinity of a channel threshold $E = E_j$ is largely determined by the phase space factor k_j/k_i. The transition amplitude $f_{j,i}$ is given by a matrix element of the form (4.283) and generally assumes a finite value at $E = E_j$. It will be essentially constant in a sufficiently small interval around E_j. An exception occurs when lower partial-waves are absent in the partial waves expansion for symmetry reasons. Let l be the lowest orbital angular momentum quantum number contributing to the integral in (4.283) in a partial-waves expansion of the plane wave $\psi^*_{\mathrm{out},j} = \exp(-\mathrm{i}\boldsymbol{k}_j\cdot\boldsymbol{r}')$. Then the dependence of the integral on the wave number k_j is given by the spherical Bessel function $j_l(k_j r')$ [cf. (4.30)] and is proportional to k_j^l [(A.49) in Appendix A.4]. The absolute square of the transition amplitude is thus proportional to k_j^{2l} and, remembering the phase space factor k_j/k_i, we obtain *Wigner's threshold law for inelastic scattering cross sections*,

$$\sigma_{i\to j}(E) \propto \left(\sqrt{E-E_j}\right)^{2l+1} \,. \tag{4.315}$$

Here l is the lowest orbital angular momentum quantum number observed in the exit channel.

Wigner's threshold law is a consequence of the long-ranged centrifugal potential in the exit channel,

$$V_{j,j}(r) \overset{r\to\infty}{\sim} \frac{\hbar^2}{2\mu}\frac{\gamma}{r^2}, \quad \gamma = l(l+1)\,. \tag{4.316}$$

Whatever happens at short distances, the particle has to penetrate the repulsive potential tail (4.316) in order to be detectable at large distances. If the potential in the exit channel is attractive at small or moderate distances, then the cross section for any process can be expected to be proportional to the probability P_{T} for transmission through the potential barrier formed by the long-ranged tail (4.316) and the shorter-ranged attractive terms in the interaction. If the potential has a WKB region on the near side of the barrier, then the problem of transmission through and reflection by the barrier can be formulated as in Sect. 1.4.2, except that on the near side of the barrier where the potential cannot be assumed to be constant, the plane waves e.g. on the left-hand side of (1.175), are replaced by appropriate WKB wave functions. Explicit expressions for the behaviour of the transmission probability through such a centrifugal barrier at energies $E = \hbar^2k^2/(2\mu)$ near threshold are given in [ME01], see also [FT04], and have the general form

$$P_{\mathrm{T}} \overset{k\to 0}{\propto} k^{2\mu_\gamma}, \quad \mu_\gamma = \sqrt{\gamma + \frac{1}{4}} = l + \frac{1}{2}\,. \tag{4.317}$$

For the special case that the shorter-ranged attractive potential is well represented by a power-law term in the barrier region,

$$V(r) \overset{r\to\infty}{\sim} V_{\gamma,m}(r) = \frac{\hbar^2}{2\mu}\left(\frac{\gamma}{r^2} - \frac{(\beta_m)^{m-2}}{r^m}\right), \quad m > 2, \tag{4.318}$$

the near-threshold behaviour of the transmission probability through the barrier is exactly given by

$$P_{\mathrm{T}} \overset{k\to 0}{\sim} P(m,\gamma)\,(k\beta_m)^{2\mu_\gamma}, \tag{4.319}$$

with the coefficient

$$P(m,\gamma) = \frac{4\pi^2/2^{2\mu_\gamma}}{(m-2)^{2\nu}\mu_\gamma\nu\left[\Gamma(\mu_\gamma)\Gamma(\nu)\right]^2}, \quad \nu = \frac{2\mu_\gamma}{m-2}\,. \tag{4.320}$$

The equations (4.319), (4.320) hold not only for positive integer values of l but for any value of the parameter γ larger than $-1/4$ (corresponding to $l > -1/2$), which is the limit below which the (attractive) inverse-square potential supports an infinite dipole series of bound states, see Sect. 3.1.5. For vanishing or weakly negative values of the strength of the inverse-square potential, $-1/4 < \gamma \leq 0$, there no longer is a barrier to tunnel through, but the potential tail still has a nonclassical quantal region separating the internal WKB region from the region of asymptotic free-particle motion, and the probability for transmission through this quantal region of the potential tail vanishes according to (4.317) near threshold. Thus Wigner's threshold law also applies to s waves, ($\gamma = 0$, $l = 0$), and it can even be formally extended to weakly attractive inverse-square potentials as long as $\gamma > -1/4$. The opening of a channel j at the threshold E_j also affects the energy dependence of inelastic cross sections to other exit channels as well as the elastic scattering cross section. The qualitative behaviour of any observable near a threshold can be understood quite generally using arguments similar to those already applied for near-threshold quantization in Sect. 3.1.2. The calculation of the value of any observable generally involves the full solution of the Schrödinger equation with contributions from all channels. Directly at an inelastic threshold $E = E_j$, the contribution of channel j involves the threshold wave function u_0 which is defined in (3.40) and tends to unity asymptotically. Immediately above E_j, i.e. at energy $E = E_j + \hbar^2(k_j)^2/(2\mu)$, the wave function in channel j is a superposition of the two fundamental solutions which are asymptotically proportional to $\exp(\pm \mathrm{i}k_j r)$ and can, to lowest order in k_j, be written as $u_0 \pm \mathrm{i}k_j u_1$, where u_1 is the threshold solution which behaves asymptotically as r, see (3.40). Immediately below threshold, $E = E_j - \hbar^2(\kappa_j)^2/(2\mu)$, the wave function in channel j is asymptotically proportional to $\exp(-\kappa_j r)$, i.e. to $u_0 - \kappa_j u_1$. The full solution of the Schrödinger equation can thus be expected to contain a contribution proportional to $k_j \propto \sqrt{E - E_j}$ just above E_j and a contribution proportional to $\kappa_j \propto \sqrt{E_j - E}$ just below E_j. The contributions from all other channels with thresholds away from E_j can be expected to be smooth functions of energy around $E \approx E_j$. When calculating, e.g., the integrated elastic scattering cross section, we expect a sudden decline just above the inelastic threshold E_j, because flux is now lost into the newly opened channel j; this is described by a leading term proportional to $\sqrt{E - E_j}$ with a negative coefficient. Just below E_j the energy dependence of the integrated elastic scattering cross section is dominated by a leading term proportional to $\sqrt{E_j - E}$. If the coefficient is also negative, then we observe a conspicuous *cusp* at threshold, as illustrated in Fig. 4.17(a); a positive

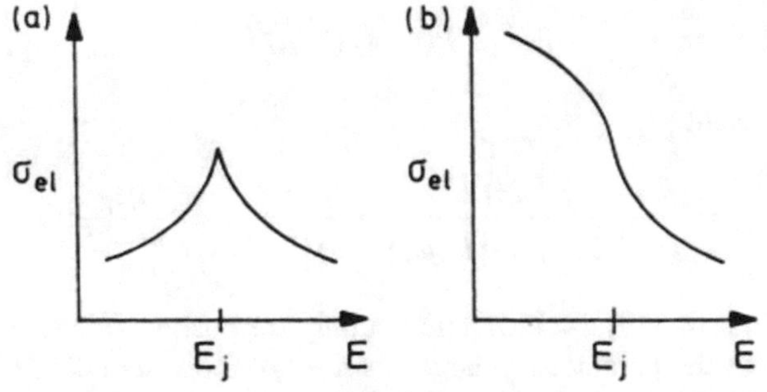

Fig. 4.17 Schematic illustration of singularities in the integrated elastic cross section at an inelastic channel threshold E_j: (**a**) cusp, (**b**) rounded step

coefficient leads to a *rounded step* as in Fig. 4.17(b). In both cases the channel threshold E_j manifests itself as a singularity with infinite gradient in the integrated elastic cross section, provided s-waves are not excluded in the exit channel j on symmetry grounds; otherwise the corresponding contributions are proportional to $(k_j)^{2l+1}$ or to $(\kappa_j)^{2l+1}$, $l \geq 1$, and they are masked by other terms depending smoothly on energy, i.e. on $(k_j)^2$ or $(\kappa_j)^2$.

It may well happen that the internal energy E_i in the entrance channel is larger than the internal energy E_j in the exit channel. This case, which corresponds to an *exothermic reaction* in chemistry, is called *superelastic scattering*. The exit channel j is then already open at the threshold E_i of the entrance channel, and the outgoing electron has an asymptotic kinetic energy which is larger by $E_i - E_j$ than the asymptotic kinetic energy of the incoming electron. At the reaction threshold E_i, the wave number k_i in the entrance channel starts at zero, but the wave number k_j in the exit channel is finite. Unless the corresponding scattering amplitudes vanish, the cross sections (4.274), (4.275) diverge at the reaction threshold for superelastic scattering.

The threshold behaviour of inelastic scattering cross sections is very different when the interaction potentials contain a very-long-ranged Coulomb term. In the matrix element for the scattering amplitude [cf. (4.283)] we now have a Coulomb wave instead of the plane wave in the exit channel. The partial-waves expansion (4.158) shows that the energy dependence of the transition amplitude $f'_{j,i}$ just above the threshold E_j is given by the regular Coulomb functions $F_l(\eta_j, k_j r)$ (divided by $k_j r$). In an attractive Coulomb potential we have according to (1.145), (1.141)

$$\frac{F_l(\eta_j, k_j r)}{k_j r} \stackrel{E \to E_j}{=} \sqrt{\frac{\pi \hbar^2}{2\mu k_j r}} \frac{1}{a_Z \sqrt{\mathcal{R}}} J_{2l+1}\left(\sqrt{\frac{8r}{a_Z}}\right), \tag{4.321}$$

so that $|f'_{j,i}|^2$ is inversely proportional zu k_j just above E_j, regardless of which angular momenta contribute to the partial waves sum. Thus the inelastic cross sections (4.294) tend smoothly to finite values at the respective thresholds, when there is an attractive Coulomb potential in the exit channel.

The differential cross section for elastic scattering in the presence of an attractive Coulomb potential behaves smoothly above an inelastic threshold, in accordance with the smooth behaviour of the inelastic cross sections—remember, the integrated elastic cross section diverges. Below an inelastic threshold however, an attractive Coulomb potential supports whole Rydberg series of Feshbach resonances. Consider the simple case that only the elastic channel i is open below the channel threshold E_j and that the electron-ion interaction can be described by a radially symmetric potential. The phase shift δ_l in each partial wave l is then given by a formula like (3.233),

$$\delta_l = \pi \mu_i - \arctan\left[\frac{|R_{i,j}|^2}{\tan\left[\pi(\nu_j + \mu_j)\right]}\right], \quad \nu_j = \sqrt{\frac{\mathcal{R}}{E_j - E}}. \tag{4.322}$$

Here μ_i, $R_{i,j}$ and μ_j are just the weakly energy-dependent MQDT parameters in the two-channel case (Sect. 3.4.1), which also depend on l. Equation (4.322) describes a Rydberg series of resonant jumps of the phase shift by π (cf. Fig. 3.18). The individual partial-wave amplitudes f_l' [see (4.160)] oscillate between zero and a maximum value of $|f_l'| = (2l + 1)/k_i$ an infinite number of times just below the threshold E_j, and this leads to increasingly narrow oscillations in the differential cross section (4.156) as we approach E_j from below. In practice these oscillations can only be resolved up to a certain energy above which the observed cross section merges into a smooth function which connects to the cross section above the threshold.

4.4.4 An Example

A comprehensive review on electron-atom scattering was written by McCarthy and Weigold in 1991 [MW91a]. Most of the theoretical investigations of inelastic electron-atom scattering have of course been performed for the hydrogen atom. Here the spectrum and the eigenstates of the target atom are known and many matrix elements can be evaluated analytically.

Detailed calculations of cross sections for inelastic electron-hydrogen scattering at comparatively low energies were performed e.g. by Callaway [Cal82, Cal88]. Williams [Wil76, Wil88] performed accurate measurements in the energy region between the first inelastic threshold $(3/4)\mathcal{R} \approx 10.20$ eV and the $n = 3$ threshold at $(8/9)\mathcal{R} \approx 12.09$ eV, see also [SS89]. In this energy region the channels in which the hydrogen-atom electron is excited to the $n = 2$ shell are open, but all higher channels are closed.

The calculations in [Cal82] are based on a close-coupling expansion. Eigenstates of the hydrogen atom up to principal quantum number $n = 3$ were included exactly; higher closed channels were approximated by pseudostates. When spin-dependent effects are neglected, the coupled channel equations reduce to blocks labelled by a good total orbital angular momentum quantum number L and a good total spin S. Different variational methods [Cal78, Nes80] were used to solve the coupled channel equations.

Figure 4.18 shows integrated inelastic scattering cross sections for energies just above the first inelastic threshold. The upper curve shows the excitation of the hydrogen atom into the $2p$ state, the lower curve shows excitation into the $2s$ state. The dots are the experimental values and the solid lines show the results of the calculations of [Cal82], which have been smoothed a little in order to simulate the finite experimental resolution. This gives the theoretical curves a finite gradient at threshold (10.20 eV), where it really should be infinite. A bit above the inelastic threshold both curves show a distinct maximum suggesting a resonance. The calculations for $L = 1$ and $S = 0$ actually do yield a resonant eigenphase in this region. Fitting its energy dependence to an analytic form similar to (1.234) gives a resonance position $E_R \approx 10.2\,\text{eV}$ and a width of $\Gamma \approx 0.02\,\text{eV}$.

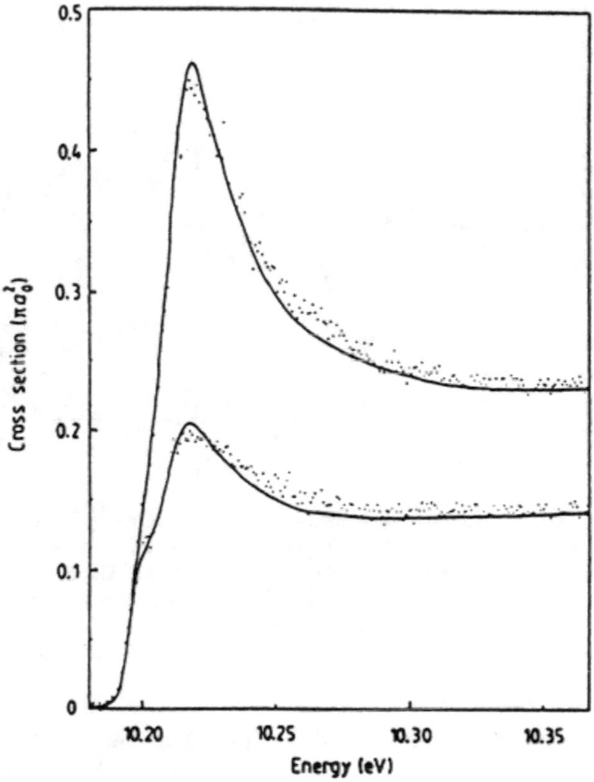

Fig. 4.18 Integrated cross sections for inelastic electron scattering by hydrogen just above the inelastic threshold (10.20 eV). The *upper curve* shows the $1s \to 2p$ excitation, the *lower curve* shows the $1s \to 2s$ excitation. The *dots* are the experimental data of Williams and the *solid lines* are the theoretical results from [Cal82], which have been smoothed a little in order to simulate finite experimental resolution (From [Wil88])

Figure 4.19 shows the integrated inelastic cross sections of Fig. 4.18 at somewhat higher energies just below the $n = 3$ threshold. Again the upper curve shows the $1s \to 2p$ transition while the lower curve shows the $1s \to 2s$ transition. The solid curves again show the (smoothed) results of the calculations [Cal82], and the dots are the data from [Wil88]. Just below the $n = 3$ threshold the barely closed $n = 3$ channels support a number of bound states which couple to and can decay into the open $n = 1$ and $n = 2$ channels and hence appear as Feshbach resonances. The positions and widths of these resonances are derived from the jumps in the eigenphases which are fitted to the analytic form (1.234) [Cal82]. The irregularly oscillating structure in the cross sections is obviously due to these resonances, the positions of which are shown as vertical lines above the abscissa. Similar structures can also be seen in differential inelastic cross sections as measured by Warner et al. [WR90].

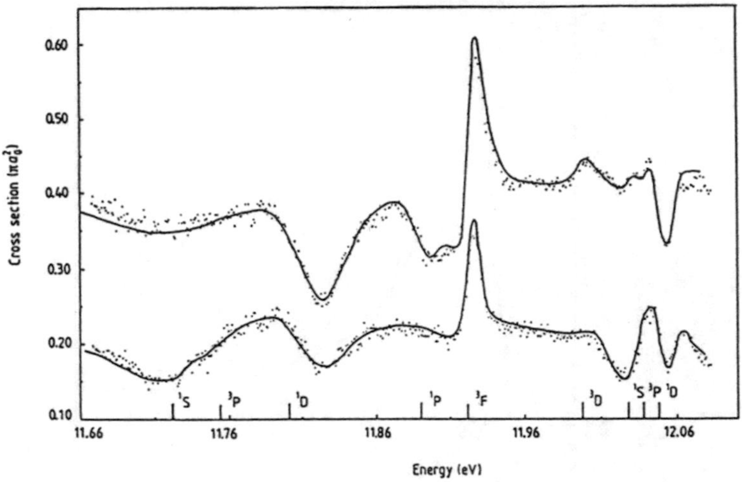

Fig. 4.19 Integrated cross sections for inelastic electron scattering by hydrogen just below the threshold for $n = 3$ excitations of the hydrogen atom (12.09 eV). The *upper curve* shows the $1s \rightarrow 2p$ excitation, the *lower curve* shows the $1s \rightarrow 2s$ excitation. The *dots* are the experimental data of Williams and the *solid lines* are the theoretical results from [Cal82], which have been smoothed a little in order to simulate finite experimental resolution. The *vertical lines* above the abscissa show the positions of a number of Feshbach resonances (From [Wil88])

Further experimental advances made high precision studies of Feshbach resonances in H^- possible [SZ95, BK96]. The hydrogen atom has the unique property that its excited energy levels include degenerate states of different parity. This means that the internal states need not be parity eigenstates, and the leading asymptotic terms in the diagonal channel potentials can contain inverse-square terms which are attractive and strong enough to support a dipole series of Feshbach resonances, see Sect. 3.1.6.. Details on recent studies of H^-—and of other negative ions—can be found in the comprehensive review by Andersen [And04].

4.5 Exit Channels with Two Unbound Electrons

The considerations of Sects. 4.1–4.4 are based on the assumption that only one of the spatial coordinates can become very large, namely the displacement vector of the incoming or scattered electron. The many-electron wave function vanishes in regions of coordinate space where the coordinates of two or more electrons are large. In these circumstances the asymptotic boundary conditions of the wave functions are easy to formulate, and an *ab initio* description of the possible elastic and inelastic scattering processes can be justified in a straightforward way, e.g. via the close-coupling ansatz (3.163) in connection with Feshbach's projection formalism.

The formulation of reaction theory becomes much more difficult if states with two or more outgoing electrons become important. This is the case if the energy of the projectile electron is sufficient to ionize the target atom or detach an electron from the target ion. This section briefly sketches and highlights some aspects of the theoretical description of such (e,2e) *reactions* with exactly two outgoing electrons in the exit channel. For a more detailed description of (e,2e) reactions see e.g. [Rud68] or the articles by Byron and Joachain [BJ89] and McCarthy and Weigold [MW91b].

4.5.1 General Formulation

For a better understanding of the general structure of the wave functions in an (e,2e) reaction we shall first replace the electrons by distinguishable particles without electric charge. The complications due to the indistinguishability of the electrons and the very-long-ranged Coulomb interactions will be discussed in Sect. 4.5.2.

The dynamics of two outgoing particles is described by continuum wave functions depending on both displacement vectors $\boldsymbol{r}_1$ and $\boldsymbol{r}_2$, i.e. on six coordinates altogether. The remaining degrees of freedom are described by bound internal wave functions $\phi_{\rm int}^{(n)}$ depending on the remaining displacement vectors $\boldsymbol{r}_3, \dots \boldsymbol{r}_N$ and all spin coordinates. They may be taken to be eigenstates of a corresponding internal Hamiltonian $\hat{H}_{\rm int}$ with the respective eigenvalues E_n. Each such eigenstate defines a *break-up channel n*.

The description of inelastic scattering in Sect. 4.4 was limited to scattering channels with one outer electron. This made it easy to reduce the equations of motion to the coupled channel equations (4.269) for the orbital wave functions of the outer electron. If both scattering channels and break-up channels are important, it is not so easy to formulate a set of coupled channel equations, because the channel wave functions are functions in different spaces: either functions of one displacement vector (scattering channels) or functions of two displacement vectors (break-up channels). We can achieve a consistent description by working in the space of wave functions of the whole N-particle system. A channel wave function is always associated to a corresponding internal wave function depending on the respective remaining degrees of freedom—$\phi_{\rm int}^{(n)}(\boldsymbol{r}_3, \dots \boldsymbol{r}_N; m_{s_1} \dots m_{s_N})$ in the break-up channels and $\psi_{\rm int}^{(j)}(\boldsymbol{r}_2, \dots \boldsymbol{r}_N; m_{s_1} \dots m_{s_N})$ in the scattering channels.

In order to study the asymptotic structure of the wave function, we again use the method of Green's functions. First we write the N-particle Hamiltonian $\hat{H}$ as a sum of the kinetic energies $\hat{t}_1 = -(\hbar^2/2\mu)\Delta_{r_1}$ and $\hat{t}_2 = -(\hbar^2/2\mu)\Delta_{r_2}$ of particle 1 and particle 2 respectively, plus an internal Hamiltonian $\hat{H}_{\rm int}$ and a residual term $\hat{V}_{\rm R}$ containing all contributions not included in the previous terms,

$$\hat{H} = \hat{t}_1 + \hat{t}_2 + \hat{H}_{\rm int} + \hat{V}_{\rm R}, \tag{4.323}$$

and we make the N-particle Schrödinger equation look like an inhomogeneous equation,

$$(E - \hat{t}_1 - \hat{t}_2 - \hat{H}_{\rm int})\Psi = \hat{V}_{\rm R}\Psi\,. \tag{4.324}$$

The Green's function $\hat{\mathbf{G}}$, which is now also an operator in the space of internal wave functions $\phi_{\rm int}^{(n)}(\boldsymbol{r}_3, \dots \boldsymbol{r}_N; \dots)$, is defined as a solution of the following equation,

$$(E - \hat{t}_1 - \hat{t}_2 - \hat{H}_{\rm int})\hat{\mathbf{G}} = \delta(\boldsymbol{r}_1 - \boldsymbol{r}'_1)\,\delta(\boldsymbol{r}_2 - \boldsymbol{r}'_2)\,\mathbf{1}\,. \tag{4.325}$$

The bold **1** on the right-hand side of (4.325) stands for the unit operator in the space of internal wave functions $\phi_{\rm int}^{(n)}$.

A formal solution of the "inhomogeneous equation" (4.324) is

$$\Psi = \hat{\mathbf{G}}\,\hat{V}_{\rm R}\,\Psi\,. \tag{4.326}$$

In contrast to the Lippmann-Schwinger equations for elastic and inelastic scattering, (4.15) and (4.280), the right-hand side of (4.326) contains no solution of the "homogeneous equation" ($\hat{V}_{\rm R} \equiv 0$) determined by the incoming boundary conditions. The reason is that the initial state contains only one free (incoming) particle, while all others are bound, and hence it is not a solution of the homogeneous equation which now describes two free particles.

We can use the integral equation (4.326) to derive the asymptotic form of the wave function in the break-up channels. Equation (4.325) can be fulfilled by a Green's function of the following structure,

$$\hat{\mathbf{G}} = \sum_n \mathcal{G}_n(\boldsymbol{r}_1, \boldsymbol{r}_2; \boldsymbol{r}'_1, \boldsymbol{r}'_2) |\,\phi_{\rm int}^{(n)}\rangle\langle\phi_{\rm int}^{(n)}\,|\,, \tag{4.327}$$

where the sum should cover a complete set of internal states $\phi_{\rm int}^{(n)}$ (and not only bound states). In the break-up channels, $E > E_n$, the dependence of the Green's function on the displacement vectors is given by the factors $\mathcal{G}_n(\boldsymbol{r}_1, \boldsymbol{r}_2; \boldsymbol{r}'_1, \boldsymbol{r}'_2)$ which fulfill the following equations:

$$(E - E_n - \hat{t}_1 - \hat{t}_2)\,\mathcal{G}_n(\boldsymbol{r}_1, \boldsymbol{r}_2; \boldsymbol{r}'_1, \boldsymbol{r}'_2) = \delta(\boldsymbol{r}_1 - \boldsymbol{r}'_1)\,\delta(\boldsymbol{r}_2 - \boldsymbol{r}'_2)\,. \tag{4.328}$$

$E - E_n$ is the asymptotic kinetic energy available to the two outgoing particles in the open break-up channel n.

For a more economical notation we collect the two displacement vectors $\boldsymbol{r}_1$ and $\boldsymbol{r}_2$ into one six-component displacement vector,

$$\boldsymbol{R} \equiv (\boldsymbol{r}_1, \boldsymbol{r}_2)\,. \tag{4.329}$$

With the abbreviations

$$E - E_n = \frac{\hbar^2}{2\mu} K_n^2\,,$$

$$\Delta_6 = \Delta_{r_1} + \Delta_{r_2} = \frac{\partial^2}{\partial x_1^2} + \frac{\partial^2}{\partial y_1^2} + \frac{\partial^2}{\partial z_1^2} + \frac{\partial^2}{\partial x_2^2} + \frac{\partial^2}{\partial y_2^2} + \frac{\partial^2}{\partial z_2^2}\,, \tag{4.330}$$

(4.328) becomes the equation defining the Green's function for the Helmholtz equation in six dimensions (except for a factor $2\mu/\hbar^2$),

$$(K_n^2 + \Delta_6)\mathcal{G}_n(\boldsymbol{R}, \boldsymbol{R}') = \frac{2\mu}{\hbar^2}\,\delta(\boldsymbol{R}, \boldsymbol{R}')\,. \tag{4.331}$$

The Green's function which fulfills (4.331) and which is appropriate for two outgoing particles in the break-up channel n, is (see Problem 4.9)

$$\mathcal{G}_n(\boldsymbol{R}, \boldsymbol{R}') = -\frac{\mu K_n^2}{8\pi^2\hbar^2} \frac{\mathrm{i}H_2^{(1)}(K_n|\boldsymbol{R} - \boldsymbol{R}'|)}{|\boldsymbol{R} - \boldsymbol{R}'|^2}\,. \tag{4.332}$$

Here $H_\nu^{(1)}$ is the Hankel function of order ν (see Appendix A.4). For small values of $|\boldsymbol{R} - \boldsymbol{R}'|$ we obtain (see (A.46) in Appendix A.4)

$$\mathcal{G}_n(\boldsymbol{R}, \boldsymbol{R}') = -\frac{\mu}{2\pi^3\hbar^2} \frac{1}{|\boldsymbol{R} - \boldsymbol{R}'|^4}\,, \quad |\boldsymbol{R} - \boldsymbol{R}'| \to 0\,; \tag{4.333}$$

for large values of $|\boldsymbol{R} - \boldsymbol{R}'|$ (see (A.45) in Appendix A.4),

$$\mathcal{G}_n(\boldsymbol{R}, \boldsymbol{R}') = \sqrt{\mathrm{i}}\,\frac{\mu}{\hbar^2}\,K_n^{3/2} \frac{\mathrm{e}^{\mathrm{i}K_n|\boldsymbol{R} - \boldsymbol{R}'|}}{(2\pi|\boldsymbol{R} - \boldsymbol{R}'|)^{5/2}}\,, \quad |\boldsymbol{R} - \boldsymbol{R}'| \to \infty\,. \tag{4.334}$$

For $R \gg R'$ we can expand in R'/R, as we did in Sect. 4.1.1 [cf. (4.16)],

$$\mathcal{G}_n(\boldsymbol{R}, \boldsymbol{R}') = \sqrt{\mathrm{i}}\,\frac{\mu}{\hbar^2}\,K_n^{3/2} \frac{\mathrm{e}^{\mathrm{i}K_n R}}{(2\pi R)^{5/2}} \left[\mathrm{e}^{-\mathrm{i}\boldsymbol{K_R}\cdot\boldsymbol{R}'} + O\left(\frac{R'}{R}\right)\right]. \tag{4.335}$$

Here $\boldsymbol{K_R}$ is the six-component wave vector of length K_n pointing in the direction of the (six-component) displacement vector $\boldsymbol{R}$.

We obtain the asymptotic form of the wave function by inserting the Green's function given by (4.327) and (4.335) into (4.326),

$$\Psi \overset{R\to\infty}{=} \sum_n \sqrt{\mathrm{i}}\,\frac{\mu}{\hbar^2}\,K_n^{3/2} \frac{\mathrm{e}^{\mathrm{i}K_n R}}{(2\pi R)^{5/2}} |\phi_{\mathrm{int}}^{(n)}\rangle\langle\phi_{\mathrm{int}}^{(n)}\psi_n^{(\boldsymbol{K_R})}|\hat{V}_{\mathrm{R}}|\Psi\rangle + \dots\,. \tag{4.336}$$

Here $\psi_n^{(\boldsymbol{K_R})}(\boldsymbol{R}') = \exp(\mathrm{i}\boldsymbol{K_R}\cdot\boldsymbol{R}')$ is a plane wave with a six-component wave vector $\boldsymbol{K_R}$ for the free motion of the two particles 1 and 2, which together have a kinetic energy $E - E_n$. The sum in (4.336) should be understood as a sum over all genuine break-up channels, for which $E > E_n$ and $\phi_{\mathrm{int}}^{(n)}$ is a bound state in the internal coordinates. Channels with $E < E_n$ do not contribute asymptotically ($R \to \infty$), and unbound internal states, which correspond to a break-up into more than two unbound particles, are hinted at by the dots on the right-hand side.

If we divide the six-component wave vector $\boldsymbol{K_R}$ into two three-component parts, $\boldsymbol{k}_1$ for the first three components and $\boldsymbol{k}_2$ for the last three components, we have

$$\psi_n^{(\boldsymbol{K_R})}(\boldsymbol{R}') = \mathrm{e}^{\mathrm{i}\boldsymbol{k}_1\cdot\boldsymbol{r}_1'}\,\mathrm{e}^{\mathrm{i}\boldsymbol{k}_2\cdot\boldsymbol{r}_2'}\,. \tag{4.337}$$

Thus $\psi_n^{(\boldsymbol{K_R})}$ is just a product of two plane waves for the independent free motion of the two outgoing particles 1 and 2.

Since $\boldsymbol{K_R}$ points in the same direction as the six-component displacement vector $\boldsymbol{R}$ in six-dimensional space, there is a common proportionality constant β, such that

$$\boldsymbol{k}_1 = \beta\boldsymbol{r}_1, \quad \boldsymbol{k}_2 = \beta\boldsymbol{r}_2\,. \tag{4.338}$$

Equation (4.338) says that the wave vector $\boldsymbol{k}_1$ points in the same direction as the displacement vector $\boldsymbol{r}_1$ in three-dimensional space and that $\boldsymbol{k}_2$ points in the same direction as $\boldsymbol{r}_2$. This amounts to four real conditions, because a direction in three-dimensional space is fixed by two angles. However, a direction in six-dimensional space is fixed by five angles. The remaining condition contained in the fact that the six-component vectors $\boldsymbol{K_R}$ and $\boldsymbol{R}$ are parallel, is

$$\frac{k_1}{k_2} = \frac{r_1}{r_2}\,. \tag{4.339}$$

The length K_n of the vector $\boldsymbol{K_R}$ is fixed by the kinetic energy available in the exit channel,

$$\frac{\hbar^2 K_n^2}{2\mu} = \frac{\hbar^2}{2\mu}(k_1^2 + k_2^2) = E - E_n\,. \tag{4.340}$$

The distribution of this kinetic energy among the two outgoing particles 1 and 2 is uniquely determined by the ratio (4.339).

The asymptotic form of the wave function Ψ in a break-up channel n as given by (4.336) is thus a product of the internal eigenstate $\phi_{\mathrm{int}}^{(n)}$ and an outgoing spherical wave in six-dimensional coordinate space, multiplied by a phase space factor $K_n^{3/2}$ and a *break-up amplitude* f_n, which depends on the direction of the (six-component) displacement vector $\boldsymbol{R}$,

$$\Psi \overset{R\to\infty}{=} \sum_n |\phi_{\mathrm{int}}^{(n)}\rangle\,\frac{\mathrm{e}^{\mathrm{i}K_n R}}{(2\pi R)^{5/2}}\,K_n^{3/2} f_n(\Omega_1, \Omega_2, \alpha) + \dots\,. \tag{4.341}$$

Here Ω_1 is the solid angle defining the direction of the vector $\boldsymbol{r}_1$ (in threedimensional space), Ω_2 is the solid angle for the direction of $\boldsymbol{r}_2$, and α is the so-called *hyper*angle; its tangent is just the ratio (4.339) which determines the distribution of the asymptotic kinetic energy among the two outgoing particles,

$$\tan\alpha = \frac{r_1}{r_2} . \tag{4.342}$$

The length R of the six-component displacement vector is often called the *hyper*radius. The hyperradius and the five angles $\Omega_1, \Omega_2, \alpha$ are the spherical coordinates of $\boldsymbol{R}$ in six-dimensional coordinate space. These six dimensional spherical coordinates are called *hyperspherical coordinates*.

Comparing (4.341) with (4.336) shows that the break-up amplitude f_n is given by a matrix element containing plane waves for free particle motion in the bra, in complete analogy to the case of elastic (4.17) or inelastic scattering (4.283), (4.284),

$$f_n(\Omega_1, \Omega_2, \alpha) = \sqrt{\mathrm{i}}\,\frac{\mu}{\hbar^2}\,\langle \phi_{\mathrm{int}}^{(n)} \psi_n^{(\boldsymbol{K_R})} | \hat{V}_{\mathrm{R}} | \Psi \rangle . \tag{4.343}$$

The operator $\hat{V}_{\mathrm{R}}$ in the matrix element in (4.343) contains all contributions to the Hamiltonian which are not already contained in the kinetic energy of the two particles 1 and 2 or in the internal Hamiltonian for the remaining degrees of freedom. The wave function Ψ in the ket is a solution of the full stationary Schrodinger equation which has the form (4.341) in the asymptotic part of six-dimensional coordinate space for finite values of $\tan\alpha$.

At this point the normalization of the total wave function Ψ and the physical dimensions of the break-up amplitude f_n are not yet determined. The reason is, that the Lippmann-Schwinger equation (4.326) has the form of a homogeneous integral equation, so that neither the equation itself nor its asymptotic form (4.336) fix the normalization of the wave function.[5]

We can fix the normalization of the total wave function by referring to the boundary conditions in the entrance channel. In the asymptotic region $R \to \infty$, the hyperangle $\alpha = \pi/2$, $\tan\alpha = r_1/r_2 = \infty$, just covers that part of configuration space in which only particle 1 is very far away. In this region the asymptotic behaviour of the wave function is thus determined by the boundary conditions in the entrance channel i and all elastic and inelastic scattering channels,

$$\Psi = \mathrm{e}^{\mathrm{i}k_i z_1} |\psi_{\mathrm{int}}^{(i)}\rangle + \sum_j \frac{\mathrm{e}^{\mathrm{i}k_j r_1}}{r_1} f_{j,i}(\Omega_1) |\psi_{\mathrm{int}}^{(j)}\rangle , \quad \frac{r_1}{r_2} \to \infty . \tag{4.344}$$

[5]One Lippmann-Schwinger equation is not sufficient to uniquely determine the total wave function in the presence of break-up channels. A detailed discussion of this problem can be found in [Glo83].

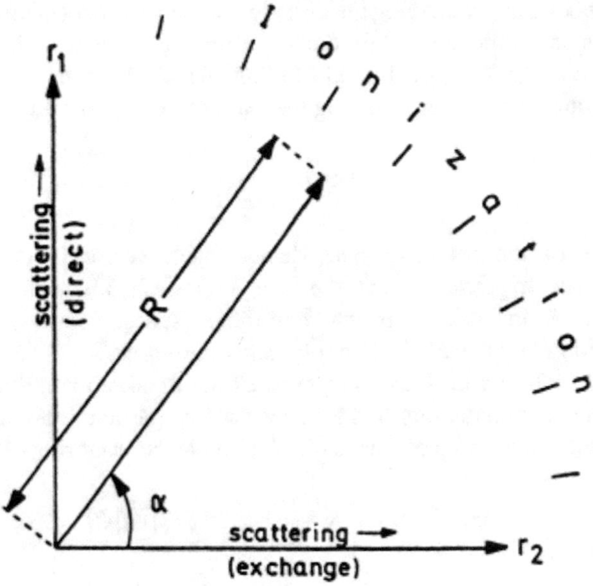

Fig. 4.20 Various asymptotic regions in six-dimensional coordinate space represented by the hyperradius R and the hyperangle α

The wave functions $\psi_{\rm int}^{(j)}$ are the internal wave functions in the scattering channels and are eigenfunctions of a corresponding internal Hamiltonian for the particles 2 to N. For $\alpha = 0$, $\tan\alpha = r_1/r_2 = 0$, the asymptotic region $R \to \infty$ covers that part of configuration space in which only particle 2 is very far away. This corresponds to elastic or inelastic scattering in conjunction with an exchange of the particles 1 and 2. Asymptotically the wave function is

$$\Psi = \sum_j \frac{{\rm e}^{{\rm i}k_j r_2}}{r_2}\, g_{j,i}(\Omega_2)|\psi_{\rm int}^{(j)}\rangle\,, \qquad \frac{r_2}{r_1} \to \infty\,. \tag{4.345}$$

Here $\psi_{\rm int}^{(j)}$ are the same internal wave functions as in (4.344), but they now describe particles 1, 3,…N. The various asymptotic regions are illustrated in Fig. 4.20 with the help of hyperradius and hyperangle.

In connection with the normalization of the wave functions we can now discuss the physical dimensions of the quantities appearing in (4.343). The total wave function Ψ in the ket has the same dimension as a dimensionless plane wave multiplied by a bound wave function, normalized to unity, for $(N-1)$ particles in three-dimensional coordinate space, i.e. $[\text{length}]^{-(3/2)(N-1)}$. On the other hand, the wave function in the bra has the dimension of a bound wave function, normalized to unity, (namely $\phi_{\rm int}^{(n)}$) for only $(N-2)$ particles, multiplied by two dimensionless plane waves (4.337); thus the dimension of the wave function in the bra is $[\text{length}]^{-(3/2)(N-2)}$. Since the integration over all $3N$ spatial coordinates contributes a dimension $[\text{length}]^{3N}$, the dimension of the matrix element in (4.343) is energy × length$^{9/2}$, and the dimension of the break-up amplitude f_n is length$^{5/2}$.

The definitions of cross sections are based on a generalization of the current density (4.4) to particle currents in six-dimensional coordinate space,

$$j_6 = \frac{\hbar}{2\mathrm{i}\mu}[\psi^*(\boldsymbol{R})\nabla_6\psi(\boldsymbol{R}) - \psi(\boldsymbol{R})\nabla_6\psi^*(\boldsymbol{R})]\,. \tag{4.346}$$

The subscript "6" refers to the six-dimensional space, as in (4.330). For a wave function of the form (4.341) with a spatial part

$$\psi(\boldsymbol{R}) \overset{R\to\infty}{=} \frac{\mathrm{e}^{\mathrm{i}K_nR}}{(2\pi R)^{5/2}} K_n^{3/2} f_n(\Omega_1, \Omega_2, \alpha), \tag{4.347}$$

we obtain an outgoing current density in six-dimensional space in complete analogy to the three-dimensional case (4.5),

$$j_6 = \frac{\hbar K_n^4}{\mu} \frac{|f_n(\Omega_\mathrm{h})|^2}{(2\pi R)^5} \frac{\boldsymbol{R}}{R} + O\left(\frac{1}{R^6}\right)\,. \tag{4.348}$$

We have abbreviated the solid angle $(\Omega_1, \Omega_2, \alpha)$ in six-dimensional space by Ω_h. The corresponding angular element is (see Problem 4.11)

$$\begin{aligned} \mathrm{d}\Omega_\mathrm{h} &= \sin^2\alpha \cos^2\alpha \,\mathrm{d}\alpha\, \mathrm{d}\Omega_1\, \mathrm{d}\Omega_2 \\ &= \sin^2\alpha \cos^2\alpha \,\mathrm{d}\alpha\ \sin\theta_1\, \mathrm{d}\theta_1\, \mathrm{d}\phi_1\ \sin\theta_2 \mathrm{d}\theta_2\, \mathrm{d}\phi_2\,. \end{aligned} \tag{4.349}$$

The quantity

$$\mathrm{d}^3\sigma_{i\to n} = \frac{|j_6| R^5\, \mathrm{d}\Omega_\mathrm{h}}{\hbar k_i/\mu} \tag{4.350}$$

is the particle flux into the solid angle $\mathrm{d}\Omega_\mathrm{h}$, divided by the incoming current density $\hbar k_i/\mu$ (of one particle) in the entrance channel i. Outgoing particle flux in the solid angle $\mathrm{d}\Omega_\mathrm{h}$ implies that particle 1 is travelling in a direction contained in $\mathrm{d}\Omega_1$, that particle 2 is travelling in a direction contained in $\mathrm{d}\Omega_2$, and that the tangent of the ratio k_1/k_2 lies between α and $\alpha + \mathrm{d}\alpha$. It is customary to express this ratio in terms of the asymptotic kinetic energy $T_1 = \hbar^2 k_1^2/(2\mu)$ of particle 1 or $T_2 = \hbar^2 k_2^2/(2\mu)$ of particle 2. These kinetic energies are related to the hyperangle α via

$$k_1 = K_n \sin\alpha\,, \quad k_2 = K_n \cos\alpha\,. \tag{4.351}$$

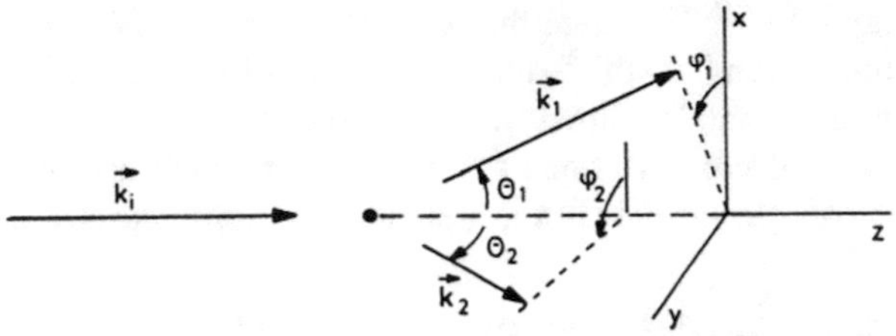

Fig. 4.21 Schematic illustration of an (e,2e) reaction. $\boldsymbol{k}_i$ is the wave vector of the incoming particle parallel to the z-axis, $\boldsymbol{k}_1$ is the wave vector of the outgoing particle 1 travelling away in the direction Ω_1, and $\boldsymbol{k}_2$ is the wave vector of the outgoing particle 2 travelling away in the direction Ω_2 with the kinetic energy $T_2 = \hbar^2 k_2^2/(2\mu)$

With

$$K_n^4 \sin^2\alpha \cos^2\alpha \,|\mathrm{d}\alpha| = k_1^2 k_2^2 \,|\mathrm{d}\alpha| = k_1 k_2^2 \,|\mathrm{d}k_2| = \frac{k_1 k_2}{2} \,|\mathrm{d}(k_2^2)| = k_1 k_2 \frac{\mu}{\hbar^2} \mathrm{d}T_2 \tag{4.352}$$

(4.350) becomes the *triple differential cross section* in its usual form,

$$\frac{\mathrm{d}^3\sigma_{i-n}}{\mathrm{d}\Omega_1 \mathrm{d}\Omega_2 \mathrm{d}T_2} = \frac{k_1 k_2}{k_i} \frac{\mu}{\hbar^2} \frac{|f_n(\Omega_1, \Omega_2, T_2)|^2}{(2\pi)^5} . \tag{4.353}$$

This is the number of reactions, normalized to the incoming current density, in which particle 1 travels away in the direction Ω_1 and particle 2 travels away with kinetic energy T_2 in the direction Ω_2, while the remaining particles stay behind in the bound eigenstate $\phi_{\mathrm{int}}^{(n)}$ of the internal Hamiltonian (see also Fig. 4.21). Since the square of the break-up amplitude has the physical dimension of a length to the fifth power (see discussion shortly after (4.345) above), the triple differential cross section (4.353) has the dimension of an area divided by an energy.

4.5.2 Application to Electrons

In order to apply the formulation of the preceding section to (e,2e) reactions, we have to take into consideration firstly the indistinguishability of the electrons and secondly their electric charge which is the origin of the very-long-ranged Coulomb interaction.

If the target atom (or ion) is a one-electron atom, then there are only two electrons whose indistinguishability must be considered. If there are more than two electrons altogether, we must also consider effects of exchange between the two continuum electrons in the break-up channels and the bound electrons left behind. Here we shall assume that these latter effects are accounted for by appropriate modifications in the definition of the interaction $\hat{V}_R$, similar to the discussion in Sect. 3.3.1 [cf. (3.176)], and we shall only treat the exchange of the two continuum electrons.

The formulation in Sect. 4.5.1 with the asymptotic equations (4.341), (4.344), (4.345) assumes that electron 1 is the incoming electron in channel i. We could just as easily have chosen electron 2 as the incoming electron. If we call the corresponding solution of the full Schrödinger equation Ψ', then the asymptotic formulae for Ψ' are obviously

$$\Psi' \stackrel{R\to\infty}{=} \sum_n |\phi_{\mathrm{int}}^{(n)}\rangle \frac{\mathrm{e}^{\mathrm{i}K_n R}}{(2\pi R)^{5/2}} K_n^{3/2}\, g_n(\Omega_1, \Omega_2, \alpha) + \dots, \tag{4.354}$$

$$\Psi' = \mathrm{e}^{\mathrm{i}k_i z_2}|\psi_{\mathrm{int}}^{(i)}\rangle + \sum_j \frac{\mathrm{e}^{\mathrm{i}k_j r_2}}{r_2} f_{j,i}(\Omega_2)|\psi_{\mathrm{int}}^{(j)}\rangle, \quad \frac{r_2}{r_1} \to \infty, \tag{4.355}$$

$$\Psi' = \sum_j \frac{\mathrm{e}^{\mathrm{i}k_j r_1}}{r_1} g_{j,i}(\Omega_1)|\psi_{\mathrm{int}}^{(j)}\rangle, \quad \frac{r_1}{r_2} \to \infty. \tag{4.356}$$

The reciprocity in the direct scattering amplitudes $f_{j,i}$ and the exchange amplitudes $g_{j,i}$ is already built into (4.355) and (4.356). The break-up amplitude g_n in (4.354) is given in analogy to (4.343) by

$$g_n(\Omega_1, \Omega_2, \alpha) = \sqrt{\mathrm{i}}\, \frac{\mu}{\hbar^2} \langle \phi_{\mathrm{int}}^{(n)} \psi_n^{(\boldsymbol{K}_R)} | \hat{V}_R | \Psi' \rangle . \tag{4.357}$$

As can be seen by permuting the labels 1 and 2, it is related to the break-up amplitude f_n by

$$\begin{aligned} g_n(\Omega_1, \Omega_2, \alpha) &= f_n(\Omega_2, \Omega_1, \frac{\pi}{2} - \alpha) \quad \text{or} \\ g_n(\Omega_1, \Omega_2, T_2) &= f_n(\Omega_2, \Omega_1, T_1) . \end{aligned} \tag{4.358}$$

The reciprocity relation (4.358) is known as the *Peterkop theorem*.

How the indistinguishability of the electrons 1 and 2 affects the triple differential cross section for final states in the break-up channel n, depends on whether the spins of the two electrons in the exit channel are coupled to a total spin zero (singlet) or 1 (triplet) [cf. Sect. 2.2.4, (2.81), (2.82)]. In the singlet case, the total wave function must be symmetric with respect to an interchange of the spatial coordinates $\boldsymbol{r}_1$ and $\boldsymbol{r}_2$ alone, because the spin part of the wave function (2.82) is antisymmetric. We obtain an appropriate solution of the full Schrödinger equation in this case, by adding the solution Ψ defined by (4.341), (4.344), (4.345) to the solution Ψ' defined by (4.354)–(4.356),

$$\Psi_{S=0} = \frac{1}{\sqrt{2}}(\Psi + \Psi') . \tag{4.359}$$

In the formula (4.353) for the triple differential cross section, this amounts to replacing the break-amplitude f_n by the sum of f_n and g_n (divided by $\sqrt{2}$). We also have to add the cross sections for Ω_1, Ω_2, T_2 and Ω_2, Ω_1, T_1 because we cannot distinguish the two electrons in the exit channel. With the help of the Peterkop theorem (4.358) we thus obtain the following result for singlet coupling of the spins of the outgoing electrons:

$$\left(\frac{\mathrm{d}^3\sigma_{i\to n}}{\mathrm{d}\Omega_1\mathrm{d}\Omega_2\mathrm{d}T_2}\right)_{S=0} = \frac{k_1k_2}{k_i}\frac{\mu}{\hbar^2}\frac{|f_n^{\mathrm{s}}(\Omega_1,\Omega_2,T_2)|^2}{(2\pi)^5},$$
$$f_n^{\mathrm{s}} = f_n + g_n\,. \tag{4.360}$$

The analogous result for triplet coupling of the spins of the outgoing electrons is

$$\left(\frac{\mathrm{d}^3\sigma_{i\to n}}{\mathrm{d}\Omega_1\mathrm{d}\Omega_2\mathrm{d}T_2}\right)_{S=1} = \frac{k_1k_2}{k_i}\frac{\mu}{\hbar^2}\frac{|f_n^{\mathrm{t}}(\Omega_1,\Omega_2,T_2)|^2}{(2\pi)^5},$$
$$f_n^{\mathrm{t}} = f_n - g_n\,. \tag{4.361}$$

The spin coupling of the outgoing electrons is not measured in general, so the observed triple differential cross section is the average of the expressions (4.360), (4.361), weighted with the multiplicity $2S+1$,

$$\frac{\mathrm{d}^3\sigma_{i\to n}}{\mathrm{d}\Omega_1 d\Omega_2\mathrm{d}T_2} = \frac{k_1k_2}{k_i}\frac{\mu}{\hbar^2}\frac{\frac{1}{4}|f_n^{\mathrm{s}}|^2+\frac{3}{4}|f_n^{\mathrm{t}}|^2}{(2\pi)^5}\,. \tag{4.362}$$

The consideration of the very-long-ranged Coulomb interactions poses more serious problems. In order to formulate an equation like (4.343) we must know the asymptotic form of the two-electron wave function (in the presence of Coulomb forces), firstly to determine the "free waves" in the bra and secondly to fix the solution Ψ of the full Schrödinger equation in the ket. The crucial difficulty is, that the continuum electrons are never really free, not even at very large distances, because they feel not only the Coulomb interaction due to the ion left behind (if it doesn't happen to be a neutral atom), but also their mutual long-ranged Coulomb repulsion.

An obvious guess for extending the formula (4.343) to charged electrons in the break-up channel consists in replacing the plane waves in the bra by Coulomb waves $\bar{\psi}_{\mathrm{C},r_1}$ and $\bar{\psi}_{\mathrm{C},r_2}$ in the field of the residual ion. $\bar{\psi}_{\mathrm{C},r_1}$ and $\bar{\psi}_{\mathrm{C},r_2}$ are the distorted waves (4.154) introduced in Sect. 4.1.13; the associated wave vector has the length k_1 or k_2 and points in the direction of the radius vector $\boldsymbol{r}_1$ or $\boldsymbol{r}_2$ respectively. The fact that the outgoing electrons do not travel independently, not even asymptotically, can be incorporated in the form of a phase ϕ. The expression for the break-up amplitude then still has the form (4.343), but the "free wave" (4.337) in the bra is replaced by

$$\bar{\psi}_{\mathrm{C}}^{(\boldsymbol{K_R})}(\boldsymbol{R}') = \bar{\psi}_{\mathrm{C},r_1}(\boldsymbol{r}_1')\,\bar{\psi}_{\mathrm{C},r_2}(\boldsymbol{r}_2')\,\mathrm{e}^{\mathrm{i}\phi}\,. \tag{4.363}$$

For a naked residual ion (no electrons) we are dealing with a pure three-particle Coulomb problem. In this case the wave function (4.363) actually is a solution asymptotically if we take ϕ to be the phase by which a Coulomb wave for the relative motion of the two electrons differs from a plane wave (with the same asymptotic wave number) [BB89]. For large separations of the two outgoing electrons we have

$$\phi = -\eta' \ln\left(kr' + \boldsymbol{k}\cdot\boldsymbol{r}'\right),$$

$$\boldsymbol{k} = \frac{1}{2}(\boldsymbol{k}_1 - \boldsymbol{k}_2), \quad \boldsymbol{r}' = \boldsymbol{r}'_1 - \boldsymbol{r}'_2, \quad \eta' = \frac{\mu' e^2}{\hbar^2 k}. \tag{4.364}$$

The Coulomb parameter η' here is the one for the repulsive electron-electron interaction (μ' is the reduced mass of the two electrons).

The wave function (4.363) solves the Schrödinger equation for two electrons in the field of a naked nucleus asymptotically, i.e. for large separations of the two electrons from the nucleus and from each other, but it becomes inaccurate for small separations of the two electrons, because their correlations are insufficiently accounted for by the phase factor $e^{i\phi}$ alone. Improvements have been engineered into the wave function, e.g. by Berakdar and collaborators [Ber96, BO97] with some success, but it remains a fact, that a globally accurate wave function for the three-body Coulomb problem is not yet available. A detailed discussion of the mathematics of the three- (and more-) particle Coulomb problem can be found in the book by Faddeev and Merkuriev [FM93].

The Coulomb waves (4.363) represent approximate solutions of a Schrödinger equation for two electrons in the field of a charged ion. If we base the derivation of the expression for the break-up amplitude on an "inhomogeneous differential equation" with an appropriate Green's function for the associated "homogeneous equation" as for uncharged particles in Sect. 4.5.1, then the potential in the "inhomogeneous term" should only contain those interactions which are not already included in the "homogeneous equation". If we include the effects of the very-long-ranged Coulomb interactions between the two outgoing electrons and the residual ion by replacing the free wave (4.337) in the formula (4.343) by the two-electron Coulomb wave (4.363), then we must at the same time leave the associated Coulomb potentials out of the residual potential $\hat{V}_R$.

Apart from the problem of finding the correct free waves for the bra in (4.343) and (4.357), we also need the exact wave functions Ψ and Ψ' for the respective ket. These are of course not available in general. We obtain an approximate formula in the spirit of the Born approximation (with Coulomb waves), if we replace the exact wave functions in the ket by Coulomb waves in the entrance channel. The break-up amplitude (4.343) thus becomes

$$f_n^{\mathrm{DWBA}}(\Omega_1, \Omega_2, \alpha) = \sqrt{\mathrm{i}}\,\frac{\mu}{\hbar^2}\,\langle \phi_{\mathrm{int}}^{(n)} \bar{\psi}_{\mathrm{C}}^{(\boldsymbol{K_R})} | \hat{V}_{\mathrm{R}} | \psi_{\mathrm{int}}^{(i)} \psi_{\mathrm{C}}(\boldsymbol{r}'_1) \rangle, \tag{4.365}$$

where ψ_{C} is the Coulomb wave (4.141) with incoming part travelling in the direction of the z-axis and wave number k_i. The Born approximation works best when the energy of the incoming electron is large. If we focus our attention on final states in which one electron has a large energy while the other electron has a much smaller energy, then exchange effects become unimportant and we can identify the fast electron with the incoming electron. Going one step further and replacing the Coulomb waves of the fast electron in bra and ket by the corresponding plane waves leads to the following customary form [Rud68, BJ89] of the break-up amplitude in Born approximation:

$$f_n^{\mathrm{B}}(\Omega_1, \Omega_2, \alpha) = \sqrt{\mathrm{i}}\,\frac{\mu}{\hbar^2}\,\langle \phi_{\mathrm{int}}^{(n)} \mathrm{e}^{\mathrm{i}\boldsymbol{k}_1\cdot\boldsymbol{r}'_1} \bar{\psi}_{\mathrm{C},r_2}(r'_2)\,|\hat{V}_{\mathrm{R}}|\psi_{\mathrm{int}}^{(i)} \mathrm{e}^{\mathrm{i}k_i z'_1}\rangle\,. \tag{4.366}$$

According to the considerations in the preceding paragraph, the residual potential $\hat{V}_{\mathrm{R}}$ in (4.366) no longer contains the Coulomb interaction between the slow electron 2 and the residual ion, but it does contain the Coulomb interaction between the fast electron and the residual ion as well as the Coulomb repulsion of the two outgoing electrons. For an (e,2e) reaction on a one-electron atom (or ion) the residual ion has no electrons at all and the residual potential to be inserted in (4.366) is simply

$$V_{\mathrm{R}}(\boldsymbol{r}'_1, \boldsymbol{r}'_2) = -\frac{Ze^2}{r'_1} + \frac{\mathrm{e}^2}{|\boldsymbol{r}'_1 - \boldsymbol{r}'_2|}\,. \tag{4.367}$$

This applies for the “post” form of the (distorted wave) Born approximation, where the residual interaction is that part of the full Hamiltonian that is not diagonalized in the bra in (4.366). It can be advantageous to work with the “prior” form of the DWBA, where the residual interaction refers to the incoming wave function in the ket.

4.5.3 Example

The interest in cross sections for (e,2e) reactions has been continuously strong for many years. Special attention has been given to the simplest such reaction,

$$\mathrm{e}^- + \mathrm{H} \rightarrow \mathrm{H}^+ + \mathrm{e}^- + \mathrm{e}^-, \tag{4.368}$$

for which experimental data have been available for some time [EK85, EJ86, SE87, CJ04]. As the residual ion H^+ has no internal degrees of freedom, there is precisely one break-up channel in this reaction and the associated internal energy is zero. Figure 4.22 shows the triple differential cross section for the reaction (4.368) as a function of the angle θ_2 of the slow electron. The other variables were fixed as follows: asymptotic kinetic energy of the incoming electron, $E = 150$ eV; kinetic energy of the slow electron after collision, $T_2 = 3$ eV; $\boldsymbol{k}_i$, $\boldsymbol{k}_1$ and $\boldsymbol{k}_2$ coplanar. The different parts of the figure correspond to different scattering angles of the fast electron, namely 4°, 10° and 16°. Due to the different magnitudes of the energies

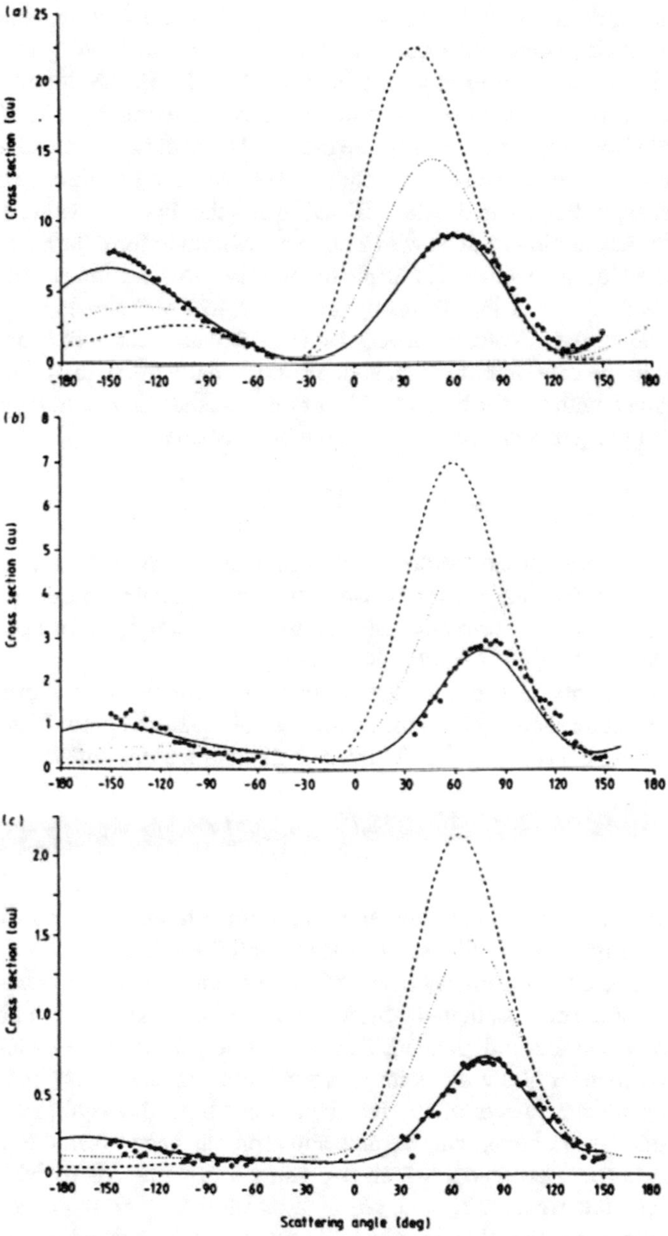

Fig. 4.22 Triple differential cross section (4.353) for the reaction (4.368) in asymmetric coplanar geometry as function of θ_2 for projectile energy $E_{\rm inc} = 150$ eV, $T_2 = 3$ eV and **(a)** $\theta_1 = 4°$, **(b)** $\theta_1 = 10°$, **(c)** $\theta_1 = 16°$. The experimental points are from [EK85] and from further measurements by Ehrhardt et al. The *dotted lines* show the results of the Born approximation (4.366). The *solid lines* were calculated using a formula similar to (4.365) with the correct asymptotic form (4.363) for the free three-particle Coulomb wave. They are normalized to the experimental data at one point in each panel. The *dashed lines* show the results of the same calculation for positron collisions (4.369) (From [BB89])

of the outgoing electrons and the plane geometry of the three wave vectors, such a choice of reaction parameters is called *asymmetric coplanar* [LM84, BJ85].

In addition to the measured points in Fig. 4.22 [EK85], the dotted lines show the calculated cross sections obtained in the Born approximation (4.366), (4.367). Although the Born approximation can be expected to be quite a good approximation at such high energies, there is still a considerable deviation from the experimental results. Brauner, Briggs and Klar [BB89] were the first to evaluate the more sophisticated expression (4.365) with a correct asymptotic form (4.363) for the free three-particle Coulomb wave. The triple differential cross section calculated in this way is shown as a solid line in each part of Fig. 4.22 and agrees very well with the experimental data. Note, however, that the calculated curve in each panel was normalized to the experimental data at one point. Finally the dashed lines show the results obtained with the formula (4.365) for the case that the incoming particle and the faster outgoing particle is not an electron but a positron:

$$\mathrm{e}^+ + \mathrm{H} \rightarrow \mathrm{H}^+ + \mathrm{e}^- + \mathrm{e}^+ \,. \tag{4.369}$$

The difference between the results for electron and positron collisions emphasizes the influence of the interaction between the two outgoing particles, which is repulsive in (4.368) and attractive in (4.369). In the simple Born approximation the cross sections for (4.368) and (4.369) are equal.

The two maxima in Fig. 4.22 are characteristic for the asymmetric coplanar geometry. It can be shown within the framework of the Born approximation [BJ89], that maxima are expected in the direction of the momentum transfer vector of the fast electron,

$$\boldsymbol{q} = \boldsymbol{k}_1 - \boldsymbol{k}_i, \tag{4.370}$$

and in the direction of $-\boldsymbol{q}$. Note that the length of the momentum transfer vector is small if the energy loss of the fast electron is small (Problem 4.12).

If we assume axial symmetry of the whole reaction around the z-axis, i.e. if we ignore polarization effects, then the triple differential cross section at a given impact energy depends on four independent variables, namely θ_1, θ_2, $\phi_1 - \phi_2$ and T_2 or T_1. Different geometries allow different approximations in the theory and illuminate different dynamical aspects of the reaction. Apart from the asymmetric coplanar geometry discussed above, considerable attention has been given e.g. to the *non-coplanar symmetric geometry,* which was studied in particular by McCarthy and collaborators. Here we have $T_1 = T_2$, $\theta_1 = \theta_2$ and $\phi_1 - \phi_2 \neq 0,\ \pi$. In the framework of the *impulse approximation,* in which the electron to be ejected is treated almost as a free electron, the triple differential cross section in non-coplanar symmetric geometry can be related to the wave function of the ejected electron before the collision [MW76, MW88].

The calculations of [BB89] reproduce the angular dependence of the ionization cross section quite well (Fig. 4.22), but they do not predict absolute cross sections. In fact, the evaluation of absolute cross sections for the reaction (4.368) has proved to be a very difficult problem over the years. The integrated or total ionization cross section,

$$\sigma_{e,2e}(E) = \int d\Omega_1 \int d\Omega_2 \int_0^E \mathrm{d}T_2 \frac{\mathrm{d}^3\sigma}{\mathrm{d}\Omega_1 \mathrm{d}\Omega_2 \mathrm{d}T_2}, \tag{4.371}$$

was measured accurately as a function of energy by Shah et al. in 1987 [SE87], and many theoretical groups have since tried to reproduce these data. The first calculation able to reproduce the absolute values and the shape of the cross section (4.371) over an energy range extending from comparatively small energies up to high energies was published by Bray and Stelbovics in 1993 [BS93]. In their method the Lippmann-Schwinger equation is solved in momentum space in the spirit of the close-coupling expansion described in Sect. 3.3.1, and the judicious choice of basis states representing the target leads to convergent results, in contrast to some other close-coupling techniques; for this reason the authors call their method the *convergent close-coupling* (CCC) method.

The performance of the CCC method in reproducing the total ionization cross section (4.371) is illustrated in Fig. 4.23. The open circles are the experimental results from [SE87] and the solid line is the calculated cross section from [BS93]. The calculation reproduces the experimental results well all the way from a bit above threshold (at 13.6 eV) to high energies where the Born approximation (4.366) works well. The fact that reproducing the shape of this curve has been no trivial matter is demonstrated by comparison with the less successful results of other quite sophisticated efforts. The long dashed line in Fig. 4.23 is from an "intermediate

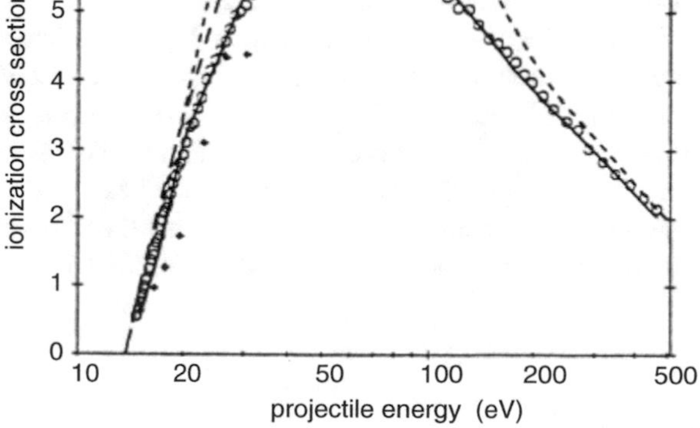

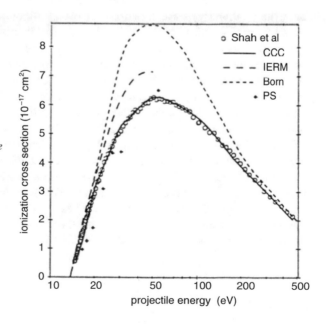

Fig. 4.23 Total ionization cross sections for electron impact on hydrogen. The *open circles* are the experimental results from [SE87] and the *solid line* is the cross section calculated via the CCC (convergent close-coupling) method [BS93]. The *short dashed line* shows the result of the Born approximation (4.366), the *long dashed line* is from the "intermediate energy R-matrix" (IERM) calculation of Scholz et al. [SW90], and the *asterisks* show the pseudo-state calculation of Callaway and Oza [CO79] (From [BS93])

energy R-matrix" (IERM) calculation by Scholz et al. [SW90], and the asterisks were obtained by Callaway and Oza [CO79] who calculated excitation probabilities of the target hydrogen atom using a pseudo-state expansion and extracted the ionization probabilities from the continuum components of the pseudo states. The short dashed line in Fig. 4.23 shows the result of the Born approximation (4.366), which becomes accurate only for energies above a few hundred eV.

The complexity of the two-electron problem in three-dimensional coordinate space has encouraged investigations of simplifying models of two electron atoms. On such model is the *s-wave model*, in which both electrons are restricted to spherical states. The coordinate space for this model is spanned by two variables, viz. the radial distances r_1 and r_2 of the electrons from the nucleus, and the potential energy is,

$$V(r_1, r_2) = -\frac{Ze^2}{r_1} - \frac{Ze^2}{r_2} + \frac{e^2}{r_>} . \tag{4.372}$$

The reduction of variables from vectors in three-dimensional space to one dimensional variables r_1, r_2 means that physical cross sections are reduced to dimensionless probabilities. In a related but not entirely equivalent picture developed by Temkin and Poet [Tem62, Poe78], the three-dimensional picture is retained, but the electron-electron interaction is truncated so as to act only for the s-wave components of the one-electron wave functions, corresponding again to the potential energy (4.372). The ionization probabilities in the s-wave model were calculated by Ihra et al. [ID95] by solving the time dependent Schrödinger equation for wave packets with a small energy spread; with this technique it is not necessary to know the (stationary) wave functions for two continuum electrons. The resulting ionization probabilities are shown in Fig. 4.24 together with the experimental ionization cross sections of [SE87]; the spin averaged probabilities of the s-wave model (solid line) are normalized to reproduce the experimental data at the maximum. Considering how hard it is, for other approximate theories to reproduce the energy dependence of the total ionization cross section (cf. Fig. 4.23), the agreement between the ionization probabilities predicted in the s-wave model and the data in Fig. 4.24 is remarkable. Since angular correlations are completely absent in the s-wave model, the good agreement in Fig. 4.24 shows, that the net effects of such angular correlations in the total ionization cross section must be negligible over a wide range of energies. Note that the ionization cross section calculated in the three-dimensional model based on the potential (4.372) contains a factor proportional to the inverse projectile energy, which describes the diminishing contribution of the s-wave to the incoming plane wave, so that the experimental energy dependence of the ionization cross section is not well reproduced in that picture.

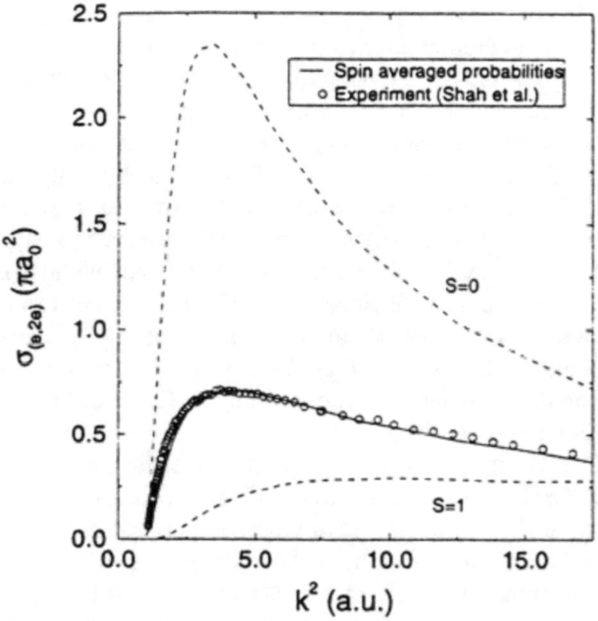

Fig. 4.24 Ionization probabilities for electron impact on hydrogen in the s-wave model. The *dashed lines* show the results for singlet and triplet symmetry and the *solid line* is their weighted average. The *open circles* are again the experimental ionization cross sections from [SE87]. The *solid line* is normalized to have the same height at maximum as the data (From [ID95])

4.5.4 *Threshold Behaviour of Ionization Cross Sections*

For total energies just above the break-up threshold E_n, both outgoing particles in a break-up process must necessarily have small energies and wave numbers, $k_1 \to 0$, $k_2 \to 0$. For short-ranged interactions the "free wave" $\psi_n^{(K_R)}(\boldsymbol{R}')$ in the break-up amplitude (4.343) is given by (4.337) and tends to a constant in this limit. The same is true for the break-up amplitude, unless the $l = 0$ components of the plane waves in (4.337) give vanishing contributions to the matrix element in (4.343), or the matrix element vanishes due to some other symmetry property. The energy dependence of the cross section (4.353) near threshold is thus generally dominated by the factors k_1 and k_2, which are both proportional to K_n according to (4.351), so the differential cross section (4.353) depends linearly on the excess energy $E - E_n$ (4.330) in the limit of small excess energies. The integrated total break-up cross section σ_n, defined in analogy to (4.371) acquires a further factor proportional to $E - E_n$ via the integration over T_2 from zero to its maximum value (which is $E - E_n$), so the energy dependence of the total break-up cross section is generally given by,

$$\sigma_n \propto (E - E_n)^2, \quad E \to E_n, \quad E > E_n, \tag{4.373}$$

as long as the forces on the outgoing particles are of shorter range. Such situations are not so common in atomic physics (electron detachment from a negative ion by a neutral projectile would be an example), but they are important in nuclear physics (e.g. neutron induced ejection of a neutron from a nucleus).

The situation is more complicated for electron impact ionization, because the "free wave" (4.337) does not contain the effect of the very-long-ranged Coulomb interaction. It is instructive to look at what could be expected, if the correlation of the two continuum electrons were neglected, and the "free wave" (4.337) were replaced by a product of two Coulomb waves, as in (4.363) but without the correlating factor $e^{i\phi}$. The low-energy behaviour ($k \to 0$) of the radial Coulomb functions in an attractive Coulomb potential can be deduced from (4.321) or from (A.78) in Appendix A.5 and is seen to be proportional to $|\eta|^{-1/2} \propto \sqrt{k}$ regardless of the value of the angular momentum quantum number. This means that $F_l(\eta, kr)/(kr)$ which enters into the partial-waves expansion of a free Coulomb wave in place of the spherical Bessel functions in the expansion of the plane wave, is proportional to $1/\sqrt{k}$ for all l. The break-up amplitude (4.343) is now proportional to $1/\sqrt{k_1 k_2}$ for small k_1, k_2, so the differential ionization cross section (4.362) becomes independent of energy near the ionization threshold. After integrating over the energy of one of the outgoing electrons this leads to the statement, that the total ionization cross section depends linearly on the excess energy near threshold, if (!) the correlations between the outgoing electrons are neglected.

How these correlations affect the threshold behaviour of ionization cross sections has been a topic of interest and controversy for more than sixty years. A pioneering study by Wannier from 1953 [Wan53] is still the valid reference today. Wannier derived a threshold law for ionization by studying the volume of classical phase space available to the two outgoing electrons. That this is reasonable can be understood when considering that the classical limit for Coulombic systems is at total energy zero, which is just the ionization threshold in a system consisting of a projectile electron and a one-electron target atom (cf. Sects. 4.1.12, 5.3.4(b)). Wannier's derivation is based on the recognition that, due to the electron-electron repulsion, the two electrons can only both escape exactly at threshold if they move away from the nucleus in opposite directions with equal velocities which tend to zero with increasing separation. For small positive energies a small volume of classical phase space opens to the ionization process, and carefully analyzing how this happens leads to the following dependence of the total ionization cross section on the excess energy E above the ionization threshold, now at $E = 0$:

$$\sigma_{e,2e}(E) \propto E^{\nu_W}, \quad \nu_W = \frac{1}{4}\left(\sqrt{\frac{100Z-9}{4Z-1}} - 1\right). \tag{4.374}$$

This is *Wannier's threshold law.* The Wannier exponent ν_W depends only on the charge Z of the residual nucleus (or ion). Its value is $1.12689\ldots$ for $Z = 1$, it is $1.05589\ldots$ for $Z = 2$, and it approaches unity for $Z \to \infty$. This is consistent with the result expected when neglecting correlations between the outgoing electrons, an approximation which should become better and better with increasing Z.

The experimental and theoretical investigation of the energy region near the two-electron threshold has long been a field of continuing and intense activity, see e.g. [GL93, BS03] and references given there. Extensions were formulated to account for difference between singlet and triplet coupling of the two electron spins [KS76, GL93] and for ionization by positron impact [Kla81]. Wannier's classical theory was challenged frequently (see [Tem91] and references given there), but it is still generally accepted as appropriate sufficiently close to threshold. Various generalizations extended the range of energies above threshold, where the ionization cross section can be fitted to a simple analytical form, both for electron induced [Fea93] and positron induced [IM97a] ionization. The break-up threshold of atoms with more than two electrons was studied in particular by Kuchiev and Ostrovsky [KO98].

One widely studied simplification of the full two-electron problem is the collinear model, in which both electrons are restricted to lie on opposite sides on a straight line through the nucleus. The coordinate space for this model is spanned by two variables, viz. the respective distances r_1 and r_2 of the electrons from the nucleus, and the potential energy is,

$$V(r_1, r_2) = -\frac{Ze^2}{r_1} - \frac{Ze^2}{r_2} + \frac{e^2}{r_1 + r_2}. \tag{4.375}$$

Classical ionization probabilities were calculated within this model by Rost [Ros94], simply by initiating classical trajectories corresponding to an incoming projectile electron and a bound target electron oscillating between the nucleus and an outer classical turning point, and counting those trajectories which asymptotically (i.e. after long times) describe two outgoing electrons. The resulting ionization probabilities for electron impact ionization of hydrogen are shown in Fig. 4.25 (solid line) together with experimental data from [MC68]. The dashed line shows the proportionality to $E^{1.127\ldots}$ expected from Wannier's threshold law (4.374). The solid line was normalized to the data at one point (5.84 eV). Figure 4.25 illustrates two points. Firstly, the threshold behaviour (4.374) is reproduced accurately for small energies, but the range where this formula is relevant is quite small, and experimental verification or falsification of Wannier's law is difficult, because its deviation from a linear behaviour is not very pronounced. [This difference is even less pronounced for nuclear charges larger than one, but it is more pronounced in positron induced ionization [IM97a].] Secondly, the collinear classical model reproduces the energy dependence of the experimental (!) data well for energies up to several eV above the ionization threshold. This indicates that the physics determining the ionization cross section is already contained in the collinear configuration, and it shows that classical dynamics determines the energy dependence of the cross section well beyond the regime where Wannier's law (4.374) is applicable.

The convincing results in Fig. 4.25 may conceal the fact that the relation between classical mechanics and quantum mechanics for Coulomb systems near the break-up threshold is enriched with unexpected subtleties. If for example we consider

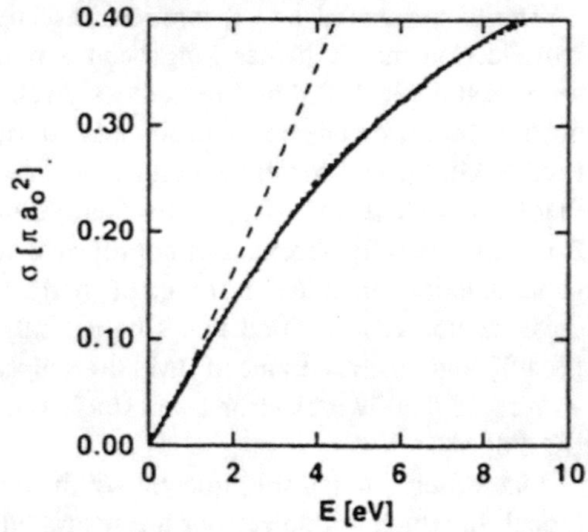

Fig. 4.25 Total ionization cross sections for electron impact on hydrogen in the near-threshold region. The *circles* are the experimental results from [MC68] and the *solid line* is the classical ionization probability calculated within the collinear model (4.375). It was normalized to the data at one point (5.84 eV). The *dashed line* shows the proportionality to $E^{1.127}$ expected from Wannier's threshold law (From [Ros94])

the unphysical case of a continuously varying nuclear charge Z smaller than one, then (4.374) shows that the Wannier exponent tends to infinity as $Z \to 1/4$.[6] Indeed, for $Z = 1/4$ two classical electrons at equal distances on opposite sides of the nucleus feel no force at all, because the attraction by the nucleus is exactly cancelled by the repulsion due to the other electron. Calculations by Ihra et al. predicted an exponential damping of the ionization cross section by a factor proportional to $\exp(-\text{const.}/E^{\nu})$ in this case, but the power ν and the constant involved are different in the classical and quantum calculations [IM97b, CI98]. A further interesting example is the s-wave model defined by the potential energy (4.372) in Sect. 4.5.3, where classical ionization is strictly forbidden in a finite energy interval above threshold [HD93]. A semiclassical treatment based on Wannier's picture predicts an exponential damping of the quantum ionization probability in this case [MI97, CI00].

Problems

4.1

a) Verify the identity

$$\lim_{a\to\infty} a \int_{-1}^{1} (1+x)f(x)\mathrm{e}^{\mathrm{i}a(1-x)}\mathrm{d}x = 2\mathrm{i}f(1).$$

[6]The unphysical case $Z = 1/4$ is however equivalent to a situation in which two particles of charge $-4Z$ move in the field of a central particle of charge $+Z$, which could be realized physically, at least in principle.

b) When we use the stationary scattering wave function (4.3) to calculate the particle flux $\oint \boldsymbol{j}\cdot \mathrm{d}\boldsymbol{s}$ through the surface of an asymptotically large sphere, we obtain a contribution I_{out} as in (4.9) and a contribution I_{interf} coming from interference terms between the incoming plane wave and the outgoing spherical wave. Use the identity a) to show that

$$I_{\text{interf}} = \frac{\hbar}{\mu} 2\pi \mathrm{i}[f(\theta = 0) - f^*(\theta = 0)],$$

which leads to the optical theorem (4.10).

4.2 Show that the free-particle Green's function in three-dimensional coordinate space,

$$\mathcal{G}(\boldsymbol{r},\boldsymbol{r}') = -\frac{\mu}{2\pi\hbar^2}\frac{\mathrm{e}^{\mathrm{i}k|\boldsymbol{r}-\boldsymbol{r}'|}}{|\boldsymbol{r}-\boldsymbol{r}'|},$$

can be approximated by the expression (4.16) for $r \gg r'$,

$$\mathcal{G}(\boldsymbol{r},\boldsymbol{r}') = -\frac{\mu}{2\pi\hbar^2}\frac{\mathrm{e}^{\mathrm{i}kr}}{r}\left[\mathrm{e}^{-\mathrm{i}\boldsymbol{k}_r\cdot\boldsymbol{r}'} + O\left(\frac{r'}{r}\right)\right], \quad \boldsymbol{k}_r = k\frac{\boldsymbol{r}}{r}.$$

4.3

a) Calculate the density of states $\varrho(E)$ for plane waves of unit amplitude in three-dimensional coordinate space, $\psi(\boldsymbol{k}) = \exp(\mathrm{i}\boldsymbol{k}\cdot\boldsymbol{r})$, $E = \hbar^2k^2/(2\mu)$. (Impose periodic boundary conditions in a cube of length L and study the limit $L \to \infty$.)
b) Use the Golden Rule (2.139) to give an expression for the transition probability per unit time from an initial state ψ_{i} into final states consisting of the plane waves above with wave vectors pointing in directions contained in the angular element $\mathrm{d}\Omega$.

Confirm the following observation: If the matrix element of the transition operator $\hat{T}$ is related to the scattering amplitude f as in (4.18), then the transition probability per unit time is just the differential scattering cross section $|f|^2$ multiplied by the incoming current density $\hbar k/\mu$.

4.4 Use the phase shifts (1.133) for elastic scattering by a hard sphere,

$$\tan\delta_l = -\frac{j_l(kr_0)}{n_l(kr_0)},$$

to discuss the dependence of the integrated scattering cross section (4.38),

$$\sigma = \frac{4\pi}{k^2}\sum_{l=0}^{\infty}(2l+1)\sin^2\delta_l,$$

on energy. Which partial waves l contribute significantly to the cross section at high energy E ?

4.5 An electron at a distance $\boldsymbol{r}$ from an atom generates an electric field $\boldsymbol{E} = e\boldsymbol{r}/r^3$ at the position of the atom. An electric field of strength $\boldsymbol{E}$ induces an electric dipole moment $\boldsymbol{d} = \alpha_{\rm d}\boldsymbol{E}$ in an atom with a dipole polarizability $\alpha_{\rm d}$. The force $\boldsymbol{F}$ which a dipole of dipole moment $\boldsymbol{d}$ exerts on an electron at a distance $\boldsymbol{r}$ is

$$\boldsymbol{F} = (e/r^3)[\boldsymbol{d} - 3\boldsymbol{r}(\boldsymbol{r}\cdot\boldsymbol{d})/r^2]\,.$$

Show that an electron, which is brought from infinity to a point at a distance r from an atom with dipole polarizability $\alpha_{\rm d}$, does the work

$$W(r) = e^2\,\frac{\alpha_{\rm d}}{2r^4}\,.$$

4.6 An electron (spin $\frac{1}{2}$) is scattered by a potential. Consider the solution ψ' of the stationary Schrödinger equation with the boundary conditions (4.248)

$$\psi' = {\rm e}^{{\rm i}kz}\begin{pmatrix}0\\1\end{pmatrix} + \frac{{\rm e}^{{\rm i}kr}}{r}\begin{pmatrix}g'(\theta,\phi)\\ f'(\theta)\end{pmatrix}, \qquad r\to\infty\,.$$

Show that the partial wave amplitudes f_l' and g_l' in the expansions

$$f'(\theta) = \sum_{l=0}^{\infty} f_l'\sqrt{\frac{4\pi}{2l+1}}\,Y_{l,0}(\theta),$$

$$g'(\theta,\phi) = \sum_{l=1}^{\infty} g_l'\sqrt{l(l+1)}\sqrt{\frac{4\pi}{2l+1}}\,Y_{l,-1}(\theta,\phi)$$

are given by formulae like (4.245),

$$f_l' = \frac{l+1}{2{\rm i}k}\left[\exp\left(2{\rm i}\delta_l^{(l+1/2)}\right)-1\right] + \frac{l}{2{\rm i}k}\left[\exp\left(2{\rm i}\delta_l^{(l-1/2)}\right)-1\right],$$

$$g_l' = \frac{1}{2{\rm i}k}\left[\exp\left(2{\rm i}\delta_l^{(l+1/2)}\right) - \exp\left(2{\rm i}\delta_l^{(l-1/2)}\right)\right].$$

Hint: Repeat the considerations following (4.239) for a z-component of the total angular momentum $m' = -1/2$.

4.7 Consider a two-component spinor normalized to unity,

$$|\chi\rangle = \begin{pmatrix}A\\B\end{pmatrix}, \qquad |A|^2 + |B|^2 = 1\,.$$

Show that the polarization vector $\boldsymbol{P} = \langle\chi|\hat{\boldsymbol{\sigma}}|\chi\rangle$ has the components given in (4.255),

$$P_x = 2\Re[A^*B], \quad P_y = 2\Im[A^*B], \quad P_z = |A|^2 - |B|^2 .$$

$\hat{\boldsymbol{\sigma}}$ is the vector of the three Pauli spin matrices,

$$\hat{\sigma}_x = \begin{pmatrix} 0 & 1 \\ 1 & 0 \end{pmatrix}, \quad \hat{\sigma}_y = \begin{pmatrix} 0 & -\mathrm{i} \\ \mathrm{i} & 0 \end{pmatrix}, \quad \hat{\sigma}_z = \begin{pmatrix} 1 & 0 \\ 0 & -1 \end{pmatrix} .$$

Show that the projection $\hat{\sigma}_P = \boldsymbol{P}\cdot\hat{\boldsymbol{\sigma}} = P_x\hat{\sigma}_x + P_y\hat{\sigma}_y + P_z\hat{\sigma}_z$ onto the direction of $\boldsymbol{P}$ is given by

$$\hat{\sigma}_P = \begin{pmatrix} |A|^2 - |B|^2 & 2AB^* \\ 2A^*B & |B|^2 - |A|^2 \end{pmatrix},$$

and that the spinor $|\chi\rangle$ is an eigenstate of $\hat{\sigma}_p$ with eigenvalue $+1$.

4.8 Consider the elastic scattering of two electrons with parallel spins (total spin $S = 1$). In the centre-of-mass system this corresponds to the scattering of a particle of reduced mass $\mu = m_e/2$ in the repulsive Coulomb potential e^2/r. The indistinguishability of the two electrons leads to a modification of the formulae for scattering amplitude and cross section.

a) Show that the Rutherford formula (4.147) for the differential cross section is replaced by the following *Mott formula*,

$$\frac{\mathrm{d}\sigma_{\mathrm{M}}^{\mathrm{t}}}{\mathrm{d}\Omega} = \frac{\eta^2}{4k^2}\left[\frac{1}{\sin^4\frac{1}{2}\theta} + \frac{1}{\cos^4\frac{1}{2}\theta} - 2\frac{\cos(\eta\ln\tan^2\frac{1}{2}\theta)}{\sin^2\frac{1}{2}\theta\,\cos^2\frac{1}{2}\theta}\right].$$

b) Which orbital angular momentum quantum numbers l contribute to the partial-waves expansion of the wave function?
c) What changes in a) and b) if we consider the scattering of two electrons whose spins are coupled to $S = 0$? Which differential cross section do we observe in the scattering of unpolarized electrons?

4.9 Show that the Green's function of the Helmholtz equation in n dimensions,

$$\mathcal{G}(\boldsymbol{x},\boldsymbol{x}') = -\left(\frac{K}{2\pi}\right)^{\nu}\frac{\mathrm{i}H_\nu^{(1)}(K|\boldsymbol{x}-\boldsymbol{x}'|)}{4|\boldsymbol{x}-\boldsymbol{x}'|^{\nu}}, \quad \nu = \frac{n}{2} - 1,$$

fulfills the defining equation

$$(K^2 + \Delta_n)\mathcal{G}(\boldsymbol{x},\boldsymbol{x}') = \delta(\boldsymbol{x}-\boldsymbol{x}') .$$

Here $H_\nu^{(1)}(\varrho)$ is the Hankel function of order ν (Appendix A.4). It is a solution of Bessel's differential equation

$$\frac{\mathrm{d}^2 w}{\mathrm{d}\varrho^2} + \frac{1}{\varrho}\frac{\mathrm{d}w}{\mathrm{d}\varrho} + \left(1 - \frac{\nu^2}{\varrho^2}\right) w = 0$$

with the boundary conditions

$$\mathrm{i}H_\nu^{(1)}(\varrho) \stackrel{\varrho\to 0}{=} \frac{\Gamma(\nu)}{\pi}\left(\frac{\varrho}{2}\right)^{-\nu}, \quad \mathrm{i}H_\nu^{(1)}(\varrho) \stackrel{\varrho\to\infty}{=} \sqrt{\frac{2\mathrm{i}}{\pi}}\frac{\mathrm{e}^{\mathrm{i}\varrho}}{\mathrm{i}^\nu\sqrt{\varrho}} .$$

4.10 Evaluate the integral

$$I_n = \int_{-\infty}^{\infty} \mathrm{d}x_1 \cdots \int_{-\infty}^{\infty} \mathrm{d}x_n, \mathrm{e}^{-x_1^2 - x_2^2 \cdots - x_n^2}$$

in two different ways: (i) as a product of n one-dimensional integrals, (ii) by transforming it into a radial integral. Show that this leads to the following formulae for the surface $S_n(R)$ and the volume $V_n(R)$ of the n-dimensional sphere of radius R:

$$S_n(R) = \frac{2\pi^{n/2}}{\Gamma\left(\frac{n}{2}\right)} R^{n-1}, \quad V_n(R) = \frac{\pi^{n/2}}{\Gamma\left(\frac{n}{2}+1\right)} R^n .$$

4.11 Two displacement vectors $\boldsymbol{r}_1$ and $\boldsymbol{r}_2$ are described in hyperspherical coordinates by the length R of the six-component vector $(\boldsymbol{r}_1, \boldsymbol{r}_2)$ and the five angles $\theta_1,\ \phi_1,\ \theta_2,\ \phi_2,\ \alpha$,

$$\begin{aligned} x_1 &= R\sin\alpha\sin\theta_1\cos\phi_1, & x_2 &= R\cos\alpha\sin\theta_2\cos\phi_2, \\ y_1 &= R\sin\alpha\sin\theta_1\sin\phi_1, & y_2 &= R\cos\alpha\sin\theta_2\sin\phi_2, \\ z_1 &= R\sin\alpha\cos\theta_1, & z_2 &= R\cos\alpha\cos\theta_2, \end{aligned}$$

where $\alpha = 0, \ldots \frac{\pi}{2}$, $\theta_i = 0, \ldots \pi$ and $\phi_i = 0, \ldots\ 2\pi$.

a) Show that the hyperspherical angular element $\mathrm{d}\Omega_\mathrm{h}$ is given by

$$\begin{aligned} \mathrm{d}\Omega_\mathrm{h} &= \sin^2\alpha\ \cos^2\alpha\ \mathrm{d}\alpha\ \mathrm{d}\Omega_1\, d\Omega_2 \\ &= \sin^2\alpha\ \cos^2\alpha\ \mathrm{d}\alpha\ \sin\theta_1\ \mathrm{d}\phi_1\ \sin\theta_2\ \mathrm{d}\theta_2\ \mathrm{d}\phi_2 . \end{aligned}$$

b) The surface S_n of an n-dimensional sphere of radius R is given by (Problem 4.10)

$$S_n = \frac{2\pi^{n/2}}{\Gamma\left(\frac{n}{2}\right)} R^{n-1} .$$

Verify that integration over the hyperspherical solid angle Ω_h gives the correct result for $n = 6$, namely π^3.

4.12

a) Determine the length and the direction of the momentum transfer vector (4.370), $\boldsymbol{q} = \boldsymbol{k}_1 - \boldsymbol{k}_i$, for the (e, 2e) reaction (4.368) in asymmetric coplanar geometry with the parameters of Fig. 4.22.
b) Determine the length and the direction of the momentum transfer vector $\boldsymbol{q}$ for the (e, 2e) reaction (4.368) in symmetric coplanar geometry ($\theta_1 = \theta_2$, $T_1 = T_2$) with incoming kinetic energy $E_{\rm inc} = 150$ eV.

References

[Adh86] S.K. Adhikari, Am. J. Phys. **54**, 362 (1986)
[AF08] F. Arnecke, H. Friedrich, P. Raab, Phys. Rev. A **78**, 052711 (2008)
[AJ77] W.O. Amrein, J.M. Jauch, K.B. Sinha, *Scattering Theory in Quantum Mechanics* (W.A. Benjamin, Reading (Mass.), 1977)
[And04] T. Andersen, Phys. Reports **394**, 157 (2004)
[AS70] M. Abramowitz, I.A. Stegun (eds.), *Handbook of Mathematical Functions* (Dover Publications, New York, 1970)
[Bar83] G. Barton, Am. J. Phys. **51**, 420 (1982)
[Bar89] K. Bartschat, Phys. Reports **180**, 1 (1989)
[Ber96] J. Berakdar, Phys. Rev. A **53**, 2314 (1996)
[Bet49] H.A. Bethe, Phys. Rev. **76**, 38 (1949)
[BB89] M. Brauner, J.S. Briggs, H. Klar, J. Phys. B **22**, 2265 (1989)
[BJ85] F.W. Byron, C.J. Joachain, B. Piraux, J. Phys. B **18**, 3203 (1985)
[BJ89] F.W. Byron, C.J. Joachain, Phys. Reports **179**, 211 (1989)
[BK96] P. Balling, P. Kristensen, H.H. Andersen, U.V. Pedersen, V.V. Petrunin, L. Præstegaard, H.K. Haugen, T. Andersen, Phys. Rev. Lett. **77**, 2905 (1996)
[Blu81] K. Blum, *Density Matrix Theory and Applications* (Plenum Press, New York, 1981)
[BM72] M.V. Berry, K.E. Mount, Rep. Prog. Phys. **35**, 315 (1972)
[BM88] K. Bartschat, R.P. McEachran, A.D. Stauffer, J. Phys. B **21**, 2789 (1988)
[BO97] J. Berakdar, P.F. O'Mahoney, F. Mota-Furtado, Z. Phys. D **39**, 41 (1997)
[BR95] J. Burgdörfer, C.O. Reinhard, J. Sternberg, J. Wang, Phys. Rev. A **51**, 1248 (1995)
[Bra83] B.H. Bransden, *Atomic Collision Theory* (Benjamin Cummings, Reading (Mass.), 1983)
[BS93] I. Bray, A.T. Stelbovics, Phys. Rev. Lett. **70**, 746 (1993)
[BS03] P.L. Bartlett, A.T. Stelbovics, I. Bray, Phys. Rev. A **68**, 030701 (2003)
[Bur77] P.G. Burke, *Potential Scattering in Atomic Physics* (Plenum Press, New York, 1977)
[Cal78] J. Callaway, Phys. Reports **45**,89 (1978)
[Cal82] J. Callaway, Phys. Rev. A **26**, 199 (1982)
[Cal88] J. Callaway, Phys. Rev. A **37**, 3692 (1988)
[CG49] G.F. Chew, M.L. Goldberger, Phys. Rev. **75**, 1637 (1949)
[CI98] P. Chocian, W. Ihra, P.F. O'Mahony, Phys. Rev. A **57**, 3583 (1998)
[CI00] P. Chocian, W. Ihra, P.F. O'Mahony, Phys. Rev. A **62**, 104704 (2000)
[CJ04] J.G. Childers, K.E. James Jr., I. Bray, M. Baertschy, M.A. Khakoo, Phys. Rev. A **69**, 022709 (2004)
[CO79] J. Callaway, D.H. Oza, Phys. Lett. **72 A**, 207 (1979)
[CP48] H.B.G. Casimir, D. Polder, Phys. Rev. **73**, 360 (1948)

[CU87] J. Callaway, K. Unnikrishnan, D.H. Oza, Phys. Rev. **A36**, 2576 (1987)
[CU89] J. Callaway, K. Unnikrishnan, Phys. Rev. Lett. **40**, 1660 (1989)
[DG65] E. Del Giudice, E. Galzenati, Nuovo Cimento **38**, 443 (1965)
[EK85] H. Ehrhardt, G. Knoth, P. Schlemmer, K. Jung, Phys. Lett. A **110**, 92 (1985)
[EJ86] H. Ehrhardt, K. Jung, G. Knoth, P. Schlemmer, Z. Phys. D **1**, 3 (1986)
[Fea93] J. Feagin, J. Phys. B **28**, 1495 (1995)
[FE12] M. Fink, J. Eiglsperger, J. Madroñero, H. Friedrich, Phys. Rev. A **85**, 040702(R) (2012)
[FM47] E. Fermi, L. Marshall, Phys. Rev. **71**, 666 (1947)
[FM93] L.D. Faddeev, S.P. Merkuriev, *Quantum Scattering Theory for Several Particle Systems* (Kluwer Academic Publisher, Dordrecht, 1993)
[Fri16] H. Friedrich, *Scattering Theory*, 2nd edn. (Springer, Heidelberg, 2016)
[FT04] H. Friedrich, J. Trost, Phys. Reports **397**, 359 (2004)
[Gao99] B. Gao, Phys. Rev. A **59**, 2778 (1999)
[GK91] A. Gellrich, J. Kessler, Phys. Rev. A **43**, 204 (1991)
[GL93] X.Q. Guo, M.S. Lubell, J. Phys. B **26**, 1221 (1993)
[Glo83] W. Glöckle, *The Quantum Mechanical Few-Body Problem* (Springer, Berlin, Heidelberg, New York, 1983)
[HD93] G. Handke, M. Draeger, W. Ihra, H. Friedrich, Phys. Rev. A **48**, 3699 (1993)
[ID95] W. Ihra, M. Draeger, G. Handke, H. Friedrich, Phys. Rev. A **52**, 3752 (1995)
[IF92] W. Ihra, H. Friedrich, Phys. Rev. A **45**, 5278 (1992)
[IM97a] W. Ihra, J.H. Macek, F. Mota-Furtado, P.F. O'Mahony, Phys. Rev. Lett. **78**, 4027 (1997)
[IM97b] W. Ihra, F. Mota-Furtado, P.F. O'Mahony, Phys. Rev. A **55**, 4263 (1997)
[Ino71] M. Inokuti, Rev. Mod. Phys. **43**, 297 (1971)
[JH76] R.H. Jansen, F.J. de Heer, J. Phys. B **9**, 213 (1976)
[Joa87] C.J. Joachain, *Quantum Collision Theory* (North-Holland, Amsterdam, 1987)
[Kes85] J. Kessler, *Polarized Electrons*, 2nd edn. (Springer, Berlin, Heidelberg, New York, 1985)
[Kes91] J. Kessler, Adv. At. Mol. Opt. Phys., vol. 27, ed. by Sir David R. Bates (1991) 81
[Kes00] J. Kessler, Phys. Essays **13**, Sp. Iss. 421 (2000)
[Kla81] H. Klar, J. Phys. B **14**, 4165 (1981)
[KM09] N.N. Khuri, A. Martin, J.-M. Richard, T.T. Wu, J. Math. Phys. **50**, 072105 (2009)
[KO98] M.Y. Kuchiev, V.N. Ostrovsky, Phys. Rev. A **58**, 321 (1998); V.N. Ostrovsky, J. Phys. B **37**, 4657 (2004)
[KS76] H. Klar, W. Schlecht, J. Phys B **9**, 1699 (1976)
[Lap82] I.R. Lapidus, Am. J. Phys. **50**, 45 (1982)
[Lev49] N. Levinson, Kgl. Danske Videnskab. Selskab, Mat.-Fys. Medd. **25**(9) (1949)
[LK63] B.R. Levy, J.B. Keller, J. Math. Phys. **4**, 54 (1963)
[LL71] L.D. Landau, L.M. Lifshitz, *Course of Theoretical Physics, vol. 1, Mechanics* (Addison-Wesley, Reading, 1971)
[LM84] B. Lohmann, I.E. McCarthy, A.T. Stelbovics, E. Weigold, Phys. Rev. A **30**, 758 (1984)
[MC68] J.W. McGowan, E.M. Clarke, Phys. Rev. **167**, 43 (1968)
[ME01] M.J. Moritz, C. Eltschka, H. Friedrich, Phys. Rev. A **63**, 042102 (2001); **64**, 022101 (2001)
[Mes70] A. Messiah, *Quantum Mechanics, vol. 1*, Chap. 10 (North Holland, Amsterdam, 1970)
[MI97] J.H. Macek, W. Ihra, Phys. Rev. A **55**, 2024 (1997)
[Mul13] T.-O. Müller, Phys. Rev. Lett. **110**, 260401 (2013)
[MW76] I.E. McCarthy, E. Weigold, Phys. Reports **27**, 275 (1976)
[MW88] I.E. McCarthy, E. Weigold, Rep. Prog. Phys. **51**, 299 (1988)
[MW91a] I.E. McCarthy, E. Weigold, Adv. At. Mol. Opt. Phys., vol. 27, ed. by Sir David R. Bates (1991) 165
[MW91b] I.E. McCarthy, E. Weigold, Adv. At. Mol. Opt. Phys., vol. 27, ed. by Sir David R. Bates (1991) 201
[Nes80] R.K. Nesbet, *Variational Methods in Electron Scattering Theory* (Plenum Press, New York, 1980)

[New82] R.G. Newton, *Scattering Theory of Waves and Particles* (Springer, Berlin, Heidelberg, New York, 1982)
[NT07] P. Naidon, E. Tiesinga, W.F. Mitchell, P. Julienne, New J. Phys. **9**, 19 (2007)
[OL83] J.K. O'Connell, N.F. Lane, Phys. Rev. A **27**, 1893 (1983)
[OS61] T.F. O'Malley, L. Spruch, L. Rosenberg, J. Math. Phys. **1**, 491 (1961)
[OR62] T.F. O'Malley, L. Rosenberg, L. Spruch, Phys. Rev. **125**, 1300 (1962)
[Poe78] R. Poet, J. Phys. B **11**, 3081 (1978)
[Ros94] J.-M. Rost, Phys. Rev. Lett. **72**, 1998 (1994)
[RT84] D.F. Register, S. Trajmar, Phys. Rev. A **29**, 1785 (1984); D.F. Register, S. Trajmar, G. Steffensen, D.C. Cartwright, Phys. Rev. A **29**, 1793 (1984)
[Rud68] M.R.H. Rudge, Rev. Mod. Phys. **40**, 564 (1968)
[SE87] M.B. Shah, D.S. Elliott, H.B. Gilbody, J. Phys. B **20**, 3501 (1987)
[Sit91] A. Sitenko, *Scattering Theory* (Springer, Berlin, Heidelberg, New York, 1991)
[SS89] M.P. Scott, T.T. Scholz, H.R.J. Walters, P.G. Burke, J. Phys. B **22**, 3055 (1989)
[SW90] T.T. Scholz, H.R.J. Walters, P.G. Burke, J. Phys. B **23**, L467 (1990)
[Swa55] P. Swan, Proc. Roy. Soc. **228**, 10 (1955)
[SZ95] A. Stintz, X.M. Zhao, C.E.M. Strauss, W.B. Ingalls, G.A. Kyrala, D.J. Funk, H.C. Bryant, Phys. Rev. Lett. **75**, 2924 (1995)
[Tay72] J.R. Taylor, *Scattering Theory: The Quantum Theory of Nonrelativistic Collisions* (Wiley, New York, 1972)
[Tem62] A. Temkin, Phys. Rev. **126**, 130 (1962)
[Tem91] A. Temkin, J. Phys. B **24**, 2147 (1991)
[Tic09] C. Ticknor, Phys. Rev. A **80**, 052702 (2009)
[TP71] R.R. Teachout, R.T. Pack, At. Data **3**, 195 (1971)
[VE84] B.J. Verhaar, P.H.W. van den Eijnde, M.A. Voermans, M.M.J. Schaffrath, J. Phys. A **17**, 595 (1984)
[Wan53] G.H. Wannier, Phys. Rev. **90**, 817 (1953)
[WC75] J.F. Williams, A. Crowe, J. Phys. B **8**, 2233 (1975)
[Wil76] J.F. Williams, J. Phys. B **9**, 1519 (1976)
[Wil79] J.F. Williams, J. Phys. B **12**, 265 (1979)
[Wil88] J.F. Williams, J. Phys. B **21**, 2107 (1988)
[WR90] C.D. Warner, R.M. Rutter, G.C. King, J. Phys. B **23**, 93 (1990)

Chapter 5
Special Topics

The last decades have seen great advances in experimental atomic physics. Exotic states of atoms can be prepared with the help of intense and short laser pulses, experiments can be performed on individual atoms and ions in electromagnetic traps and the dependence of their properties on their environment can be investigated, and high resolution laser spectroscopy has made precision studies of the finest details of complicated atomic spectra possible. The experimental advances have confronted the theory with new challenges. It has become apparent that intricate and interesting effects can occur even in seemingly simple systems with only few degrees of freedom, and that their theoretical description often is by no means easy. Complementary to high precision spectroscopy, the availability of ultra-short light pulses on the femtosecond time scale and below has made it possible to study highly localized excitations in atoms and molecules and to follow the evolution of wave packets on an atomic scale. The availability of ultra-cold atoms has made experimental tests of fundamental postulates of quantum mechanics possible, and it has led to the realization in the laboratory of degenerate condensates of gases of bosonic and of fermionic atoms. The new experimental techniques enable active manipulation of ultra-cold atoms in the extremely quantum mechanical regime.

In order to describe multiphoton processes, which typically occur in intense light fields, it is necessary to go beyond traditional perturbative treatments of the interaction of atoms with light. This is the subject of Sect. 5.1. The power of classical and semiclassical methods in understanding and describing structure and dynamics on an atomic scale has become increasingly apparent since the mid-1980's. Section 5.2 presents a brief discussion of how far the concept of coherent wave packets moving along classical trajectories can be formulated in a quantum mechanically consistent way, and Sect. 5.3 describes recent advances of our understanding of the relation between classical and quantum dynamics, in particular for the interesting case that the classical motion is chaotic. Section 5.4 is devoted to the subject of Bose-Einstein condensates of atomic gases, which were

H. Friedrich, *Theoretical Atomic Physics*, Graduate Texts in Physics,
DOI 10.1007/978-3-319-47769-5_5

prepared and observed for the first time in 1995 and have since proved to be an abounding source of exciting new physics.

In the study of Bose-Einstein condensates in particular, and of systems of ultracold atoms (or molecules) in general, it is important to understand the properties of states with extremely low energy. In a diatomic system these are the states very close to the continuum threshold in an atom-atom potential, which typically supports a number of bound states and falls off faster than $1/r^2$ for large values of the atom-atom separation r. Near-threshold bound states in such shorter-ranged potentials were treated in considerable detail in Sect. 3.1.2, and Sect. 5.5 shows how to connect the near-threshold bound states below the continuum threshold to the continuum of scattering states with energies just above the continuum threshold.

A crucial parameter influencing the properties of bound and continuum states near the continuum threshold of an atom-atom system, i.e. the dissociation threshold of the diatomic molecule, is the atom-atom scattering length, which is related to the threshold quantum number via parameters depending only on the properties of the tail of the shorter-ranged atom-atom potential. The value of the scattering length, resp. of the threshold quantum number, can be manipulated with the tool of Feshbach resonances, tuned to lie near the dissociation threshold. Section 5.6 is devoted to the description of the properties and influence of near-threshold Feshbach resonances.

Finally, Sect. 5.7 contains a brief introduction to some aspects of atom optics, where the guiding and trapping of atom waves is the focus of attention.

5.1 Multiphoton Absorption

The description of electromagnetic transitions in Sect. 2.4 is based on the assumption that the interaction of the electromagnetic field with an atom can be regarded as a small perturbation. This justifies applying first-order perturbation theory in the form of the Golden Rule and yields probabilities for transitions in which one photon is absorbed or emitted (Sect. 2.4.4). Transitions in which two or more photons are absorbed or emitted simultaneously only become important in very strong fields. Such strong fields can be produced by very intense lasers, and the investigation of atomic processes in the presence of a laser field, in particular of multiphoton processes, is a very important subfield of atomic physics and optics. A summary of experimental and theoretical work up to the early eighties is contained in [CL84]. For comprehensive monographs see [DK85, Fai86]. Further developments are summarized in [SK88]; see also [NC90, Gav92, DK94, DF00].

5.1.1 *Experimental Observations on Multiphoton Ionization*

If the energy of a single photon is smaller than the ionization potential of an atom (in a given initial state), then photoionization can only proceed via the absorption of several photons. The intensity of the laser determines how much electromagnetic field energy is available in the immediate vicinity of the atom (see Problem 5.1). Laser powers well beyond 10^{12} W/cm^2 with pulses lasting for nanoseconds have been available for many years. Early experiments on multiphoton ionization involved just counting the ions created by a strong laser pulse. An example is shown in Fig. 5.1, where strontium atoms were exposed to the pulses of a Nd:YAG laser (=neodymium:yttrium-aluminium-garnet). The wave length of the laser light is 1.064 μm corresponding to a photon energy of $\hbar\omega = 1.165$ eV. At

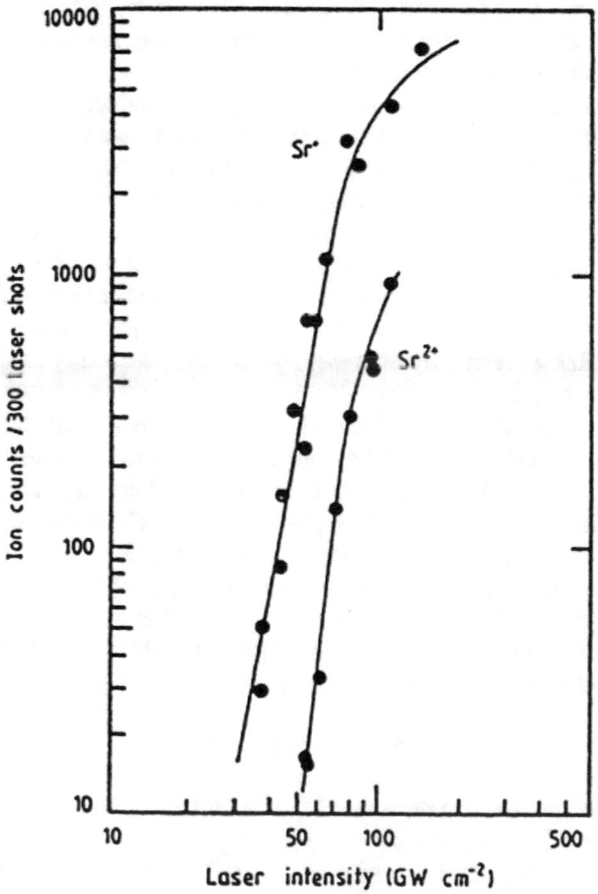

Fig. 5.1 Numbers of Sr^+ and Sr^{++} ions observed in multiphoton ionization by a Nd:YAG-Laser ($\hbar\omega = 1.165$ eV) as functions of the laser intensity (from [FK82])

least five photons are needed to ionize a strontium atom; at least fifteen photons are needed to eject two electrons [FK84].

The number of ions as a function of the laser intensity I follows a straight line over large stretches in the doubly logarithmic representation of Fig. 5.1, which indicates a power law. Extending the perturbation theory of Sect. 2.4 to higher orders gives the probability $P(n)$ for absorbing n photons in lowest non-vanishing order as

$$P(n) \propto I^n \,. \tag{5.1}$$

The expected proportionality to I^5 for singly ionized strontium is well fulfilled in Fig. 5.1, but the probability for double ionization rises more slowly than the minimum number (fifteen) of photons would suggest. The deviations from the straight lines at higher intensities in Fig. 5.1 can be attributed to a saturation effect which occurs when all atoms in the region hit by the laser pulse are ionized. The applicability of lowest-order perturbation theory is limited to non-resonant absorption. Resonance effects involving appropriate intermediate states can make the picture much more complicated [TL89].

The general interest in multiphoton ionization grew rapidly after first investigations of the ejected electrons revealed that these could have kinetic energies much larger than expected for absorption of the minimum number of photons necessary. A first explanation of these observations was, that an electron already excited into the continuum could acquire a higher final kinetic energy by the further absorption of photons. This picture corresponds to ionizing an atom out of a continuum state and led to the rather unfortunate name of *above-threshold ionisation* (ATI). A more appropriate name is *excess-photon ionization* EPI, which merely expresses the observed fact that electrons absorb more photons than necessary for ionization and refrains from further interpretation.

Figure 5.2 shows ATI or EPI spectra for the ionization of xenon by photons from a Nd:YAG-Laser ($\hbar\omega = 1.165$ eV) at four different laser intensities. The minimum number of photons needed depends on whether the Xe^+ ion is left behind in one or the other of two states energetically separated by 1.31 eV. If the ion is left behind in the lower $P_{3/2}$ state, which corresponds to the ejection of an electron from a $5p_{3/2}$ state, then at least eleven photons are needed; for a Xe^+ ion in the $P_{1/2}$ state corresponding to ejection of a $5p_{1/2}$ electron we need at least twelve photons. The asymptotic kinetic energy of an electron after absorption of n photons is just the difference of $n\hbar\omega$ and the ionization potential I_P,

$$E_{\rm kin}(n) = n\hbar\omega - I_{\rm P} \,. \tag{5.2}$$

These energies are shown at the top of Fig. 5.2 for the two ionization channels. The maxima in Fig. 5.2 show appreciable absorption of up to eight excess photons. The figure also shows features which were established as characteristic in the course of many further experiments. Amongst these are the observation that the relative probability for absorbing a larger number of excess photons increases with increasing

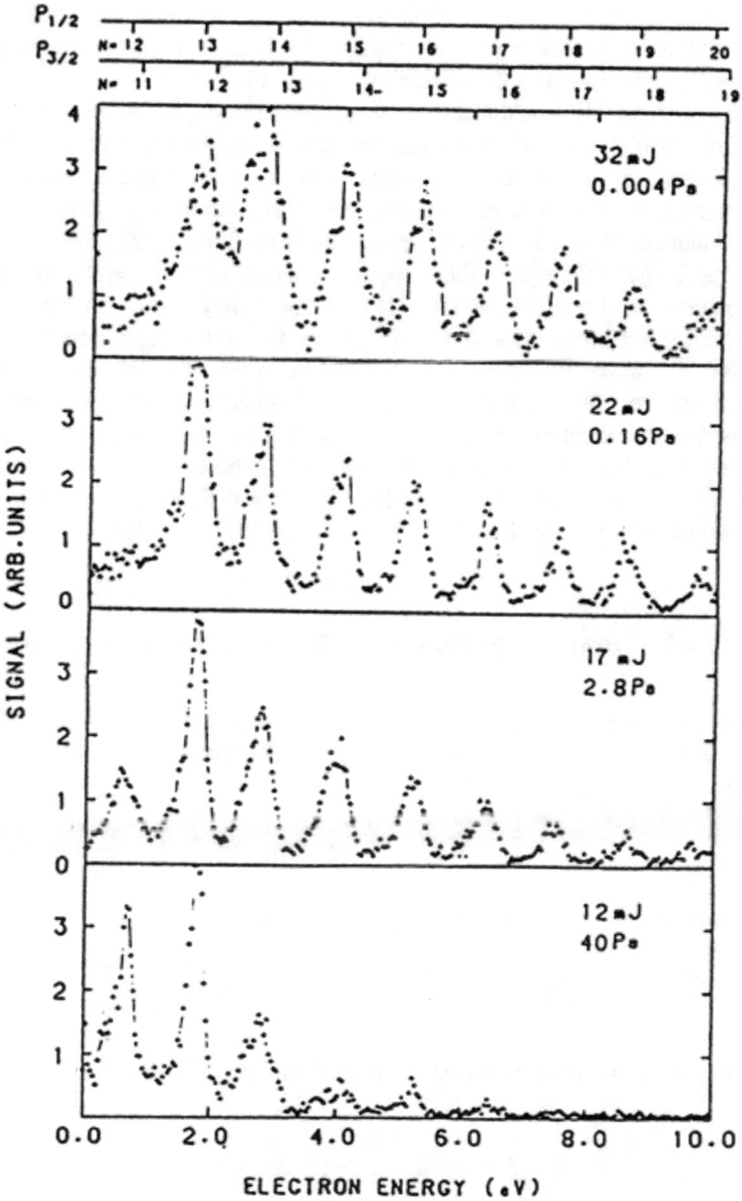

Fig. 5.2 Energy spectra of electrons ejected in the multiphoton ionization of xenon by a Nd:YAG laser ($\hbar\omega = 1.165$ eV) for various intensities ($\approx$ numbers shown as mJ $\times 2 \times 10^{12}$ $[\mathrm{W/cm^2}]$) and pressures. The asymptotic kinetic energy expected according to (5.2) for electrons having absorbed n photons is shown for the two ionization channels at the top edge of the figure (From [KK83])

laser intensity, and that the probability for absorbing no or only one excess photon is smaller than the probability for absorption of a larger number of excess photons at sufficiently high intensity (see also Fig. 5.3 below).

Whereas perturbative methods may be applied to multiphoton ionization as long as the field strengths are not too high, they are not appropriate for describing the nonmonotonic dependence of the heights of the absorption peaks on the number of excess photons. (For a discussion of perturbative methods see [Cra87, Kar91].) The explanation of simple-looking spectra such as those in Fig. 5.2 is already a serious challenge to theory. Further experimental data such as angular distributions of the ejected electrons are available [FW88], and they should enable us to sort out the merits of various theoretical approaches. The following two sections briefly sketch two examples for a non-perturbative description of multiphoton ionization. Both sections treat the example of an atom in a spatially constant monochromatic field. Further complications arise when considering the finite temporal duration of a light pulse and the rise and fall of its intensity explicitly. Large scale numerical calculations which directly solve the time-dependent Schrödinger equation have been quite successful in such situations, see e.g. [KS97] and references given there.

5.1.2 Calculating Ionization Probabilities via Volkov States

Consider a one-electron atom in an oscillating electromagnetic field described by a vector potential $\boldsymbol{A}$. In the radiation gauge (2.150) the vector potential for light polarized linearly in the x-direction is

$$\boldsymbol{A}(\boldsymbol{r},t) = -A_0 \boldsymbol{e}_x \sin\omega t\,. \tag{5.3}$$

For right or left circular polarization around the z-axis we have

$$\boldsymbol{A}(\boldsymbol{r},t) = -\frac{A_0}{\sqrt{2}}(\boldsymbol{e}_x \sin\omega t \mp \boldsymbol{e}_y \cos\omega t)\,. \tag{5.4}$$

According to (2.148) the associated electric field $\boldsymbol{E}$ for linear or circular polarization is

$$\begin{aligned} \boldsymbol{E}(\boldsymbol{r},t) &= \boldsymbol{E}_0\, \boldsymbol{e}_x \cos\omega t\,, \quad \text{or} \\ \boldsymbol{E}(\boldsymbol{r},t) &= \frac{\boldsymbol{E}_0}{\sqrt{2}}\,(\boldsymbol{e}_x \cos\omega t \pm \boldsymbol{e}_y \sin\omega t)\,. \end{aligned} \tag{5.5}$$

In both cases the amplitudes E_0 and A_0 are related by

$$E_0 = \frac{\omega}{c} A_0\,. \tag{5.6}$$

Since the wave length of the laser light is much larger than typical spatial dimensions of the atom, we can assume a spatially homogeneous field, i.e. constant amplitudes E_0, A_0 (dipole approximation). The Hamiltonian is (cf. (2.151))

$$\hat{H} = \frac{[\hat{\boldsymbol{p}} + (e/c)\boldsymbol{A}(\boldsymbol{r},t)]^2}{2\mu} + V(\boldsymbol{r})\,. \tag{5.7}$$

Apart from the vector potential $\boldsymbol{A}$ it also contains the static potential $V(\boldsymbol{r})$ describing the interaction of the electron with the residual ion in the absence of a laser field.

If we decompose the Hamiltonian (5.7) into an atomic part $\hat{\boldsymbol{p}}^2/(2\mu) + V$ and an additional term $\hat{H}_1$ due to the laser field, then

$$\hat{H}_1 = \frac{e}{\mu c}\boldsymbol{A}\cdot\hat{\boldsymbol{p}} + \frac{e^2}{2\mu c^2}\boldsymbol{A}^2\,. \tag{5.8}$$

The technique of using a Green's function to formally solve a Schrödinger equation, which was repeatedly demonstrated in Chap. 4, can be generalized to the time-dependent Schrödinger equation

$$\mathrm{i}\hbar\frac{\partial}{\partial t}\psi(\boldsymbol{r},t) = \hat{H}\psi(\boldsymbol{r},t) \tag{5.9}$$

(see e.g. Appendix A of [Rei80]). This yields an implicit expression for the probability amplitude a_{fi} describing a transition caused by the time-dependent interaction (5.8), in which an initial atomic state $\psi_{\mathrm{i}}(\boldsymbol{r},t) = \phi_{\mathrm{i}}(\boldsymbol{r})\exp[-(\mathrm{i}/\hbar)E_{\mathrm{i}}t]$ evolves into a final state $\psi_{\mathrm{f}}(\boldsymbol{r},t)$, which is a solution of the full Schrödinger equation (5.9),

$$a_{\mathrm{fi}} = \frac{1}{\mathrm{i}\hbar}\int_{-\infty}^{\infty}\langle\psi_{\mathrm{f}}|\hat{H}_1|\psi_{\mathrm{i}}\rangle\,\mathrm{d}t\,. \tag{5.10}$$

If the ionization limit of the field-free atom is at $E = 0$ then the (negative) energy eigenvalue E_{i} of the initial bound state is just minus the (positive) ionization potential I_{P}, which has to be overcome for ionization out of this state.

The formula (5.10) looks similar to the expression (2.134) for transition amplitudes in time-dependent perturbation theory. In contrast to this expression however, (5.10) is exact (like analogous formulae (4.17), (4.283) in time-independent scattering theory), provided the final state wave function ψ_{f} really is an exact solution of the Schrödinger equation.

In an approximation originally due to Keldysch and developed by Reiss [Rei80], the exact solution ψ_{f} in (5.10) is replaced by solutions of the Schrödinger equation for a free electron in a laser field. The ket of the matrix element in (5.10) then contains a solution of the (time-dependent) Schrödinger equation including the atomic potential but without a laser field, while the bra contains a solution of the Schrödinger equation containing the laser field but no atomic potential. For a spatially homogeneous monochromatic laser field these latter solutions are known analytically and are called *Volkov states*.

In the absence of an atomic potential the Hamiltonian (5.7) is

$$\hat{H}_0 = \frac{[\hat{\boldsymbol{p}} + (e/c)\boldsymbol{A}(\boldsymbol{r},t)]^2}{2\mu} . \tag{5.11}$$

For linearly polarized light (5.3) we have

$$\hat{H}_0 = \frac{\hat{\boldsymbol{p}}^2}{2\mu} - \frac{eA_0}{\mu c}\,\hat{p}_x \sin\omega t + \frac{e^2A_0^2}{2\mu c^2}\sin^2\omega t , \tag{5.12}$$

and it is straightforward to verify that the following Volkov states are solutions of the time-dependent Schrödinger equation:

$$\psi_{\rm V}(\boldsymbol{r},t) = \exp\left[\mathrm{i}\boldsymbol{k}\cdot\boldsymbol{r} - \mathrm{i}\frac{\hbar k^2}{2\mu}t - \mathrm{i}k_x\frac{eA_0}{\omega\mu c}\cos\omega t \right. \\ \left. - \frac{\mathrm{i}}{\hbar}\frac{e^2A_0^2}{2\mu c^2}\left(\frac{t}{2} - \frac{1}{4\omega}\sin 2\omega t\right)\right] . \tag{5.13}$$

For circular polarization (5.4) we have

$$\hat{H}_0 = \frac{\hat{\boldsymbol{p}}^2}{2\mu} - \frac{eA_0}{\sqrt{2}\,\mu c}(\hat{p}_x \sin\omega t \mp \hat{p}_y\cos\omega t) + \frac{e^2A_0^2}{4\mu c^2} , \tag{5.14}$$

and the corresponding Volkov states are

$$\psi_{\rm V}(\boldsymbol{r},t) = \exp\left[\mathrm{i}\boldsymbol{k}\cdot\boldsymbol{r} - \mathrm{i}\frac{\hbar k^2}{2\mu}t \right. \\ \left. - \mathrm{i}\frac{eA_0}{\sqrt{2}\,\omega\mu c}(k_x\cos\omega t \pm k_y \sin\omega t) - \frac{\mathrm{i}}{\hbar}\frac{e^2A_0^2}{4\mu c^2}t\right] . \tag{5.15}$$

The Volkov states (5.13), (5.15) look like ordinary plane waves with an additional oscillating phase,

$$\psi_{\rm V} = \exp[\mathrm{i}\boldsymbol{k}\cdot\boldsymbol{r} - (\mathrm{i}/\hbar)E_{\rm V}t + \delta_{\rm osc}] . \tag{5.16}$$

The oscillating phase describes the *wiggling* of the electron in the oscillating field. In the energy there is an additional term which is constant in space and time and depends quadratically on the amplitude of the field. It is called the *ponderomotive energy* $E_{\rm P}$,

$$E_{\rm V} = \frac{\hbar^2k^2}{2\mu} + E_{\rm P} , \quad E_{\rm P} = \frac{e^2A_0^2}{4\mu c^2} = \frac{e^2E_0^2}{4\mu\omega^2} . \tag{5.17}$$

The Keldysch approximation allows an analytic evaluation of the integral in (5.10). In the case of circular polarization (5.15) we obtain the following expression for the probability per unit time that an electron is ejected into the solid angle $\mathrm{d}\Omega$:

$$\frac{\mathrm{d}P}{\mathrm{d}\Omega} \propto \sum_{n=n_0}^{\infty} \left(n - \frac{E_{\mathrm{P}}}{\hbar\omega}\right)^2 \sqrt{n-\varepsilon}\, |\tilde{\phi}_{\mathrm{i}}(\boldsymbol{k})|^2 J_n^2 \left(2 \sin\theta \sqrt{\frac{E_{\mathrm{P}}}{\hbar\omega}} \sqrt{n-\varepsilon}\right), \tag{5.18}$$

where J_n is the ordinary Bessel function (Appendix A.4). The quantity ε in (5.18) stands for the sum of the ionization potential and the ponderomotive energy in units of the photon energy $\hbar\omega$,

$$\varepsilon = \frac{I_{\mathrm{P}} + E_{\mathrm{P}}}{\hbar\omega}. \tag{5.19}$$

$\tilde{\phi}_{\mathrm{i}}(\boldsymbol{k})$ is the Fourier transform of the spatial part of the initial wave function $\phi_{\mathrm{i}}(\boldsymbol{r})$, and θ is the angle between the wave vector $\boldsymbol{k}$ and the z-axis. The right-hand side of (5.18) depends only on the direction of the outgoing wave vector $\boldsymbol{k}$ (more precisely: only on the polar angle θ); the length of $\boldsymbol{k}$ is fixed by energy conservation,

$$\frac{\hbar^2 k^2}{2\mu} = n\hbar\omega - (I_{\mathrm{P}} + E_{\mathrm{P}}) = (n - \varepsilon)\hbar\omega. \tag{5.20}$$

The summation index n in (5.18) stands for the number of photons absorbed in the ionization process. The summation begins with the smallest number n_0 for which $n - \varepsilon$ is positive. Note that the energy to be overcome consists of the ionization potential I_{P} plus the ponderomotive energy E_{P}. More energy is needed to ionize the atom in the presence of the electromagnetic field.

A formula like (5.18) can also be derived for linearly polarized light (see [Rei80]). Expressions similar to (5.18) were already found in 1973 by Faisal [Fai73].

The Keldysch approximation is quite successful if the atomic potential V is very short-ranged [BM89]. In realistic situations the Keldysch-Faisal-Reiss theory (KFR) is not always so successful in describing the multiphoton ionization data quantitatively [Buc89]. This may be due to the fact that the effect of the static very-long-ranged Coulomb potential between the ejected electron and the residual ion is not included. Furthermore, the consequences of the Keldysch approximation are not gauge invariant. The KFR theory nevertheless is able to reproduce some of the qualitative features of the energy spectra of the ejected electrons. As an example Fig. 5.3 shows ionization probabilities (5.18) integrated over all angles, in comparison with experimental spectra from the multiphoton ionization of xenon by circularly polarized pulses from a Nd:YAG laser. The calculated ionization probabilities have been decomposed into contributions from various photon numbers n which are related to the energy of the ejected electron via (5.2).

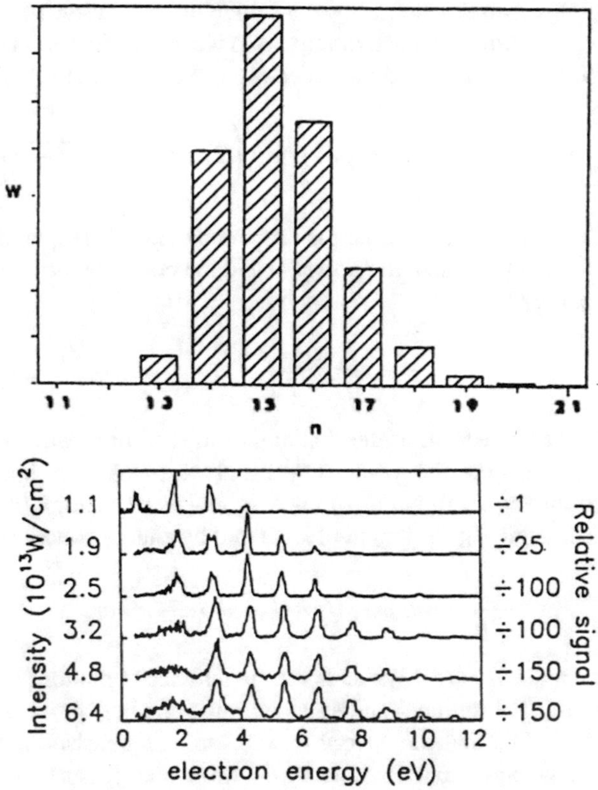

Fig. 5.3 The upper picture shows angle integrated ionization probabilities (5.18) decomposed into contributions from various photon numbers n. The parameters correspond to the ionization of xenon by photons with $\hbar\omega = 1.165$ eV and a field strength characterized by a ponderomotive energy of $E_P/\hbar\omega = 1$ (from [Rei87]). The lower picture shows the energy spectra of photoelectrons from the ionization of xenon by circularly polarized pulses from a Nd:YAG laser ($\hbar\omega = 1.165$ eV) at various intensities (from [MB87])

If the duration of the laser pulses is not too short, the energy of the photoelectrons registered in the detector is given by (5.2) and the ponderomotive energy need not be subtracted. The reason lies in the fact that the field strength and hence the ponderomotive energy, which are regarded as constant over a few wave lengths of the laser, fall off from their respective maximum values to zero over a distance corresponding to the spatial extension of the pulse. The resulting gradient of the ponderomotive energy exerts a force on the electron, the *ponderomotive force*.

After absorbing n photons the electron leaves the atom with a kinetic energy given by (5.20). The ponderomotive force then accelerates the electron away from the centre of the pulse so that it reaches the detector with the asymptotic kinetic energy given by (5.2). Such acceleration due to the ponderomotive force can also be observed in different contexts, e.g. in the scattering of free electrons by a strong laser pulse [Buc89]. For very short laser pulses (shorter than picoseconds) the laser field has subsided before the acceleration by the ponderomotive force becomes effective, and the energy shifts due to the ponderomotive energy in (5.20), which can also be interpreted as ac-Stark shifts of the bound-state energies (see Sect. 3.5.3), are observed in the detectors [RW90a, DP90].

5.1.3 Calculating Ionization Probabilities via Floquet States

This section briefly sketches the use of the theory of Floquet states introduced in Sect. 3.5.3 for the nonperturbative treatment of multiphoton ionization. For more details the reader is referred to an article on this subject by Potvliege and Shakeshaft [PS92].

In the field gauge (3.344) the Hamiltonian $\hat{H}$ for an atom in a spatially constant and monochromatic field is the sum of the time-independent Hamiltonian $\hat{H}_A$ for the field-free atom and an additional potential oscillating with the circular frequency ω. As discussed in Sect. 3.5.3 we can use the ansatz

$$\psi = \exp[-(\mathrm{i}/\hbar)\varepsilon t]\Phi_\varepsilon(t)\,, \quad \Phi_\varepsilon(t + 2\pi/\omega) = \Phi_\varepsilon(t) \tag{5.21}$$

to reduce the time-dependent Schrödinger equation to an eigenvalue equation for the generalized Hamiltonian

$$\hat{\mathcal{H}} = \hat{H} - \mathrm{i}\hbar\frac{\partial}{\partial t} \tag{5.22}$$

(cf. (3.346), (3.347)). The eigenvalues of (5.22) are the quasi-energies ε, and the associated solutions (5.21) are the Floquet states or quasi-energy states. In a monochromatic field the Hamiltonian including the atom-field interaction has the general form

$$\hat{H} = \hat{H}_A + \hat{W}\,\mathrm{e}^{\mathrm{i}\omega t} + \hat{W}^\dagger\,\mathrm{e}^{-\mathrm{i}\omega t}\,, \tag{5.23}$$

and the precise nature of the time-independent coupling operator $\hat{W}$ depends on polarization and gauge. If we express the periodic time dependence of the Φ_ε in terms of a Fourier series,

$$\Phi_\varepsilon = \sum_n \mathrm{e}^{-in\omega t} \psi_{\varepsilon,n} , \tag{5.24}$$

then the eigenvalue equation for the generalized Hamiltonian (5.22) becomes a set of time-independent coupled equations for the Fourier components $\psi_{\varepsilon,n}$,

$$\hat{H}_\mathrm{A} \psi_{\varepsilon,n} + \hat{W} \psi_{\varepsilon,n+1} + \hat{W}^\dagger \psi_{\varepsilon,n-1} = (\varepsilon - n\hbar\omega)\psi_{\varepsilon,n} . \tag{5.25}$$

Potvliege and Shakeshaft solved the coupled equations (5.25) numerically for the case that $\hat{H}_\mathrm{A}$ describes a hydrogen atom [PS89]. This involves the consideration of asymptotic ($r \to \infty$) boundary conditions whose explicit form depends on the choice of gauge. The calculations yield complex eigenvalues

$$\varepsilon_i = E_i + \Delta_i - \mathrm{i}\frac{\Gamma_i}{2} , \tag{5.26}$$

where E_i are the energy eigenvalues of the field-free hydrogen atom, and Δ_i are real energy shifts which should become the ac-Stark shifts (3.354) in the weak-field limit. The origin of the imaginary part in (5.26) is that each initially bound state can couple to and decay into continuum states for sufficiently large n, i.e. by coupling to a sufficient number of photons. As a consequence the absolute square of the wave function of the Floquet state decreases proportional to $\exp[-\Gamma_i t/\hbar]$ corresponding to an ionization rate per unit time of $\Gamma_i/\hbar$ (see also [PS90]). Figure 5.4 shows $\Gamma_i/\hbar$ for ionization from the $1s$ ground state of the hydrogen atom by a linearly polarized

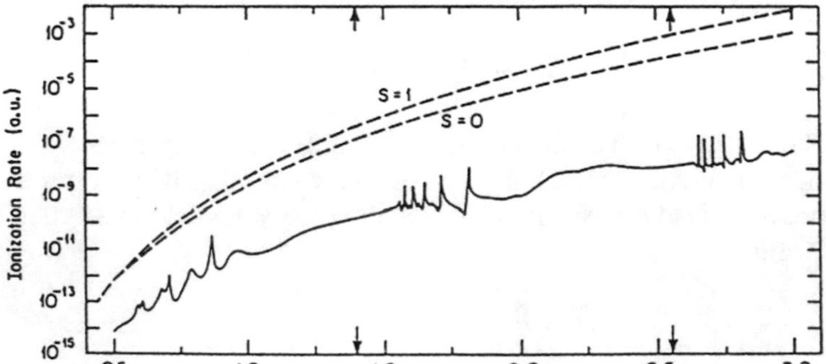

Fig. 5.4 Probability per unit time for ionization of a hydrogen atom from its $1s$ ground state by a strong laser field with the Nd:YAG frequency ($\hbar\omega = 1.165eV$) as a function of the laser intensity. The *dashed lines* show the results of perturbation theory in lowest non-vanishing order (5.1) for the absorption of $n = 12$ ($S = 0$) or $n = 13$ ($S = 1$) photons (from [PS89, PS92])

Nd:YAG laser ($\hbar\omega = 1.165$ eV) as a function of the laser intensity. The dashed lines show for comparison the results of lowest non-vanishing order perturbation theory for ionization by $n = 12$ or $n = 13$ photons. The resonance-like structures in the non-perturbative curve occur when the quasi-energy of the Floquet state which corresponds to the $1s$ state of the H atom in the field-free limit crosses or almost crosses the quasi-energies of other states as the laser intensity is varied.

One remarkable feature of Fig. 5.4 is that the non-perturbative result, which includes ionization by an arbitrary number (at least twelve) of photons, lies substantially lower than the perturbative ionization probabilities for exactly twelve or exactly thirteen photons. The authors of [PS89] conclude that perturbative treatments can overestimate the probability for ionizing an atom in a strong laser field by orders of magnitude. Perturbative methods have, on the other hand, been successful in reproducing the angular distributions of the photo-electrons. Figure 5.5 shows angular distributions of electrons ejected in the multiphoton ionization of hydrogen by photons with an energy of 3.5 eV. The minimum number of photons needed for ionization is four. The various parts of the figure correspond to absorption of up to three excess photons. The perturbative calculation reproduces the measured angular distributions quite well in all cases.

The investigation of atoms under the influence of intense laser pulses has been refined considerably in the last few decades. Spectra of the type shown in Fig. 5.2 with dozens of excess-photon peaks have been observed, revealing a not necessarily monotonic dependence of the intensities of the higher-order peaks on their order. Similar structures are also observed in *photon* spectra emitted by atoms under the influence of intense laser pulses, the remarkable feature here being the occurrence of higher-energy photons corresponding to odd harmonics of the frequency of the original laser pulse. This generation of higher harmonics is a useful mechanism for creating light pulses of very short wave lengths. The strengths of higher-order peaks both in above-threshold (excess-photon) ionization and in higher-harmonic generation show characteristic plateaus in their dependence on order, meaning that the expected decline of intensity with order is attenuated over several peaks. The non-trivial dependence of the peak intensity on order has been explained very successfully on the basis of a classical picture in which the electron which is ejected from the atom by the intense field is accelerated back towards the residual ion when the oscillating field of the laser changes direction. The subsequent "re-scattering" of the electron by the ion strongly influences the observed peak intensities in the spectra of the (higher-harmonic) photons or of the photo-electrons. For details see the comprehensive review by Becker et al. [BG02].

5.2 Classical Trajectories and Wave Packets

Many effects which are correctly and satisfactorily described by quantum mechanics can already be largely explained within the framework of classical mechanics, which is frequently considered easier to understand and visualize. It thus makes sense to

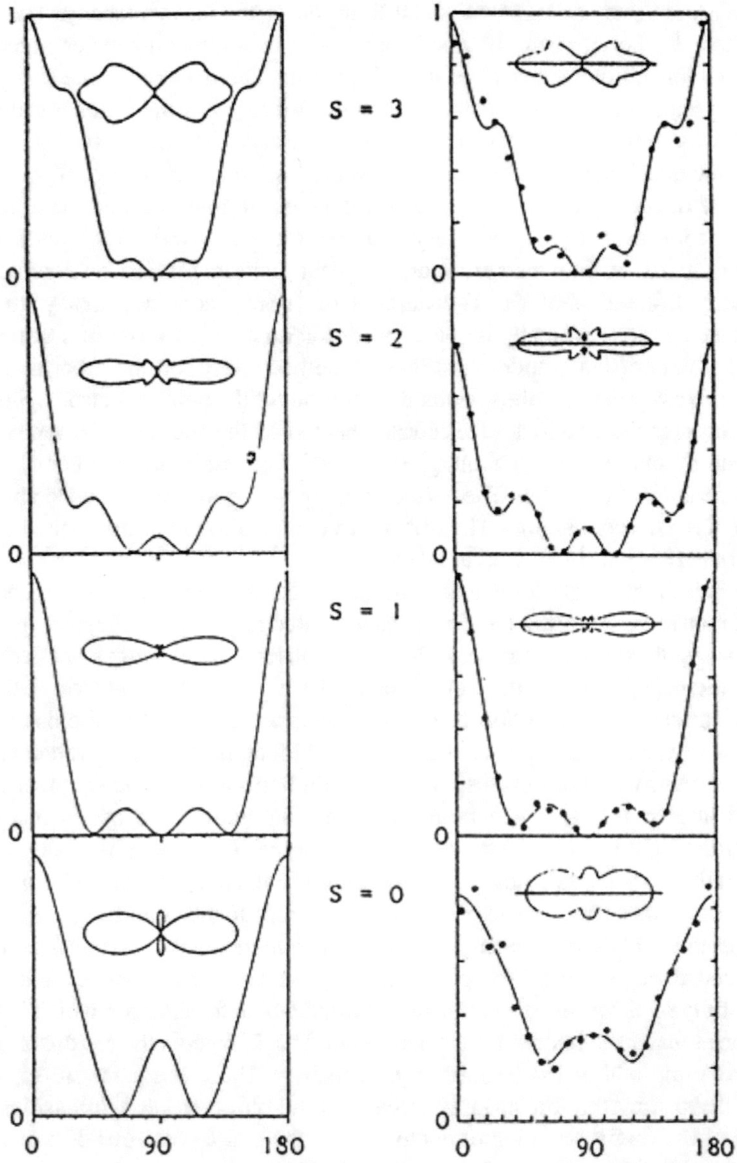

Fig. 5.5 Angular distributions of photo-electrons observed in the multiphoton ionization of hydrogen by photons with an energy of 3.5 eV. The various parts of the figure correspond to the absorption of $S = 0$ to $S = 3$ excess photons. The left half of the picture shows the results of a perturbative calculation, the right half shows the measured distributions whose absolute heights were fitted to the calculated curves (from [KM88])

compare the two theories, to establish the correspondence of classical and quantum mechanical descriptions and to highlight the genuine quantum effects which cannot be explained classically.

5.2.1 Phase Space Densities

In classical mechanics a physical system with f degrees of freedom is described by a *Hamiltonian function* $H(q_1, \dots q_f; p_1, \dots p_f; t)$ which depends on the f coordinates q_i, on f canonically conjugate momenta p_i and perhaps also on the time (see textbooks on mechanics, e.g. [Gol80, LL71] or [Sch90]). The temporal evolution of the system is described by a *trajectory* $(q_i(t), p_i(t))$ in *phase space*. The trajectory is a solution of the following system of $2f$ coupled ordinary differential equations:

$$\dot{q}_i = \frac{\partial H}{\partial p_i}, \quad \dot{p}_i = -\frac{\partial H}{\partial q_i}. \tag{5.27}$$

These are the *canonical equations* of classical mechanics. The initial conditions $q_i(t_0), p_i(t_0)$ uniquely determine the evolution of the system for all times.

If we don't know the state of the system at time t_0 exactly, we can describe it by a *classical phase space density* $\varrho_{\text{cl}}(q_i, p_i; t_0)$. It is the probability density for finding the system in the state q_i, p_i at time t_0. Being a probability density, ϱ_{cl} cannot be negative, and its integral over all possible states in phase space must be unity at all times,

$$\int \mathrm{d}^f q_i \int \mathrm{d}^f p_i \, \varrho_{\text{cl}}(q_i, p_i; t) = 1 \,. \tag{5.28}$$

We obtain an equation of motion for the classical phase space density by realizing that the probability for a state of the system cannot change along a trajectory in phase space, because this just describes the evolution of the system. This means that $\varrho_{\text{cl}}\left(q_i\left(t\right), p_i\left(t\right); t\right)$ must be constant in time if $q_i(t)$, $p_i(t)$ are solutions of the canonical equations (5.27),

$$\frac{\mathrm{d}}{\mathrm{d}t}\varrho_{\text{cl}}(q_i(t), p_i(t); t) = \sum_{i=1}^{N}\left(\dot{q}_i\frac{\partial \varrho_{\text{cl}}}{\partial q_i} + \dot{p}_i\frac{\partial \varrho_{\text{cl}}}{\partial p_i}\right) + \frac{\partial \varrho_{\text{cl}}}{\partial t} = 0 \,. \tag{5.29}$$

Inserting the expressions given by the canonical equations (5.27) for $\dot{q}_i$ and $\dot{p}_i$ into (5.29) and writing the resulting sum as a *Poisson bracket*,

$$\{H, \varrho_{\text{cl}}\} \stackrel{\text{def}}{=} \sum_{i=1}^{N}\left(\frac{\partial H}{\partial p_i}\frac{\partial \varrho_{\text{cl}}}{\partial q_i} - \frac{\partial H}{\partial q_i}\frac{\partial \varrho_{\text{cl}}}{\partial p_i}\right), \tag{5.30}$$

reduces (5.29) to the compact form

$$\frac{\partial \varrho_{\text{cl}}}{\partial t} = -\{H, \varrho_{\text{cl}}\} . \tag{5.31}$$

Equation (5.31) is the equation of motion for the classical phase space density in a system described by the Hamiltonian function H, and it is called the *Liouville equation*.

For simplicity we now consider a system consisting of a particle in a conservative potential in three spatial dimensions. The Hamiltonian function is

$$H(\boldsymbol{r},\boldsymbol{p}) = \frac{\boldsymbol{p}^2}{2\mu} + V(\boldsymbol{r}) , \tag{5.32}$$

and the Liouville equation has the form

$$\frac{\partial}{\partial t}\varrho_{\text{cl}}(\boldsymbol{r},\boldsymbol{p};t) = -\frac{\boldsymbol{p}}{\mu}\cdot\nabla_{\boldsymbol{r}}\,\varrho_{\text{cl}} - \boldsymbol{F}(\boldsymbol{r})\cdot\nabla_{\boldsymbol{p}}\,\varrho_{\text{cl}} , \quad \boldsymbol{F}(\boldsymbol{r}) = -\nabla_{\boldsymbol{r}} V(\boldsymbol{r}) . \tag{5.33}$$

It describes the flow of ϱ_{cl} in phase space under the influence of the inertial term (the first term on the right-hand side) and a field of force $\boldsymbol{F}$.

In quantum mechanics we describe the state of a system by a wave function $|\psi(t)\rangle$ which (in coordinate representation) is a function of the displacement vector, $\psi(\boldsymbol{r},t)$, and should be normalized to unity. The time evolution of $|\psi\rangle$ is determined by the Hamiltonian operator $\hat{H}$ and is described by the time-dependent Schrödinger equation,

$$\mathrm{i}\hbar\frac{\partial}{\partial t}|\psi\rangle = \hat{H}|\psi\rangle . \tag{5.34}$$

We can alternatively describe a pure state $|\psi\rangle$ by the associated density operator

$$\hat{\varrho}(t) = |\psi(t)\rangle\langle\psi(t)| \tag{5.35}$$

(cf. Sect. 4.3.3). In coordinate representation the density operator is an integral operator with the integral kernel

$$\varrho(\boldsymbol{r},\boldsymbol{r}';t) = \psi(\boldsymbol{r},t)\,\psi^*(\boldsymbol{r}',t) . \tag{5.36}$$

The quantum mechanical wave function in *momentum representation* is a function $\tilde{\psi}$ depending on the momentum variable $\boldsymbol{p}$, and it is related to the wave function $\psi(\boldsymbol{r},t)$ in coordinate representation by a Fourier transformation:

$$\tilde{\psi}(\boldsymbol{p},t) = \frac{1}{(2\pi\hbar)^{3/2}}\int \mathrm{e}^{-\mathrm{i}\boldsymbol{p}\cdot\boldsymbol{r}/\hbar}\,\psi(\boldsymbol{r},t)\,\mathrm{d}\boldsymbol{r} . \tag{5.37}$$

In momentum representation the density operator for the pure state (5.35) has the form

$$\tilde{\varrho}(\boldsymbol{p},\boldsymbol{p}';t) = \tilde{\psi}(\boldsymbol{p},t)\,\tilde{\psi}^*(\boldsymbol{p}',t)\,. \tag{5.38}$$

A mixed quantum mechanical state is described by an incoherent superposition of pure states with (non-negative) probabilities w_n (see Sect. 4.3.3),

$$\hat{\varrho}(t) = \sum_n w_n|\psi_n(t)\rangle\langle\psi_n(t)|\,, \tag{5.39}$$

and its coordinate and momentum representations are corresponding generalizations of (5.36) and (5.38) respectively.

If the wave function $|\psi(t)\rangle$ of a pure state (5.35) or the wave functions $|\psi_n(t)\rangle$ of the mixed state (5.39) obey the time-dependent Schrödinger equation (5.34), then the associated density matrix obeys the *von Neumann equation*

$$\frac{\partial\hat{\varrho}}{\partial t} = -\frac{\mathrm{i}}{\hbar}[\hat{H},\hat{\varrho}]\,. \tag{5.40}$$

Here $[\hat{H},\hat{\varrho}] = \hat{H}\hat{\varrho} - \hat{\varrho}\hat{H}$ is the commutator of $\hat{H}$ and $\hat{\varrho}$ as usual.

The von Neumann equation (5.40) is more flexible than the Schrödinger equation, in particular because it can be generalized to describe dissipative effects, see e.g. [Haa73]. If an initial (pure) state with a given energy E is subject to dissipative effects due to coupling to other degrees of freedom such as internal excitations of the particle, then the density matrix will generally evolve into a mixed state containing contributions corresponding to other (lower) energies than E. This can be described in by an additional dissipative term on the right-hand side of (5.40); the structure of such a dissipative term is more complicated than a simple commutator of a given operator with $\hat{\rho}$.

The von Neumann equation (5.40) for the quantum mechanical density operator has the same form as the Liouville equation (5.31) for the classical phase space density if we identify the Poisson bracket in the classical equation with $(\mathrm{i}/\hbar)$ times the commutator of quantum mechanics. The similarity between classical and quantum mechanics becomes more apparent if we represent the density operator, which depends on two displacement vectors in coordinate representation and on two momentum vectors in momentum representation, by its *Wigner function* $\varrho_{\mathrm{W}}(\boldsymbol{R},\boldsymbol{P};t)$, which depends on one displacement vector and one momentum vector. We obtain the Wigner function of $\hat{\varrho}$ either from the coordinate representation $\varrho\,(\boldsymbol{r},\boldsymbol{r}';t)$ by a Fourier transformation with respect to the variable $\boldsymbol{r}-\boldsymbol{r}'$, or from the momentum representation $\tilde{\varrho}(\boldsymbol{p},\boldsymbol{p}';t)$ by a Fourier transformation with respect to the variable $\boldsymbol{p}-\boldsymbol{p}'$,

$$\begin{aligned}\varrho_{\mathrm{W}}(\boldsymbol{R},\boldsymbol{P};t) &= \frac{1}{(2\pi\hbar)^3}\int \mathrm{e}^{-\mathrm{i}\boldsymbol{P}\cdot\boldsymbol{s}/\hbar}\varrho\left(\boldsymbol{R}+\frac{\boldsymbol{s}}{2},\boldsymbol{R}-\frac{\boldsymbol{s}}{2};t\right)\mathrm{d}\boldsymbol{s}\\ &= \frac{1}{(2\pi\hbar)^3}\int \mathrm{e}^{+\mathrm{i}\boldsymbol{R}\cdot\boldsymbol{q}/\hbar}\tilde{\varrho}\left(\boldsymbol{P}+\frac{\boldsymbol{q}}{2},\boldsymbol{P}-\frac{\boldsymbol{q}}{2};t\right)\mathrm{d}\boldsymbol{q}\,. \qquad (5.41)\end{aligned}$$

The Wigner function (5.41) is real, because the density operator is Hermitian. The coordinate or momentum representation of the density operator can be recovered from the Wigner function by inverting the corresponding Fourier transformation in (5.41).

The Wigner function has several properties reminiscent of a classical phase space density. Integrating over the momentum variables yields the (quantum mechanical) probability density in coordinate space, e.g. for the pure state (5.35),

$$\int \varrho_{\mathrm{W}}(\boldsymbol{R},\boldsymbol{P};t)\,\mathrm{d}\boldsymbol{P} = \varrho(\boldsymbol{R},\boldsymbol{R};t) = |\psi(\boldsymbol{R},t)|^2\,. \qquad (5.42)$$

Conversely, integrating over the spatial variables yields the quantum mechanical probability density in momentum space,

$$\int \varrho_{\mathrm{W}}(\boldsymbol{R},\boldsymbol{P};t)\,\mathrm{d}\boldsymbol{R} = \tilde{\varrho}(\boldsymbol{P},\boldsymbol{P};t) = |\tilde{\psi}(\boldsymbol{P},t)|^2\,. \qquad (5.43)$$

Integrating the Wigner function over the whole of phase space we obtain the conservation of total probability [cf. (5.28)],

$$\int \varrho_{\mathrm{W}}(\boldsymbol{R},\boldsymbol{P};t)\,\mathrm{d}\boldsymbol{R}\mathrm{d}\boldsymbol{P} = \int |\psi(\boldsymbol{R},t)|^2\mathrm{d}\boldsymbol{R} = \int |\tilde{\psi}(\boldsymbol{P},t)|^2\mathrm{d}\boldsymbol{P} = 1\,. \qquad (5.44)$$

However, the Wigner function is also different from a classical phase space density in some crucial aspects. In particular, the values of the function can be negative, and it is only after integrations such as in (5.42)–(5.44) that genuine probability interpretations become possible.

We obtain an equation of motion for the Wigner function by formulating the von Neumann equation (5.40) in the Wigner representation. We assume a Hamiltonian operator

$$\hat{H} = \hat{T} + \hat{V}\,, \quad \hat{T} = \frac{\hat{\boldsymbol{p}}^2}{2\mu}\,, \quad \hat{V} \equiv V(\boldsymbol{r})\,. \qquad (5.45)$$

The Wigner function $[\hat{T},\hat{\varrho}]_W$ of the commutator of $\hat{T}$ and $\hat{\varrho}$ is most easily calculated by Fourier transformation from the momentum representation [lower line (5.41)],

$$[\hat{T},\hat{\varrho}]_W = \frac{1}{(2\pi\hbar)^3}\int d\boldsymbol{q}\,e^{+i\boldsymbol{R}\cdot\boldsymbol{q}/\hbar}\,\frac{1}{2\mu}\left[\left(\boldsymbol{P}+\frac{\boldsymbol{q}}{2}\right)^2-\left(\boldsymbol{P}-\frac{\boldsymbol{q}}{2}\right)^2\right]$$
$$\times\,\tilde{\varrho}\left(\boldsymbol{P}+\frac{\boldsymbol{q}}{2},\boldsymbol{P}-\frac{\boldsymbol{q}}{2};t\right) = \frac{\hbar}{i}\frac{\boldsymbol{P}}{\mu}\,\nabla_{\boldsymbol{R}}\,\varrho_W(\boldsymbol{R},\boldsymbol{P};t)\,. \tag{5.46}$$

The Wigner function for the commutator of the potential energy and $\hat{\varrho}$ is more easily calculated by Fourier transformation from the coordinate representation (upper line (5.41)),

$$[\hat{V},\hat{\varrho}]_W = \frac{1}{(2\pi\hbar)^3}\int d\boldsymbol{s}\,e^{-i\boldsymbol{P}\cdot\boldsymbol{s}/\hbar}\left[V\left(\boldsymbol{R}+\frac{\boldsymbol{s}}{2}\right)-V\left(\boldsymbol{R}-\frac{\boldsymbol{s}}{2}\right)\right]$$
$$\times\varrho\left(\boldsymbol{R}+\frac{\boldsymbol{s}}{2},\boldsymbol{R}-\frac{\boldsymbol{s}}{2};t\right). \tag{5.47}$$

The Wigner representation of the von Neumann equation now reads

$$\frac{\partial}{\partial t}\varrho_W = -\frac{i}{\hbar}\left([\hat{T},\hat{\varrho}]_W + [\hat{V},\hat{\varrho}]_W\right), \tag{5.48}$$

with the two terms $[\hat{T},\hat{\varrho}]_W$ and $[\hat{V},\hat{\varrho}]_W$ given by (5.46) and (5.47) respectively. The kinetic energy term given by (5.46) has the same structure as the inertial term in the classical Liouville equation (5.33). The potential energy term acquires the same structure as the contribution due to the force-field in (5.33), if the potential $V(\boldsymbol{R}\pm\boldsymbol{s}/2)$ in (5.47) is expanded to second order in a Taylor series about $V(\boldsymbol{R})$,

$$V\left(\boldsymbol{R}\pm\frac{\boldsymbol{s}}{2}\right) = V(\boldsymbol{R}) \pm \frac{1}{2}\boldsymbol{s}\cdot\nabla_{\boldsymbol{R}}V(\boldsymbol{R}) + \frac{1}{8}\sum_{i,j}s_i s_i\frac{\partial^2 V}{\partial R_i\partial R_j} \pm \dots\,. \tag{5.49}$$

If we insert the expansion (5.49) into (5.47) the even terms vanish and the Wigner representation of the von Neumann equation becomes

$$\frac{\partial}{\partial t}\varrho_W(\boldsymbol{R},\boldsymbol{P};t) = -\frac{\boldsymbol{P}}{\mu}\cdot\nabla_{\boldsymbol{R}}\,\varrho_W + \nabla_{\boldsymbol{R}}\,V(\boldsymbol{R})\cdot\nabla_{\boldsymbol{P}}\,\varrho_W + \dots\,. \tag{5.50}$$

The dots on the right-hand side of (5.50) stand for contributions from cubic and higher terms in the expansion of the potential (5.49).

For potentials depending at most quadratically on the coordinates, the expansion (5.49) is exact and the quantum mechanical von Neumann equation in Wigner representation is identical to the classical Liouville equation (5.33). A Wigner function given at a certain time t_0 will thus evolve in phase space exactly as if it were a classical phase space density obeying the Liouville equation, provided the potential contains no anharmonic terms. Many phenomena which are often taught as typical examples of quantum mechanical behaviour, such as the *spreading of a wave packet* describing free-particle motion, can be completely understood classically. If, in the case of a free particle, an initial phase space density containing a distribution of momenta spreads in coordinate space in the course of time, then this is not a quantum mechanical effect, just think e.g. of a 100-metre race with athletes running at different speeds. The uncertainty relation of quantum mechanics does however forbid an initial state with a finite uncertainty in coordinate space together with a sharply defined momentum as would be necessary—both classically and in quantum mechanics—to avoid dispersion of the probability distribution in coordinate space. (See Problem 5.2.)

5.2.2 Coherent States

The concept of coherent states is useful for the description of the time-dependent motion of wave packets, in particular if the Hamiltonian is the Hamiltonian of a harmonic oscillator. To keep formulae simple we restrict the discussion in this section to a one-dimensional harmonic oscillator,

$$\hat{H} = \frac{\hat{p}^2}{2\mu} + \frac{\mu}{2}\omega^2 x^2 = -\frac{\hbar^2}{2\mu}\frac{\partial^2}{\partial x^2} + \frac{\mu}{2}\omega^2 x^2 . \tag{5.51}$$

(For a discussion of coherent states of a three-dimensional harmonic oscillator see [AB91].)

The eigenvalues of the Hamiltonian (5.51) are $E_n = \left(n + \frac{1}{2}\right)\hbar\omega$, $n = 0, 1, 2 \ldots$ The associated eigenstates (normalized to unity) are $|n\rangle$, and in coordinate representation they are polynomials of degree n multiplied by a Gaussian. The ground state wave function consists of this Gaussian alone,

$$|0\rangle \equiv \psi_0(x) = \left(\beta\sqrt{\pi}\right)^{-1/2} \mathrm{e}^{-x^2/(2\beta^2)} . \tag{5.52}$$

According to (1.83) the natural oscillator width β is related to the oscillator frequency ω by

$$\beta = \sqrt{\frac{\hbar}{\mu\omega}} . \tag{5.53}$$

In momentum representation the ground state wave function is also a Gaussian (cf. (5.37)),

$$\tilde{\psi}_0\,(p) = \frac{1}{\sqrt{2\pi\hbar}} \int_{-\infty}^{\infty} \mathrm{e}^{-\mathrm{i}px/\hbar}\,\psi_0(x)\mathrm{d}x = \left(\sqrt{\pi}\hbar/\beta\right)^{-1/2} \mathrm{e}^{-p^2\beta^2/(2\hbar^2)} . \tag{5.54}$$

We define the operators

$$\hat{b} = \frac{\mu\omega x + \mathrm{i}\hat{p}}{\sqrt{2\mu\hbar\omega}}, \quad \hat{b}^\dagger = \frac{\mu\omega x - \mathrm{i}\hat{p}}{\sqrt{2\mu\hbar\omega}}. \tag{5.55}$$

The commutation relations for $\hat{b}$ and $\hat{b}^\dagger$ follow from the commutation relations (1.33) for position and momentum,

$$[\hat{b}, \hat{b}^\dagger] = 1. \tag{5.56}$$

The Hamiltonian (5.51) has a very simple form if it is expressed in terms of the operators $\hat{b}^\dagger$, $\hat{b}$:

$$\hat{H} = \hbar\omega\left(\hat{b}^\dagger\hat{b} + 1/2\right). \tag{5.57}$$

From (5.56), (5.57) we obtain the commutation relations between $\hat{b}^\dagger$ or $\hat{b}$ and $\hat{H}$,

$$[\hat{H}, \hat{b}^\dagger] = \hbar\omega\hat{b}^\dagger, \quad [\hat{H}, \hat{b}] = -\hbar\omega\hat{b}. \tag{5.58}$$

It follows from the first equation (5.58) and the commutation relation $[\hat{b}, \hat{b}^\dagger] = 1$ that the operator $\hat{b}^\dagger$ transforms the eigenstate $|n\rangle$ of $\hat{H}$ into the eigenstate $|n+1\rangle$ (except for a normalization constant), i.e. $\hat{b}^\dagger$ is a *quantum creation operator*. In the same way it follows from the second commutation relation (5.58) that $\hat{b}$ is a *quantum annihilation operator* which transforms the eigenstate $|n\rangle$ into an eigenstate with $n-1$ quanta. Together with the correct normalization and phase convention we have

$$\hat{b}\,|n\rangle = \sqrt{n}\,|n-1\rangle, \quad \hat{b}^\dagger\,|n\rangle = \sqrt{n+1}\,|n+1\rangle \tag{5.59}$$

(see also Problem 2.6). $\hat{b}^\dagger\hat{b}$ is an operator which just counts the number of oscillator quanta excited in the eigenstates of the Hamiltonian (5.51) or (5.57),

$$\hat{b}^\dagger\hat{b}\,|n\rangle = n\,|n\rangle. \tag{5.60}$$

The *coherent states* $|z\rangle$ are defined as superpositions of eigenstates of the Hamiltonian (5.51),

$$|z\rangle = \mathrm{e}^{-zz^*/2}\sum_{n=0}^{\infty}\frac{(z^*)^n}{\sqrt{n!}}|n\rangle = \mathrm{e}^{-|z|^2/2}\mathrm{e}^{z^*\hat{b}^\dagger}\,|0\rangle, \tag{5.61}$$

where z is an arbitrary complex number. The states (5.61) are normalized to unity,

$$\langle z|z\rangle = \mathrm{e}^{-|z|^2}\sum_{n,n'}\frac{z^n(z^*)^{n'}}{\sqrt{n!\,n'!}}\langle n|n'\rangle = \mathrm{e}^{-|z|^2}\sum_{n=0}^{\infty}\frac{(|z|^2)^n}{n!} = 1, \tag{5.62}$$

but they are not orthogonal. The mean number of quanta excited in the coherent state $|z\rangle$ is

$$\langle z|\hat{b}^\dagger\hat{b}\,|z\rangle = |z|^2 \tag{5.63}$$

(see Problem 5.3).

In order to obtain the wave function of the coherent state $|z\rangle$ in coordinate representation, we start from the second equation (5.61). We can factorize the operator $\exp(z^*\hat{b}^\dagger)$ into a product

$$\begin{aligned} \mathrm{e}^{z^*\hat{b}^\dagger} &= \exp\left(\frac{z^*(\mu\omega x - \mathrm{i}\hat{p})}{\sqrt{2\mu\hbar\omega}}\right) \\ &= \exp\left(\frac{(z^*)^2}{4}\right)\exp\left(\frac{-\mathrm{i}z^*\hat{p}}{\sqrt{2\mu\hbar\omega}}\right)\exp\left(\frac{z^*x}{\sqrt{2}\beta}\right). \end{aligned} \tag{5.64}$$

In doing so we have used the relation

$$\mathrm{e}^{(\hat{A}+\hat{B})} = \mathrm{e}^{\hat{A}}\mathrm{e}^{\hat{B}}\mathrm{e}^{-[\hat{A},\hat{B}]/2} \tag{5.65}$$

for the operators $\hat{A} = -\mathrm{i}z^*\hat{p}/\sqrt{2\mu\hbar\omega}$ and $\hat{B} = z^*x\sqrt{\mu\omega/(2\hbar)} = z^*x/\left(\sqrt{2}\beta\right)$. The relation (5.65) is a special case of the *Baker-Campbell-Hausdorff relation* which applies when the commutator $[\hat{A},\hat{B}]$ (here it is the constant $-(z^*)^2/2$) commutes both with $\hat{A}$ and with $\hat{B}$ (see Problem 5.4). Before applying (5.64) we recall that the action of an operator of the form $\exp(\mathrm{i}a\hat{p})$ on an arbitrary wave function $\psi(x)$ merely consists in shifting the argument by $a\hbar$ [cf. (1.67)],

$$\mathrm{e}^{\mathrm{i}a\hat{p}}\psi(x) = \mathrm{e}^{a\hbar\,\partial/\partial x}\psi(x) = \sum_{n=0}^{\infty}\frac{(a\hbar)^n}{n!}\frac{\partial^n}{\partial x^n}\psi(x) = \psi(x+a\hbar)\,. \tag{5.66}$$

The coordinate representation of $|z\rangle$ is thus

$$\begin{aligned} |z\rangle &\equiv \psi_z(x) \\ &= \mathrm{e}^{-\left[|z|^2-(z^*)^2\right]/2}\left(\sqrt{\pi}\beta\right)^{-1/2}\exp\left(-\frac{\left(x-z^*\sqrt{2}\beta\right)^2}{2\beta^2}\right). \end{aligned} \tag{5.67}$$

The coherent state $|z\rangle$ is just a Gaussian wave packet which is shifted in position and momentum from the harmonic oscillator ground state (5.52), (5.54). To see this we construct the associated Wigner function according to (5.41),

$$
\begin{aligned}
\varrho_W(X,P) &= \frac{1}{2\pi\hbar}\int_{-\infty}^{\infty} \mathrm{e}^{-\mathrm{i}Ps/\hbar}\,\psi_z(X+s/2)\psi_z^*(X-s/2)\,\mathrm{d}s \\
&= \mathrm{e}^{(z-z^*)^2/2}\,\frac{(\sqrt{\pi}\beta)^{-1}}{2\pi\hbar}\int_{-\infty}^{\infty}\mathrm{e}^{-\mathrm{i}Ps/\hbar}\exp\left(-\frac{(X+\frac{s}{2}-z^*\sqrt{2}\beta)^2}{2\beta^2}\right) \\
&\quad\times\exp\left(-\frac{\left(X-\frac{s}{2}-z\sqrt{2}\beta\right)^2}{2\beta^2}\right)\mathrm{d}s \\
&= \frac{1}{\pi\hbar}\,\mathrm{e}^{-(X-X_z)^2/\beta^2}\,\mathrm{e}^{-(P-P_z)^2\beta^2/\hbar^2}\,.
\end{aligned}
\tag{5.68}
$$

The shifts X_z, P_z in position and momentum are

$$
\begin{aligned}
X_z &= \sqrt{2}\beta\,\Re(z) = \frac{\beta}{\sqrt{2}}(z+z^*)\,, \\
P_z &= -\sqrt{2}\frac{\hbar}{\beta}\,\Im(z) = \frac{\mathrm{i}\hbar}{\sqrt{2}\beta}(z-z^*)\,.
\end{aligned}
\tag{5.69}
$$

The Wigner function (5.68) of a coherent state is positive everywhere so there exists a corresponding classical system described by a numerically identical phase space density.

Coherent states evolve in time in a particularly simple way. Consider a coherent state $|z_0\rangle$ which is characterized at time t_0 by the complex number z_0. In order to apply the time evolution operator $\exp[-(\mathrm{i}/\hbar)\hat{H}(t-t_0)]$ cf. (1.41), to the first form (5.61) of $|z_0\rangle$, we only have to multiply the eigenstates $|n\rangle$ of $\hat{H}$ by the respective phase factors $\exp[-\mathrm{i}(n+1/2)\omega\,(t-t_0)]$,

$$
\begin{aligned}
\exp\left[-\tfrac{\mathrm{i}}{\hbar}\hat{H}(t-t_0)\right]|z_0\rangle &= \mathrm{e}^{-|z_0|^2/2}\sum_{n=0}^{\infty}\frac{(z_0^*)^n}{\sqrt{n!}}\,\mathrm{e}^{-\mathrm{i}(n+1/2)\omega(t-t_0)}\,|n\rangle \\
&= \mathrm{e}^{-\mathrm{i}\omega(t-t_0)/2}\mathrm{e}^{-|z|^2/2}\sum_{n=0}^{\infty}\frac{(z(t)^*)^n}{\sqrt{n!}}|n\rangle \\
&= \mathrm{e}^{-\mathrm{i}\omega(t-t_0)/2}|z(t)\rangle\,,
\end{aligned}
\tag{5.70}
$$

where $|z(t)\rangle$ again is a coherent state, namely the one characterized by the complex number

$$
z\,(t) = \mathrm{e}^{\mathrm{i}\omega(t-t_0)}z_0\,.
\tag{5.71}
$$

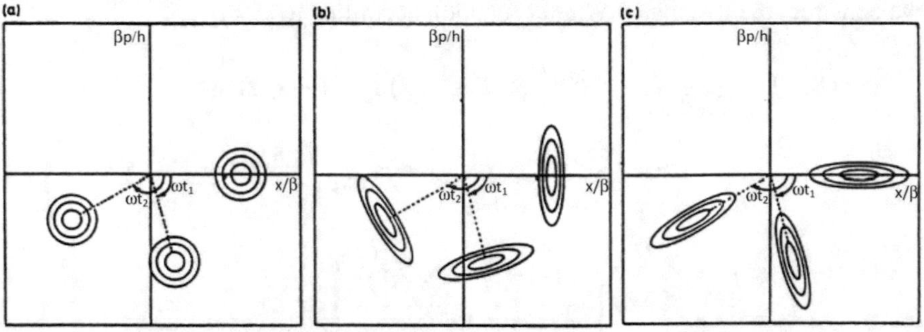

Fig. 5.6 Time evolution of minimal wave packets in phase space under the influence of a harmonic oscillator potential with the natural oscillator width β. The wave packet starts at its maximum (positive) displacement at time $t = 0$. Each part of the picture shows contour lines of the Wigner function (corresponding from the inside outwards to $\varrho_W/(\varrho_W)_{\max} = 0.9,\ 0.7,\ 0.5$) at $t = 0$, $\omega t_1 = 75°$ and $\omega t_2 = 150°$. (**a**) shows the coherent state, (**b**) a state squeezed in amplitude and (**c**) a state squeezed in phase

Except for a phase factor $\exp[-i\omega(t - t_0)/2]$, which doesn't affect probabilities, the time evolution of a coherent state is simply given by a rotation of the characteristic number z in the complex plane. Thus both real and imaginary part of z oscillate with the oscillator frequency ω, and the coherent wave packet $|z(t)\rangle$ oscillates in position and momentum without changing its Gaussian shape or its widths, see Fig. 5.6(a).

The coherent state (5.61) represents a *minimal wave packet* in which the product of position uncertainty $\Delta_x = \beta/\sqrt{2}$ and momentum uncertainty $\Delta_p = \hbar/(\sqrt{2}\beta)$ takes on the minimum value $\hbar/2$ allowed by the uncertainty relation (1.34). This minimal property is a property of any Gaussian wave packet. Consider for example a Gaussian wave packet of the form (5.67) but with a different width β' in place of the natural oscillator width β of (5.53). Now the position uncertainty is $\Delta_x = \beta'/\sqrt{2}$ and the momentum uncertainty is $\Delta_p = \hbar/\left(\sqrt{2}\beta'\right)$. If β' is smaller than the natural oscillator width β given by (5.53), then the wave packet is *squeezed* in coordinate space in comparison with the coherent states; the momentum space distribution is correspondingly broader. If β' is larger than the natural oscillator width β, then the momentum distribution is narrower than for the coherent states; the wave packet is squeezed in momentum space.

The time evolution of squeezed states is not quite as simple as for the coherent states, but almost. The Wigner function of any Gaussian wave packet has the form (5.68) (with the appropriate width parameter) and is non-negative. Its time evolution follows the quantum mechanical von Neumann equation and is exactly the same as the time evolution of a numerically identical classical phase space density according to the Liouville equation, because the potential is harmonic. So the Wigner function follows the classical trajectories in phase space, and these are concentric circles which are traversed uniformly with a period $2\pi/\omega$. The Wigner function thus executes a circular motion in phase space, during which it keeps its shape but changes its orientation with respect to the position and momentum axes as illustrated in Figs. 5.6(b) and (c). (Note that all Wigner functions, and not only Gaussian wave packets, evolve in this way as long as the potential is harmonic.) Figure 5.6(b) shows the time evolution of a minimal wave packet which is squeezed in position ($\beta' = \beta/2$) and starts at its maximum (positive) displacement at time $t = 0$. After one quarter of a period, $\omega t = \pi/2$, it has moved to $x = 0$ and is now squeezed in momentum, after half a period it has moved to its maximum negative displacement and is again squeezed in position, and so it goes on until it returns to the original state after a whole period. Figure 5.6(c) on the other hand shows the time evolution of a minimal wave packet which is squeezed in momentum at $t = 0$, after a quarter of a period it is squeezed in position, etc., etc.. A time-independent way of classifying the squeezed nature of the states is to call the wave packet in Fig. 5.6(b) *squeezed in amplitude*, $(p^2/\mu + \mu\omega^2 x^2)^{1/2}$, and the wave packet in Fig. 5.6(c) *squeezed in phase*, $\arctan(p/\mu\omega x)$.

In the quantum mechanical description of the electromagnetic field in Sect. 2.4.2 we treated the photons in a given mode as quanta of a harmonic oscillator. For a single mode λ the equations (2.159), (2.160) become

$$\begin{aligned} \boldsymbol{A} &= \frac{\boldsymbol{\pi}_\lambda}{L^{3/2}}(q_\lambda \mathrm{e}^{-\mathrm{i}\omega_\lambda t} + q_\lambda^* \mathrm{e}^{+\mathrm{i}\omega_\lambda t}), \\ \boldsymbol{E} &= \frac{\boldsymbol{\pi}_\lambda}{L^{3/2}} \frac{\mathrm{i}\omega_\lambda}{c}(q_\lambda \mathrm{e}^{-\mathrm{i}\omega_\lambda t} - q_\lambda^* \mathrm{e}^{+\mathrm{i}\omega_\lambda t}), \\ \boldsymbol{B} &= \frac{\mathrm{i}\boldsymbol{k}_\lambda \times \boldsymbol{\pi}_\lambda}{L^{3/2}}(q_\lambda \mathrm{e}^{-\mathrm{i}\omega_\lambda t} - q_\lambda^* \mathrm{e}^{+\mathrm{i}\omega_\lambda t}). \end{aligned} \tag{5.72}$$

We have invoked the dipole approximation ($\exp(i\boldsymbol{k}_\lambda \cdot \boldsymbol{r}) \approx 1$), because this keeps formulae simple and we are not concerned with the spatial structure of the fields at the moment. If we replace the amplitudes q_λ and q_λ^* by position and momentum variables according to (2.162) and drop the factors $\exp(\pm\mathrm{i}\omega_\lambda t)$ in order to move from the Heisenberg representation to the Schrödinger representation as suggested by (2.170), then we obtain the following relations connecting the electromagnetic field operators to the position and momentum operators $\hat{x}_\lambda$ and $\hat{p}_\lambda$ of the harmonic oscillator associated with the mode λ (in the Schrödinger representation):

$$\begin{aligned} \hat{\boldsymbol{A}} &= \frac{\boldsymbol{\pi}_\lambda}{L^{3/2}}\sqrt{4\pi c^2}\,\hat{x}_\lambda, \\ \hat{\boldsymbol{E}} &= -\frac{\boldsymbol{\pi}_\lambda}{L^{3/2}}\sqrt{4\pi}\,\hat{p}_\lambda, \qquad \hat{\boldsymbol{B}} = -\frac{\boldsymbol{k}_\lambda \times \boldsymbol{\pi}_\lambda}{|\boldsymbol{k}_\lambda| L^{3/2}}\sqrt{4\pi}\,\hat{p}_\lambda. \end{aligned} \tag{5.73}$$

In a given mode λ the vector potential together with the electric or the magnetic field strength thus play the role of conjugate position and momentum variables for the harmonic oscillator describing this mode. (See also Problem 5.3.)

Coherent states play an important role in the investigation of the statistical properties of light in the framework of quantum optics. States of the electromagnetic field are usually called "classical" if they can be written as a superposition of coherent states $|z\rangle$ with a regular, non-negative amplitude function $P(z)$. A coherent state $|z_0\rangle$ itself would correspond to $P(z) = \delta(z - z_0)$, which would be at the edge of the classical regime defined in this way. A state of the field in which the photon number distribution is more sharply peaked than in a coherent state can in general not be represented in terms of superpositions of coherent states with regular, non-negative amplitude $P(z)$. This is the regime of "nonclassical light". An eigenstate of the field with a fixed finite number n_λ of photons in a given mode λ is an example of nonclassical light. The Wigner function (5.41) of such a state takes on negative values and hence cannot be interpreted as a classical phase space density, see Problem 5.3(c).

The time evolution of coherent states reflects the classical dynamics. The finite widths of their position and momentum distributions satisfy the requirements of the uncertainty relation. The creation and observation of squeezed states of light has been a subject of considerable interest for many years. The popularity of squeezed states stems from the fact that their uncertainty (in amplitude or phase) lies below the natural quantum mechanical uncertainty (of the coherent state), and this makes it possible to overcome limits to resolution due to natural quantum fluctuations in sensitive measurement processes [MS83]. For a detailed treatment of the quantum theory of light see e.g. [KS68, MS90, Lou00]. A special illumination of quantum optics from the point of view of phase space representations is given in [Sch01].

After all that has been said in this section we must not forget that the simple picture of a wave packet evolving along classical trajectories without changing its shape is bound to the harmonic nature of the Hamiltonian. This makes the classical oscillation frequency independent of the amplitude and the quantum mechanical energy levels equidistant. Life isn't always so simple, as can already be seen in the example of wave-packet spreading for a free particle. The concept of coherent states can however be used with advantage in other physical systems, e.g. in a space of angular momentum eigenstates. The eigenvalues of the z component of angular momentum are actually equidistant, but the spectrum for a given angular momentum quantum number l is bounded from above and below (1.58). For a general description of coherent states in systems characterized by various symmetry groups see e.g. [Per86, Hec87, ZF90].

5.2.3 Coherent Wave Packets in Real Systems

The harmonic oscillator treated in the preceding section is untypical for the dynamical evolution of wave packets in as far as two important results cannot be transferred to more general systems. Firstly, the evolution of the classical and the quantum mechanical phase space distributions is no longer the same if the potential contains anharmonic terms. Secondly, phase space distributions with finite uncertainties in position and momentum usually spread in coordinate space, even classically. A wave packet for a particle moving in a general potential may follow a classical trajectory in an average way, but beyond this there usually is dispersion, which can be understood classically, and there are genuine quantum mechanical effects resulting from terms indicated by the dots on the right-hand sides of equations (5.49), (5.50).

Considerable effort has gone into the search for coherent wave packets which are exact solutions of the Schrödinger equation and at the same time expose the correspondence to classical mechanics more clearly than the usual stationary eigenstates [Nau89, GD89, DS90, YM90]. The behaviour of wave packets in a Coulomb potential (1.134) is obviously of special interest in atomic physics. In a pure Coulomb potential the energy eigenvalues $E_n = -\mathcal{R}/n^2$ are highly degenerate. For each eigenvalue there are (without spin) n^2 eigenstates which can be labelled by the angular momentum quantum number $l = 0, 1, \dots\ n-1$ and the azimuthal quantum number $m = -l, \dots\ l$. In a pure Coulomb potential there is a further constant of motion in the form of the *Runge-Lenz vector*

$$\hat{\boldsymbol{M}} = \frac{1}{2\mu}(\hat{\boldsymbol{p}}\times\hat{\boldsymbol{L}} - \hat{\boldsymbol{L}}\times\hat{\boldsymbol{p}}) - e^2\frac{\boldsymbol{r}}{r}\,. \tag{5.74}$$

Classically its length is a measure for the excentricity of the closed Kepler ellipses, and it points in the direction of the larger principal axis. Using the components of the angular momentum $\hat{\boldsymbol{L}}$ and the Runge-Lenz vector (5.74) Nauenberg [Nau89] and Gay et al. [GD89] constructed a generalized angular momentum in two and three spatial dimensions respectively and searched for solutions of the Schrödinger equation with a minimum uncertainty in appropriate components of this generalized angular momentum. Superposing degenerate eigenstates with a given principal quantum number n in this way leads to a stationary solution of the Schrödinger equation which is no longer characterized by good angular momentum quantum numbers l and m, but which is optimally localized around a classical Kepler ellipse (see Fig. 5.7).

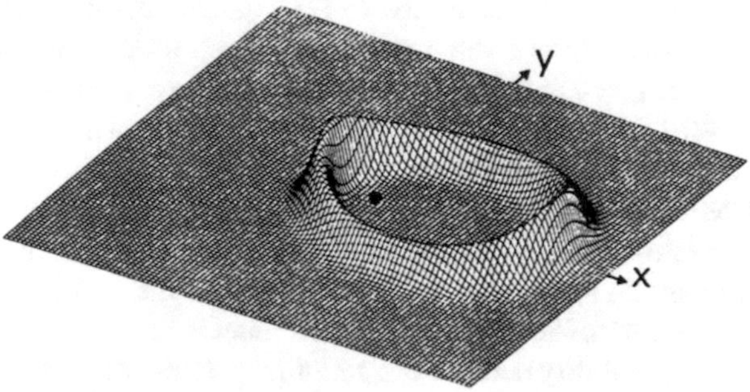

Fig. 5.7 Probability density $|\psi(\boldsymbol{r})|^2$ for a stationary solution of the Schrödinger equation in a pure Coulomb potential showing optimal localization around a Kepler ellipse of given excentricity (0.6 in this case) (From [GD89])

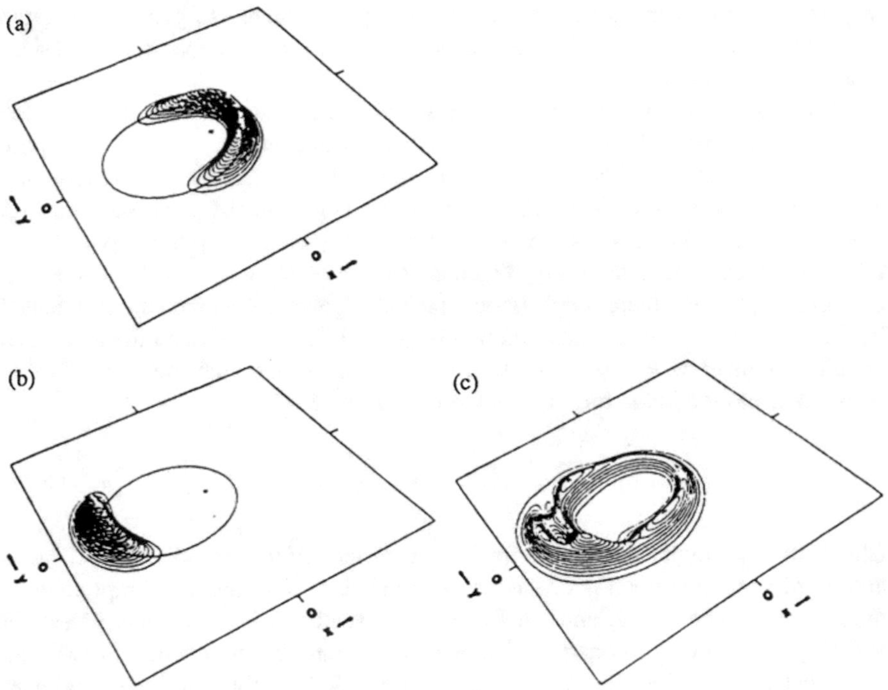

Fig. 5.8 Probability density for an initial wave packet which is localized around the perihelion of a Kepler ellipse (**a**). After half a revolution it is localized around the aphelion (**b**). In the course of time, spreading and interference effects become noticeable, as can already be seen after two revolutions (**c**) (From [Nau89])

In order to construct a non-stationary wave packet to simulate classical motion along a Kepler ellipse we have to superpose eigenstates with different principal quantum numbers n. The time evolution of a Gaussian superposition is shown in Fig. 5.8. Figure 5.8(a) shows a wave packet localized around the perihelion of a Kepler ellipse at time $t = 0$. After half a revolution the wave packet has arrived at the aphelion, Fig. 5.8(b). Localization along the trajectory is even a little narrower here. This is due to the slower speed near the aphelion and simply illustrates congestion. As time goes on the wave packet actually spreads. After two revolutions it has already spread out over the whole Kepler ellipse, Fig. 5.8(c). Figure 5.8(c) also shows signs of quantum mechanical interference where the faster head of the wave packet has caught up with the slower tail. These interference effects, which lead to oscillations in the probability density, are genuine quantum effects which cannot be described classically.

Coherent wave packets which are sharply localized and move along classical trajectories must be superpositions of stationary states involving different energies. Such wave packets can only be produced in the laboratory by perturbations of the

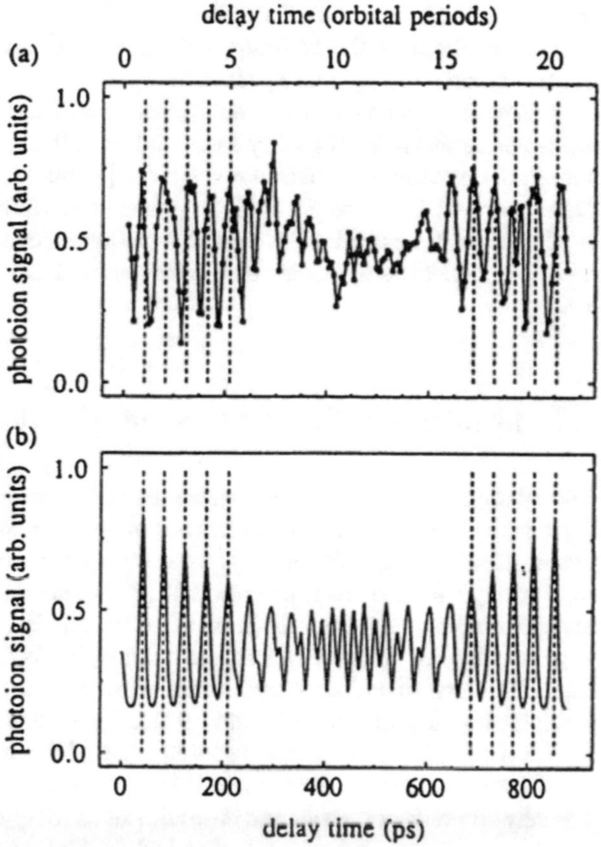

Fig. 5.9 Photoionization probability for $n \approx 65$ Rydberg states of a potassium atom which were excited by a 15 ps laser pulse. The abscissa shows the time delay of the second, the ionizing, pulse. (**a**) Experiment, (**b**) theoretical calculation (From [YM90])

Hamiltonian which are strongly localized both in space and in time. This can be achieved with intense laser pulses of durations of the order of picoseconds.

Figure 5.9 shows the results of an experiment in which Rydberg states around $n = 65$ in potassium were excited by a laser pulse of 15 picoseconds. At the corresponding energy the period of revolution for a classical Kepler ellipse is near 40 ps. The potassium atom is ionized by a second time-delayed laser pulse. Most of the time the excited electron is far away from the K^+ ion and, similar to a free electron, cannot absorb energy from the laser field (cf. Problem 5.5). There is an appreciable probability for ionization only if the electron is close to the K^+ ion, which happens every 40 picoseconds. The observed photoionization rate as a function of the time delay of the second laser pulse indeed shows maxima corresponding to this period. The signal is washed out after several periods due to spreading of the wave packet. A little later we observe a *revival* to a more or less coherent wave packet with oscillations again corresponding to the period of the classical revolution. The reason for this revival is that the time evolution of a state consisting of a superposition of a finite number of energy eigenstates always is quasiperiodic (or periodic). The coherence of the various interfering contributions is maintained during the evolution and enables the regeneration (to a large extent) of the original localized wave packet.

Review articles on electronic wave packets in Rydberg atoms were published by Alber and Zoller [AZ91] and by Jones and Noordham [JN98]. For a review on the subject of quantum wave packet revivals see [Rob04]. With continuing progress on the experimental side, very short laser pulses on the femtosecond timescale became available, and time resolved studies of wave-packet evolution are being extensively used to analyze the dynamics of atomic and molecular systems, see e.g. [EK99, TA16].

5.3 Regular and Chaotic Dynamics in Atoms

The relation between classical mechanics and quantum mechanics is understood reasonably well for systems which are *integrable*, meaning essentially that the classical motion is quasiperiodic and corresponds, in an appropriate representation, to a superposition of one-dimensional oscillations. Integrability is, however, the exception rather than the rule in classical mechanics, even for seemingly simple systems with few degrees of freedom. Although this has been known in principle since the work of Poincaré and others more than a hundred years ago, the far-reaching implications only became generally realized and accepted in the late 1970's [LL83, SJ05]. A tangible consequence of this realization was the explosive development field of the field of nonlinear dynamics, "chaos", which permeated into virtually all fields of physics and beyond. The continuing progress in understanding the rich and diverse behaviour in classical dynamics made it urgently desirable to understand if and how the nonlinearity of classical evolution survives the transition to strictly linear quantum mechanics [Haa01]. Simple atoms provided important examples of naturally occurring and experimentally accessible systems in which the quantum manifestations of classical chaos can be studied. The study of simple atoms, with or without the presence of external fields, led to important and exciting advances in our understanding of the relation between classical and quantum dynamics [GG89, Gay91, CK97, SS98]. A collection of articles by some of the most prominent researchers in this field is contained in [FE97]. A monograph on the subject was written by Blümel and Reinhardt [BR09].

5.3.1 Chaos in Classical Mechanics

The trajectories $(q_i(t), p_i(t))$ describing the evolution of a system with f degrees of freedom are solutions of the canonical equations (5.27) and, for given initial conditions $q_i(t_0)$, $p_i(t_0)$, they determine the state of the system for all later times. It is helpful to collect the $2f$ components $q_1, \ldots, q_f; p_1, \ldots p_f$ of a point in phase space in one symbol x. How regular or "chaotic" the classical motion is depends on how rapidly a small deviation Δx from a given trajectory $x(t)$ can grow in time. We generally regard a system as chaotic if a small deviation can increase exponentially in time which means that neighbouring trajectories *diverge exponentially.*

In order to formulate this statement more precisely we consider a given trajectory $x(t)$ and a small deviation $\Delta x(t_0)$ at time t_0. At a later time t_1 the trajectory which started at $x(t_0) + \Delta x(t_0)$ will deviate from the original trajectory by a separation $\Delta x(t_1)$. In the limit of infinitesimal deviations there is a linear relation connecting the deviations at time t_0 and at time t_1. Since the phase space points as well as the deviations Δx are quantities with $2f$ components, this linear relation is mediated by $2f \times 2f$ matrix which is called the *stability matrix* $\mathbf{M}(t_1, t_0)$:

$$\Delta x(t_1) = \mathbf{M}(t_1, t_0)\Delta x(t_0)\,. \tag{5.75}$$

Since Δx has several components, an initial deviation in one direction in phase space may grow strongly in the course of time, while an initial deviation in a different direction might increase at a slower rate or even become smaller. In a *conservative system* the Hamiltonian function H does not depend explicitly on time, and it follows from the special structure of the canonical equations (5.27) that the stability matrix is a *symplectic matrix*, which means,;

$$\mathbf{MJM}^\dagger = \mathbf{J}, \quad \mathbf{J} = \begin{pmatrix} \mathbf{0} & \mathbf{1} \\ -\mathbf{1} & \mathbf{0} \end{pmatrix}; \tag{5.76}$$

here $\mathbf{0}$ is the $f \times f$ matrix full of zeros and $\mathbf{1}$ is the $f \times f$ unit matrix. If λ_1 is an eigenvalue of $\mathbf{M}$, so are λ_1^*, $1/\lambda_1$ and $1/\lambda_1^*$. The $2f$ eigenvalues of the stability matrix occur in quartets or, if they are real or have unit modulus, in pairs. Their product is unity, expressing the fact that the total phase space volume of a set of initial conditions doesn't change in the course of the dynamical evolution in a conservative system *(Liouville's theorem).*

The definition of chaos is based on the fastest growing deviation from a given trajectory, and the growth rate is related to the *matrix norm* of the stability matrix. A matrix norm $\|\mathbf{M}\|$ is non-negative and can e.g. be defined as the square root of the largest eigenvalue of the Hermitian matrix $\mathbf{M}^\dagger\mathbf{M}$ [HJ85]. The dynamics is *unstable* in the point x in phase space if the norm of the stability matrix increases exponentially along the trajectory beginning with $x(t_0)$—more precisely, if the *Liapunov exponent*

$$\lambda \stackrel{\text{def}}{=} \lim_{t-t_0\to\infty} \frac{\ln \|\mathbf{M}(t, t_0)\|}{t - t_0}\,, \tag{5.77}$$

which is defined in the long-time limit, does not vanish but is positive. Roughly speaking this says that neighbouring trajectories diverge exponentially, and the Liapunov exponent (5.77) is the factor in the exponent which determines the rate of divergence (see Fig. 5.10).

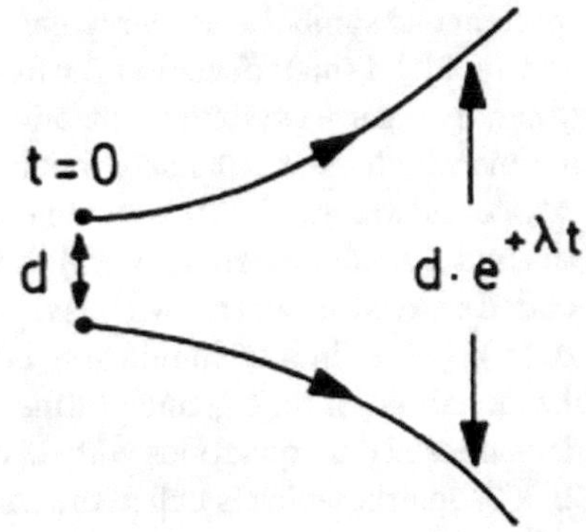

Fig. 5.10 Schematic illustration of the exponential divergence of neighbouring trajectories in phase space

The Liapunov exponent is a property of the classical trajectory; all phase space points along one trajectory have the same Liapunov exponent (see Problem 5.6). Every trajectory is either stable (if its Liapunov exponent vanishes), or unstable (if its Liapunov exponent is positive). An unstable trajectory need not be very complicated. Simple periodic trajectories (periodic orbits) can be stable or unstable. The Liapunov exponent of a periodic orbit of period T can be defined via the eigenvalues of the stability matrix over one period—the *monodromy matrix* $\mathbf{M}(T,0)$. If Λ is the largest modulus of an eigenvalue of $\mathbf{M}(T,0)$, then the norm of $\mathbf{M}(T,0)^n$ becomes equal to Λ^n for large n [HJ85] and the Liapunov exponent is given by,

$$\lambda = \lim_{n\to\infty} \frac{\ln(\Lambda^n)}{nT} = \frac{\ln\Lambda}{T} . \tag{5.78}$$

Instability of a periodic orbit means that infinitesimally small deviations lead to trajectories which move away from the periodic orbit at an exponential rate and hence cannot themselves be periodic (see Problem 5.7).

A region in phase space is chaotic if all trajectories are unstable. Chaos can already occur in a system with only one spatial degree of freedom if the Hamiltonian function depends explicitly on time. A periodic time dependence as caused by an oscillating external field is an important example. The simplest example of a mechanical system driven by a periodic force is the *periodically kicked rotor,* which has been studied in considerable detail in particular by Blümel and collaborators [BR09]. The Hamiltonian function is

$$H(\theta;p;t) = \frac{p^2}{2} + k\cos\theta \sum_n \delta(t-nT) . \tag{5.79}$$

The coordinate θ describes the rotation around a fixed axis and p is the associated canonically conjugate angular momentum (the moment of inertia is unity). At the end of each period T the rotor gets a kick, the strength of which is determined by the coefficient k and the momentary angle θ (see Fig. 5.11). The kick changes the angular momentum by $k\sin\theta$. Between two kicks the rotor rotates freely so that the

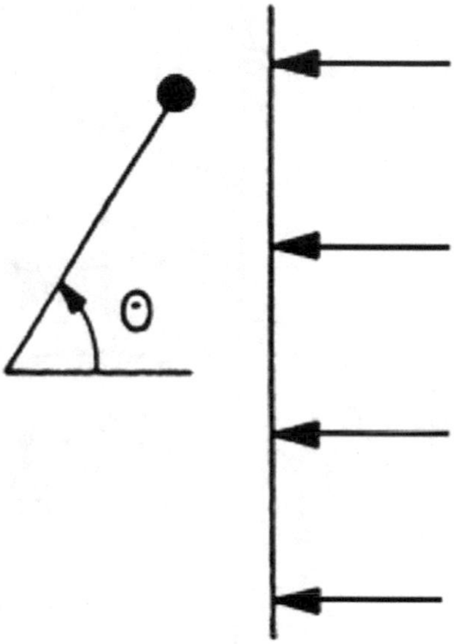

Fig. 5.11 The periodically kicked rotor. At time nT it experiences a torque $k \sin\theta\, \delta(t - nT)$. Whether a kick accelerates or decelerates the rotational motion depends on the sense of rotation and the angle θ at the time of the kick

angle increases by pT in a period. The angle θ_{n+1} and the angular momentum p_{n+1} after $n + 1$ periods can thus be expressed by the following recursion relation:

$$p_{n+1} = p_n + k \sin\theta_n\,, \quad \theta_{n+1} = \theta_n + p_{n+1}T\,. \tag{5.80}$$

This equation describes the entire dynamics of the kicked rotor as a mapping of the two-dimensional phase space into itself. Because of its fundamental importance it is known as the *standard mapping*. A trajectory which begins at $\theta = \theta_0, p = p_0$ at time $t = 0$ is completely described by the sequence of points (θ_n, p_n), $n = 0, 1, 2, \ldots$.

The dynamics described by the standard mapping (5.80) can be quite complicated, as can be seen by studying the sequence of points (θ_n, p_n) in phase space. In the *integrable limit* $k = 0$ we have uniform rotation, the angular momentum p is constant and the angle θ increases by pT each period. The points (θ_n, p_n) of a trajectory in phase space all lie on the straight line p = const.. Obviously a small deviation in initial conditions can only grow linearly in time in this case. For finite values of k—actually kT is the relevant quantity—we observe more structure in phase space. Figure 5.12 shows the sequence of points (θ_n, p_n) generated by five different sets of initial conditions for $kT = 0.97$. We can clearly distinguish two different types of trajectories: regular trajectories for which all points lie on a one-dimensional curve, and irregular trajectories whose points (θ_n, p_n) are spattered more or less uniformly over a finite area in phase space. The two regular trajectories in Fig. 5.12 describe *quasiperiodic* motion and the associated curves in phase space form boundaries which cannot be crossed by other trajectories and hence divide phase space into separated regions. Detailed numerical calculations by Greene [Gre79] and others have shown that the share of irregular or chaotic trajectories increases with increasing values of the parameter kT. For large values

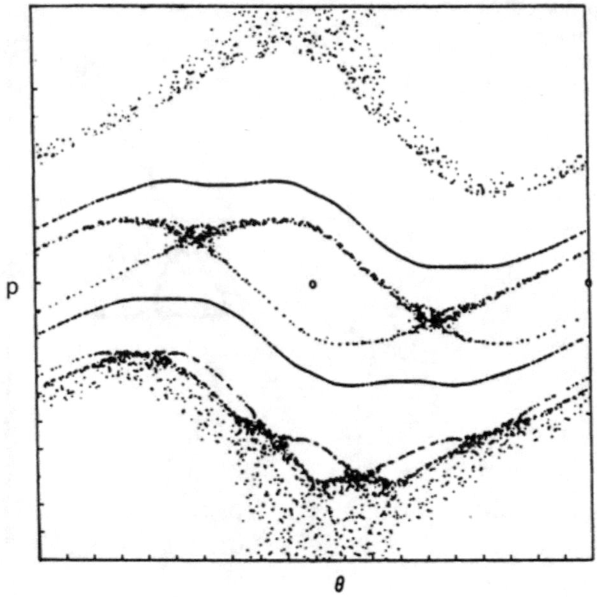

Fig. 5.12 Trajectories of the periodically kicked rotor (5.79), (5.80) in phase space for a coupling constant $kT = 0.97$ (from [Gre79])

of kT the boundary curves break up and the irregular trajectories can explore the whole of phase space. Numerical calculations also show that the distribution $P(p)$ of angular momenta becomes a Gaussian after a large number n of kicks, provided kT is sufficiently large, and that the square of the width of this Gaussian grows linearly with n as in ordinary diffusion or random walk processes. After n periods we have [CF86]

$$P(p) \approx \left(kT\sqrt{n\pi}\right)^{-1} \mathrm{e}^{-p^2/[n(kT)^2]} ,$$
$$\langle p^2 \rangle = \int p^2 P(p)\, \mathrm{d}p \approx \frac{1}{2} n (kT)^2 . \tag{5.81}$$

As p^2 is proportional to the kinetic energy of the system, (5.81) implies that the energy distribution of the system broadens as in diffusion.

In a conservative system the Hamiltonian function H does not depend explicitly on time, the energy $H(q_1(t), \dots\ q_f(t); p_1(t) \dots\ p_f(t))$ of the system is always an *integral of motion* and all trajectories with the same energy move on a $(2f-1)$-dimensional subspace of phase space called the *energy shell*. In a one-dimensional conservative system any bound motion is a (not usually harmonic) oscillation between two classical turning points and hence is periodic. The trajectories are closed curves in the two-dimensional phase space, see Figs. 5.13(a) and (b). A small deviation from a given trajectory leads to a slightly different trajectory which again

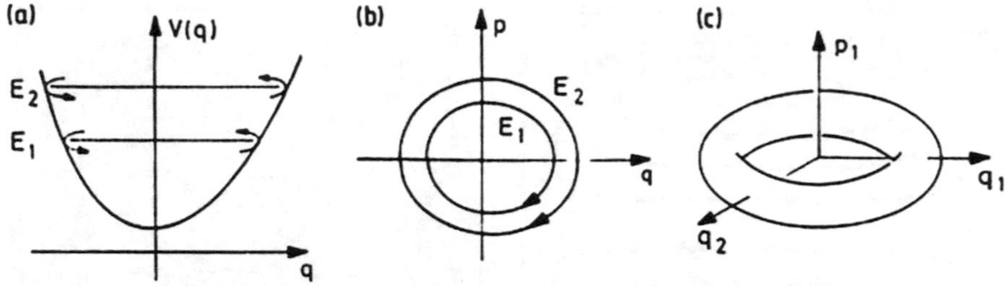

Fig. 5.13 (**a**) Bound motion in a one-dimensional conservative system, $H(q,p) = \frac{1}{2}p^2 + V(q)$. (**b**) Periodic trajectories of the one-dimensional conservative system in phase space. (**c**) Two-dimensional torus in the three-dimensional energy shell of a conservative system with $f = 2$ degrees of freedom

is periodic and the separation of two trajectories can only grow linearly in time. Such a system has no chaos (although there may be isolated unstable points).

The simplest conservative potentials capable of being chaotic have $f = 2$ degrees of freedom. Chaos is possible if the system is not integrable, i.e. if there is no further integral of motion. Otherwise the motion of the system is usually periodic or quasiperiodic. In a two-dimensional system with two independent integrals of motion a trajectory in four-dimensional phase space is confined to a two-dimensional surface which usually has the topology of a *torus*. The parameters of the torus are determined by the energy and the second integral of motion, see Fig. 5.13(c). More generally: a mechanical system with f degrees of freedom is called *integrable* if its Hamiltonian function can be written as a function of f independent integrals of motion and no longer depends on the associated canonically conjugate variables [Gol80]. In an integrable system all Liapunov exponents vanish [Mey86]. The f integrals of motion confine the trajectories in $2f$-dimensional phase space to f-dimensional subspaces which are also called "tori" if $f > 2$.

Two anharmonically coupled oscillators already provide an example of a two-dimensional conservative system which isn't integrable. To be specific let's consider the Hamiltonian function

$$H = \frac{1}{2}\left(p_1^2 + p_2^2 + q_1^4 + q_2^4 + \gamma q_1^2 q_2^2\right) . \tag{5.82}$$

The potential energy $V = (q_1^4 + q_2^4 + \gamma q_1^2 q_2^2)/2$ in (5.82) is a homogeneous function of the coordinates, $V(\sigma q_1, \sigma q_2) = \sigma^d V(q_1, q_2)$, with $d = 4$. Hence the dynamics is essentially independent of energy, see Sect. 5.3.4. The properties of the dynamics are determined by the coupling parameter γ. In the integrable limit $\gamma = 0$ the motion factorizes into two independent periodic oscillations in the variables q_1 and q_2.

We can visualize the dynamics in a conservative system with $f = 2$ degrees of freedom in a way similar to Fig. 5.12 if we look at a two-dimensional surface of section of the three-dimensional energy shell and register the points at which a trajectory pierces this surface (perhaps subject to a condition concerning the

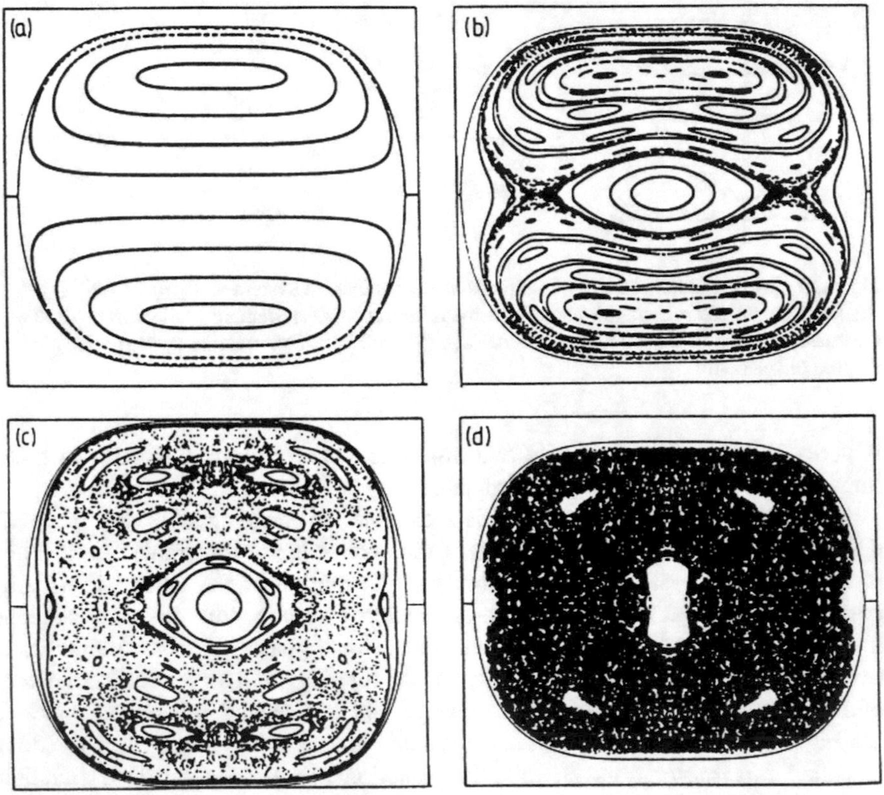

Fig. 5.14 Poincaré surfaces of section for the system (5.82) with the following values of the coupling parameter γ: 6 (**a**), 7 (**b**), 8 (**c**) and 12 (**d**). The surface of section is the p_1q_1-plane at $q_2 = 0$ (From [Eck88])

direction of the motion normal to the surface). The resulting figure is called a *Poincaré surface of section.* A periodic trajectory appears on a Poincaré surface of section as a single point or a finite (small) number of points. A quasiperiodic trajectory running on a two-dimensional torus in the energy shell appears as a one-dimensional curve on the Poincaré surface of section, similar to Fig. 5.12. An irregular or chaotic trajectory, which densely fills a finite three-dimensional volume in the energy shell, covers a finite area of the Poincaré surface of section with more or less uniformly spattered points. Figure 5.14 shows Poincaré surfaces of section for the system (5.82) at four different values of the coupling constant γ. At $\gamma = 6$ the motion is still largely on regular tori. With increasing values of the coupling constant the share of phase space filled with irregular trajectories becomes bigger and bigger. At $\gamma = 12$ the whole of phase space is filled with irregular trajectories, except for small islands of regularity. For a numerical calculation of the Liapunov exponents of the trajectories in this example see [Mey86].

5.3.2 *Traces of Chaos in Quantum Mechanics*

Both the concept of Liapunov exponents and the picture of Poincaré surfaces of section are defined via classical trajectories and cannot be transferred to quantum mechanics in an obvious way. We shall not enter here into the frequently controversial discussion on how to define "quantum chaos" or whether or not this concept makes sense at all. Instead we shall turn to the more modest question of how the fact that a classical system is chaotic affects the corresponding quantum mechanical system.

The quantum mechanical version of the periodically kicked rotor (5.79) is described by the Hamiltonian operator

$$\hat{H} = -\frac{\hbar^2}{2}\frac{\partial^2}{\partial\theta^2} + k\cos\theta \sum_n \delta(t - nT)\,. \tag{5.83}$$

Solutions of the time-independent Schrödinger equation (1.38) can easily be constructed with the help of the time evolution operator (1.40). To this end we expand the wave function $\psi(\theta, t)$ in a Fourier series in the angle θ,

$$\psi(\theta, t) = \sum_{\nu=-\infty}^{\infty} c_\nu\, \mathrm{e}^{\mathrm{i}\nu\theta}\,, \tag{5.84}$$

which is the same as expanding in eigenstates of the free rotor ($k = 0$). The Hamiltonian is time-independent between two kicks so that the time evolution (1.41) simply amounts to multiplication of the basis functions $\exp(\mathrm{i}\nu\theta)$ by the respective factors $\exp[-\mathrm{i}(\hbar/2)\nu^2 T]$. In the infinitesimally short time between t_- immediately before and t_+ immediately after a kick the Hamiltonian depends explicitly on time and we have to replace the product $\hat{H}(t_+ - t_-)$ in the time evolution operator by the integral $\int_{t_-}^{t_+} \hat{H}(t)\mathrm{d}t$. Thus the wave function ψ is just multiplied by $\exp(-\mathrm{i}k\cos\theta/\hbar)$ during a kick.

If $\psi_n(\theta) = \sum_\nu c_\nu(n)\exp(\mathrm{i}\nu\theta)$ is the wave function immediately after the nth kick, then the wave function one period later is

$$\psi_{n+1}(\theta) = \mathrm{e}^{-\mathrm{i}k\cos\theta/\hbar} \sum_{\nu=-\infty}^{\infty} c_\nu(n)\, \mathrm{e}^{\mathrm{i}(\nu\theta - \hbar T\nu^2/2)}\,, \tag{5.85}$$

and its expansion in a Fourier series defines a new set of coefficients $c_\nu(n+1)$ (see e.g. [Eck88]).

The search for traces of chaos led to the question, whether the quantum mechanical evolution according to (5.85) involves diffusive behaviour and a linear increase of the kinetic energy in time or in number of kicks as in (5.81). If the period T is an integral multiple of $4\pi/\hbar$, then the wave function is simply multiplied by a factor $\exp(-\mathrm{i}k\cos\theta/\hbar)$ each period. In case of such a *resonance* the kinetic energy even increases quadratically with the number of kicks. According to [IS79] such resonances, for which there is no classical equivalent, occur whenever the period T is a rational multiple of $\pi/\hbar$. The time evolution (5.85) away from resonances was investigated numerically by Casati et al. [CF86]. This led to the following picture: For small times an initially localized distribution with only one or few non-vanishing coefficients c_ν spreads diffusively at first, but with a smaller diffusion constant than in the corresponding classical case. After a certain time t_S a saturation is reached, the diffusive spreading ceases and we have quasiperiodic motion in phase space. The time t_S is larger if $\hbar$ is smaller. Thus classical chaos is suppressed in quantum mechanics by the finite value of $\hbar$ [Cas90]. For more details on the classical and quantum dynamics of the kicked rotor the reader is referred to [Blu97, Haa01, BR09].

A conservative quantum mechanical system is primarily characterized by its spectrum of energy eigenvalues. In a bound system the spectrum is discrete. A state in a bounded energy interval is always a superposition of a finite number of energy eigenstates and so its time evolution must be (at least) quasiperiodic. At sufficiently high excitation energies and level densities the spectrum may nevertheless be very complicated, and the investigation of statistical properties of spectra has revealed connections to the regular or chaotic nature of the corresponding classical dynamics. Some of the more basic results are presented below; more details are contained e.g. in the monograph by Haake [Haa01].

The opposite of a (classically) chaotic system is an integrable system with a Hamiltonian function which can be expressed in terms of integrals of motion. The corresponding quantum mechanical Hamiltonian operator should then be a corresponding function of constants of motion so that the energy eigenvalues depend on several independent good quantum numbers. The eigenvalues e.g. of a separable Hamiltonian of the form

$$\hat{H} = \hat{H}_1 + \hat{H}_2 + \cdots + \hat{H}_N \tag{5.86}$$

are just sums of the eigenvalues E_{n_i} of the operators $\hat{H}_i$,

$$E_{n_1,n_2,\cdots n_N} = E_{n_1} + E_{n_2} + \cdots + E_{n_N} \,. \tag{5.87}$$

If the individual eigenvalue sequences E_{n_i}, $n_i = 1, 2, \ldots$ are not correlated, then the summation in (5.87) produces a rather irregular sequence of eigenvalues for the whole system, somewhat similar to a sequence of randomly distributed numbers. Such a random spectrum is called a *Poisson Spectrum*.

If the classical system is chaotic it will probably not be possible to label the energy eigenvalues of the corresponding quantum mechanical system by good quantum numbers in a straightforward way. The energy eigenvalues are eigenvalues of a Hermitian matrix. When there are no good quantum numbers at all (apart from the energy) one tries to understand the spectrum by studying the spectra generated by *random matrices* [GM98]; these are matrices whose elements are distributed randomly subject to certain restrictions. One generally considers a whole ensemble of Hermitian $N \times N$ matrices, whose matrix elements are individually and independently randomly distributed. One property of the eigenvalue distribution in the limit $N \to \infty$ is the concentration on a semicircle of unit radius (with appropriate choice of energy scale) centred around a mean value. The probability of realization of a particular matrix is given by

$$P(H) \propto \exp\left[-\frac{N}{\lambda\,\sigma^2}\mathrm{Tr}(H^2)\right] \tag{5.88}$$

with $\lambda = 2$; the parameter σ defines the energy scale. The probability (5.88) is invariant under a unitary transformation of the matrix H,

$$H'_{i,j} = \langle\psi'_i|\hat{H}|\psi'_j\rangle = \sum_{k,l} U^*_{k,i}H_{k,l}U_{l,j} \quad \text{or} \quad H' = U^\dagger H U\,, \tag{5.89}$$

i.e., it does not depend on the choice of basis. The corresponding ensemble of random matrices is called a *Gaussian unitary ensemble* GUE.

In some cases, e.g. for the coupled oscillators (5.82), we can assume that the matrix of the quantum mechanical Hamiltonian is not only Hermitian but real and symmetric. It is then reasonable to replace the requirement of invariance under unitary transformations by the requirement that the probability for a given real and symmetric random matrix be invariant under *orthogonal transformations*; these are transformations of the form (5.89) except that the unitary matrix U is replaced by an orthogonal matrix O (whose transposed matrix is equal to its inverse). The ensemble of random matrices is now called a *Gaussian orthogonal ensemble* GOE, and the parameter entering in the definition (5.88) is $\lambda = 4$.

Although exact proofs are scarce, the results of many numerical experiments indicate that a quantum mechanical spectrum shows similarities to a random or Poisson spectrum if the corresponding classical system is regular, and to the spectrum of random matrices (GOE or GUE) if the corresponding classical system is chaotic.

In order to formulate these statements more quantitatively we consider a spectrum $E_1 \le E_2 \le \ldots \le E_n \le \ldots$. Such a spectrum can be expressed in terms of the *mode number*

$$N(E) = \sum_n \Theta(E - E_n)\,. \tag{5.90}$$

The step function $\Theta(x)$ vanishes for $x < 0$ and is unity for $x \ge 0$, so that $N(E)$ is just the number of eigenstates with energies up to (and including) E. The mode number $N(E)$ is a staircase function; it fluctuates around the *mean mode number* $\tilde{N}(E)$, which can be obtained by dividing the classically allowed region in $2f$-dimensional phase space by the fth power of $2\pi\hbar$. An example for $N(E)$ and $\tilde{N}(E)$ is illustrated in Fig. 5.15. The derivative of the mode number with respect to energy is the *level*

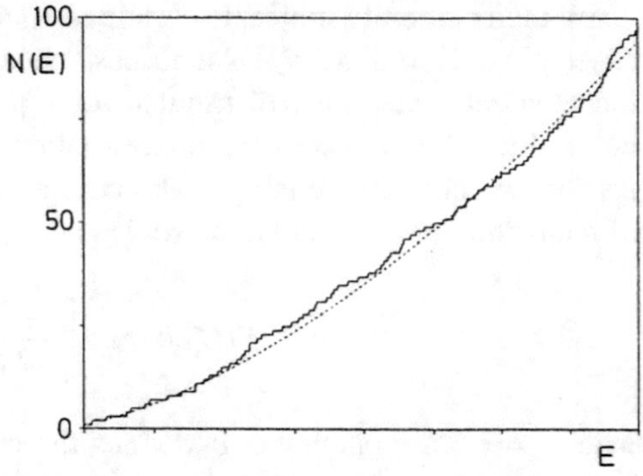

Fig. 5.15 Examples for the mode number $N(E)$ and the mean mode number $\widetilde{N}(E)$ (*dashed*)

density, and the derivative of the mean mode number defines the *mean level density*,

$$d(E) = \frac{\mathrm{d}N(E)}{\mathrm{d}E} = \sum_n \delta(E - E_n)\,, \quad \tilde{d}(E) = \frac{\mathrm{d}\tilde{N}(E)}{\mathrm{d}E}\,. \tag{5.91}$$

The statistical properties of a spectrum can best be studied if the weakly energy-dependent effects reflecting the mean level density are normalized away. This can be achieved by replacing the spectrum E_n by the sequence of numbers

$$\varepsilon_n = \tilde{N}(E_n) = \int_{E_1}^{E_n} \tilde{d}(E)\mathrm{d}E\,, \tag{5.92}$$

which has all the fluctuation properties of the original spectrum but corresponds to a mean level density of unity.

A frequently studied property of spectra is the distribution of the separations between neighbouring levels, $E_{n+1}-E_n$ or $\varepsilon_{n+1}-\varepsilon_n$, the so-called *nearest neighbour spacings* NNS. It is relatively straightforward to show that the NNS of a Poisson spectrum follow an exponential distribution (see Problem 5.8). For a mean level density unity the probability density $P(s)$ for a separation s of neighbouring levels is given by

$$P(s) = \mathrm{e}^{-s}\,. \tag{5.93}$$

The high probability for small separations of neighbouring levels expresses the fact that degeneracies or near degeneracies are not unusual if there are other good quantum numbers beside the energy, as is the case when the corresponding classical system is regular. On the other hand, if there are no further good quantum numbers, the residual interaction leads to repulsion of close lying states and hinders degeneracies (see Problem 1.6). It can actually be shown [Eck88, Haa01] that the NNS distribution $P(s)$ for the eigenvalue spectra of random matrices is proportional to s for small separations in the GOE case and to s^2 in the GUE case. The NNS

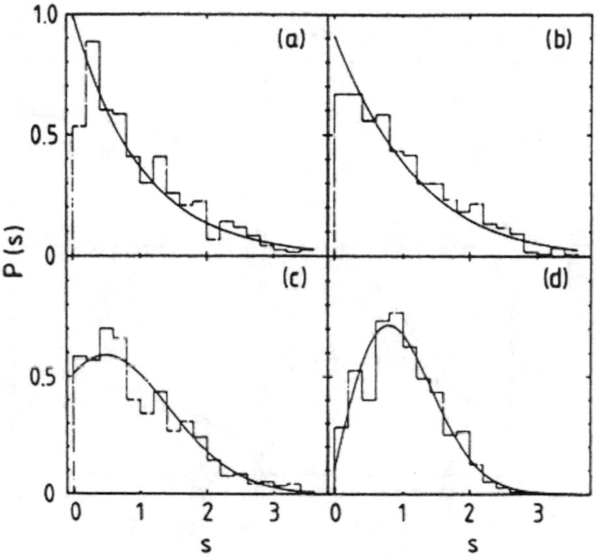

Fig. 5.16 NNS distributions of the quantum mechanical energy spectrum for the coupled oscillators (5.82). The four parts of the picture belong to the same values of the coupling parameter γ as in Fig. 5.14. The curve in (**a**) is the Poisson distribution (5.93). The curve in (**d**) is the Wigner distribution (5.94) (From [Eck88])

distribution in the GOE case is quite well approximated by a *Wigner distribution*

$$P(s) = \frac{\pi}{2} s \, \mathrm{e}^{\;(\pi/4)s^2} \,. \tag{5.94}$$

Figure 5.16 shows the NNS distributions for energy spectra of the Hamiltonian operator corresponding to the Hamiltonian function (5.82). The four parts of the picture belong to the same four values of the coupling parameter as in Fig. 5.14. Note the transition from a Poisson distribution (5.93) at $\gamma = 6$ (a), where the classical dynamics is still largely regular, to the Wigner distribution (5.94) at $\gamma = 12$ (d), where the classical dynamics is largely irregular.

Higher correlations of the spectrum can be studied via various statistical measures [BG84, BH85]. One popular measure is the *spectral rigidity* $\Delta_3(L)$ which measures the deviation of the mode number from a straight line over a stretch of spectrum of length L,

$$\Delta_3(L) = \frac{1}{L} \min_{A,B} \int_x^{x+L} [N(\varepsilon) - A\varepsilon - B]^2 \mathrm{d}\varepsilon \,. \tag{5.95}$$

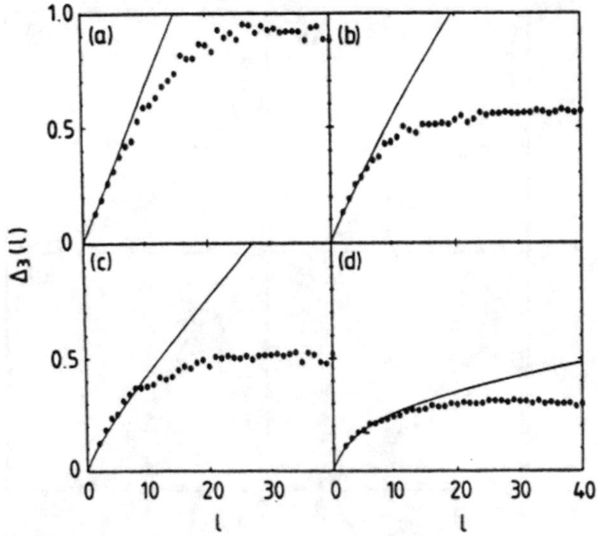

Fig. 5.17 Spectral rigidity of the quantum mechanical energy spectrum of the coupled oscillators (5.82). The four parts of the figure correspond to the same values of the coupling parameter γ as in Figs. 5.14 and 5.16. The straight line in (**a**) is the expectation (5.96) for a Poisson spectrum. The curve in (**d**) is the function (5.97) expected for a GOE spectrum (From [Eck88])

Δ_3 is on the average independent of the starting point x in the special cases discussed above. The dependence of Δ_3 on L is linear for a Poisson spectrum,

$$\Delta_3(L) = \frac{L}{15}, \tag{5.96}$$

and approximately logarithmic for a GOE spectrum,

$$\Delta_3 \approx \frac{1}{\pi^2} \ln(L) - 0.007, \quad L \gg 1. \tag{5.97}$$

(See e.g. [BG84] for further details.) Figure 5.17 shows the spectral rigidity for the coupled oscillators (5.82). The four parts of the figure again correspond to the same four values of the coupling constant γ as in Figs. 5.14 and 5.16.

Figures 5.14, 5.16 and 5.17 clearly show that the transition from regularity to chaos in the classical system is accompanied by a simultaneous transition in the statistical properties of the energy spectrum of the corresponding quantum mechanical system. The NNS distribution and the spectral rigidity (and further statistical measures—see e.g. [BH85]) correspond to the expectations for a Poisson spectrum in the classically regular regime and to the expectations for an ensemble of random matrices in the classically chaotic regime. Beware of over-interpretations of this statement! It does not mean that the quantum mechanical spectrum of a classically chaotic system is identical in detail to a random matrix spectrum. All the eigenvalues of a Hamiltonian together contain much more information than a small number of statistical measures. The identity of the physical system itself is still contained in the spectrum and can be extracted e.g. by analysing long-ranged spectral correlations, as discussed in the following section. It is equally obvious that the spectrum of a classically regular system won't be identical in detail to a Poisson spectrum, even if the NNS distribution and other statistical measures agree with the corresponding expectations.

This section concludes with a further warning, namely that there are individual physical systems whose behaviour deviates from the *generic* behaviour described above. Consider e.g. a system of harmonic oscillators which is always integrable and can be characterized by its normal modes. If the frequencies are commensurable, then all energy eigenvalues (without zero-point energy) are integral multiples of a smallest energy. There are many exact degeneracies but no level spacings between zero and this smallest energy. The NNS distribution will never approach the Poisson distribution (5.93) no matter how many states are included in the statistical analysis.

5.3.3 Semiclassical Periodic Orbit Theory

The use of classical trajectories and in particular of periodic orbits in the analysis of quantum mechanical spectra has a long history [Gut97], and it has become an important instrument for understanding and describing the quantum mechanics of systems whose corresponding classical dynamics may be integrable or not integrable [Cha92, FE97, BR09]. Elements of the theory are sketched here for the case of a conservative system with f degrees of freedom. A detailed elaboration is contained in the book by Brack and Bhaduri [BB97a].

The starting point is the quantum mechanical propagator or Green's function $G(q_a, t_a; q_b, t_b)$, which describes the time evolution of a quantum mechanical wave function in coordinate space,

$$\psi(q_b, t_b) = \int G(q_a, t_a; q_b, t_b)\psi(q_a, t_a)\, \mathrm{d}q_a\,, \tag{5.98}$$

and is just the coordinate representation of the time evolution operator introduced in Sect. 1.1.3,

$$G(q_a, t_a; q_b, t_b) = \langle q_b|\hat{U}(t_b, t_a)|q_a\rangle\,. \tag{5.99}$$

In Feynman's *path integral* formulation of quantum mechanics, the propagator is written as

$$G(q_a,t_a;q_b,t_b) = \int \mathcal{D}[q] \exp\left(\frac{\mathrm{i}}{\hbar}\int_{t_a}^{t_b} L(q,\dot{q})\,\mathrm{d}t\right), \tag{5.100}$$

where $L(q_1,\dots q_f;\dot{q}_1,\dots \dot{q}_f)$ is the classical *Lagrangian*, which is related to the Hamiltonian function $H(q_1,\dots q_f;p_1\dots p_f)$ by

$$L(q_1,\dots q_f;\dot{q}_1,\dots \dot{q}_f) = \sum_{i=1}^{f} p_i\dot{q}_i - H(q_1,\dots q_f;p_1\dots p_f)\,. \tag{5.101}$$

The symbol $\int \mathcal{D}[q]$ in (5.100) stands for a mathematically non-trivial integration over all paths in coordinate space connecting the initial coordinate q_a at time t_a to the final coordinate q_b at time t_b.

A semiclassical approximation of the propagator is derived using the stationary phase approximation in much the same way as it was used to derive a semiclassical approximation to the integral representation of the scattering amplitude in Sect. 4.1.11. The condition of stationary phase for the integrand in (5.100) selects those paths between (q_a,t_a) and (q_b,t_b) for which the Lagrangian action,

$$W(q_a,q_b;t_b-t_a) = \int_{t_a}^{t_b} L(q,\dot{q})\,\mathrm{d}t\,, \tag{5.102}$$

is stationary under infinitesimal variations of path, and these are just those paths which fulfill the classical equations of motion, i.e. the classical trajectories [LL71]. The resulting semiclassical expression for the propagator is,

$$G_{\mathrm{sc}}(q_a,t_a;q_b,t_b) = (2\pi\mathrm{i}\hbar)^{-f/2} \times \sum_{\mathrm{cl.traj}} \sqrt{\left|\det\frac{\partial^2 W}{\partial q_a\partial q_b}\right|}\,\mathrm{e}^{-\mathrm{i}\kappa_{\mathrm{traj}}\pi/2} \exp\left[\frac{\mathrm{i}}{\hbar}W(q_a,q_b;t_b-t_a)\right]. \tag{5.103}$$

The significance of the various contributions to the expression on the right hand side of (5.103) can be appreciated by recalling the expression (4.138) for the semiclassical scattering amplitude in Sect. 4.1.11. Each term in the sum corresponds to a classical trajectory and carries a phase given by the action along the trajectory. The weight of each contribution is related to the density of trajectories and is given by the square root of the determinant of the $f\times f$ matrix of second derivatives of the Lagrangian action, which is called the *van Vleck determinant* and becomes singular at focal points. Each term also contains an additional phase $\kappa_{\mathrm{traj}}\pi/2$ where κ_{traj} counts the number of focal points along the trajectory.

A connection to the quantum mechanical energy spectrum can be made by realizing that the Fourier transform of the time evolution operator, $\hat{U}(t_b, t_a) = \exp\left[-\frac{\mathrm{i}}{\hbar}\hat{H}(t_b - t_a)\right]$, is,

$$\int_0^\infty \mathrm{e}^{(\mathrm{i}/\hbar)Et}\,\hat{U}(t,0)\,\mathrm{d}t = \int_0^\infty \mathrm{e}^{(\mathrm{i}/\hbar)(E-\hat{H})t}\mathrm{d}t$$
$$= \left.\frac{\hbar\,\mathrm{e}^{(\mathrm{i}/\hbar)(E-\hat{H})t}}{\mathrm{i}(E-\hat{H})}\right|_0^\infty = \frac{\mathrm{i}\hbar}{E-\hat{H}}, \tag{5.104}$$

where the contribution at $t = \infty$ is argued to vanish via an infinitesimal positive imaginary contribution to the energy E. An analogous Fourier transform of the Green's function (5.99) is just the coordinate representation of the term on the right of lower line of (5.104),

$$\tilde{G}(q_a, q_b; E) \stackrel{\text{def}}{=} \frac{1}{\mathrm{i}\hbar}\int_0^\infty \mathrm{e}^{(\mathrm{i}/\hbar)Et} G(q_a, 0; q_b, t)\,\mathrm{d}t$$
$$= \langle q_b \,|\frac{1}{E-\hat{H}}|\, q_a\rangle = \sum_n \frac{\psi_n(q_b)\,\psi_n(q_a)^*}{E-E_n}. \tag{5.105}$$

The expression on the right of the lower line of (5.105) is obtained by inserting the unit operator expressed via a complete set of energy eigenfunctions $\psi_n(q)$ with eigenvalues E_n according to (1.22). The Green's function (5.105) is thus a sum over pole terms, one for each eigenstate, and the respective residua are defined by the product of the eigenfunctions' values at q_a and q_b. Taking the trace eliminates the dependence on the wave functions,

$$\mathrm{Tr}[\tilde{G}(E)] = \int \tilde{G}(q, q; E)\,\mathrm{d}q = \sum_n \frac{1}{E-E_n}. \tag{5.106}$$

The pole terms $1/(E-E_n)$ consist of a real principle value singularity at $E = E_n$ plus an imaginary component proportional to $\delta(E-E_n)$, which can be traced back to the infinitesimal imaginary contribution to the energy mentioned above,

$$\lim_{\varepsilon\to 0}\frac{1}{E+\mathrm{i}\varepsilon-E_n} = \mathcal{P}\left(\frac{1}{E-E_n}\right) - \mathrm{i}\pi\delta(E-E_n), \tag{5.107}$$

so the imaginary part of the trace of the Green's function is directly proportional to the energy level density (5.91),

$$d(E) = -\frac{1}{\pi}\Im\left\{\mathrm{Tr}\left[\tilde{G}(E)\right]\right\}. \tag{5.108}$$

A semiclassical approximation to the energy-level density can thus be obtained by subjecting the semiclassical approximation (5.103) of the time Green's function (propagator) to the Fourier transformation (5.105) and inserting the trace of the result into (5.108). The Fourier transformation introduces an integral over time, so the Fourier transformed Green's function contains contributions from all classical trajectories which travel from q_a to q_b in any time t. Approximating the time integrals with the help of the stationary phase approximation selects only those trajectories whose conserved energy is equal to the energy E in the argument of the Fourier transformed (approximate) Green's function, and the result is,

$$\tilde{G}_{\rm sc}\,(q_a, q_b; E) \\ = \frac{2\pi}{(2\pi \mathrm{i}\hbar)^{(f+1)/2}} \sum_{\rm cl.traj.} \sqrt{|D|}\, \exp\left[\frac{\mathrm{i}}{\hbar} S\,(q_a, q_b; E) - \mathrm{i}\mu_{\rm traj}\frac{\pi}{2}\right] . \quad (5.109)$$

Now the phase in the contribution of each trajectory is dominantly determined by the action,

$$S\,(q_a, q_b; E) = \int_{q_a}^{q_b} p\,\mathrm{d}q\,, \quad (5.110)$$

which resembles the action integral introduced in Sect. 1.6.3 and is often referred to as the *reduced action* in order to distinguish it from the Lagrangian action (5.102). The amplitude factor $\sqrt{|D|}$ now involves the determinant of an $(f+1)\times(f+1)$ matrix,

$$D = \det\begin{pmatrix} \frac{\partial^2 S}{\partial q_a \partial q_b} & \frac{\partial^2 S}{\partial q_a \partial E} \\ \frac{\partial^2 S}{\partial E \partial q_b} & \frac{\partial^2 S}{\partial E^2} \end{pmatrix}, \quad (5.111)$$

and the index $\mu_{\rm traj}$ counts the number of focal points along the trajectory.

Taking the trace over the semiclassical Green's function (5.109) leads to a sum of integrals over all coordinates q involving classical trajectories which begin and end at q, $q_a = q_b = q$. The f coordinates are reexpressed locally as one coordinate in the instantaneous direction of the respective trajectory and $f-1$ coordinates transverse to the instantaneous motion, and the integral over these latter $f-1$ coordinates is performed via the stationary phase approximation. This selects just those trajectories for which also $p(q_a) = p(q_b)$, so the sum over trajectories closed in coordinate space is now reduced to the sum over trajectories closed in phase space, i.e. the periodic orbits. The integration over the coordinate along the trajectory is performed explicitly, yielding a factor proportional to period of the orbit. The resulting approximation of the expression (5.108) for the energy-level density eventually is,

$$d(E) = \tilde{d}(E) - \frac{1}{\pi\hbar}\Re\left\{\sum_{\mathrm{ppo}} T_{\mathrm{ppo}} \sum_{n_{\mathrm{p}}=1}^{\infty} \frac{\exp\left[\mathrm{i}\left(\frac{S_{\mathrm{ppo}}}{\hbar} - \nu_{\mathrm{ppo}}\frac{\pi}{2}\right)n_{\mathrm{p}}\right]}{\sqrt{\left|\det\left(\mathbf{M}_{\mathrm{ppo}}{}^{n_{\mathrm{p}}} - \mathbf{1}\right)\right|}}\right\}. \tag{5.112}$$

The sums in (5.112) are over all "primitive periodic orbits", i.e. periodic orbits run around just once, and over all numbers n_{p} of passages around each "ppo". S_{ppo} is the action (5.110) integrated over one passage of the ppo,

$$S_{\mathrm{ppo}} = \oint_{\mathrm{ppo}} p\, \mathrm{d}q\,, \tag{5.113}$$

and T_{ppo} is its period. $\mathbf{M}_{\mathrm{ppo}}$ stands for the $2(f-1) \times 2(f-1)$ *reduced monodromy matrix* over one period of the orbit; it involves only the $f-1$ coordinates and conjugate momenta transverse to the orbit. The topologically invariant phase index ν_{ppo} is a generalized Maslov index which counts the focal points along the trajectory and contains additional contributions arising from the evaluation of integrals via the stationary phase approximation—for more details see [BB97a]. Finally, the first term $\tilde{d}(E)$ on the right-hand side of (5.112) is a smoothly energy-dependent term due to the contribution of the trajectories of zero length ($q_b \to q_a$ with no detours) to the trace of the semiclassical Green's function. It is identified with the mean level density introduced in (5.91).

The formula (5.112) connects the fluctuating part $d(E) - \tilde{d}(E)$ of the quantum mechanical energy-level density to the periodic orbits of the corresponding classical system and is known as *Gutzwiller's trace formula* [Gut97]. In the form given here it assumes that the periodic orbits are isolated, but extensions to more general situations have been formulated [BB97a]. The beauty of the trace formula is, that it is applicable, irrespective of whether the classical system is regular with stable orbits or chaotic with unstable periodic orbits. The information on the stability or instability of an orbit is contained in the amplitude factors $1/\sqrt{\left|\det(\mathbf{M}_{\mathrm{ppo}}{}^{n_{\mathrm{p}}} - \mathbf{1})\right|}$ and their dependence on the number of passages n_{p}.

The amplitude factors have a particularly simple form in the case of two degrees of freedom, where there is only one coordinate transverse to the orbit and the reduced monodromy matrix has just two eigenvalues, Λ_1 and Λ_2. The two further eigenvalues of the full monodromy matrix are both unity. Because of the symplectic property of the monodromy matrix, there are only two essentially different possibilities. If Λ_1 and Λ_2 are complex, they must have unit modulus and

$$\Lambda_{1,2} = \mathrm{e}^{\pm 2\pi \mathrm{i} w} \Rightarrow \det(\mathbf{M}_{\mathrm{ppo}}{}^{n_{\mathrm{p}}} - \mathbf{1}) = 4\sin^2(\pi n_{\mathrm{p}} w)\,. \tag{5.114}$$

In this case the orbit is stable. Such orbits are called *elliptic periodic orbits* and are characterised by the *winding number* w in (5.114). If the eigenvalues Λ_1, Λ_2 of the reduced monodromy matrix have moduli different from unity, then they must be real and

$$\Lambda_1 = \pm \mathrm{e}^{\lambda T_{\text{ppo}}}, \quad \Lambda_2 = \pm \mathrm{e}^{-\lambda T_{\text{ppo}}}$$
$$\Rightarrow \det\left(\mathbf{M}_{\text{ppo}}{}^{n_{\text{p}}} - \mathbf{1}\right) = \begin{cases} -4\sinh^2(n_{\text{p}}\lambda T_{\text{ppo}}/2) \\ 4\cosh^2(n_{\text{p}}\lambda T_{\text{ppo}}/2) \end{cases}. \tag{5.115}$$

In this case the orbit is unstable and its Liapunov exponent is $|\lambda|$ according to (5.78). Such an orbit is called a *hyperbolic orbit* for the "+" version of $\pm$ signs in (5.115), and it is called an *inverse hyperbolic orbit* for the "−" version.

The contributions corresponding to several passages of a ppo in the trace formula (5.112) interfere constructively at energies fulfilling

$$S_{\text{ppo}} = \oint_{\text{ppo}} p\,\mathrm{d}q = \left(n + \frac{\nu_{\text{ppo}}}{4}\right) 2\pi\hbar\,. \tag{5.116}$$

This equation strongly resembles a quantization condition, cf. (1.308), but it must now be interpreted differently. E.g. for unstable periodic orbits the amplitudes in (5.112) fall off exponentially with n_{p} and the leading terms will produce smooth maxima of constructive interference at energies fulfilling (5.116). Equation (5.116) is thus a *resonance condition* describing the positions of modulation peaks due to constructive interference of phases of multiple passages of the periodic orbit.

The modulation frequency due to a periodic orbit is the inverse of the separation of successive peaks given by (5.116). From the definitions (5.101), (5.102) and (5.110) it follows [LL71], that $S(q_a, q_b; E) = W(q_a, q_b; t) + Et$ and that $\partial S/\partial E = t$, where t is time a classical trajectory takes to travel from q_a to q_b with (conserved) energy E. For the primitive periodic orbits this implies

$$\frac{\mathrm{d}}{\mathrm{d}E} S_{\text{ppo}}(E) = T_{\text{ppo}}\,, \tag{5.117}$$

so the separation of successive energies fulfilling (5.116) is approximately $2\pi\hbar/T_{\text{ppo}}$. The modulations due to a ppo thus appear as prominent peaks in the Fourier transformed spectra at times corresponding to the period of the orbit. The classical periodic orbits with the shortest periods are responsible for the longest ranged modulations in the quantum mechanical energy spectra.

Gutzwiller's trace formula underlines the importance of the periodic orbits for the phase space structure of a mechanical system. The periodic orbits may form a subset of measure zero in the set of all classical trajectories, but it is a dense subset, because any (bound) trajectory can be well approximated for a given time by a periodic orbit of sufficiently long period. The sum over all periodic orbits in the trace formula is extremely divergent, and its mathematically safe evaluation has been the subject of extensive work by many authors, see e.g. [Cha92, FE97]. Terms due to individual primitive periodic orbits tend to diverge at points of bifurcation, and Main [Mai97] discussed techniques for smoothly bridging such points; they are based on connection procedures similar to the uniform approximation of WKB wave functions near classical turning points. As a semiclassical expression Gutzwiller's trace formula contains just contributions of leading order in $\hbar$. Only few authors have so far addressed the question of higher order corrections [GH97]. Diffractive corrections related to orbits "creeping" along the edge of the classically allowed region were discussed in particular by Wirzba [Wir92, Wir93]. It may also be worth mentioning, that allowing nonintegral Maslov indices in the trace formula to account for finite wave length effects at reflections and focal points could be one possibility of improving results without too much additional effort [FT96, BB97b].

After Gutzwiller derived the trace formula around 1970 [Gut97], it was all but ignored for one and a half decades. Its first application to spectra of a real physical system was the case of a hydrogen atom in a uniform magnetic field [Win87a, FW89]. As a practical aid for understanding gross features of quantum spectra on the basis of simple classical orbits it was thereafter remarkably successful in describing such diverse phenomena as the magic numbers of shell structure observed in alkali metal clusters [BB97a] and conductance fluctuations in semiconductor microstructures [RU96, DS97].

5.3.4 Scaling Properties for Atoms in External Fields

One important advantage of studying atoms (or molecules) in external fields is, that the field parameters are tunable variables, and investigation of the properties of the atom as function of these variables provides a much richer body of information than can be observed in the isolated specimen. Due to scaling properties, the classical dynamics of an atom in external fields depends on some combinations of field parameters in a trivial way. This section summarizes these scaling properties for the case of an external electric or magnetic constant or time-dependent field, or any superposition thereof, see also [Fri98].

(a) Classical Mechanics

We start by discussing the concept of *mechanical similarity* for a conservative system Σ with a finite number of degrees of freedom. Such a system is described by a kinetic energy,

$$T = \frac{m}{2}\left(\frac{\mathrm{d}\boldsymbol{r}}{\mathrm{d}t}\right)^2 , \tag{5.118}$$

and a potential energy $U(\boldsymbol{r})$. [The mass m can be different for the various degrees of freedom, but this is irrelevant for the following.] The *similarity transformation*,

$$\boldsymbol{r}' = \sigma \boldsymbol{r}\,, \quad t' = \tau t\,, \quad \frac{\mathrm{d}\boldsymbol{r}'}{\mathrm{d}t'} = \frac{\sigma}{\tau}\frac{\mathrm{d}\boldsymbol{r}}{\mathrm{d}t}\,, \tag{5.119}$$

with the two positive constants σ and τ, transforms the system Σ into a system Σ', whose kinetic energy T' is related to the kinetic energy T in (5.118) by

$$T' = \left(\frac{\sigma}{\tau}\right)^2 T\,. \tag{5.120}$$

Suppose the potential energy in the system Σ is given by a homogeneous function $V(\boldsymbol{r})$ of degree d, i.e.

$$V(\sigma \boldsymbol{r}) = \sigma^d V(\boldsymbol{r})\,, \tag{5.121}$$

multiplied by a parameter F, which gives us a handle on the potential strength, $U(\boldsymbol{r}) = FV(\boldsymbol{r})$. Let the potential energy U' in the system Σ' be given by the same (homogeneous) function V, multiplied by a strength parameter F', $U'(\boldsymbol{r}') = F'V(\boldsymbol{r}')$. Because of homogeneity (5.121), the potential energy U' is related to the potential energy U in Σ by

$$U'(\boldsymbol{r}') = \frac{F'}{F}\sigma^d U(\boldsymbol{r})\,. \tag{5.122}$$

If and only if the field strengths fulfill the relation

$$\sigma^d F' = \left(\frac{\sigma}{\tau}\right)^2 F\,, \tag{5.123}$$

then the potential energies are related by the same multiplicative factor $(\sigma/\tau)^2$ as the kinetic energies (5.120). The classical Lagrangian $T' - U'$ in the system Σ' is then just a multiple of the Lagrangian $L = T - U$ in Σ, and the equations of motion in both systems are the same [LL71]. The coordinate space trajectory $\boldsymbol{r}(t)$ is a solution of the equations of motion in Σ if and only if the trajectory $\boldsymbol{r}'(t')$, which is related to $\boldsymbol{r}(t)$ by the similarity transformation (5.119), is a solution of the equations of motion in Σ'. This is the property of mechanical similarity of the systems Σ and Σ' and the condition for mechanical similarity is, that the field strengths obey (5.123). The (conserved) energy $E = \frac{1}{2}m\dot{\boldsymbol{r}}(t)^2 + U(\boldsymbol{r}(t))$ of motion along the trajectory $\boldsymbol{r}(t)$ in Σ is related to the associated energy E' in Σ' via

$$E' = \left(\frac{\sigma}{\tau}\right)^2 E\,. \tag{5.124}$$

The condition (5.123) contains two parameters σ and τ and can always be fulfilled for any values of the field strengths F and F'. Together with the relation (5.124) we can, for any field strengths F and F' and energies E and E' uniquely determine the constants σ and τ defining the similarity transformation (5.119) connecting the trajectory $\boldsymbol{r}(t)$ in Σ with the trajectory $\boldsymbol{r}'(t)$ in Σ',

$$\sigma = \left(\frac{F}{F'}\frac{E'}{E}\right)^{1/d} , \qquad \tau = \sigma\sqrt{\frac{E}{E'}} . \tag{5.125}$$

[It shall be taken for granted throughout, that potential strengths and energies have the same sign in Σ' as in Σ.] From (5.125) we see e.g., that trajectories at different energies E, E' for one and the same potential strength, $F' = F$, are related by a stretching factor $\sigma = (E'/E)^{1/d}$ in coordinate space, whereas the traversal times are stretched by the factor $\sigma\sqrt{E/E'}$.

The considerations above are readily generalized to a potential which can be written as a sum of n homogeneous terms of degree d_i, $i = 1, \ldots, n$. The potential U in the system Σ is now

$$U(\boldsymbol{r}) = \sum_{i=1}^{n} F_i V_i(\boldsymbol{r}) ; \quad V_i(\sigma\boldsymbol{r}) = \sigma^{d_i} V_i(\boldsymbol{r}) , \quad i = 1, \ldots, n , \tag{5.126}$$

and the potential U' in the system Σ' differs only through different potential strengths,

$$U'(\boldsymbol{r}') = \sum_{i=1}^{n} F'_i V_i(\boldsymbol{r}') . \tag{5.127}$$

The systems Σ and Σ' are mechanically similar, if U' is just U multiplied by $(\sigma/\tau)^2$ when $\boldsymbol{r}'$ and t' are related to $\boldsymbol{r}$ and t via (5.119). The condition (5.123) must now be fulfilled for each of the n terms independently, and the first equation (5.125) is replaced by the n equations,

$$\sigma = \left[\left(\frac{\sigma}{\tau}\right)^2 \frac{F_i}{F'_i}\right]^{1/d_i} = \left(\frac{E'}{E}\frac{F_i}{F'_i}\right)^{1/d_i} , \quad i = 1, \ldots, n . \tag{5.128}$$

The relation between the total energies E and E' is again given by (5.124).

Equating the right-hand sides of (5.128) for two different terms i and j and collecting unprimed and primed quantities on separate sides leads to the condition

$$\frac{|F_i/E|^{d_j}}{|F_j/E|^{d_i}} = \frac{|F'_i/E'|^{d_j}}{|F'_j/E'|^{d_i}} . \tag{5.129}$$

If we consider an ensemble of systems Σ corresponding to different field strengths F_i and energies E (excluding changes of sign), then (5.129) shows that the classical dynamics within the ensemble is invariant within mechanical similarity if

$$\frac{|F_j/E|^{d_j}}{|F_j/E|^{d_i}} = \text{const.} \tag{5.130}$$

for each pair of labels i, j. For $n > 2$ these conditions are not independent. The $n+1$ parameters E, F_i $(i = 1, \ldots, n)$ are effectively subjected to $n-1$ independent conditions, because σ and τ generate a two-parameter manifold of mechanically similar systems.

The Coulomb potential describing the forces in an atom (or molecule or ion) is homogeneous of degree $d_1 = -1$, and the corresponding strength parameter F_1 may be assumed to be constant for a given specimen. This fixes the scaling parameters,

$$\sigma = \frac{E}{E'}, \quad \tau = \left(\frac{E}{E'}\right)^{3/2}, \tag{5.131}$$

according to (5.124), (5.128). In the presence of $n-1$ homogeneous external fields of degree d_j $(j = 2, \ldots, n)$ the conditions (5.130) reduce to

$$F_j/|E|^{d_j+1} = \text{const.}\ , \quad j = 2, \ldots n\,, \tag{5.132}$$

when inserting $d_1 = -1$, $F_1 = \text{const.}$ for $i = 1$. The $n-1$ conditions for mechanical similarity are thus, that the *scaled field strengths* $\tilde{F}_j$, defined by

$$\tilde{F}_j = F_j/|E|^{d_j+1}\,, \tag{5.133}$$

be constant. The values of these $n-1$ scaled field strengths determine the properties of the classical dynamics which are invariant to within similarity transformations (5.119). For each set of values of the scaled field strengths there is now a one-parameter family of mechanically similar systems and not a two-parameter family, because the field strength F_1 is kept fixed.

If $j = 2$ labels a homogeneous external electric field, the potential V_2 is homogeneous of degree one, and F_2 is the electric field strength f. The scaled electric field strength is

$$\tilde{f} = f/E^2\,, \tag{5.134}$$

and all systems with the same value of $\tilde{f}$ (and the same sign of E) are mechanically similar.

A homogeneous external magnetic field with field strength γ [see (3.330)] is studied more conveniently by directly subjecting the equations of motion for a charged particle in such a magnetic field to the similarity transformation (5.119).

The equations of motion in the systems Σ and Σ' are seen to be equivalent if the respective magnetic field strengths γ and γ' are related by

$$\gamma' = \gamma/\tau\,. \tag{5.135}$$

Comparing with (5.123) shows that this corresponds to the behaviour of a homogeneous potential of degree two, and the square of the magnetic field strength γ plays the role of "field strength" F. For an atom ($d_1 = -1$, $F_1 =$ const.) in a constant homogeneous magnetic field of strength γ, the scaled magnetic field strength $\tilde{\gamma}$ is thus defined via (5.133) with $F_j = \gamma^2$, $d_j = 2$, as

$$\tilde{\gamma}^2 = \gamma^2/|E|^3, \quad \tilde{\gamma} = \gamma/|E|^{3/2}. \tag{5.136}$$

The conditions for invariant classical dynamics of an atom in an external electric or magnetic field are conventionally stated as the condition of constant *scaled energy,* which is $E/\sqrt{f}$ for the electric field and $E/\gamma^{2/3}$ for the magnetic field. The nomenclature evolved historically [FW89], and has probably been a mistake from the pedagogical point of view. This becomes clear when we consider an atom in a superposition of homogeneous electric and magnetic fields. We are then confronted with two different definitions of scaled energy, and usually the conditions of mechanical similarity are expressed as requiring one of these scaled energies and the ratio f^3/γ^4 to be constant. The more natural statement of the conditions for mechanical similarity for an atom in a superposition of electric and magnetic fields is surely that both scaled field strengths, $\tilde{f}$ and $\tilde{\gamma}$, be constant. This of course implies the constance of the above mentioned scaled energies and of the ratio $\tilde{f}^3/\tilde{\gamma}^4$, which is equal to f^3/γ^4 and is independent of energy.

In the presence of a time-dependent external field the expression (5.126) for the potential energy of the system Σ must be generalized, e.g. to

$$U(\boldsymbol{r},t) = \sum_{i=1}^{n} F_i V_i(\boldsymbol{r}) + F_0 V_0(\boldsymbol{r}) \Phi(\omega t)\,;$$
$$V_i(\sigma \boldsymbol{r}) = \sigma^{d_i} V_i(\boldsymbol{r})\,, \quad i = 0, 1, \ldots, n\,, \tag{5.137}$$

where we have added a homogeneous potential V_0 with strength F_0 multiplied by a time-dependent function $\Phi(\omega t)$, which is usually, but not necessarily, a harmonic function (sine, cosine or exp $(\pm \mathrm{i}\omega t)$). The time function Φ need not even be periodic, but the parameter $\omega (> 0)$ is included explicitly to give us a handle on the time scale. The corresponding potential energy in the system Σ' is

$$U'(\boldsymbol{r}',t') = \sum_{i=1}^{n} F_i' V_i(\boldsymbol{r}') + F_0' V_0(\boldsymbol{r}') \Phi(\omega' t')\,. \tag{5.138}$$

Again we study the effect of the similarity transformation (5.119) on the kinetic and potential energy. The systems Σ and Σ' are mechanically similar, if kinetic and potential energies in Σ' differ from those in Σ by the same multiplicative factor. The time function Φ is generally assumed to be bounded, so it cannot be a homogeneous function. Hence we have no freedom to choose the parameter τ connecting the times t and t'; if U' is to be proportional to U there is no choice but to set

$$\tau = \frac{\omega}{\omega'}\,, \quad \text{so that } \omega' t' = \omega t\,. \tag{5.139}$$

The time scale parameter ω replaces the energy of the time-independent case as additional parameter (beside the field strengths) determining the classical dynamics of the system. Whereas (5.124) fixes the ratio σ/τ in the time-independent case, (5.139) fixes the time stretching parameter τ in the time-dependent case. This leaves one free parameter σ and the $n+1$ conditions,

$$\sigma^{d_i} F_i' = \left(\sigma \frac{\omega'}{\omega}\right)^2 F_i\,, \quad i = 0, 1, \ldots, n\,. \tag{5.140}$$

Resolving for σ now yields

$$\sigma = \left[\left(\frac{\omega'}{\omega}\right)^2 \frac{F_i}{F_i'}\right]^{\frac{1}{d_i-2}} \quad \text{for all} \quad i = 0, 1, \ldots, n\,. \tag{5.141}$$

For any pair (i, j) of labels this implies

$$\frac{|F_i/\omega^2|^{d_j-2}}{|F_j/\omega^2|^{d_i-2}} = \frac{|F_i'/\omega'^2|^{d_j-2}}{|F_j'/\omega'^2|^{d_i-2}}\,, \tag{5.142}$$

in other words, mechanical similarity is given if

$$\frac{|F_i/\omega^2|^{d_j-2}}{|F_j/\omega^2|^{d_i-2}} = \text{const.} \tag{5.143}$$

The potential (5.137) may contain more than one time-dependent contribution. As long as the dependence of each contribution on the coordinates is homogeneous, the results derived for the label $i = 0$ above are easily generalized to a finite number of time-dependent terms. Note, however, that only one time scale parameter ω can be accommodated, because there can be only one time stretching factor τ, see (5.139). If the potential contains e.g. a superposition of several harmonic terms with different frequencies, then the mutual ratios of these frequencies have to be the same in all mechanically similar systems, so that there is effectively only one parameter defining the time scale.

For an atom (or ion) in a time-dependent field and $n-1$ external static fields we again assume the label $i=1$ to describe the constant ($F_1=$ const.) Coulomb field ($d_1=-1$) of the atom, and this fixes the stretching parameter σ via (5.141),

$$\sigma=\left(\frac{\omega}{\omega'}\right)^{2/3}. \tag{5.144}$$

The conditions (5.143) now suggest the following definition for the scaled field strengths:

$$\tilde{F}_j \stackrel{\text{def}}{=} F_j/\omega^{\frac{2}{3}(d_j+1)}. \tag{5.145}$$

With these definitions the n conditions for mechanical similarity can be expressed as the requirement

$$\tilde{F}_j=\text{const.}\quad j=0,2,3\ldots,n. \tag{5.146}$$

For an atom described by a constant Coulomb field ($i=1$) in a superposition of one ($j=0$) time-dependent and $n-1$ ($j=2,\ldots,n$) static external fields, the classical dynamics is determined to within mechanical similarity by the values (5.146) of these n scaled field strengths.

The time-dependent field is very often the oscillating electric field of microwave or laser radiation, so $F_0=f_{\text{rad}}$ is the amplitude of the oscillating field of circular frequency ω, and $d_0=1$. The corresponding scaled field strength $\tilde{f}_{\text{rad}}$, which is constant under the conditions of mechanical similarity, is

$$\tilde{f}_{\text{rad}}=f_{\text{rad}}/\omega^{4/3} \tag{5.147}$$

according to (5.145). For an external static electric field of strength f, the scaled field strength $\tilde{f}$ is analogously given by $\tilde{f}=f/\omega^{4/3}$. For an additional magnetic field of strength γ (with F_j corresponding to γ^2), the scaled field strength $\tilde{\gamma}$ is given by

$$\tilde{\gamma}^2=\gamma^2/\omega^2,\quad \tilde{\gamma}=\gamma/\omega, \tag{5.148}$$

according to (5.145). Under the conditions of mechanical similarity, $\tilde{f}_{\text{rad}}$, $\tilde{f}$ and $\tilde{\gamma}$ are constant, and so are $f_{\text{rad}}^3/\gamma^4$ and f^3/γ^4 as in the time-independent case.

(b) Quantum Mechanics

The quantum mechanical system corresponding to the classical system Σ introduced above is described by the Schrodinger equation,

$$-\frac{\hbar^2}{2m}\Delta\psi(\boldsymbol{r},t)+U(\boldsymbol{r},t)\psi(\boldsymbol{r},t)=\mathrm{i}\hbar\frac{\partial}{\partial t}\psi(\boldsymbol{r},t), \tag{5.149}$$

and can be obtained by quantization via the *canonical commutation relations* between the coordinates r_i and the momenta $p_j = m\,\mathrm{d}r_j/\mathrm{d}t$

$$[r_i, p_j] = \mathrm{i}\hbar\delta_{i,j}\,. \tag{5.150}$$

When the classical dynamics of the system Σ is related to the classical dynamics of the system Σ' via the *non-canonical* similarity transformation (5.119), the coordinates transform as $r_i' = \sigma r_i$ and the momenta as $p_j' = (\sigma/\tau)p_j$. [The latter also holds if the momentum p_j contains a term proportional to a vector potential describing a homogeneous magnetic field, because the vector potential must be proportional to a product of the magnetic field strength, transforming according to (5.135), and a linear function of the coordinates.] The same quantum mechanics is thus obtained by quantization of the system Σ' via the *non-canonical commutation relations,*

$$\left[r_i', p_j'\right] = \mathrm{i}\hbar'\delta_{i,j}\,, \tag{5.151}$$

where $\hbar'$ is an *effective Planck's constant*,

$$\hbar' = \frac{\sigma^2}{\tau}\hbar\,. \tag{5.152}$$

If the field strengths F_i in Σ are varied under the conditions of mechanical similarity, then canonical quantization in the system Σ leads to the same quantum mechanics as non-canonical quantization in the mechanically similar "scaled system" system Σ' according to (5.151), with a variable effective Planck's constant (5.152).

For a system with one time-independent homogeneous potential of degree d, $U(\boldsymbol{r}) = FV(r)$, the constants σ and τ are given by (5.125), and the effective Planck's constant in the scaled system Σ' is

$$\hbar' = \frac{\sigma^2}{\tau}\hbar = \left(\frac{F}{F'}\right)^{\frac{1}{d}}\left(\frac{E'}{E}\right)^{\frac{1}{d}+\frac{1}{2}}\hbar\,. \tag{5.153}$$

For given energy E' and field strength F' in the scaled system, the semiclassical limit can now be defined as the limit $\hbar' \to 0$, and the anticlassical or extreme quantum limit is $\hbar' \to \infty$, see also Sect. 1.6.3 in Chap. 1. Equation (5.153) determines which combination of energy E and field strength F corresponds to the semiclassical limit $\hbar' \to 0$. This obviously depends in the following way on the degree d of homogeneity of the potential:

$$0 < d: \quad F \to 0 \quad \text{or} \quad |E| \to \infty \tag{5.154}$$

$$-2 < d < 0: \quad |F| \to \infty \quad \text{or} \quad E \to 0 \tag{5.155}$$

$$d = -2: \quad |F| \to \infty \quad \text{and} \quad E \quad \text{arbitrary} \tag{5.156}$$

$$d < -2: \quad |F| \to \infty \quad \text{or} \quad |E| \to \infty\,. \tag{5.157}$$

When $d > 0$ or $d < -2$, the semiclassical limit for a given field strength F corresponds to the high energy limit $|E| \to \infty$. However, if the degree d of homogeneity of the potential lies between zero and -2, then the semiclassical limit of the Schrödinger equation for a given field strength corresponds to the limit of vanishing values of the energy E. This applies in particular to all Coulomb systems, where $d = -1$, and it is perhaps not surprising when remembering that the energies of the bound states of a one-electron atom vanish in the (semiclassical) limit of large quantum numbers. It is, however, not trivial and not widely appreciated, that large energies, $E \to \infty$, actually correspond to the anticlassical or extreme quantum limit in Coulombic systems, see, e.g., Fig. 4.15 in Sect. 4.2.6.

Now consider a potential U in (5.149) consisting of n contributions, $U(\boldsymbol{r}) = \sum_{i=1}^{n} F_i V_i(\boldsymbol{r})$ where V_i is a (time-independent) homogeneous potential of degree d_i. The equivalence of the canonical Schrödinger equation for energy E and field strengths F_i with the non-canonical Schrödinger equation containing the modified Planck's constant (5.152) is maintained, as long as energy and field strengths are varied under the conditions of mechanical similarity described above. This implies

$$\hbar' = \frac{\sigma^2}{\tau}\hbar = \left(\frac{F_i}{F_i'}\right)^{\frac{1}{d_i}} \left(\frac{E'}{E}\right)^{\frac{1}{d_i}+\frac{1}{2}} \hbar \quad \text{for all } i\,. \tag{5.158}$$

The conditions of the semiclassical limit correspond in each contribution i to the limiting behaviour (5.154)–(5.157), depending on the degree d_i of homogeneity of the respective term. These conditions are compatible in the case of mechanical similarity (5.129). For example, if the label $i = 1$ describes the fixed Coulomb potential in an atom, then the condition of constant scaled field strengths (5.133) implies

$$F_j \propto |E|^{1+d_j}\,. \tag{5.159}$$

The semiclassical limit $\hbar' \to 0$ corresponds to $E \to 0$. For any further contributions with a positive degree of homogeneity, e.g. an external electric field with $d_j = 1$, or an external magnetic field with $d_j = 2$, the strengths F_j must tend to zero as prescribed by (5.159) in the semiclassical limit. Note in particular, that a fixed strength of the Coulomb potential and a non-vanishing external electric and/or magnetic field are incompatible with the conditions of the semiclassical limit.

(c) Scaled-Fields Spectroscopy

The energy and the $n-1$ strengths of the static external fields in which an atom is placed have $n-1$ conditions to fulfill for mechanical similarity to hold, e.g. that the scaled field strengths (5.133) be constant. When the field strength of the Coulombic forces describing the atom is kept fixed, there remains one continuous parameter, which can be varied without changing the classical dynamics, except to within a similarity transformation (5.119). This makes it possible to study the variations of the quantum system corresponding to different values of the effective Planck's constant without changing the classical dynamics. Although the energy itself or any one of the external field strengths could be chosen as the variable parameter, a prudent choice is

$$\chi = \frac{\tau}{\sigma^2}\frac{1}{\hbar} = \frac{1}{\hbar'}, \tag{5.160}$$

which has the dimensions of an inverse action and is just the inverse of the effective Planck's constant $\hbar'$

A justification for this choice can be found by looking at Gutzwiller's trace formula (5.112) or variations thereof [Cha92, FE97], which typically express the energy-level density or some other quantum mechanical property in terms containing the actions $S_{\rm traj}$ along classical trajectories,

$$\text{property of qm spectrum} = \text{function}\left[\exp\left(\frac{\rm i}{\hbar}S_{\rm traj}\right)\right],$$
$$S_{\rm traj} = \int_{\rm traj} \boldsymbol{p}\cdot {\rm d}\boldsymbol{r}\,. \tag{5.161}$$

Regarding both sides of the upper line of (5.161) as functions of the variable χ defined by (5.160) leads to the following form of this general equation:

$$\text{property of qm spectrum}(\chi) = \text{function}\left(\exp\left[{\rm i}\chi S'_{\rm traj}\right]\right), \tag{5.162}$$

where we have expressed the actions $S_{\rm traj}$ through the "scaled actions"

$$S'_{\rm traj} = \int_{\rm traj} \boldsymbol{p}'\cdot {\rm d}\boldsymbol{r}' = \frac{\sigma^2}{\tau}S_{\rm traj} = \frac{\hbar'}{\hbar}S_{\rm traj}\,. \tag{5.163}$$

The scaled classical actions (5.163) depend only on the fixed energy E', which defines the energy at which the effective Planck's constant $\hbar'$ assumes its physical value $\hbar$, and on the $n-1$ values of the scaled field strengths (5.133), which determine the classical dynamics. In the general formula (5.162) these scaled actions appear as Fourier conjugates to the variable χ. Applying a Fourier transform to (5.162) will thus reveal structures associated with classical trajectories at values of the conjugate variable corresponding to the scaled actions of the trajectories. An example is given in Sect. 5.3.5 (b).

For an atom in external static fields the scaling parameters σ and τ are given by (5.131) and the natural variable (5.160) is,

$$\chi = \frac{1}{\hbar}\sqrt{\frac{E'}{E}} \propto \frac{1}{\hbar\sqrt{|E|}}\,. \tag{5.164}$$

The definition of the natural variable χ depends on which field strength we are keeping constant, and not on which external fields (of variable strength) are present; the constance of the strength F_1 of the Coulombic ($d_1 = -1$) potential describing the atom leads to the simple result (5.164), $\chi \propto 1/\sqrt{|E|}$. For an external magnetic field of variable strength γ, this corresponds to $\chi \propto \gamma^{-1/3}$ when the scaled field strength $\tilde{\gamma}$ is kept constant, cf. (5.136). For an external static electric field of variable strength, $\chi \propto 1/\sqrt{|E|}$ corresponds to $\chi \propto f^{-1/4}$ when the scaled field strength $\tilde{f}$ is kept fixed, cf. (5.134). In a superposition of electric and magnetic fields both relations apply, which is consistent because f^3/γ^4 is constant under the conditions of mechanical similarity.

The technique of scaled-fields spectroscopy is well established for the example of atoms in external electric and magnetic fields and has been called "scaled-energy spectroscopy" [MW91] and also "recurrence spectroscopy" [MM97, DS97], because of the dominating role which periodic and recurring classical orbits play in appropriately Fourier transformed spectra.

(d) Time-dependent Potentials

The Schrödinger equation (5.149) with the time-dependent potential (5.137) is equivalent to a non-canonical Schrödinger equation containing the effective Planck's constant (5.152) with the scaled potential (5.138) as long as the frequency parameter and the potential strengths obey the conditions (5.139) and (5.141) for mechanical similarity. If the label $i = 1$ describes the fixed ($F_1 = F_1'$) Coulomb potential ($d_1 = -1$) of an atom, the stretching parameter σ is given by (5.144), and the conditions for mechanical similarity reduce to the requirement that the scaled field strengths (5.145) be constant.

For a concrete experiment with a one-electron atom in a time-dependent field, the initial (unperturbed) state of the atom is described by a quantum number n_0, and $n_0 2\pi\hbar$ is the classical action S of the electron on the corresponding orbit. The similarity transformation (5.119) transforms actions as

$$S' = \frac{\sigma^2}{\tau} S = \left(\frac{\omega}{\omega'}\right)^{1/3} S\,, \tag{5.165}$$

according to (5.139) and (5.144), hence $n_0\omega^{1/3}$ is the corresponding scaled quantum number which remains constant under the conditions of mechanical similarity. The cube of the scaled quantum number, $n_0^3\omega$ is naturally called the scaled frequency.

Using the initial quantum number n_0 as reference rather than the frequency parameter ω leads to

$$\tilde{f} = fn_0^4 \tag{5.166}$$

as an alternative definition [instead of (5.147)] for the scaled strengths of the time-dependent or static electric fields [Koc92, Ric97]. The corresponding alternative to (5.148) for the scaled strength of an external magnetic field is

$$\tilde{\gamma} = \gamma n_0^3 \,. \tag{5.167}$$

With σ and τ given by (5.144) and (5.139), the effective Planck's constant is

$$\hbar' = \left(\frac{\omega}{\omega'}\right)^{1/3} \hbar \,, \tag{5.168}$$

and the semiclassical limit $\hbar' \to 0$ corresponds to $\omega \to 0$. Note that a finite time scale for the time-dependent part of the potential is incompatible with the semiclassical limit under the conditions of mechanical similarity. For fixed field strength of the Coulomb potential describing the atom, the semiclassical limit for an atom in external time-dependent and/or time-independent electric and magnetic fields corresponds to the static limit according to (5.168) and to vanishing field strengths according to (5.145), (5.159).

5.3.5 *Examples*

(a) Ionization of the Hydrogen Atom in a Microwave Field

General interest in simple Hamiltonians with a periodic time dependence received a great boost after Bayfield and Koch observed the ionization of hydrogen atoms in a microwave field in 1974 [BK74, BG77]. Hydrogen atoms in an initial state with principal quantum number $n_0 = 66$ were ionized in a microwave field of about 10 GHz. This corresponds to a photon energy of $\hbar\omega \approx 4 \cdot 10^{-5}$eV, so that more than 70 photons would have to be absorbed to ionize a H atom (from the $n_0 = 66$ level). The perturbative approach, which may be useful at least for relatively weak intensities and which was discussed in connection with multiphoton ionization in Sect. 5.1, is not practicable when so many photons are absorbed. Consequently intensified efforts were undertaken to solve the time-dependent Schrödinger equation directly for this case.

There are experimental grounds (e.g. strong polarization of the H atom prepared in an additional electric field) which may justify treating the problem in only one spatial dimension. This can of course only work if the microwave field is linearly polarized in the direction of this one spatial coordinate. The Hamiltonian is then (in atomic units)

$$\hat{H} = -\frac{1}{2}\frac{\partial^2}{\partial z^2} - \frac{1}{z} + fz\cos\omega t\,, \quad z > 0\,, \tag{5.169}$$

where f is the strength of the oscillating electric field. This Hamiltonian is somewhat similar to the Hamiltonian (5.83) of the kicked rotor. In the corresponding classical system the periodic trajectories in the field-free case ($f = 0$) are just straight-line oscillations between the position of the nucleus ($z = 0$) and a maximal displacement which depends on the energy. The similarity to a free rotor becomes most apparent when we perform a canonical transformation from the variables p, z to the appropriate *action-angle variables* I, θ. Here $I = S/(2\pi\hbar) = [\oint p\,\mathrm{d}z]/(2\pi\hbar)$ is the action in units of $2\pi\hbar$ and is the classical counterpart of the principal quantum number, and θ is the canonically conjugate angle variable, which varies from zero to 2π during a period of oscillation starting at the nucleus and ending with the return to the nucleus [Jen84]. In the field-free case the trajectories in phase space are simply straight lines $I =$ const. as for the rotor. The influence of a microwave field can be seen in Fig. 5.18 showing trajectories in phase space for a microwave frequency of 7.11 GHz and a field strength of 9.1 V/cm.

Figure 5.18 shows that most classical trajectories are quasiperiodic for actions smaller than 65 to 70, while irregular trajectories dominate at higher actions. These irregular trajectories, along which the action can grow to arbitrarily large values as in the case of the kicked rotor, are interpreted as *ionizing trajectories*. Thus the phase space picture Fig. 5.18 is interpreted as indicating that initial states with an action (i.e. principal quantum number) up to about 65 remain localized in quantum number (and hence bound) in a microwave field of the corresponding frequency and strength, while initial states above $n_0 \approx 68$ are ionized. The threshold above which ionization is possible depends on the field strength and the frequency of the microwave field. For increasing frequency and/or field strength ionization becomes possible for smaller and smaller quantum numbers of the initial state. For a given microwave frequency ω and a given initial quantum number n_0 there is a *critical field strength* or threshold f_{cr} above which ionization begins. According to the scaling properties of an atom in a time oscillating field, cf. (5.165), (5.166), we expect this (classical) condition for ionization to relate the scaled quantum number $n_0\omega^{1/3}$ to the scaled electric field strength fn_0^4. Casati et al. [CC87] derived the estimate $f_{\mathrm{cr}}n_0^4 \approx 1/(50n_0\omega^{1/3})$ for the threshold for ionization.

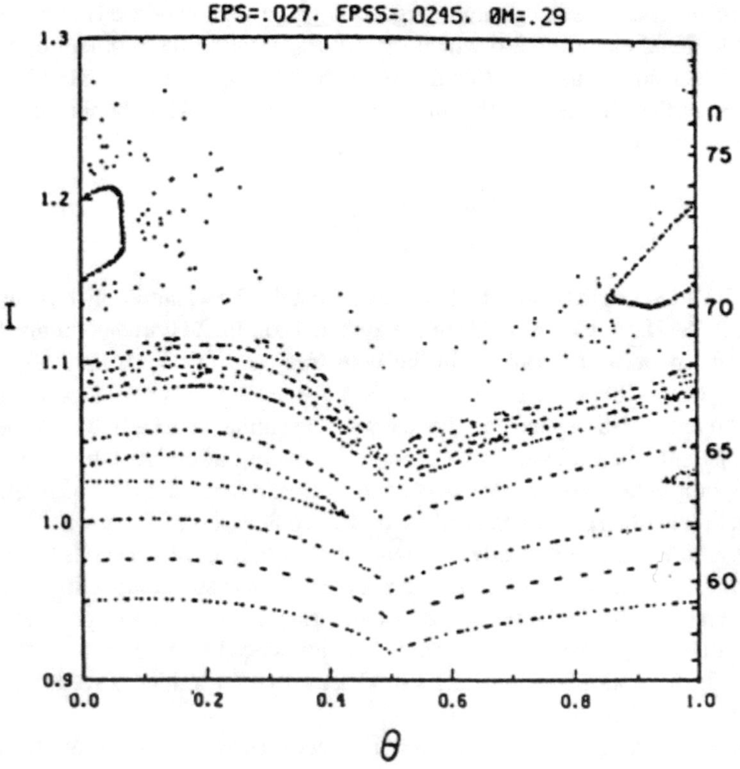

Fig. 5.18 Classical Trajectories as calculated by Jensen for the one-dimensional hydrogen atom in a microwave field of 7.11 GHz and a field strength of 9.1 V/cm (From [Bay86])

More sophisticated calculations going beyond the one-dimensional model (5.169) can and have been performed. Figure 5.19 shows a comparison of experimental ionization thresholds with the results of a full three-dimensional classical calculation. The scaled electric field strengths at which the ionization probability reaches 10% and 90% are plotted as functions of the scaled frequency $n_0^3\,\omega$ and include initial quantum numbers between $n_0 = 32$ and $n_0 = 90$ for a microwave frequency of 9.923 GHz [KL95]. The classical calculations are due to Rath and Richards and include the effect of switching on and switching off the microwave field. The classical calculations reproduce the non-trivial structure of the experimental threshold fields well for scaled frequencies below about 0.8 atomic units, except perhaps near simple fractions, 1/2, 1/3, etc. These discrepancies are attributed to quantum mechanical resonance effects, because they occur at scaled frequencies at which just two, three, etc. photons of energy ω (in atomic units) are needed to excite the initial state with quantum number n_0 to the next excited state with quantum number $n_0 + 1$.

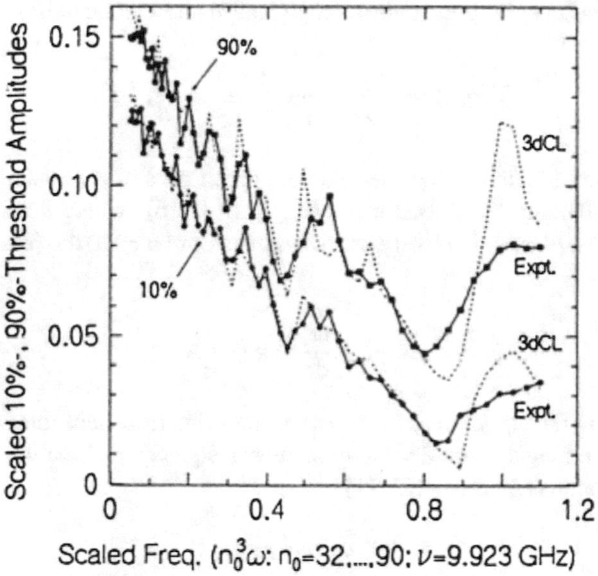

Fig. 5.19 Experimental scaled field strengths at which the probability for ionizing a hydrogen atom in a microwave field of 9.923 GHz reaches 10% (*dots*) and 90% (*squares*) as functions of the scaled frequency $n_0^3\,\omega$. The *dotted lines* show the results of a classical calculation due to Rath and Richards (From [KL95])

Further work on atoms in oscillating electromagnetic fields includes the study of ionization by circularly or elliptically polarized microwaves and the use of alkali atoms in place of hydrogen [Ric97, DZ97, BR09]. Progress continued with shorter wavelengths and higher intensities [PK97]. Amongst the many interesting properties exhibited by a Rydberg atom in a temporally oscillating field, one which received particular attention is the occurrence of non-dispersing wave packets which are well localized and follow a classical periodic orbit without spreading [BD02].

(b) Hydrogen Atom in a Uniform Magnetic Field

The hydrogen atom in a uniform magnetic field has become one of the most widely studied if not *the* most widely studied example for a conservative Hamiltonian system with chaotic classical dynamics [TN89, FW89, HR89, Gay91, RW94, Mai97, SS98, PK14]. Its popularity is mainly due to the fact that it is a real system for which observed spectra and the results of quantum mechanical calculations agree down to the finest detail (see Fig. 3.34 in Sect. 3.5.2). The system corresponds very accurately to a point particle moving in a two-dimensional potential (see (3.332) and Fig. 3.31). For a given value L_z of the z-component of the orbital angular momentum this potential is (in cylindrical coordinates (3.293) and atomic units),

$$V(\varrho, z) = \frac{L_z^2}{2\varrho^2} - \frac{1}{\sqrt{\varrho^2 + z^2}} + \frac{1}{8}\gamma^2\varrho^2 . \tag{5.170}$$

The Hamiltonian describing the quantum mechanics of the system contains the potential (5.170) and the operator $\hat{\boldsymbol{p}}^2/(2\mu)$ [cf.(3.316)] where $\hat{\boldsymbol{p}}$ is the canonical momentum. The classical velocity $\mathrm{d}\boldsymbol{r}/\mathrm{d}t$ is however related to the (classical) kinetic momentum,

$$\mu\boldsymbol{v} = \mu\frac{\mathrm{d}\boldsymbol{r}}{\mathrm{d}t} = \boldsymbol{p} + \frac{e}{c}\boldsymbol{A} . \tag{5.171}$$

If we transform the equations of motion to a coordinate system rotating around the direction of the magnetic field with an angular frequency ω, then the velocity $\boldsymbol{v}'$ in the rotating frame is given by [LL71]

$$\boldsymbol{v}' = \boldsymbol{v} + \boldsymbol{r}\times\boldsymbol{\omega} , \tag{5.172}$$

where $\boldsymbol{\omega}$ is the vector of length ω pointing in the direction of the magnetic field. The canonical momentum $\boldsymbol{p}$ in the inertial frame is,

$$\boldsymbol{p} = \mu\boldsymbol{v} - \frac{e}{c}\boldsymbol{A} = \mu\left(\boldsymbol{v} + \frac{e}{2\mu c}\boldsymbol{r}\times\boldsymbol{B}\right) = \mu\boldsymbol{v}', \quad \text{when} \quad \boldsymbol{\omega} = \frac{e\boldsymbol{B}}{2\mu c} , \tag{5.173}$$

where we have used the definition $\boldsymbol{A} = -\boldsymbol{r}\times\boldsymbol{B}/2$ of the symmetric gauge, on which the derivation of the potential (5.170) was based. The canonical momentum in the inertial frame thus corresponds to the kinetic momentum in the frame of reference which rotates around the z-axis pointing in the direction of the magnetic field with the constant rotational frequency ω equal to half the cyclotron frequency.

According to Sect. 5.3.4 the classical dynamics of the system depends not on the energy E and the field strength γ independently, but only on the *scaled field strength* $\tilde{\gamma} = \gamma|E|^{-3/2}$ or on the *scaled energy* $\varepsilon = E\gamma^{-2/3} = \pm\tilde{\gamma}^{-2/3}$. In the bound regime (negative energies) the separable limit corresponding to a hydrogen atom without an external field is given by $\varepsilon = -\infty$, $\tilde{\gamma} = 0$. The "field-free threshold" $E = 0$ corresponds to $\varepsilon = 0$, ($|\tilde{\gamma}| = \infty$) and is identical to the classical ionization threshold. Because of the finite zero-point energy of the electron's motion perpendicular to the field the (quantum mechanical) ionization threshold actually lies higher (see (3.335)).

Numerical solutions of the classical equations of motion were obtained already in the 1980's by various authors [Rob81, RF82, HH83, DK84]. Figure 5.20 (a–d) shows Poincaré surfaces of section for four different values of the scaled energy and $L_z = 0$. The surface of section is the ϱp_ϱ-plane at $z = 0$. Similar to Fig. 5.14 we clearly see an increasing share of phase space filled with irregular trajectories

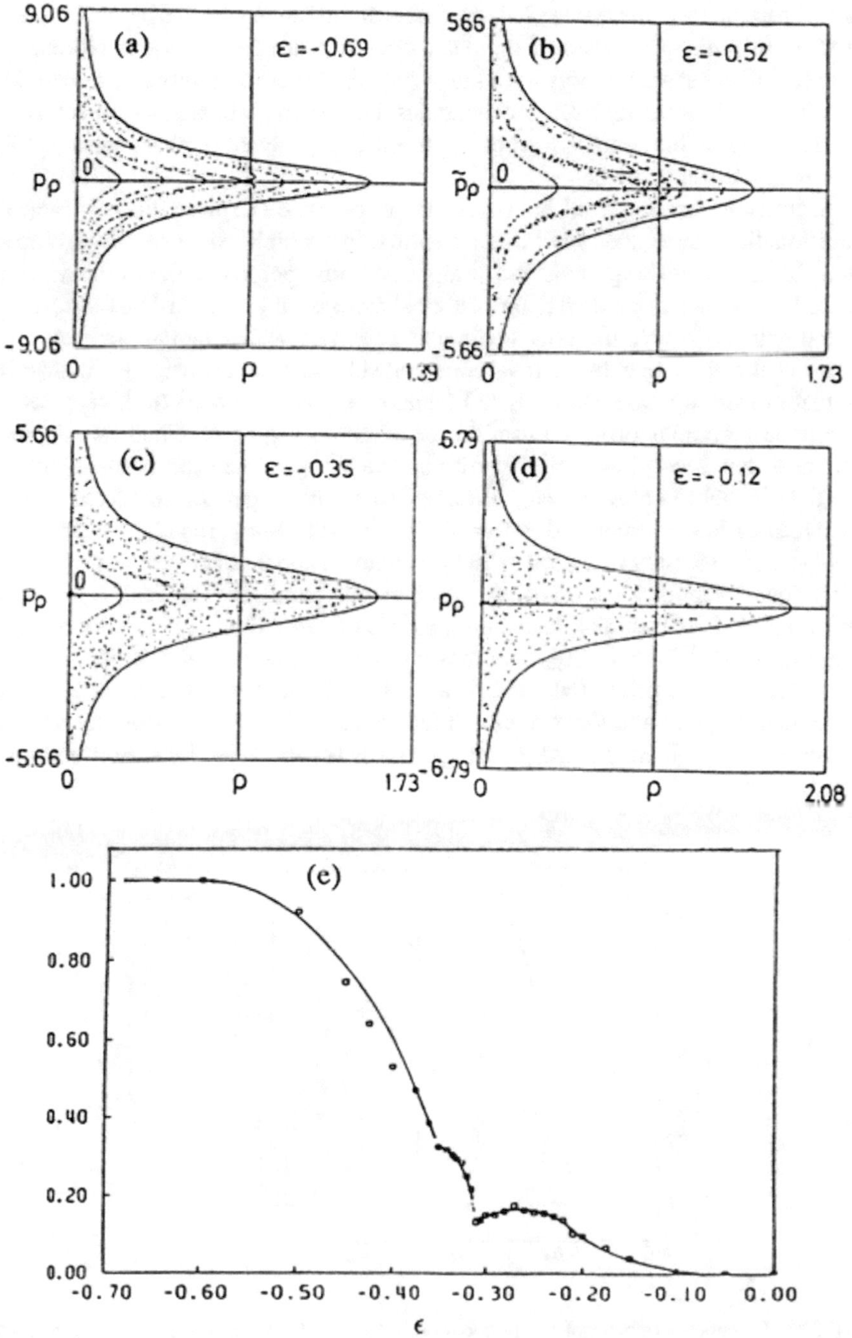

Fig. 5.20 Poincaré surfaces of section for $L_z = 0$ and four different values of the scaled energy ε (**a–d**). The surface of section is the $\varrho\, p_\varrho$-plane at $z = 0$. The bottom panel is taken from [SN93] and shows a measure for the share of regular orbits in phase space as function of the scaled energy

as the parameter ε increases. This is demonstrated again in the bottom panel of Fig 5.20 in which the share of regular trajectories in phase space is plotted as a function of the scaled energy. Around $\varepsilon \approx -0.35$ there is a more or less sudden transition to dominantly irregular dynamics, but the share of regular trajectories is not a monotonic function of ε. Above $\gamma \approx -0.1$ virtually all of phase space is filled with irregular trajectories.

In the field-free case, all bound orbits are periodic (Kepler ellipses). Near the field-free limit there are only three periodic orbits which exist even for arbitrarily weak but non-vanishing fields: the straight-line orbit perpendicular to the direction of the field (which is labelled I_1 for historical reasons), the straight-line orbit parallel to the field (I_∞) and the almost circular orbit (C) which merges into an exact circle in the field-free limit. It is comparatively easy to investigate the stability of these orbits by calculating their Liapunov exponents [Win87b, SN88, SN93]. The almost circular orbit is unstable for all finite values of ε and its Liapunov exponent increases monotonically with ε. The straight-line orbit perpendicular to the field is stable below $\varepsilon_0 = -0.127268612$. The larger dip in Fig. 5.20 (e) is attributed to the confluence of an unstable orbit with the perpendicular orbit I_1 at $\varepsilon = -0.316186$ [SN93]. Above ε_0 the Liapunov exponent of I_1 grows proportional to the square root of $\gamma - \gamma_0$. The straight-line orbit parallel to the field, I_∞, is stable up to $\gamma = -0.391300824$, and then intervals of instability and stability alternate (see Fig. 5.21). Whenever I_∞ becomes unstable, a new periodic orbit is born by *bifurcation*. These orbits (labelled $I_2, I_3, \ldots$) are initially stable but soon become unstable at higher values of ε at which further periodic orbits are born by renewed bifurcation. As ε increases the growing chaos is accompanied by a proliferation of periodic orbits.

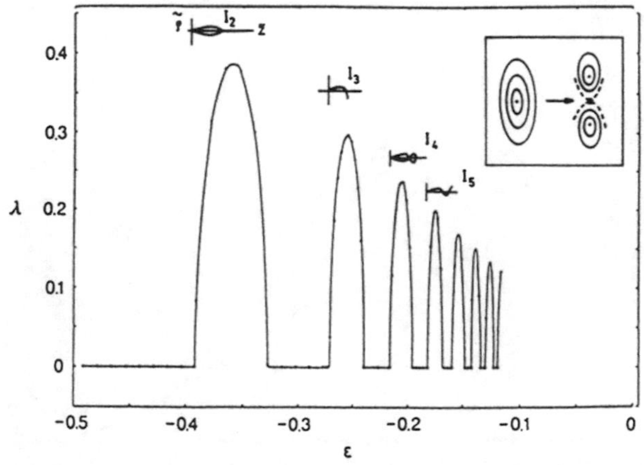

Fig. 5.21 Liapunov exponent of the periodic orbit I_∞ parallel to the direction of the magnetic field. Whenever I_∞ becomes unstable a further initially stable orbit $I_2, I_3, \ldots$ is born by bifurcation. The inset demonstrates schematically how such a bifurcation shows up in the Poincaré surfaces of section

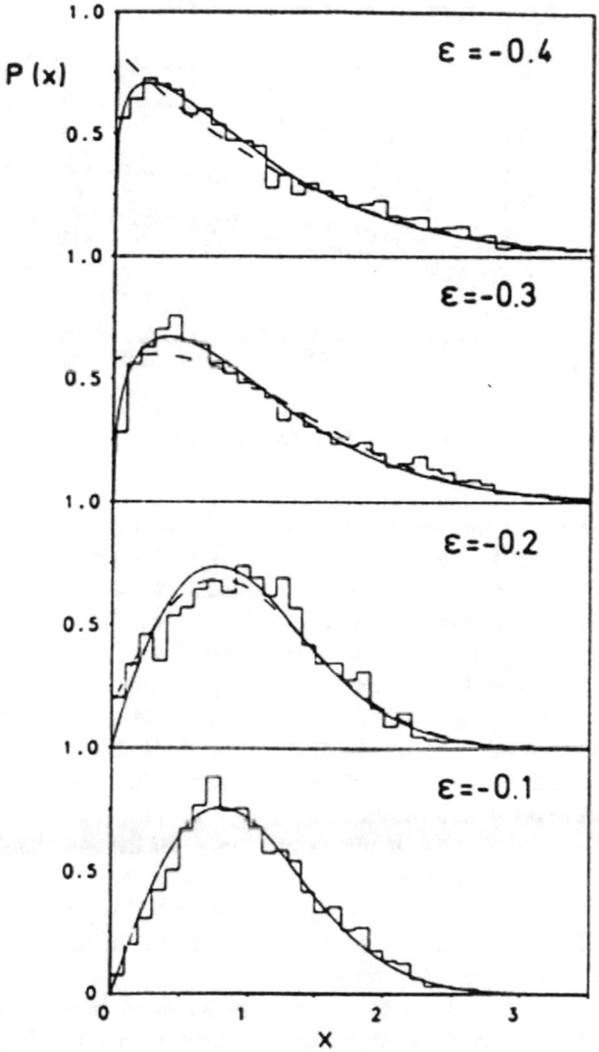

Fig. 5.22 NNS distributions for scaled energies between −0.4 and −0.1. (The *dashed and solid lines* show attempts to fit analytic formula to the distributions in the transition region between regularity and chaos (see [FW89]))

The transition to chaos manifests itself in the statistical properties of the quantum mechanical spectrum of the hydrogen atom in a uniform magnetic field, just as it does for the coupled harmonic oscillators (cf. Figs. 5.16, 5.17). This was shown almost simultaneously in 1986 in [WF86], [DG86] and [WW86]. Figure 5.22 shows e.g. the NNS distributions for four different values of the scaled energy ε. The transition from a distribution close to a Poisson distribution (5.93) at $\gamma = -0.4$ to a Wigner distribution (5.94) at $\gamma = -0.1$ is apparent.

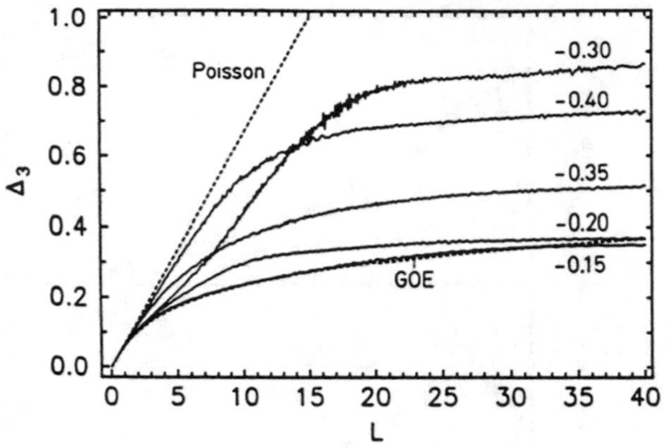

Fig. 5.23 Spectral rigidity (5.95) for various scaled energies ε

Figure 5.23 shows the spectral rigidity (5.95) for values of ε between -0.4 and -0.15. The "odd curve out" at $\varepsilon = -0.30$ clearly reveals what can also be observed by closer inspection for other statistical measures: the transition from Poisson statistics in the regular regime to GOE statistics in the chaotic regime is not monotonic. This is due to non-universal properties of the dynamics which are specific to the system under investigation. Attempts to find simple universal laws or rules for the statistical properties of energy spectra in the transition region between regularity and chaos were only moderately successful [PR94].

Statistical measures such as NNS distributions and the spectral rigidity describe correlations of short and medium range in the spectrum and show a universal behaviour in the regular or classically chaotic limits. On the other hand, long-ranged correlations of the spectrum generally reflect specific properties of the physical system under consideration. This is expressed quantitatively in Gutzwiller's trace formula (5.112) which relates the fluctuating part of the quantum mechanical level density to the classical periodic orbits.

As discussed in Sect. 5.3.4 (c), spectra of an atom in a uniform magnetic field of strength γ are most appropriately recorded for fixed scaled field strength (corresponding to fixed scaled energy) as functions of the natural variable $\chi = \gamma^{-1/3}$, which is proportional to the inverse of the effective Planck's constant. The Fourier transformed spectra then reveal prominent structures at values of the conjugate variable corresponding to the scaled actions of the periodic orbits. This is illustrated in Fig. 5.24 showing the absolute square of the Fourier transformed spectrum in the $m^{\pi_z} = 2^+$ and the $m^{\pi_z} = 2^-$ subspaces for $\varepsilon = -0.2$. The maxima in the Fourier transformed spectra can uniquely be related to simple classical periodic orbits; the corresponding orbits are shown in the right half of the figure.

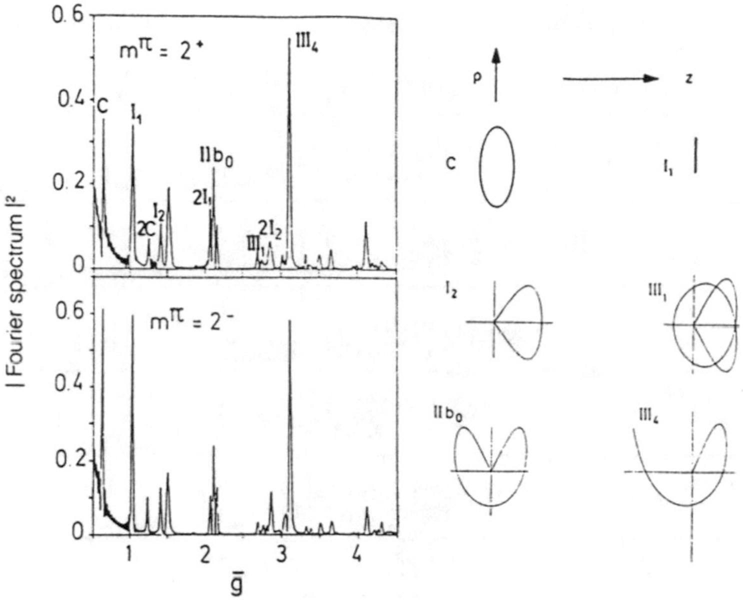

Fig. 5.24 Absolute square of the Fourier transformed spectrum as function of the variable $\bar{g}$, which is conjugate to $\gamma^{-1/3}$, in the $m^{\pi_z} = 2^+$ and $m^{\pi_z} = 2^-$ subspaces at $\gamma = -0.2$. The positions of the peaks are numerically equal to the scaled actions of the classical periodic orbits shown in the right half of the figure (from [Fri90])

The relation between simple periodic classical orbits and modulations in quantum mechanical spectra can also be extended to other observables such as e.g. photoabsorption spectra. Figure 5.25 shows the famous photoabsorption cross sections for barium atoms as measured by Garton and Tomkins in 1969 [GT69]. Near the field-free threshold $E = 0$ we notice modulation peaks separated by about 1.5 times the energy separation of the Landau states of free electrons in a magnetic field. It was soon noticed that these modulation peaks, which are called *quasi-Landau resonances*, can be connected to the classical periodic orbit perpendicular to the field by a relation like (5.116). Later investigations of the photoabsorption cross sections revealed whole series of modulations which can be related to classical periodic orbits in essentially the same way as the modulations in the energy spectra (Fig. 5.24) [HM90]. (The difference between barium and hydrogen is not so important in the present context, where we are dealing with highly excited states extending over large regions in coordinate space, because it only affects the potential $V(\varrho, z)$ in a very small region around the origin.) The quasi-Landau modulations are a very instructive experimental example for how unstable periodic classical orbits in a classically chaotic system manifest themselves in quantum mechanical spectra.

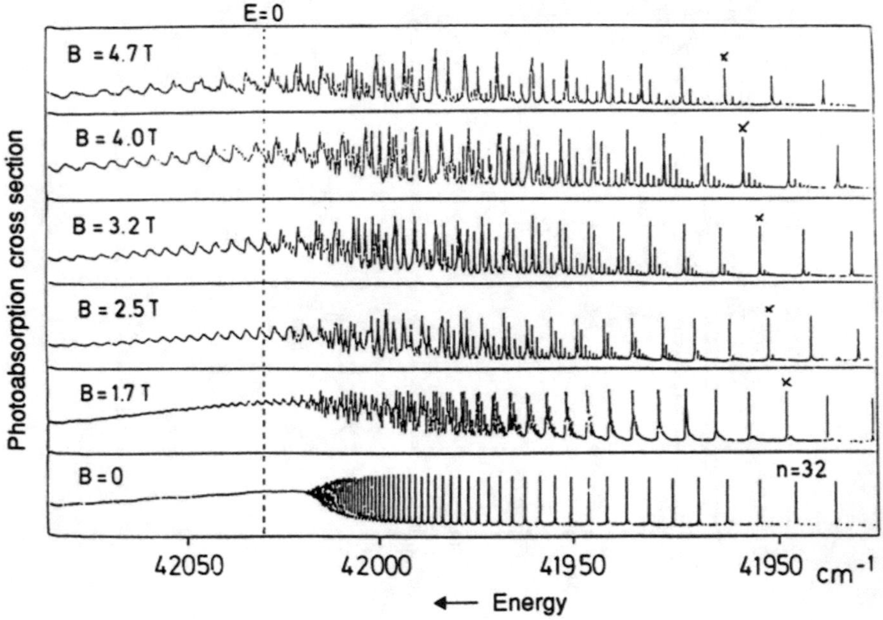

Fig. 5.25 Photoabsorption spectra of barium atoms in a uniform magnetic field (from [GT69])

The role of periodic classical orbits in shaping the structure of the quantum spectra of atoms in a magnetic field is continuing to be a subject of considerable interest. Further advances were achieved in understanding the influence of the non-Coulombic core of the potential in atoms other than hydrogen [O'M89, DM95a, HM95] and in incorporating "ghost orbits" into the periodic orbit theory. Ghost orbits occur close to points of bifurcation and are periodic solutions of the classical equations of motion in complex phase space, which become real periodic orbits after bifurcation. For an overview of related developments see [Mai97, BM99, FM05].

The problem of one electron moving in an attractive Coulomb field and a uniform magnetic field becomes substantially more complicated when an additional electric field is applied. One reason for studying this problem is, that the real hydrogen atom is a two-body system, and its motion in a magnetic field effectively induces an electric field in the Hamiltonian describing its internal motion, see Sect. 3.5.2. All features of regular and chaotic motion and their manifestations in quantum spectra are of course present for an atom in a superposition of electric and magnetic fields, and the richness and diversity of effects is enormous, see, e.g. [MU97, MS98] and references therein.

(c) The Helium Atom

The successful description of the spectrum of the hydrogen atom on the basis of the Bohr-Sommerfeld quantization condition (see Sect. 3.2.1) in the early days of quantum mechanics brought forth numerous attempts to describe the energy spectrum of the helium atom in a similar way [Bor25]. These attempts were unsuccessful for more than sixty years, because a two-electron atom or ion is a nonintegrable system, and the relation between classical mechanics and quantum mechanics was not at all well understood for such systems. Atoms (or ions) with at least two electrons are essentially different from one-electron atoms, because they are, at any total energy, classically unstable for most initial conditions. This is so, because one of the electrons can approach the nucleus arbitrarily closely and so acquire an arbitrarily low energy which leaves enough energy to be transferred to another electron for it to be excited into the continuum. In phase space, a thin skeleton of periodic orbits and nonperiodic trapped orbits remains bound, but most trajectories lead to ionization. Various periodic orbits of the classical helium atom had been known early on [Bor25], but naive applications of the Bohr-Sommerfeld quantization condition had failed to reproduce the energy eigenvalues of low-lying states which were known accurately from experiment and from approximate solutions of the Schrödinger equation.

A satisfactory semiclassical approximation of the energy levels in helium was achieved in 1991 on the basis of periodic orbit theory by Ezra, Richter, Tanner and Wintgen [ER91, WR92]. The method is based on approximating the so-called "dynamical zeta function", whose logarithmic derivative with respect to energy is just the trace of the Green's function (5.106). Individual energy levels are identified with the zeros of the dynamical zeta function, which correspond to the poles of the trace of the Green's function. The dynamical zeta function can be approximated by a product of terms associated with classical periodic orbits such that the logarithmic derivative of this product yields the semiclassical approximation to the trace of the Green's function as summarized in Gutzwiller's trace formula (5.112). If the periodic orbits can be classified by a digital code, then the product over all orbits can be expanded in terms of the lengths of the codes and the expansion truncated after a certain length. This method is known as *cycle expansion* technique [CE89]. The zeros of the approximate dynamical zeta function obtained in this way provide approximations to the energy levels of the system.

The analysis in [ER91] was based on the collinear model of helium, in which both electrons are restricted to lie on different sides on a straight line through the nucleus, see (4.375). In this model there are two spatial coordinates, namely the separations r_1, r_2 of the two electrons from the nucleus, and its four-dimensional classical phase space is a genuine subspace of the phase space of the full two-electron problem in three dimensions. The periodic orbits within this model are easily classified in a systematic way by registering collisions of each of the electrons with the nucleus. All orbits of the collinear model are unstable, i.e. have positive Liapunov exponent, but the collinear motion is seen to be stable against bending

Table 5.1 Energies (in atomic units) of some (n_1s, n_2s) states of the helium atom. The quantum mechanical energy $E_{\rm qm}$ is compared with the semiclassical approximation $E_{\rm sc}$ based on the cycle expansion and with the energies $E_{\rm as}$ obtained via modified Bohr-Sommerfeld quantization of the asymmetric stretch vibration. (From [ER91])

n_1, n_2	$E_{\rm qm}$	$E_{\rm sc}$	$E_{\rm as}$
1, 1	−2.904	−2.932	−3.100
2, 2	−0.778	−0.778	−0.804
2, 3	−0.590	−0.585	
3, 3	−0.354	−0.353	−0.362
3, 4	−0.281	−0.282	
4, 4	−0.201	−0.199	−0.205
4, 5	−0.166	−0.166	
4, 6	−0.151	−0.151	
5, 5	−0.129	−0.129	−0.132
5, 6	−0.110	−0.109	
5, 7	−0.100	−0.101	
6, 6	−0.0902	−0.0895	−0.0917

away from the straight line. The symmetric vibration of both electrons has an infinite Liapunov exponent due to the highly singular triple collision when both electrons meet at the nucleus simultaneously. This so-called "Wannier" mode plays an important role for the ionization process, as discussed in Sect. 4.5.4, but its relevance for the level spectrum well below the break-up threshold is not so clear [Ros94]. The next simplest periodic motion of two electrons in collinear helium is the asymmetric stretch vibration in which both electrons alternately collide with the nucleus and are reflected at their outer turning point. In the application in [ER91] the cycle expansion was truncated so as to include the contributions of all primitive periodic orbits with up to six collisions of one of the electrons with the nucleus during one period. The energies obtained in this way are compared in Table 5.1 with the results of exact quantum mechanical calculations for some of the (n_1s, n_2s) states with similar quantum numbers n_1, n_2. The results show that the energies of several low-lying states of helium can be approximated in the framework of semiclassical periodic orbit theory and the collinear model with an accuracy of a few per cent of the level spacing.

The simplest version of the cycle expansion includes only one periodic orbit, viz. the asymmetric stretch vibration mentioned above, and it corresponds to a modified Bohr-Sommerfeld quantization of this periodic orbit,

$$S_{\rm as}(E_n) = \oint_{\rm as} p\,{\rm d}q = 2\pi\hbar(n + \mu_{\rm d})\,. \tag{5.174}$$

The constant $\mu_{\rm d}$ on the right-hand side plays the role of a negative quantum defect, i.e. a quantum excess, and contains the Maslov index divided by four together with a winding number correction accounting for the zero point motion of the (stable) bending mode. Because the potential is homogeneous of degree −1, the energy dependence of the action on the left-hand side of (5.174) is [cf. (5.163), (5.164) in Sect. 5.3.4 (c)],

$$S_{as}(E) = \frac{S_{as}(-1)}{\sqrt{-E}}\,, \tag{5.175}$$

and the quantization condition (5.174) yields a Rydberg formula,

$$E_n = -\frac{[S_{as}(-1)/(2\pi\hbar)]^2}{(n+\mu_d)^2}\,. \tag{5.176}$$

The quantum mechanical energies of the symmetrically excited (ns, ns) states in helium actually follow such a Rydberg formula quite well, and the data correspond to a value of 1.8205 for $S_{as}(-1)/(2\pi\hbar)$ and a quantum excess $\mu_d = 0.0597$. The deviation of the value 1.8205 from the value two, which one would expect for two non-interacting electrons in the field of the $Z = 2$ nucleus, is attributed to mutual screening of the nucleus by the partner electron. The action of the asymmetric stretch vibration is $S_{as}(-1)/(2\pi\hbar) = 1.8290$. Our experience with quantization of the one-dimensional Coulomb problem in Sect. 3.2.1, (3.131), indicates that a contribution 3 rather than 1 to the Maslov index is appropriate for reflection at an attractive Coulomb singularity. In any case, the two reflections during one period of the asymmetric stretch vibration lead to a half-integral contribution to μ_d, and together with the winding number correction the quantum excess μ_d acquires a theoretical value 0.039 modulo unity [ER91]. The resulting energies (5.175) for the symmetrically excited states are shown in the column E_{as} in Table 5.1. Modified Bohr-Sommerfeld quantization of the asymmetric stretch vibration thus gives a fair account of the energies of the symmetrically excited states.

The interpretation that symmetrically excited states in helium are strongly influenced by the asymmetric stretch vibration of the collinear configuration is supported by quantum mechanical calculations. In the subspace defined by total orbital angular momentum zero there are three independent coordinates, viz. r_1, r_2 and $r_{12} = |\boldsymbol{r}_1 - \boldsymbol{r}_2|$. Figure 5.26 shows the probability distribution $|\psi(r_1, r_2, r_{12})|^2$ of the eigenfunction with $n_1 = 6, n_2 = 6$ on the section of coordinate space defined by $r_{12} = r_1 + r_2$, corresponding to the collinear configuration. The localization of the wave function along the asymmetric stretch orbit, shown as a solid line, is quite apparent.

The analysis of the classical dynamics of two-electron atoms and ions has revealed some interesting and curious results. The so-called "Langmuir orbit", which corresponds to a maximal vibrational bending of the symmetric electron-nucleus-electron configuration, has been shown to be stable for nuclear charge $Z = 2$ [RW90b]. Further stable orbits exist in the "frozen planet" configuration in which both electrons are on the same side of the nucleus at very different separations, and the farther electron ("frozen planet") moves slowly in a limited region of coordinate space, while the nearer electron oscillates rapidly to and from the nucleus [RW90c]. Although most classical trajectories are unstable, the

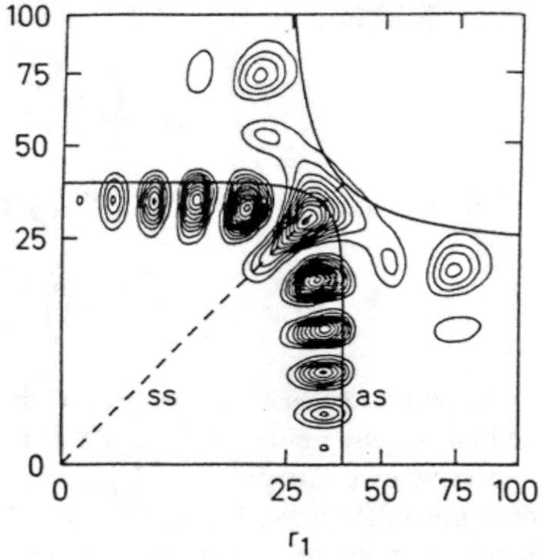

Fig. 5.26 Probability distribution $|\psi(r_1, r_2, r_{12})|^2$ of the (6*s*, 6*s*) helium atom eigenfunction on the section of coordinate space defined by $r_{12} = r_1 + r_2$. The *solid line* labelled 'AS' shows the asymmetric stretch orbit, the *dashed line* is the Wannier orbit, $r_1 = r_2$ (From [WR92])

existence of such stable orbits means, that the classical dynamics of the helium atom is not fully ergodic.

In highly asymmetric configurations of a two-electron atom or ion, one electron can move for a long time on a very extended Kepler orbit, while the other electron stays near the nucleus. Even though such orbits are generally unstable, their Liapunov exponent can be arbitrarily small, because the motion of the two electrons is almost independent, being on individual and only slightly perturbed Kepler ellipses for an arbitrarily long time. This phenomenon of long intervals of regularity on trajectories which are in fact unstable is called "intermittency". The quantum analogue in two-electron atoms is provided by the highly excited states in Rydberg series, where one electron is excited to very high quantum numbers, while the other electron is in a state of low or medium quantum number. Adaptations of periodic orbit theory were quite successful in establishing the link between classical and quantum dynamics in these situations [RT97, BQ97, Ros98]. An extensive review on two-electron atoms is contained in [TR00]. Several partial successes have opened windows on special features of two-electron atoms, but a comprehensive understanding of the structure of the Hilbert space and the energy spectrum, the coexistence Rydberg-like series of narrow levels and broad, overlapping resonances generating Ericson fluctuations [BR09], has not been achieved, not even qualitatively. It remains true, that highly doubly excited two-electron atoms below the break-up threshold constitute one of the most fundamental still unsolved problems of quantum mechanics.

The developments around and after 1990 reinstated classical mechanics as a relevant theory, even in the atomic domain. It remains undisputed, that quantum mechanics is the formalism for a correct quantitative description of atomic phenomena. It is also clear that the uncertainty principle holds and that it would be wrong to picture the electrons in an atom as point particles moving on classical trajectories with well defined positions and momenta. Through the advances described in this section it has however become apparent, that the properties of a classical system, in particular of its periodic orbits, are visible in spectra of the corresponding quantum system, and that we can understand and sometimes quantitatively describe features of the quantum mechanical observables on the basis of our knowledge of the classical orbits.

5.4 Bose-Einstein Condensation in Atomic Gases

5.4.1 Quantum Statistics of Fermions and Bosons

Consider a large number of independent identical particles, each described by the same one-body Hamiltonian with eigenstates $|\nu\rangle$ and eigenvalues ε_ν, $\nu = 1, 2, 3, \ldots$. We can construct a basis of eigenstates of the many-body system from the products of the one-body eigenstates, which should be antisymmetrized or symmetrized if the particles are fermions or bosons respectively. Due to the indistinguishability of the particles, a many-body state depends only on the numbers n_ν of particles occupying the various one-body eigenstates $|\nu\rangle$, and we shall collect all these numbers $n_1, n_2, \ldots, n_\nu, \ldots$ in one label r. The total energy E_r in the many-body state r is,

$$E_r = \sum_{\nu=1}^{\infty} n_\nu \varepsilon_\nu \,, \tag{5.177}$$

and the total number N_r of particles is,

$$N_r = \sum_{\nu=1}^{\infty} n_\nu \,. \tag{5.178}$$

The standard procedure for describing such a system in the framework of statistical mechanics is to imagine an ensemble of systems corresponding to all possible realizations of the many body state, and the values we deduce for observable physical quantities depend on the probability with which the various possibilities are realized. In the *grand canonical ensemble* the probabilities are determined by the temperature T and the chemical potential μ, and are proportional to $\exp\left[-(E_r - \mu N_r)/(k_B T)\right]$, where k_B is Boltzmann's constant. This is generally expressed with the help of the grand canonical *partition function,*

$$Y = \sum_r \mathrm{e}^{-\beta(E_r - \mu N_r)} \,, \quad \beta = 1/(k_B T) \,, \tag{5.179}$$

so the probability P_r for realizing an individual state r of the whole many- body system is,

$$P_r = \frac{1}{Y}\,e^{-\beta(E_r-\mu N_r)} = \frac{1}{Y}\,e^{-\beta\sum_{\nu=1}^{\infty} n_\nu(\varepsilon_\nu-\mu)} = \frac{1}{Y}\prod_{\nu=1}^{\infty} e^{-\beta(\varepsilon_\nu-\mu)n_\nu}\,. \tag{5.180}$$

The full many-body partition function (5.179) can be rearranged to a product,

$$Y = \prod_{\nu=1}^{\infty} Y_\nu\,, \quad Y_\nu = \sum_n e^{-\beta(\varepsilon_\nu-\mu)n}\,, \tag{5.181}$$

and each factor Y_ν is actually a one-state partition function for a grand canonical ensemble of one-state systems, in which the particles can only occupy the one single-particle quantum state $|\nu\rangle$. For fermions, each state $|\nu\rangle$ can only be occupied by $n = 0$ or $n = 1$ particles because of the Pauli principle, and the summation over n is easily performed, $Y_\nu = 1 + \exp[-\beta(\varepsilon_\nu - \mu)]$. The probability for the state $|\nu\rangle$ being unoccupied is $P_0 = 1/Y_\nu$ and the probability for being occupied is $P_1 = \exp[-\beta(\varepsilon_\nu - \mu)]/Y_\nu$, so the average occupation number $\langle n_\nu\rangle$ is,

$$\langle n_\nu\rangle = \sum_{n=0,1} nP_n = \frac{\exp[-\beta(\varepsilon_\nu-\mu)]}{1+\exp[-\beta(\varepsilon_\nu-\mu)]} = \frac{1}{\exp[\beta(\varepsilon_\nu-\mu)]+1}\,. \tag{5.182}$$

For bosons there is no restriction on the number of particles which can occupy a given single particle state $|\nu\rangle$, and Y_ν is a geometric series which sums to $Y_\nu = 1/(1-\exp[-\beta(\varepsilon_\nu-\mu)])$, provided $\varepsilon_\nu > \mu$. The average occupation number in the state $|\nu\rangle$ is now

$$\langle n_\nu\rangle = \sum_{n=0}^{\infty} nP_n = \left(1 - e^{-\beta(\varepsilon_\nu-\mu)}\right)\sum_{n=1}^{\infty} n\,e^{-\beta(\varepsilon_\nu-\mu)n}\,. \tag{5.183}$$

The right-hand side of (5.183) can be evaluated by writing the sum as $1/\beta$ times the derivative of $\sum_{n=0}^{\infty}\exp[-\beta(\varepsilon_\nu-\mu)n] = 1/(1-\exp[-\beta(\varepsilon_\nu-\mu)])$ with respect to μ, and this yields

$$\langle n_\nu\rangle = \frac{1}{\exp[\beta(\varepsilon_\nu-\mu)]-1}\,. \tag{5.184}$$

At energies much larger than the chemical potential, $\varepsilon_\nu-\mu \gg k_{\rm B}T$, the difference between fermions (5.182) and bosons (5.184) disappears and the (small) occupation probabilities approach an exponential behaviour, $\langle n_\nu\rangle = \exp[-\beta(\varepsilon_\nu-\mu)]$, typical for classical particles. At low temperatures, $\beta \to \infty$, the occupation probability (5.182) for fermions degenerates to $1 - \Theta(\varepsilon_\nu - \mu)$, i.e. the chemical potential corresponds to the Fermi energy [cf. (2.102)] up to which all single-particle states are filled, while all higher-lying single-particle states are unoccupied. This is the extreme case of a degenerate Fermi gas. For bosons (5.184) the chemical potential must be smaller than the lowest single particle energy, and the occupation probability tends to infinity when $\varepsilon_\nu \to \mu$. The condensation to a degenerate Bose gas, i.e. Bose-Einstein condensation, is conveniently illustrated for the example of an ideal gas.

An ideal gas can be considered to be a system of free particles of mass m moving in a large cube of side length L. The single-particle states can be labelled by three positive integers $(\nu_x, \nu_y, \nu_z) \equiv \boldsymbol{\nu}$, and the corresponding single-particle energies are (cf. Problem 2.4),

$$\varepsilon_{\boldsymbol{\nu}} = \frac{\hbar^2\pi^2}{2mL^2}(\nu_x^2+\nu_y^2+\nu_z^2) = C\left(\frac{\nu}{L}\right)^2, \quad \text{with} \quad C = \frac{\hbar^2\pi^2}{2m}. \tag{5.185}$$

The average total number of particles $\langle N\rangle$ is,

$$\langle N\rangle = \sum_{\nu_x,\nu_y,\nu_z} \langle n_{\boldsymbol{\nu}}\rangle \approx \frac{4\pi}{8}L^3 \int_0^\infty \tilde{\nu}^2\,\mathrm{d}\tilde{\nu}\,\frac{1}{\exp\left[\beta(C\tilde{\nu}^2-\mu')\right]-1}, \tag{5.186}$$

where the sum over the discrete lattice ν_x, ν_y, ν_z has been replaced by an integral over the vector $\tilde{\boldsymbol{\nu}} = (\nu_x-1, \nu_y-1, \nu_z-1)/L$ in the octant, $\tilde{\nu}_i \geq 0$, and $\mu' = \mu - \varepsilon_{\rm gs}$ is the chemical potential relative to the energy of the non-degenerate ground state, $\tilde{\boldsymbol{\nu}} = 0$.[1] The integral on the right-hand side of (5.186) can be evaluated by decomposing the quotient into a geometric series, yielding,

$$\frac{\langle N\rangle}{L^3} = \frac{\pi}{2}\sum_{l=1}^{\infty}\int_0^\infty \tilde{\nu}^2\,\mathrm{e}^{-l\beta(C\tilde{\nu}^2-\mu')}\mathrm{d}\tilde{\nu} = \frac{\pi}{2}\sum_{l=1}^{\infty}\mathrm{e}^{l\beta\mu'}\frac{\sqrt{\pi}}{4(l\beta C)^{3/2}}. \tag{5.187}$$

Inserting the expression for C as given in (5.185) yields

$$\frac{\langle N\rangle}{L^3} = \left(\frac{mk_{\rm B}T}{2\pi\hbar^2}\right)^{3/2}\sum_{l=1}^{\infty}\frac{\mathrm{e}^{l\beta\mu'}}{l^{3/2}}. \tag{5.188}$$

Equation (5.188) defines the temperature dependence of the chemical potential μ'. As T is decreased from some high value, μ' must increase from some large negative value, if the total average particle number $\langle N\rangle$, or number density $\langle N\rangle/L^3$, is to remain constant. At a critical temperature $T_{\rm c}$, the value of μ' reaches zero. As the temperature is reduced below $T_{\rm c}$, μ' remains zero. The formula (5.186) now only accounts for the particles in excited states, because its derivation

[1] Terms of order $\tilde{\nu}/L$ are neglected in the exponent on the right-hand side of (5.186), but reference to the energy of the non degenerate ground state, $\varepsilon_{\rm gs} = O(1/L^2)$ is retained for pedagogical reasons.

relied on the condition $\mu < \varepsilon_\nu$. The critical temperature is defined by inserting $\mu' = 0$ in (5.188),

$$\frac{\langle N\rangle}{L^3} = \left(\frac{mk_\mathrm{B}T_c}{2\pi\hbar^2}\right)^{3/2}\sum_{l=1}^{\infty}\frac{1}{l^{3/2}}\,. \tag{5.189}$$

The sum on the right-hand side of (5.189) is just the value of the zeta function, $\zeta(x) = \sum_{l=1}^{\infty} l^{-x}$, for argument $x = 3/2$, $\zeta(3/2) = 2.612\ldots$. The critical temperature T_c is reached when the number density $\varrho = \langle N\rangle/L^3$ is, except for the factor $\zeta(3/2) = 2.612\ldots$, equal to the inverse cube of the *thermal wave length* $\lambda(T) = 2\pi\hbar/\sqrt{2\pi mk_\mathrm{B}T}$,

$$\varrho = \frac{2.612\ldots}{\lambda(T_\mathrm{c})^3}\,. \tag{5.190}$$

The thermal wave length $\lambda(T)$ is the de Broglie wave length $2\pi\hbar/p$ of a particle whose kinetic energy $p^2/(2m)$ is equal to $\pi k_\mathrm{B}T$. At the critical temperature the thermal wave length becomes so large that it is of the order of the linear dimensions of the volume ϱ^{-1} available to each particle.

As the temperature is reduced below T_c, the chemical potential μ' remains zero and the number N_exc of particles in excited states is given by,

$$\frac{N_\mathrm{exc}}{L^3} = \left(\frac{mk_\mathrm{B}T}{2\pi\hbar^2}\right)^{3/2}\sum_{l=1}^{\infty}\frac{1}{l^{3/2}} = \left(\frac{T}{T_\mathrm{c}}\right)^{3/2}\frac{\langle N\rangle}{L^3}\,, \quad T \le T_\mathrm{c}\,. \tag{5.191}$$

The number N_0 of particles which has condensed into the non-degenerate ground state is,

$$N_0 = \langle N\rangle\left[1 - \left(\frac{T}{T_\mathrm{c}}\right)^{\frac{3}{2}}\right]\,, \quad T \le T_\mathrm{c}\,. \tag{5.192}$$

For $T \to 0$ all particles condense into the ground state. This is the extreme case of a degenerate Bose gas.

The condensation of a significant fraction of the Bose gas into its ground state has dramatic consequences for its thermodynamical properties. The condensed particles don't contribute to the pressure of the gas, and they don't participate in the transfer of heat. Below T_c the specific heat of the gas falls off with diminishing temperature due to the diminishing fraction of particles participating.

Atoms as a whole behave like fermions if their total number of nucleons (neutrons and protons) and electrons is odd, and like bosons if it is even. The standard text-book example for Bose-Einstein condensation was, for many years, liquid ^{4}He, which shows a phase transition to superfluidity at a temperature of 2.17°K.

Homogeneous Bose-Einstein condensates have been a topic of continuing study and interest in the field of condensed matter physics for many years. The condensation of atoms trapped in an external potential produces spatially confined Bose-Einstein condensates which have a finite volume and a surface and hence exhibit new and interesting features not present in the homogeneous case. The successful preparation of such condensates of atomic gases in 1995 greatly stimulated interest in their theoretical description. A representative introduction to the theory of non-homogeneous Bose-Einstein condensates is contained in the papers collected in [BE96].

5.4.2 The Effect of Interactions in Bose-Einstein Condensates

The Hamiltonian for a system of identical particles of mass m in a common external potential $V(\boldsymbol{r})$, which interact via a two-body potential $W(\boldsymbol{r}_i - \boldsymbol{r}_j)$ is [cf. (2.53)],

$$\hat{H} = \sum_{i=1}^{N} \frac{\hat{\boldsymbol{p}}_i^2}{2m} + \sum_{i=1}^{N} V(\boldsymbol{r}_i) + \sum_{i<j} W(\boldsymbol{r}_i - \boldsymbol{r}_j) \,. \tag{5.193}$$

The Hartree-Fock method described in Sect. 2.3.1 treats a system of interacting fermions on the basis of Slater determinants, so that the independent particle picture is formally kept, but a part of the interaction between the particles is taken into account in the form of a mean field. An analogous ansatz for bosons is to start with a many-body wave function Ψ consisting of a product of single-particle wave functions,

$$\Psi(\boldsymbol{r}_1, \ldots, \boldsymbol{r}_N) = \prod_{i=1}^{N} \psi_i(\boldsymbol{r}_i) \,. \tag{5.194}$$

The right-hand side of (5.194) should in principle be symmetrized with respect to the particle labels. In a product ansatz for the ground state of the many-boson system we assume all particles to occupy the same single-particle state, $\psi_i(\boldsymbol{r}) = \psi(\boldsymbol{r}), i = 1, \ldots N$, so the symmetry requirement is fulfilled automatically. Minimizing the expectation value of the Hamiltonian (5.193) with respect to variations of the single-particle wave functions leads to an equation for ψ. The calculations are now simpler than for the fermion case in Sect. 2.3.1, in particular, there is no exchange potential. Instead of the Hartree-Fock equations (2.88) we obtain a "Schrödinger equation" with the one-body Hamiltonian,

$$\hat{h}_\Psi = \frac{\hat{\boldsymbol{p}}^2}{2m} + V(\boldsymbol{r}) + W_{\rm mf}(\boldsymbol{r}) \,, \tag{5.195}$$

and the mean-field contribution is [cf. (2.90), (2.91)],

$$W_{\rm mf}(\boldsymbol{r}) = \int {\rm d}\boldsymbol{r}' \sum_{i=1}^{N} |\psi_i(\boldsymbol{r}')|^2 W(\boldsymbol{r}-\boldsymbol{r}') \\ = \int {\rm d}\boldsymbol{r}' N |\psi(\boldsymbol{r}')|^2 W(\boldsymbol{r}-\boldsymbol{r}') \,. \tag{5.196}$$

The resulting "Schrödinger equation" is usually formulated for the renormalized wave function,

$$\psi_N(\boldsymbol{r}) = \sqrt{N}\psi(\boldsymbol{r}) \,, \tag{5.197}$$

and its time-dependent version reads [Gro63],

$$\mathrm{i}\hbar \frac{\partial \psi_N}{\partial t} = \left(-\frac{\hbar^2}{2m}\Delta + V(\boldsymbol{r})\right) \psi_N(\boldsymbol{r},t) \\ + \psi_N(\boldsymbol{r},t) \int |\psi_N(\boldsymbol{r}',t)|^2 \, W(\boldsymbol{r}-\boldsymbol{r}') \, {\rm d}\boldsymbol{r}' \,. \tag{5.198}$$

Since this equation is nonlinear in ψ_N, it is necessary to specify the normalization condition,

$$\int |\psi_N(\boldsymbol{r},t)|^2 {\rm d}\boldsymbol{r} = N. \tag{5.199}$$

Equation (5.198) is known as the *Gross-Pitaevskii equation* or also as the *nonlinear Schrödinger equation*. Its time-independent version reads,

$$\left(-\frac{\hbar^2}{2m}\Delta + V(\boldsymbol{r})\right) \psi_N(\boldsymbol{r}) + \psi_N(\boldsymbol{r}) \int |\psi_N(\boldsymbol{r}')|^2 \, W(\boldsymbol{r}-\boldsymbol{r}') \, {\rm d}\boldsymbol{r}' \\ = \mu \psi_N(\boldsymbol{r}) \,, \tag{5.200}$$

where we have written the chemical potential μ for the energy $\varepsilon_{\rm gs}$ of the single-particle ground state, in accordance with the conditions for condensation described in Sect. 5.4.1.

The two-body potential W may be expected to disturb the independent-particle picture only weakly, if its range is short compared to the spatial extension of the condensate wave function ψ_N. In this case we may approximate $\int |\psi_N(\boldsymbol{r}')|^2 W(\boldsymbol{r}-\boldsymbol{r}')\,{\rm d}\boldsymbol{r}'$ by $|\psi_N(\boldsymbol{r})|^2 \int W(\boldsymbol{r}')\,{\rm d}\boldsymbol{r}'$ in (5.198) and (5.200). According to (4.19), the spatial integral over the potential W is, except for a constant, identical to the low-energy limit of the Born approximation $f^{\rm B}$ to the amplitude for particle-particle scattering under the influence of the two-body potential W,

$$\int W(\mathbf{r}')\,\mathrm{d}\mathbf{r}' = -\lim_{k\to 0}\frac{4\pi\hbar^2}{m}f^{\mathrm{B}}\,. \tag{5.201}$$

[Remember that the reduced mass of relative motion of two particles of mass m is $m/2$.] In the low-energy limit the scattering amplitude (4.31) reduces to the partial wave amplitude $f_{l=0}$ for the s-wave, which in turn can be expressed via the scattering length a according to (4.34), (4.59), $f_{l=0} = (1/k)\sin\delta_{l=0} + \ldots = -a + O(k)$ for $k \to 0$. If the effect of the interaction is sufficiently weak, we may identify f^{B} with $-a$ and obtain the following generally used forms [DG97] of the time-dependent and time-independent Gross-Pitaevskii equation:

$$\mathrm{i}\hbar\frac{\partial\psi_N}{\partial t} = \left(-\frac{\hbar 2}{2m}\Delta + V(\mathbf{r}) + \frac{4\pi\hbar^2}{m}a|\psi_N(\mathbf{r},t)|^2\right)\psi_N(\mathbf{r},t)\,, \tag{5.202}$$

$$\left(-\frac{\hbar^2}{2m}\Delta + V(\mathbf{r}) + \frac{4\pi\hbar^2}{m}a|\psi_N(\mathbf{r})|^2\right)\psi_N(\mathbf{r}) = \mu\,\psi_N(\mathbf{r})\,. \tag{5.203}$$

The effect of two-body interactions on the condensate wave function is thus, in a first approximation, controlled by the scattering length of the two-body potential. The importance of atom-atom collisions for the understanding of Bose-Einstein condensates rekindled interest in quantum and semiclassical analyses of the atom-atom interaction, in particular in the regime of extremely low energies [Jul96, MW96, CH96, TE98, EM00]. In certain cases it is actually possible to tune the scattering length of the atom-atom interaction by varying the strength of an external magnetic field [CC00]. This works via a near-threshold Feshbach resonance in an inelastic channel of the atom-atom system. The energy E_{R} of such a Feshbach resonance relative to the elastic-channel threshold generally depends on the strength of the magnetic field, so it can be tuned to any small positive or negative value simply by adjusting the field strength appropriately. Tunable near-threshold Feshbach resonances are discussed in detail in Sect. 5.6.

The condensate wave function ψ_N can be obtained by numerical solution of the Gross-Pitaevskii equation. This was done by by Dalfovo et al. [DP96] for the case of a cylindrically symmetric harmonic external potential,

$$V(\mathbf{r}) = \frac{m}{2}\omega_\perp^2(x^2 + y^2 + \lambda^2 z^2)\,. \tag{5.204}$$

Calculations were performed for a perpendicular oscillator width, $\beta_\perp = \sqrt{\hbar/(m\omega_\perp)}$ cf. (1.83), of 1.222×10^{-4} cm corresponding to about 23000 Bohr radii, and the frequency ratio λ was chosen as $\sqrt{8}$. The physical parameters of the particles correspond to ^{87}Rb atoms; the scattering length a was taken to be 100 Bohr radii and positive, corresponding to a repulsive atom-atom potential; the number of atoms in the condensate was assumed to be 5000. The resulting wave function along

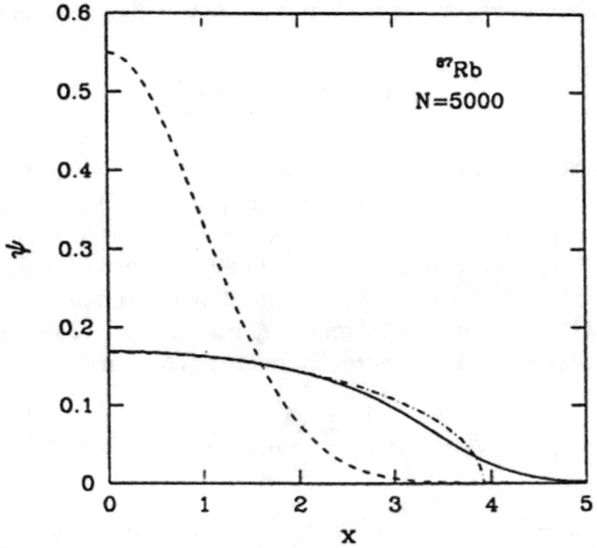

Fig. 5.27 Ground state wave function $\psi_N(x, 0, 0)$ (in arbitrary units) for a Bose-Einstein condensate of 5000 ^{87}Rb atoms in the external potential (5.204). The length labelling the abscissa is in units of the oscillator width $\beta_\perp$. The *solid line* shows the solution of the Gross-Pitaevskii equation, the *dashed line* is the Gaussian ground state harmonic oscillator wave function describing the non-interacting case, and the *dot-dashed line* is the result of the Thomas-Fermi approximation, $\psi_N(\boldsymbol{r}) = \sqrt{[m/(4\hbar^2 a)][\mu - V(\boldsymbol{r})]}$ (From [DP96])

the x-axis perpendicular to the axis of symmetry is shown in Fig. 5.27. The dashed line shows the x-dependence of the wave function in the non-interacting case, where it is just the Gaussian for the harmonic oscillator ground state. The solid line shows the result of numerically solving the Gross-Pitaevskii equation (5.203). The repulsive atom-atom interaction clearly stretches and flattens the profile of the wave function. The dash-dotted line in Fig. 5.27 shows the result corresponding to $|\psi_N(\boldsymbol{r})|^2 = [m/(4\pi\hbar^2 a)][\mu - V(\boldsymbol{r})]$, which is obtained simply by neglecting the kinetic energy term $-[\hbar^2/(2m)]\Delta\psi_N$ in the Gross-Pitaevskii equation (5.203). This so-called "Thomas-Fermi approximation" implies a large product of scattering length and density; it describes the profile of the condensate wave function in Fig. 5.27 quite well in the interior but poorly near the surface.

One example of differences between homogeneous and confined condensates is provided by work [OS97, DG97] on the excitation spectrum of a trapped condensate. Excitations of the condensate wave function are described in the framework of the Hartree-Fock approximation as one-particle-one-hole excitations, in which one of the particles occupies an excited single particle state rather than the ground state ψ, cf. Sect. 2.3.1. A further reaching theory due to Bogoliubov is based on the concept of *quasiparticles,* which is more general, because a single-quasiparticle state involves a superposition of occupied and unoccupied particle states, see e.g. [ED96]. Bogoliubov's theory has a long history of successful applications to the description of superfluidity in condensed matter and nuclear physics. It is able to describe collective excitations such as the phonons in a homogeneous Bose-Einstein condensate, but as a generalization of the Hartree-Fock approximation it can also account for excitations dominantly of a single-particle nature.

Dalfovo et al. [DG97] have recently used Bogoliubov theory to calculate the excitation spectrum of a condensate of 10000 Rubidium atoms in a spherical external harmonic oscillator potential of oscillator width 0.791×10^{-4} cm corresponding to about 15000 Bohr radii; the (positive) scattering length is 110 Bohr radii. The excitation spectrum obtained in this way is shown in the top half of Fig. 5.28. The bottom half shows the spectrum obtained in the Hartree-Fock approximation, in

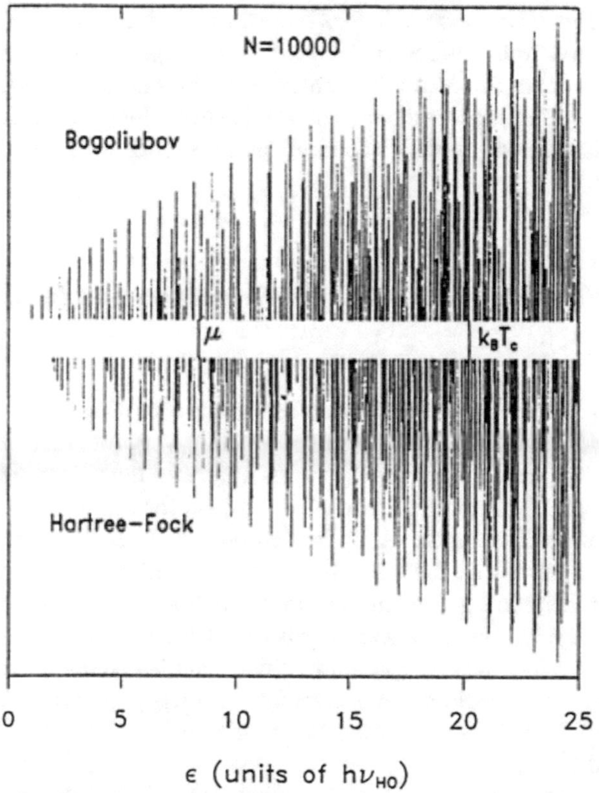

Fig. 5.28 Excitation spectrum of a Bose-Einstein condensate of 10000 rubidium atoms in a spherically symmetric harmonic potential with frequency parameter ω. The top half shows the results derived using Bogoliubov theory, the bottom half is based on simple single-particle excitations. The lengths of the lines are proportional to $2l + 1$, where l is the orbital angular momentum quantum number. μ labels the chemical potential obtained from the solution of the Gross-Pitaevskii equation for the ground state wave function, $k_B T_c$ denotes the critical temperature for non-interacting bosons in the external harmonic oscillator potential. All energies are in units of $\hbar\omega = h\nu_{HO}$ (From [DG97])

which the quasiparticles of Bogoliubov theory reduce to single particle excitations. The lengths of the lines in Fig. 5.28 are proportional two $2l+1$, where l is the orbital angular momentum quantum number of the respective state. Also shown in Fig. 5.28 are the chemical potential μ, which follows from the solution of the Gross-Pitaevskii equation (5.203) for the ground state of the condensate, and the critical temperature $k_B T_c$ which would apply in the case of non-interacting particles in the given external potential. In the non-interacting case the chemical potential is equal to the energy $\frac{3}{2}\hbar\omega$ of the non-degenerate single-particle ground state, and the excitation spectrum consists of positive integral multiples of $\hbar\omega$.

The lowest few excitations shown in the top half of Fig. 5.28 correspond to collective phononic excitations and cannot be accounted for in the simple Hartree-Fock approach based on single-particle excitations. Apart from these very low states, the Hartree-Fock approach does however reproduce the general structure of the excitation spectrum well, even for excitation energies lower than the chemical potential. The occurrence of single-particle excitations at low energies is attributed to the existence of a surface region where the density is low, cf. Fig. 5.27, and hence is a characteristic feature in which confined systems differ from homogeneous Bose gases [DG97].

5.4.3 Realization of Bose-Einstein Condensation in Atomic Gases

In order to experimentally realize Bose-Einstein condensation in an atomic gas, it is necessary to accumulate a large number of atoms at very low temperature. Neutral atoms can be trapped in an inhomogeneous magnetic field, provided they have a substantial magnetic dipole moment, as is the case for alkali atoms. Deceleration of moving atoms can be achieved by irradiation with laser light which is tuned to be selectively absorbed by the faster atoms. Modern procedures are quite intricate and subtle and involve e.g. the intelligent exploitation of the hyperfine structure of the atomic sublevels, which are temporarily populated in order to shield the coldest atoms from emission and absorption of photons and hence optimize their survival rates. Progress in the development of techniques for trapping and cooling atoms was rewarded in 1997 by the award of the Noble Prize in Physics to S. Chu, C. Cohen-Tannoudji and W. Phillips.

A further process, viz. evaporative cooling, proved vital in achieving temperatures low enough at densities high enough to enable condensation. A radio-frequency magnetic field causes a spin-flip in the faster atoms near the edge of the sample, these are no longer trapped and evaporate, thus cooling the sample. The radio frequency is continuously reduced, thus peeling off layer after layer of comparatively faster atoms. The final frequency ν_{evap} is a measure for the temperature of the atoms remaining in the sample.

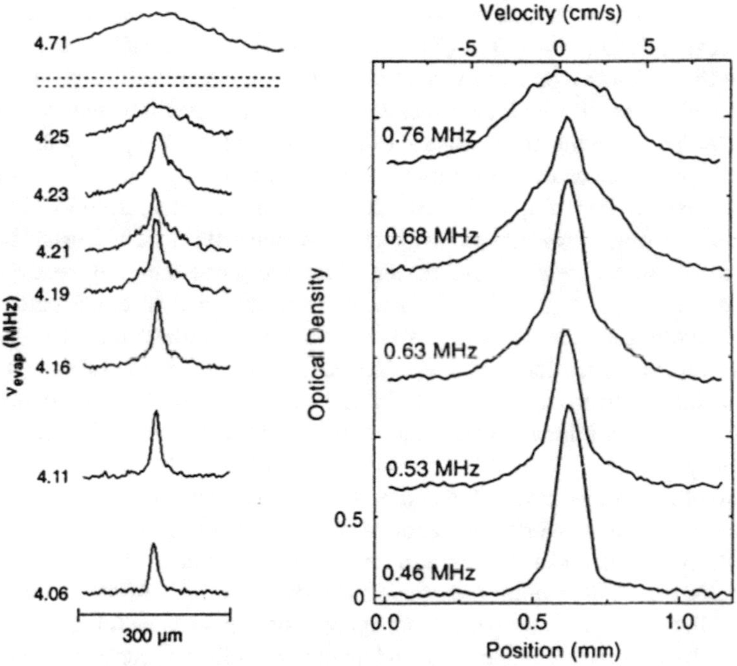

Fig. 5.29 Velocity distributions of atoms released from a magneto-optical trap after being evaporatively cooled. The various curves correspond to different final radio frequencies ν_{evap} which steer the evaporative cooling process and are a measure for the temperature of the sample. The left-hand part of the figure shows the results of Anderson et al. for samples of ^{87}Rb atoms (from [AE95]), the right hand part of the figure shows the results of Davis et al. for samples of sodium atoms (from [DM95b])

The procedure sketched above was applied in 1995 by Anderson et al. [AE95] at the Joint Institute for Labotoratory Astrophysics (JILA) in Boulder, Colorado, and by Davis et al. [DM95b] at the Massachussetts Institute of Technology (MIT) to cool trapped samples of alkali atoms. The velocity distribution of the atoms was then determined by time-of-flight measurements after the confining trap potential had been switched off. Resulting velocity distributions are shown in Fig. (5.29) for different values of the radio frequency ν_{evap} which steers the evaporative cooling process and determines the final temperature of the sample. The left-hand part of the figure shows the results of Anderson et al. [AE95] who cooled a vapour of ^{87}Rb atoms. As ν_{evap} falls below 4.25 MHz, an increasing fraction of the sample belongs to a sharp peak around velocity zero. This is seen as evidence of condensation of this fraction of atoms to the lowest quantum state in the trap potential. The sample at 4.25 MHz, where the transition begins, has a temperature of 1.7×10^{-7} °K and contains 2×10^4 atoms at a number density of 2.6×10^{12} per cubic centimetre; this corresponds to $\varrho = 0.3/\lambda(T)^3$, cf. (5.190). Near 4.1 MHz the sample still contains 2000 atoms which are virtually all in the condensate. The right-hand part of Fig. (5.29) shows the analogous results of Davis et al. [DM95b], who worked with sodium atoms. Here the condensation of atoms sets in at a frequency of 0.7 MHz, where the temperature of the sample is estimated to be $2 \times 10^{-6}\,^\circ$K. Davis et al. observed condensates of up to 5×10^5 atoms at number densities up to 10^{14} cm^{-3}.

The pioneering experimental work at JILA and MIT in 1995 established the existence of Bose-Einstein condensates of atomic gases and was rewarded with the Nobel Prize in Physics in 2001, which was awarded in equal parts to Eric Cornell, Wolfgang Ketterle and Carl Wieman. It is remarkable that, after the 1997 Nobel Prize mentioned above, a further Prize was awarded for such closely related achievements only four years later. This shows that the importance of the new developments related to cold atoms were well appreciated in the academic community.

Many other groups have since succeeded in manufacturing Bose-Einstein condensates. Subsequent work concentrated on understanding the properties of this new state of matter and focussed e.g. on the internal energy and the specific heat of condensates, on the stability of condensates of atoms with attractive interactions, on the collective and single-particle excitations of condensates (cf. Sect. 5.4.2), on collisions between two condensates, and on the possibility of constructing intensive coherent atomic beams analogous to laser beams [Ket97].

Further breakthroughs were made in the observation of Bose-Einstein condensates of diatomic molecules [JB03], which are of particular interest when the individual atoms are fermionic [RG04]. In the latter reference, Regal et al. describe a system of cold fermionic atoms (^{40}K) which is subjected to an external magnetic field in order to tune the atom-atom scattering length a via an appropriate Feshbach resonance, as described in Sect. 5.6. When the energy of the Feshbach resonance is marginally below the elastic threshold of the atom-atom system, the atoms pair off in weakly bound diatomic molecules which are bosonic and form a molecular Bose-Einstein condensate. When the energy of the Feshbach resonance is marginally above the threshold, it no longer supports such weakly bound molecules. On this "attractive side of the Feshbach resonance", Regal et al. infer a condensation of fermionic atom pairs similar to the formation of Cooper pairs in the BCS theory of superfluids [DL05]. (See also [CS05].)

The field of cold and ultracold atoms and molecules has now become a prominent subfield of atomic and molecular physics in general. In 2013, a new book series on the subject was launched [MW13].

5.5 Near-Threshold Quantization and Scattering for Deep Shorter-Ranged Potentials

For potentials with long-ranged (attractive) Coulombic tails proportional to $1/r$, the quasicontinuum of bound states below the continuum threshold merges smoothly into the continuum of scattering states above threshold, as described elegantly in the framework of quantum defect theory, see Sects. 3.2–3.4.

For shorter-ranged potentials falling off faster than $1/r^2$ asymptotically, the situation is very different, e.g., because there is at most a finite number of bound states and hence no quasicontinuum below threshold. As already shown (for vanishing angular momentum) in Sect. 3.1.2, the near-threshold bound-state energies and wave functions in this case depend sensitively on the threshold quantum number $\upsilon_{\rm D}$, which, for potentials falling off faster than $1/r^3$, is related to the s-wave scattering length a according to (3.59). In this section we illuminate the connection to scattering states above the continuum threshold.

5.5.1 Connecting Near-Threshold Quantization and Near-Threshold Scattering

Near-threshold quantization, discussed in Sect. 3.1.2, involved matching the regular solution of the radial Schrödinger equation with the full potential to a solution of the radial Schrödinger equation (3.17),

$$-\frac{\hbar^2}{2\mu}\frac{{\rm d}^2u}{{\rm d}r^2}+V_{\rm tail}(r)\,u(r)=E\,u(r)\;, \tag{5.205}$$

obeying bound-state boundary conditions. The potential in (5.205) is the attractive reference potential $V_{\rm tail}(r)$, which is more singular than $1/r^2$ at small distances, is a good approximation of the full potential at large distances and falls off faster than $1/r^2$ for $r\to\infty$. The influence of the potential tail was contained in one single quantization function (3.22), constructed at each energy E with the help of the small-r behaviour of the asymptotically bound solution of (5.205), which is accurately given by its WKB representation for $r\to 0$.

At positive energies, there are two linearly independent physically meaningful solutions of (5.205) for each energy E, and the small-r behaviour of each solution is determined by an amplitude and a phase, e.g. in the WKB representation of this solution for $r\to 0$. One overall normalization constant is always arbitrary, so the quantum mechanical properties of the reference potential are manifest not in one tail function, as in subthreshold quantization, but in three tail functions at positive energies.

One choice of two linearly independent solutions of (5.205) is provided by the wave functions obeying the following large-r boundary conditions [MK11]:

$$u_{\rm s}(r) \overset{r\to\infty}{\sim} \sin(kr)\,, \quad u_{\rm c}(r) \overset{r\to\infty}{\sim} \cos(kr)\,. \tag{5.206}$$

Beyond the short-range deviations of the full interaction from the reference potential $V_{\rm tail}(r)$, the regular solution $u_{\rm reg}(r)$ of the full problem is a superposition of the two solutions of (5.205),

$$u_{\rm reg}(r) \overset{r\text{ large}}{\propto} \cos\delta_0\, u_{\rm s}(r) + \sin\delta_0\, u_{\rm c}(r)\,. \tag{5.207}$$

The properties of the reference potential $V_{\rm tail}(r)$ are contained in the amplitudes and phases of the WKB representations of $u_{\rm s}(r)$ and $u_{\rm c}(r)$ for $r \to 0$, where these representation become exact. The explicit expressions for the WKB representations contain the lower integration limit in the action integrals as point of reference. In the presence of a classical turning point, this turning point is a natural choice, but for the singular, attractive reference potential $V_{\rm tail}(r)$, there is no classical turning point at positive energy. One conspicuous point is the distance r_E at which the potential $V(r_E)$ is equal to minus the energy E,

$$V_{\rm tail}(r_E) = -|E|\,; \tag{5.208}$$

it lies in the heart of the nonclassical region of $V_{\rm tail}(r)$, see also (5.324) in Sect. 5.7.3. With this choice, the WKB representations of the two solutions of (5.205) defined by the boundary conditions (5.206) can be written as

$$\begin{aligned} u_{\rm s}(r) &\overset{r\to 0}{\sim} \frac{A_{\rm s}}{\sqrt{p_{\rm tail}(E;r)}} \sin\left(\frac{1}{\hbar}\int_{r_E}^{r} p_{\rm tail}(E;r')\,{\rm d}r' - \phi_{\rm s}\right)\,, \\ u_{\rm c}(r) &\overset{r\to 0}{\sim} \frac{A_{\rm c}}{\sqrt{p_{\rm tail}(E;r)}} \cos\left(\frac{1}{\hbar}\int_{r_E}^{r} p_{\rm tail}(E;r')\,{\rm d}r' - \phi_{\rm c}\right)\,, \end{aligned} \tag{5.209}$$

with the local classical momentum $p_{\rm tail}(E;r) = \sqrt{2\mu\,[E - V_{\rm tail}(r)]}$, which is real and positive in the whole range $0 < r < \infty$. Equation (5.209) defines the amplitudes $A_{\rm s,c}$ which are real and taken to be positive, and the phases $\phi_{\rm s,c}$, which are real. These amplitudes and phases are tail functions determined entirely by the reference potential $V_{\rm tail}(r)$. They are functions of energy, but for simplicity in notation this is not explicitly written in the formulae below. Note that the lower limit r_E of the integrals in (5.209) is larger than the upper limit r when $r \to 0$.

At distances r which are small enough for the WKB representations (5.209) of $u_{\rm s}(r)$ and $u_{\rm c}(r)$ to be valid, and at the same time large enough so that the reference potential $V_{\rm tail}(r)$ is a good approximation of the full interaction, the regular solution with the asymptotic behaviour (5.207) can be written as

$$u_{\text{reg}}(r) \propto \frac{1}{\sqrt{p_{\text{tail}}(E;r)}} \sin\left(\frac{1}{\hbar}\int_{r_E}^{r} p_{\text{tail}}(E;r')\,\mathrm{d}r' - \phi_{\text{sr}}(E)\right) . \tag{5.210}$$

The position r in (5.210) lies beyond the short-range deviations of the full interaction from the reference potential $V_{\text{tail}}(r)$, and the inner boundary condition $u_{\text{reg}}(0) = 0$ is carried over in terms of the phase $\phi_{\text{sr}}(E)$. From (5.207) and (5.209) it follows, that $\phi_{\text{sr}}(E)$ is related to the scattering phase shift δ_0 by

$$\tan\delta_0 = \frac{A_{\text{s}}}{A_{\text{c}}}\frac{\sin(\phi_{\text{s}} - \phi_{\text{sr}}(E))}{\cos(\phi_{\text{c}} - \phi_{\text{sr}}(E))} . \tag{5.211}$$

The choice of the reference point r_E in (5.210) may seem unconventional, but it allows the WKB expression to be written in terms of $p_{\text{tail}}(E;r')$ rather than $p(E;r') = \sqrt{2\mu\,[E - V(r')]}$, which involves the full interaction. A more conventional WKB representation for $u_{\text{reg}}(r)$ is,

$$u_{\text{reg}}(r) \propto \frac{1}{\sqrt{p(E;r)}} \cos\left(\frac{1}{\hbar}\int_{r_{\text{in}}(E)}^{r} p(E;r')\,\mathrm{d}r' - \frac{\phi_{\text{in}}(E)}{2}\right) , \tag{5.212}$$

which defines the inner reflection phase $\phi_{\text{in}}(E)$, compare the upper line of (3.15). For distances r beyond the short-range deviations of the full interaction from the reference potential $V_{\text{tail}}(r)$, $p_{\text{tail}}(E;r)$ and $p(E;r)$ are essentially equal, so the factors in front of the sine in (5.210) and cosine (5.212) are the same. Equating the sine and cosine parts relates $\phi_{\text{in}}(E)$ to $\phi_{\text{sr}}(E)$:

$$\begin{aligned}\phi_{\text{sr}}(E) &= \frac{\phi_{\text{in}}(E)}{2} - \frac{\pi}{2} - \frac{1}{\hbar}\int_{r_{\text{in}}(E)}^{r} p(E;r')\,\mathrm{d}r' - \frac{1}{\hbar}\int_{r}^{r_E} p_{\text{tail}}(E;r')\,\mathrm{d}r' \\ &= \frac{\phi_{\text{in}}(E)}{2} - \frac{\pi}{2} - \frac{1}{\hbar}\int_{r_{\text{in}}(E)}^{r_E} p(E;r')\,\mathrm{d}r' .\end{aligned} \tag{5.213}$$

Since the range of integration in the second integral in the top line of (5.213) is beyond the short-range deviations, the momentum $p_{\text{tail}}(E;r')$ can be replaced by $p(E;r')$ in this integral, which leads to the expression in the bottom line. With the quantization condition at threshold, (3.10) in Sect. 3.1.2, the phase $\phi_{\text{sr}}(E)$ can be related to the threshold quantum number υ_{D},

$$\begin{aligned}\phi_{\text{sr}}(E) = {}& -\upsilon_{\text{D}}\pi - \frac{\phi_{\text{out}}(0)}{2} - \frac{\pi}{2} - \frac{\phi_{\text{in}}(0) - \phi_{\text{in}}(E)}{2} \\ &+ \frac{1}{\hbar}\int_{r_E}^{\infty} p(0;r)\,\mathrm{d}r + \frac{1}{\hbar}\int_{r_{\text{in}}(0)}^{r_E} p(0;r)\,\mathrm{d}r - \frac{1}{\hbar}\int_{r_{\text{in}}(E)}^{r_E} p(E;r)\,\mathrm{d}r .\end{aligned} \tag{5.214}$$

The difference $\phi_{\rm in}(0)-\phi_{\rm in}(E)$ of the inner reflection phases in (5.215) is a smooth function of energy and vanishes at $E = 0$. The leading near-threshold energy dependence of the right-hand side of (5.215) comes from the difference of action integrals in the lower line. Replacing the momenta $p(0;r)$ and $p(E;r)$ in the second and third integrals, i.e. in those with upper limit r_E, by $p_{\rm tail}(0;r)$ and $p_{\rm tail}(E;r)$ introduces an error of order E at most. This is because the difference between p and $p_{\rm tail}$ is limited to short distances and hence a smooth function of E, while the difference of the two integrals clearly vanishes at $E = 0$. In the first integral, covering the range r_E to infinity, $p(0;r)$ can be replaced by $p_{\rm tail}(0;r)$, because r is always beyond the range of the short-range deviations. With the abbreviation

$$\xi = \frac{1}{\hbar}\int_{r_E}^{\infty} p_{\rm tail}(0;r)\,{\rm d}r + \frac{1}{\hbar}\int_0^{r_E} [p_{\rm tail}(0;r) - p_{\rm tail}(E;r)]\,{\rm d}r - \frac{\phi_{\rm out}(0)}{2} - \frac{\pi}{2}\,, \tag{5.215}$$

we can rewrite (5.215) as

$$\phi_{\rm sr}(E) = -\upsilon_{\rm D}\pi + \xi + \pi f_{\rm sr}(E)\,, \tag{5.216}$$

where $f_{\rm sr}(E)$ is a smooth function of energy which vanishes at threshold and accounts for all residual short-range effects. The expression (5.211) thus becomes

$$\tan\delta_0 = \frac{A_{\rm s}}{A_{\rm c}}\,\frac{\sin\left([\upsilon_{\rm D} - f_{\rm sr}(E)]\pi - \xi + \phi_{\rm s}\right)}{\cos\left([\upsilon_{\rm D} - f_{\rm sr}(E)]\pi - \xi + \phi_{\rm c}\right)}\,. \tag{5.217}$$

The influence of the reference potential $V_{\rm tail}(r)$ on the low-energy behaviour of the scattering phase shift δ_0 is expressed through the three tail functions, $A_{\rm s}/A_{\rm c}$, $\phi_{\rm s}$ and $\phi_{\rm c}$. The auxiliary tail function ξ defined in (5.215) is needed to compensate the effects of choosing the lower integration limit in the action integrals to be r_E rather than some energy independent value. Such a choice would, however, introduce an unnecessary element of arbitrariness in the formulation.

Towards threshold, the solutions $u_{\rm s}(r)$ and $u_{\rm c}(r)$ of (5.205), defined by their asymptotic behaviour (5.206), approach the threshold solutions $u_1^{(0)}(r)$ and $u_0^{(0)}(r)$, which were introduced in Sect. 3.1.2 and are defined by the asymptotic behaviour (3.40),

$$u_{\rm s}(r) \overset{k\to 0}{\sim} k\,u_1^{(0)}(r)\,, \quad u_{\rm c}(r) \overset{k\to 0}{\sim} u_0^{(0)}(r)\,. \tag{5.218}$$

Consequently, the threshold limits of the tail functions can be expressed in terms of the amplitudes $D_{0,1}$ and phases $\phi_{0,1}$ defining the WKB representations (3.41) of

$u_1^{(0)}(r)$ and $u_0^{(0)}(r)$, and the threshold value of ξ follows from (5.215):

$$\frac{A_s}{A_c} \overset{k\to 0}{\sim} k\frac{D_1}{D_0}\,, \quad \phi_s \overset{k\to 0}{\longrightarrow} -\frac{\pi}{2}-\frac{\phi_1}{2}\,, \quad \phi_c \overset{k\to 0}{\longrightarrow} -\frac{\phi_0}{2}\,, \quad \xi \overset{k\to 0}{\longrightarrow} -\frac{\pi}{2}-\frac{\phi_0}{2}\,. \tag{5.219}$$

With $f_{sr}(E=0) = 0$, the near-threshold limit of (5.217) is seen to be

$$\tan\delta_0 \overset{k\to 0}{\sim} -k\frac{D_1}{D_0}\left[\cos\left(\frac{\phi_0-\phi_1}{2}\right)+\frac{\sin\left(\frac{\phi_0-\phi_1}{2}\right)}{\tan(\upsilon_D\pi)}\right] = -k\left(\bar{a}+\frac{b}{\tan(\upsilon_D\pi)}\right)\,. \tag{5.220}$$

The threshold length b and the mean scattering length $\bar{a}$ are as already defined in (3.46) and (3.48) in Sect. 3.1.2, so (5.220) is consistent with the expression (3.59) for the scattering length a. Remember that a finite value for the mean scattering length $\bar{a}$ exists only for reference potentials $V_{tail}(r)$ falling off faster than $1/r^3$ at large distances.

Equation (5.217) transparently exposes how the energy dependence of the scattering phase shift δ_0 is influenced by the reference potential $V_{tail}(r)$. As for near-threshold quantization discussed in Sect. 3.1.2, the threshold quantum number υ_D, more precisely the remainder $\Delta_D = \upsilon_D - \lfloor \upsilon_D \rfloor$, crucially determines the leading energy dependence of δ_0. For reference potentials $V_{tail}(r)$ falling off faster than $1/r^3$ at large distances, the leading proportionality of $\tan\delta_0$ to k comes from the prefactor A_s/A_c in front of the quotient of sine and cosine, and the actual value of the scattering length a depends sensitively on Δ_D, as seen in (5.220) and in (3.59) in Sect. 3.1.2.

At large energies, the prefactor A_s/A_c in (5.217) approaches unity exponentially, as is most easily seen when analyzing the transmission and *quantum reflection* through the nonclassical part of V_{tail} in terms of the wave functions (5.209), see (5.344), (5.361) in Sect. 5.7.3 below. This means that the arguments of sine and cosine in the quotient become identical and equal to δ_0 itself,

$$\delta_0 \overset{k\to\infty}{\sim} [\upsilon_D - f_{sr}(E)]\,\pi - \xi + \phi_s\,. \tag{5.221}$$

In this semiclassical regime, the threshold quantum number υ_D affects the scattering phase shift only as an additive constant. Further effects due to the short-range deviation of the full interaction from the reference potential $V_{tail}(r)$ enter via the correction term $f_{sr}(E)$, which is a smooth function of energy, in particular at threshold, and vanishes at $E = 0$:

$$f_{sr}(E) = \gamma_{sr}E + O\left(E^2\right)\,. \tag{5.222}$$

Again, the description above is particularly useful for single-power tails (3.63), for which the tail properties depend not on energy and potential strength independently, but only on the dimensionless product $k\beta_\alpha$ of the wave number k and the quantum length β_α. The point of reference in units of β_α is

$$\frac{r_E}{\beta_\alpha} = (k\beta_\alpha)^{-2/\alpha} \tag{5.223}$$

according to (5.208), and the auxiliary function (5.215) is given by [MK11]

$$\xi = -\left(\frac{3}{4} + \frac{\nu}{2}\right)\pi + 2\nu\eta_\alpha(k\beta_\alpha)^{1-2/\alpha}\,, \quad \text{with} \quad \nu = \frac{1}{\alpha - 2} \quad \text{and} \tag{5.224}$$

$$\eta_\alpha = \sqrt{2} - \frac{\alpha}{\alpha+2}\,{}_2F_1\left(\frac{1}{2}, \frac{1}{2} + \frac{1}{\alpha}; \frac{3}{2} + \frac{1}{\alpha}; -1\right)\,; \tag{5.225}$$

${}_2F_1$ stands for the hypergeometric function defined by (A.72) in Appendix A.5. The leading near-threshold behaviour of the tail functions $A_\mathrm{s}/A_\mathrm{c}$, ϕ_s and ϕ_c is, for any $\alpha > 3$ [MK11],

$$\frac{A_\mathrm{s}}{A_\mathrm{c}} \overset{k\to 0}{\sim} \nu^{2\nu}\frac{\Gamma(1-\nu)}{\Gamma(1+\nu)}k\beta_\alpha = k\sqrt{\bar{a}^2 + b^2}\,, \tag{5.226}$$

$$\phi_{\mathrm{s/c}} \overset{k\to 0}{\sim} \left(-\frac{1}{2} \pm \frac{\nu - \frac{1}{2}}{2}\right)\pi + 2\nu\eta_\alpha(k\beta_\alpha)^{1-2/\alpha}\,. \tag{5.227}$$

The leading near-threshold behaviour of $\tan\delta_0$ is as given in (5.220), with $\bar{a}$ and b as given in (3.67) and Table 3.1 in Sect. 3.1.2. In the semiclassical limit of large k, the prefactor $A_\mathrm{s}/A_\mathrm{c}$ approaches unity exponentially, compare (5.361) in Sect. 5.7.3 below, and

$$\phi_{\mathrm{s/c}} \overset{k\to\infty}{\sim} -\rho_\alpha(k\beta_\alpha)^{1-2/\alpha}\,, \quad \rho_\alpha = \sqrt{2} - \frac{\alpha/2}{\alpha-1}\,{}_2F_1\left(\frac{1}{2}, 1 - \frac{1}{\alpha}; 2 - \frac{1}{\alpha}; -1\right) \tag{5.228}$$

for $\alpha > 2$. The high-k behaviour of the phase shift is thus

$$\delta_0 \overset{k\to\infty}{\sim} \left(\upsilon_\mathrm{D} - f_\mathrm{sr}(E) + \frac{3}{4} + \frac{\nu}{2}\right)\pi - (\rho_\alpha + 2\nu\eta_\alpha)(k\beta_\alpha)^{1-2/\alpha}\,, \tag{5.229}$$

as already given in [FG99]. Numerical values of the dimensionless parameters η_α and ρ_α are listed in Table 5.2

For a single-power tail (3.63), the quantum length β_α can be related to an energy E_{β_α},

$$E_{\beta_\alpha} = \frac{\hbar^2}{2\mu\beta_\alpha^2}\,, \tag{5.230}$$

Table 5.2 Numerical values of dimensionless parameters η_α and ρ_α as defined in (5.225) and (5.228), respectively

α	3	4	5	6	7	$\alpha \to \infty$
η_α	0.908797	0.847213	0.802904	0.769516	0.743463	0.532840
ρ_α	0.769516	0.847213	0.885769	0.908797	0.924102	1

which defines a scale separating the extreme quantum region immediately near threshold from the regime of somewhat larger energies, where the influence of the reference potential can be described semiclassically. (See also (3.77) in Sect. 3.1.2.) For $E \ll E_{\beta_\alpha}$ corresponding to $k\beta_\alpha \ll 1$, the near-threshold expansions (5.226), (5.227) apply and the phase shift may be expressed via the scattering length according to (5.220); for $\alpha = 3$ the near-threshold expansion of the phase shift is expressed via the remainder Δ_D according to (4.113). As the energy increases beyond E_{β_α} corresponding to $k\beta_\alpha$ growing beyond unity, the semiclassical expression (5.229) becomes increasingly accurate.

As specific examples consider single-power reference potentials (3.63) with $\alpha = 6$ and $\alpha = 4$. The auxiliary function (5.215) is given according to (5.224) in these cases by

$$\xi = -\frac{7}{8}\pi + \frac{1}{2}\eta_6(k\beta_6)^{2/3} \quad \text{for } \alpha = 6 \quad \text{and} \tag{5.231}$$

$$\xi = -\pi + \frac{1}{2}\eta_4(k\beta_4)^{1/2} \quad \text{for } \alpha = 4\,. \tag{5.232}$$

The tail functions A_s/A_c, ϕ_s and ϕ_c are shown for both powers in Fig. 5.30.

The scattering phase shifts that follow via (5.217) are shown for various values of the remainder Δ_D in Fig. 5.31. The leading linear behaviour near threshold, which is in accordance with Wigner's threshold law, is restricted to the extreme quantum regime $k\beta_\alpha \ll 1$ corresponding to $E \ll E_{\beta_\alpha}$. The scattering length a depends sensitively on the remainder Δ_D according to (3.59) and for large $|a|$, the linear regime is restricted even further by the condition $k|a| < 1$. The dot-dashed lines in Fig. 5.31 show the cases of vanishing scattering length, which are achieved with $\Delta_\text{D} = \frac{3}{4}$ for $\alpha = 6$ and $\Delta_\text{D} = \frac{1}{2}$ for $\alpha = 4$. In these cases, the versions (4.100) or (4.107) of the effective-range expansion don't work, but the corresponding expansions for δ_0, e.g. (4.101) for potentials falling off faster than $1/r^5$ at large distances, are applicable. See Sects. 4.1.7 and 4.1.8 in Chapter 4.

Since the quantum lengths β_α are very large in realistic systems, typically hundreds or even many thousands of atomic units (Bohr radii), the truly quantum mechanical near-threshold regime $k\beta_\alpha \ll 1$ is tiny, as already observed for near-threshold quantization in Sect. 3.1.2. In contrast to the bound regime below threshold however, the energy spectrum above threshold is continuous and any ever so small range of energies near threshold accommodates physically meaningful wave functions.

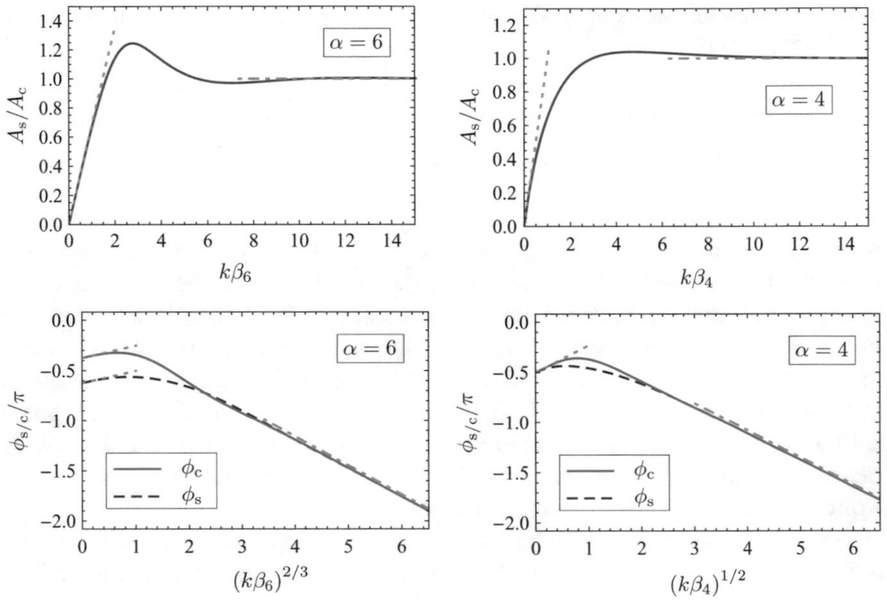

Fig. 5.30 Tail functions for a single-power reference potential (3.63) with $\alpha = 6$ (left-hand panels) and $\alpha = 4$ (right-hand panels). The upper panels show the ratios A_s/A_c of the amplitudes defined by the WKB representations (5.209) of the wave functions $u_s(r)$ and $u_c(r)$ in the limit $r \to 0$, as functions of $k\beta_\alpha$; the lower panels show the phases ϕ_s and ϕ_c as functions of $(k\beta_\alpha)^{1-2/\alpha}$ (from [MK11])

The phase shifts shown in Fig. 5.31 were obtained via (5.217) without considering possible short-range corrections due to the deviation of the full interaction from the reference potential $V_{\rm tail}(r)$ at small distances, i.e. assuming $f_{\rm sr} \equiv 0$. The characteristic length scale for such short-range corrections is typically of the order of a few atomic units (Bohr radii), associated with a characteristic energy much larger than E_{β_α}. In the energy range covered in Fig. 5.31, the effect of the short-range correction term $f_{\rm sr}$ in (5.217) is negligibly small in a sufficiently deep Lennard-Jones type potential where the potential tail is well described by the single-power form (3.63) [MK11].

Consider again the Lennard-Jones potential (3.74) with $B_{\rm LJ} = 10^4$, which was studied as Example 1 in Sect. 3.1.2. The short-range correction function $f_{\rm sr}(E)$ was derived from the exact numerically calculated phase shifts by resolving (5.217), and $\upsilon_{\rm D} - f_{\rm sr}(E)$ is shown as the solid black curve in the right-hand part ($E > 0$) of Fig. 5.32. The left-hand part ($E < 0$) of the figure repeats the plot in the right-hand part of Fig. 3.4, where $\upsilon + F_6(\kappa_\upsilon \beta_6)$ is plotted as function of energy for the highest five bound states $\upsilon = 19, \dots 23$. Note that the energy is now given in the units of E_{β_6} as defined in (3.77). It is related to the depth $\mathcal{E}$ of the Lennard-Jones potential by $E_{\beta_6}/\mathcal{E} = (B_{\rm LJ})^{-3/2}/\sqrt{2}$, which in the present case

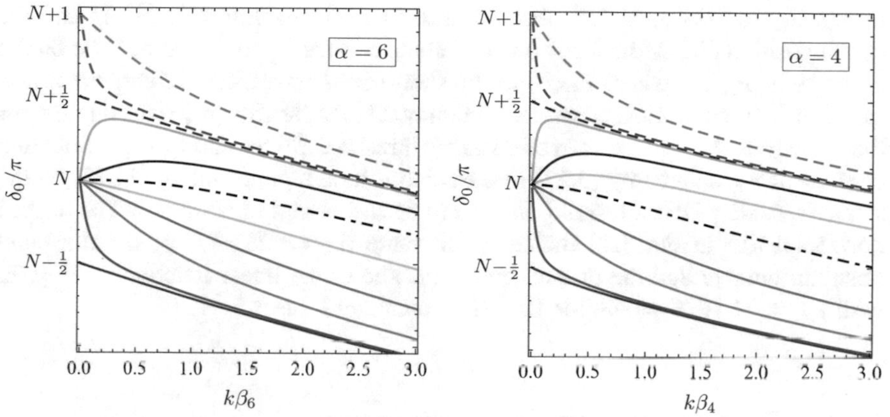

Fig. 5.31 *s*-wave phase shifts as given by (5.217) for a potential with a single-power tail (3.63) for various values of the remainder Δ_D. The additional short-range correction given through $f_{sr}(E)$ is taken to be zero. The *solid lines* show (from bottom to top) $\Delta_D = 0,\ 0.01,\ 0.1, 0.25, 0.5, 0.9, 0.99$ for $\alpha = 6$ and $\Delta_D = 0,\ 0.01,\ 0.1, 0.25, 0.75, 0.9, 0.99$ for $\alpha = 4$. For the lowest three values of Δ_D, the plots are repeated with a shift of π, which would correspond to one additional bound state in a potential well. The *dot-dashed lines* show the respective phase shift for the value of Δ_D for which the scattering length vanishes, $\Delta_D = \frac{3}{4}$ for $\alpha = 6$ and $\Delta_D = \frac{1}{2}$ for $\alpha = 4$ (adapted from [MK11])

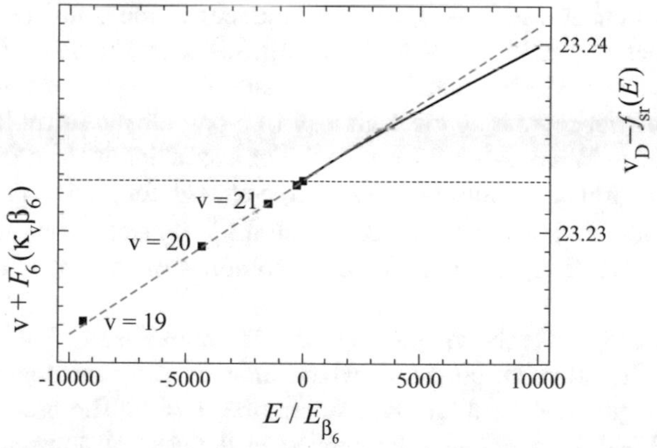

Fig. 5.32 For the Lennard-Jones potential (3.74) with $B_{LJ} = 10^4$, the left-hand part ($E < 0$) shows $\upsilon + F_6(\kappa_\upsilon \beta_6)$ as function of energy for the highest five bound states $\upsilon = 19, \ldots 23$ (*solid squares*). The right-hand part ($E > 0$) shows $\upsilon_D - f_{sr}(E)$, derived from the exact numerically calculated phase shifts by resolving (5.217). The *dashed horizontal line* indicates the value $\upsilon_D = 23.2327$ of the threshold quantum number; the *dashed green line* shows the linear function $\upsilon_D - \gamma_{sr} E$, with $\gamma_{sr} = -1.16/\mathcal{E} = -8.2 \times 10^{-7}/E_{\beta_6}$, compare Table 3.5

means $E_{\beta_6} \approx 0.7 \times 10^{-6}\mathcal{E}$. According to the quantization rule (3.11) and the decomposition (3.24), the black squares in the left-hand part of Fig. 5.32 lie on the curve $\upsilon_\mathrm{D} - F_\mathrm{sr}(E)$, where $F_\mathrm{sr}(E)$ is the short-range correction to the quantization function. This curve clearly merges smoothly into the function $f_\mathrm{sr}(E)$ accounting for the analogous short-range correction above threshold. So the short-range correction coefficient γ_sr, defined by (3.25) in the subthreshold regime and by (5.222) on the scattering side of the threshold, is seen to be the same in both cases. The dashed horizontal line in Fig. 5.32 indicates the value $\upsilon_\mathrm{D} = 23.2327$ of the threshold quantum number and the dashed green line shows the linear function $\upsilon_\mathrm{D} - \gamma_\mathrm{sr}E$, with $\gamma_\mathrm{sr} = -1.16/\mathcal{E} = -8.2 \times 10^{-7}/E_{\beta_6}$, compare Table 3.5.

5.5.2 Nonvanishing Angular Momentum

For nonvanishing angular momentum quantum number l, the radial Schrödinger equation (5.205) with the reference potential $V_\mathrm{tail}(r)$ becomes

$$-\frac{\hbar^2}{2\mu}\frac{\mathrm{d}^2u}{\mathrm{d}r^2} + V_\mathrm{tail}^{(l)}(r)u(r) = E\,u(r)\,, \quad V_\mathrm{tail}^{(l)}(r) = V_\mathrm{tail}(r) + \frac{l(l+1)\hbar^2}{2\mu r^2}\,. \tag{5.233}$$

Since $V_\mathrm{tail}(r)$ is more singular than $1/r^2$ at small distances, its influence becomes increasingly dominant for $r \to 0$, and the influence of the centrifugal potential in (5.233) becomes negligible in this limit. At large distances, however, the centrifugal term dominates over $V_\mathrm{tail}(r)$, which falls off faster than $1/r^2$, and this gives rise to a centrifugal barrier separating the regime of free-particle motion at large distances from the region of WKB validity for $r \to 0$. For a sufficiently deep full interaction, there still is a region of r values where r is large enough for the full interaction to be accurately represented by the reference potential $V_\mathrm{tail}^{(0)}(r)$ and at the same time small enough for the WKB representations of the solutions of (5.233) to be sufficiently accurate.

As example, Fig. 5.33 shows the tail of the potential already featured in Fig. 3.1 together with the effective potential, which includes the centrifugal potential, in this case for angular momentum quantum number $l = 8$. The procedure outlined in Sects. 3.1.2 and 5.5.1 can also be applied in the case of nonvanishing angular momentum. In the bound-state regime, the outer classical turning point $r_\mathrm{out}(E)$ does not go to infinity for $E \to 0$, but assumes a finite value $r_{E=0}$ corresponding to the inner base point of the centrifugal barrier. With this in mind, the tail contribution $F_\mathrm{tail}(E)$ to the quantization function is still defined by (3.22) in Sect. 3.1.2, but the local classical momentum $p_\mathrm{tail}(r')$ in the action integrals is now replaced by

$$p_\mathrm{tail}^{(l)}(r') = \sqrt{2\mu\left[E - V_\mathrm{tail}^{(l)}(r')\right]}\,. \tag{5.234}$$

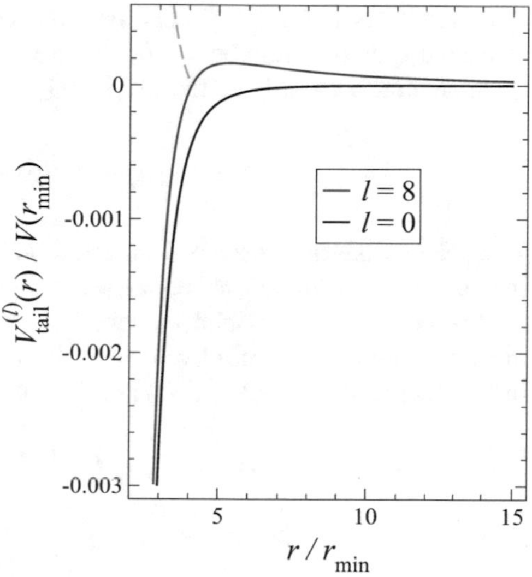

Fig. 5.33 Tail of the deep potential already featured in Fig. 3.1 (*solid black line*), together with the effective potential $V^{(l)}_{\rm tail}(r)$ as defined in (5.233), for angular momentum quantum number $l = 8$ (*solid blue line*). The *dashed orange line* shows the location of the reference point r_E which is defined for positive energies by (5.239)

For small noninteger values of l in the range $-\frac{1}{2} < l < +\frac{1}{2}$, the leading near-threshold behaviour of $F_{\rm tail}(E)$ was derived in [ME01] for single power tails (3.63),[2]

$$F^{(l)}_{\alpha}(\kappa\beta_\alpha) \overset{\kappa\to 0}{\sim} \frac{\pi\;\nu(0)^{2\nu(l)}}{\sin\left[\left(l+\frac{1}{2}\right)\pi\right]\left(l+\frac{1}{2}\right)\nu(l)\left[\Gamma\left(l+\frac{1}{2}\right)\Gamma\left(\nu(l)\right)\right]^2}\left(\frac{\kappa\beta_\alpha}{2}\right)^{2l+1}$$
$$+\,O\left((\kappa\beta_\alpha)^{4l+2}\right) + O(E)\,, \quad -\frac{1}{2} < l < +\frac{1}{2}\,; \tag{5.235}$$

here $\nu(l)$ is a generalization of $\nu \equiv \nu(0)$ as defined in (5.224),

$$\nu(l) = \frac{2l+1}{\alpha-2}\,. \tag{5.236}$$

At the upper end of the interval given in (5.235), i.e., $l = \frac{1}{2}$, the energy dependence $(\kappa\beta_\alpha)^{2l+1}$ is already of order E. For all higher l-values, in particular for all positive integers, the leading energy dependence of the tail contribution to the quantization function $F^{(l)}_{\rm tail}(E)$ is of order E. A separation of tail effects from the influence of short-range deviations of the full interaction from the reference potential $V_{\rm tail}(r)$ is still possible for $l > 0$. As in Figs. 3.4 and 3.6 in Sect. 3.1.2, a plot of $\upsilon + F^{(l)}_{\rm tail}(E_\upsilon)$ against E_υ approaches a straight-line behaviour towards threshold, from which

[2]Noninteger values of l are not merely of academic interest. They can describe the effects of inverse-square potentials of other origin than the centrifugal term. In two-dimensional scattering described in Sect. 4.2, the radial Schrödinger equation with integer angular momentum quantum number m resembles that of the 3D case when $l = |m| - \frac{1}{2}$.

the parameters $\upsilon_\mathrm{D}(l)$ and γ_sr can be extracted. For inverse-power tails (3.63), the threshold quantum number $\upsilon_\mathrm{D}(l)$ for nonvanishing l is related to the threshold quantum number $\upsilon_\mathrm{D}(0)$ by [ME01, FT04],

$$\upsilon_\mathrm{D}(l) = \upsilon_\mathrm{D}(0) - \frac{l}{\alpha - 2} . \tag{5.237}$$

This relation has been used by Lemeshko and B. Friedrich [LF09, LF10] to estimate the number of ro-vibrational bound states in diatomic molecules and molecular ions.

The procedure described in Sect. 5.5.1 can easily be extended to the case of nonvanishing angular momentum quantum number l. For $l \neq 0$, the two linearly independent solutions of (5.233) are chosen to be those behaving asymptotically as

$$\begin{aligned} u_\mathrm{s}^{(l)}(r) &\overset{r\to\infty}{\sim} kr\,j_l(kr) \overset{r\to\infty}{\sim} \sin\left(kr - l\frac{\pi}{2}\right) , \\ u_\mathrm{c}^{(l)}(r) &\overset{r\to\infty}{\sim} -kr\,y_l(kr) \overset{r\to\infty}{\sim} \cos\left(kr - l\frac{\pi}{2}\right) . \end{aligned} \tag{5.238}$$

The amplitudes $A_\mathrm{s,c}$ and phases $\phi_\mathrm{s,c}$ are defined via the WKB representations of these wave functions for $r \to 0$, as in (5.209) for the case $l = 0$, but the local classical momentum p_tail is replaced by $p_\mathrm{tail}^{(l)}$ given by (5.234). The point of reference r_E is now chosen as the classical turning point of $-V_\mathrm{tail}^{(l)}(r)$,

$$V_\mathrm{tail}^{(l)}(r_E) = V_\mathrm{tail}(r_E) + \frac{l(l+1)\hbar^2}{2\mu(r_E)^2} = -E < 0 . \tag{5.239}$$

The dashed orange line in Fig. 5.33 shows the location of reference point r_E for each positive energy E. At threshold, $r_E \equiv r_0$ coincides with the inner base point of the centrifugal barrier, which is also the limit of the outer classical turning point $r_\mathrm{out}(E)$ when the threshold is approached from below. The auxiliary tail function (5.215) is, for $l > 0$, defined by

$$\xi^{(l)} = \frac{1}{\hbar}\int_{r_E}^{r_0} p_\mathrm{tail}^{(l)}(0;r)\,\mathrm{d}r + \frac{1}{\hbar}\int_0^{r_E}\left[p_\mathrm{tail}^{(l)}(0;r) - p_\mathrm{tail}^{(l)}(E;r)\right]\mathrm{d}r - \frac{\phi_\mathrm{out}(0)}{2} - \frac{\pi}{2} . \tag{5.240}$$

The theory described above, including nonvanishing angular momenta, has been shown to work well in a realistic application to near-threshold bound and continuum states of the $^{88}Sr_2$ molecule in [KM11].

5.5.3 Summary

For a deep potential with an attractive tail falling off faster than $1/r^2$ at large distances, tail effects and short-range effects are most effectively identified by defining a reference potential $V_\mathrm{tail}(r)$, which describes the full interaction accurately at large distances and tends to $-\infty$ more rapidly than $-1/r^2$ at small distances. The influence of the reference potential is contained in a few tail functions, which are functions of energy that are determined solely by $V_{\rm tail}(r)$. They are related to the amplitudes and phases in the WKB representation of exact solutions of the Schrödinger equation, with $V_{\rm tail}(r)$, in the limit $r \to 0$. Since the WKB approximation is exact for $r \to 0$ in this case, referring to the WKB representation does *not* imply a semiclassical approximation.

The near-threshold bound-state energies and scattering phase shifts are significantly influenced by the threshold quantum number $\upsilon_{\rm D}$, or rather by its remainder $\Delta_{\rm D} = \upsilon_{\rm D} - \lfloor \upsilon_{\rm D} \rfloor$, which is a property of the full interaction and tells us how close this is to supporting a bound state exactly at threshold. Further effects of the short-range deviation of the full interaction from $V_{\rm tail}(r)$ enter via a smooth function of energy which vanishes at threshold. We called it $F_{\rm sr}(E)$ below threshold and $f_{\rm sr}(E)$ above threshold, but both functions merge smoothly with a common gradient at $E = 0$:

$$F_{\rm sr}(E) = \gamma_{\rm sr} E + O\left(E^2\right) \;\text{ for } E < 0\,, \quad f_{\rm sr}(E) = \gamma_{\rm sr} E + O\left(E^2\right) \;\text{ for } E > 0\,. \tag{5.241}$$

The short-range correction (5.241) vanishes in the limit that the range of the deviations of the full interaction from the reference potential $V_{\rm tail}(r)$ is small compared to the characteristic length scales of $V_{\rm tail}(r)$.

The positions of the near-threshold energy levels are determined by the quantization rule (3.11), which can be written as (3.26) when the quantization function is written as a sum of $F_{\rm tail}(E)$ and the short-range correction $F_{\rm sr}(E)$. The contribution $F_{\rm tail}(E)$ is a tail function depending only on the properties of the reference potential $V_{\rm tail}(r)$. The immediate near-threshold behaviour of the quantization function $F(E)$ and of the quantization rule (3.11) is universal for all potentials falling off faster than $1/r^2$ at large distances,

$$F(E) \overset{\kappa \to 0}{\sim} \frac{b\kappa}{\pi} + O(E)\,, \quad \upsilon_{\rm D} - \upsilon \overset{\kappa_\upsilon \to 0}{\sim} \frac{b\kappa_\upsilon}{\pi} + O(E)\,, \tag{5.242}$$

where b is the threshold length. It is a property of $V_{\rm tail}(r)$ alone and is defined by (3.46).

At above-threshold energies, the s-wave scattering phase shift is given by (5.217). The ratio $A_{\rm s}/A_{\rm c}$, the angles $\phi_{\rm s}$, and $\phi_{\rm c}$, as well as the auxiliary function ξ are tail functions depending only on the reference potential $V_{\rm tail}(r)$.

The immediate near-threshold behaviour of the phase shift depends sensitively on the remainder $\Delta_{\rm D} = \upsilon_{\rm D} - \lfloor \upsilon_{\rm D} \rfloor$. For potentials falling off faster than $1/r^3$ at large distances, we have $\tan\delta_0 \overset{k \to 0}{\sim} -ka$ and the scattering length a is related to the remainder $\Delta_{\rm D}$ by (3.59), i.e.

$$a = \bar{a} + \frac{b}{\tan(\Delta_{\rm D}\pi)}\,, \tag{5.243}$$

where $\bar{a}$ is the mean scattering length defined in (3.48). The relation (5.243) follows from the immediate near-threshold behavior (5.219) of the tail functions occurring in (5.217). For potentials falling off as $-1/r^3$ asymptotically, the near-threshold behaviour of the tail functions yields [Mul13] the behaviour (4.113).

The semiclassical limit is approached away from threshold, both for positive and negative energies, i.e. for large $|E|$. The behaviour of the scattering phase shift is given in the high-k limit by (5.221), and the influence of the threshold quantum number reduces to a simple additive constant in this limit.

The theory described in this section is particularly elegant for potential tails that are well described by a single-power reference potential (3.63). In this case, all tail functions depend only on $\kappa\beta_\alpha$ (below threshold) or $k\beta_\alpha$ (above threshold). The transition between the immediate near threshold quantum regime and the semiclassical regime away from threshold occurs when $\kappa\beta_\alpha$ or $k\beta_\alpha$ is of the order of unity. The range of the quantum regime is tiny when compared with typical potential depths, because the length scale of the reference potential is very large (in atomic units) for typical atomic or molecular interactions.

5.5.4 Relation to Other Approaches

Deep potentials typically occurring in atomic and molecular physics have been studied by many researchers over the years. Inspired by the success of quantum-defect theory for Coulombic potentials, i.e. modified Coulomb potentials with short-range deviations from the pure $1/r$ behaviour, Greene et al. [GF79, GR82] and Giusti [Giu80] formulated an adaptation of quantum-defect theory to more general situations, in particular to potentials falling off faster than $1/r^2$ at large distances. This approach was applied to elastic and inelastic scattering by several authors [Mie84, MJ84, Gao98, GT05, Gao10]. The description of scattering in these references is essentially equivalent to the theory described Sects. 3.1.2, 5.5.1 and 5.5.2 in that it attempts to separate the effects due to the singular reference potential from the short-range effects due to the deviation of the full interaction from the reference potential at small distances. For a compact review of this line of work see the description beginning on p. 4962 of [QJ12]. Although the applications of this "generalized quantum-defect theory" have been very successful, the use of the language of quantum-defect theory in connection with potentials falling off faster than $1/r^2$ at large distances has been and remains unfortunate.

The term "quantum defect" was introduced for systems described by modified Coulomb potentials to account for the shift of energy levels relative to the levels in a pure Coulomb potential, which serves as reference potential. Towards the continuum threshold, the quantum defects merge into a quantum defect function which (with a factor π) corresponds to the scattering phase shift above threshold, relative to the phase of the regular wave functions in the reference potential, the pure Coulomb potential, see Sects. 3.2 and 3.4 in Chapter 3.

For potentials falling off faster than $1/r^2$ at large distances, the reference potentials generally in use are too singular to supply a reference spectrum of bound states or a definite phase of scattering states, relative to which a "defect" or additional phase shift could be defined. Other marked differences are the number of bound states, which is infinite for Coulombic potentials and finite for potentials falling off faster than $1/r^2$ at large distances, and the semiclassical limit, which is at $E \to 0$ for Coulombic potentials and $|E| \to \infty$ for potentials falling off faster than $1/r^2$.

Samuel Johnson once wrote: "Language is the dress of thought" [Joh81]. For the treatment of potentials which fall off faster than $1/r^2$ at large distances, the language of quantum-defect theory is more of a disguise. Interpreting potentials that fall off faster than $1/r^2$ as a generalization of Coulombic potentials tends to obscure the fundamental differences between these two types of interaction. This is potentially confusing and can promote misconceptions. One example is provided by the observation made by Gao in 1999, that for single-power potential tails proportional to $-1/r^6$ or to $-1/r^3$ conventional WKB quantization leads to poorer results towards the dissociation threshold [Gao99]. Although the failure of conventional WKB quantization at threshold for such potentials was long well known [PK83], the observation in [Gao99] was celebrated as sensational evidence for the "breakdown" of Bohr's correspondence principle, according to which the behaviour of a quantized system is expected to become increasingly (semi-)classical as the quantum number tends to infinity. This alleged breakdown of Bohr's correspondence principle was spotlighted in two key media, *Physical Review Focus* [PR99] and *Nature*'s "News" [Bal99]. Apart from the fact that the limit of infinite quantum number cannot be reached in a system with a finite number of bound states, it was textbook knowledge at the time, that for homogeneous potential tails proportional to $1/r^{\alpha}$, the semiclassical limit is for $|E| \to \infty$ when $\alpha > 2$, and this means $E \to -\infty$ in the bound-state regime, see e.g. discussion involving (5.153)–(5.156) in the Second Edition of *Theoretical Atomic Physics*, published in 1998. "Large quantum numbers" means not large υ, but large $\upsilon_{\rm D} - \upsilon$, and the semiclassical limit is approached not towards threshold but towards increasing binding energy, at least as far as the finite depth of any realistic potential well permits. Deep potentials falling off faster than $1/r^2$ at large distances thus show conformity with Bohr's correspondence principle and not its breakdown. Appropriate refutations of [Gao99] were published in 2001 [EF01, BA01]. In order to avoid accidents such as the one documented by [Gao99, Bal99, PR99], it is important to have a proper appreciation of the differences between potentials with a Coulombic tail and those falling off faster than $1/r^2$ at large distances.

A further difference to Coulombic potentials is, that realistic atomic potentials falling off faster than $1/r^2$ are often not so well represented at large distance by the leading asymptotic inverse-power term alone, at least not in an energy range encompassing more than one or two of the most weakly bound states. The universality of the theory for single-power reference potentials (3.63), where the universal tail functions depending on $\kappa\beta_\alpha$ below and on $k\beta_\alpha$ above threshold apply to all potentials with a given power α, regardless of strength, is lost when a more sophisticated reference potential is used. The tail functions must then be calculated independently for each specific system, and the question arises, whether it may not be worthwhile to simply solve the radial Schrödinger equation directly to obtain bound-state energies and scattering phase shifts.

A pragmatic approach to describe near-threshold states of deep potentials is based on defining a (analytical) model potential $V_{\mathrm{mod}}(r)$, which is a good approximation of the potential tail at large distances, where it is well known, and is nonsingular at small distances, where the exact interaction is often not so well known. Being regular at the origin, the model potential supports a finite number of bound states below threshold and well defined scattering states above threshold. The lesser known short-range part of the potential can be equipped with a small number of model parameters to be fitted in order to reproduce known benchmarks of problem under investigation, e.g. bound-state energy levels and the scattering length. For the bound and continuum states in a relatively narrow energy range around threshold, the behaviour of the wave functions at short distances is essentially independent of energy, and their behaviour at large distances can be obtained by solving the radial Schrödinger equation. Near-threshold effects depending on the potential tail can be described accurately in this way, because the model potential accurately represents the exact interaction at large distances. This approach is very flexible and easily extended to multi-channel scattering situations. It has been followed successfully in recent years, in particular by Tiemann and collaborators [ST00, LT02, DT06, SK08, SK12, SS12].

5.6 Near-Threshold Feshbach Resonances

5.6.1 Motivation

In a first approximation, a condensate of N indistinguishable bosonic particles is described by a completely symmetric many-body wave function, in which each individual boson occupies the same single-particle quantum state, $\psi_N(\mathbf{r})$. In a mean-field treatment of the interparticle interactions, this single-particle wave function is determined via the Gross-Pitaevskii equation, (5.203). The two-particle interaction between the bosons (e.g. bosonic alkali atoms) is accounted for by the scattering length a in the term which contains $|\psi_N(\mathbf{r})|^2$ and makes the equation nonlinear. Clearly, the magnitude and the sign of the scattering length have a dominating influence of the solution of (5.203) and on whether or not a Bose-Einstein condensate can form at all.

As discussed on several occasions in this book, the scattering length depends sensitively on how close the highest bound state in a potential well is to the continuum threshold, which in an atom-atom system is the dissociation threshold, see e.g. (4.84) in Sect. 4.1.7 and (3.61) in Sect. 3.1.2; it acquires large positive values for bound states very close to threshold and large negative values if the potential just fails to support a further bound state, see e.g. Fig. 4.4 in Sect. 4.1.7. As shown below, this general behaviour of the scattering length also holds when the weakly or almost bound state involved originates from an inelastic channel, i.e., when there is a Feshbach resonance at an energy very near to the threshold of the elastic channel. In diatomic systems, elastic and inelastic channels can have different magnetic properties (e.g. magnetic moments of the individual atoms), so the bound and continuum states in the elastic and in inelastic channels can acquire different shifts in the presence of an external magnetic field. This makes it possible to tune the position of a Feshbach resonance relative to the threshold of the elastic channel by varying the strength of the external field, and thus offers a practical way of manipulating and controlling Bose Einstein condensates through the corresponding variations of the scattering length. A comprehensive review on Feshbach resonances as a tool to control the interaction in gases of ultracold atoms was published in 2010 by Chin et al. [CG10].

Consider the two-channel situation illustrated schematically in Fig. 5.34. In the presence of an external magnetic field of strength B, the channel thresholds are separated by $\Delta\mu B$ due to the difference $\Delta\mu$ in the relevant magnetic moments. The upper channel is closed for energies near the threshold of the lower channel, which we call "incident channel" for want of a better word. In the absence of channel coupling, the closed channel supports a bound state at an energy E_0 near the threshold of the incident channel, and the coupling of this state to the incident-channel wave functions appears as a Feshbach resonance in the incident channel.

Close to the threshold of the incident channel, which we take to be at $E = 0$, the behaviour of the incident-channel phase shift δ is determined by the scattering length a: $\delta \stackrel{k\to 0}{\sim} -ak$. As the position of the Feshbach resonance is tuned to pass the threshold of the incident channel, a pole singularity of the scattering length is observed at a given strength B_0 of the magnetic field. This is generally empirically

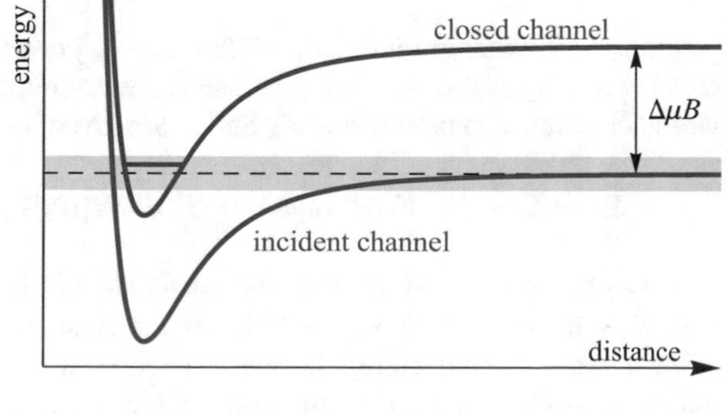

Fig. 5.34 Schematic illustration of atom-atom potentials in a two-channel situation. The closed channel (*red curve*) acquires a shift $\Delta\mu\, B$ relative to the lower, the "incident" channel (*blue curve*) due to different effects of a magnetic field of strength B. The closed channel supports a bound state close to the threshold of the incident channel

parametrized as [MV95, CG10]

$$a = a_{\rm bg}\left(1 + \frac{\Delta B}{B - B_0}\right) , \tag{5.244}$$

where $a_{\rm bg}$ is the background scattering length for the incident channel in the absence of channel coupling. It has become customary in the cold-atoms community to use the term "magnetic Feshbach resonance" to describe such a pole in the scattering length. This can be confusing to anyone with a broader education in scattering theory, because Feshbach resonances are a much more general phenomenon and not restricted to energies near a threshold.

The empirical formula (5.244) satisfactorily describes the pole of the scattering length that occurs when a Feshbach resonance crosses the threshold of the incident channel, but it does not reveal the physical origin of the parameters involved nor their interdependencies. The theory described in this section aims to provide a physically motivated parametrization of a Feshbach resonance near threshold which transparently reveals its influence on scattering properties and on the bound-state spectrum. The treatment is based on the extension into the continuum of the theory of near-threshold quantization for deep shorter-ranged potentials typical for diatomic systems, as described above in Sect. 5.5.1.

5.6.2 *Threshold-Insensitive Parametrization of a Feshbach Resonance*

The influence of a single isolated Feshbach resonance on the scattering phase shift of the incident channel was given in Sect. 1.5.2,

$$\delta = \delta_{\rm bg} + \delta_{\rm res} , \quad \tan\delta_{\rm res} = -\frac{\Gamma/2}{E - E_{\rm R}} , \tag{5.245}$$

where $\delta_{\rm bg}$ is the background phase shift due to the potential in the uncoupled incident channel and $\delta_{\rm res}$ is the resonant phase shift due to coupling to the bound state in the closed channel. The parameters $E_{\rm R}$ and Γ are given by

$$E_{\rm R} = E_{\rm c} + \langle u_{\rm c}|V_{\rm c,i}\hat{G}V_{\rm i,c}|u_{\rm c}\rangle , \quad \Gamma = 2\pi|\langle u_{\rm c}|V_{\rm c,i}|\bar{\rm u}_{\rm i}^{\rm (reg)}\rangle|^2 , \tag{5.246}$$

where $u_{\rm c}$ is the wave function of the bound state in the uncoupled closed channel (called ϕ_0 in Sect. 1.5.2), $V_{\rm c,i}$ and $V_{\rm i,c}$ are the channel-coupling potentials, $\bar{\rm u}_{\rm i}^{\rm (reg)}(r)$ is the energy-normalized regular wave function in the uncoupled incident channel (called $\phi_{\rm reg}$ in Sect. 1.5.2) and the operator $\hat{G}$ is the propagator (Green's operator) in the uncoupled incident channel; its kernel is the Green's function

$$G(r,r') = -\pi\,\bar{\mathrm{u}}_{\mathrm{i}}^{(\mathrm{reg})}(r_<)\,\bar{\mathrm{u}}_{\mathrm{i}}^{(\mathrm{irr})}(r_>)\,. \tag{5.247}$$

The pole of $\tan\delta_{\mathrm{res}}$ defines the resonance energy, i.e. the position E_{R} of the resonance, which differs from the bound-state energy E_{c} in the uncoupled closed channel by a index given by the matrix element containing the incident-channel propagator. When E_{R} is far from the incident-channel threshold and the channel coupling is not too strong, the energy dependence of Γ is weak and its value at $E = E_{\mathrm{R}}$ defines the width of the resonance. This straightforward interpretation breaks down towards the incident-channel threshold. The matrix element describing the shift between E_{c} and E_{R} goes smoothly through a constant value at threshold, but the energy dependence of the parameter Γ poses a more serious problem.

The behaviour of $\bar{\mathrm{u}}_{\mathrm{i}}^{(\mathrm{reg})}(r)$ is, beyond the range of the incident-channel potential, given by

$$\bar{\mathrm{u}}_{\mathrm{i}}^{(\mathrm{reg})}(r) = \sqrt{\frac{2\mu}{\pi\hbar^2 k}}\,\sin[k(r+\delta_{\mathrm{bg}}/k)] \overset{k\to 0}{\sim} \sqrt{\frac{2\mu k}{\pi\hbar^2}}\,(r-a_{\mathrm{bg}})\,, \tag{5.248}$$

compare (1.223) in Sect. 1.5.2. Remember that the near-threshold behaviour of the phase shift δ_{bg} in the uncoupled incident channel is $\delta_{\mathrm{bg}} \overset{k\to 0}{\sim} -a_{\mathrm{bg}}k$. From (5.248) it follows, that $\bar{\mathrm{u}}_{\mathrm{i}}^{(\mathrm{reg})}(r)$ can be written as

$$\bar{\mathrm{u}}_{\mathrm{i}}^{(\mathrm{reg})}(r) = \sqrt{\frac{2\mu k}{\pi\hbar^2}}\,\tilde{u}_{\mathrm{i}}^{(\mathrm{reg})}(r) \quad \text{with} \quad \tilde{u}_{\mathrm{i}}^{(\mathrm{reg})}(r) \overset{r\to\infty,\;k\to 0}{\sim} (r-a_{\mathrm{bg}})\,. \tag{5.249}$$

The irregular radial wave $\bar{\mathrm{u}}_{\mathrm{i}}^{(\mathrm{irr})}(r)$ behaves, beyond the range of the incident-channel potential, as

$$\bar{\mathrm{u}}_{\mathrm{i}}^{(\mathrm{irr})}(r) = \sqrt{\frac{2\mu}{\pi\hbar^2 k}}\,\cos\left[k(r+\delta_{\mathrm{bg}}/k)\right] \overset{k\to 0}{\sim} \sqrt{\frac{2\mu}{\pi\hbar^2 k}}\,\cos\left[k(r-a_{\mathrm{bg}})\right]\,, \tag{5.250}$$

compare (1.227) in Sect. 1.5.2, and can thus be written as

$$\bar{\mathrm{u}}_{\mathrm{i}}^{(\mathrm{irr})}(r) = \sqrt{\frac{2\mu}{\pi\hbar^2 k}}\,\tilde{u}_{\mathrm{i}}^{(\mathrm{irr})}(r) \quad \text{with} \quad \tilde{u}_{\mathrm{i}}^{(\mathrm{irr})}(r) \overset{r\to\infty,\;k\to 0}{\longrightarrow} 1\,; \tag{5.251}$$

the wave function $\tilde{u}_{\mathrm{i}}^{(\mathrm{irr})}(r)$ converges to a k-independent function of r at threshold. In a product of $\bar{\mathrm{u}}_{\mathrm{i}}^{(\mathrm{reg})}(r)$ and $\bar{\mathrm{u}}_{\mathrm{i}}^{(\mathrm{irr})}(r)$, the near-threshold dependencies on k cancel, so the Green's function (5.247) and the matrix element defining the energy shift in the first equation (5.246) tend to finite limits at threshold. On the other hand, the parameter Γ as defined in (5.246) vanishes proportional to k, which makes (5.245) less easy to interpret near threshold.

This problem can be solved by formulating a threshold-insensitive description of the Feshbach resonance, which is possible when the incident-channel potential is deep in the spirit of Sects. 3.1.2, 5.5.1 and well described at large distances by a singular reference potential $V_{\rm tail}(r)$ [SM12]. If channel-coupling effects are of sufficiently short range, then the regular wave function in the incident channel can be written in the form (5.210) in a range of r-values, which are large enough so that the wave function already contains all the effects due to the deviation of the full interaction (including channel coupling) from the uncoupled reference potential $V_{\rm tail}(r)$, and at the same time small enough for the WKB representation of the wave in the reference potential $V_{\rm tail}(r)$ to be sufficiently accurate. As elaborated in [SM12], the effect of the Feshbach resonance on the phase of the regular wave under the influence of $V_{\rm tail}(r)$ can be obtained in a way similar to the derivation of (5.245) and (5.246) above, except that the (energy-normalized) continuum wave functions of the incident channel are replaced by incident-channel wave functions $u_{\rm i}^{\rm (reg)}(r)$ which, in the range of r values referred to above, have the form (5.210) with the phase $\phi_{\rm sr}$ given by (5.216),

$$u_{\rm i}^{\rm (reg)}(r) = \sqrt{\frac{2\mu}{\pi\hbar}}\frac{1}{\sqrt{p_{\rm tail}(E;r)}}\sin\left(\frac{1}{\hbar}\int_{r_E}^{r} p_{\rm tail}(E;r')\,{\rm d}r' - \phi_{\rm sr}(E)\right) . \tag{5.252}$$

[Remember that, in the range of r values considered here, the upper limit r of the integral in (5.252) is smaller than the lower limit r_E.] The effect of channel coupling on the incident-channel wave is the same as in the standard treatment leading to (5.245) and (5.246). The regular solution acquires an additional resonant phase

$$\phi_{\rm sr}(E) \longrightarrow \phi_{\rm sr}(E) + \arctan\left(\frac{\bar{\Gamma}/2}{E-E_{\rm R}}\right) , \tag{5.253}$$

and the width $\bar{\Gamma}$ is given by

$$\bar{\Gamma} = 2\pi|\langle u_{\rm c}|V_{\rm c,i}|u_{\rm i}^{\rm (reg)}\rangle|^2 , \tag{5.254}$$

where the wave function $u_{\rm i}^{\rm (reg)}(r)$ is as defined in connection with (5.252). As long as the range of r values, where both the bound-state wave function $u_{\rm c}(r)$ in the uncoupled closed channel and the coupling potential $V_{\rm c,i}$ are significantly nonvanishing, is small, the matrix element in (5.254) is essentially independent of energy in the near-threshold regime, because the regular wave function, which behaves as (5.252) at small distances, only becomes sensitive to the threshold at large distances. The width $\bar{\Gamma}$ defined by (5.254) is thus threshold-insensitive. At energies far above the incident-channel threshold, the wave function (5.252) becomes equal to the energy-normalized regular wave function $\bar{\rm u}_{\rm i}^{\rm (reg)}(r)$, so

$$\Gamma \stackrel{E\ \rm large}{\longrightarrow} \bar{\Gamma} . \tag{5.255}$$

With the appropriate choice of the irregular radial wave function $u_{\rm i}^{(\rm irr)}(r)$, to replace $\bar{u}_{\rm i}^{(\rm irr)}(r)$ in (5.247), the product of $u_{\rm i}^{(\rm reg)}$ and $u_{\rm i}^{(\rm irr)}$ converges to a well-defined function at $E = 0$. The matrix element defining the small shift between the position $E_{\rm R}$ of the Feshbach resonance and the energy $E_{\rm c}$ of the bound state in the uncoupled closed channel is threshold-insensitive.

The determination of the scattering phase shift in the incident channel follows as already described in Sect. 5.5.1 after (5.210). The result is

$$\tan\delta = \frac{A_{\rm s}}{A_{\rm c}}\,\frac{\sin\left([\Delta_{\rm D} - f_{\rm sr}(E)]\pi + \bar{\delta}_{\rm res} - \xi + \phi_{\rm s}\right)}{\cos\left([\Delta_{\rm D} - f_{\rm sr}(E)]\pi + \bar{\delta}_{\rm res} - \xi + \phi_{\rm c}\right)}\,, \tag{5.256}$$

with the threshold-insensitive resonant phase shift,

$$\bar{\delta}_{\rm res} = -\arctan\left(\frac{\bar{\Gamma}/2}{E - E_{\rm R}}\right)\,. \tag{5.257}$$

In (5.256), $\Delta_{\rm D} = \upsilon_{\rm D} - \lfloor \upsilon_{\rm D} \rfloor$ is the noninteger remainder of the threshold quantum number $\upsilon_{\rm D}$, and the functions $A_{\rm s}/A_{\rm c}$, $\phi_{\rm s}$ and $\phi_{\rm c}$ as well as the auxiliary function ξ are tail functions depending only on the reference potential $V_{\rm tail}(r)$ in the incident channel, as defined through (5.206) and (5.209) in Sect. 5.5.1; $f_{\rm sr}(E)$ is a smooth function of E which vanishes at threshold and accounts for residual corrections due to the deviation of the full interaction in the uncoupled incident channel from the reference potential $V_{\rm tail}(R)$ at small distances.

Since the resonance is a short-range effect, it makes sense to amalgamate the threshold-insensitive resonant phase and the uncoupled, single channel remainder $\Delta_{\rm D}$ to an "extended remainder",

$$\bar{\Delta}_{\rm D}(E) = \Delta_{\rm D} - \frac{1}{\pi}\arctan\left(\frac{\bar{\Gamma}/2}{E - E_{\rm R}}\right)\,. \tag{5.258}$$

With the definition (5.258) of the extended remainder the formula (5.256) becomes,

$$\tan\delta = \frac{A_{\rm s}}{A_{\rm c}}\,\frac{\sin\left(\left[\bar{\Delta}_{\rm D}(E) - f_{\rm sr}(E)\right]\pi - \xi + \phi_{\rm s}\right)}{\cos\left(\left[\bar{\Delta}_{\rm D}(E) - f_{\rm sr}(E)\right]\pi - \xi + \phi_{\rm c}\right)}\,. \tag{5.259}$$

At energies sufficiently far above the incident-channel threshold, the ratio $A_{\rm s}/A_{\rm c}$ tends to unity and the phases $\phi_{\rm s}$ and $\phi_{\rm c}$ become equal. Hence the arguments of sine and cosine in the quotient on the right-hand side of (5.259) become the same and equal to the phase δ on the left-hand side, but instead of (5.221) in Sect. 5.5.1 we now have

$$\delta \overset{E\ \text{large}}{\approx} \left[\bar{\Delta}_{\rm D}(E) - f_{\rm sr}(E)\right]\pi - \xi + \phi_{\rm s} \;=\; \delta_{\rm bg} + \delta_{\rm res} \tag{5.260}$$

$$\text{with}\quad \delta_{\rm bg} = [\Delta_{\rm D} - f_{\rm sr}(E)]\,\pi - \xi + \phi_{\rm s} \ \ \text{and}\ \ \delta_{\rm res} = -\arctan\left(\frac{\bar{\Gamma}/2}{E - E_{\rm R}}\right)\,;$$

this is consistent with (5.245), (5.255) above.

5.6.3 Influence on the Scattering Length

We now assume, that the potential falls off faster than $1/r^3$ asymptotically, so that a well defined scattering length exists. Towards threshold, an additive decomposition of the scattering phase shift δ into a background contribution and a resonant term, as in (5.260), is no longer possible. The behaviour $A_{\rm s}/A_{\rm c} \overset{k\to 0}{\propto} k$, as given in the first equation (5.219) in Sect. 5.5.1, ensures the behaviour $\delta \overset{k\to 0}{\sim} -ak$ for the scattering phase shift, and the value of the scattering length is obtained by the same steps that led to the far right-hand side of (5.220),

$$\tan\delta \overset{k\to 0}{\sim} -k\left(\bar{a} + \frac{b}{\tan(\bar{\upsilon}_{\rm D}(E=0)\pi)}\right) = -k\left(\bar{a} + \frac{b}{\tan(\bar{\Delta}_{\rm D}(E=0)\pi)}\right). \tag{5.261}$$

The essential difference between (5.261) and (5.220) is that, in place of the threshold quantum number $\upsilon_{\rm D}$, (5.261) contains the threshold value of the "extended threshold quantum number",

$$\bar{\upsilon}_{\rm D}(E) = \upsilon_{\rm D} - \frac{1}{\pi}\arctan\left(\frac{\bar{\Gamma}/2}{E - E_{\rm R}}\right), \tag{5.262}$$

or, equivalently, the extended remainder (5.258). Equation (5.261) shows that, even in the presence of a near-threshold Fesbach resonance, the phase shift $\delta(k)$ is nailed down to be an integer multiple of π at threshold, which precludes the existence of a resonance feature of finite width in the scattering phase shift straddling the threshold, as observed for the additional phase shifts in potentials with an attractive Coulombic tail, see Fig. 3.17 in Sect. 3.3.4.

The scattering length following from (5.261) is the term in the big round brackets on the right-hand sides,

$$a = \bar{a} + \frac{b}{\tan[\bar{\Delta}_{\rm D}(E=0)\pi]} = \bar{a} + \frac{b}{\tan\left[\Delta_{\rm D}\pi + \arctan\left(\bar{\Gamma}/(2E_{\rm R})\right)\right]}. \tag{5.263}$$

In the absence of channel coupling, the incident-channel phase shift is the background phase shift $\delta_{\rm bg}$, and its leading near-threshold behaviour is $\delta_{\rm bg} \stackrel{k\to 0}{\sim} -a_{\rm bg}k$, which defines the background scattering length $a_{\rm bg}$. It is related to the single-channel remainder, i.e. the remainder $\Delta_{\rm D}$ in the uncoupled incident channel by (3.59),

$$a_{\rm bg} = \bar{a} + \frac{b}{\tan(\Delta_{\rm D}\pi)} \quad \Longrightarrow \quad \Delta_{\rm D}\pi = \arctan\left(\frac{b}{a_{\rm bg} - \bar{a}}\right) . \tag{5.264}$$

Inserting the expression on the far right of (5.264) for $\Delta_{\rm D}\pi$ in (5.263) gives

$$a = \left[a_{\rm bg} + \frac{\bar{\Gamma}/2}{E_{\rm R}}\left(\bar{a}\,\frac{a_{\rm bg} - \bar{a}}{b} - b\right)\right]\left[1 + \frac{\bar{\Gamma}/2}{E_{\rm R}}\left(\frac{a_{\rm bg} - \bar{a}}{b}\right)\right]^{-1} . \tag{5.265}$$

Equation (5.265) is a universally valid formula for the scattering length a as function of the position $E_{\rm R}$ of a Feshbach resonance, which may be tuned, e.g. as a function of the strength of an external field, from values above threshold, $E_{\rm R} > 0$, to values below threshold $E_{\rm R} < 0$. On the right-hand side of (5.265), $a_{\rm bg}$ is the background scattering length due to the potential in the uncoupled incident channel and $\bar{\Gamma}$ is the threshold-insensitive width (5.254). The lengths $\bar{a}$ and b are the mean scattering length and the threshold length of the singular reference potential $V_{\rm tail}(r)$; they are properties of the $V_{\rm tail}(r)$ only and independent of the position and width of the Feshbach resonance. For a given reference potential describing the large-distance behaviour of the potential in the incident channel, the value of the scattering length depends on two quantities with a clear physical interpretation: the background scattering length $a_{\rm bg}$ and the ratio of the threshold-insensitive width $\bar{\Gamma}$ to the position $E_{\rm R}$ of the Feshbach resonance relative to the threshold.

If the distance $E_{\rm R}$ of the Feshbach resonance from threshold is much larger than its width, then the scattering length a is barely affected by the channel coupling,

$$\frac{\bar{\Gamma}}{E_{\rm R}} \to 0 \Longrightarrow a \to a_{\rm bg} . \tag{5.266}$$

If the uncoupled incident channel supports a bound state exactly at threshold, then $|a_{\rm bg}| \to \infty$. From (5.265) we deduce,

$$|a_{\rm bg}| \to \infty \quad \Longrightarrow \quad a = \bar{a} + b\,\frac{E_{\rm R}}{\bar{\Gamma}/2} . \tag{5.267}$$

In this case, the scattering length a is a linear function of $E_{\rm R}$ and there is no pole.

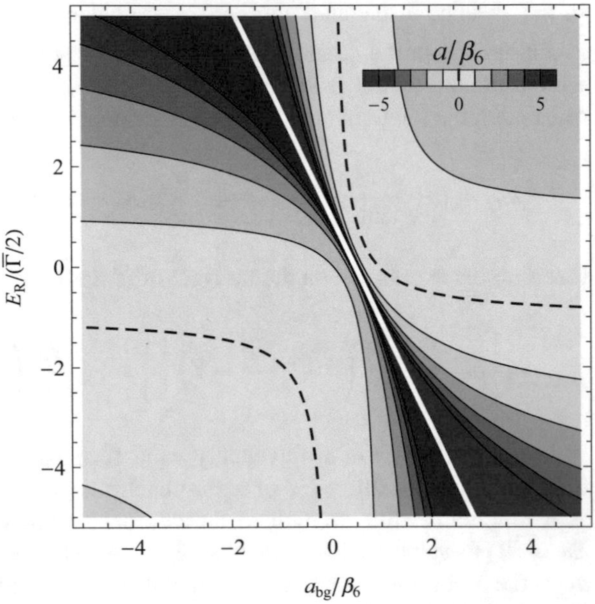

Fig. 5.35 For a single-power reference potential (3.63) with $\alpha = 6$, the figure shows values of the scattering length a given by (5.265) as function of the background scattering length a_{bg} (in units of β_6) and the position E_R of a Feshbach resonance (in units of half its threshold-insensitive width, i.e. of $\bar{\Gamma}/2$). *Dark red areas* indicate large positive, *dark blue areas* large negative values. The *white diagonal* shows the pole E_{Rpole} as given by (5.268). Vanishing values of a occur along the dashed lines (From [SM12])

For $|a_{bg}| < \infty$, the pole of the scattering length, which is customarily called *the* (magnetic) Feshbach resonance in the cold-atoms community, generally occurs for a nonvanishing value of E_R:

$$|a| \to \infty \quad \text{for} \quad E_R = E_{\text{Rpole}}\,, \quad E_{\text{Rpole}} = \frac{\bar{\Gamma}}{2}\,\frac{(\bar{a} - a_{bg})}{b} = -\frac{\bar{\Gamma}/2}{\tan(\Delta_D \pi)}\,. \tag{5.268}$$

Whether the value of E_{Rpole} is above or below threshold depends on the sign of $\bar{a} - a_{bg}$, which in turn depends on whether the (single-channel) remainder Δ_D is smaller or larger than $\frac{1}{2}$. If the background scattering length a_{bg} is smaller than the mean scattering length of the reference potential $V_{\text{tail}}(r)$, then $\tan(\Delta_D \pi)$ is negative, corresponding to $\frac{1}{2} < \Delta_D < 1$, and $E_{\text{Rpole}} > 0$; if $a_{bg} > \bar{a}$, then $\tan(\Delta_D \pi)$ is positive, corresponding to $0 < \Delta_D < \frac{1}{2}$, and $E_{\text{Rpole}} < 0$.

A plot of the scattering length (5.265) as function of a_{bg} and $E_R/(\bar{\Gamma}/2)$ is shown in Fig. 5.35 for an inverse-power tail (3.63) with $\alpha = 6$. Dark red areas indicate large positive, dark blue areas large negative values. The white diagonal shows the position of the pole of a as given by (5.268). It crosses the vertical axis $a_{bg} = 0$ at $E_R/(\bar{\Gamma}/2) = 1$, because the two tail parameters $\bar{a}$ and b are equal in this case, compare (3.67) and Table 3.1 in Sect. 3.1.2.

5.6.4 Influence on the Bound-State Spectrum

The derivation of (5.256) was based on the influence of the Feshbach resonance on the regular incident-channel wave function (5.252), and this influence consists of an additional resonant phase in the argument of the sine on the right-hand side, see (5.253). The distances r where the representation (5.252) of the regular radial wave function is valid lie in the WKB regime where the potential is deep and where the wave functions are insensitive to the position of the threshold. The derivation can thus be continued to the bound-state regime at negative energies, which leads to a simple modification of the quantization rule (3.11)

$$\upsilon_{\rm D} - \frac{1}{\pi}\arctan\left(\frac{\bar{\Gamma}/2}{E_\upsilon - E_{\rm R}}\right) - \upsilon = F(E_\upsilon)\;, \tag{5.269}$$

i.e., the threshold quantum number $\upsilon_{\rm D}$ is simply replaced by the extended threshold quantum number (5.262),

$$\bar{\upsilon}_{\rm D}(E_\upsilon) - \upsilon = F(E_\upsilon) = F_{\rm tail}(E_\upsilon) + F_{\rm sr}(E_\upsilon)\;, \tag{5.270}$$

where the expression on the far right contains the decomposition (3.24) of the quantization function $F(E)$ into the tail contribution $F_{\rm tail}(E)$, as defined by (3.22) in Sect. 3.1.2, and the short-range correction $F_{\rm sr}(E)$, which is a smooth function of energy and vanishes at $E = 0$. Since the quantization functions in (5.270) vanish for $E_\upsilon = 0$, the condition for the existence of a bound state exactly at threshold is now, that the threshold value of the extended threshold quantum number $\bar{\upsilon}_{\rm D}(E = 0)$ be an integer, i.e. that the threshold value of the extended remainder be zero:

$$\bar{\Delta}_{\rm D}(E=0) = \Delta_{\rm D} + \frac{1}{\pi}\arctan\left(\frac{\bar{\Gamma}/2}{E_{\rm R}}\right) = 0\;. \tag{5.271}$$

[Remember that the branch of the arcus-tangent is chosen such that $\arctan(1/x)$ varies smoothly from zero to $-\pi$ as x varies from $-\infty$ to ∞.]

If the position $E_{\rm R}$ of the Feshbach resonance lies somewhat above threshold, then its influence on the bound-state spectrum is small. If it lies below threshold, $E_{\rm R} < 0$, then the quantization rule (5.270) produces one additional bound state, an *intruder* or *perturber state* in the vicinity of $E_{\rm R}$, compared to the "unperturbed" spectrum of the uncoupled incident channel. [We keep the term "incident" channel at subthreshold energies, even though there can be no genuine incident waves when the channel is closed.]

The exact position of the intruder state, i.e. of the perturber, depends on the position and width of the Feshbach "resonance" and on the unperturbed spectrum. Near the threshold of a deep incident-channel potential, the unperturbed spectrum is essentially determined by the singular reference potential $V_{\rm tail}(r)$ and

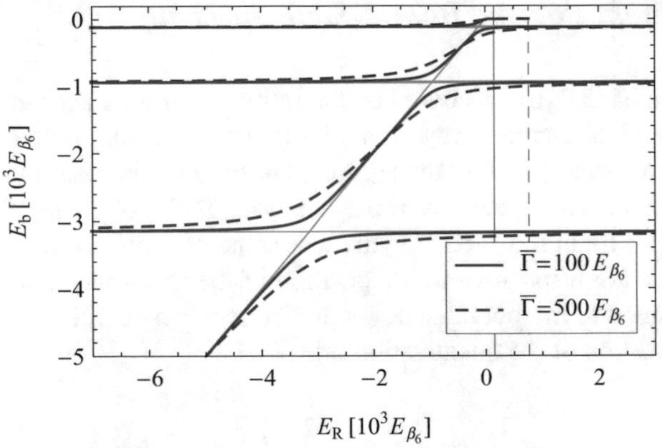

Fig. 5.36 For a deep incident-channel potential with a single-power tail (3.63) with $\alpha = 6$ and a remainder $\Delta_D = 0.9$, the highest three bound-state energies following from (5.269) are shown as functions of the position E_R of a Feshbach resonance. The *solid blue (dashed red) lines* correspond to a threshold-insensitive width $\bar{\Gamma} = 100\,E_{\beta_6}$ ($\bar{\Gamma} = 500\,E_{\beta_6}$). The short-range correction term $F_{sr}(E)$ is neglected. The unit of energy is $E_{\beta_6} = \hbar^2/\left[2\mu(\beta_6)^2\right]$. The *straight horizontal lines* show the unperturbed bound-state energies and the straight diagonal line corresponds to $E_b = E_R$. The *straight vertical lines* indicate the respective values of E_R at which the scattering length diverges according to (5.268) (Adapted from [SM12])

the remainder Δ_D, as discussed in Sect. 3.1.2. Figure 5.36 shows the dependence on E_R of the energies of the highest three states, as given by (5.269), in a deep potential with an inverse-power tail (3.63) with $\alpha = 6$ for a value $\Delta_D = 0.9$ of the (single-channel) remainder. The straight horizontal lines in Fig. 5.36 show the unperturbed bound-state energies; the solid blue and dashed red lines show the perturbed bound-state energies corresponding, respectively, to the values $\bar{\Gamma} = 100\,E_{\beta_6}$ and $\bar{\Gamma} = 500\,E_{\beta_6}$ of the threshold-insensitive width. The short-range correction $F_{sr}(E)$ is neglected here.

Without channel coupling, the spectrum would consist of the unperturbed levels in the incident channel (straight horizontal lines in Fig. 5.36) plus the intruder at $E_b = E_R$ (straight diagonal line in Fig. 5.36). Channel coupling leads to avoided crossings between the unperturbed levels and the intruder state. The value of E_R for which the least bound state is exactly at threshold defines the position $E_{R\text{pole}}$ of the pole of the scattering length as given by (5.268). The straight vertical lines in Fig. 5.36 indicate the values of E_R at which this pole occurs for the respective choice of $\bar{\Gamma}$. According to (5.268), the pole occurs at $E_R = -\bar{\Gamma}/[2\,\tan(0.9\pi)] \approx 1.54 \times \bar{\Gamma}$ in the present case(s).

The bound state at threshold is a two-component wave function with contributions from the incident channel and the closed channel. Its composition can be understood in a physically appealing way as a consequence of level repulsion between the Feshbach resonance at E_R, which comes from the closed-channel bound state, and a weakly bound incident-channel state just below threshold or a state just above threshold, which is only marginally unbound. A small value of the single-channel remainder Δ_D implies that the uncoupled incident channel supports a bound state close to threshold, which can be pushed to threshold by level repulsion from a lower-lying Feshbach resonance. A single-channel remainder Δ_D close to unity suggests a marginally unbound state just above threshold, which can be pushed down to threshold from a higher-lying Feshbach resonance. (This is the situation depicted in Fig. 5.36.) In both cases, the bound state at threshold is close to the uncoupled incident channel wave function with a small contribution due to coupling from the closed channel. If Δ_D is close to $\frac{1}{2}$, then the uncoupled incident channel is as far as possible from supporting a bound state at threshold. The two-channel wave function of the bound state at threshold is then strongly influenced by the Feshbach resonance from the closed-channel and it occurs at a value E_{Rpole} close to zero. If Δ_D is a little below $\frac{1}{2}$, then $E_{\text{Rpole}} < 0$; a Feshbach resonance just below threshold is pushed up to threshold by the highest bound state of the incident channel. When Δ_D is a little above $\frac{1}{2}$, a Feshbach resonance just above threshold is pushed down by coupling to the incident channel; $E_{\text{Rpole}} > 0$ in this case.

A relation connecting the scattering length as given by (5.263) with the asymptotic inverse penetration length κ_b of a bound state very near threshold can be found, as in the derivation of (3.61) in Sect. 3.1.2, by exploiting (5.269)–(5.271). The low-energy expansion of the quantization function (3.22) (multiplied by π) gives [cf. (3.24), (3.25), (3.47)]

$$\pi F(E_b) \overset{\kappa_b \to 0}{\sim} b\kappa_b - \frac{1}{2}(d\kappa_b)^2 + \pi\gamma_{sr}E_b \,. \tag{5.272}$$

From (5.269) we have

$$\Delta_D\pi = \pi F(E_b) + \arctan\left(\frac{\bar{\Gamma}/2}{E_b - E_R}\right) \quad (\text{mod } \pi) \,; \tag{5.273}$$

inserting this expression for $\Delta_D\pi$ in the argument of the tangent on the far right-hand side of (5.263) leads to

$$a \overset{\kappa_b \to 0}{\sim} \frac{1}{\kappa_b} + \rho_{\text{eff}} + \frac{\hbar^2}{2\mu\, b}\left[\pi\gamma_{sr} - \frac{\bar{\Gamma}/2}{(E_R)^2 + (\bar{\Gamma}/2)^2}\right] + O(\kappa_b) \,. \tag{5.274}$$

Equation (5.274) shows that the universal leading behaviour, already formulated as (4.84) in Sect. 4.1.7, namely $a \overset{\kappa_b \to 0}{\sim} 1/\kappa_b + O\left(\kappa_b^0\right)$, also holds when the near-threshold bound state is generated by the coupling of the incident channel to a near-threshold Feshbach resonance. A different result given at the end of Sect. 4.1.3 in the Third Edition of *Theoretical Atomic Physics* is incorrect.

5.6.5 Relation to the Empirical Formula (5.244)

In a typical experiment involving a Feshbach resonance whose position is tuned passed an incident channel's threshold, the quintessential observation is the pole of the scattering length, which occurs when the energy E_{R} of the Feshbach resonance assumes the value E_{Rpole}, as given in (5.268). Expressing E_{R} as $E_{\mathrm{Rpole}} + E_{\mathrm{R}} - E_{\mathrm{Rpole}}$ and exploiting (5.264) and (5.268), we can rewrite (5.263) as

$$a = a_{\mathrm{bg}} - \frac{b}{\sin^2(\Delta_{\mathrm{D}}\pi)} \frac{\bar{\Gamma}/2}{E_{\mathrm{R}} - E_{\mathrm{Rpole}}} . \tag{5.275}$$

In order to connect to the empirical formula (5.244), let's assume that the energy E_{R} of the Feshbach resonance depends linearly on the strength B of an external magnetic (or other) field,

$$E_{\mathrm{R}} = E_{\mathrm{Rpole}} + \Delta\mu(B - B_0) , \tag{5.276}$$

where B_0 is the field strength of the pole and $\Delta\mu$ is a constant with physical dimension energy per field strength. This choice of notation is consistent with the label $\Delta\mu\, B$ for the variable energy in Fig. 5.34. As function of the field strength B, the scattering length (5.275) is

$$a = a_{\mathrm{bg}} - \frac{b}{\sin^2(\Delta_{\mathrm{D}}\pi)} \frac{\bar{\Gamma}/2}{\Delta\mu(B - B_0)} = a_{\mathrm{bg}} \left[1 - \frac{b/a_{\mathrm{bg}}}{\sin^2(\Delta_{\mathrm{D}}\pi)} \frac{\bar{\Gamma}/2}{\Delta\mu(B - B_0)} \right] , \tag{5.277}$$

so the width ΔB, introduced as an empirical parameter in (5.244), is explicitly given as

$$\Delta B = -\frac{b}{a_{\mathrm{bg}}} \frac{1}{\sin^2(\Delta_{\mathrm{D}}\pi)} \frac{\bar{\Gamma}}{2\Delta\mu} . \tag{5.278}$$

Expressing $\sin^2(\Delta_{\mathrm{D}}\pi)$ in terms of a_{bg} according to (5.264) gives an expression for ΔB in terms of a_{bg} and the tail parameters $\bar{a}$ and b:

$$\Delta B = -\frac{\bar{\Gamma}}{2\Delta\mu} \frac{1}{b} \left[\frac{\bar{a}^2 + b^2}{a_{\mathrm{bg}}} - 2\bar{a} + a_{\mathrm{bg}} \right] . \tag{5.279}$$

Equations (5.278), (5.279) show that the width ΔB of a "magnetic Feshbach resonance", as observed in a typical experiment, reflects not only the strength of the coupling between the bound state in the closed channel and the incident-channel wave functions, which is expressed in the threshold-insensitive width $\bar{\Gamma}$. It also depends sensitively on the properties of the uncoupled incident channel, as expressed in the background phase shift a_{bg}. If the uncoupled incident channel supports a bound state (or if there is a virtual state) very near threshold, a_{bg} becomes very large and the empirical formula (5.244) is no longer applicable, as discussed in connection with (5.267) above. Another interesting situation is $a_{\text{bg}} \to 0$, corresponding to little or no interaction in the absence of channel coupling. In this case, the width ΔB as defined via (5.244) diverges, and a more appropriate empirical formula would be,

$$a = a_{\text{bg}} + \frac{\Delta_B}{B - B_0} \quad \text{with} \quad \Delta_B \equiv a_{\text{bg}} \Delta B = -\frac{\bar{\Gamma}}{2\Delta\mu} \frac{1}{b} \left(\bar{a}^2 + b^2 - 2a_{\text{bg}}\bar{a} + a_{\text{bg}}^2 \right) . \tag{5.280}$$

The width Δ_B defined in this way has the physical dimension of a length times field strength. In the limit of vanishing background phase shift, $a_{\text{bg}} \to 0$, it converges to a finite value determined by the threshold-insensitive width $\bar{\Gamma}$ of the Feshbach resonance and the tail parameters $\bar{a}$ and b.

5.7 Some Aspects of Atom Optics

When experimenting with ultra-cold atoms under extremely quantum mechanical conditions it is helpful to be able to guide and manipulate the atomic matter waves in much the same way as electromagnetic waves can be guided and manipulated in optical devices. One obvious difference between atom waves and light is the rich internal structure of an atom which allows a large variety of inelastic processes in addition to conventional reflection and refraction. Beside this, there are several similarities but also essential differences in the properties of matter waves and light waves.

For stationary states of a particle of mass M moving with the energy $E = \hbar^2k^2/(2M)$ under the influence of a potential $V(\boldsymbol{r}) = \upsilon(\boldsymbol{r})\hbar^2/(2M)$, the time-independent Schrödinger equation is,

$$\left(\Delta - \upsilon(\boldsymbol{r}) + k^2\right) \psi(\boldsymbol{r}) = 0 , \tag{5.281}$$

which has essentially the same structure as the wave equation for light with a spatially varying index of refraction proportional to $\sqrt{k^2 - \upsilon(\boldsymbol{r})}$. Hence some results of conventional wave optics can be transferred to the atom-wave situation. However, typical potentials occurring in atomic systems do not necessarily correspond to the behaviour of the index of refraction for typical optical systems, so many problems arising in atom optics have not received the corresponding attention in the optical community.

The time-dependent wave equations for massive particles and for light contain an essential difference in the terms involving time derivatives. The time-dependent Schrödinger equation (1.155) contains the first derivative with respect to time, whereas the wave equation (2.152) for electromagnetic waves contains the second derivative. Whenever time evolution is important, the behaviour of the quantum mechanical matter wave can be expected to be different from the behaviour of an electromagnetic wave. E.g. for a plane monochromatic wave whose amplitude is a function of $\boldsymbol{k}\cdot\boldsymbol{r} - \omega t$, the frequency ω and wave vector $\boldsymbol{k}$ are connected via the dispersion relation. For a particle wave described by the Schrödinger equation this is given by (1.162), whereas for an electromagnetic wave the dispersion relation is

$$\omega = ck = c\sqrt{\boldsymbol{k}\cdot\boldsymbol{k}}\,. \tag{5.282}$$

As a consequence, the wave packet of electromagnetic waves in the vacuum, or in a dielectric medium with constant index of refraction, does not show the spreading described in Sect. 1.4.1 for matter waves.

When constructing wave guides for atoms or other atom-optical devices it is desirable to keep the atoms away from material surfaces in order to avoid unwanted inelastic reactions and adsorption ("sticking"). This is a non-trivial problem, because atom-wall interactions generally feature long-ranged attractive potential tails as described in Sect. 5.7.1. One technique of keeping atoms away from surfaces is based on evanescent-wave mirrors, which exploit forces generated by laser light as explained in Sect. 5.7.2. Finally, Sect. 5.7.3 describes the phenomenon of quantum reflection, through which atoms can be reflected by the nonclassical region of the attractive tail of an atom-surface potential before they come close to the surface. This section gives only a brief introduction to these few aspects of the interesting and highly topical field of atom optics. For a comprehensive introduction to the field the reader is referred to the book by Meystre [Mey01].

5.7.1 Atom-Wall Interactions

In order to understand or construct an atom-optical device it is important to understand the interaction of the atom waves with the surfaces defining the device. At close distances of the order of a few atomic units, the atom-surface interaction is strongly influenced by the forces between the individual electrons in the atom and the electrons and ions in the surface, and it is quite complicated. Beyond this "close" region of a few atomic units, the atom-surface interaction is well described by a simple local potential.

Let us first consider the interaction of a neutral polarizable particle with a perfectly conducting plane wall. Assume that the wall lies in the half-space $z \leq 0$ and that the particle is located at a distance $z > 0$ from the surface, which lies in the xy-plane. The presence of the particle leads to induced charges on the surface of the wall, and these induced charges generate an electric field which seems to come from a mirror-image particle located at a distance z behind the surface. Since the particle is electrically neutral, the leading contributions come from its electric dipole moment $\boldsymbol{d}$ which is subjected to the influence of the apparent image dipole

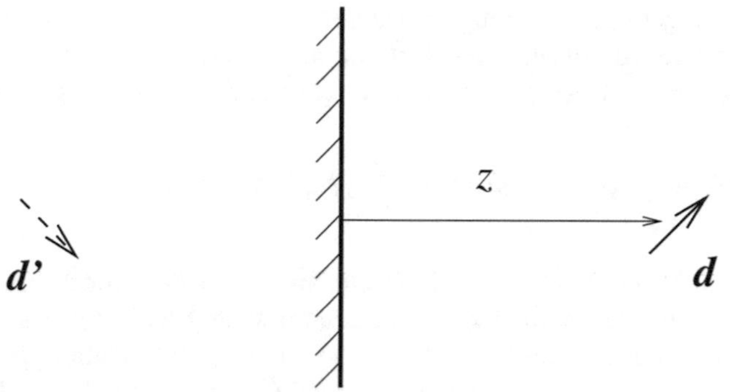

Fig. 5.37 Schematic illustration of a dipole in front of a conducting wall together with its image dipole

$\boldsymbol{d}'$ as illustrated in Fig. 5.37. The potential energy of such a system of two dipoles is the dipole-dipole interaction [Jac75]

$$V = \frac{1}{D^3}[\boldsymbol{d}\cdot\boldsymbol{d}' - 3(\boldsymbol{e}\cdot\boldsymbol{d})(\boldsymbol{e}\cdot\boldsymbol{d}')] \, , \tag{5.283}$$

where D is the spatial separation of the two dipoles and $\boldsymbol{e}$ is the unit vector pointing from one to the other. In the situation illustrated in Fig. 5.37, $\boldsymbol{e}$ is the unit vector in z direction, i.e. normal to the surface of the wall, the separation D is $2z$ and the dipole moments are related by $d'_\perp = d_\perp$ and $d'_{||} = -d_{||}$. (The subscript $\perp$ denotes the component normal to the surface and the subscript $||$ denotes the projection of the vector onto the xy-plane parallel to the surface.) The fact that the interaction is between two *induced* dipoles leads to a further factor $\frac{1}{2}$ on the right-hand side of (5.283), so the electrostatic *van der Waals potential* between a neutral particle and a conducting wall is,

$$V^{\mathrm{vdW}}(z) = -\frac{1}{16z^3}[(\boldsymbol{d}_{||})^2 + 2(d_\perp)^2] \, . \tag{5.284}$$

When the polarizable particle is a neutral atom in a quantum mechanical stationary eigenstate ψ_0, (5.284) is readily adapted to

$$V^{\mathrm{vdW}}(z) = -\frac{C_3}{z^3}, \quad C_3 = \frac{1}{16}\langle\psi_0|(\hat{\boldsymbol{d}}_{||})^2 + 2(\hat{d}_\perp)^2|\psi_0\rangle \, ; \tag{5.285}$$

now $\hat{\boldsymbol{d}}$ stands for the dipole operator (2.186) of the atomic electrons,

$$\hat{\boldsymbol{d}} = -e\sum_{i=1}^{Z}\boldsymbol{r}_i \, , \tag{5.286}$$

where Z is the total number of electrons in the (neutral) atom. Note that the components of the dipole operator enter quadratically on the right-hand side of

(5.285), so we get non-vanishing contributions even when ψ_0 is a parity eigenstate with no permanent dipole moment. If the atom is in a spherical state, the expectation value of $(\hat{\boldsymbol{d}}_{||})^2 = (\hat{d}_x)^2 + (\hat{d}_y)^2$ is twice the expectation value $(\hat{d}_\perp)^2 = (\hat{d}_z)^2$, so

$$C_3 = \frac{1}{12}\langle\psi_0|\hat{\boldsymbol{d}}^2|\psi_0\rangle = \frac{1}{12}\sum_n \langle\psi_0|\hat{\boldsymbol{d}}|\psi_n\rangle\langle\psi_n|\hat{\boldsymbol{d}}|\psi_0\rangle\,. \tag{5.287}$$

The far right-hand side of (5.287) contains the sum over a complete set of eigenstates of the atom (including continuum states) and exposes the potential strength as a sum of contributions corresponding to dipole transitions from the initial state to all possible states of the atom. In 1948 Casimir and Polder [CP48] pointed out that the electrostatic formula (5.285) only applies for distances z smaller than the wavelengths of all non-vanishing transition matrix elements contributing to the sum,

$$z \ll \bar{\lambda}_{0,n}\,, \quad \bar{\lambda}_{0,n} = \frac{\lambda_{0,n}}{2\pi} = \frac{\hbar c}{|E_n - E_0|}\,; \tag{5.288}$$

E_n is the energy eigenvalue of the atomic eigenstate ψ_n. At distances larger than the relevant transition wavelengths of the atom, the corresponding transition time becomes shorter than the time a light signal needs to travel between the atom and the wall. We can no longer ignore relativistic effects ("retardation") and radiative corrections accounting for the modification of the electromagnetic vacuum through the presence of the atom and the wall. These effects are discussed in detail in [CP48, Bar74, Har90], and they depend crucially on whether the atom is in its ground state or in a metastable state, or if there are non-vanishing dipole transition matrix elements to lower-lying states. If the wall is not perfectly conducting, then a more sophisticated theory is necessary to take this into account.

If the deviations from perfect conductivity are accurately described by a fixed dielectric constant ε, then the interaction between an atom in a spherical ground state or a metastable state and the dielectric wall can be compactly written (in atomic units) as [TS93, YD97]

$$V_\varepsilon(z) = -\frac{(\alpha_{\rm fs})^3}{2\pi}\int_0^\infty \alpha_{\rm d}({\rm i}\omega)\omega^3\int_1^\infty \exp(-2\omega z p\,\alpha_{\rm fs})\,h(p,\varepsilon){\rm d}p\,{\rm d}\omega\,, \tag{5.289}$$

where

$$h(p,\varepsilon) = \frac{s-p}{s+p} + (1-2p^2)\frac{s-\varepsilon p}{s+\varepsilon p} \quad \text{with} \quad s = \sqrt{\varepsilon - 1 + p^2}\,; \tag{5.290}$$

$\alpha_{\rm fs} \equiv 1/c = 0.007297353\ldots$ is the fine-structure constant and ε is the dielectric constant of the wall; $\alpha_{\rm d}$ is the frequency-dependent dipole polarizability of the projectile atom in its eigenstate ψ_0 see (3.355) in Sect. 3.5.3,

$$\alpha_d(i\omega) = \sum_n 2(E_n - E_0) \frac{|\langle\psi_0|\sum_{j=1}^Z x_j|\psi_n\rangle|^2}{(E_n - E_0)^2 + \omega^2} . \tag{5.291}$$

For a perfectly conducting surface, a simpler formula is obtained by taking $\varepsilon \to \infty$ in (5.290) and integrating over p in (5.289),

$$\begin{aligned} V_\infty(z) &= -\frac{1}{4\pi z^3} \int_0^\infty \alpha_d(i\omega)[1 + 2\alpha_{fs}\omega z + 2(\alpha_{fs}\omega z)^2] \exp(-2\alpha_{fs}\omega z)\, d\omega \\ &= -\frac{1}{4\pi\alpha_{fs} z^4} \int_0^\infty \alpha_d\left(i\frac{x}{\alpha_{fs} z}\right) [1 + 2x + 2x^2] \exp(-2x)\, dx . \end{aligned} \tag{5.292}$$

For small z values, we can put $z = 0$ in the upper line of (5.292) and obtain the van der Waals potential between the atom and a conducting surface,

$$V_\infty^{vdW}(z) = -\frac{C_3(\infty)}{z^3} , \quad C_3(\infty) = \frac{1}{4\pi} \int_0^\infty \alpha_d(i\omega)\, d\omega . \tag{5.293}$$

Inserting the expression (5.291) for $\alpha_d(i\omega)$ and using $\int_0^\infty d\omega/(\eta^2 + \omega^2) = \pi/(2|\eta|)$ brings us back to (5.287). For finite values of the dielectric constant ε, the derivation of the small-z behaviour of the potential is a bit more subtle, but the result is quite simple [TS93, YB98],

$$V_\varepsilon^{vdW}(z) = -\frac{C_3(\varepsilon)}{z^3} , \quad C_3(\varepsilon) = \frac{\varepsilon - 1}{\varepsilon + 1} C_3(\infty) . \tag{5.294}$$

For large z values, we can assume the argument of α_d in the lower line of (5.292) to be zero and perform the integral over x. This gives the *highly retarded limit* of the Casimir-Polder potential between the atom and a conducting surface,

$$V_\infty^{ret}(z) = -\frac{C_4(\infty)}{z^4} , \quad C_4(\infty) = \frac{3}{8\pi} \frac{\alpha_d(0)}{\alpha_{fs}} . \tag{5.295}$$

For finite values of the dielectric constant ε, we have [YD97, YB98]

$$V_\varepsilon^{ret}(z) = -\frac{C_4(\varepsilon)}{z^4} , \quad C_4(\varepsilon) = \frac{\varepsilon - 1}{\varepsilon + 1} \phi(\varepsilon) C_4(\infty) , \tag{5.296}$$

where $\phi(\varepsilon) = \frac{1}{2}\frac{\varepsilon+1}{\varepsilon-1} \int_0^\infty h(p+1, \varepsilon)(p+1)^{-4}\, dp$ is a well defined smooth function which increases monotonically from the value $\frac{23}{30}$ for $\varepsilon = 1$ to unity for $\varepsilon \to \infty$. Explicit expressions for $\phi(\varepsilon)$ and a table of values are given in [YD97].

Table 5.3 Parameters determining the "short"-range behaviour (5.293), (5.294) and the long-range behaviour (5.295), (5.296) of the atom-surface potentials calculated by Marinescu et al. [MD97] for hydrogen and by Yan and Babb [YB98] for metastable helium. The length L is the distance (5.297) separating the regime of "small" distances from the regime of large distances; $\rho_{\rm qr} = \beta_3/\beta_4$ is the parameter determining the relative importance of the "small"-distance regime and the large-distance regime for quantum reflection, see (5.366) in Sect. 5.7.3. All quantities are in atomic units.

Atom	H	He(2^1 S)			He(2^3 S)		
ε	∞	∞	2.295	2.123	∞	2.295	2.123
C_3	0.25	2.6712	1.0498	0.9605	1.9009	0.7471	0.6836
C_4	73.61	13091	3918	3582	5163	1545	1413
β_3	919	38980	15320	14017	27740	10902	9975
β_4	520	13820	7561	7230	8680	4748	4540
L	294	4901	3732	3729	2716	2068	2067
$\rho_{\rm qr}$	1.77	2.82	2.03	1.94	3.20	2.30	2.20

The atom-surface potential behaves as $-C_3/z^3$ for "small" distances [(5.293), (5.294)] and as $-C_4/z^3$ for large distances [(5.295), (5.296)]. The ratio

$$L = \frac{C_4}{C_3} = \frac{(\beta_4)^2}{\beta_3} \tag{5.297}$$

defines a length scale separating the regime of "small" z values, $z \ll L$, from the regime of large z values, $z \gg L$. In (5.297) we have introduced the parameters β_3 and β_4 which express the potential strength in the respective limit in terms of a length, as for the homogeneous potentials (3.1) discussed in Sect. 3.1.

The expressions (5.289) and (5.292) were evaluated for the interaction of a hydrogen atom with a conducting surface by Marinescu et al. [MD97] and for the interaction of metastable helium $2\,^1S$ and $2\,^3S$ atoms with a conducting surface ($\varepsilon = \infty$) and with BK-7 glass ($\varepsilon = 2.295$, $\phi(\varepsilon) = 0.761425$) and fused silica ($\varepsilon = 2.123$, $\phi(\varepsilon) = 0.760757$) surfaces by Yan and Babb [YB98]. A list of the potential parameters determining the "short"-range and the long-range parts of the respective potentials is given in Table 5.3.

The lengths β_3 and β_4 are natural length scales corresponding to typical distances where quantum effects associated with the "short"- or long-range part of the potential are important. These distances are of the order of hundreds or thousands or even tens of thousands of atomic units. The words "small" or "short" refer to lengths which are small compared to these very large distances, a few tens of atomic units, say, but larger than the close distances of a few atomic units, where more intricate details of the atom-surface interaction involving the microscopic structure of the atom and of the surface become important.

For the potential (5.292) between the atom and a conducting surface, we can also make some general statements about the next-to-leading terms at large and "small" distances. For large distances we can exploit the fact that the dipole polarizability

(5.291) is an even function of the imaginary part of its argument, so $V_\infty(z)$ as given in the second line of (5.292) is an even function of $1/z$ and the next term in the large-distance expression (5.295) must fall off at least as $1/z^6$,

$$V_\infty(z) \stackrel{z\to\infty}{\sim} -\frac{C_4}{z^4} + O\left(\frac{1}{z^6}\right) . \tag{5.298}$$

For small distances z we can calculate a correction to the expression (5.293) via a Taylor expansion of the integral in the first line of (5.292),

$$\begin{aligned} V_\infty(z) &= -\frac{I(z)}{z^3} \stackrel{z\to 0}{\sim} -\frac{1}{z^3}\left(I(0) + z\left.\frac{dI}{dz}\right|_{z=0}\right) \quad \text{with} \quad I(0) = C_3(\infty) \quad \text{and} \\ I(z) &= \frac{1}{4\pi}\int_0^\infty \alpha_d(i\omega)\left[1 + 2\alpha_{fs}\,\omega z + 2(\alpha_{fs}\,\omega z)^2\right]\exp\left(-2\alpha_{fs}\,\omega z\right)d\omega\,, \\ \frac{dI}{dz} &= -\frac{1}{\pi}\int_0^\infty \alpha_d(i\omega)(\alpha_{fs}\,\omega)^3\,z^2\,\exp(-2\alpha_{fs}\,\omega z)\,d\omega \\ &= -\frac{\alpha_{fs}}{4\pi}\int_0^\infty \alpha_d\left(\frac{iy}{2\alpha_{fs}z}\right)\left[\frac{y}{2\alpha_{fs}z}\right]^2 y\,\exp(-y)\,dy\,. \end{aligned} \tag{5.299}$$

The last line in (5.299) follows from the second-last line via a change of variable, $y = 2\alpha_{fs}\omega z$, $\omega = y/(2\alpha_{fs}z)$. The limit of small z corresponds to the limit of large ω and with (5.291) we have

$$\lim_{\omega\to\infty} \alpha_d(i\omega)\omega^2 = \sum_n 2(E_n - E_0)|\langle\psi_0|\sum_{j=1}^{Z} x_j|\psi_n\rangle|^2 = Z\,. \tag{5.300}$$

The fact that the sum over n in (5.300) reduces to the total number Z of electrons in the atom is just the sum rule (2.220) formulated in Sect. 2.4.6. In the limit $z \to 0$, the product of the dipole polarizability and the square of the square bracket in the integrand in the last line of (5.299) can thus be replaced by Z,

$$\left.\frac{dI}{dz}\right|_{z=0} = -Z\frac{\alpha_{fs}}{4\pi}\int_0^\infty y\,\exp\left(-y\right)dy = -\frac{Z\alpha_{fs}}{4\pi}\,. \tag{5.301}$$

The leading and next-to-leading contributions to the potential between the atom and the conducting wall at "small" distances are thus,

$$V_\infty(z) \stackrel{z\to 0}{\sim} -\frac{C_3(\infty)}{z^3} + \frac{Z\alpha_{fs}}{4\pi}\frac{1}{z^2}\,. \tag{5.302}$$

Remember that "small" means small compared to the lengths listed in Table 5.3, but this can still be quite large in atomic units. The second term on the right-hand side of (5.302), i.e. the leading retardation correction to the van der Waals potential between an atom and a conducting surface at "small" distances, was first derived by Barton for one-electron atoms in 1974 [Bar74]. An intriguing feature of this correction is, that it is universal: it depends only on the number Z of electrons in the atom and not on its eigenstate ψ_0.

If we factor the van der Waals term out of the potential,

$$V(z) = -\frac{C_3}{z^3}\, v\,, \tag{5.303}$$

then the transition from the "small"-distance regime to the large-distance regime is contained in the *shape function v,* which tends to unity for "small" distances and behaves as $(C_4/C_3)/z = (z/L)^{-1}$ at large distances. A simple rational approximation which fulfills these boundary conditions is

$$v\left(\frac{z}{L}\right) = \frac{1+\xi\, z/L}{1+\eta\, z/L+\xi\,(z/L)^2}\,, \tag{5.304}$$

containing two parameters η and ξ. For the simplest case of a ground-state hydrogen atom in front of a conducting wall, the static dipole polarizability which determines the coefficient of the asymptotic $-1/z^4$ part of the potential according to (5.295) is known, $\alpha_d(0) = 9/2$ a.u., see Problem 3.9. Also, the expectation value of r^2 which determines the van der Waals coefficient according to (5.286), (5.287) is known [BS77] to be $\langle\psi_0|r^2|\psi_0\rangle = 3$ a.u., so in this case,

$$C_3 = \frac{\langle\psi_0|r^2|\psi_0\rangle}{12} = 0.25\,\text{a.u.}\,, \quad C_4 = \frac{3\alpha_d(0)}{8\pi\alpha_{fs}} \approx 73.61\,\text{a.u.} \tag{5.305}$$

In the rational approximation (5.304) for the shape function, the parameter η must be unity in order to reproduce the next-to-leading behaviour (5.298) at large distances, and the parameter ξ must be chosen as

$$\xi = 1 - \frac{Z\alpha_{fs}}{4\pi}\frac{C_4}{(C_3)^2} \tag{5.306}$$

in order to reproduce the universal next-to-leading correction at "small" distances (5.302). For the hydrogen atom in front of a conducting wall we have $Z = 1$ and the values (5.305) giving $\xi = 0.31608\ldots$. The rational approximation (5.304) thus leads to the following atom-surface potential (in atomic units),

$$V_H(z) = -\frac{C_3}{z^3}\left[\frac{1+\xi z/L}{1+z/L+\xi(z/L)^2}\right],$$
$$C_3 = 0.25\,\text{a.u.}\,, \quad \xi = 0.31608\ldots\,, \quad L = C_4/C_3 \approx 294\,\text{a.u.} \tag{5.307}$$

This expression does in fact approximate the exact potential between a ground-state hydrogen atom and a conducting wall very well as illustrated in Fig. 5.38, where it is

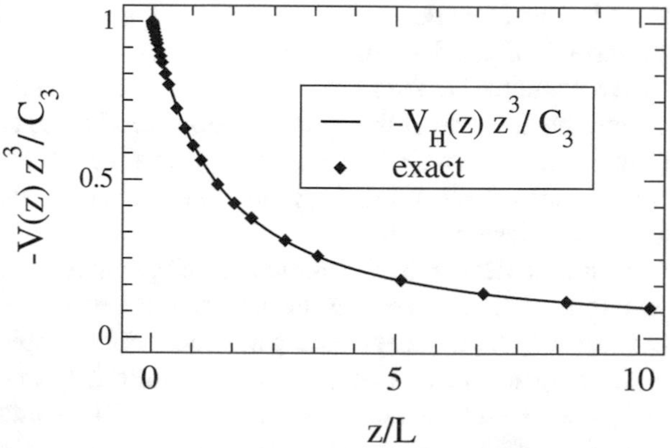

Fig. 5.38 Shape function for the potential between a ground-state hydrogen atom and a conducting wall. The *solid line* shows the quotient in the square bracket on the right-hand side of (5.307); the *filled diamonds* show the exact numerical results calculated by Marinescu et al. [MD97]

compared to the numerical results calculated and tabulated in [MD97]. The rational approximation (5.307) actually reproduces the numerical values to within a relative error of 0.6% in the whole range of z values.

The large-distance behaviour of the atom-surface potential becomes more complicated, when the atom is not in its ground state or a metastable state, but rather in an excited state with non-vanishing dipole matrix elements to lower-lying states. For more detailed discussions see [Bar74] and [Har90]. For the case of a conducting surface, the asymptotic behaviour of the atom-surface potential is (in atomic units) [Bar74],

$$V(z) \overset{z\to\infty}{\sim} -\frac{C_4}{z^4} + \sum_n \frac{\Theta(E_0 - E_n)}{|\bar{\lambda}_{0,n}|^3} \times \qquad (5.308)$$

$$\left[|\langle\psi_n \left| \hat{\boldsymbol{d}}_{\parallel} \right| \psi_0\rangle|^2 \left(\frac{\cos(\zeta_n)}{\zeta_n} - \frac{\sin(\zeta_n)}{(\zeta_n)^2} - \frac{\cos(\zeta_n)}{(\zeta_n)^3} \right) \right.$$

$$\left. - 2|\langle\psi_n|\hat{d}_{\perp}|\psi_0\rangle|^2 \left(\frac{\sin(\zeta_n)}{(\zeta_n)^2} + \frac{\cos(\zeta_n)}{(\zeta_n)^3} \right) \right] + O\left(\frac{1}{(\zeta_n)^6}\right).$$

Here C_4 is as defined in (5.295), $\hat{\boldsymbol{d}}_{\parallel}$ and $\hat{d}_{\perp}$ are the parallel and normal parts of the dipole operator (5.286) and $\zeta_n = 2z/\bar{\lambda}_{0,n}$ is the n-dependent ratio of the round-trip distance from the atom to the wall and back and the transition wavelength $\bar{\lambda}_{0,n}$ connecting the initial state ψ_0 to the respective lower-lying state ψ_n, cf. (5.288). The first term $-C_4/z^4$ on the right-hand side of (5.309) only represents the leading asymptotic behaviour of the atom-surface potential at large distances if there are no lower-lying states with non-vanishing dipole transition matrix elements. Otherwise the sum over n contributes terms of longer range with coefficients that oscillate as functions of z. The theta function in the sum ensures that only lower-lying states contribute to this term. The wavelengths of the oscillations are just half the transition wavelengths $\lambda_{0,n}$ (without bar) to the lower-lying states. An interesting special case arises for the metastable $2S$ state of the hydrogen atom[FJ05]; it is connected via a non-vanishing dipole matrix element to the $2P$ state, with which it can be considered degenerate as long as the effects of relativity and quantum electrodynamics (Lamb shift) on the atomic structure are negligible.

The effect of further details of the structure of the atom, the surface and the electromagnetic field in between on the interaction between an atom and a surface have recently been receiving increasing attention. For example, Al-Amri and Babiker [AB04] investigated the influence of replacing the empty space in front of a conducting wall by a dielectric medium, and Shresta et al. [SH03] studied the effects of the movement of the atom on the various corrections to the atom-wall potential. Babb et al. [BK04] studied the joint effect of the dynamic polarizability of the atom, finite conductivity of the wall metal and nonzero temperature of the system. A rough estimate of where a finite surface temperature T might affect the derivation of the atom-surface potentials above can be obtained by comparing the thermal energy $k_B T$ with the corresponding photon energy $\hbar\omega = 2\pi\hbar c/\lambda$. One degree Kelvin corresponds to roughly 3×10^{-6} a.u. and a wavelength of roughly 3×10^{8} a.u. A wall at room temperature, $T \approx 300$ K, can thus be expected to substantially modify the results derived for zero temperature at distances near 10^6 a.u. and larger.

5.7.2 Evanescent-Wave Mirrors

A neutral polarizable atom in an electric field acquires an induced electric dipole moment and, if the field is non-homogeneous, it exerts a force on the induced dipole. This is one way of understanding the polarization potential (4.119) between an atom and a charged particle. Electromagnetic light fields can also exert forces on polarizable atoms if they are strong enough, and this is the case for sufficiently intense lasers. The forces which an intense light field exerts on an atom can be understood on the basis of the simplest possible non-trivial model of the atom, namely the "two-level atom" which has only two internal stationary eigenstates, the ground state $|g\rangle$ and the excited state $|e\rangle$.

The Hamiltonian $\hat{H}_A$ describing the centre-of-mass motion and the internal structure of such a two-level atom of mass M is,

$$\hat{H}_A = \frac{\hat{p}^2}{2M} + \hbar\omega_0 |e\rangle\,\langle e|\,. \tag{5.309}$$

We have written the excitation energy of the excited state in terms of the frequency ω_0, so the wave functions solving the time-dependent Schrödinger equation with the

Hamiltonian (5.309) are

$$|g\rangle \quad \text{or} \quad |e\rangle\, \mathrm{e}^{-\mathrm{i}\omega_0 t} \tag{5.310}$$

multiplied by a wave function $\psi(\boldsymbol{r},t)$ describing the free-particle motion of the centre of mass of the atom.

The dipole operator describing the internal dipole moment of the two-level atom is introduced as

$$\hat{\boldsymbol{d}} = d\,\boldsymbol{e}_{\mathrm{d}}(|e\rangle\langle g| + |g\rangle\langle e|)\,. \tag{5.311}$$

As for realistic atoms, where the expectation value of the dipole operator vanishes in eigenstates of given parity (see Sect. 2.4.5), the expectation value of the operator (5.311) vanishes in both states of the two-level atom, but the transition matrix element connecting the two states is finite,

$$\langle g|\hat{\boldsymbol{d}}|g\rangle = \langle e|\hat{\boldsymbol{d}}|e\rangle = 0, \quad \langle g|\hat{\boldsymbol{d}}|e\rangle = \langle e|\hat{\boldsymbol{d}}|g\rangle = d\,\boldsymbol{e}_{\mathrm{d}}. \tag{5.312}$$

The (real) parameter d describes the strength of the dipole transition and the unit vector $\boldsymbol{e}_{\mathrm{d}}$ describes the orientation of the dipole; such a vector of orientation has to be included explicitly, because the two-level atom has no internal spatial coordinates.

Let us now look at the effect on the atom of a light field oscillating with a frequency ω. The difference between this frequency and the resonance frequency ω_0 of the two-level atom is the *detuning*

$$\delta = \omega - \omega_0\,. \tag{5.313}$$

The light field is described classically, and the electric field at the position $\boldsymbol{r}$ of the atom is written as

$$\boldsymbol{E}(\boldsymbol{r}) = \boldsymbol{\epsilon}\,\mathcal{E}(\boldsymbol{r})\,\cos\left[\omega t + \Phi(\boldsymbol{r})\right]\,, \tag{5.314}$$

where $\boldsymbol{\epsilon}$ is a vector describing the direction (polarization) of the field, $\mathcal{E}(\boldsymbol{r})$ is a slowly varying amplitude factor and $\Phi(\boldsymbol{r})$ is a spatially varying phase which, e.g. for a monochromatic wave with wave vector $\boldsymbol{k}$, is simply $-\boldsymbol{k}\cdot\boldsymbol{r}$. The interaction energy of the dipole (5.311) with the electric field (5.314) is

$$\begin{aligned} \hat{H}_{\mathrm{AL}} &= -\hat{\boldsymbol{d}}\cdot\boldsymbol{E}(\boldsymbol{r}) = \hbar\Omega(\boldsymbol{r})\,\cos(\omega t + \Phi)(|e\rangle\langle g| + |g\rangle\langle e|) \quad \text{with} \\ \Omega(\boldsymbol{r}) &= -d\,(\boldsymbol{\epsilon}\cdot\boldsymbol{e}_{\mathrm{d}})\,\mathcal{E}(\boldsymbol{r})/\hbar\,. \end{aligned} \tag{5.315}$$

Here $\Omega(\boldsymbol{r})$ is the *Rabi frequency*; when $\boldsymbol{\epsilon}$, $\mathcal{E}$ and Φ are all independent of $\boldsymbol{r}$, the time-dependent Schrödinger equation with the Hamiltonian $\hat{H}_{\mathrm{A}} + \hat{H}_{\mathrm{AL}}$ has approximate solutions in which the internal state of the atom oscillates between $|g\rangle$ and $|e\rangle$ with a frequency near Ω.

The influence of the dipole coupling term (5.315) on the centre-of-mass motion of the atom depends on the internal state of the atom. Transitions between the ground state and the excited state are forced by the external light field, but additionally the excited state can decay via spontaneous emission with a rate given by (2.192), (2.193) in Sect. 2.4.4,

$$P = \frac{4}{3}\frac{d^2\omega_0^3}{\hbar c^3}\,. \tag{5.316}$$

The decay rate (5.316) corresponds to a width $\Gamma = \hbar P$ of the excited state of the free (i.e. without external field) two-level atom, see Sect. 2.4.1. Spontaneous emission brings a statistical element into the internal dynamics of the atom, so it is appropriate to describe its internal state via the von Neumann equation (5.40) for the density operator

$$\hat{\rho}_{\rm int} = \rho_{gg}|g\rangle\langle g| + \rho_{ge}|g\rangle\langle e| + \rho_{eg}|e\rangle\langle g| + \rho_{ee}|e\rangle\langle e|\,. \tag{5.317}$$

In this way, it is also possible to describe dissipative effects, as are exploited in "laser cooling". Here an atom moving upstream in a laser beam absorbs photons (and their momentum) and spontaneously re-emits them in arbitrary directions, which leads to a net loss of momentum.

The time evolution of the density matrix (5.317) is influenced by the *saturation parameter*

$$s = \frac{1}{2}\frac{\Omega(\boldsymbol{r})^2}{\delta^2 + (P/2)^2}\,. \tag{5.318}$$

For large values of the saturation parameter, the internal state of the atom evolves into a steady configuration in which both the ground state and the excited state are almost equally populated. For small values of s it evolves into a steady configuration in which the population of the ground state is significantly higher than the population of the excited state. Detailed analysis of the equations of motion [Ash78, CD92, Mey01, For01] reveals that the effect of the light field on the centre-of-mass motion of the atom due to the dipole coupling term (5.315) contains a conservative and a dissipative component, and that the conservative component is well described by the effective potential

$$V_{\rm dip}(\boldsymbol{r}) = \frac{\hbar\delta}{2}\ln\left(1 + \frac{\Omega(\boldsymbol{r})^2/2}{\delta^2 + (P/2)^2}\right)\,; \tag{5.319}$$

the associated force $-\boldsymbol{\nabla} V_{\rm dip}$ is called the *dipole force*.

The sign of the potential (5.319) depends on the sign of the detuning (5.313). For *blue detuning*, i.e. for $\omega > \omega_0$, $\delta > 0$, the potential is positive and the atom is attracted to regions of small field intensities, it is "weak-field seeking". For *red detuning* on the other hand, i.e. for $\omega < \omega_0$, $\delta < 0$, the potential is negative and the atom is attracted to regions of large field intensities, it is "strong-field seeking". The possibility of exerting mechanical forces on neutral atoms through light has paved the way to many new fascinating experiments. One example is the trapping and guiding of atoms in an "optical lattice", which is a spatially periodic electric field due to the standing waves generated by appropriately adjusted counter-propagating lasers, see e.g. [Blo04] and references therein. In this section we focus on another example with direct practical use, namely the evanescent wave mirror.

When light in a dielectric medium is incident on a surface to an optically less dense outside, it undergoes total internal reflection if the angle of incidence θ_i is large enough: $\sin(\theta_i) > 1/n_m$, where n_m is the refractive index of the dielectric medium relative to the outside. Some light does penetrate into the outside as a decaying, "evanescent" wave characterized by a finite penetration depth. To be precise, assume that the surface is the xy plane and the dielectric medium is the half-space $z < 0$. A monochromatic plane wave is totally reflected at the surface as sketched in Fig. 5.39. Outside the medium, i.e. for positive z values, there is an electric field (5.314) oscillating with the frequency ω and propagating parallel to the surface. In the normal direction, the amplitude $\mathcal{E}$ decays with a penetration depth $1/\kappa$,

$$\mathcal{E}(z) = \mathcal{E}_0\,\mathrm{e}^{-\kappa z}\,, \quad \kappa = k\sqrt{(n_m)^2\sin^2(\theta_i) - 1}\,; \tag{5.320}$$

here k is the wave number outside the medium, which is connected to the frequency ω by the dispersion relation (5.282). This translates into the following

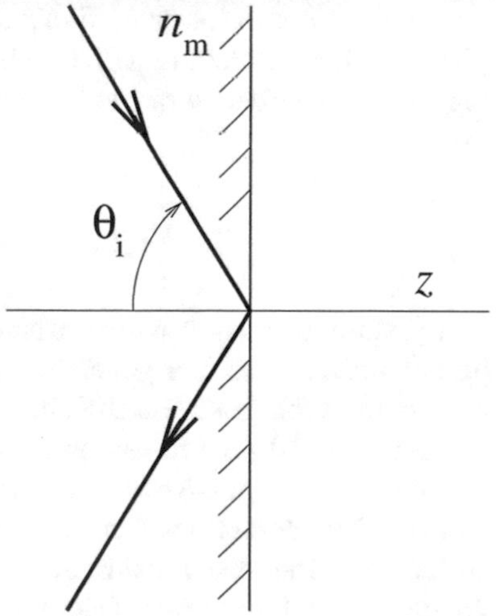

Fig. 5.39 Total internal reflection of a plane wave in a medium—with refractive index (relative to the outside) n_m—at the surface in the xy plane

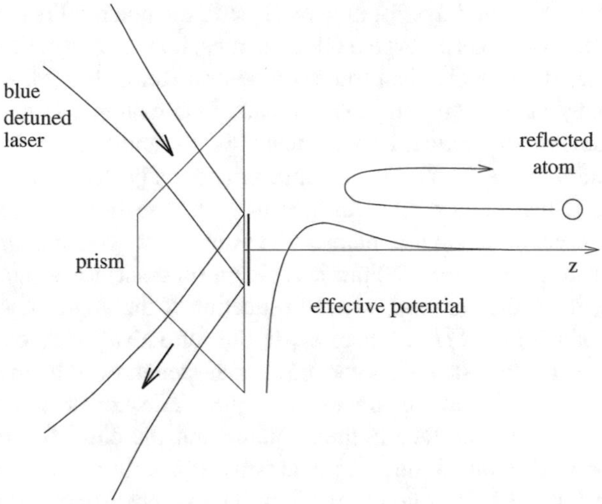

Fig. 5.40 Schematic illustration of an evanescent-wave atomic mirror. The blue detuned laser light incident on the vertical surface from the left generates a repulsive dipole potential (5.322) which, together with the attractive atom-surface potential discussed in Sect. 5.7.1 produces an effective potential with a barrier (from [CS98], courtesy of Robin Côté)

behaviour of the square of the Rabi frequency,

$$\Omega(z)^2 = (\Omega_0)^2\,\mathrm{e}^{-2\kappa z}\,, \quad (\Omega_0)^2 = \frac{d^2(\mathcal{E}_0)^2}{\hbar^2}(\boldsymbol{\epsilon}\cdot\boldsymbol{e}_\mathrm{d})^2\,. \tag{5.321}$$

For small values of the saturation parameter (5.318) we can expand the logarithm in (5.319); when the detuning (5.313) is so large that we can neglect the contribution of the spontaneous decay rate in the denominator $\delta^2 + (P/2)^2$, the potential simplifies to

$$V_\mathrm{dip}(z) = \frac{\hbar\Omega(z)^2}{4\delta} = \frac{\hbar(\Omega_0)^2}{4\delta}\,\mathrm{e}^{-2\kappa z}\,. \tag{5.322}$$

By shining a laser into a prism so that it is totally reflected by one of the prism surfaces, we can generate a repulsive or attractive dipole force for atoms approaching the prism on the other side. If the laser is blue detuned with respect to the relevant dipole transition in the approaching atoms ($\delta > 0$), then the atoms are subject to a repulsive force due to the evanescent light wave in front of the prism surface. The interaction of an atom with the surface also contains the attractive atom-surface potential discussed in Sect. 5.7.1. If the repulsive dipole potential is strong enough, the total atom-surface potential has a barrier as sketched in Fig. 5.40. If the energy of the atom is lower than the barrier height, it is reflected at the outer classical turning point of the barrier. Thus the dipole force of the evanescent light wave helps to make a mirror which reflects sufficiently cold approaching atoms with near to 100% efficiency.

The strength of the evanescent-wave potential (5.322) depends on the intensity $(\mathcal{E}_0)^2/2$ of the electric field at the surface, on the strength d of the dipole transition matrix element (5.312), and on a factor of order unity related to the polarization of the electric field and the orientation vector $\boldsymbol{e}_\mathrm{d}$ of the dipole transition. The electric field intensity at the surface is related to the power of the laser light and other circumstances [For01]. For a rough estimate we can refer to Problem 5.1, from which it follows that a laser power of 10^{17} W/cm^2 corresponds to an energy density of the order of an atomic unit. The inverse penetration depth κ is necessarily less than the wave number $n_\mathrm{m}k$ of the incoming wave in the prism and depends on the angle of incidence according to (5.320). Towards the critical angle, $n_\mathrm{m}\sin(\theta_\mathrm{i}) \to 1$, κ tends to zero corresponding to infinitely large values of the penetration depth.

For practical applications it is convenient to work with atoms which approximately fulfill the requirements of the two-level model with a level separation in the range of available laser frequencies. One popular choice is the metastable $2\,^3S$ state of helium which is connected by a dipole matrix element to the higher-lying $2\,^3P_2$ state. The transition wavelength is $\lambda = 1083$ nm $= 20466$ a.u., so $\omega_0 = 0.04207$ a.u. The upper $2\,^3P_2$ state can only decay via spontaneous emission to the lower $2\,^3S$ state and its lifetime is 98 ns= 4.05×10^9 a.u., so the spontaneous decay rate is $P = 0.247 \times 10^{-9}$ a.u.; this corresponds to a dipole strength $d^2 = 6.4$ a.u. according to (5.316). Dall et al. used evanescent light fields to guide such metastable helium atoms through hollow optical fibres consisting of fused silica capillaries [DH99]. This is just one example of how evanescent light fields can be used to construct atom-optical devices.

5.7.3 *Quantum Reflection*

The transmission through and reflection by a potential barrier were described in Sect. 1.4.2. For an incident particle with an energy greater than the barrier height (i.e. the maximum of the potential), there is no classical turning point; reflection of the particle is classically forbidden and is a purely quantum mechanical effect. This "quantum reflection" is the above-barrier analog of below-barrier tunnelling through the classically forbidden region in coordinate space. Quantum reflection can also occur in the absence of a barrier in a purely attractive potential. The only condition is, that there be a quantal region of coordinate space in which the quantality function (1.298) is significantly non-vanishing and that to either side there be semiclassical regions in which the WKB approximation is accurate, so that we can construct solutions of the Schrödinger equation which can unambiguously be classified as leftward travelling or rightward travelling. A simple example is the sharp step potential, see (4.80) in Sect. 4.1.7. Here the quantal region reduces to the single point at which the potential is discontinuous, while the WKB approximation is exact on either side of this discontinuity.

This section focusses on singular attractive potential tails $V_{\rm tail}(r)$ which tend to zero faster than $1/r^2$ at large distances $r \to \infty$, and to $-\infty$ faster than $-1/r^2$ for $r \to 0$. At a given positive energy,

$$E = \frac{\hbar^2 k^2}{2M} > 0\,, \qquad (5.323)$$

semiclassical approximations are good at large distances, where the Schrödinger equation essentially describes free-particle motion, and again at small distances $r \to 0$. In between, there is a nonclassical, quantal region giving rise to quantum reflection.

Quantum reflection by a step potential or an attractive potential tail is always important towards threshold, $E \to 0$, because the reflection probability approaches unity in this limit, see (5.335) and Fig. 5.42 below. In contrast to reflection by a potential barrier, however, the reflection remains classically forbidden all the way down to threshold. For a potential barrier, the classical reflection probability is unity below the barrier and zero above, and the contribution of quantum mechanics is merely a smoothing of the edges of this step function, see top half of Fig. 5.41. For a potential step or a purely attractive potential tail, reflection is classically forbidden at all energies (above threshold) and all reflection is a purely quantum mechanical phenomenon, see bottom half of the figure.

For the singular attractive potential tail $V_{\rm tail}(r)$ at energy $E > 0$, a characteristic distance is provided by the point r_E already introduced in (5.208) in Sect. 5.5.1. It corresponds to the classical turning point of the repulsive potential $-V_{\rm tail}(r)$, i.e. the point where the absolute value of $V_{\rm tail}(r)$ is equal to the total energy of the particle,

$$|V_{\rm tail}(r_E)| = E\,. \qquad (5.324)$$

A typical classical action is provided the product of r_E and the asymptotic momentum $\hbar k$, corresponding in units of $\hbar$ to $k\,r_E$. Thus $k\,r_E$ is a generalization of the concept of the reduced classical turning point introduced after (3.18) in Sect. 3.1.2. For the singular attractive potential $V_{\rm tail}(r)$ falling off faster than $1/r^2$, the high-energy limit $k \to \infty$ implies $k\,r_E \to \infty$ and corresponds to the semiclassical limit of the Schrödinger equation (5.205), while the threshold limit $k \to 0$ implies $k\,r_E \to 0$ and corresponds to the anticlassical limit.

The local classical momentum $p_{\rm tail}(E;r) = \sqrt{2\mu[E - V_{\rm tail}(r)]}$ is real and positive for all distances $0 < r < \infty$. At distances noticeably smaller than r_E, $p_{\rm tail}(E;r)$ is dominated by the contribution from $V_{\rm tail}(r)$ and becomes independent of energy. The quantality function (1.298) becomes insensitive to the energy and vanishes for $r \to 0$, so the WKB representations of the solutions of the Schrödinger equation with the potential $V_{\rm tail}(r)$, (5.205), become exact in the limit $r \to 0$. This implies that the solutions of (5.205) can, for any energy E, be unambiguously decomposed into

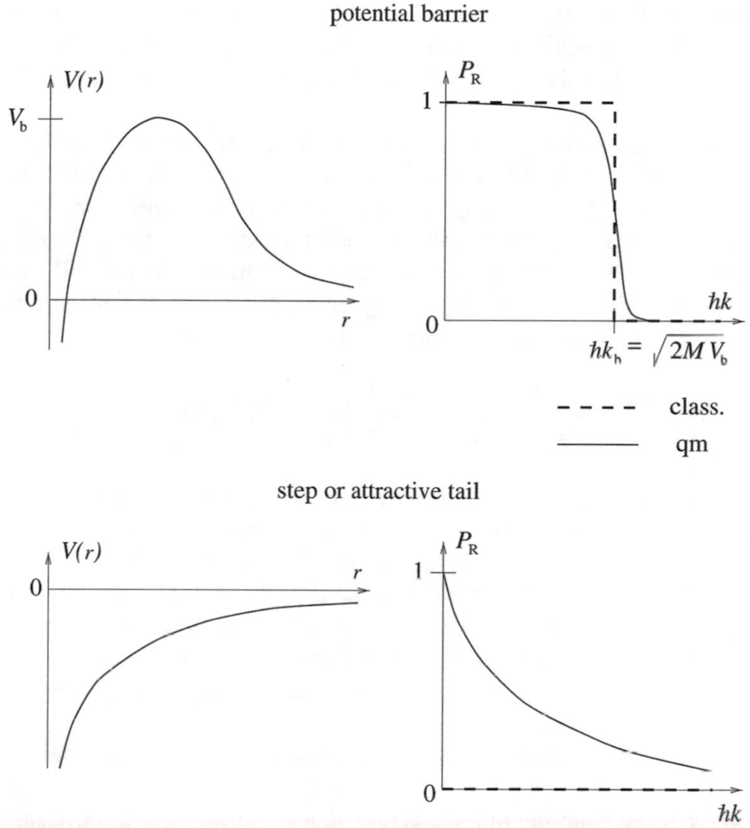

Fig. 5.41 Schematic illustration of the qualitative behaviour of the reflection probability P_R in a one-dimensional potential. For reflection by a potential barrier of height V_b, the contribution of quantum mechanics is merely to smooth out the step function describing the classical reflection probability (top half). For a potential step or a purely attractive potential tail, reflection is classically forbidden at all energies (above threshold) and all reflection is a purely quantum mechanical phenomenon (bottom half)

incoming and outgoing radial waves at small distances. At distances much larger than r_E, the potential $V_{tail}(r)$ is only a small correction to the dominant, constant part $\hbar k$ of $p_{tail}(E; r)$, and the Schrödinger equation (5.205) becomes that for free-particle motion. For $r \gg r_E$, the wave function essentially describes free-particle motion and can also be decomposed into incoming and outgoing waves. In between the near-origin regime $r \to 0$ and the large-distance regime $r \gg r_E$, there is the *nonclassical region* of the reference potential $V_{tail}(r)$, with distances of the order of the generalized reduced classical turning point r_E, where the condition (1.297) is not well fulfilled—at least at low energies. This nonclassical, quantal region of the potential tail is the source of quantum reflection.

The fact that the quantal region of $V_{\rm tail}$ is centred around r_E can be shown analytically for single-power attractive potential tails (3.63), see Problem 1.10, and it was shown numerically in [FJ02] for retarded van der Waals potentials of the type (5.303).

For each energy E, i.e. for each wave number k, there are two linearly independent solutions of (5.205), and the physically relevant linear combination of these two solutions is chosen by defining appropriate boundary conditions at small distances. For ordinary scattering problems, this boundary condition is chosen to ensure that the regular solution of the radial Schrödinger equation with the full interaction matches to the solution of (5.205) at large distances. Other choices are, however, possible. Choosing incoming boundary conditions at $r \to 0$,

$$u(r) \overset{r\to 0}{\sim} \frac{T}{\sqrt{p_{\rm tail}(E;r)}} \exp\left(-\frac{\rm i}{\hbar}\int_{r_0}^{r} p_{\rm tail}(E;r')\,{\rm d}r'\right) , \tag{5.325}$$

corresponds to assuming that all incoming flux which is transmitted through the nonclassical region of the potential tail to small distances is absorbed. Note that, for sufficiently small r, the upper integration limit r is smaller than the lower integration limit r_0 limit in the integral in (5.325), so the integral itself is negative. Writing the argument of the WKB wave function as upper limit in the action integral has the advantage, that wave functions containing $\exp\left(-\frac{\rm i}{\hbar}\int^r\cdots\right)$ are easily identified as inward-travelling waves, whereas wave functions containing $\exp\left(+\frac{\rm i}{\hbar}\int^r\cdots\right)$ are outward-travelling waves.

Starting with the incoming boundary conditions (5.325), the Schrödinger equation (5.205) can be integrated outwards, which yields a well defined solution that can be decomposed into incoming and outgoing radial waves at large distances,

$$u(r) \overset{r\to\infty}{\sim} \frac{1}{\sqrt{\hbar k}}\left({\rm e}^{-{\rm i}kr} + R\,{\rm e}^{+{\rm i}kr}\right) . \tag{5.326}$$

Since the potential $V_{\rm tail}(r)$ is strongly r-dependent for $r \to 0$, the right-hand side of (5.325) necessarily contains the prefactor $1/\sqrt{p_{\rm tail}(E;r)}$. The factor $1/\sqrt{\hbar k}$ on the right-hand side of (5.326) is included for consistency. The transmission coefficient T in (5.325) can be chosen such that there is no further proportionality constant in front of the incoming wave in (5.326). The phase of T also depends on the choice of the lower integration limit r_0 in the action integral. Equation (5.326) defines the *quantum reflection amplitude* R. Comparing (5.326) with (4.48) and (4.49) in Sect. 4.1.5 shows that the reflection amplitude R can be interpreted as minus the s-wave S-matrix,

$$R \equiv -S_{l=0} = -{\rm e}^{2{\rm i}\delta_0} , \tag{5.327}$$

with an s-wave scattering phase shift δ_0. Incoming boundary conditions imply absorption, so the S-matrix is no longer unitary, which is expressed through a complex phase shift δ_0.

The immediate near-threshold behaviour of the quantum reflection amplitude can be easily derived [FT04] on the basis of the two threshold ($E = 0$) solutions $u_0^{(0)}(r)$ and $u_1^{(0)}(r)$ of the radial Schrödinger equation (5.205), which are defined by their asymptotic behaviour (3.40). From their small-r behaviour (3.41), it follows that the linear combination

$$\begin{aligned} u(r) &= \frac{e^{i\phi_0/2}}{D_1}\, u_1^{(0)}(r) - \frac{e^{i\phi_1/2}}{D_0}\, u_0^{(0)}(r) \\ &\overset{r\to 0}{\propto} \frac{1}{\sqrt{p_{\rm tail}(0;r)}} \exp\left(-\frac{i}{\hbar}\int_{\infty}^{r} p_{\rm tail}(0;r')\,dr'\right) \end{aligned} \tag{5.328}$$

obeys incoming boundary conditions for $r \to 0$. At large distances, the superposition (5.328) behaves as

$$u(r) \overset{r\to\infty}{\sim} -\frac{e^{i\phi_1/2}}{D_0} + \frac{e^{i\phi_0/2}}{D_1}\, r\,, \tag{5.329}$$

which is to be compared with

$$\frac{1}{\sqrt{\hbar k}}\left(e^{-ikr} + R\,e^{+ikr}\right) \overset{kr\to 0}{\propto} 1 + R \;-\; ik(1-R)\,r\,. \tag{5.330}$$

Since the ratio of the constant term and the coefficient of r must be the same in (5.329) and (5.330), we obtain

$$\frac{D_0}{D_1}\,e^{i(\phi_0-\phi_1)/2} = \frac{ik(1-R)}{1+R} \implies R \overset{k\to 0}{\sim} -\frac{1 - ik\,e^{i(\phi_0-\phi_1)/2} D_1/D_0}{1 + ik\,e^{-i(\phi_0-\phi_1)/2} D_1/D_0}\,, \tag{5.331}$$

and, with the threshold length b and mean scattering length $\bar{a}$ as defined in (3.46), (3.48),

$$\begin{aligned} R &\overset{k\to 0}{\sim} -\left[1 - 2k\frac{D_1}{D_0}\left[\sin\left(\frac{\phi_0-\phi_1}{2}\right) + i\cos\left(\frac{\phi_0-\phi_1}{2}\right)\right]\right] \\ &= -\left[1 - 2i(\bar{a} - ib)k\right]\,. \end{aligned} \tag{5.332}$$

Expressing R in terms of the complex phase shift $\boldsymbol{\delta}_0$ according to (5.327) reveals the following near-threshold behaviour of $\boldsymbol{\delta}_0$,

$$\boldsymbol{\delta}_0 \overset{k\to 0}{\sim} -(\bar{a} - i\,b)k = -\boldsymbol{a}\,k\,. \tag{5.333}$$

Thus the mean scattering length $\bar{a}$ and the threshold length b, introduced in Sect. 3.1.2 as tail parameters of a singular reference potential $V_{\rm tail}(r)$, appear as the real part and minus the imaginary part of the *complex scattering length* [VF05, AF06],

$$\boldsymbol{a} = \bar{a} - \mathrm{i}\,b\,, \tag{5.334}$$

which describes the leading near-threshold behaviour of the quantum reflection amplitude. The mean scattering length is well defined only for potentials falling off faster than $1/r^3$ at large distances, but the threshold length b is well defined for potentials falling off faster than $1/r^2$. The leading near-threshold behaviour of the modulus of the quantum reflection amplitude is determined by the threshold length b,

$$|R| \overset{k\to 0}{\sim} 1 - 2bk + O(k^2) = \mathrm{e}^{-2bk} + O(k^2)\,. \tag{5.335}$$

Note that the probability $|R|^2$ for quantum reflection approaches unity at threshold, so quantum reflection always becomes dominant at sufficiently low energies.

The effective-range expansion, described for the phase shifts of ordinary scattering in Sect. 4.1.7, can be adapted for the complex phase shifts of quantum reflection, as described in [AF06]. Equation (4.99) becomes

$$\begin{aligned} k\cot\boldsymbol{\delta}_0 &\overset{k\to 0}{\sim} -\frac{1}{\bar{a}-\mathrm{i}\,b} + \frac{1}{2}\boldsymbol{r}_{\mathrm{eff}}\,k^2\,,\\ \boldsymbol{r}_{\mathrm{eff}} &= 2\int_0^\infty \left(\left[w^{(0)}(r)\right]^2 - \left[u^{(0)}(r)\right]^2\right)\,\mathrm{d}r\,, \end{aligned} \tag{5.336}$$

but the radial wave function $u^{(0)}(r)$ is now defined as the solution of (5.205) which obeys incoming boundary conditions for $r \to 0$ and the following boundary conditions for large r:

$$u^{(0)}(r) \overset{r\to\infty}{\sim} 1 - \frac{r}{\bar{a}-\mathrm{i}\,b}\,. \tag{5.337}$$

The wave function $w^{(0)}(r)$ in (5.336) assumes the form on the right-hand side of (5.337) in the whole range of r-values, from the origin to infinity,

$$w^{(0)}(r) = 1 - \frac{r}{\bar{a}-\mathrm{i}\,b}\,. \tag{5.338}$$

The parameter $\boldsymbol{r}_{\mathrm{eff}}$ in (5.336) is the complex effective range. As for the real effective range in ordinary scattering, it is well defined for potentials $V_{\mathrm{tail}}(r)$ falling off faster than $1/r^5$ at large distances.

The tail parameters of attractive single-power tails (3.63) can be related in a very elegant way to corresponding parameters of the repulsive inverse-power potentials (4.74) discussed in Sect. 4.1.7. To see this, observe that the repulsive inverse-power potential (4.74) becomes the attractive inverse-power potential (3.63) by an appropriate transformation of the quantum length β_α. With $\nu = 1/(\alpha-2)$:

$$\beta_\alpha \to \beta_\alpha^{-\mathrm{i}\pi\nu} \quad \Longrightarrow \quad \frac{(\beta_\alpha)^{\alpha-2}}{r^\alpha} \to -\frac{(\beta_\alpha)^{\alpha-2}}{r^\alpha} \,. \tag{5.339}$$

The same transformation, $\beta_\alpha \to \beta_\alpha^{-\mathrm{i}\pi\nu}$, transforms the purely imaginary local classical momentum under the repulsive inverse-power potential to a real local classical momentum in the attractive inverse-power potential. The radial wave function which is exactly equal to its WKB representation in the limit $r \to 0$ for inverse-power tails with $\alpha > 2$, is transformed from the regular solution which vanishes monotonically for $r \to 0$ in the repulsive case to the oscillating solution obeying incoming boundary conditions in the attractive case. All properties which depend on the quantum length β_α carry over from the repulsive to the attractive case via the transformation (5.339). The scattering length, which is given by (4.78) for the repulsive inverse-power potential (4.74), transforms according to

$$a = \nu^{2\nu}\frac{\Gamma(1-\nu)}{\Gamma(1+\nu)}\beta_\alpha \longrightarrow \nu^{2\nu}\frac{\Gamma(1-\nu)}{\Gamma(1+\nu)}\beta_\alpha\,[\cos(\pi\nu) - \mathrm{i}\sin(\pi\nu)] \;=\; \bar{a} - \mathrm{i}\,b \tag{5.340}$$

to the complex scattering length $\boldsymbol{a} = \bar{a} - \mathrm{i}\,b$; the expressions following for the mean scattering length $\bar{a}$ and the threshold length b according to (5.340) are those already given in (3.67) . Similarly, the complex effective range $\boldsymbol{r}_{\mathrm{eff}}$ appearing in (5.336) is, for attractive single-power potentials (3.63) with $\alpha > 5$, just $\mathrm{e}^{-\mathrm{i}\pi\nu}$ times the real effective range r_{eff} of the corresponding repulsive inverse-power potential (3.63) with the same quantum length β_α [AF06]. The straightforward relationship between repulsive and attractive inverse-power potentials makes it possible to adapt the extensive results on the near-threshold behaviour of phase shifts which were derived in [DG65] for repulsive inverse-power potentials to the description of quantum reflection by attractive inverse-power potentials.

In the limit of large energies, we may use a semiclassical expression for the reflection amplitudes which was derived by Pokrovskii et al. [PS58, PU58]. We use the reciprocity relation (1.176) to adapt the formula of [PS58, PU58] to the reflection amplitude R (corresponding to R_{r}), defined via the boundary conditions (5.325), (5.326),

$$R(k)^* \overset{k\to\infty}{\sim} \mathrm{i}\exp\left(\frac{2\mathrm{i}}{\hbar}\int^{r_{\mathrm{t}}} p(r)\,\mathrm{d}r\right) . \tag{5.341}$$

Here r_{t} is the complex turning point with the smallest (positive) imaginary part. For a single-power potential (3.63) it can be written as

$$r_{\mathrm{t}} = (-1)^{1/\alpha} r_E = \mathrm{e}^{\mathrm{i}\pi/\alpha} r_E \,, \tag{5.342}$$

where r_E as defined by (5.324) is (5.223),

$$r_E = \beta_\alpha (k\beta_\alpha)^{-2/\alpha} . \tag{5.343}$$

Real values of the momentum $p(r)$ only contribute to the phase of the right-hand side of (5.341), so $|R|$ is unaffected by a shift of the lower integration point anywhere along the real axis. Integrating along the path $r/r_E = \cos(\pi/\alpha) + \mathrm{i}\xi \sin(\pi/\alpha)$ with $\xi = 0 \to 1$ gives the result [FJ02]

$$|R| \overset{k\to\infty}{\sim} \exp(-B_\alpha\, kr_E) = \exp[-B_\alpha (k\beta_\alpha)^{1-2/\alpha}] ,$$
$$B_\alpha = 2\sin\left(\frac{\pi}{\alpha}\right) \Re \left\{ \int_0^1 \sqrt{1 + \left[\cos\left(\frac{\pi}{\alpha}\right) + \mathrm{i}\xi \sin\left(\frac{\pi}{\alpha}\right)\right]^{-\alpha}}\, \mathrm{d}\xi \right\} . \tag{5.344}$$

In terms of the energy E, the particle mass μ and the strength parameter C_α of the potential (3.63), the energy-dependent factor in the exponent is

$$(k\beta_\alpha)^{1-2/\alpha} = \frac{1}{\hbar} E^{\frac{1}{2}-\frac{1}{\alpha}} (C_\alpha)^{1/\alpha} \sqrt{2\mu} = \frac{p_{\mathrm{as}} r_E}{\hbar} , \tag{5.345}$$

where $p_{\mathrm{as}} = \hbar k$ is the asymptotic ($r \to \infty$) classical momentum. The high-energy behaviour (5.344) of the reflectivity as function of $\hbar$ is an exponential decrease typically expected for an analytical potential which is continuously differentiable to all orders, see [Ber82]. Numerical values of the coefficients B_α were derived in [FJ02] are listed in bottom row of Table 3.1 in Sect. 3.1.2.

Plots of $\ln|R|$, as function both of $k\beta_\alpha$ and of $(k\beta_\alpha)^{1-2/\alpha}$, are shown in Fig. 5.42. The linear initial fall-off of the various curves in the left-hand part of the figure is in agreement with (5.335), and the gradients $-2b/\beta_\alpha$ reflect the respective threshold

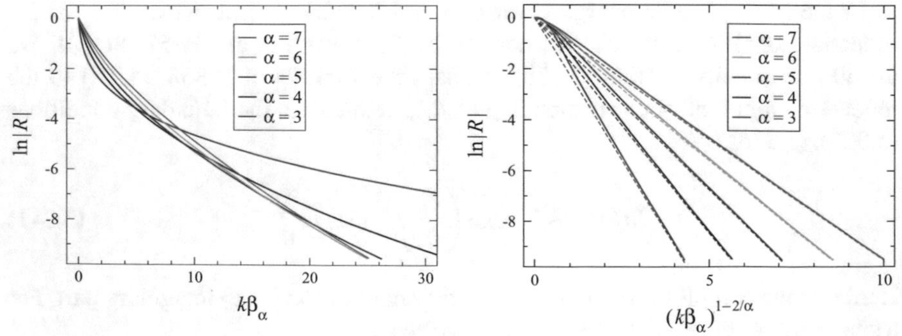

Fig. 5.42 Logarithmic plot of the modulus $|R|$ of the quantum reflection amplitude for attractive inverse-power potentials (3.63) for $\alpha = 3, \ldots 7$ as functions of $k\beta_\alpha$ (left-hand part) and of $(k\beta_\alpha)^{1-2/\alpha}$ (right-hand part). The *straight dashed lines* in the right-hand part show the functions $-B_\text{o}\,(k\beta_\alpha)^{1-2/\alpha}$ with the coefficients B_o given in the bottom row of Table 3.1 (adapted from [FJ02])

lengths b as already given in (3.67) and Table 3.1. In the right-hand part of the figure, the fall-off at large values of $(k\beta_\alpha)^{1-2/\alpha}$ is in agreement with (5.344); the straight dashed lines show $-B_\alpha(k\beta_\alpha)^{1-2/\alpha}$ with the values B_α as given in the bottom row of Table 3.1. With increasing power α, the exponent $B_\alpha\,(k\beta_\alpha)^{1-2/\alpha}$ describing the high-energy behaviour of $|R|$ approaches the exponent $-2bk$ describing its low-energy behaviour, see the corresponding entries in the last column of Table 3.1. Thus the low- and high-energy behaviour of $|R|$ merges into a single exponential form for single-power tails (3.63) with large power α,

$$|R| \overset{\alpha\to\infty}{\sim} \mathrm{e}^{-2\pi k\beta_\alpha/\alpha}\,. \tag{5.346}$$

for all energies.

For the phase of the quantum reflection amplitude, the near-threshold behaviour follows from (5.332),

$$\arg(R) \overset{k\to 0}{\sim} \pi - 2k\bar{a}\,. \tag{5.347}$$

The mean scattering length $\bar{a}$ is only defined for potentials falling off faster than $1/r^3$ at large distances. For a potential proportional to $-1/r^3$,

$$V_3^{(-)}(r) = -\frac{C_3}{r^3} = -\frac{\hbar^2}{2\mu}\frac{\beta_3}{r^3}\,, \tag{5.348}$$

the inward travelling wave is proportional to $H_1^{(1)}(\zeta)/\zeta$ with $\zeta = 2\sqrt{\beta_3/r}$ in the semiclassical region $r \to 0$, and matching to the asymptotic waves (5.330) gives

$$\arg(R) \overset{k\to 0}{\sim} \pi - 2k\beta_3\ln(k\beta_3)\,. \tag{5.349}$$

Note that the formula (5.335) for the near-threshold behavior of $|R|$ holds for all potentials falling off faster than $-1/r^2$, even for those such as (5.348), where the phase of the reflection amplitude diverges at threshold.

Looking at the high-energy limit, the phase of the right-hand side of (5.341) depends more sensitively on the choice of lower integration limit, which is not specified in [PS58, PU58]. The k-dependence of the integral in the exponent is determined by the complex classical turning point (5.342), $r_\mathrm{t} = r_E[\cos(\pi/\alpha) + \mathrm{i}\,\sin(\pi/\alpha)]$. If we assume that the real part of the integral becomes proportional to $\hbar k \times \Re(r_\mathrm{t}) = \hbar k r_E\cos(\pi/\alpha)$ for large k, then the high-energy behaviour of the phase of the reflection amplitude is

$$\arg R \overset{k\to\infty}{\sim} c - c_0 k r_E = c - c_0(k\beta_\alpha)^{1-2/\alpha} \tag{5.350}$$

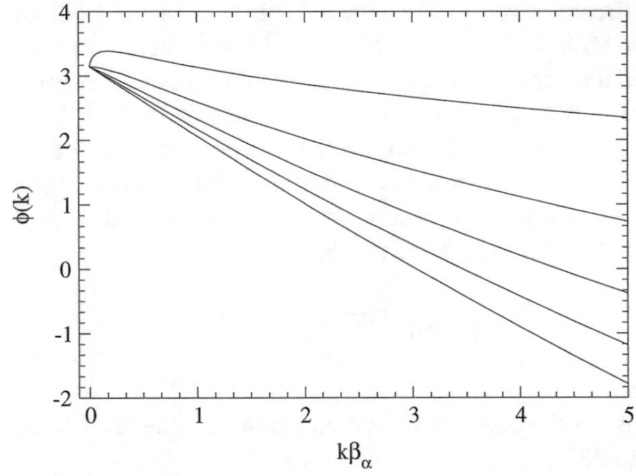

Fig. 5.43 Phase $\phi = \arg R$ of the quantum reflection amplitude for a homogeneous attractive potential (3.63) as function of $k\beta_\alpha$. From top to bottom the *curves* show the results for α = 3, 4, 5, 6 and 7 (from [FJ04])

with real constants c, c_0. This conjecture is supported by numerical calculations as demonstrated in [FJ04], see Fig. 5.43.

The energy dependence of the phase of the reflection amplitude can be related to the time gain or delay of a wave packet during reflection as described in Sect. 1.4.3. If the momentum distribution of the incoming wave packet is sharply peaked around a mean momentum $\hbar k_0$, then the shape of the reflected wave packet is essentially the same as for the incident wave packet. The derivative of $\arg[R(k)]$ with respect to k, taken at k_0, describes an apparent shift Δr in the point of reflection,

$$\Delta r = -\frac{1}{2}\frac{\mathrm{d}}{\mathrm{d}k}[\arg(R)]_{k=k_0}\,. \tag{5.351}$$

[Note that this is $-1/2$ times the shift (1.201) which corresponded to twice the distance by which the apparent point of reflection lay behind the origin.] The time evolution of the reflected wave packet corresponds to reflection of a free wave at the point $r = \Delta r$ rather than at $r = 0$. For a free particle moving with the constant velocity $v_0 = \hbar k_0/\mu$ this implies a time gain [cf. (1.202)]

$$\begin{aligned}\Delta t = \frac{2\Delta r}{v_0} &= -\frac{\mu}{\hbar k_0}\frac{\mathrm{d}}{\mathrm{d}k}[\arg(R)]_{k=k_0}\\ &= -\hbar\frac{\mathrm{d}}{\mathrm{d}E}[\arg(R)]_{E=\hbar^2k_0^2/(2\mu)}\,.\end{aligned} \tag{5.352}$$

For a positive (negative) value of Δr the reflected wave packet thus experiences a time gain (delay) relative to a free particle (with the same asymptotic velocity v_0) travelling to $r = 0$ and back. Note however, that the classical particle moving under the accelerating influence of the attractive potential is faster than the free particle; the quantum reflected wave packet may experience a time gain with respect to a free particle but nevertheless be delayed relative to the classical particle moving in the same potential (see (5.355) and Fig. 5.45 below).

Equation (5.347) implies that the near-threshold behaviour of the space shift (5.351) and of the time shift (5.352) is

$$\Delta r \overset{k_0 \to 0}{\sim} \bar{a}\,, \quad \Delta t \overset{k_0 \to 0}{\sim} \frac{2\mu}{\hbar k_0}\bar{a}\,. \tag{5.353}$$

The near-threshold behaviour of the time shift due to reflection for a wave packet with a narrow momentum distribution is determined by the mean scattering length $\bar{a}$. Near threshold, the quantum reflected wave packet evolves as for a free particle reflected at $r = \bar{a}$.

For energies above the near-threshold region, analytical solutions of the Schrödinger equation are not available (except for $\alpha = 4$), and the reflection amplitudes have to be obtained numerically. Equation (5.350) implies that the space shift (5.351) is given for large energies by

$$\Delta r \overset{k_0 \to \infty}{\sim} \frac{c_0}{2}\left(1 - \frac{2}{\alpha}\right) r_E\,. \tag{5.354}$$

The space shifts (5.351) obtained from the numerical solutions of the Schrödinger equation are plotted in Fig. 5.44 as functions of $k\beta_\alpha$ for α =3, 4, 5, 6 and 7. Except for $\alpha = 3$ and values of $k\beta_3$ less than about 0.15, the space shifts are always positive: according to (5.352) this corresponds to time gains relative to the

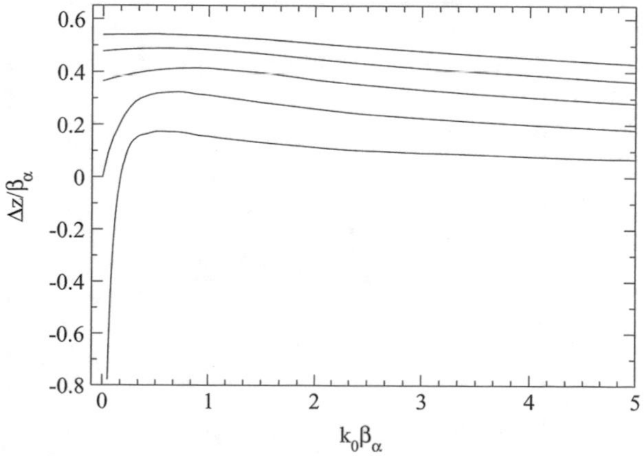

Fig. 5.44 Space shift (5.351) for quantum reflection by the single-power potential (3.63) as function of $k_0\beta_\alpha$. From bottom to top the *curves* show the results for $\alpha = 3$, 4, 5, 6 and 7 (from [FJ04])

free particle reflected at $z = 0$. For $\alpha = 3$ and energies close to threshold there are significant time delays. Note, however, that the classical particle accelerated under the influence of the attractive potential is faster than the free particle [with the same asymptotic velocity $v_0 = \hbar k_0/\mu$], and its time gain is

$$(\Delta t)_{\rm cl} = 2\mu \int_0^\infty \left(\frac{1}{\hbar k_0} - \frac{1}{p(r)} \right) {\rm d}r = \frac{2\mu}{\hbar k_0}\,\tau(\alpha)\, r_E\,, \tag{5.355}$$

where $\tau(\alpha)$ depends only on α

$$\tau(\alpha) = \frac{1}{\sqrt{\pi}} \Gamma\left(\frac{1}{2} + \frac{1}{\alpha}\right) \Gamma\left(1 - \frac{1}{\alpha}\right). \tag{5.356}$$

Numerical values of $\tau(\alpha)$ are given in Table 5.4.

The time gain (5.355) corresponds to the space shift

$$(\Delta r)_{\rm cl} = \frac{v_0 (\Delta t)_{\rm cl}}{2} = \tau(\alpha) r_E\,; \tag{5.357}$$

the classical particle which is accelerated in the potential and reflected at $r = 0$ eventually returns at the same time as a free particle reflected at $(\Delta r)_{\rm cl}$. The classical space shifts (5.357) are generally larger than the space shifts of the quantum reflected wave, as illustrated in Fig. 5.45 for the example $\alpha = 4$. At high energies

Table 5.4 Numerical values of $\tau(\alpha)$ as defined in (5.356)

α	3	4	5	6	7	8	$\alpha \to \infty$
$\tau(\alpha)$	0.862370	0.847213	0.852623	0.862370	0.872491	0.881900	1

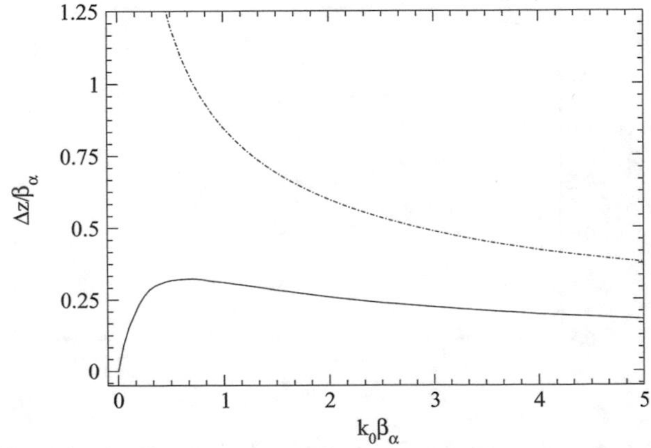

Fig. 5.45 Space shift (5.351) for quantum reflection by the single-power potential (3.63) with $\alpha = 4$ as function of $k_0\beta_4$. The *solid line* shows the space shift of the quantum reflected wave whereas the *dot-dashed line* shows the classical space shift (5.354) (from [FJ04])

both the classical space shifts (5.357) and the quantum space shift (5.354) show the same dependence on $k_0\beta_\alpha$, i.e., proportionality to r_E, but the coefficient $\tau(\alpha)$ in the classical case is larger than the corresponding coefficient in the quantum case. At small energies, the classical space shift diverges as r_E, see (5.343), whereas the quantum space shift remains bounded by a positive distance of the order of the quantum length β_α, see Figs. 5.44, 5.45. Although the quantum reflected wave may experience a time gain relative to the free particle reflected at $r = 0$, it is always delayed relative to the classical particle which is accelerated in the attractive potential [FJ04].

5.7.3.1 Rephrasing (5.217) in Terms of the Amplitudes for Transmission and Quantum Reflection

Equation (5.217) in Sect. 5.5.1 contains three tail functions, A_s/A_c, ϕ_s and ϕ_c, which are defined by the low-r behaviour (5.209) of the solutions $u_s(r)$ and $u_c(r)$ of the radial Schrödinger equation (5.205); the solutions u_s and u_c of (5.205) are defined by their asymptotic behaviour (5.206).

As an alternative choice, the parameters of quantum reflection by the nonclassical part the reference potential $V_{\rm tail}(r)$ can also serve as appropriate tail functions to describe the influence of $V_{\rm tail}(r)$ on the scattering phase shifts [MK11a]. To see this, consider the solution $u_{\rm inc}(r)$ of (5.205) which obeys incoming boundary conditions (5.325) for $r \to 0$ and behaves as (5.326) for $r \to \infty$. In terms of the solutions $u_s(r)$ and $u_s(r)$, with the asymptotic behaviour (5.206) we have

$$u_{\rm inc}(r) = -\frac{\rm i}{\sqrt{\hbar k}}(1-R)u_s(r) + \frac{1}{\sqrt{\hbar k}}(1+R)u_c(r) \,. \tag{5.358}$$

From (5.209) the small-r behaviour of this wave function is

$$\begin{aligned} u_{\rm inc}(r) \overset{r\to 0}{\sim} & \frac{{\rm e}^{-{\rm i}I}}{2\sqrt{\hbar k\, p_{\rm tail}(E;r)}}\left[(1-R)A_s{\rm e}^{{\rm i}\phi_s} + (1+R)A_c{\rm e}^{{\rm i}\phi_c}\right] \\ & + \frac{{\rm e}^{+{\rm i}I}}{2\sqrt{\hbar k\, p_{\rm tail}(E;r)}}\left[(1+R)A_c{\rm e}^{-{\rm i}\phi_c} - (1-R)A_s{\rm e}^{-{\rm i}\phi_s}\right] \,, \end{aligned} \tag{5.359}$$

with $I = \frac{1}{\hbar}\int_{r_E}^{r} p_{\rm tail}(E;r')\,{\rm d}r'$. Since $u_{\rm inc}(r)$ is required to obey incoming boundary conditions for $r \to 0$, the content of the square bracket in the lower line of (5.359) must vanish,

$$(1+R)A_c{\rm e}^{-{\rm i}\phi_c} = (1-R)A_s{\rm e}^{-{\rm i}\phi_s} \,. \tag{5.360}$$

The quotient A_s/A_c of the real and positive amplitudes defined by (5.209) is thus related to the quantum reflection amplitude R by

$$\frac{A_s}{A_c} = \left|\frac{1+R}{1-R}\right| . \tag{5.361}$$

The phase of the square bracket on the right-hand side of the upper line of (5.359) can be deduced by exploiting (5.360) to replace either $(1-R)A_s$ by $(1+R)A_c e^{i(\phi_c-\phi_s)}$ or $(1+R)A_c$ by $(1-R)A_s e^{i(\phi_s-\phi_c)}$. This phase represents the argument of the transmission coefficient T as defined by (5.325), provided that the lower limit r_0 in the action integral is taken as r_E. With this definition of T,

$$\arg T = \phi_s + \arg(1+R) = \phi_c + \arg(1-R) . \tag{5.362}$$

In terms of the amplitudes for reflection by and transmission through the nonclassical region of the reference potential $V_{tail}(r)$, (5.217) reads

$$\tan\delta_0 = \left|\frac{1+R}{1-R}\right| \frac{\sin\left([\upsilon_D - f_{sr}(E)]\pi - \xi + \arg T - \arg(1+R)\right)}{\cos\left([\upsilon_D - f_{sr}(E)]\pi - \xi + \arg T - \arg(1-R)\right)} . \tag{5.363}$$

5.7.3.2 Observation of Quantum Reflection

Quantum reflection is observable in collisions of ultracold atoms with surfaces. At *large* distances, the projectile interacts with a plane surface via electrostatic van der Waals forces, which are modified at *very large* distances due to retardation [CP48]. Such "Casimir-Polder potentials" have all the properties assumed for the reference potential $V_{tail}(r)$ in this section. Due to translational invariance parallel to the surface, the motion normal to the surface is decoupled from the parallel motion, and it is governed by a one-dimensional Schrödinger equation equivalent to the s-wave radial equation of scattering in three-dimensional space. Very low normal velocities can be achieved with grazing incidence of very slow projectiles. Atoms which are transmitted through the nonclassical region of the potential are accelerated towards the surface and are likely to transfer at least some small fraction of their kinetic energy to the surface, which leads to trapping of the atom at the surface if its total energy falls below zero. Such "sticking" is classically expected to become dominant at very low velocities, but early experiments with liquid helium surfaces indicated a suppression of sticking probabilities towards threshold, which was confirmed in quantum mechanical calculations [Bre80, BB82]. The quenched sticking probabilities are due to quantum reflection in the potential tail, whereby only a fraction of the incident atoms actually penetrates through to the deep attractive part of the atom-surface potential [CK92, YD93]. Quantitative measurements of quantum reflection probabilities for ultracold atoms scattering off solid surfaces have since been performed by several groups, e.g. [Shi01, DD03, PS04], and the growing activity in the field of ultracold atoms and molecules has drawn particular attention to this phenomenon [CH96, MH01, DM03, CS03, FJ04, OK05, MF07, ZM11].

As discussed in Sect. 5.7.1, the van der Waals potential for a neutral atom at a distance z from and a plane conducting or dielectric surface is $-C_3/z^3$, but at very large distances it becomes equal to $-C_4/z^4$ due to retardation effects [CP48]. The quotient $L = C_4/C_3$ has the dimensions of a length and roughly defines a transition range separating the nonretarded van der Waals regime $z \ll L$ from the highly retarded regime $z \gg L$. At very small distances of a few atomic units or so, the atom-surface potential is rather complicated, but this "close region" is not important when considering quantum reflection with incoming boundary conditions. Beyond the close region, the singular, attractive atom-surface potential can be written as $V_{\rm tail}(r) = (C_3/z^3)\,\upsilon(z/L)$, cf. (5.303). The shape function $\upsilon(x)$ interpolates between the $-C_3/z^3$ behaviour for $z \ll L$ and the $-C_4/z^4$ behaviour for $z \gg L$.

In order to explain the quantum reflection probabilities that he observed in his pioneering experiments involving metastable neon atoms and solid surfaces, Shimizu [Shi01] modelled the atom-surface potential with a very simple shape function,

$$\upsilon_1(x) = \frac{1}{1+x} \Longrightarrow V_{\rm tail}(z) = -\frac{\hbar^2}{2\mu}\left[\frac{z}{\beta_3} + \frac{z^2}{(\beta_4)^2}\right]^{-1} . \tag{5.364}$$

The lengths β_3 and β_4 are the quantum lengths for the single-power forms (3.63), which the potential (5.364) approaches in the limits $z \to 0$ and $z \to \infty$, respectively. An alternative interpolation is guided by the exact potential for a hydrogen atom interacting with a perfectly conducting surface, which was calculated numerically in [MD97] and is well approximated by the rational function (5.307), the shape function in this case is

$$\upsilon_{\rm H}(x) = \frac{1+\xi x}{1+x+\xi x^2} , \quad \xi = 0.31608 . \tag{5.365}$$

As shown in [FJ02], which part of the atom-wall potential dominantly influences quantum reflection depends on the ratio ρ of the quantum lengths characterizing the single-power limits at small and large distances,

$$\rho = \frac{\sqrt{2\mu}\, C_3}{\hbar\sqrt{C_4}} = \frac{\beta_3}{\beta_4} , \tag{5.366}$$

see last row of Table 5.3 in Sect. 5.7.1. For $\rho < 1$, the energy dependence of $|R|$ is largely determined by the nonretarded van der Waals part of the potential; for $\rho > 1$, the retarded $-C_4/z^4$ part is dominant. Thus the smaller of the two quantum lengths is the one belonging to the dominant term. This observation may be counter-intuitive, but it is understandable when looking at the expression for the atom-wall potential that is given on the far right of (5.364).

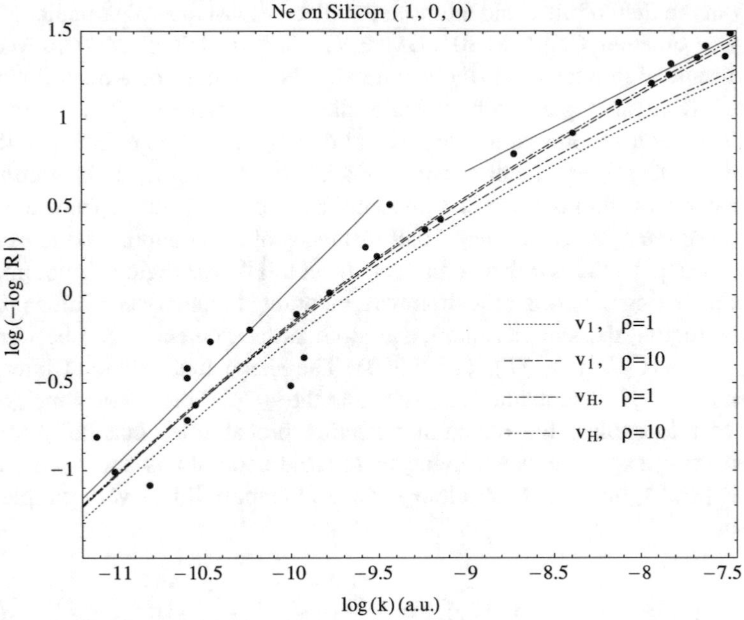

Fig. 5.46 Modulus of the quantum reflection amplitude, as observed in the scattering of metastable neon atoms off a silicon surface [Shi01]. The figure shows $\ln(-\ln|R|)$ as function of $\ln(k)$ (natural logarithms) with k measured in atomic units, i.e. in units of the inverse Bohr radius. The *curves* show the results obtained by numerically solving the Schrödinger equation (5.205) with potentials (5.303) constructed with the shape functions υ_1 and $\upsilon_{\rm H}$. The quantum length β_4 associated with the strength C_4 of the potential in the highly retarded limit is $\beta_4 = 11\,400$ a.u. in all cases. For the $-C_3/z^3$ van der Waals limit of the potential, the quantum length is $\beta_3 = 11\,400$ a.u. for $\rho = 1$ and $\beta_3 = 114\,000$ a.u. for $\rho = 10$. The *straight red line* in the bottom left corner shows the behaviour $\ln|R| \sim -2\,\beta_4 k$ expected in the low-k regime. The *straight red line* in the top right corner shows the behaviour $\ln|R| \propto -\sqrt{\beta_4 k}$ expected in the high-k regime for a single-power $1/z^4$ potential (From [FJ02])

The transition from the leading linear behaviour (5.335) of $|R|$ near threshold to the high-k behaviour (5.344) can be exposed by studying $\ln(-\ln|R|)$ as a function of $\ln k$,

$$|R| = \mathrm{e}^{-B\,k^C} \quad \Longrightarrow \quad \ln(-\ln|R|) = \ln(B) + C\,\ln(k)\,. \tag{5.367}$$

A plot of $\ln(-\ln|R|)$ against $\ln(k)$ is shown in Fig. 5.46 for the quantum reflection of metastable neon atoms by a silicon surface, as studied by Shimizu in [Shi01]. The dots are the experimental data and the curves are the results obtained by numerically solving the Schrödinger equation (5.205) with potentials (5.303) constructed with the shape functions υ_1 (5.364) and $\upsilon_{\rm H}$ (5.365). The quantum length corresponding to the highly retarded $-C_4/z^4$ part of the potential was $\beta_4 = 11\,400$ a.u. in all four cases. The value of β_3 was chosen to be equal to β_4, corresponding to $\rho = 1$, or to be ten times larger, corresponding to $\rho = 10$. The straight red line in the bottom left of the figure has unit gradient, corresponding to the universal near-threshold behaviour (5.335). The results obtained with all potentials in Fig. 5.46 approach such behaviour in the low-k limit, and the data are consistent, albeit with a very large scatter. Towards large k, the gradients of the curves in Fig. 5.46 decrease gradually. The experimental points are well fitted by the two curves with $\rho = 10$, i.e. with $\beta_3 = 114\,000$ a.u.. They are essentially the same for both shape functions, (5.364) and (5.365), and they are also independent of β_3 as long as β_3 is significantly larger than β_4. Essentially the same result is obtained with a single-power $-1/z^4$ potential with the appropriate quantum length $\beta_4 = 11\,400$ a.u. The straight red line in the top right corner of the figure shows the large-k behaviour expected in this case according to (5.344), with $B_\alpha k r_E = B_\alpha (k\beta_4)^{1/2}$; its gradient is $\frac{1}{2}$. In contrast, the large-k behaviour of the two curves with $\rho = 1$ is closer to the expectation of a $-1/z^3$ potential, where the asymptotic gradient is $\frac{1}{3}$. One expects the nonretarded $-1/z^3$ part of the potential at moderate distances to have increasing influence at higher energies, but at the energies where this happens, the quantum reflection yields are very small.

As already pointed out by Shimizu in [Shi01], the highly retarded part of the neon-surface interaction is essentially responsible for quantum reflection observed in the experiment. Also for other atom-wall systems, involving e.g. bosonic alkali atoms, hydrogen or metastable helium, the crucial parameter β_3/β_4 is generally significantly larger than unity [FJ02, FT04]. Quantum reflection is well described on the basis of the highly retarded, single-power $-1/z^4$ potential in all these cases.

Druzhinina and DeKieviet [DD03] measured the probability for quantum reflection of (ground-state) ^{3}He atoms by a rough quartz surface, and they measured up to energies high enough to detect significant deviations from the results expected for the $-C_4/r^4$ potential alone. Shimizu and collaborators have also investigated reflection from surfaces structured with roof- or wall-like ridges [SF02, OK05, OT05]. This can significantly enhance the reflection probability and be used in the development of atom-optical imaging devices.

A significant advance towards lower temperatures and higher reflection probabilities was reported by Pasquini et al. [PS04], who collided a Bose-Einstein condensate of sodium atoms with a silicon surface at temperatures of the order of nano-Kelvins. Pasquini et al. observed evidence for quantum reflection probabilities above 50% at *normal incidence,* i.e. without resorting to near-grazing angles to reduce the normal component of the incident velocity. The demonstrated possibility of achieving such high quantum reflection probabilities irrespective of angle of incidence raises the question of whether one might base the construction of atom wave guides or traps on the phenomenon quantum reflection alone [Jur05]. Quantum reflection is a very universal and insensitive mechanism independent of auxiliary requirements such as the laser fields needed to generate the repulsive potential of an evanescent-wave mirror. It requires the atoms to be cooled to extremely low temperatures, but the walls of the device need not be cold.

It is worth emphasizing, that all characteristic lengths, including the transition length (5.297) are very large, typically several hundreds or thousands of atomic units (Bohr radii) [FJ02, FT04]. Quantum reflection is generated at really large atom-surface distances. The same applies for the quantum reflection of ultracold molecules, as was impressively demonstrated in a recent experiment by Zhao et al. who scattered helium dimers off a solid diffraction grating at very low energies corresponding to normal incident velocities near 10 cm/s, translating to a kinetic energy near 0.6 neV ($\approx 2 \times 10^{-11}$ a.u.) in the normal direction. The very fragile helium dimer, with a binding energy of only 4×10^{-8} a.u. and a bond length of almost 100 a.u. (Bohr radii), is expected to fragment while being accelerated under the influence of the attractive molecule-surface potential with a well depth near 2×10^{-4} a.u. However, a noticeable fraction of the incident dimers is spared this fate due to quantum reflection, which occurs "tens of nanometers above the actual surface where the … forces are still too feeble to break up even the fragile He_2 bond" [ZM11].

5.7.3.3 Nonplanar Surfaces

For atoms scattering off an absorbing sphere, the radius of the sphere enters as a further length in the problem. As shown in [AF07], the nonclassical region of the potential tail moves to smaller atom-sphere separations r when the radius of the sphere is decreased, but the transition region between nonretarded van der Waals regime and the highly retarded regime is essentially independent of this radius and roughly the same as for an atom in front of a plane surface. The sensitivity of quantum reflection to the nonretarded part of the atom-surface potential thus becomes increasingly noticeable for smaller spheres.

It is interesting to consider the threshold limits of the cross sections for elastic scattering and for absorption of atoms interacting with an absorbing sphere. The electrostatic van der Waals potential is proportional to $1/r^6$, but at very large distances the atom-sphere potential is proportional to $1/r^7$ due to retardation effects [CP48]. Towards threshold, the scattering amplitude is dominated by the s-wave ($l = 0$), and the complex scattering phase shift is determined by the complex scattering length $\boldsymbol{a} = \bar{a} - \mathrm{i}b$. With (4.34) in Sect. 4.1.3 and (5.333), (5.334) above,

$$f(\theta) \overset{k\to 0}{\sim} \frac{1}{k}\boldsymbol{\delta}_0 \overset{k\to 0}{\sim} -\boldsymbol{a} = -\bar{a} + \mathrm{i}b \,. \tag{5.368}$$

The elastic scattering cross section $|f(\theta)|^2$ remains finite, the square of the real scattering length in the nonabsorbing case is simply replaced by the absolute square of the complex scattering length in the presence of absorption,

$$\frac{\mathrm{d}\sigma_{\mathrm{el}}}{\mathrm{d}\Omega} \overset{k\to 0}{\sim} |\boldsymbol{a}|^2 = \bar{a}^2 + b^2 \,, \quad \sigma_{\mathrm{el}} \overset{k\to 0}{\sim} 4\pi(\bar{a}^2 + b^2) \,. \tag{5.369}$$

In contrast, the absorption cross section, as given by (4.171) in Sect. 4.1.14, behaves as follows towards threshold:

$$\sigma_{\rm abs} \overset{k\to 0}{\sim} \frac{\pi}{k^2}\left(1-\left|e^{2i\delta_0}\right|^2\right) \overset{k\to 0}{\sim} \frac{\pi}{k^2}\left(1-|1-2kb-2i\bar{a}k|^2\right) \overset{k\to 0}{\sim} \frac{4\pi b}{k}\,. \tag{5.370}$$

This is consistent with the optical theorem (4.10), according to which

$$\sigma_{\rm tot} = \frac{4\pi}{k}\,\Im[f(\theta=0)] \overset{k\to 0}{\sim} \frac{4\pi}{k}\,\Im[-\bar{a}+ib]\,; \tag{5.371}$$

the total cross section $\sigma_{\rm tot} = \sigma_{\rm el} + \sigma_{\rm abs}$ is dominated towards threshold by the diverging contribution of the absorption cross section (5.370).

The absorption cross section, which is related to the probability for transmission through the nonclassical region of the potential tail, can be used to calculate the rate for a reaction that occurs when projectile and target meet [Dic07]. Since this involves an average over the product of $\sigma_{\rm abs}$ and the asymptotic relative velocity $\hbar k/\mu$, reaction rates following from absorption cross sections that diverge as in (5.370) tend to finite limits at threshold.

The description of an atom interacting with cylinder is more difficult than for an atom interacting with a plane wall or with a sphere. One reason is, that the atom-cylinder interaction is much more complicated, see e.g. [KB06]. Furthermore, due to translational invariance along the direction parallel to the cylinder axis, the scattering problem is actually two-dimensional, and quantum mechanical scattering theory in two dimensions is somewhat more subtle than in the one- and three-dimensional cases, in particular near threshold, see Sect. 4.2. For an atom interacting with a perfectly conducting cylinder, the nonclassical region of the potential tail is not so sensitive to the radius of the cylinder. As in the case of the plane wall, the highly retarded part of the atom-cylinder potential is important for quantum reflection in realistic cases [FE12]. For an atom interacting with a dielectric cylinder, however, the nonretarded part of the interaction is more likely to play a role [Fin13].

5.7.4 *Quantum Reflection and Near-Threshold Quantization in Two Spatial Dimensions*

A theory in two spatial dimensions is needed to describe a system restricted to a plane, or as in the atom-cylinder example mentioned above, for a higher-dimensional system in which degrees of freedom beyond two decouple from the two degrees in focus. Scattering theory in two dimensions was discussed in detail in Sect. 4.2, with special emphasis on the case of s-waves in 2D, see Sect. 4.2.5. This section complements the discussion of Sect. 4.2 by summarizing the features introduced through two-dimensionality for quantum reflection and near-threshold quantization.

For a reference potential $V_{\rm tail}(r)$, which is attractive and more singular than $1/r^2$ at short distances and falls off faster than $1/r^2$ at large distances, the radial Schrödinger equation for s-waves in two dimensions (4.210),

$$\left[-\frac{\hbar^2}{2\mu}\frac{\mathrm{d}^2}{\mathrm{d}r^2}-\frac{1}{4}\frac{\hbar^2}{2\mu\,r^2}+V_{\rm tail}(r)\right]u_{m=0}(r)=E\,u_{m=0}(r)\,, \tag{5.372}$$

can be solved with incoming boundary conditions, which describes absorption in the close region $r \to 0$. At large distances, the radial wave function still has the form given in the bottom line of (4.200), but the phase shift is now complex. With $m = 0$,

$$u(r) \overset{r\to\infty}{\propto} \mathrm{e}^{-\mathrm{i}kr}-\mathrm{i}\,\mathrm{e}^{2\mathrm{i}\boldsymbol{\delta}}\mathrm{e}^{\mathrm{i}kr} \propto \mathrm{e}^{-\mathrm{i}\left(kr+\frac{\pi}{4}\right)}-\mathrm{e}^{2\mathrm{i}\boldsymbol{\delta}}\mathrm{e}^{\mathrm{i}\left(kr+\frac{\pi}{4}\right)}\,. \tag{5.373}$$

The right-hand side(s) of (5.373) represent an incoming radial wave together with an outgoing radial wave, which is generated by quantum reflection in the nonclassical part of coordinate space. Defining the coefficient of $\mathrm{e}^{\mathrm{i}(kr+\pi/4)}$ as the quantum reflection amplitude gives

$$R=-\mathrm{e}^{2\mathrm{i}\boldsymbol{\delta}}\,, \tag{5.374}$$

similar to (5.327) for s-waves in 3D.

The leading near-threshold behaviour of the complex phase shift $\boldsymbol{\delta}$ is given by a formula similar to (4.214), except that the real scattering length a is replaced by a complex scattering length $\boldsymbol{a}$, which is defined through the zero-energy solution $u^{(0)}(r)$ of (5.372) obeying incoming boundary conditions for $r \to 0$:

$$u^{(0)}(r) \overset{r\to\infty}{\propto} -\sqrt{r}\,\ln\left(\frac{r}{\boldsymbol{a}}\right)=-\sqrt{r}\,\ln\left(\frac{r}{|\boldsymbol{a}|}\right)+\sqrt{r}\,\mathrm{i}\arg(\boldsymbol{a})\,. \tag{5.375}$$

For the complex phase shift $\boldsymbol{\delta}$ we have

$$\cot\boldsymbol{\delta} \overset{k\to 0}{\sim} \frac{2}{\pi}\left[\ln\left(\frac{k\boldsymbol{a}}{2}\right)+\gamma_{\rm E}\right], \tag{5.376}$$

which, for the quantum reflection amplitude (5.374), implies

$$R \overset{k\to 0}{\sim} -1-\frac{\mathrm{i}\,\pi}{\ln\left(\frac{k\boldsymbol{a}}{2}\right)+\gamma_{\rm E}+\mathrm{i}\left(\arg(\boldsymbol{a})-\frac{\pi}{2}\right)}\,. \tag{5.377}$$

The results (5.376) and (5.377) are derived in [AF08], where further terms up to and including $O(k^2)$ are also given. (Note that the quantum reflection amplitude in [AF08] is i times the amplitude R defined above.)

For *near-threshold quantization* in a deep potential which is well described at large distances by the singular reference potential $V_{\rm tail}(r)$, the quantization rule

$\upsilon_D - \upsilon = F(E)$ is determined by the quantization function $F(E)$, and the universal near-threshold behaviour of this quantization function for s-states in 2D is

$$F(E) \overset{\kappa\to 0}{\sim} \frac{1}{\pi} \arctan\left(\frac{\arg \boldsymbol{a}}{\ln\left(\frac{k|\boldsymbol{a}|}{2}\right) + \gamma_E}\right) + O\left(\kappa^2\right) . \tag{5.378}$$

The complex scattering length $\boldsymbol{a}$ is as defined in (5.375), and it is a property of the reference potential $V_{\text{tail}}(r)$. The relation connecting the threshold quantum number υ_D with the scattering length a reads

$$a = |\boldsymbol{a}| \exp\left(-\frac{\arg(\boldsymbol{a})}{\tan(\upsilon_D \pi)}\right) , \tag{5.379}$$

so, for a bound-sate energy $E_b = -\hbar^2\kappa_b^2/(2\mu)$ very close to threshold,

$$a \overset{\kappa_b\to 0}{\sim} \frac{2\,\exp(-\gamma_E)}{\kappa_b} + O(\kappa_b) . \tag{5.380}$$

For further details, see [AF08].

Problems

5.1 Consider an atom of radius $n^2 a_0$, a_0 being the Bohr radius. Give an estimate for the power in W/cm^2 which a laser must have, if the electromagnetic field energy in the volume occupied by the atom is to be roughly as big as the binding energy $\mathcal{R}/n^2$ ($\mathcal{R}$ is the Rydberg energy).

5.2

a) Consider a free particle of mass μ in one spatial dimension. At time $t = 0$ it is described by a minimal Gaussian wave packet of width β moving with the mean velocity $v_0 = \hbar k_0/\mu$ in the direction of the positive x-axis,

$$\psi(x, t = 0) = (\sqrt{\pi}\beta)^{-1/2}\, e^{-x^2/(2\beta^2)} e^{ik_0 x} .$$

Calculate the wave functions $\psi(x, t)$ in coordinate space and $\tilde{\psi}(p, t)$ in momentum space as well as the associated probability densities $|\psi(x, t)|^2$ and $|\tilde{\psi}(p, t)|^2$ at a later time t.

b) Calculate the density matrix ϱ and the Wigner function ϱ_W for the pure state described by the wave function $\psi(x, t)$ in a).

c) Classically the free particle may be described at time $t = 0$ by an initial phase space density with finite uncertainty in position and momentum,

$$\varrho_{\rm cl}(x, p; t = 0) = \frac{1}{\alpha\beta\pi}\,\mathrm{e}^{-x^2/\beta^2}\,\mathrm{e}^{-(p-p_0)^2/\alpha^2}\,.$$

Use the classical trajectories $p(t) = p(0), \quad x(t) = x(0) + (p/\mu)t$ and the form (5.29) of the Liouville equation,

$$\frac{\mathrm{d}}{\mathrm{d}t}\varrho_{\rm cl}(x(t), p(t); t) = 0\,,$$

to calculate the phase space density at a later time t. Compare the resulting probability densities in position and momentum with the quantum mechanical results.

5.3

a) Show that the coherent state (5.61) is an eigenstate of the quantum annihilation operator with eigenvalue z^*,

$$\hat{b}\,|z\rangle = z^*|\,z\rangle\,,$$

and use this result to calculate the expectation value of the number operator $\hat{b}^\dagger\hat{b}$.

b) Use (5.73) to calculate the time average of the expectation value of the energy $(\hat{\boldsymbol{E}}^2 + \hat{\boldsymbol{B}}^2)L^3/(8\pi)$ of a monochromatic field in the coherent state $|z\rangle = \left|z_0\mathrm{e}^{\mathrm{i}\omega t}\right\rangle$. Compare the result with the quantity $\hbar\omega\langle z|\hat{b}^\dagger\hat{b} + 1/2|z\rangle$ following from a).

c) Calculate the Wigner function (5.41) for the ground state (5.52) of the one-dimensional harmonic oscillator and for the first excited state,

$$\psi_1(x) = \hat{b}^\dagger\psi_0(x) = (\beta\sqrt{\pi})^{-1/2}\frac{2x}{\beta\sqrt{2}}\,\mathrm{e}^{-x^2/(2\beta^2)}\,.$$

5.4 Verify the special form (5.65) of the Baker-Campbell-Hausdorff relation for two operators $\hat{A}$ and $\hat{B}$, which both commute with their commutator, $[\hat{A}, [\hat{A}, \hat{B}]] = [\hat{B}, [\hat{A}, \hat{B}]] = 0$,

$$\mathrm{e}^{\hat{A}+\hat{B}} = \mathrm{e}^{\hat{A}}\mathrm{e}^{\hat{B}}\mathrm{e}^{-[\hat{A},\hat{B}]/2}\,.$$

Hint: Study the derivative of the function $\hat{f}(\lambda) = \mathrm{e}^{\lambda\hat{A}}\mathrm{e}^{\lambda\hat{B}}$ with respect to λ.

5.5 A photon (rest mass zero) behaves like a particle with energy $E = \hbar\omega$ and momentum $p = \hbar\omega/c$. Show that a free electron cannot absorb or emit a photon without violating energy and momentum conservation.

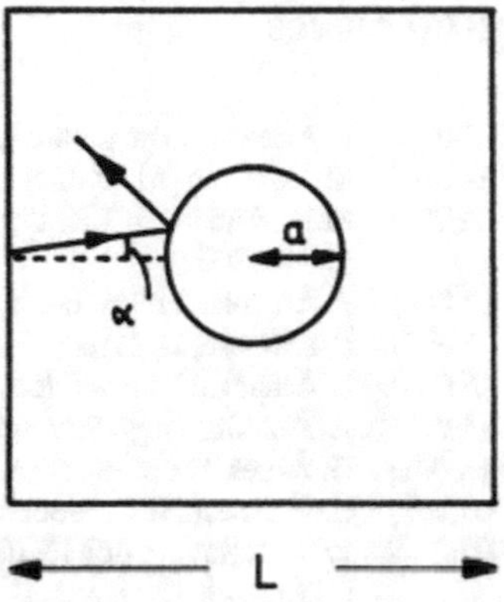

Fig. 5.47 Realization of *Sinai's billiard* [Sin70]. The parameters in Problem 5.7 were chosen to correspond roughly to the dimensions in real billiards

5.6 Show that the stability matrix defined by (5.75) for motion following a classical trajectory $x(t)$ obeys a chain rule of the form

$$\mathbf{M}(t_2, t_0) = \mathbf{M}(t_2, t_1)\mathbf{M}(t_1, t_0)\,,$$

and conclude that the Liapunov exponent defined by (5.77) is the same for all phase space points on the trajectory.

Hint: Matrix norms fulfill the triangle inequality

$$\|\mathbf{M}_1\mathbf{M}_2\| \leq \|\mathbf{M}_1\| \cdot \|\mathbf{M}_2\|\,.$$

5.7 Consider a square of length L. In the centre of the square there is a circular disc of radius a. A point particle travels from the middle of one side of the square towards the disc at an angle α (see Fig. 5.47). It is reflected by the sides of the square and the edge of the disc. Determine the direction of motion of the particle after up to five collisions with the disc for $L = 2$ m, $a = 5$ cm and an initial angle of $\alpha = 0.3°$, $0.0003°$, $0.0000003°$, $0.0000000003°$. (Follow the trajectory only as long as collisions with the initial side of the square and the disc alternate.) Estimate the Liapunov exponent for the periodic orbit $\alpha = 0$.

5.8 Start with a number x from a randomly distributed set of numbers (Poisson spectrum) and choose N further numbers y in the interval $x < y < x + L$. How big is the probability that none of the numbers y lies in the interval $(x, x + s)$? Consider the limit $N \to \infty$, $L \to \infty$ at constant level density $d = N/L$, and show that the probability density $P(s)$ for a nearest neighbour spacing s is given by

$$P(s) = d\,\mathrm{e}^{-ds}\,.$$

References

[AB91] F. Arickx, J. Broeckhove, P. Van Leuven, Phys. Rev. A **43**, 1211 (1991)

[AB04] M. Al-Amri, M. Babiker, Phys. Rev. A **69**, 065801 (2004)

[AE95] M.H. Anderson, J.R. Ensher, M.R. Matthews, C.E. Wieman, E.A. Cornell, Science **269**, 198 (1995)

[AF06] F. Arnecke, H. Friedrich, J. Madroñero, Phys. Rev. A **74**, 062702 (2006)

[AF07] F. Arnecke, H. Friedrich, J. Madroñero, Phys. Rev. A **75**, 042903 (2007)

[AF08] F. Arnecke, H. Friedrich, P. Raab, Phys. Rev. A **78**, 052711 (2008)

[Ash78] A. Ashkin, Phys. Rev. Lett. **40**, 729 (1978)

[AZ91] G. Alber, P. Zoller, Phys. Reports **199**, 231 (1991)

[BA01] C. Boisseau, E. Audouard, J. Vigué, Phys. Rev. Lett. **86**, 2694 (2001)

[Bal99] P. Ball, Nature, doi:10.1038/news991202-2, http://www.nature.com/news/1999/991126/full/news991202-2.html

[Bar74] G. Barton, J. Phys. B **7**, 2134 (1974)

[Bay86] J. Bayfield, in *Fundamental Aspects of Quantum Theory*, ed. by V. Gorini, A. Frigerio (Plenum Press, New York, 1986), p. 183

[BB82] J. Böheim, W. Brenig, J. Stutzki, Z. Phys. B **48**, 43 (1982); Erratum: Z. Phys. B **49**, 362 (1982)

[BB97a] M. Brack, R.K. Bhaduri, *Semiclassical Physics* (Addison-Wesley, Reading, 1997)

[BB97b] J. Blaschke, M. Brack, Phys. Rev. A **56**, 182 (1997)

[BD02] A. Buchleitner, D. Delande, J. Zakrzewski, Phys. Reports **368**, 409 (2002)

[BE96] K. Burnett, M. Edwards, C.W. Clark (eds.), J. Res. Natl. Inst. Stand. Technol. **101** (1996). Special Issue on Bose-Einstein Condensation

[Ber82] M.V. Berry, J. Phys. A **15**, 3693 (1982)

[BG77] J. Bayfield, L.D. Gardner, P.M. Koch, Phys. Rev. Lett. **39**, **76** (1977)

[BG84] O. Bohigas, M.-J. Giannoni, in *Mathematical and Computational Methods in Nuclear Physics,* Lecture Notes in Physics, vol. 209, ed. by J.S. Dehesa, J.M.G. Gomez, A. Polls (Springer, Berlin, Heidelberg, New York, 1984)

[BG02] W. Becker, F. Grasborn, R. Kopold, D.B. Milošević, G.G. Paulus, H. Walther, Adv. At. Mol. Opt. Phys. **48**, 35 (2002)

[BH85] O. Bohigas, R.U. Haq, A. Panday, Phys. Rev. Lett. **54**, 1645 (1985)

[BK74] J. Bayfield, P.M. Koch, Phys. Rev. Lett. **33**, 258 (1974)

[BK04] J.F. Babb, G.L. Klimchitskaya, V.M. Mostepanenko, Phys. Rev. A **70**, 042901 (2004)

[Blo04] I. Bloch, Phys. World **17**(4), 25 (2004)

[Blu97] R. Blümel, in *Classical, Semiclassical and Quantum Dynamics in Atoms*, Lecture Notes in Physics, vol. 485, ed. by H. Friedrich, B. Eckhardt (Springer, Berlin, 1997), p. 154

[BM89] W. Becker, J.K. McIver, M. Confer, Phys. Rev. A **40**, 6904 (1989)

[BM99] T. Bartsch, J. Main, G. Wunner, Ann. Phys. (NY) **277**, 19 (1999)

[Bor25] M. Born, *Vorlesungen über Atommechanik* (Springer, Berlin, 1925)

[BQ97] J. Burgdörfer, Y. Qiu and J. Müller, in *Classical, Semiclassical and Quantum Dynamics in Atoms,* Lecture Notes in Physics, vol. 485, ed. by H. Friedrich, B. Eckhardt (Springer, Berlin, 1997), p. 304

[BR09] R. Blümel, W.P. Reinhardt, *Chaos in Atomic Physics* (Cambridge University Press, Cambridge, 2009)

[Bre80] W. Brenig, Z. Phys. B **36**, 227 (1980)

[BS77] H.A. Bethe, E. Salpeter, *Quantum Mechanics of One- and Two- Electron Atoms* (Plenum Publishing, New York, 1977)

[Buc89] P.H. Bucksbaum, in *Atomic Spectra and Collisions in External Fields*, ed. by K.T. Taylor, M.H. Nayfeh, C.W. Clark (Plenum Press, New York, 1989), p. 359

[Cas90] G. Casati, in *Atoms in Strong Fields,* ed. by C.A. Nicolaides, C.W. Clark, M.H. Nayfeh (Plenum Press, New York, 1990), p. 231

[CC87] G. Casati, B.V. Chirikov, I. Guarneri, D.L. Shepelyansky, Phys. Reports **154**, 77 (1987)

[CC00] S.L. Cornish, N.R. Claussen, J.L. Roberts, E.A. Cornell, C.E. Wieman, Phys. Rev. Lett. **85**, 1795 (2000)
[CD92] C. Cohen-Tannoudj, J. Dupont-Roc, G. Grynberg, *Atom-Photon Interactions* (Wiley, New York, 1992)
[CE89] P. Cvitanović, B. Eckhardt, Phys. Rev. Lett. **63**, 823 (1989)
[CF86] G. Casati, J. Ford, I. Guarneri, F. Vivaldi, Phys. Rev. **A 34**, 1413 (1986)
[CG10] C. Chin, R. Grimm, P. Julienne, E. Tiesinga, Rev. Mod. Phys. **82**, 1225 (2010)
[Cha92] Chaos **2** (1992), special focus issue on *Periodic Orbit Theory*
[CK92] D.P. Clougherty, W. Kohn, Phys. Rev. B **46**, 4921 (1992)
[CK97] L.S. Cederbaum, K.C. Kulander, N.H. March (eds.), *Atoms and Molecules in Intense Fields*, Structure and Bonding, vol. 86 (Springer, Berlin, 1997)
[CL84] S.L. Chin, P. Lambropoulos (eds.), *Multiphoton Ionization of Atoms* (Academic Press, New York, 1984)
[CH96] R. Côté, E.J. Heller, A. Dalgarno, Phys. Rev. A **53**, 234 (1996)
[CP48] H.B.G. Casimir, D. Polder, Phys. Rev. **73**, 360 (1948)
[Cra87] M. Crance, Phys. Reports **144**, 117 (1987)
[CS98] R. Côté, B. Segev, M.G. Raizen, Phys. Rev. A **58**, 3999 (1998)
[CS03] R. Côté, B. Segev, Phys. Rev. A **67**, 041604(R) (2003)
[CS05] Q. Chen, J. Stajic, S. Tan, K. Levin, Phys. Reports **412**, 1 (2005)
[DD03] V.V. Druzhinina, M. DeKieviet, Phys. Rev. Lett. **91**, 193202 (2003)
[DF00] L.F. DiMauro, R.R. Freeman, K. Kulander (eds.), *Multiphoton Processes Conference* (Springer, New York, 2000)
[DG86] D. Delande, J.-C. Gay, Phys. Rev. Lett. **57**, 2006 (1986)
[DG65] E. Del Giudice, E. Galzenati, Nuovo Cimento **38**, 435 (1965)
[DG97] F. Dalfovo, S. Giorgini, M. Guilleumas, L. Pitaevskii, S. Stringari, Phys. Rev. A **56**, 3840 (1997)
[DH99] R.G. Dall, M.D. Hoogerland, K.G.H. Baldwin, S.J. Buckman, J. Opt. B **1**, 396 (1999)
[Dic07] A.S. Dickinson, J. Phys. B **40**, F237 (2007)
[DK84] J.B. Delos, S.K. Knudson, D.W. Noid, Phys. Rev. A **30**, 1208 (1984)
[DK85] N.B. Delone, V.P. Krainov, *Atoms in Strong Light Fields* (Springer, Berlin, Heidelberg, New York, 1985)
[DK94] N.B. Delone, V.P. Krainov, *Multiphoton Processes in Atoms* (Springer, Berlin, Heidelberg, New York, 1994)
[DL05] J. Dziarmaga, M. Lewenstein, Phys. Rev. Lett. **94**, 090403 (2005)
[DM95a] P.A. Dando, T.S. Monteiro, D. Delande, K.T. Taylor, Phys. Rev. Lett. **74**, 1099 (1995)
[DM95b] K.B. Davis, M.-O. Mewes, M.R. Andrews, N.J. van Druten, D.S. Durfee, D.M. Kurn, W. Ketterle, Phys. Rev. Lett. **75**, 3969 (1995)
[DM03] E.I. Dashevskaya, A.I. Maergoiz, J. Troe, I. Litvin, E.E. Nikitin, J. Chem. Phys. **118**, 7313 (2003)
[DP90] M. Dörr, R.M. Potvliege, R. Shakeshaft, Phys. Rev. A **41**, 558 (1990)
[DP96] F. Dalfovo, L. Pitaevskii, S. Stringari, J. Res. Natl. Inst. Stand. Technol. **101**, 537 (1996)
[DS90] Z. Daĉić Gaeta, C.R. Stroud Jr., Phys. Rev. A **42**, 6308 (1990)
[DS97] J.B. Delos, C.D. Schwieters, in *Classical, Semiclassical and Quantum, Dynamics in Atoms*, Lecture Notes in Physics, vol. 485, ed. by H. Friedrich, B. Eckhardt (Springer, Berlin, 1997), p. 233
[DT06] O. Docenko, M. Tamanis, J. Zaharova, R. Ferber, A. Pashov, H. Knöckel, E. Tiemann, J. Phys. B **39**, S929 (2006)
[DZ97] D. Delande, J. Zakrzewski, in *Classical, Semiclassical and Quantum, Dynamics in Atoms*, Lecture Notes in Physics, vol. 485, ed. by H. Friedrich, B. Eckhardt (Springer, Berlin, 1997), p. 205
[Eck88] B. Eckhardt, Phys. Reports **163**, 205 (1988)
[ED96] M. Edwards, R.J. Dodd, C.W. Clark, K. Burnett, J. Res. Natl. Inst. Stand. Technol. **101**, 553 (1996)
[EF01] C. Eltschka, H. Friedrich, M.J. Moritz, Phys. Rev. Lett. **86**, 2693 (2001)

[EK99] V.A. Ermoshin, A.K. Kazansky, V. Engel, J. Chem. Phys. **111**, 7807 (1999)
[EM00] C. Eltschka, M.J. Moritz, H. Friedrich, J. Phys. B **33**, 4033 (2000)
[ER91] G. Ezra, K. Richter, G. Tanner, D. Wintgen, J. Phys. B **24**, L413 (1991)
[Fai73] F.H.M. Faisal, J. Phys. **6**, L89 (1973)
[Fai86] F.H.M. Faisal, *Theory of Multiphoton Processes* (Plenum Press, New York, London, 1986)
[FE97] H. Friedrich, B. Eckhardt (eds.), *Classical, Semiclassical and Quantum, Dynamics in Atoms*, Lecture Notes in Physics, vol. 485 (Springer, Berlin, 1997)
[FE12] M. Fink, J. Eiglsperger, J. Madroñero, H. Friedrich, Phys. Rev. A **85**, 040702(R) (2012)
[FG99] V.V. Flambaum, G. Gribakin, C. Harabati, Phys. Rev. A **59**, 1998 (1999)
[Fin13] M. Fink, Doctoral thesis, Technical University Munich, 2013. mediatum.ub.tum.de/doc/1141600.pdf
[FJ02] H. Friedrich, G. Jacoby, C.G. Meister, Phys. Rev. A **65**, 032902 (2002)
[FJ04] H. Friedrich, A. Jurisch, Phys. Rev. Lett. **92**, 103202 (2004)
[FJ05] H. Friedrich, A. Jurisch, Phys. Lett. A **335**, 43 (2005)
[FK82] D. Feldmann, J. Krautwald, S.L. Chin, A. von Hellfeld, K.H. Welge, J. Phys B **15**, 1663 (1982)
[FK84] D. Feldmann, J. Krautwald, K.H. Welge, in *Multiphoton Ionization of Atoms*, ed. by S.L. Chin, P. Lambropoulos (Academic Press, New York, 1984), p. 233
[FM05] T. Fabičič, J. Main, T. Bartsch, G. Wunner, J. Phys. B **38**, S219 (2005)
[For01] F. de Fornel, *Evanescent Waves From Newtonian Optics to Atomic Optics* (Springer, Berlin, 2001)
[Fri90] H. Friedrich, in *Atoms in Strong Fields*, ed. by C.A. Nicolaides, C.W. Clark, M. Nayfeh (Plenum Press, New York, 1990), p. 247
[Fri98] H. Friedrich, in *Atoms and Molecules in Strong External Fields*, Proceedings of the 172nd WE-Heraeus Seminar (Bad Honnef, 1997), ed. by P. Schmelcher, W. Schweizer (Plenum Publishing, New York, 1998)
[FT96] H. Friedrich, J. Trost, Phys. Rev. A **54**, 1136 (1996)
[FT04] H. Friedrich, J. Trost, Phys. Reports **397**, 359 (2004)
[FW88] D. Feldmann, B. Wolff, M. Wemhöner, K.H. Welge, in *Multiphoton, Processes*, ed. by S.J. Smith, P.L. Knight (Cambridge University Press, Cambridge, 1988), p. 35
[FW89] H. Friedrich, D. Wintgen, Phys. Reports **183**, 37 (1989)
[Gao98] B. Gao, Phys. Rev. A **58**, 4222 (1998)
[Gao99] B. Gao, Phys. Rev. Lett. **83**, 4225 (1999)
[Gao10] B. Gao, Phys. Rev. Lett. **104**, 231201 (2010)
[Gav92] M. Gavrila (ed.), *Atoms in Intense Laser Fields* (Academic Press, Boston, 1992)
[Gay91] J.-C. Gay (Guest Editor), Comm. At. Mol. Phys., vol. XXV, Special Volume on *Irregular Atomic Systems and Quantum Chaos*, 1991
[GD89] J.-C. Gay, D. Delande, A. Bommier, Phys. Rev. A **39**, 6587 (1989)
[GF79] C. Greene, U. Fano, G. Strinati, Phys. Rev. A **19**, 1485 (1979)
[GG89] T. Grozdanov, P. Grujic, P. Krstic (eds.), *Classical Dynamics in Atomic and Molecular Physics* (World Scientific Publishers, Singapore, 1989)
[GH97] P. Gerwinski, F. Haake, in *Classical, Semiclassical and Quantum Dynamics in Atoms*, Lecture Notes in Physics, vol. 485, ed. by H. Friedrich, B. Eckhardt (Springer, Berlin, 1997), p. 112
[Giu80] A. Giusti, J. Phys. B **13**, 3867 (1980)
[GM98] T. Guhr, A. Müller-Groeling, H.A. Weidenmüller, Phys. Reports **299**, 189 (1998)
[Gol80] H. Goldstein, *Classical Mechanics* (Addison-Wesley, Reading, 1980)
[GR82] C.H. Greene, A.R.P. Rau, Phys. Rev. A **26**, 2441 (1982)
[Gre79] J.M. Greene, J. Math. Phys. **20**, 1183 (1979)
[Gro63] E.P. Gross, J. Math. Phys. **4**, 195 (1963)
[GT69] W.R.S. Garton, F.S. Tomkins, Astrophys. J. **158**, 839 (1969)
[GT05] B. Gao, E. Tiesinga, C.J. Williams, P.S. Julienne, Phys. Rev. A **72**, 042719 (2005)

[Gut97] M.C. Gutzwiller, in *Classical, Semiclassical and Quantum Dynamics in Atoms*, Lecture Notes in Physics, vol. 485, ed. by H. Friedrich, B. Eckhardt (Springer, Berlin, 1997), p. 8

[Haa73] F. Haake, Springer Tracts Mod. Phys. **66**, 98 (1973)

[Haa01] F. Haake, *Quantum Signatures of Chaos*, 2nd. edn. (Springer, Berlin, 2001)

[Har90] S. Haroche, in *Les Houches LIII*, Course 13 (1990), p. 767

[Hec87] K.T. Hecht, *The Vector Coherent State Method and Its Application to Problems of Higher Symmetry*, Lecture Notes in Physics, vol. 290 (Springer, Berlin, Heidelberg, New York, 1987)

[HH83] A. Harada, H. Hasegawa, J. Phys. A **16**, L259 (1983)

[HJ85] R.A. Horn, C.A. Johnson, *Matrix Analysis* (Cambridge University Press, Cambridge, 1985)

[HM90] A. Holle, J. Main, G. Wiebusch, H. Rottke, K.H. Welge, in *Atoms in Strong Fields*, ed. by C.A. Nicolaides, C.W. Clark, M. Nayfeh (Plenum Press, New York, 1990), p. 175

[HM95] B. Hüpper, J. Main, G. Wunner, Phys. Rev. Lett. **74**, 744 (1995)

[HR89] H. Hasegawa, M. Robnik, G. Wunner, Prog. Theor. Phys. Suppl. No. 98, 198 (1989)

[IS79] F.M. Izraelev, D.L. Shepelyansky, Dokl. Akad. Nauk. SSSR **249**, 1103 (1979) (Engl. transl. Sov. Phys. Dokl. **24**, 996 (1979))

[Jac75] J.D. Jackson, *Classical Electrodynamics*, 2nd edn. (Wiley, New York, 1975), p. 143

[JB03] S. Jochim, M. Bartenstein, A. Altmeyer, G. Hendl, S. Riedl, C. Chin, J. Hecker Denschlag, R. Grimm, Science **302**, 2101 (2003)

[Jen84] R.V. Jensen, Phys. Rev. A **30**, 386 (1984)

[JN98] R.R. Jones, L.D. Noordham, Adv. At. Mol. Opt. Phys. **38**, 1 (1998)

[Joh81] S. Johnson, *Life of Abraham Cowley*, in *The Lives of the Most Eminent English Poets*, ed. by R. Lonsdale (Oxford University Press, Oxford, 2006). (First published 1781)

[Jul96] P.S. Julienne, J. Res. Natl. Inst. Stand. Technol. **101**, 487 (1996)

[Jur05] A. Jurisch, Doctoral thesis, Technical University Munich, 2005. mediatum.ub.tum.de/doc/603101/603101.pdf

[Kar91] E. Karule, Adv. At. Mol. Opt. Phys., vol. 27, ed. by Sir David Bates, B. Bederson (1991) 265

[KB06] G.L. Klimchitskaya, E.V. Blagov, V.M. Mostepanenko, J. Phys. A **39**, 6481 (2006)

[Ket97] W. Ketterle, Phys. Bl. **53**, 677 (1997)

[KK83] P. Kruit, J. Kimman, H.G. Muller, M.J. van der Wiel, Phys. Rev. A **28**, 248 (1983)

[KL95] P.M. Koch, K.A.H. van Leeuwen, Phys. Reports **255**, 289 (1995)

[KM88] G. Kracke, H. Marxer, J.T. Broad, J.S. Briggs, Z. Phys. D **8**, 103 (1988)

[KM11] A. Kaiser, T.-O. Müller, H. Friedrich, J. Chem. Phys. **135**, 214302 (2011)

[Koc92] P.M. Koch, Chaos **2**, 131 (1992)

[KS68] J.R. Klauder, E.C.G. Sudarshan, *Fundamentals of Quantum Optics* (W.A. Benjamin, New York, 1968)

[KS97] K.C. Kulander, K.J. Schafer, in *Atoms and Molecules in Intense Fields*, ed. by L.S. Cederbaum, K.C. Kulander, N.H. March, Structure and Bonding, vol. 86, 149 (1997)

[LL71] L.D. Landau, L.M. Lifschitz, *Course of Theoretical Physics, vol. 1, Mechanics* (Addison-Wesley, Reading, 1971)

[LL83] A.J. Lichtenberg, M.A. Liberman, *Regular and Stochastic Motion* (Springer, Berlin, Heidelberg, New York, 1983)

[LF09] M. Lemeshko, B. Friedrich, Phys. Rev. A **79**, 050501 (2009)

[LF10] M. Lemeshko, B. Friedrich, J. At. Mol. Sci. **1**, 39 (2010)

[LT02] T. Laue, E. Tiesinga, C. Samuelis, H. Knöckel, E. Tiemann, Phys. Rev. A **65**, 023412 (2002)

[Lou00] R. Loudon, *The Quantum Theory of Light*, 3rd. edn. (Oxford University Press, Oxford, 2000)

[Mai97] J. Main, in *Classical, Semiclassical and Quantum Dynamics in Atoms*, Lecture Notes in Physics, vol. 485, ed. by H. Friedrich, B. Eckhardt (Springer, Berlin, 1997), p. 248

[MB87] T.J. McIlrath, P.H. Bucksbaum, R.R. Freeman, M. Bashkansky, Phys. Rev. A **35**, 4611 (1987)

[MD97] M. Marinescu, A. Dalgarno, J.F. Babb, Phys. Rev. A **55**, 1530 (1997)

[Mey86] H.-D. Meyer, J. Chem. Phys. **84**, 3147 (1986)

[Mey01] P. Meystre, *Atom Optics* (Springer, New York, 2001)

[ME01] M.J. Moritz, C. Eltschka, H. Friedrich, Phys. Rev. A **64**, 022101 (2001)

[MF07] J. Madroñero, H. Friedrich, Phys. Rev. A **75**, 022902 (2007)

[MH01] A. Mody, M. Haggerty, J.M. Doyle, E.J. Heller, Phys. Rev. B **64**, 085418 (2001)

[Mie84] F. Mies, J. Chem. Phys. **80**, 2514 (1984)

[MJ84] F. Mies, P.S. Julienne, J. Chem. Phys. **80**, 2526 (1984)

[MK11] T.-O. Müller, A. Kaiser, H. Friedrich, Phys. Rev. A **84**, 032701 (2011)

[MK11a] T.-O. Müller, A. Kaiser, H. Friedrich, Phys. Rev. A **84**, 052702 (2011)

[MM97] J. Main, V.A. Mandelshtam, H.S. Taylor, Phys. Rev. Lett. **78**, 4351 (1997)

[MS83] P. Meystre, M.O. Scully (eds.), *Quantum Optics, Experimental Gravitation and Measurement Theory* (Plenum Press, New York, 1983)

[MS90] P. Meystre, M. Sargent III, *Elements of Quantum Optics* (Springer, Berlin, Heidelberg, New York, 1990)

[MS98] J. Main, M. Schwacke, G. Wunner, Phys. Rev. A **57**, 1149 (1998)

[MU97] J. von Milczewski, T. Uzer, Phys. Rev. A **55**, 6540 (1997)

[Mul13] T.-O. Müller, Phys. Rev. Lett. **110**, 260401 (2013)

[MV95] A.J. Moerdijk, B.J. Verhaar, A. Axelsson, Phys. Rev. A **51**, 4852 (1995)

[MW91] J. Main, G. Wiebusch, K.H. Welge, Comments At. Mol. Phys. **XXV**, 233 (1991)

[MW96] F.H. Mies, C.J. Williams, P.S. Julienne, M. Krauss, J. Res. Natl. Inst. Stand. Technol. **101**, 521 (1996)

[MW13] K.W. Madison, Y. Wang, A.M. Rey, K. Bongs, (eds.), *Annual Review of Cold Atoms and Molecules*, vol. 1 (World Scientific, Singapore, 2013)

[Nau89] M. Nauenberg, Phys. Rev. A **40**, 1133 (1989)

[NC90] C.A. Nicolaides, C.W. Clark, M.H. Nayfeh (eds.), *Atoms in Strong Fields* (Plenum Press, New York, 1990)

[OK05] H. Oberst, D. Kouznetsov, K. Shimizu, J. Fujita, F. Shimizu, Phys. Rev. Lett. **94**, 013203 (2005)

[O'M89] P.F. O'Mahony, Phys. Rev. Lett. **63**, 2653 (1989)

[OS97] P. Öhberg, E.L. Surkov, I. Tittonen, S. Stenholm, M. Wilkens, G.V. Shlyapnikov, Phys. Rev. A **56**, R3346 (1997)

[OT05] H. Oberst, Y. Tashiro, K. Shimizu, F. Shimizu, Phys. Rev. A **71**, 052901 (2005)

[Per86] A.M. Peremolov, *Generalized Coherent States* (Springer, Berlin, 1986)

[PK83] R. Paulsson, F. Karlsson, R.J. LeRoy, J. Chem. Phys. **79**, 4346 (1983)

[PK97] M. Protopapas, C.H. Keitel, P.L. Knight, Rep. Prog. Phys. **60**, 389 (1997)

[PK14] V.S. Popov, B.M.Karnakov, Phys.-Uspekhi **57**, 257 (2014)

[PR94] T. Prosen, M. Robnik, J. Phys. A **27**, 8059 (1994)

[PR99] Physical Review Focus: http://physics.aps.org/story/v4/st26

[PS58] V.L. Pokrovskii, S.K. Savvinykh, F.K. Ulinich, Sov. Phys. JETP **34**(7), 879 (1958)

[PS89] R.M. Potvliege, R. Shakeshaft, Phys. Rev. A **40**, 3061 (1989)

[PS90] M. Pont, R. Shakeshaft, R.M. Potvliege, Phys. Rev. A **42**, 6969 (1990)

[PS92] R.M. Potvliege, R. Shakeshaft, in *Atoms in Intense Laser Fields*, ed. by M. Gavrila (Academic Press, Boston, 1992), p. 373

[PS04] T.A. Pasquini, Y. Shin, C. Sanner, M. Saba, A. Schirotzek, D.E. Pritchard, W.E. Ketterle, Phys. Rev. Lett. **93**, 223201 (2004)

[PU58] V.L. Pokrovskii, F.K. Ulinich, S.K. Savvinykh, Sov. Phys. JETP **34**(7), 1119 (1958)

[QJ12] G. Quémener, P.S. Julienne, Chem. Rev. **112**, 4949 (2012)

[Rei80] H.R. Reiss, Phys. Rev. A **22**, 1786 (1980)

[Rei87] H.R. Reiss, in *Photons and Continuum States of Atoms and Molecules*, ed. by N.K. Rahman, C. Guidotti, M. Allegrini (Springer, Berlin, Heidelberg, New York, 1987), p. 98

[RF82] W.P. Reinhardt, D. Farelly, J. de Physique (Paris) **43** (Coll. C2, suppl. 11), C2–29 (1982)
[RG04] C.A. Regal, M. Greiner, D.S. Jin, Phys. Rev. Lett. **92**, 040403 (2004)
[Ric97] D. Richards, in *Classical, Semiclassical and Quantum Dynamics in Atoms*, Lecture Notes in Physics, vol. 485, ed. by H. Friedrich, B. Eckhardt (Springer, Berlin, 1997), p. 172
[Rob81] M. Robnik, J. Phys. A **14**, 3195 (1981)
[Rob04] R.W. Robinett, Phys. Reports **392**, 1 (2004)
[Ros94] J.-M. Rost, J. Phys. B **27**, 5923 (1994)
[Ros98] J.-M. Rost, Phys. Reports **297**, 271 (1998)
[RT97] J.-M. Rost, G. Tanner, in *Classical, Semiclassical and Quantum Dynamics in Atoms*, Lecture Notes in Physics, vol. 485, ed. by H. Friedrich, B. Eckhardt (Springer, Berlin, 1997), p. 273
[RU96] K. Richter, D. Ullmo, R.A. Jalabert, Phys. Reports **276**, 1 (1996)
[RW90a] H. Rottke, B. Wolff, M. Brickwedde, D. Feldmann, K.H. Welge, Phys. Rev. Lett. **64**, 404 (1990)
[RW90b] K. Richter, D. Wintgen, J. Phys. A **23**, L197 (1990)
[RW90c] K. Richter, D. Wintgen, Phys. Rev. Lett. **65**, 1965 (1990)
[RW94] H. Ruder, G. Wunner, H. Herold, F. Geyer, *Atoms in Strong Magnetic Fields* (Springer, Heidelberg, 1994)
[Sch90] F. Scheck, *Mechanics—From Newton's Laws to Deterministic Chaos* (Springer, Berlin, Heidelberg, New York 1990)
[Sch01] W.P. Schleich, *Quantum Optics in Phase Space* (Wiley-VCH, Weinheim, 2001)
[SF02] F. Shimizu, J. Fujita, Phys. Rev. Lett. **88**, 123201 (2002)
[Shi01] F. Shimizu, Phys. Rev. Lett. **86**, 987 (2001)
[SH03] S. Shresta, B.L. Hu, N.G. Phillips, Phys. Rev. A **68**, 062101 (2003)
[Sin70] Y. Sinai, Russ. Math. Surv. **25**, 137 (1970)
[SJ05] H.G. Schuster, W. Just, *Deterministic Chaos—An Introduction*, 4th. edn. (Wiley-VCH, Weinheim, 2005)
[SK88] S.J. Smith, P.L. Knight (eds.), *Multiphoton Processes* (Cambridge University Press, Cambridge, 1988)
[SK08] A. Stein, H. Knöckel, E. Tiemann, Phys. Rev. A **78**, 042508 (2008)
[SK12] M. Steinke, H. Knöckel, E. Tiemann, Phys. Rev. A **85**, 042720 (2012)
[SN88] W. Schweizer, R. Niemeier, H. Friedrich, G. Wunner, H. Ruder, Phys. Rev A **38**, 1724 (1988)
[SN93] W. Schweizer, R. Niemeier, G. Wunner, H. Ruder, Z. Phys. D **25**, 95 (1993)
[SM12] F. Schwarz, T.-O. Müller, H. Friedrich, Phys. Rev. A **85**, 052703 (2012)
[SS98] P. Schmelcher, W. Schweizer (eds.), *Atoms and Molecules in Strong External Fields*, Proceedings of the 172nd WE-Heraeus Seminar (Bad Honnef, 1997) (Plenum Publishing, New York, 1998)
[SS12] T. Schuster, R. Scelle, A. Trautmann, S. Knoop, M.K. Oberthaler, M.M. Haverhals, M.R. Goosen, S.J.J.M.F. Kokkelmans, E. Tiemann, Phys. Rev. A **85**, 042721 (2012)
[ST00] C. Samuelis, E. Tiesinga, E. Laue, T. Elbs, H. Knöckel, E. Tiemann, Phys. Rev. A **63**, 012710 (2000)
[TA16] I.F. Tenney, M. Artamonov, T. Seideman, P.H. Bucksbaum, Phys. Rev. A **93**, 013421 (2016)
[TE98] J. Trost, C. Eltschka, H. Friedrich, J. Phys. B **31**, 361 (1998); Europhys. Lett. **43**, 230 (1998)
[TL89] X. Tang, P. Lambropoulos, A. L'Huillier, S.N. Dixit, Phys. Rev. A **40**, 7026 (1989)
[TN89] K.T. Taylor, M.H. Nayfeh, C.W. Clark (eds.), *Atomic Spectra and, Collisions in External Fields* (Plenum Press, New York, 1989)
[TR00] G. Tanner, K. Richter, J.M. Rost, Rev. Mod. Phys. **72**, 497 (2000)
[TS93] Y. Tikochinsky, L. Spruch, Phys. Rev. A **48**, 4223 (1993)
[VF05] A.Y. Voronin, P. Froelich, J. Phys. B **38**, L301 (2005); A.Y. Voronin, P. Froelich, B. Zygelman, Phys. Rev. A **72**, 062903 (2005)

[WF86] D. Wintgen, H. Friedrich, Phys. Rev. Lett. **57**, 571 (1986)
[Win87a] D. Wintgen, Phys. Rev. Lett. **58**, 1589 (1987)
[Win87b] D. Wintgen, J. Phys. B **20**, L511 (1987)
[Wir92] A. Wirzba, Chaos **2**, 77 (1992)
[Wir93] A. Wirzba, Nucl. Phys. A **560**, 136 (1993)
[WR92] D. Wintgen, K. Richter, G. Tanner, Chaos **2**, 19 (1992)
[WW86] G. Wunner, U. Woelk, I. Zech, G. Zeller, T. Ertl, F. Geyer, W. Schweizer, H. Ruder, Phys. Rev. Lett. **57**, 3261 (1986)
[YB98] Z.-C. Yan, J.F. Babb, Phys. Rev. A **58**, 1247 (1998)
[YD93] I.A. Yu, J.M. Doyle, J.C. Sandberg, C.L. Cesar, D. Kleppner, T.J. Greytak, Phys. Rev. Lett. **71**, 1589 (1993)
[YD97] Z.-C. Yan, A. Dalgarno, J.F. Babb, Phys. Rev. A **55**, 2882 (1997)
[YM90] J.A. Yeazell, M. Mallalieu, C.R. Stroud Jr., Phys. Rev. Lett. **64**, 2007 (1990)
[ZF90] W.-M. Zhang, D.H. Feng, R. Gilmore, Rev. Mod. Phys. **62**, 867 (1990)
[ZM11] B.S. Zhao, G. Meijer, W. Schöllkopf, Science **331**, 892 (2011)

Solutions to the Problems

1.1. Bound states only exist for energies

$$E \equiv -\frac{\hbar^2\kappa^2}{2\mu} < 0\,, \quad E + V_0 = \frac{\hbar^2 k^2}{2\mu} > 0\,, \quad \kappa^2 + k^2 = \frac{2\mu}{\hbar^2}V_0\,.$$

The solution of the radial Schrödinger equation for $l = 0$ is

$$\phi(r) \propto \sin kr \quad \text{for} \quad r \leq r_0\,; \quad \phi(r) \propto \mathrm{e}^{-\kappa r} \quad \text{for} \quad r \geq r_0\,.$$

The matching condition (1.92) implies

$$\cot kr_0 = -\kappa/k = -\sqrt{\frac{2\mu V_0}{\hbar^2 k^2} - 1}\,. \tag{1}$$

Each intersection of the left-hand side and the right-hand side of (1) (as functions of k) yields a bound state. The right-hand side varies from $-\infty$ at $k = 0$ to zero at $k_{\max} = (1/\hbar)\sqrt{2\mu V_0}$. The number of branches of $\cot kr$ which intersect the right-hand side is given by the largest number n for which $(n - \frac{1}{2})\pi/r_0 < k_{\max}$, thus the number of bound states is near $r_0\sqrt{2\mu V_0}/(\pi\hbar)$. Note that there is no bound state if $(2\mu r_0^2/\hbar^2)V_0 < (\pi/2)^2$.

1.2.

a)

$$\langle\phi|\phi_n\rangle = \sqrt{2}\left(\frac{2\beta b}{\beta^2 + b^2}\right)^{\frac{3}{2}}\left(\frac{\beta^2 - b^2}{\beta^2 + b^2}\right)^n\left[\frac{\frac{1}{2}\cdot\frac{3}{2}\cdots(n + \frac{1}{2})}{n!}\right]^{1/2}$$

H. Friedrich, *Theoretical Atomic Physics*, Graduate Texts in Physics,
DOI 10.1007/978-3-319-47769-5

b)

$$\langle\phi|\phi_n\rangle = \left(\frac{na}{b}\right)^{\frac{3}{2}} \left(s - \frac{n+1}{2}\right) (s-1)^{n-2} s^{-(n+2)}\,, \quad s = \frac{na+b}{2b}$$

c) Harmonic oscillator, $b = \beta/2$:

n	0	1	2	3	4	5
$\lvert\langle\phi\lvert\phi_n\rangle\rvert^2$	0.5120	0.2765	0.1244	0.0523	0.0212	0.0084
$\sum^n \lvert\langle\phi\lvert\phi_\nu\rangle\rvert^2$	0.5120	0.7885	0.9129	0.9652	0.9864	0.9948

Coulomb potential, $b = a/2$:

n	1	2	3	4	5
$\lvert\langle\phi\lvert\phi_n\rangle\rvert^2$	0.7023	0.0419	0.0110	0.0045	0.0022
$\sum^n \lvert\langle\phi\lvert\phi_\nu\rangle\rvert^2$	0.7023	0.7442	0.7552	0.7597	0.7619

d) Coulomb potential, $b = a$ (use orthonormality relations rather than formula b) above)

n	1	2	3	4	5
$\lvert\langle\phi\lvert\phi_n\rangle\rvert^2$	1	0	0	0	0
$\sum^n \lvert\langle\phi\lvert\phi_\nu\rangle\rvert^2$	1	1	1	1	1

Coulomb potential, $b = 2a$:

n	1	2	3	4	5
$\lvert\langle\phi\lvert\phi_n\rangle\rvert^2$	0.7023	0.25	0.0127	0.0039	0.0017
$\sum^n \lvert\langle\phi\lvert\phi_\nu\rangle\rvert^2$	0.7023	0.9523	0.9650	0.9689	0.9706

1.3. From (1.139) and abbreviating $2r/(na)$ as x, we have

$$\langle\phi_{n,l}|r|\phi_{n,l}\rangle = \frac{(n-l-1)!}{4(n+1)!}\, a \int_0^\infty x^{2l+1}[xL_{n-l-1}^{2l+1}(x)]^2\, \mathrm{e}^{-x}\, \mathrm{d}x\,.$$

From (A.16) in Appendix A.2 we have

$$xL_{n-l-1}^{2l+1}(x) = 2nL_{n-l-1}^{2l+1}(x) - (n-l)L_{n-l}^{2l+1}(x) - (n+l)L_{n-l-2}^{2l+1}(x)\,,$$

and, exploiting (A.15), we obtain

$$\begin{aligned}
&\langle\phi_{n,l}|r|\phi_{n,l}\rangle \\
&= \frac{(n-l-1)!}{4(n+l)!}\,a\left[(2n)^2\,\frac{(n+l)!}{(n-l-1)!}\right. \\
&\left. +(n-l)^2\frac{(n+l+1)!}{(n-l)!}+(n+l)^2\,\frac{(n-l-1)!}{(n-l-2)!}\right] \\
&= \frac{a}{4}\left[4n^2+(n+l+1)(n-l)+(n+l)(n-l-1)\right] \\
&= \frac{a}{2}\left[3n^2-l(l+1)\right] .
\end{aligned}$$

1.4.

$$\begin{aligned}
&\tilde{\psi}(k,0) \\
&= \frac{1}{\pi^{3/4}\sqrt{2\beta}}\int_{-\infty}^{\infty}\exp\left\{-\frac{1}{2\beta^2}[x^2-2x\,(x_0-\mathrm{i}\beta^2(k-k_0))+(x_0)^2]\right\}\mathrm{d}x \\
&= \frac{1}{\pi^{3/4}\sqrt{2\beta}}\int_{-\infty}^{\infty}\exp\left\{-\frac{1}{2\beta^2}[x-(x_0-\mathrm{i}\beta^2(k-k_0))]^2\right\}\mathrm{d}x \\
&\qquad\times\exp\left\{-\mathrm{i}(k-k_0)x_0-\frac{\beta^2}{2}(k-k_0)^2\right\} .
\end{aligned}$$

The integral over x is the Gaussian integral, $\int_{-\infty}^{\infty}\exp\left\{-[x-\ldots]^2/(2\beta^2)\right\}\mathrm{d}x = \beta\sqrt{2\pi}$, so

$$\tilde{\psi}(k,0)=(\beta/\sqrt{\pi})^{1/2}\,\mathrm{e}^{-\mathrm{i}(k-k_0)x_0}\,\mathrm{e}^{-(k-k_0)^2\beta^2/2} .$$

In momentum representation, the Hamiltonian for the free particle simply acts as a multiplication by $p^2/(2\mu)=\hbar^2k^2/(2\mu)$, so the time evolution operator (1.41) acts by multiplying the wave function $\tilde{\psi}(k)$ with $\exp\left[-\mathrm{i}\hbar k^2t/(2\mu)\right]$,

$$\tilde{\psi}(k,t)=(\beta/\sqrt{\pi})^{1/2}\,\mathrm{e}^{-\mathrm{i}(k-k_0)x_0}\,\mathrm{e}^{-\mathrm{i}k^2a(t)^2/2}\,\mathrm{e}^{-(k-k_0)^2\beta^2/2} ;$$

here we have introduced the abbreviation $a(t)=\sqrt{\hbar t/\mu}$. Note that the time evolution does not affect the probability distribution in momentum space,

$$|\tilde{\psi}(k,t)|^2=\frac{\beta}{\sqrt{\pi}}\,\mathrm{e}^{-\beta^2(k-k_0)^2} .$$

Transforming back to coordinate space gives

$$\psi(x,t) = \frac{1}{\sqrt{2\pi}} \int_{-\infty}^{\infty} e^{ikx}\tilde{\psi}(k,t)\,dk$$
$$= \frac{\sqrt{\beta/2}}{\pi^{3/4}} \int_{-\infty}^{\infty} \exp\left\{-i(k-k_0)x_0 - \frac{\beta^2}{2}(k-k_0)^2 - i\frac{a^2}{2}k^2 + ikx\right\} dk\,.$$

The integrand above can be written as

$$\exp\left\{-\frac{\beta^2 + ia^2}{2}\left(k - \frac{\beta^2 k_0 + i(x-x_0)}{\beta^2 + ia^2}\right)^2\right\} \quad \times$$
$$\exp\left\{\frac{ik_0\beta^2(2x - k_0a^2) - (x-x_0)^2 - 2k_0a^2x_0}{2(\beta^2 + ia^2)}\right\} \quad .$$

The second factor is independent of the integration variable k and the integral over the first factor is a Gaussian integral with value $\sqrt{2\pi/(\beta^2 + ia^2)}$, so the time dependent wave function in coordinate space is

$$\psi(x,t) = \frac{(\beta\sqrt{\pi})^{-1/2}}{\sqrt{1 + ia^2/\beta^2}} \exp\left[\frac{-(x-x_0)^2 - 2k_0a^2x_0 + ik_0\beta^2(2x - k_0a^2)}{2(\beta^2 + ia^2)}\right].$$

The structure of this wave packet is easier to appreciate if we look at the corresponding probability density $|\psi(x,t)|^2 = \psi^*(x,t)\psi(x,t)$,

$$|\psi(x,t)|^2 = \frac{1}{b(t)\sqrt{\pi}} \exp\left[-\frac{(x - x_0 - k_0a^2)^2}{b(t)^2}\right], \quad b(t) = \beta\sqrt{1 + \frac{a^4}{\beta^4}}\,.$$

Recalling the abbreviation above, $a(t)^2 = \hbar t/\mu$ and $k_0a^2 = \upsilon_0 t$, brings us to the result (1.168).

For a normalized Gaussian distribution

$$P(y) = \frac{1}{\sigma\sqrt{\pi}} e^{-(y-y_0)^2/\sigma^2}, \qquad \int_{\infty}^{\infty} P(y)\,dy = 1\,,$$

with mean value y_0 and width paramater σ, the square of the uncertainty (variance, fluctuation) Δy can be written as

$$(\Delta y)^2 = \int_{\infty}^{\infty} y^2 P(y)\,dy - (y_0)^2 = \frac{1}{\sigma\sqrt{\pi}} \int_{\infty}^{\infty} (y-y_0)^2\, e^{-(y-y_0)^2/\sigma^2} dy = \frac{\sigma^2}{2}$$

giving $\Delta y = \sigma/\sqrt{2}$. For the uncertainty Δx of the wave packet in coordinate space, replace y by x, $P(y)$ by $|\psi(x,t)|^2$ and σ by $b(t)$, so

$$\Delta x = \frac{1}{\sqrt{2}} b(t) = \frac{\beta}{\sqrt{2}} \sqrt{1 + \frac{\hbar^2 t^2}{\mu^2 \beta^4}} \, .$$

For the uncertainty in momentum space, replace y by k, $P(y)$ by $|\tilde{\psi}(k,t)|^2$ and σ by $1/\beta$, so

$$\Delta k = \frac{1}{\beta\sqrt{2}}, \quad \Delta p = \hbar \Delta k = \frac{\hbar}{\beta\sqrt{2}} \, .$$

Note that the momentum probability distribution remains unchanged during the time evolution of the free-particle wave function, whereas the wave packet spreads in coordinate space. This shows that the uncertainty relation (1.34) is an inequality in general. In the present case

$$\Delta x \Delta p = \frac{\hbar}{2} \sqrt{1 + \frac{\hbar^2 t^2}{\mu^2 \beta^4}} \geq \frac{\hbar}{2} \, .$$

Wave packets fulfilling the equality, $\Delta x \Delta p = \hbar/2$, are called *minimum uncertainty wave packets*. The initial Gaussian wave packet (1.167) is such a minimum uncertainty wave packet.

1.5. One way is to show that

$$\int_0^\infty \left[\left(E + \frac{\hbar^2}{2\mu} \frac{\mathrm{d}^2}{\mathrm{d}r^2} \right) G_0(r, r') \right] f(r) \, \mathrm{d}r = f(r') \tag{1}$$

for sufficiently well behaved square integrable trial functions f. For $r \neq r'$ the integrand vanishes, because $\frac{\hbar^2}{2\mu} \frac{\mathrm{d}^2}{\mathrm{d}r^2} G_0(r, r')$ always equals $-E G_0(r, r')$. Thus showing (1) reduces to showing that

$$\lim_{\varepsilon \to 0} \int_{r'-\varepsilon}^{r'+\varepsilon} \left[\left(k^2 + \frac{\mathrm{d}^2}{\mathrm{d}r^2} \right) \sin(kr_<) \cos(kr_>) \right] f(r) \, \mathrm{d}r = -k f(r') \, . \tag{2}$$

The contribution proportional to k^2 on the left-hand side of (2) vanishes in the limit $\varepsilon \to 0$. For the remaining contribution we integrate by parts twice and obtain (for finite and positive ε)

$$\text{l.h.s.} = \left[\left(\frac{\mathrm{d}}{\mathrm{d}r}\sin(kr_<)\cos(kr_>)\right)f(r)\right]_{r'-\varepsilon}^{r'+\varepsilon} \tag{3}$$
$$-\left[\sin(kr_<)\cos(kr_>)\frac{\mathrm{d}f}{\mathrm{d}r}\right]_{r'-\varepsilon}^{r'+\varepsilon} + \int_{r'-\varepsilon}^{r'+\varepsilon}\sin(kr_<)\cos(kr_>)\frac{\mathrm{d}^2f}{\mathrm{d}r^2}\,\mathrm{d}r\,.$$

The latter two terms in (3) vanish in the limit $\varepsilon \to 0$ and the remaining term yields

$$\text{l.h.s} = \{-k\sin[k(r'+\varepsilon)]\sin[k(r'-\varepsilon)]\}f(r'+\varepsilon)$$
$$-\{k\cos[k(r'-\varepsilon)]\cos[k(r'+\varepsilon)]\}f(r'-\varepsilon)\,,$$

which clearly becomes equal to the right-hand side of (2) in the limit $\varepsilon \to 0$.

1.6. In this Hilbert space the state vectors are two-component vectors $\binom{a_1}{a_2}$, and the eigenstates of $\hat{H}_0$ are $\psi_1^{(0)} = \binom{1}{0}$ and $\psi_0^{(2)} = \binom{0}{1}$ with (unperturbed) eigenvalues ε_1 and ε_2 respectively.

a) In lowest non-vanishing order perturbation theory (1.253) yields

$$\psi_1 = \psi_1^{(0)} + \psi_1^{(1)} = \psi_1^{(0)} + \frac{w}{\varepsilon_1 - \varepsilon_2}\psi_2^{(0)}\,,$$
$$\psi_2 = \psi_2^{(0)} + \psi_2^{(1)} = \psi_2^{(0)} + \frac{w}{\varepsilon_2 - \varepsilon_1}\psi_1^{(0)}\,,$$

and (1.255) yields

$$E_1 = E_1^{(0)} + E_1^{(2)} = \varepsilon_1 + \frac{w^2}{\varepsilon_1 - \varepsilon_2}\,,$$
$$E_2 = E_2^{(0)} + E_2^{(2)} = \varepsilon_2 - \frac{w^2}{\varepsilon_1 - \varepsilon_2}\,.$$

b) To diagonalize $\hat{H}$ in this case we first solve the secular equation (cf. (1.259), (1.279))

$$\det\begin{pmatrix} \varepsilon_1 - E & w \\ w & \varepsilon_2 - E \end{pmatrix} = (\varepsilon_1 - E)(\varepsilon_2 - E) - w^2 = 0\,,$$

yielding the exact eigenvalues

$$E_\pm = \frac{\varepsilon_1 + \varepsilon_2}{2} \pm \sqrt{w^2 + \left(\frac{\varepsilon_1 - \varepsilon_2}{2}\right)^2}\,.$$

The corresponding eigenstates $\binom{a_1}{a_2}$ follow from solving the simultaneous equations

$$(E - \varepsilon_1)a_1 = wa_2\,, \quad (E - \varepsilon_2)a_2 = wa_1\,,$$

for the respective eigenvalue. The eigenstates can be characterized by the ratios

$$\frac{a_1}{a_2} = \frac{\varepsilon_1 - \varepsilon_2}{2w} \pm \frac{w}{|w|}\sqrt{1 + \left(\frac{\varepsilon_1 - \varepsilon_2}{2w}\right)^2}\,,$$

and a_1, a_2 are of course only defined to within a common arbitrary factor. The perturbative results are good for $|\varepsilon_1 - \varepsilon_2| \gg |w|$, but they give poor results for $|\varepsilon_1 - \varepsilon_2| \ll |w|$.

1.7.

a) For energy E the classical turning points b and $a = -b$ are given by $b = \sqrt{2E/(\mu\omega^2)}$. From (1.308) we have

$$\left(n + \frac{\mu_\phi}{4}\right)\pi\hbar = \int_{-b}^{b} \mu\omega\sqrt{b^2 - x^2}\,\mathrm{d}x = \mu\omega\frac{\pi}{2}b^2 = \frac{\pi E}{\omega}\,,$$

which yields $E_n = (n + \mu_\phi/4)\hbar\omega$ in agreement with the exact result, when the Maslov index μ_ϕ is taken to be two, corresponding to a phase loss of $\pi/2$ of the WKB wave at each turning point.

b) The classical turning points are $a = 0$ and $b = L$ independent of E. With $p = \sqrt{2\mu E}$ we have $\int_a^b p\,\mathrm{d}x = \sqrt{2\mu E}L$, and the quantization condition (1.308) yields the exact quantum mechanical result $E_n = (\pi\hbar)^2(n+1)^2/(2\mu L^2)$, $n = 0, 1, 2, \ldots$, *provided* the Maslov index μ_ϕ is taken to be four, corresponding to a phase loss of π at each turning point.

c) For $x > L$ we have $|p(x)| = \hbar\kappa = \sqrt{2\mu(V_0 - E)} = \text{const.}$, and the WKB wave function,

$$\psi_{\mathrm{WKB}}(x) = \frac{N}{\sqrt{\hbar\kappa}}\,\mathrm{e}^{-\kappa(x-L)}\,, \quad x > L\,, \tag{1}$$

exactly solves the Schrödinger equation. For $x < L$ we have $p(x) = \hbar k = \sqrt{2\mu E} = \text{const.}'$, and the (real) WKB wave function

$$\psi_{\mathrm{WKB}}(x) = \frac{2}{\sqrt{\hbar k}}\cos\left(k(L - x) - \frac{\phi}{2}\right)\,, \quad x < L\,, \tag{2}$$

is an exact solution of the Schrödinger equation in this region as well. Matching these (exact) wave functions and their derivatives at $x = L$ fixes the two constants

N and ϕ in (1) and (2),

$$\phi = 2\arctan(\kappa/k)\,, \quad N = 2\sqrt{\kappa k/(\kappa^2 + k^2)}\,. \tag{3}$$

The exact wave function constructed in this way coincides with the WKB wave function, except at the classical turning point L, where the WKB wave function is not defined.
For the particle in the well bounded by two steps, the WKB wave functions represent exact solutions of the Schrödinger equation in the regions $x < 0$, $0 < x < L$ and $x > L$. The (exact) wave functions decay as $\exp[-\kappa|x|]$ for $x < 0$ and as $\pm\exp[-\kappa(x - L)]$ for $x > L$; the "+" and "−" signs refer to solutions which are symmetric or antisymmetric with respect to reflection at $x = L/2$. Matching the WKB wave functions at each turning point is the same as matching the exact solutions; it leads to two expressions for the wave function in the classically allowed region, and the condition that these be equal is,

$$\cos\left(kx - \frac{\phi}{2}\right) = \pm\cos\left(k(L-x) - \frac{\phi}{2}\right), \quad 0 < x < L\,, \tag{4}$$

which is fulfilled if and only if $kL = \phi + n\pi$, i.e. $\hbar kL = \pi\hbar(n + \mu_\phi/4)$ with $\mu_\phi = 2\phi/(\pi/2)$. This is just the quantization condition (1.308), (1.309) with the phase loss ϕ at each turning point as given by (3). With the Maslov index corresponding to the correct reflection phase(s) ϕ the quantization condition (1.308) gives the exact energy eigenvalues. When matched with the correct phase ϕ and amplitude factor N (3), the WKB wave function in the regions $x < 0$, $0 < x < L$ and $x > L$ is equal to the corresponding continuous exact wave function with continuous derivative.
[Note that the quantization condition for the ground state ($n = 0$) is $\tan(kL/2) = \kappa/k = \sqrt{2\mu V_0/(\hbar^2 k^2) - 1}$ and always has a solution, no matter how small L and V_0 are. This is in contrast to the potential step in the radial Schrödinger equation of three-dimensional space (see Problems 1.1 and 1.8).]

1.8. For the kinetic energy $\hat{T} = -\frac{\hbar^2}{2\mu}\frac{\mathrm{d}^2}{\mathrm{d}x^2}$ we have

$$\begin{aligned}\left\langle\psi|\hat{T}|\psi\right\rangle &= \left(\sqrt{\pi}b\right)^{-1}\frac{\hbar^2}{2\mu}\int_{-\infty}^{\infty}\left(\frac{\mathrm{d}}{\mathrm{d}x}\,\mathrm{e}^{-x^2/(2b^2)}\right)^2 \mathrm{d}x \\ &= \left(\sqrt{\pi}b\right)^{-1}\frac{\hbar^2}{2\mu}\frac{1}{b^4}\int_{-\infty}^{\infty} x^2\,\mathrm{e}^{-x^2/b^2}\,\mathrm{d}x = \frac{\hbar^2}{4\mu b^2}\,,\end{aligned}$$

which tends to zero as $1/b^2$ when $b \to \infty$.
For any potential $V(x)$ the expectation value

$$\langle\psi|V|\psi\rangle = \left(\sqrt{\pi}b\right)^{-1}\int_{-\infty}^{\infty} V(x)\,,\mathrm{e}^{-x^2/b^2}\,\mathrm{d}x$$

approaches $1/(\sqrt{\pi}b)$ times the constant $\int_{-\infty}^{\infty} V(x)\,\mathrm{d}x$ as $b \to \infty$. If this constant is negative, then the more slowly vanishing negative contribution of the potential energy will outweigh the more rapidly vanishing positive contribution of the kinetic energy for sufficiently large b, giving in sum a negative energy expectation value, which in turn must be larger than the lowest energy eigenvalue due to (1.272).

The same reasoning cannot be applied in three dimensions, because there the normalized Gaussian is $\left(\sqrt{\pi}b\right)^{-3/2}\mathrm{e}^{-x^2/(2b^2)}$ and the potential energy expectation value vanishes as b^{-3} for large b. Indeed, from Problem 1.1 we see that the attractive spherical square well has no bound state if V_0 is too small. In two dimensions the Gaussian trial function does not lead to conclusive results, but an alternative choice of trial functions can be used to prove the existence of at least one bound state in a dominantly attractive potential (see e.g. Perez, Malta and Coutinho, *Am. J. Phys.* **58** (1990) 519).

1.9. For energies $E = -|E|$ sufficiently close to threshold the outer classical turning point $b(E)$ is given by

$$E = -\frac{C_l}{b^2}, \quad b\,(E) = \sqrt{\frac{C_l}{|E|}} > r_0\,, \quad C_l = C - \frac{\hbar^2}{2\mu}\left(l + \frac{1}{2}\right)^2 . \tag{1}$$

From (1.308) we have

$$\left(n + \frac{\mu_\phi}{4}\right)\pi\hbar = \int_a^b p(r)\,\mathrm{d}r = \int_a^{r_0} p(r)\,\mathrm{d}r + \int_{r_0}^b p(r)\,\mathrm{d}r\,. \tag{2}$$

As E approaches zero the first term on the far right of (2) tends to a constant, but the second term grows beyond all bounds:

$$\begin{aligned}
\int_{r_0}^{b(E)} p(r)\,\mathrm{d}r &= \int_{r_0}^{b} \sqrt{2\mu\left(\frac{C_l}{r^2} - |E|\right)}\,\mathrm{d}r = \frac{\sqrt{2\mu C_l}}{b}\int_{r_0}^{b}\frac{\sqrt{b^2-r^2}}{r}\,\mathrm{d}r \\
&= \frac{\sqrt{2\mu C_l}}{b}\left[\sqrt{b^2-r^2} - b\ln\left(\frac{b+\sqrt{b^2-r^2}}{r}\right)\right]_{r_0}^{b} \\
&= \sqrt{2\mu C_l}\left[\ln\left(\frac{b+\sqrt{b^2-r_0^2}}{r_0}\right) - \sqrt{1-\frac{r_0^2}{b^2}}\right] \\
&\overset{b\to\infty}{\to} \sqrt{2\mu C_l}\ln b + \mathrm{const.}
\end{aligned} \tag{3}$$

For $E \to 0$, which implies $b \to \infty$ and $n \to \infty$, we have

$$b \propto \exp\left[\frac{(n+\mu_\phi/4)\pi\hbar}{\sqrt{2\mu C_l}}\right] \quad \text{or} \quad E = -\frac{C_l}{b^2} = -c_1\,\mathrm{e}^{-c_2(l)n}$$

$$\text{with} \quad c_2(l) = \frac{2\pi\hbar}{\sqrt{2\mu C_l}} = \frac{2\pi\hbar}{\sqrt{2\mu C - (l+1/2)^2\hbar^2}}\,. \tag{4}$$

The magnitudes of the energies are determined by the constant c_1 which depends on the constants entering in (2) and (3). These in turn depend crucially on the parameter r_0 and the nature of the potential inside r_0. An infinite sequence of bound states exists as long as $C_l = C - (l+1/2)^2\hbar^2/(2\mu)$ is positive. The ratio E_n/E_{n+1} of successive binding energies is $\exp[c_2(l)]$. If $C_l \le 0$ there is at most a finite number of bound states. Although these statements were derived using the WKB approximation including the Langer modification, they agree with the exact quantum mechanical results (see Morse and Feshbach, *Methods of Theoretical Physics* Part II, (McGraw-Hill, New York, 1953) p. 1665), Sect. 3.1.5.

1.10. The quantality function is,

$$Q(r) = \frac{5\alpha^2(\beta_\alpha)^{2\alpha-4}}{16r^{2\alpha+2}\left[k^2+(\beta_\alpha)^{\alpha-2}/r^\alpha\right]^3} - \frac{\alpha(\alpha+1)(\beta_\alpha)^{\alpha-2}}{4r^{\alpha+2}\left[k^2+(\beta_\alpha)^{\alpha-2}/r^\alpha\right]^2}$$

and maxima of $|Q(r)|$ occur when

$$k^2 = F_\alpha\frac{(\beta_\alpha)^{\alpha-2}}{r^\alpha}\,,$$

where

$$F_\alpha = \frac{5}{4} - \frac{9}{2\alpha+4} \pm \frac{9\alpha}{4\alpha+8}\sqrt{1-\frac{20}{27}\left(\frac{\alpha+2}{\alpha+1}\right)}\,. \tag{1}$$

The positions of these maxima are

$$r_{\max} = [F_\alpha]^{1/\alpha} r_E\,,$$

with r_E as given by (1.324). For $\alpha > 4$, the function $Q(r)$ has a zero at $r = \frac{1}{4}[1-5/(\alpha+1)]^{1/4}\,r_E$; there is a larger maximum of $|Q(r)|$ above [corresponding to the plus sign in (1)] and a smaller one below [corresponding to the minus sign in (1)] this zero. For $\alpha = 3, 4$, only the plus sign in (1) yields a positive value for F_α. For the plus sign in (1), the values of $[F_\alpha]^{1/\alpha}$ are:

α	3	4	5	6	7	8	9	10
$[F_\alpha]^{1/\alpha}$	0.8952	1	1.0370	1.0511	1.0560	1.0569	1.0560	1.0543

The fact that the maximum of $|Q(r)|$ lies close to the point r_E defined by $|V(r_E)| = E$ has also been demonstrated for more general attractive potential tails, e.g. those of the Casimir-van der Waals type which behave as $-1/r^3$ for small and as $-1/r^4$ for large distances [FJ02]. This can, however, not be a general theorem. Sharp or smooth step potentials, for which $|V(r)|$ never exceeds a given depth (or height) V_0 of the step, provide a counter-example, because r_E cannot be defined for $E > V_0$.

1.11.

$$[\hat{p}^2, r] = -\hbar^2[\Delta, r] = -\hbar^2\left[\frac{\partial^2}{\partial r^2} + \frac{2}{r}\frac{\partial}{\partial r}, r\right].$$

The first identity follows immediately once we realize that

$$\frac{\partial^2}{\partial r^2} r\phi - r\frac{\partial^2 \phi}{\partial r^2} = 2\frac{\partial \phi}{\partial r},$$

and that

$$\frac{2}{r}\frac{\partial}{\partial r} r\phi - 2\frac{\partial \phi}{\partial r} = \frac{2}{r}\phi \quad \text{for all} \quad \phi(r).$$

Second identity:

$$\begin{aligned}
[\hat{p}^2, r^2] &= [\hat{p}^2, r]r + r[\hat{p}^2, r] \\
&= -2\hbar^2\left(\frac{\partial}{\partial r} r + 1 + r\frac{\partial}{\partial r} + 1\right) = -2\hbar^2\left(2r\frac{\partial}{\partial r} + 3\right).
\end{aligned}$$

2.1. Using the properties (1.350), (1.352) we obtain,

$$\begin{aligned}
(\hat{\boldsymbol{\sigma}}\cdot A)(\hat{\boldsymbol{\sigma}}\cdot B) &= (\hat{\sigma}_x A_x + \hat{\sigma}_y A_y + \hat{\sigma}_z A_z)(\hat{\sigma}_x B_x + \hat{\sigma}_y B_y + \hat{\sigma}_z B_z) \\
&= \hat{\sigma}_x^2 A_x B_x + \hat{\sigma}_y^2 A_y B_y + \hat{\sigma}_z^2 A_z B_z \\
&\quad + \hat{\sigma}_x\hat{\sigma}_y A_x B_y + \hat{\sigma}_y\hat{\sigma}_z A_y B_z + \hat{\sigma}_z\hat{\sigma}_x A_z B_x \\
&\quad + \hat{\sigma}_y\hat{\sigma}_x A_y B_x + \hat{\sigma}_z\hat{\sigma}_y A_z B_y + \hat{\sigma}_x\hat{\sigma}_z A_x B_z \\
&= A\cdot B + \mathrm{i}\hat{\sigma}_x(A_y B_z - A_z B_y) \\
&\quad + \mathrm{i}\hat{\sigma}_y(A_z B_x - A_x B_z) + \mathrm{i}\hat{\sigma}_z(A_x B_y - A_y B_x) \\
&= A\cdot B + \mathrm{i}\hat{\boldsymbol{\sigma}}\cdot(A\times B).
\end{aligned}$$

For $A = r$ and $B = \hat{p}$ we have

$$(\hat{\boldsymbol{\sigma}}\cdot r)(\hat{\boldsymbol{\sigma}}\cdot\hat{p}) = r\cdot\hat{p} + \mathrm{i}\hat{\boldsymbol{\sigma}}\cdot(r\times\hat{p}) = \frac{\hbar}{\mathrm{i}} r\frac{\partial}{\partial r} + \mathrm{i}\hat{\boldsymbol{\sigma}}\cdot\hat{L}.$$

For $A = B = r$ we have

$$(\hat{\sigma}\cdot r)\,(\hat{\sigma}\cdot r) = r^2,$$

hence

$$\begin{aligned}(\hat{\sigma}\cdot\hat{p}) &= \frac{1}{r^2}(\hat{\sigma}\cdot r)(\hat{\sigma}\cdot r)(\hat{\sigma}\cdot\hat{p})\\ &= \frac{1}{r^2}(\hat{\sigma}\cdot r)\left(\frac{\hbar}{\mathrm{i}}r\frac{\partial}{\partial r} + \mathrm{i}\hat{\sigma}\cdot\hat{L}\right).\end{aligned}$$

2.2. The unperturbed eigenfunctions of the hydrogen atom ($Z = 1$) or a hydrogenic ion ($Z > 1$) for fixed quantum numbers n and l and arbitrary quantum numbers j and m are degenerate with respect to the unperturbed Hamiltonian (2.13). All three relativistic corrections are diagonal in j and m, so we choose the unperturbed eigenfunctions as

$$\Phi_{n,j,m,l} \equiv \frac{\phi_{n,l}(r)}{r}\mathcal{Y}_{j,m,l}\,, \tag{1}$$

where $\phi_{n,l}(r)$ are the radial eigenfunctions (1.139) (with Bohr radius (2.15)) and $\mathcal{Y}_{j,m,l}$ are the generalized spherical harmonics (1.358).

For the *spin-orbit* term we obtain for $l > 0$:

$$\langle\Phi_{n,j,m,l}|\frac{1}{r^3}\hat{L}\cdot\hat{S}|\Phi_{n,j,m,l}\rangle = \int_0^\infty \frac{[\phi_{n,l}(r)]^2}{r^3}\,\mathrm{d}r \times \frac{\hbar^2}{2}F(j,l)\,,$$

where $F(j,l)$ is the factor from (1.362) which is equal to l for $j = l + 1/2$ and equal to $-(l+1)$ for $j = l - 1/2$. Using the substitution $x = 2r/(na_Z)$, the energy shift in first-order perturbation theory is thus

$$\Delta E_{LS} = \frac{Ze^2}{2m_0^2c^2}\frac{4(n-l-1)!}{n^4a_Z^3(n+l)!}\int_0^\infty x^{2l-1}\left[L_{n-l-1}^{2l+1}(x)\right]^2\mathrm{e}^{-x}\,\mathrm{d}x\,\frac{\hbar}{2}F(j,l)\,. \tag{2}$$

The only non-vanishing case up to $n = 2$ is that of the quantum numbers $n = 2$, $l = 1$, for which the Laguerre polynomials are unity and the integral in (2) can be easily calculated. A more general formula can be obtained using the expectation value of $1/r^3$ as given by Bethe and Salpeter [BS77]:

$$\begin{aligned}\Delta E_{LS} &= \frac{Z^4e^2}{2m_0^2c^2}\frac{1}{a_0^3n^3(l+1)\left(l+\frac{1}{2}\right)l}\frac{\hbar^2}{2}F(j,l)\\ &= \frac{1}{4}m_0c^2(Z\alpha_{\mathrm{fs}})^4\frac{F(j,l)}{(l+1)\left(l+\frac{1}{2}\right)l}\,,\end{aligned} \tag{3}$$

where we have inserted $a_0/Z = \hbar^2/(Zm_0e^2)$ for a_Z, and $\alpha_{\rm fs} = e^2/(\hbar c) \approx 1/137$ is the fine structure constant. For $l = 1$ the factor $F(j,l)$ is unity for $j = 3/2$ and -2 for $j = 1/2$.

The *Darwin term* contributes only for $l = 0$ implying $j = 1/2$, and we have

$$\begin{aligned}\langle \Phi_{n,j,m,l=0}|\hat{H}_{\rm D}|\Phi_{n,j,m,l=0}\rangle &= \frac{\pi\hbar^2 Ze^2}{2m_0^2c^2}|\Phi_{n,j,m,l=0}(r=0)|^2 \\ &= \frac{\hbar^2 Z^4 e^2}{2m_0c^2}\frac{1}{(na_0)^3} = \frac{1}{2n^3}m_0c^2(Z\alpha)^4\,,\end{aligned} \tag{4}$$

where we have again written a_0/Z for a_Z.

Combining the formula (3) for $l \neq 0$ and the formula (4) for $l = 0$ we have

$$\begin{aligned}&\langle \Phi_{n,j,m,l}|\hat{H}_{LS} + \hat{H}_{\rm D}|\Phi_{n,j,m,l}\rangle \\ &= \frac{1}{4n^3}m_0c^2(Z\alpha)^4 \begin{cases} \left[(l+1)\left(l+\frac{1}{2}\right)\right]^{-1} & \text{for } j = l+1/2 \\ -\left[\left(l+\frac{1}{2}\right)l\right]^{-1} & \text{for } j = l-1/2 \end{cases}.\end{aligned} \tag{5}$$

The *kinetic energy correction* can be written as

$$-\frac{\hat{p}^2\hat{p}^2}{8m_0^3c^2} = -\frac{1}{2m_0c^2}\left(\hat{H}_Z + \frac{Ze^2}{r}\right)\left(\hat{H}_Z + \frac{Ze^2}{r}\right),$$

where $\hat{H}_Z$ is the unperturbed Hamiltonian (2.13). Hence

$$\begin{aligned}\langle \Phi_{n,j,m,l}|\hat{H}_{\rm ke}|\Phi_{n,j,m,l}\rangle &= \frac{1}{2m_0c^2}\langle \Phi_{n,j,m,l}|\left(\hat{H}_Z + \frac{Ze^2}{r}\right)^2|\Phi_{n,j,m,l}\rangle \\ &= -\frac{1}{2m_0c^2}\times \\ &\left[\left(\frac{\mathcal{R}_Z}{n^2}\right)^2 - 2\frac{\mathcal{R}_Z}{n^2}\langle \Phi_{n,j,m,l}|\frac{Ze^2}{r}|\Phi_{n,j,m,l}\rangle + \langle \Phi_{n,j,m,l}|\frac{Z^2e^4}{r^2}|\Phi_{n,j,m,l}\rangle\right].\end{aligned} \tag{6}$$

The expectation value of the potential energy $-Ze^2/r$ in the unperturbed eigenstates is just twice the total unperturbed energy $-\mathcal{R}_Z/n^2$ by the virial theorem. For the last term in the big square bracket in (6) we need to calculate an integral as in (2) above, but with x^{2l} instead of x^{2-l} in the integrand. This is easy to do directly for $n \leq 2$. A more general formula can be derived using the expression for the expectation value of $1/r^2$ as given by Bethe and Salpeter [BS77]:

$$\langle \Phi_{n,j,m,l}|\frac{1}{r^2}|\Phi_{n,j,m,l}\rangle = \frac{1}{n^3(l+\frac{1}{2})a_Z^2} = \frac{Z^2}{n^3(l+\frac{1}{2})a_0^2}\,. \tag{7}$$

Combining (6) and (7) gives

$$\langle \Phi_{n,j,m,l}|\hat{\mathrm{H}}_{\mathrm{ke}}|\Phi_{n,j,m,l}\rangle = \frac{m_0c^2}{2}(Z\alpha)^4\left[\frac{3}{4n^4} - \frac{1}{n^3\left(l+\frac{1}{2}\right)}\right] . \tag{8}$$

Note that the sum (5) + (8) of the relativistic corrections in first-order perturbation theory agrees with the leading terms of the expansion of the exact eigenvalues according to (2.36).

2.3. The wave function $\psi(r)$ is the normalized 1s eigenfunction of the single-particle hydrogenic Hamiltonian corresponding to a charge number Z', defined such that the Bohr radius $\hbar^2/(Z'\mu e^2)$ coincides with β,

$$\frac{\hbar^2}{Z'\mu e^2} = \beta\,, \quad Z'e^2 = \frac{\hbar^2}{\mu\beta}\,. \tag{1}$$

The two-electron wave function Ψ is essentially the symmetric product $\psi(r_1)\psi(r_2)$ of the spatial one-electron wave functions; the antisymmetric spin-parts give trivial factors unity in all matrix elements.

The expectation value of the one-body part of the Hamiltonian is given according to (2.73), and the one-electron matrix elements can be calculated in a straightforward way. It is more elegant to exploit the virial theorem by which the expectation value of the one-electron kinetic energy in ψ is just minus the one-electron total energy $-\mathcal{R}_{Z'} = -(Z')^2\mathcal{R}$ of the hydrogenic Hamiltonian corresponding to charge number Z'. Similarly, the expectation value of the one-electron potential energy $-Z'e^2/r$ is twice the total energy and hence the expectation value of $-Ze^2/r$ is $-2ZZ'\mathcal{R}$. Summing the contributions for the two electrons we obtain the following result for the expectation value of the one-body terms in the Hamiltonian $\hat{H}$:

$$\langle\Psi|\sum_{i=1,2}\left(\frac{\hat{p}_i^2}{2\mu} - \frac{Ze^2}{r_i}\right)|\Psi\rangle = [2(Z')^2 - 4ZZ']\mathcal{R}\,. \tag{2}$$

For the expectation value of the interaction term we exploit the hint and obtain

$$\langle\Psi|\frac{e^2}{|r_1 - r_2|}|\Psi\rangle = \frac{4\pi e^2}{\pi^2\beta^6} \times$$

$$\sum_{l,m}\int \mathrm{d}r_1 \int \mathrm{d}r_2\, \mathrm{e}^{-2r_1/\beta}\mathrm{e}^{-2r_2/\beta}\frac{r_<^l}{(2l+1)r_>^{l+1}}Y_{l,m}^*(\Omega_1)Y_{l,m}(\Omega_2) \tag{3}$$

$$= \frac{\mathrm{e}^2}{2\beta}\int_0^\infty \mathrm{d}x_1 \int_0^\infty \mathrm{d}x_2 \frac{x_1^2x_2^2}{x_>}\,\mathrm{e}^{-(x_1+x_2)}\,,$$

where we have used the substitutions $x_i = 2r_i/\beta$, and $x_>$ is the larger of x_1, x_2. The integral can be evaluated with elementary means,

$$\int_0^\infty \mathrm{d}x_1 x_1^2\,\mathrm{e}^{-x_1}\left[\frac{1}{x_1}\int_0^{x_1} x_2^2\,\mathrm{e}^{-x_2}\mathrm{d}x_2 + \int_{x_1}^\infty x_2\,\mathrm{e}^{-x_2}\mathrm{d}x_2\right] = \frac{5}{4},$$

and hence

$$\langle\Psi|\frac{e^2}{|r_1 - r_2|}|\Psi\rangle = \frac{5}{4}\frac{e^2}{2\beta} = \frac{5}{4}Z'\mathcal{R}. \tag{4}$$

Thus the total energy expectation value is

$$\langle E\rangle = \left\langle\Psi|\hat{H}|\Psi\right\rangle = \left[2(Z')^2 - 4ZZ' + \frac{5}{4}Z'\right]\mathcal{R}.$$

The minimum of $\langle E\rangle$ corresponding to $\mathrm{d}\langle E\rangle/\mathrm{d}Z' = 0$ occurs at

$$Z' = Z - \frac{5}{16} \quad \text{corresponding to} \quad \beta = \frac{\hbar^2}{\mu e^2(Z - 5/16)}, \tag{5}$$

and the minimum energy is

$$\langle E\rangle_{\min} = \left(-2Z^2 + \frac{5}{4}Z - \frac{25}{128}\right)\mathcal{R}.$$

For charge numbers up to $Z = 10$ we obtain the following energies (in atomic units, $2\mathcal{R}$) which compare quite favourably with the results of Hartree-Fock calculations as listed in Table 2.1 in Sect. 2.3.2:

Z	1	2	3	4	5
$\langle E\rangle_{\min}$	−0.473	−2.848	−7.223	−13.598	−21.973
Z	6	7	8	9	10
$\langle E\rangle_{\min}$	−32.348	−44.723	−59.098	−75.473	−93.848

For the H^- ion ($Z = 1$), the energy −0.473 a.u. lies, as does the (restricted) Hartree-Fock energy −0.4879 ... a.u. listed in Table 2.1, above the energy −0.5 a.u. of a hydrogen atom plus a free electron at at rest. Chandrasekhar [Cha44] showed that introducing correlations via the simple ansatz

$$\psi(r_1, r_2) = \frac{u(r_1, r_2)}{r_1 r_2}, \quad u(r_1, r_2) \propto \mathrm{e}^{-\alpha r_1 - \beta r_2}\,\mathrm{e}^{-\alpha r_2 - \beta r_1}$$

already leads to a variational minimum energy of −0.513303 a.u. The radial wave function $u(r_1, r_2)$ lies within the space defined by the s-wave model, cf. Sect. 4.5.3, and the full wave function $\psi(r_1, r_2)$ contains no dependence at all on the angular variables. The calculation in [Cha44] shows than radial correlations alone already account for a large part of the binding energy of the H^- ion, see also [ME00].

For part b) of the problem we need the 1s wave function of Problem 2.3 a) and the 2p one-electron wave functions $\psi_{p,m}$:

$$\psi_s(r) = \frac{\phi_1(r)}{r} Y_{0,0}(\Omega)\,, \quad \psi_{p,m}(r) = \frac{\phi_2(r)}{r} Y_{l=1,m}(\Omega)\,. \tag{6}$$

Note that both one-electron wave functions correspond to the Bohr radius β defined by (1). The two-electron singlet and triplet wave functions in LS coupling are

$$\begin{aligned}\Psi_s &= \frac{1}{\sqrt{2}}[\psi_s(r_1)\psi_{p,m}(r_2) + \psi_s(r_2)\psi_{p,m}(r_1)]\chi(S=0)\,,\\ \Psi_t &= \frac{1}{\sqrt{2}}[\psi_s(r_1)\psi_{p,m}(r_2) - \psi_s(r_2)\psi_{p,m}(r_1)]\chi(S=1)\,.\end{aligned} \tag{7}$$

The symbol χ stands for the antisymmetric ($S = 0$) or symmetric ($S = 1$) spin part of the two-electron wave function. The subscript m in (7) labels the azimuthal quantum number of the one-electron p-orbital and is at the same time the quantum number of the z-component of the total orbital angular momentum.

The expectation value of the one-body part of $\hat{H}$ can be calculated as in part a) above, except that the total one-body energy of the hydrogenic Hamiltonian corresponding to charge number Z' now is $-Z'^2\mathcal{R}/4$ in the second single-particle state. Thus equation (2) above is modified to

$$\langle\Psi|\sum_{i=1,2}\left(\frac{\hat{p}_i^2}{2\mu} - \frac{Ze^2}{r_i}\right)|\Psi\rangle = \left[\frac{5}{4}(Z')^2 - \frac{5}{2}ZZ'\right]\mathcal{R}\,, \tag{8}$$

and this holds for both singlet and triplet states (7).

The difference between singlet and triplet states shows up in the expectation value of the interaction term. For example for the singlet state we have

$$\begin{aligned}&\langle\psi_s|\frac{e^2}{|r_1 - r_2|}|\Psi_s\rangle\\ &= \frac{1}{2}\Bigg[\langle\psi_s(r_1)\psi_{p,m}(r_2)|\frac{e^2}{|r_1 - r_2|}|\psi_s(r_1)\psi_{p,m}(r_2)\rangle\\ &\quad +\langle\psi_{p,m}(r_1)\psi_s(r_2)|\frac{e^2}{|r_1 - r_2|}|\psi_{p,m}(r_1)\psi_s(r_2)\rangle\end{aligned} \tag{9}$$

$$+\langle\psi_s(r_1)\psi_{p,m}(r_2)|\frac{e^2}{|r_1-r_2|}|\psi_{p,m}(r_1)\psi_s(r_2)\rangle$$
$$+\langle\psi_{p,m}(r_1)\psi_s(r_2)|\frac{e^2}{|r_1-r_2|}|\psi_s(r_1)\psi_{p,m}(r_2)\rangle\Big]$$
$$= E_\mathrm{d} + E_\mathrm{ex} ,$$

where we have introduced the abbreviations

$$\begin{aligned} E_\mathrm{d} &= \langle\psi_s(r_1)\psi_{p,m}(r_2)|\frac{e^2}{|r_1-r_2|}|\psi_s(r_1)\psi_{p,m}(r_2)\rangle \\ &= \langle\psi_{p,m}(r_1)\psi_s(r_2)|\frac{e^2}{|r_1-r_2|}|\psi_{p,m}(r_1)\psi_s(r_2)\rangle , \\ E_\mathrm{ex} &= \langle\psi_s(r_1)\psi_{p,m}(r_2)|\frac{e^2}{|r_1-r_2|}|\psi_{p,m}(r_1)\psi_s(r_2)\rangle \\ &= \langle\psi_{p,m}(r_1)\psi_s(r_2)|\frac{e^2}{|r_1-r_2|}|\psi_s(r_1)\psi_{p,m}(r_2)\rangle . \end{aligned} \tag{10}$$

For the triplet state (9) is replaced by

$$\langle\Psi_\mathrm{t}|\frac{e^2}{|r_1-r_2|}|\Psi_\mathrm{t}\rangle = \mathrm{E}_\mathrm{d} - E_\mathrm{ex} . \tag{11}$$

The task now is to calculate the direct and exchange parts of the interaction energy as defined by (10).

For the direct part we have

$$\begin{aligned} E_\mathrm{d} &= \sum_{l',m'} \frac{e^2}{2l'+1} \times \\ &\int \mathrm{d}r_1 \int \mathrm{d}r_2 \, \frac{[\phi_1(r_1)\phi_2(r_2)]^2}{r_1^2 r_2^2} |Y_{1,m}(\Omega_2)|^2 \frac{r_<^{l'}}{r_>^{l'+1}} Y^*_{l',m'}(\Omega_1) Y_{l',m'}(\Omega_2) \\ &= e^2 \int_0^\infty \mathrm{d}r_1 \int_0^\infty \mathrm{d}r_2 \, \frac{[\phi_1(r_1)\phi_2(r_2)]^2}{r_>} . \end{aligned} \tag{12}$$

The last line in (12) follows from the fact that the integral over the angles Ω_1 reduces the sum over l' and m' to the single term $l' = 0$. Inserting the explicit forms of the radial wave functions,

$$\phi_1(r) = \frac{2r}{\beta^{3/2}} \, e^{-r/\beta} , \quad \phi_2(r) = \frac{r^2}{2\sqrt{6}\beta^{5/2}} \, e^{-r/(2\beta)} , \tag{13}$$

and implementing the substitutions $x_i = r_i/\beta$ leads to

$$E_\mathrm{d} = \frac{\mathrm{e}^2}{6\beta}\int_0^\infty \mathrm{d}x_1 \int_0^\infty \mathrm{d}x_2\, \frac{x_1^2 x_2^4}{x_>}\,\mathrm{e}^{-2x_1}\,\mathrm{e}^{-x_2} = \frac{\mathrm{e}^2}{6\beta}\frac{118}{81} = \frac{118}{243} Z'\mathcal{R}\,.$$

For the exchange part we have

$$E_\mathrm{ex} = \sum_{l',m'} \frac{\mathrm{e}^2}{2l'+1}\int \mathrm{d}r_1 \int \mathrm{d}r_2\, \frac{\phi_1^*(r_1)\phi_2^*(r_2)\phi_2(r_1)\phi_1(r_2)}{r_1^2 r_2^2} \times$$
$$Y_{1,m}^*(\Omega_2)Y_{1,m}(\Omega_1)\frac{r_<^{l'}}{r_>^{l'+1}}Y_{l',m'}^*(\Omega_1)Y_{l',m'}(\Omega_2) \qquad (14)$$
$$= \frac{\mathrm{e}^2}{3}\int_0^\infty \mathrm{d}r_1 \int_0^\infty \mathrm{d}r_2\, \phi_1(r_1)\phi_2(r_2)\phi_2(r_1)\phi_1(r_2)\frac{r_<}{r_>^2}\,.$$

The last line in (14) follows from the fact that the integral over the angles Ω_1 and Ω_2 reduces the sum over l' and m' to the single term corresponding to $l' = 1$ and $m' = m$. Note that the result does not depend on the azimuthal quantum number m of the trial functions (7). Inserting the explicit forms of the radial wave functions (13) and implementing the substitutions $x_i = 3r_i/(2\beta)$ leads to

$$E_\mathrm{ex} = \frac{\mathrm{e}^2}{18\beta}\left(\frac{2}{3}\right)^7 \int_0^\infty \mathrm{d}x_1 \int_0^\infty \mathrm{d}x_2\, \frac{x_1^3 x_2^3 x_<}{x_>^2}\,\mathrm{e}^{-x_1}\,\mathrm{e}^{-x_2}$$
$$= \frac{\mathrm{e}^2}{18\beta}\left(\frac{2}{3}\right)^7 \frac{21}{4} = \frac{224}{6561} Z'\mathcal{R}\,.$$

2.4. The wave functions obeying the correct boundary conditions, namely $\psi(x,y,z) = 0$ if $x = 0$, $y = 0$ or $z = 0$, or if $x = L$, $y = L$ or $z = L$, are $\psi \propto \sin(k_x x)\sin(k_y y)\sin(k_z z)$,

$$k_z = \frac{\pi}{L}n_z\,,\, k_y = \frac{\pi}{L}n_y\,,\, k_z = \frac{\pi}{L}n_z\,, \quad n_i = 1,2,\ldots \quad i \equiv x,\, y,\, z\,.$$

Note that only positive k's count. Changing the sign of one of the wave numbers merely multiplies the total wave function by -1. The energy eigenvalues are

$$E_{n_x,n_y,n_z} = \frac{\hbar^2\pi^2}{2\mu L^2}(n_x^2 + n_y^2 + n_z^2)\,.$$

The number of states with energy up to E_F corresponds to the number of cubes of side length π/L which fit into the octant $k_x > 0$, $k_y > 0$, $k_z > 0$ of the sphere of radius k_F, and hence the number of states including spin is

$$N = 2 \times \frac{1}{8} \times \frac{4}{3}\pi k_F^3 \left(\frac{L}{\pi}\right)^3 = \frac{V}{3\pi^2} k_F^3 ,$$

in agreement with (2.103).

2.5. The eigenfunctions $\psi(x)$ obeying the correct boundary condition $\psi(0) = \psi(L) = 0$ are

$$\psi_n(x) = \sqrt{\frac{2}{L}} \sin k_n x , \quad k_n = \frac{n\pi}{L} , \quad n = 1, 2, \ldots ;$$

the factor $\sqrt{2/L}$ ensures normalization to unity. The number ρ_k of eigenstates per unit wave number k is the reciprocal of the separation of k-values and is equal to L/π. With $E = \hbar^2 k^2/(2\mu)$ the number ρ_E of eigenstates per unit energy is

$$\rho_E = \rho_k \left(\frac{dE}{dk}\right)^{-1} = \frac{L}{2\pi} \sqrt{\frac{2\mu}{\hbar^2 E}} .$$

2.6. Using $[\hat{A}^2, \hat{B}] = \hat{A}[\hat{A}, \hat{B}] + [\hat{A}, \hat{B}]\hat{A}$ and remembering (1.33) we have

$$\left[\hat{H}, \hat{b}^\dagger\right] = \frac{1}{\sqrt{2\hbar\omega}} \left(\frac{\omega}{2} \left[\hat{p}^2, x\right] - \frac{i\omega^2}{2} \left[x^2, \hat{p}\right] \right) = \hbar\omega \hat{b}^\dagger ,$$
$$\left[\hat{H}, \hat{b}\right] = \frac{1}{\sqrt{2\hbar\omega}} \left(\frac{\omega}{2} \left[\hat{p}^2, x\right] + \frac{i\omega^2}{2} \left[x^2, \hat{p}\right] \right) = -\hbar\omega \hat{b} . \qquad (1)$$

Hence

$$\hat{H}\left(\hat{b}^\dagger \psi_n\right) = \left(n + \frac{3}{2}\right) \hbar\omega \left(\hat{b}^\dagger \psi_n\right) , \quad \hat{H}\left(\hat{b} \psi_n\right) = \left(n - \frac{1}{2}\right) \hbar\omega \left(\hat{b} \psi_n\right) , \qquad (2)$$

showing that $\hat{b}^\dagger \psi_n$ is, within a factor, ψ_{n+1} and that $\hat{b}\psi_n$ is, within a factor, ψ_{n-1}. Each ψ_n is an eigenstates of $\hat{b}^\dagger \hat{b} = \left(\hat{H} - \frac{1}{2}\hbar\omega\right) / (\hbar\omega)$ with eigenvalue n. Let $\hat{b}\psi_n = c_n \psi_{n-1}$. Then

$$\langle \psi_n | \hat{b}^\dagger \hat{b} | \psi_n \rangle = n = |c_n|^2 \langle \psi_{n-1} | \psi_{n-1} \rangle = |c_n|^2 .$$

Except, possibly, for a phase, c_n must be equal to $\sqrt{n}$, and this also holds for $n = 0$. If $\hat{b}^\dagger \psi_n = d_n \psi_{n+1}$, then

$$\langle \psi_{n+1} | \hat{b}^\dagger | \psi_n \rangle = d_n = \langle \psi_n | \hat{b} | \psi_{n+1} \rangle^* = c_{n+1}^* ,$$

hence d_n must be equal to $\sqrt{n+1}$ (except, possibly, for a phase).

2.7. Since the transition matrix element in (2.192) does not depend on spin, we ignore the spin degrees of freedom and take the initial state to be

$$\Psi_{\rm i} = \frac{\phi_2(r)}{r} Y_{l=1,m}(\Omega) \,, \quad \phi_2(r) = \frac{r^2}{2\sqrt{6}a^{5/2}} \, {\rm e}^{-r/(2a)} \,, \tag{1}$$

where a is the Bohr radius. The only final state to which $\Psi_{\rm i}$ can decay is

$$\Psi_{\rm f} = \frac{\phi_1(r)}{r} Y_{0,0}(\Omega) \,, \quad \phi_1(r) = \frac{2r}{a^{3/2}} \, {\rm e}^{-r/a} \,. \tag{2}$$

If we express the vector r in spherical components as in (2.204), (2.206), then

$$\begin{aligned} |r_{\rm fi}|^2 &= |\langle \Psi_{\rm f}|r|\Psi_{\rm i}\rangle|^2 = \sum_{\nu=-1}^{1} |\langle \Psi_{\rm f}|r^{(\nu)}|\Psi_{\rm i}\rangle|^2 \\ &= \left(\int_0^\infty \phi_1(r)\phi_2(r) r \, {\rm d}r \right)^2 \times ({\rm CG})^2 \,, \end{aligned} \tag{3}$$

with the Clebsch-Gordan coefficient

$$({\rm CG}) = \langle 00|1,-m,1,m\rangle = \pm \frac{1}{\sqrt{3}} \,.$$

The last line can be obtained e.g. by exploiting (A.11), (A.12) in Appendix A.1.

The total decay probability per unit time is given by (2.192), (2.193) in Appendix A.1 and is

$$\begin{aligned} P_{\rm i} &= \frac{4}{3}\frac{{\rm e}^2\omega^3}{\hbar c^3}|r_{\rm fi}|^2 = \frac{4}{9}\frac{{\rm e}^2\omega^3}{\hbar c^3}\left(\frac{1}{\sqrt{6}a^4} \int_0^\infty r^4 \, {\rm e}^{-3r/(2a)} {\rm d}r \right)^2 \\ &= \frac{4}{9}\frac{{\rm e}^2\omega^3}{\hbar c^3}\frac{a^2}{6}\left(\frac{2}{3}\right)^{10} \left(\int_0^\infty x^4 \, {\rm e}^{-x} {\rm d}x \right)^2 = \frac{4}{9}\frac{{\rm e}^2\omega^3}{\hbar c^3} 96 a^2 \left(\frac{2}{3}\right)^{10} \\ &= \frac{4}{9} 96 \left(\frac{2}{3}\right)^{10} \frac{(\alpha_{\rm fs})^3}{4} \omega \left(\frac{\hbar\omega}{\mathcal{R}}\right)^2 = 6.268 \times 10^8 \, {\rm s}^{-1} \,, \end{aligned}$$

and so the lifetime of the state is $\tau = 1/P_{\rm i} = 1.595 \times 10^{-9}$ seconds. Note that the decay probability does not depend on the m quantum number of the initial state, so averaging over the three degenerate initial p-states states doesn't affect the result.

2.8. In this case we have

$$\mu \frac{\rm i}{\hbar} \left[\hat{H}_{\rm A}, r_i \right] = \hat{p}_i + \frac{\mu}{2m_{\rm nuc}} \sum_{k \neq l} [\hat{p}_k \cdot \hat{p}_l, r_i] \,.$$

Using $\left[\hat{A}\hat{B},\hat{C}\right]=\hat{A}\left[\hat{B},\hat{C}\right]+\left[\hat{A},\hat{C}\right]\hat{B}$ and remembering (1.33) we have

$$[\hat{p}_k\cdot\hat{p}_l, r_i] = \frac{\hbar}{\mathrm{i}}\left(\hat{p}_k\delta_{l,i}+\hat{p}_l\delta_{k,i}\right)\,,$$

and so

$$\mu\frac{\mathrm{i}}{\hbar}\left[\hat{\mathrm{H}}_{\mathrm{A}},r_i\right]=\hat{p}_i+\frac{\mu}{m_{\mathrm{nuc}}}\sum_{k\neq i}\hat{p}_k\,. \tag{1}$$

In obtaining (2.189) we inserted $\sum_{i=1}^{N}\mu\frac{\mathrm{i}}{\hbar}\left[\hat{H}_{\mathrm{A}},r_i\right]$ for $\sum_{i=1}^{N}\hat{p}_i$, whereas (1) shows that

$$\sum_{i=1}^{N}\mu\frac{\mathrm{i}}{\hbar}\left[\hat{H}_{\mathrm{A}},r_i\right]=\sum_{i=1}^{N}\hat{p}_i+(N-1)\frac{\mu}{m_{\mathrm{nuc}}}\sum_{i=1}^{N}\hat{p}_i\,,$$

i.e. we should have inserted

$$\frac{m_{\mathrm{nuc}}}{m_{\mathrm{nuc}}+(N-1)\,\mu}\sum_{i=1}^{N}\mu\frac{\mathrm{i}}{\hbar}\left[\hat{H}_{\mathrm{A}},r_i\right]\quad\text{for}\quad\sum_{i=1}^{N}\hat{p}_i\,.$$

Formula (2.189) is modified in that the right-hand side acquires an additional factor $m_{\mathrm{nuc}}^2/[m_{\mathrm{nuc}}+(N-1)\mu]^2$. In formula (2.220) the right-hand side acquires an additional factor $m_{\mathrm{nuc}}/[m_{\mathrm{nuc}}+(N-1)\mu]$.

3.1. The contributions of the two integrals in (3.135) in the region $r > r_0$ cancel, provided the (common) outer classical turning point lies beyond r_0, which is the case for sufficiently large n, i.e. sufficiently close to the threshold $E = 0$. The energy E can be neglected in the remaining finite integrals, giving

$$\pi\,\mu_{n,\ell}=\int_{a_V}^{r_0}\sqrt{\frac{2}{a_Z\,r}-\frac{(l+1/2)^2}{r^2}}\,\mathrm{d}r-\int_{a_C}^{r_0}\sqrt{\frac{2}{ar}-\frac{(l+1/2)^2}{r^2}}\,\mathrm{d}r \tag{1}$$

in the limit $n\to\infty$. In (1), $a=\hbar^2/(\mu e^2)$ is the Bohr radius (corresponding to charge number unity), $a_Z=a/Z$ is the Bohr radius corresponding to charge number Z, a_V is the inner classical turning point of the full potential given by

$$a_V=\frac{(l+1/2)^2}{2}\frac{a}{Z} \tag{2}$$

and a_C is the inner classical turning point of the pure Coulomb potential,

$$a_C=\frac{(l+1/2)^2}{2}\,a\,. \tag{3}$$

We have made use of the Langer modification and replaced $l(l+1)$ by $\left(l+\frac{1}{2}\right)^2$.

For sufficiently small l-values analytic integration of (1) gives

$$\pi\mu_{n,l} \overset{n\to\infty}{=} (2l+1)\left[\lambda_Z - \arctan\lambda_Z - (\lambda - \arctan\lambda)\right] , \qquad (4)$$

with the l-dependent parameters

$$\lambda = \sqrt{\frac{2r_0}{a(l+1/2)^2} - 1} , \quad \lambda_Z = \sqrt{\frac{2r_0}{a_Z(l+1/2)^2} - 1} = \sqrt{\frac{2Zr_0}{a(l+1/2)^2} - 1} . \qquad (5)$$

If $l+\frac{1}{2} > \sqrt{2r_0/a}$, then the inner turning point in the pure Coulomb potential lies beyond r_0 and the terms containing λ in (4) don't contribute. If $l+\frac{1}{2} > \sqrt{2Zr_0/a}$, then the inner turning point in the full potential also lies beyond r_0 and the (semi-classical) quantum defect vanishes.

Taking $r_0 = a/3$ and $Z = 19$ as a rough model for potassium yields the following quantum defects according to (4): 1.667 for $l = 0$, 0.970 for $l = 1$, 0.352 for $l = 2$, 0.005 for $l = 3$ and zero for higher l-values. (Compare Fig. 3.9.)

3.2. The sign of an oscillator strength $f_{n_\mathrm{f},n_\mathrm{i}}$ (or mean oscillator strength $\bar{f}_{n_\mathrm{f},n_\mathrm{i}}$) is determined by the sign of the transition energy $\hbar\omega = \varepsilon_\mathrm{f} - \varepsilon_\mathrm{i}$ (see Sect. 2.4.6). Oscillator strengths are negative for emission, $\varepsilon_\mathrm{f} < \varepsilon_\mathrm{i}$, and positive for absorption, $\varepsilon_\mathrm{f} > \varepsilon_\mathrm{i}$, from a given initial state Φ_i. The inequalities in energy can be replaced by inequalities in the principal quantum number n, because ε depends monotonically on n. For transitions in which the orbital angular momentum quantum number l increases by unity, the upper equation (3.155) says that the sum of all mean oscillator strengths is positive, i.e. the sum of all (positive) oscillator strengths corresponding to an increase of the n must outweigh the sum of all (negative) oscillator strengths corresponding to a decrease of n. Conversely, the sum of all oscillator strengths for transitions in which l decreases by unity is negative according to the lower equation (3.155), and hence the oscillator strengths in which n decreases dominate.

According to Table 1.4 the three radial wave functions relevant for the concrete example are

$$\phi_{2p}(r) = \frac{r^2}{2\sqrt{6}a^{5/2}}\,\mathrm{e}^{-r/(2a)} , \quad \phi_{3d}(r) = \frac{4}{81\sqrt{30}}\frac{r^3}{a^{7/2}}\,\mathrm{e}^{-r/(3a)} ,$$

$$\phi_{3s}(r) = \frac{r}{9\sqrt{3}a^{3/2}}\left(6 - 4\frac{r}{a} + \frac{4}{9}\frac{r^2}{a^2}\right)\mathrm{e}^{-r/(3a)} ,$$

and the corresponding radial integrals are

$$\int_0^\infty \phi_{3s}(r)\, r\, \phi_{2p}(r)\,\mathrm{d}r = \frac{1}{54\sqrt{2}}\int_0^\infty \left(\frac{r}{a}\right)^4 \left(6-4\frac{r}{a}+\frac{4}{9}\frac{r^2}{a^2}\right) \mathrm{e}^{-5r/(6a)}\,\mathrm{d}r$$
$$= \frac{4a}{9\sqrt{2}}\left(\frac{6}{5}\right)^6 ,$$
$$\int_0^\infty \phi_{3d}(r)\, r\, \phi_{2p}(r)\,\mathrm{d}r = \frac{1}{243\sqrt{5}}\int_0^\infty \left(\frac{r}{a}\right)^6 \mathrm{e}^{-5r/(6a)}\,\mathrm{d}r = \frac{2^{11}3^4}{5^6\sqrt{5}}\,a\,.$$

With $\hbar\omega = \left(\frac{1}{4}-\frac{1}{9}\right)\mathcal{R} = \frac{5}{72}\hbar^2/(\mu a^2)$ we have

$$\bar{f}_{3s,2p} = \frac{5}{108}\frac{1}{3a^2}\left(\int_0^\infty \phi_{3s}(r)\, r\, \phi_{2p}(r)\,\mathrm{d}r\right)^2 = \frac{2^{13}3^4}{5^{11}} = 0.0136\,,$$
$$\bar{f}_{3d,2p} = \frac{5}{108}\frac{2}{3a^2}\left(\int_0^\infty \phi_{3d}(r)\, r\, \phi_{2p}(r)\,\mathrm{d}r\right)^2 = \frac{2^{21}3^4}{5^{12}} = 0.6958\,.$$

Transitions from the $2p$ state to d-states must have non-negative oscillator strengths, because the $n = 1$ shell only contains s-states. According to the upper equation (3.155) the sum of all these oscillator strengths is 10/9. As shown above, the transition to the $3d$ state already exhausts more than 60 % of the sum.

3.3. Part a) of the problem is essentially the same as Problem 1.6b) in Chap. 1. Writing $\langle\phi_{02}|V_{2,3}|\phi_{03}\rangle$ as $W_{2,3}$, the energy eigenvalues are

$$E_\pm = \frac{E_{02}+E_{03}}{2} \pm \sqrt{W_{2,3}^2 + \left(\frac{E_{02}-E_{03}}{2}\right)^2}\,, \tag{1}$$

and the (normalized) eigenstates can be expressed as

$$a_2 = \frac{1}{\sqrt{2}}\sqrt{1+\frac{\alpha}{\sqrt{1+\alpha^2}}}\,,\quad a_3 = \frac{1}{\sqrt{2}}\sqrt{1-\frac{\alpha}{\sqrt{1+\alpha^2}}}\,,$$
$$b_2 = \frac{1}{\sqrt{2}}\sqrt{1-\frac{\alpha}{\sqrt{1+\alpha^2}}}\,,\quad b_3 = -\frac{1}{\sqrt{2}}\sqrt{1+\frac{\alpha}{\sqrt{1+\alpha^2}}}\,. \tag{2}$$

We have assumed phases chosen such that $W_{2,3} \geq 0$ and used the abbreviation

$$\alpha = \frac{E_{02}-E_{03}}{2W_{2,3}}\,. \tag{3}$$

In the spirit of the Golden Rule, the decay width due to decay to the energy normalized regular wave function ϕ_{reg} in the open channel 1 is (cf. (1.232), (2.144))

$$\Gamma_\pm = 2\pi\left|\langle\psi_\pm|V|\phi_{\mathrm{reg}}\rangle\right|^2 .$$

Inserting the wave function ψ_+ given by the (real) coefficients a_2, a_3 in (2) we obtain

$$\begin{aligned} \Gamma_+ &= 2\pi |a_2 \langle \phi_{02}|V_{2,1}|\phi_{\text{reg}}\rangle + a_3 \langle \phi_{03}|V_{3,1}|\phi_{\text{reg}}\rangle|^2 \\ &= \pi (W_{2,1}^2 + W_{3,1}^2) + \pi \frac{\frac{E_{02}-E_{03}}{2}\left(W_{2,1}^2 - W_{3,1}^2\right) + 2W_{2,1}W_{3,1}W_{2,3}}{\sqrt{\left(\frac{E_{02}-E_{03}}{2}\right)^2 + W_{2,3}^2}}, \end{aligned} \tag{4}$$

where we have written $W_{i,1}$ for $\langle \phi_{0i}|V_{i,1}|\phi_{\text{reg}}\rangle$, $i = 2, 3$. The same calculation for the second solution ψ_- gives

$$\Gamma_- = \pi (W_{2,1}^2 + W_{3,1}^2) - \pi \frac{\frac{E_{02}-E_{03}}{2}\left(W_{2,1}^2 - W_{3,1}^2\right) + 2W_{2,1}W_{3,1}W_{2,3}}{\sqrt{\frac{(E_{02}-E_{03})^2}{2} + W_{2,3}^2}}. \tag{5}$$

Equations (4), (5) have the same structure as equation (3.213), which was obtained via the exact solution of the Schrödinger equation in the space spanned by the whole continuum channel 1 and the two isolated bound states in the channels 2 and 3. Also equation (1) above has the same structure as equation (3.208). The perturbative treatment in this problem misses the Green's function corrections to the resonance energies, cf. (3.203), and to the direct coupling matrix element, cf. (3.204).

3.4. The structure of the oscillator strength function becomes clearer if we write

$$\frac{\mathrm{d}f_{E\mathrm{i}}}{\mathrm{d}E} = \frac{2\mu}{\hbar}\omega d_1^2 \times Q, \quad Q = \frac{\left(1 + \frac{d_2}{d_1}\frac{W_{2,1}}{E-\varepsilon_2} + \frac{d_3}{d_1}\frac{W_{3,1}}{E-\varepsilon_3}\right)^2}{1 + \left(\frac{\pi W_{2,1}^2}{E-\varepsilon_2} + \frac{\pi W_{3,1}^2}{E-\varepsilon_3}\right)^2}. \tag{1}$$

The quotient Q can formally be written as a Beutler-Fano function,

$$Q = \frac{(q+\varepsilon)^2}{1+\varepsilon^2}, \tag{2}$$

with the energy dependent parameter q and the "reduced energy" ε given by

$$\begin{aligned} q &= \frac{\frac{d_2}{d_1}W_{2,1}(E-\varepsilon_3) + \frac{d_3}{d_1}W_{3,1}(E-\varepsilon_2)}{\pi W_{2,1}^2 (E-\varepsilon_3) + \pi W_{3,1}^2 (E-\varepsilon_2)}, \\ \varepsilon &= \left(\frac{\pi W_{2,1}^2}{E-\varepsilon_2} + \frac{\pi W_{3,1}^2}{E-\varepsilon_3}\right)^{-1}. \end{aligned} \tag{3}$$

The reduced energy ε has a pole at the energy

$$E_{\mathrm{P}} = \frac{W_{2,1}^2\varepsilon_3 + W_{3,1}^2\varepsilon_2}{W_{2,1}^2 + W_{3,1}^2}, \tag{4}$$

which lies between ε_2 and ε_3. The reduced energy varies from $\varepsilon = -\infty$ at $E = -\infty$ to $\varepsilon = +\infty$ at $E = E_P$, and again from $\varepsilon = -\infty$ at $E = E_P$ to $\varepsilon = +\infty$ at $E = +\infty$. Near ε_2 the reduced energy ε is approximately $(E-\varepsilon_2)/(\pi W_{2,1}^2)$, near ε_3 it is approximately $(E-\varepsilon_3)/(\pi W_{3,1}^2)$. Thus we expect two Beutler-Fano type resonances whose low-energy or high-energy tails are contracted into the region above or below E_P respectively. If the widths $2\pi W_{2,1}^2$ and $2\pi W_{3,1}^2$ are sufficiently small, then the parameter q in (3) is roughly constant over the width of a resonance and we can assign shape parameters

$$q_2 = \frac{d_2/d_1}{\pi W_{2,1}}, \quad q_3 = \frac{d_3/d_1}{\pi W_{3,1}} \tag{5}$$

to the resonances around ε_2 and ε_3 respectively. The zeros of $\mathrm{d}f_{Ei}/\mathrm{d}E$ lie at $\varepsilon = -q_2$, $\varepsilon = -q_3$ and the maxima at $\varepsilon = 1/q_2$ an $\varepsilon = 1/q_3$. For small magnitudes of d_2/d_1, d_3/d_1 (compared with the magnitudes of $W_{2,1}$ and $W_{3,1}$) we expect window resonances, for large magnitudes we expect pronounced peaks (cf. Fig. 3.13).

The above considerations assume weak energy dependence of the parameter q in (3) over the width of a resonance. The zeros $Z_\pm$ of $\mathrm{d}f_{Ei}/\mathrm{d}E$ are given exactly as the zeros of the quadratic form

$$(E-\varepsilon_2)(E-\varepsilon_3) + \frac{d_2}{d_1}W_{2,1}(E-\varepsilon_3) + \frac{d_3}{d_1}W_{3,1}(E-\varepsilon_2)$$

and are

$$Z_\pm = \frac{\varepsilon_2+\varepsilon_3}{2} - \frac{1}{2}\left(\frac{d_2}{d_1}W_{2,1} + \frac{d_3}{d_1}W_{3,1}\right) \pm\frac{1}{2}\sqrt{\left[\varepsilon_2-\varepsilon_3-\left(\frac{d_2}{d_1}W_{2,1}-\frac{d_3}{d_1}W_{3,1}\right)\right]^2 + 4\frac{d_2d_3}{d_1^2}W_{2,1}W_{3,1}}\,. \tag{6}$$

Note that $W_{2,1}d_2/d_1$ and $W_{3,1}d_3/d_1$ have the dimensions of an energy. If the magnitudes of these numbers are small compared to $|\varepsilon_2-\varepsilon_3|$, then we can neglect the term proportional to d_2d_3/d_1^2 under the square root in (6) and obtain the two zeros

$$Z = \varepsilon_2 - \frac{d_2}{d_1}W_{2,1} \quad \text{and} \quad Z = \varepsilon_3 - \frac{d_3}{d_1}W_{3,1}\,. \tag{7}$$

This result agress with the result following from zeros of the reduced energy $\varepsilon \approx (E-\varepsilon_2)/(\pi W_{2,1}^2)$ at $-q_2$ or of $\varepsilon \approx (E-\varepsilon_3)/(\pi W_{3,1}^2)$ at $-q_3$ as obtained above. If $W_{2,1}d_2/d_1$ and $W_{3,1}d_3/d_1$ have large magnitudes compared to $|\varepsilon_2-\varepsilon_3|$, then we can neglect the epsilons under the square root in (6) and obtain one zero near the average energy $(\varepsilon_2+\varepsilon_3)/2$ and one zero shifted by $-[(d_2/d_1)W_{2,1} + (d_3/d_1)W_{3,1}]$.

3.5. It is more accurate to first read off the quantum defects μ_n and then to calculate the energies via $E_n/\mathcal{R} = -1/(n-\mu_n)^2$. Results for $\Gamma = 0.01\mathcal{R}$ are (approximately):

n	3	4	5	6	7	8	9	10
$-E_n/\mathcal{R}$	0.1126	0.0647	0.0445	0.0349	0.0267	0.0199	0.0154	0.0122
μ_n	0.02	0.07	0.26	0.72	0.88	0.92	0.93	0.94

Results for $\Gamma = 0.001\ \mathcal{R}$ are (approximately):

n	3	4	5	6	7	8	9	10
$-E_n/\mathcal{R}$	0.1111	0.0628	0.0415	0.0386	0.0277	0.0204	0.0156	0.0123
μ_n	0	0.01	0.09	0.91	0.99	0.995	1.0	1.0

For $\Gamma \to 0$ the perturber only affects the $n = 5$ state at $E = -0.04\ \mathcal{R}$. For small but finite Γ there is one energy just below and one just above $-0.04\ \mathcal{R}$. For vanishing Γ this energy becomes degenerate. Explicitly we have

n	3	4	5	6	7	8	9	10
$-E_n/\mathcal{R}$	0.1111	0.0625	0.04	0.04	0.0277	0.0204	0.0156	0.0123
μ_n	0	0	0	1	1	1	1	1

3.6. The quantum defects (modulo unity) of the bound states in Fig. 3.19 can be read off to be: –0.07, 0.04, 0.21, 0.48, 0.68, 0.76, 0.80[2] and 0.83. The first dot with an energy near $-0.25\mathcal{R}$ must correspond to an effective quantum number near 2, so we know where to start counting. The effective quantum numbers of the first eight states are thus 2.07, 2.96, 3.79, 4.52, 5.32, 6.24, 7.20 and 8.17, and the corresponding binding energies ($-E$) are (in Rydbergs): 0.233, 0.114, 0.070, 0.049, 0.035, 0.026, 0.019 and 0.015.

The energy of the perturber is the point of maximum gradient of $\mu(E)$ which lies near $E = -0.05\mathcal{R}$. The width can be estimated according to (1.236) as $2/\pi$ divided by this maximum gradient which has a numerical value of at least $14.3/\mathcal{R}$. The background quantum defect is the amount by which the value of $\mu(E)$ differs from 1/2 at the energy of the perturber. In the present example the parameters $E_\mathrm{R} = -0.052\mathcal{R}$, $\Gamma = 0.035\mathcal{R}$ and a background quantum defect of –0.06, inserted in the formula (3.222), give quantum defects which differ by less than 0.02 from the values in Fig. 3.19 (except for the lowest and the highest energy when the difference is 0.04 and 0.03 respectively). The energy of the perturber relative to the series limit of the second channel ($\varepsilon_2 = 0$ in Fig. 3.19) is $E - I_2 = E_R - 0.125\mathcal{R} = -0.177\mathcal{R}$, which corresponds to an effective quantum number (in channel 2) $\nu_2 = \sqrt{\mathcal{R}/(I_2 - E)} = 2.38$. At the energy of the perturber $\nu_2 + \mu_2$ should be an integer, so μ_2 is 0.62

(modulo unity). From the width formula (3.256) we derive $R_{1,2}^2 = \pi \nu_2^3 \Gamma / 4\mathcal{R} = 0.371$.

Summary: $|R_{1,2}| = 0.61$, $\mu_1 = -0.062$, $\mu_2 = 0.62$.

3.7. Near a perturber the quantum defects lie on a curve (3.222)

$$\mu_n = \mu_0 - \frac{1}{\pi} \arctan \frac{\Gamma/2}{E - E_R} . \tag{1}$$

The closest approach of two adjacent levels n and $n+1$ can be expected when the state n is on the low-energy tail and the state $n+1$ is on the high energy tail of the arcus-tangent curve. Appropriate expansions of the arcus-tangent yield

$$2\pi(\mu_n - \mu_0) = -\frac{\Gamma}{E_n - E_R} , \quad 2\pi(\mu_{n+1} - \mu_0 - 1) = -\frac{\Gamma}{E_{n+1} - E_R} . \tag{2}$$

It is convenient to introduce the parameter α describing the ratio of the distance of E_{n+1} from E_R to the separation of E_n and E_{n+1}:

$$E_{n+1} - E_R = \alpha(E_{n+1} - E_n) , \quad E_n - E_R = (\alpha - 1)(E_{n+1} - E_n) . \tag{3}$$

From the energy formula in the (perturbed) Rydberg series we have

$$\frac{1}{E_n} - \frac{1}{E_{n+1}} = -\frac{1}{\mathcal{R}} \left[2n(\mu_{n+1} - 1 - \mu_n) + \mu_n^2 - (\mu_{n+1} - 1)^2 \right] . \tag{4}$$

We neglect the small (compared with n) quantities μ_n^2 and $(\mu_{n+1} - 1)^2$ on the right-hand side of (4) and replace the difference $\mu_{n+1} - 1 - \mu_n$ according to (2), (3):

$$\frac{(E_{n+1} - E_n)^2}{E_{n+1} E_n} = \frac{n\Gamma}{\pi \mathcal{R}} \left(\frac{1}{\alpha} + \frac{1}{1-\alpha} \right) . \tag{5}$$

We replace the product $E_n E_{n+1}$ by $\mathcal{R}^2/(n^*)^4$, where n^* is an effective quantum number corresponding to an energy between E_n and E_{n+1}. This gives:

$$\frac{(E_{n+1} - E_n)^2}{4\mathcal{R}^2/(n^*)^6} = \frac{\Gamma (n^*)^3}{\pi \mathcal{R}} \frac{1}{4} \left(\frac{1}{\alpha} + \frac{1}{1-\alpha} \right) , \tag{6}$$

where we have neglected the difference between the quantum number n in (5) and the effective quantum number n^*. The left-hand side of (6) is just the square of the energy difference relative to the unperturbed energy difference $2\mathcal{R}/(n^*)^3$. The expression in the big brackets on the right-hand side has its minimum at $\alpha = 0.5$ and the minimum value is four, hence the minimum of the energy difference relative to the unperturbed energy difference is $\sqrt{\Gamma (n^*)^3/(\pi \mathcal{R})}$. (See also [FW85].)

3.8. Since z is the $\nu = 0$ spherical component of the vector r we must have $m' = m$. The triangle condition and parity demand $l' = l \pm 1$. Hence the only non-vanishing matrix elements are between

$$\psi_{0,0} = \frac{\mathrm{e}^{-r/(2a)}}{a^{3/2}2\sqrt{2\pi}}\left(1 - \frac{r}{2a}\right) \quad \text{and} \quad \psi_{1,0} = \frac{r\,\mathrm{e}^{-r/(2a)}}{2\sqrt{6}a^{5/2}}Y_{1,0}(\theta)\,.$$

When calculating the matrix element of $eE_z z$ between these states we can exploit the fact that z is $\sqrt{4\pi/3r}$ times the (real) function $Y_{1,0}$ and that the angle integral over $Y_{1,0}^2$ gives unity. Thus

$$\begin{aligned}\langle\psi_{0,0}\,|eE_z z|\,\psi_{1,0}\rangle &= \frac{eE_z}{12a^4}\int_0^\infty\left(1 - \frac{r}{2a}\right)r^2\,\mathrm{e}^{-r/a}r^2\mathrm{d}r \\ &= \frac{eE_z a}{12}\int_0^\infty\left(x^4 - \frac{1}{2}x^5\right)\mathrm{e}^{-x}\mathrm{d}x = -3eE_z a\,.\end{aligned}$$

The matrix W of the perturbing operator is thus

$$\mathrm{W} = \begin{pmatrix} 0 & -3eE_z a \\ -3eE_z a & 0 \end{pmatrix}.$$

Its eigenvalues follow from the secular equation (cf. (1.259)), which in this case reads $E^2 = (3eE_z a)^2$, yielding

$$E_\pm = \pm 3eE_z a\,.$$

The corresponding (normalized) eigenstates are

$$\psi_+ = \frac{1}{\sqrt{2}}\left(\psi_{0,0} - \psi_{1,0}\right), \quad \psi_- = \frac{1}{\sqrt{2}}\left(\psi_{0,0} + \psi_{1,0}\right).$$

(See Fig. 3.24).

3.9. Remembering that $[\hat{A}\hat{B}, \hat{C}] = \hat{A}[\hat{B}, \hat{C}] + [\hat{A}, \hat{C}]\hat{B}$ and exploiting (1.33) and Problem 1.11 gives

$$[\hat{H}_0, \hat{b}] = -\frac{a\hbar^2}{\mu}\left[\frac{\partial}{\partial z}\left(a + \frac{r}{2}\right) + \frac{z}{2}\left(\frac{\partial}{\partial r} + \frac{1}{r}\right)\right].$$

Remembering that $\partial/\partial z = (z/r)\partial/\partial r$ we can verify the identity by straight-forward differentiation.

Now

$$\frac{|\langle\psi_m|z|\psi_0\rangle|^2}{E_m - E_0} = \frac{\mu}{\hbar^2}\frac{\langle\psi_0|z|\psi_m\rangle\langle\psi_m|[\hat{H}_0, \hat{b}]|\psi_0\rangle}{E_m - E_0}$$
$$= \frac{\mu}{\hbar^2}\langle\psi_0|z|\psi_m\rangle\langle\psi_m|\hat{b}|\psi_0\rangle\,.$$

Summing over all m gives $(\mu/\hbar^2)\langle\psi_0|z\hat{b}|\psi_0\rangle$ via the completeness relation, so the expression for the dipole polarizability becomes

$$\alpha_{\rm d} = \frac{2\mu e^2}{\hbar^2}\left\langle\psi_0|z\hat{b}|\psi_0\right\rangle = \frac{2\mu e^2}{\hbar^2}\left[a^2\left\langle\psi_0|z^2|\psi_0\right\rangle + \frac{a}{2}\left\langle\psi_0|z^2 r|\psi_0\right\rangle\right]$$
$$= \frac{2\mu e^2}{\hbar^2}\left(a^4 + \frac{5}{4}a^4\right) = \frac{9}{2}a^3\,.$$

3.10.

$$A_{\rm s}(r) = A_{\rm L}(r) + \frac{1}{2}\begin{pmatrix} y \\ x \\ 0 \end{pmatrix} B_z = A_{\rm L}(r) + \nabla\left(\frac{xy}{2}B_z\right)\,. \qquad (1)$$

From

$$\hat{p}\psi_{\rm s} = \frac{\hbar}{\rm i}\nabla\left({\rm e}^{-\frac{{\rm i}e}{\hbar c}f}\psi_{\rm L}\right) = -\frac{e}{c}\nabla f\,{\rm e}^{-\frac{{\rm i}e}{\hbar c}f}\psi_{\rm L} + {\rm e}^{-\frac{{\rm i}e}{\hbar c}f}\frac{\hbar}{\rm i}\nabla\psi_{\rm L}$$

we deduce

$$\left(\hat{p} + \frac{e}{c}(A_{\rm L} + \nabla f)\right){\rm e}^{-\frac{{\rm i}e}{\hbar c}f}\psi_{\rm L} = {\rm e}^{-\frac{{\rm i}e}{\hbar c}f}\left(\hat{p} + \frac{e}{c}A_{\rm L}\right)\psi_{\rm L}\,. \qquad (2)$$

Applying the big bracket on the left-hand side a second time yields an expression similar to the right-hand side of (2), but with $(\hat{p} + (e/c)A_{\rm L})\psi_{\rm L}$ taking the place of ψ_L. Thus

$$\left(\hat{p} + \frac{e}{c}(A_{\rm L} + \nabla f)\right)^2{\rm e}^{-\frac{{\rm i}e}{\hbar c}f}\psi_{\rm L} = {\rm e}^{-\frac{{\rm i}e}{\hbar c}f}\left(\hat{p} + \frac{e}{c}A_{\rm L}\right)^2\psi_{\rm L}$$
$$= {\rm e}^{-\frac{{\rm i}e}{\hbar c}f}2\mu E\psi_{\rm L}\,, \qquad (3)$$

where the lower line follows from the Schrödinger equation for ψ_L. Except for the factor 2μ, equation (3) is just the Schrödinger equation for $\psi_{\rm s} = {\rm e}^{-\frac{{\rm i}e}{\hbar c}f}\psi_{\rm L}$ with the vector potential $A_{\rm s} = A_{\rm L} + \nabla f$.

In the symmetric gauge the Hamiltonian is (cf. (3.316)):

$$\hat{H}_{\mathrm{s}} = \frac{\hat{p}^2}{2\mu} + \omega\hat{L}_z + \frac{\mu}{2}\omega^2(x^2+y^2)\,, \tag{4}$$

where $\omega = eB_z/(2\mu c) = \omega_c/2$ is half the cyclotron frequency. The azimuthal quantum number m is a good quantum number, and the motion of the electron parallel to the z-axis is that of a free particle. The motion perpendicular to the z-axis is that of a two-dimensional harmonic oscillator. A discussion of the two-dimensional harmonic oscillator (which is frequently neglected in the shadow of detailed treatments of the one- and the three-dimensional case) can be found e.g. in: J.D. Talman, *Nuclear Physics* vol. **A141** (1970) p. 273. In polar coordinates ($\rho = \sqrt{x^2+y^2}$, $\tan\phi = y/x$) the eigenfunctions of the two-dimensional oscillator are

$$\Psi_{N,m}(\rho,\phi) = \mathrm{e}^{im\phi}\,\psi_{N,m}(\rho)\,.$$

$\psi_{N,m}$ are the radial eigenfunctions

$$\psi_{N,m}(\rho) = \left(b\sqrt{\pi}\right)^{-1}\left[\frac{N!}{(N+|m|)!}\right]^{1/2}\left(\frac{\rho}{b}\right)^{|m|} L_N^{|m|}\left(\frac{\rho^2}{b^2}\right)\mathrm{e}^{-\rho^2/(2b^2)}\,,$$

where $b = \sqrt{\hbar(\mu\omega)}$ is the oscillator width and $L_N^{|m|}$ stands for the Laguerre polynomials. The corresponding eigenvalues of the two-dimensional oscillator part of the Hamiltonian are $(2N+|m|+1)\hbar\omega$.

The full wave functions are thus characterized by the good quantum numbers N and m for the motion perpendicular to the field and by the wave number k_z for the free motion parallel to the field:

$$\Psi_{k_z,N,m} = \mathrm{e}^{\mathrm{i}k_z z}\mathrm{e}^{im\phi}\,\psi_{N,m}(\rho)\,. \tag{5}$$

The total energy eigenvalues, including $\hbar^2k_z^2(2\mu)$ from the motion parallel to the field and $m\hbar\omega$ from the normal Zeeman term $\omega\hat{L}_z$, are:

$$E_{k_z,N,m} = \frac{\hbar^2k_z^2}{2\mu} + (2N+|m|+m+1)\hbar\omega\,, \tag{6}$$

with $-\infty < k_z < \infty$, $m = 0, \pm1, \pm2, \ldots$ and $N = 0, 1, 2, \ldots$.

In the Landau gauge the Hamiltonian is

$$\hat{H}_{\mathrm{L}} = \frac{\mu}{2}\omega_{\mathrm{c}}^2\left(y - \frac{\hat{p}_x}{\mu\omega_{\mathrm{c}}}\right)^2 + \frac{\hat{p}_y^2}{2\mu} + \frac{\hat{p}_z^2}{2\mu}\,. \tag{7}$$

The electron moves freely in the z-direction. Its momentum in x-direction is also a good quantum number, and the value of the x-momentum fixes the centre of the harmonic oscillator motion in y-direction. Note that the oscillator frequency for this one-dimensional vibratory motion now is the full cyclotron frequency ω_c.

The eigenfunctions in the Landau gauge are characterized by the wave numbers k_x and k_z for the good momenta in x- and z-directions and by the oscillator quantum number n for the one-dimensional oscillator motion in y-direction:

$$\Phi_{k_x,k_z,n} = \mathrm{e}^{\mathrm{i}k_x x}\mathrm{e}^{\mathrm{i}k_z z}\psi_n(y)\,, \tag{8}$$

where $\psi_n(y)$ are the eigenstates of the one-dimensional harmonic oscillator (cf. Sect. 5.2.2). The corresponding energy eigenvalues are

$$E_{k_x,k_z,n} = \frac{\hbar^2 k_z^2}{2\mu} + \left(n + \frac{1}{2}\right)\hbar\omega_c\,. \tag{9}$$

In both the symmetric gauge and the Landau gauge the energy contains a continuous term $\hbar^2 k_z^2(2\mu)$ for the free motion of the electron parallel to the field, as well as a discrete part consisting of odd multiples of $\frac{1}{2}\hbar\omega_c(=\hbar\omega)$ for the so-called *Landau states* describing the motion perpendicular to the field. All eigenvalues are highly degenerate. For given values of k_z and n in the Landau gauge, all values of k_x yield the same total energy, the corresponding wave functions differ by the reference point $y_0 = hk_x/\mu\omega_c$ around which the oscillatory motion is centered. For given values of k_z and $E_{\mathrm{osc}} = (2N_{\mathrm{max}} + 1)\hbar\omega$ in the symmetric gauge, all wave functions (5) with $m = 1, N = N_{\mathrm{max}} - 1, m = 2, N = N_{\mathrm{max}} - 2, \ldots\; m = N_{\mathrm{max}}, N = 0$, as well as all eigenfunctions with $N = N_{\mathrm{max}}, m \leq 0$ belong to the same energy (6).

From parts a) and b) we know that eigenstates in the different gauges are related by

$$\psi_s(r) = \exp\left(-\frac{\mathrm{i}}{\hbar}\mu\omega xy\right)\psi_L(r) = \mathrm{e}^{\frac{-\mathrm{i}xy}{b^2}}\psi_L(r)\,, \tag{10}$$

where $b = \sqrt{\hbar(\mu\omega)}$ is the oscillator width associated with half the cyclotron frequency. Because of the degeneracies, (10) doesn't imply a one-to-one relation between the eigenstates (5) and (8). If, for example, we wish to relate the eigenstate

$$\Psi_{0,0,0} = \frac{1}{b\sqrt{\pi}}\,\mathrm{e}^{-(x^2+y^2)/(2b^2)} \tag{11}$$

in the symmetric gauge to eigenstates of the same total energy $E = \hbar\omega = \frac{1}{2}\hbar\omega_c$ in the Landau gauge, we must allow superpositions of eigenstates $\Phi_{k_x,k_z=0,n=0}$ with various wave numbers k_x i.e.

$$\Psi_{0,0,0} = \mathrm{e}^{\frac{-\mathrm{i}xy}{b^2}} \int_{-\infty}^{\infty} a(k_x)\mathrm{e}^{\mathrm{i}k_x x}\left(\frac{\sqrt{2}}{\sqrt{\pi}b}\right)^{1/2} \exp\left[-\frac{1}{b^2}\left(y - \frac{\hbar k_x}{2\mu\omega}\right)^2\right] \mathrm{d}k_x \, ,$$

with an appropriate amplitude $a(k_x)$. (Note that the oscillator width in the Landau gauge is $b_\mathrm{c} = b\sqrt{2}$.) The choice

$$a(k_x) = \frac{\sqrt{b}}{(2\pi)^{3/4}}\, \mathrm{e}^{-b^2k_x^2/4}$$

does indeed produce the wave function (11).
3.11. Using the hint we obtain the approximate expression

$$\psi_B = \frac{1}{2m_0c}\, \hat{\sigma}\cdot\left(\hat{p} + \frac{e}{c}A\right) ,$$

which we insert into the upper equation to obtain

$$\frac{1}{2m_0}\left[\hat{\sigma}\cdot\left(\hat{p} + \frac{e}{c}A\right)\right]\left[\hat{\sigma}\cdot\left(\hat{p} + \frac{e}{c}A\right)\right]\psi_A = (E + e\Phi - m_0c^2)\psi_A \, .$$

With the help of the identity in Problem 2.1 we obtain

$$\left[\frac{1}{2m_0}\left(\hat{p} + \frac{e}{c}A\right)^2 + \frac{e}{2m_0c}\mathrm{i}\hat{\sigma}\cdot(\hat{p}\times A + A\times\hat{p}) - e\Phi\right]\psi_A = (E - m_0c^2)\psi_A \, .$$

The term $(\hat{p}\times A + A\times\hat{p})$ does not vanish, because $\hat{p}$ does not commute with $A(r)$. An operator $\frac{\partial}{\partial x}A_y$ actually means $\frac{\partial A_y}{\partial x} + A_y\frac{\partial}{\partial x}$ by virtue of the product rule, so $(\hat{p}\times A + A\times \hat{p}) = (\hbar/\mathrm{i})\nabla\times A = (\hbar/\mathrm{i})B$ and the corresponding contribution to the Hamiltonian is

$$\frac{e\hbar}{2m_0c}\hat{\sigma}\cdot B = \frac{e}{m_0c}\hat{S}\cdot B \, .$$

This is the spin contribution to the Hamiltonian (3.320). Note that the factor 2 in front of the the spin in (3.320) follows without further assumption from the Dirac equation.
4.1. Integrating the left-hand side of the asserted identity by parts we obtain, for the left-hand side,

$$\frac{-a}{\mathrm{i}a}\, \mathrm{e}^{\mathrm{i}a(1-x)}(1 + x)f(x)\Big|_{-1}^{1} + \lim_{a\to\infty}\frac{a}{\mathrm{i}a}\int_{-1}^{1} \mathrm{e}^{\mathrm{i}a(1-x)}\frac{\mathrm{d}}{\mathrm{d}x}[(1 + x)f(x)]\mathrm{d}x \, .$$

The first term above is independent of a and equals $2\mathrm{i}f(1)$ as required for the identity. The second term vanishes in the limit $a \to \infty$ because of the increasing frequency of the oscillating factor in the integrand.

Inserting the asymptotic form (4.3) for the wave function into the definition (4.4) of the current density gives

$$j \overset{r\to\infty}{=} \frac{\hbar k}{\mu} e_z + j_{\mathrm{interf}} + j_{\mathrm{out}} \,, \tag{1}$$

where j_{out} is the current density (4.5), and

$$\begin{aligned} j_{\mathrm{interf}} &= \frac{\hbar k}{2\mu}\left[\mathrm{e}^{\mathrm{i}kz} f^*(\theta,\phi)\frac{\mathrm{e}^{-\mathrm{i}kr}}{r}(e_r+e_z) + \mathrm{e}^{-\mathrm{i}kz} f(\theta,\phi)\frac{\mathrm{e}^{\mathrm{i}kr}}{r}(e_r+e_z)\right] \\ &= \frac{\hbar k}{2\mu r}\left[f(\theta,\phi)\,\mathrm{e}^{\mathrm{i}kr(1-\cos\theta)}(e_r+e_z) + f^*(\theta,\phi)\,\mathrm{e}^{-\mathrm{i}kr(1-\cos\theta)}(e_r+e_z)\right]. \end{aligned}$$

Introducing $x = \cos\theta$ and writing j_r for the radial component of j_{interf}, i.e. $j_r = j_{\mathrm{interf}} \cdot r/r$, we have

$$\begin{aligned} I_{\mathrm{interf}} &= \oint j_{\mathrm{interf}} \cdot \mathrm{d}s = r^2 \int_{-1}^{1} \mathrm{d}x \int_0^{2\pi} \mathrm{d}\phi \, j_r \\ &= \frac{\hbar k r}{2\mu}\left[\int_{-1}^{1} \mathrm{d}x\,\mathrm{e}^{\mathrm{i}kr(1-x)}(1+x)\int_0^{2\pi} f(\theta,\phi)\,\mathrm{d}\phi \right. \\ &\quad \left. + \int_{-1}^{1} \mathrm{d}x\,\mathrm{e}^{-\mathrm{i}kr(1-x)}(1+x)\int_0^{2\pi} f^*(\theta,\phi)\,\mathrm{d}\phi\right]. \end{aligned}$$

In the limit $kr \to \infty$ the first integral on the right-hand side contributes $(\mathrm{i}\hbar/\mu) \times 2\pi f(\theta = 0)$, because of the identity a). Note that $f(\theta,\phi)$ is independent of ϕ at $\theta = 0$. The corresponding identity for $-kr \to -\infty$ shows us that the second integral gives a contribution $-(\mathrm{i}\hbar/\mu) \times 2\pi f^*(\theta = 0)$. Thus

$$I_{\mathrm{interf}} = -\frac{\hbar}{\mu} 4\pi\, \Im f(\theta = 0) \quad \text{for} \quad r \to \infty \,. \tag{2}$$

The optical theorem follows from the observation that, since the first term in (1) doesn't contribute to the net flux on symmetry grounds, the sum of the fluxes I_{out} from (4.9) and I_{interf} from (2) above must vanish according to (4.8).

4.2. For $r' \ll r$ we have

$$|r - r'| = r\sqrt{1 - 2\frac{r\cdot r'}{r}} \approx r - \frac{r\cdot r'}{r} + O\left(\frac{r'^2}{r}\right),$$

and so the exponential can be approximated by

$$e^{ik|r-r'|} = e^{ikr}e^{-ik_r \cdot r'}\left[1 + O\left(kr'\frac{r'}{r}\right)\right].$$

Furthermore we have

$$\frac{1}{|r-r'|} = \frac{1}{r\sqrt{1-2\frac{r\cdot r'}{r^2}+\left(\frac{r'}{r}\right)^2}} \approx \frac{1}{r}\left[1 + O\left(\frac{r'}{r}\right)\right],$$

showing that the corrections to the leading term in the expression for $\mathcal{G}$ are smaller by a factor of the order r'/r.

4.3. In a cube of length L periodic boundary conditions are fulfilled for wave vectors

$$k = \begin{pmatrix} k_x \\ k_y \\ k_z \end{pmatrix}, \quad \text{with} \quad k_x = \frac{2\pi}{L}n_x\,, \quad k_y = \frac{2\pi}{L}n_y\,, \quad k_z = \frac{2\pi}{L}n_z\,,$$

$$n_z = 0\,, \pm 1, \pm 2, \ldots\,, \quad n_y = 0, \pm 1, \pm 2, \ldots\,, \quad n_z = 0, \pm 1, \pm 2, \ldots\,.$$

In k-space there is one normalizable state for each cube of volume $(2\pi/L)^3$, hence the density of states is $(L/2\pi)^3$. In order to obtain the density of states with respect the modulus k of the wave vector, we write the volume element in k-space as $\Delta V_k = 4\pi k^2 \Delta k$, giving

$$\frac{\Delta N}{\Delta V_k} = \left(\frac{L}{2\pi}\right)^3 = \frac{\Delta N}{4\pi k^2 \Delta k}.$$

Hence we have

$$\rho_k = \frac{\Delta N}{\Delta k} = 4\pi k^2 \frac{L^3}{8\pi^3} = \frac{k^2 L^3}{2\pi^2}.$$

For the density of states with respect to the energy $E = \hbar^2 k^2/(2\mu)$ we obtain

$$\frac{\Delta N}{\Delta E} = \rho_k \left(\frac{dE}{dk}\right)^{-1} = \frac{L^3 \mu k}{2\pi^2 \hbar^2}.$$

States normalized to unity carry an amplitude factor $1/L^{3/2}$. When working with bound states of unit amplitude, the density of states must absorb the factor $1/L^3$ so that products such as occur in the Golden Rule remain independent of the choice of amplitude. Thus the density of states for plane waves with unit amplitude is

$$\rho_E = \frac{\mu k}{2\pi^2\hbar^2} = \frac{\mu^{3/2}}{\hbar^3}\frac{\sqrt{2E}}{2\pi^2} .$$

If we now visualize the scattering process as a transition from incoming plane waves ψ_{i} (travelling in the direction of the z-axis) to final plane waves ψ_{f} (travelling in the direction $\mathrm{d}\Omega$), then the transition probability per unit time is, according to the Golden Rule,

$$\mathrm{d}P(\theta,\phi) = \frac{2\pi}{\hbar}\,|\langle\psi_{\mathrm{f}}|\hat{T}|\psi_{\mathrm{i}}\rangle|^2\rho_E \times \frac{\mathrm{d}\Omega}{4\pi} = \frac{k\mu}{4\pi^2\hbar^3}|\langle\psi_{\mathrm{f}}|\hat{T}|\psi_{\mathrm{i}}\rangle|^2\,\mathrm{d}\Omega .$$

The perturbing operator causing the transition is called $\hat{T}$. If we relate the matrix element of the transition operator $\hat{T}$ to the scattering amplitude $f(\theta,\phi)$ as suggested by (4.18), then we obtain

$$\frac{\mathrm{d}P(\theta,\phi)}{\mathrm{d}\Omega} = |f(\theta,\phi)|^2\frac{\hbar k}{\mu} .$$

By dividing the transition rate per unit time for scattering into the solid angle $\mathrm{d}\Omega$ by the incoming current density, we return to the original definition (4.6) for the differential scattering cross section.

4.4. The integrated cross section can be written as a sum of contributions σ_l, which originate from the partial waves l and vary between zero (for $\delta_l = n\pi$) and maximum values of $4\pi(2l+1)/k^2$ (for $\delta_l = \left(n+\frac{1}{2}\right)\pi$),

$$\sigma = \sum_{l=0}^{\infty}\sigma_l , \quad \sigma_l = \frac{4\pi}{k^2}(2l+1)\sin^2\delta_l . \tag{1}$$

For a given partial wave l we have

$$\tan\delta_l \overset{k\to\infty}{=} -\frac{\sin(kr_0 - l\pi/2)}{\cos(kr_0 - l\pi/2)} = -\tan(kr_0 - l\pi/2) ,$$

and hence

$$\delta_l = \left(n + \frac{l}{2}\right)\pi - kr_0 , \quad \text{for} \quad k \to \infty .$$

The oscillatory asymptotic ($kr_0 \to \infty$) behaviour of $j_l(kr_0)$, $n_l(kr_0)$ turns to a monotonic decrease of j_l/n_l to zero as the argument kr_0 goes to zero (cf. (A.49) in Appendix A.4). An estimate of where this turnover takes place can be obtained by looking at the wave number k_l, where the classical radial kinetic energy at r_0,

$$E - \frac{l(l+1)\hbar^2}{2\mu r_0^2} \approx \frac{\hbar^2}{2\mu}\left(k^2 - \frac{\left(l+\frac{1}{2}\right)^2}{r_0^2}\right),$$

vanishes, and this happens at

$$k_l r_0 = l + \frac{1}{2}. \tag{2}$$

Note that we have utilized the Langer modification replacing $l\,(l+1)$ by $(l+1/2)^2$. For large values of l we have (see e.g. Ch. 9.3 in [AS70] quoted in the Appendix)

$$\frac{j_l(kr_0)}{n_l(kr_0)} \stackrel{l\to\infty}{=} \frac{1}{2}\left(\frac{e\,kr_0}{2l+1}\right)^{2l+1}. \tag{3}$$

For a given energy, i.e. for a given wave number k, partial waves up to $l_{\rm max} \approx kr_0$ contribute significantly to the cross section, contributions from higher partial waves fall off rapidly according to (3). An approximate value for the total cross section is

$$\begin{aligned}
\sigma &\approx \frac{4\pi}{k^2}\sum_{l=0}^{l_{\rm max}}(2l+1)\sin^2\left(kr_0 - \frac{1}{2}\pi\right) \\
&= \frac{4\pi}{k^2}\left[\sum_{l=1}^{l_{\rm max}} l\left[\sin^2\left(kr_0 - \frac{1}{2}\pi\right) + \sin^2\left(kr_0 - \frac{l-1}{2}\pi\right)\right]\right. \\
&\qquad \left. + \,(l_{\rm max}+1)\sin^2\left(kr_0 - \frac{l_{\rm max}}{2}\pi\right)\right] \\
&\approx \frac{4\pi}{k^2}\sum_{l=0}^{l_{\rm max}} l = \frac{4\pi}{k^2}\frac{l_{\rm max}(l_{\rm max}+1)}{2} \approx \frac{4\pi}{k^2}\frac{(kr_0)^2}{2} = 2\pi r_0^2.
\end{aligned}$$

For scattering by a finite potential of depth (or height) $V_0 = \hbar^2 k_0^2/(2\mu)$ and range r_0, the phase shifts will fall off rapidly for values of l above $k_0 r_0$, so the upper limit to the sum over contributing partial waves no longer depends on k. An approximate upper bound for σ is

$$\sigma_{\rm max} \approx \frac{4\pi}{k^2}\sum_{0\le l<k_0 r_0}(2l+1) \approx \frac{4\pi}{k^2}\times(k_0 r_0)^2 = 4\pi\frac{k_0^2 r_0^2}{k^2}.$$

At high energies $E \to \infty$ we expect the integrated scattering cross section to decrease at least as fast as $1/E$.

4.5. The work $\mathrm{d}W$ done in going from r to $r - \mathrm{d}r$ is

$$\mathrm{d}W = \boldsymbol{F}\cdot\mathrm{d}\boldsymbol{r} = \frac{e\alpha_\mathrm{d}}{r^3}\left[\frac{e\boldsymbol{r}}{r^3} - 3\frac{\boldsymbol{r}}{r^2}\frac{er^2}{r^3}\right]\cdot\mathrm{d}\boldsymbol{r} = -2\frac{e^2\alpha_\mathrm{d}}{r^5}\,\mathrm{d}r\,,$$

where we have assumed the differential displacement to be in the radial direction, $\mathrm{d}\boldsymbol{r} = (\boldsymbol{r}/r)\,\mathrm{d}r$. The work done in coming from infinity to a finite position r is

$$W = -2e^2\alpha_\mathrm{d}\int_\infty^r \frac{1}{r'^5}\,\mathrm{d}r' = \frac{e^2\alpha_\mathrm{d}}{2r^4}\,.$$

4.6. In the special case $m = -1/2$ (1.358) becomes

$$\mathcal{Y}_{l+\frac{1}{2},m,l} = \frac{1}{\sqrt{2l+1}}\begin{pmatrix}\sqrt{l}\,Y_{l,-1}(\theta,\phi)\\ \sqrt{l+1}\,Y_{l,0}(\theta)\end{pmatrix},$$

$$\mathcal{Y}_{l-\frac{1}{2},m,l} = \frac{1}{\sqrt{2l+1}}\begin{pmatrix}-\sqrt{l+1}\,Y_{l,-1}(\theta,\phi)\\ \sqrt{l}\,Y_{l,0}(\theta)\end{pmatrix}.$$

These relations can be inverted,

$$\begin{pmatrix}Y_{l,-1}(\theta,\phi)\\ 0\end{pmatrix} = \sqrt{\frac{l}{2l+1}}\,\mathcal{Y}_{l+\frac{1}{2},m,l} - \sqrt{\frac{l+1}{2l+1}}\,\mathcal{Y}_{l-\frac{1}{2},m,l}\,,$$

$$\begin{pmatrix}0\\ Y_{l,0}(\theta)\end{pmatrix} = \sqrt{\frac{l+1}{2l+1}}\,\mathcal{Y}_{l+\frac{1}{2},m,l} + \sqrt{\frac{l}{2l+1}}\,\mathcal{Y}_{l-\frac{1}{2},m,l}\,. \qquad (1)$$

Expanding the spatial part of the plane wave according to (4.30) and using the upper equation (1) yields

$$\mathrm{e}^{\mathrm{i}kz}\chi_- = \sqrt{4\pi}\sum_{l=0}^{\infty}\sqrt{2l+1}\,\mathrm{i}^l j_l(kr)\begin{pmatrix}0\\ Y_{l,0}(\theta)\end{pmatrix}$$

$$= \sqrt{4\pi}\sum_{l=0}^{\infty}\mathrm{i}^l j_l(kr)\left(\sqrt{l+1}\,\mathcal{Y}_{l+\frac{1}{2},m,l} + \sqrt{l}\,\mathcal{Y}_{l-\frac{1}{2},m,l}\right).$$

We can use (1) and decompose the outgoing spherical wave into components with good j, m and l,

$$\begin{pmatrix}g'(\theta,\phi)\\ f'(\theta)\end{pmatrix} = \sum_{l=0}^{\infty}\frac{\sqrt{4\pi}}{2l+1}\Big[\left(f'_l + lg'_l\right)\sqrt{l+1}\,\mathcal{Y}_{l+\frac{1}{2},m,l}$$

$$+ \quad \left[f'_l - (l+1)g'_l\right]\sqrt{l}\,\mathcal{Y}_{l-\frac{1}{2},m,l}\Big]\,.$$

If we now collect the radial parts of the incoming plane wave and the outgoing spherical wave for given values of l and j, we obtain expressions which look like the big square bracket in the upper line of (4.32), except that the coefficient f_l in (4.32) is now replaced by different linear combinations of f'_l and g'_l, namely $f'_l + lg'_l$ for $j = l + 1/2$ and $f'_l - (l+1)g'_l$ for $j = 1 - 1/2$. The same steps which led from (4.32) to (4.34) now give

$$f'_l + lg'_l = \frac{2l+1}{2ik}\left[\exp\left(2i\delta_l^{(l+1/2)}\right) - 1\right],$$

$$f'_l - (l+1)g'_l = \frac{2l+1}{2ik}\left[\exp\left(2i\delta_l^{(l-1/2)}\right) - 1\right].$$

Resolving for the partial wave amplitudes f'_l and g'_l yields

$$f'_l = \frac{l+1}{2ik}\left[\exp\left(2i\delta_l^{(l+1/2)}\right) - 1\right] + \frac{l}{2ik}\left[\exp\left(2i\delta_l^{(l-1/2)}\right) - 1\right],$$

$$g'_l = \frac{1}{2ik}\left[\exp\left(2i\delta_l^{(l+1/2)}\right) - \exp\left(2i\delta_l^{(l-1/2)}\right)\right].$$

4.7.

$$P_x = \begin{pmatrix} A^* \\ B^* \end{pmatrix} \cdot \left[\begin{pmatrix} 0 & 1 \\ 1 & 0 \end{pmatrix}\begin{pmatrix} A \\ B \end{pmatrix}\right] = \begin{pmatrix} A^* \\ B^* \end{pmatrix} \cdot \begin{pmatrix} B \\ A \end{pmatrix}$$

$$= A^*B + B^*A = 2\Re[A^*B],$$

$$P_y = \begin{pmatrix} A^* \\ B^* \end{pmatrix} \cdot \left[\begin{pmatrix} 0 & -i \\ i & 0 \end{pmatrix}\begin{pmatrix} A \\ B \end{pmatrix}\right] = -i\begin{pmatrix} A^* \\ B^* \end{pmatrix} \cdot \begin{pmatrix} B \\ -A \end{pmatrix}$$

$$= \frac{1}{i}(A^*B - B^*A) = 2\Im[A^*B],$$

$$P_z = \begin{pmatrix} A^* \\ B^* \end{pmatrix} \cdot \left[\begin{pmatrix} 1 & 0 \\ 0 & -1 \end{pmatrix}\begin{pmatrix} A \\ B \end{pmatrix}\right] = \begin{pmatrix} A^* \\ B^* \end{pmatrix} \cdot \begin{pmatrix} A \\ -B \end{pmatrix} = |A|^2 - |B|^2.$$

$$\hat{\sigma}_P = P_x\hat{\sigma}_x + P_y\hat{\sigma}_y + P_z\hat{\sigma}_z = \begin{pmatrix} P_z & P_x - iP_y \\ P_x + iP_y & -P_z \end{pmatrix}$$

$$= \begin{pmatrix} |A|^2 - |B|^2 & 2\left[\Re(A^*B) - i\Im(A^*B)\right] \\ 2\left[\Re(A^*B) + i\Im(A^*B)\right] & |B|^2 - |A|^2 \end{pmatrix}$$

$$= \begin{pmatrix} |A|^2 - |B|^2 & 2AB^* \\ 2A^*B & |B|^2 - |A|^2 \end{pmatrix}.$$

Operating on the spinor $|\chi\rangle = \begin{pmatrix} A \\ B \end{pmatrix}$ with $\hat{\sigma}_P$ and recalling that $|A|^2 + |B|^2$ is unity yields

$$\hat{\sigma}_P \begin{pmatrix} A \\ B \end{pmatrix} = \begin{pmatrix} (|A|^2 - |B|^2)A + 2A|B|^2 \\ 2|A|^2 B + (|B|^2 - |A|^2)B \end{pmatrix} = \begin{pmatrix} (|A|^2 + |B|^2)A \\ (|A|^2 + |B|^2)B \end{pmatrix} = \begin{pmatrix} A \\ B \end{pmatrix} .$$

4.8. In order to describe triplet scattering, we must work with solutions of the Schrödinger equation which are antisymmetric with respect to the interchange of the spatial coordinates r_1 and r_2 of the two electrons, i.e. the wave functions must have negative parity with respect to the reflection of the relative distance coordinate $r = r_1 - r_2$ at the origin, $r \to -r$. Such solutions are readily constructed from the wave functions (4.143),

$$\psi_\mathrm{t}(r) = \psi_\mathrm{C}(r) - \psi_\mathrm{C}(-r) . \tag{1}$$

For all spatial directions excluding $\theta = 0$ and $\theta = \pi$ the asymptotic form of the wave function (1) is

$$\psi_\mathrm{t}(r) = \mathrm{e}^{\mathrm{i}[kz+\eta \ln k(r-z)]} - \mathrm{e}^{-\mathrm{i}[kz-\eta \ln k(r+z)]} + [f_\mathrm{C}(\theta) - f_\mathrm{C}(\pi-\theta)] \frac{\mathrm{e}^{\mathrm{i}[kr-\eta \ln 2kr]}}{r} . \tag{2}$$

The differential scattering cross section is as usual defined as the outgoing particle flux divided by the incoming current density which is given by the $\mathrm{e}^{+\mathrm{i}[kz\ldots]}$ term in (2):

$$\frac{\mathrm{d}\sigma_\mathrm{M}^\mathrm{t}}{\mathrm{d}\Omega} = |f_\mathrm{C}(\theta) - f_\mathrm{C}(\pi-\theta)|^2 . \tag{3}$$

Noting that $\sin((\pi-\theta)/2) = \cos(\theta/2)$ and that $\ln\sin^2(\theta/2) - \ln\cos^2(\theta/2) = \ln\tan^2(\theta/2)$ we use the expression (4.144) for the Coulomb scattering amplitude to obtain

$$\begin{aligned} \frac{\mathrm{d}\sigma_\mathrm{M}^\mathrm{t}}{\mathrm{d}\Omega} &= \frac{\eta^2}{4k^2}\left|\frac{\mathrm{e}^{-\mathrm{i}\eta\ln\sin^2(\theta/2)}}{\sin^2(\theta/2)} - \frac{\mathrm{e}^{-\mathrm{i}\eta\ln\cos^2(\theta/2)}}{\cos^2(\theta/2)}\right|^2 \\ &= \frac{\eta^2}{4k^2}\left[\frac{1}{\sin^4(\theta/2)} + \frac{1}{\cos^2(\theta/2)} - \frac{\mathrm{e}^{-\mathrm{i}\eta\ln\tan^2(\theta/2)} + \mathrm{e}^{-\mathrm{i}\eta\ln\tan^2(\theta/2)}}{\sin^2(\theta/2)\cos^2(\theta/2)}\right] \\ &= \frac{\eta^2}{4k^2}\left[\frac{1}{\sin^4(\theta/2)} + \frac{1}{\cos^4(\theta/2)} - 2\frac{\cos[\eta\ln\tan^2(\theta/2)]}{\sin^2(\theta/2)\cos^2(\theta/2)}\right] . \end{aligned} \tag{4}$$

Only odd angular momentum quantum numbers l contribute to the partial waves expansion, because the even partial waves have positive parity and drop out in the superposition (1).

In singlet scattering the spatial wave function must be symmetric and the difference (1) is replaced by a sum. The corresponding formula for the differential scattering cross section becomes

$$\begin{aligned}\frac{d\sigma_M^s}{d\Omega} &= |f_C(\theta) + f_C(\pi - \theta)|^2 \\ &= \frac{n^2}{4k^2}\left[\frac{1}{\sin^4(\theta/2)} + \frac{1}{\cos^4(\theta/2)} + 2\frac{\cos\left[\eta \ln \tan^2(\theta/2)\right]}{\sin^2(\theta/2)\cos^2(\theta/2)}\right]. \end{aligned} \tag{5}$$

In the scattering of unpolarized electrons (with no measurement of spin in the final states) we observe a mean of the singlet and triplet cross sections, weighted with the respective multiplicity $2S+1$ which is unity for $S = 0$ and three for $S = 1$:

$$\begin{aligned}\frac{d\sigma_M}{d\Omega} &= \frac{1}{4}\left(\frac{d\sigma_M^s}{d\Omega} + 3\frac{d\sigma_M^t}{d\Omega}\right) \\ &= \frac{n^2}{4k^2}\left[\frac{1}{\sin^4(\theta/2)} + \frac{1}{\cos^4(\theta/2)} - \frac{\cos\left[\eta \ln \tan^2(\theta/2)\right]}{\sin^2(\theta/2)\cos^2(\theta/2)}\right].\end{aligned}$$

4.9. $\mathcal{G}(x, x')$ and the delta function in the defining equation depend only on the difference $x - x' \stackrel{\text{def}}{=} \boldsymbol{\rho}$ of the two coordinates, and for fixed x' we can replace the derivatives with respect to the components of x by the derivatives with respect to the corresponding components of $\boldsymbol{\rho}$. Thus we have to show that the function

$$G(\boldsymbol{\rho}) = -\left(\frac{K}{2\pi}\right)^{\nu} \frac{\mathrm{i}H_{\nu}^{(1)}(K|\boldsymbol{\rho}|)}{4|\boldsymbol{\rho}|^{\nu}} \tag{1}$$

fulfills the equation

$$(K^2 + \Delta_n)G(\boldsymbol{\rho}) = \delta(\boldsymbol{\rho}). \tag{2}$$

Since G depends only on $\rho = |\boldsymbol{\rho}|$, the n-dimensional gradient is

$$\nabla G(\rho) = \frac{\boldsymbol{\rho}}{\rho}\frac{dG}{d\rho},$$

and the corresponding Laplacian applied to G is

$$\begin{aligned}
\Delta_n G(\rho) &= \nabla \cdot \left(\frac{\boldsymbol{\rho}}{\rho}\frac{\mathrm{d}G}{\mathrm{d}\rho}\right) = \sum_{i=1}^{n} \frac{\partial}{\partial \rho_i}\left(\frac{\rho_i}{\rho}\frac{\mathrm{d}G}{\mathrm{d}\rho}\right) \\
&= \sum_{i=1}^{n}\left[\frac{1}{\rho}\frac{\partial G}{\partial \rho} + \rho_i \frac{\partial}{\partial \rho_i}\left(\frac{1}{\rho}\frac{\mathrm{d}G}{\mathrm{d}\rho}\right)\right] = \frac{n}{\rho}\frac{\mathrm{d}G}{\mathrm{d}\rho} + \sum_{i=1}^{n}\rho_i \frac{\mathrm{d}}{\mathrm{d}\rho}\left(\frac{1}{\rho}\frac{\mathrm{d}G}{\mathrm{d}\rho}\right)\frac{\partial \rho}{\partial \rho_i} \\
&= \frac{n}{\rho}\frac{\mathrm{d}G}{\mathrm{d}\rho} + \sum_{i=1}^{n}\rho_i\left(-\frac{1}{\rho^2}\frac{\mathrm{d}G}{\mathrm{d}\rho} + \frac{1}{\rho}\frac{\mathrm{d}^2G}{\mathrm{d}\rho^2}\right)\frac{\rho_i}{\rho} \\
&= \frac{n}{\rho}\frac{\mathrm{d}G}{\mathrm{d}\rho} + \sum_{i=1}^{n}\frac{\rho_i^2}{\rho}\left(-\frac{1}{\rho^2}\frac{\mathrm{d}G}{\mathrm{d}\rho} + \frac{1}{\rho}\frac{\mathrm{d}^2G}{\mathrm{d}\rho^2}\right) = \frac{\mathrm{d}^2G}{\mathrm{d}\rho^2} + \frac{n-1}{\rho}\frac{\mathrm{d}G}{\mathrm{d}\rho}\,.
\end{aligned}$$

Now G is equal to a constant (namely $-(\mathrm{i}/4)(K/2\pi)^\nu$) times $H_\nu^{(1)}(K\rho)/\rho^\nu$, and writing $2\nu + 1$ for $n - 1$ we have

$$\begin{aligned}
\Delta_n \frac{H_\nu^{(1)}(K\rho)}{\rho^\nu} &= \frac{\mathrm{d}^2}{\mathrm{d}\rho^2}\left(\frac{H_\nu^{(1)}(K\rho)}{\rho^\nu}\right) + \frac{2\nu+1}{\rho}\frac{\mathrm{d}}{\mathrm{d}\rho}\left(\frac{H_\nu^{(1)}(K\rho)}{\rho^\nu}\right) \\
&= \frac{1}{\rho^\nu}\left(\frac{\mathrm{d}^2H_\nu^{(1)}(K\rho)}{\mathrm{d}\rho^2} + \frac{1}{\rho}\frac{\mathrm{d}H_\nu^{(1)}(K\rho)}{\mathrm{d}\rho} - \frac{\nu^2}{\rho^2}H_\nu^{(1)}(K\rho)\right). \qquad (3)
\end{aligned}$$

Bessels differential equation for $H_\nu^{(1)}(K\rho)$ tells us that

$$\frac{\mathrm{d}^2H_\nu^{(1)}(K\rho)}{\mathrm{d}(K\rho)^2} + \frac{1}{K\rho}\frac{\mathrm{d}H_\nu^{(1)}(K\rho)}{\mathrm{d}(K\rho)} - \frac{\nu^2}{K^2\rho^2}H_\nu^{(1)}(K\rho) = -H_\nu^{(1)}(K\rho)\,,$$

and so (3) amounts to

$$\Delta_n \frac{H_\nu^{(1)}(K\rho)}{\rho^\nu} = -K^2\frac{H_\nu^{(1)}(K\rho)}{\rho^\nu}\,,$$

showing that $(K^2 + \Delta_n)G(\rho)$ must vanish as long as ρ is not the singular point, $\rho \neq 0$.

To complete the proof that $G(\rho)$ fulfills (2) we show that

$$\int_V f(\rho)(K^2 + \Delta_n)G(\rho)\,\mathrm{d}\boldsymbol{\rho} = f(0) \qquad (4)$$

for a small n-dimensional volume V enclosing the singular point $\rho = 0$. Equation (4) should hold for any appropriately well behaved trial function f.

Since we are now operating in a small volume around $\rho = 0$ we may use the leading term in the appropriate expansion for $H_\nu^{(1)}(K\rho)$ and obtain

$$G(\rho) \stackrel{\rho\to 0}{=} -\frac{\Gamma(\nu)}{4\pi^{\nu+1}\rho^{2\nu}}\,. \tag{5}$$

As the radius of the small volume decreases, its volume will decrease as the $n = (2\nu + 2)$th power in the radius, and the surface of the volume will decrease as the $(2\nu + 1)$th power of the radius. The integral over fK^2G vanishes in the limit of vanishing volume V as long as f remains bounded in the vicinity of $\rho = 0$. The remaining contribution to the integral on the left-hand side of (4) can be rewritten using Green's theorem,

$$\int_V f(\rho)\Delta_n G(\rho)\,\mathrm{d}\boldsymbol{\rho} = \int_V G(\rho)\Delta_n f(\rho)\,\mathrm{d}\boldsymbol{\rho} + \oint_{S(V)} [f\,(\nabla G) - G\,(\nabla f)]\cdot \mathrm{d}o\,. \tag{6}$$

In the limit of vanishing volume V and vanishing surface $S(V)$ of the volume, the volume integral on the right-hand side of (6) and the second term in the surface integral vanish as long as ∇f and Δf remain bounded in the vicinity of $\rho = 0$. The only non-vanishing contribution on the right-hand side of (6) is thus

$$\oint_{S(V)} f\nabla G\cdot \mathrm{d}o \stackrel{S(V)\to 0}{=} f(0)\left(\frac{\nu\Gamma(\nu)}{2\pi^{\nu+1}}\right)\oint_{S(V)} \frac{1}{\rho^{2\nu+1}}\frac{\boldsymbol{\rho}}{\rho}\cdot \mathrm{d}o\,. \tag{7}$$

For a small sphere of radius $\rho = |\boldsymbol{\rho}|$ the surface integral on the right-hand side of (7) is just $1/\rho^{2\nu+1}$ times the surface of the sphere, which is $2\pi^{n/2}\rho^{n-1}/\Gamma(n/2)$ according to Problem 4.10. Recalling that $n = 2\nu+2$ this amounts to $2\pi^{\nu+1}/\Gamma(\nu+1)$, so that the right-hand side of (7) reduces to $f(0)$.

4.10. As a product of n one-dimensional integrals we have

$$I_n = \left(\sqrt{\pi}\right)^n\,. \tag{1}$$

Transforming to a radial integral yields

$$I_n = \int_0^\infty \mathrm{e}^{-R^2} S_n(R)\,\mathrm{d}R\,, \tag{2}$$

where $S_n(R) = S_n(1)R^{n-1}$ is the surface of a sphere of radius R in n dimensions; $S_n(1)$ is the surface of the unit sphere. Equation (2) can be integrated,

$$I_n = S_n(1)\int_0^\infty R^{n-1}\,\mathrm{e}^{-R^2}\,\mathrm{d}R = S_n(1)\frac{\Gamma(n/2)}{2}\,,$$

and equating this result to the right-hand side of (1) gives

$$S(1) = \frac{2\pi^{n/2}}{\Gamma(n/2)}, \quad S(R) = \frac{2\pi^{n/2}}{\Gamma(n/2)} R^{n-1} . \tag{3}$$

The volume of the n-dimensional sphere is obtained by integrating the surface (3):

$$V_n(R) = \int_0^R S_n(r)\,\mathrm{d}r = \frac{2\pi^{n/2}}{\Gamma(n/2)} \int_0^R r^{n-1}\,\mathrm{d}r = \frac{\pi^{n/2}}{\Gamma(\frac{n}{2}+1)} R^n .$$

4.11. In ordinary spherical coordinates the six-dimensional volume element is

$$\mathrm{d}\tau = r_1^2 \mathrm{d}r_1 r_2^2 \mathrm{d}r_2 \mathrm{d}\Omega_1 \mathrm{d}\Omega_2 = r_1^2 \mathrm{d}r_1 r_2^2 \mathrm{d}r_2 \sin\theta_1 \mathrm{d}\theta_1 \mathrm{d}\phi_1 \sin\theta_2 \mathrm{d}\theta_2 \mathrm{d}\phi_2 . \tag{1}$$

Transformation to hyperspherical coordinates only affects the coordinates r_1 and r_2. The corresponding differential $\mathrm{d}r_1 \mathrm{d}r_2$ transforms as

$$\begin{aligned} \mathrm{d}r_1 \mathrm{d}r_2 &= \begin{vmatrix} \frac{\partial r_1}{\partial \alpha} & \frac{\partial r_1}{\partial R} \\ \frac{\partial r_2}{\partial \alpha} & \frac{\partial r_2}{\partial R} \end{vmatrix} \mathrm{d}R\,\mathrm{d}\alpha \\ &= \begin{vmatrix} R\cos\alpha & \sin\alpha \\ -R\sin\alpha & \cos\alpha \end{vmatrix} \mathrm{d}R\,\mathrm{d}\alpha = R\,\mathrm{d}R\,\mathrm{d}\alpha . \end{aligned}$$

Inserting this result into (1) and remembering that $r_1^2 = R^2 \sin^2\alpha$, $r_2^2 = R^2 \cos^2\alpha$ gives

$$\begin{aligned} &\mathrm{d}\tau = R^5 \mathrm{d}R \sin^2\alpha \cos^2\alpha\, \mathrm{d}\alpha\, \mathrm{d}\Omega_1\, \mathrm{d}\Omega_2 = R^5\, \mathrm{d}R\, \mathrm{d}\Omega_\mathrm{h} \\ &\text{with} \quad \Omega_\mathrm{h} = \sin^2\alpha \cos^2\alpha\, \mathrm{d}\alpha\, \mathrm{d}\Omega_1\, \mathrm{d}\Omega_2 . \end{aligned}$$

Integrating over the hyperspherical solid angle gives

$$\begin{aligned} \int \mathrm{d}\Omega_\mathrm{h} &= \int_0^{\pi/2} \sin^2\alpha \cos^2\alpha\, \mathrm{d}\alpha \int_0^{\pi} \sin\theta_1\, \mathrm{d}\theta_1 \int_0^{2\pi} \mathrm{d}\phi_1 \int_0^{\pi} \sin\theta_2\, \mathrm{d}\theta_2 \int_0^{2\pi} \mathrm{d}\phi_2 \\ &= (4\pi)^2 \int_0^{\pi/2} \sin^2\alpha \cos^2\alpha\, \mathrm{d}\alpha = (4\pi)^2 \frac{\pi}{16} = \pi^3 . \end{aligned}$$

4.12. It is convenient to work in atomic units, where energies are given in units of 2 Rydbergs $\approx$ 27.21 eV and wave numbers are in units of the inverse Bohr radius $\approx 1.89 \times 10^8 \mathrm{cm}^{-1}$. k_i is a vector pointing in the direction of the momentum of the incoming electron (the z-axis), and its length follows from $E_\mathrm{inc} = k_i^2/2$: $k_i = 3.32$. In the asymmetric coplanar geometry we have

$$T_1 = \frac{k_1^2}{2} = E_\mathrm{inc} - 0.5 - T_2 = 4.90 , \quad k_1 = 3.13 .$$

Length and direction of the momentum transfer vector q can be derived by applying elementary geometry to the triangle formed by the vectors k_1, k_i and q; θ_1 is the angle between k_1 and k_i.

By the cosine rule

$$q^2 = k_1^2 + k_i^2 - 2k_1 k_i \cos\theta_1\,, \quad q = \begin{cases} 0.32 & \text{for } \theta_1 = 4^\circ \\ 0.61 & \text{for } \theta_1 = 10^\circ \\ 0.92 & \text{for } \theta_1 = 16^\circ \end{cases} .$$

The angle θ_q through which q is turned from the direction of $-k_i$ (i.e., from the negative z-axis) is given by the sine rule:

$$\sin\theta_q = \frac{k_1}{q}\sin\theta_1\,, \quad \theta_q = \begin{cases} 43^\circ & \text{for } \theta_1 = 4^\circ \\ 63^\circ & \text{for } \theta_1 = 10^\circ \\ 70^\circ & \text{for } \theta_1 = 16^\circ \end{cases} .$$

In symmetric geometry

$$T_1 + T_2 = E_{\text{inc}} - 0.5 = 5.01\,, \quad T_1 = T_2 = 2.51\,, \quad k_1 = k_2 = 2.24\,.$$

The length of the momentum transfer vector depends on $\theta_1 = \theta_2$ and is given as above by the cosine rule

$$q^2 = k_1^2 + k_i^2 - 2k_1 k_i \cos\theta_1\,, \quad q = \begin{cases} 1.18 & \text{for } \theta_1 = 10^\circ \\ 2.35 & \text{for } \theta_1 = 45^\circ \\ 3.66 & \text{for } \theta_1 = 80^\circ \end{cases} .$$

The angle θ_q is again given by the sine rule:

$$\sin\theta_q = \frac{k_1}{q}\sin\theta_1\,, \quad \theta_q = \begin{cases} 19^\circ & \text{for } \theta_1 = 10^\circ \\ 42^\circ & \text{for } \theta_1 = 45^\circ \\ 37^\circ & \text{for } \theta_1 = 80^\circ \end{cases} .$$

Note that θ_q reaches a maximum when k_1 is orthogonal to q. In the right-angled triangle formed by k_1, k_i and q we then see that $\sin\left(\theta_{q_{\max}}\right) = k_1/k_i$.

5.1. The power P of a laser in Watt per cm^2 can be expressed as the energy density ρ (in Joule per cm^3) times the speed of light c (in cm per second). The total energy in the volume occupied by the atom is simply the product of the energy density (assumed to be constant) times the volume,

$$E = \frac{4\pi}{3}(n^2 a_0)^3 \rho = \frac{4\pi}{3}(n^2 a_0)^3 \frac{P}{c}\,,$$

and the ratio Q of E to the binding energy is

$$Q = \frac{4\pi}{3}(n^2 a_0)^3 \frac{P\,n^2}{c\,\mathcal{R}} = \frac{8\pi}{3}\frac{(n^2 a_0)^4}{\alpha_{\rm fs}\hbar c^2}P\,,$$

where $\alpha_{\rm fs} \approx 1/137$ is the fine structure constant. For a ratio $Q \approx 1$ we have

$$P \approx \frac{3}{8\pi}\frac{\alpha_{\rm fs}\hbar c^2}{n^8 a_0^4} \approx \frac{10^{17}}{n^8}\frac{W}{\rm cm^2}\,.$$

5.2. The initial wave packet is the same as in Problem 1.4 in Chap. 1 for the special case $x_0 = 0$, so the time-dependent wave function $\tilde{\psi}(p,t)$ in momentum representation ($p \equiv \hbar k$) is,

$$\tilde{\psi}(p,t) = \left(\frac{\beta}{\sqrt{\pi}\hbar}\right)^{1/2} \exp\left[-\frac{\mathrm{i}}{\hbar}\frac{p^2}{2\mu}t - \frac{\beta^2}{2\hbar^2}(p-\hbar k_0)^2\right]. \tag{1}$$

The corresponding wave function in coordinate representation is

$$\psi(x,t) = \left[\sqrt{\pi}\beta\left(1+\frac{\mathrm{i}\hbar t}{\mu\beta^2}\right)\right]^{-1/2} \mathrm{e}^{-\beta^2 k_0^2/2} \exp\left[-\frac{(x-\mathrm{i}k_0\beta^2)^2}{2\beta^2\left(1+\frac{\mathrm{i}\hbar t}{\mu\beta^2}\right)}\right]. \tag{2}$$

The probability density in coordinate space is

$$|\psi(x,t)|^2 = \frac{1}{\sqrt{\pi}B}\exp\left[-\frac{1}{B^2}\left(x-\frac{\hbar k_0}{\mu}t\right)^2\right], \quad B = \beta\sqrt{1+\frac{\hbar^2 t^2}{\mu^2\beta^4}}\,. \tag{3}$$

The probability density in momentum space does not depend on time,

$$|\tilde{\psi}(p,t)|^2 = \frac{\beta}{\sqrt{\pi}\hbar}\,\mathrm{e}^{-(p-\hbar k_0)^2\beta^2/\hbar^2}\,. \tag{4}$$

The expression for the density matrix is a little simpler in momentum representation:

$$\tilde{\rho}(p,p';t) = \frac{\beta}{\sqrt{\pi}\hbar}\exp\left[-\frac{\mathrm{i}}{\hbar}\frac{p^2-p'^2}{2\mu}t\right] \times \exp\left\{-\frac{\beta^2}{2\hbar^2}[(p-\hbar k_0)^2+(p'-\hbar k_0)^2]\right\}. \tag{5}$$

Introducing sum and difference variables, $P = (p + p')/2$, $q = p - p'$, and reorganising the exponents in (5) gives

$$\tilde{\rho}\left(P + \frac{q}{2}, P - \frac{q}{2}; t\right) = \frac{\beta}{\sqrt{\pi}\hbar}\, \mathrm{e}^{-\beta^2 k_0^2} \exp\left[-\frac{\beta^2}{4\hbar^2}\left(q - 2\frac{\mathrm{i}\hbar t}{\mu\beta^2}\right)^2\right] \times$$
$$\exp\left[-\frac{B^2}{\hbar^2}P^2 + \frac{2\beta^2}{\hbar}Pk_0\right],$$

where B is as defined in (3).

The Wigner function is given by the lower line of (5.41), adapted to the one-dimensional situation:

$$\rho_{\mathrm{w}}(X, P; t) = \frac{1}{2\pi\hbar}\int_{-\infty}^{\infty} \mathrm{e}^{\mathrm{i}Xq/\hbar}\, \tilde{\rho}\left(P + \frac{q}{2}, P - \frac{q}{2}; t\right) \mathrm{d}q$$
$$= \frac{1}{\pi\hbar}\, \mathrm{e}^{-(X - Pt/\mu)^2/\beta^2}\, \mathrm{e}^{-(P - \hbar k_0)^2 \beta^2/\hbar^2}. \qquad (6)$$

Equation (6) already looks very much like classical evolution in phase space. Indeed, the evolution of the classical phase space density in part c) can be formulated by exploiting the fact that ρ_{cl} is constant along the classical trajectories, because $\mathrm{d}\rho_{\mathrm{cl}}/\mathrm{d}t = 0$. The trajectory going through the point (x, p) at time t started at the point $(x - pt/\mu,\ p)$ at time zero, hence

$$\rho_{\mathrm{cl}}(x, p; t) = \rho_{\mathrm{cl}}\left(x - \frac{p}{\mu}t, p; 0\right) = \frac{1}{\alpha\beta\pi}\, \mathrm{e}^{-(x - pt/\mu)^2/\beta^2}\, \mathrm{e}^{-(p - p_0)^2/\alpha^2}. \qquad (7)$$

This is quantitatively equal to the quantum mechanical result (6), if we choose the width α describing the initial (and time-independent) spread in momentum according to $\alpha = \hbar/\beta$.

5.3. Applying $\hat{b}$ according to (5.59) we have

$$\hat{b}|z\rangle = \mathrm{e}^{-zz^*/2}\sum_{n=0}^{\infty}\frac{(z^*)^n}{\sqrt{n!}}\sqrt{n}|n-1\rangle = \mathrm{e}^{-zz^*/2}\sum_{n=1}^{\infty}\frac{(z^*)^n}{\sqrt{(n-1)!}}|n-1\rangle$$
$$= z^*\mathrm{e}^{-zz^*/2}\sum_{n=0}^{\infty}\frac{(z^*)^n}{\sqrt{n!}}|n\rangle = z^*|z\rangle\,.$$

The conjugate equation is $\langle z|\hat{b}^\dagger = \langle z|z$ and hence

$$\langle z|\hat{b}^\dagger\hat{b}|z\rangle = \langle z|zz^*|z\rangle = |z|^2\,. \qquad (1)$$

We use (5.73) to express the electric and magnetic field strengths in terms of the momentum operator $\hat{p}$ and obtain

$$\frac{L^3}{8\pi}\langle z|\hat{E}^2+\hat{B}^2|z\rangle = \langle z|\frac{\hat{p}^2}{2}+\frac{\hat{p}^2}{2}|z\rangle = \langle z|\hat{p}^2|z\rangle\,. \tag{2}$$

The expectation value of $\hat{p}^2$ is related to the uncertainty Δ_p (which is equal to $\hbar/(\sqrt{2}\beta)$ in the present case) and the expectation value of $\hat{p}$ according to (1.35):

$$\Delta_p^2 = \frac{\hbar^2}{2\beta^2} = \langle\hat{p}^2\rangle - \langle\hat{p}\rangle^2\,,\quad \langle z|\hat{p}^2|z\rangle = \frac{\hbar^2}{2\beta^2} + \langle z|\hat{p}|z\rangle^2\,. \tag{3}$$

From (5.68) we expect that the expectation value of $\hat{p}$ in the coherent state $|z\rangle$ is P_z as given in the lower line of (5.69). This is in fact the case and can be verified by calculating the expectation value in momentum representation,

$$\langle z|\hat{p}|z\rangle = \int_{-\infty}^{\infty} p\,|\tilde{\psi}_z(p)|^2 \mathrm{d}z\,.$$

Note that the absolute square of the momentum wave function above can be derived directly from the Wigner function (see (5.43))

$$|\tilde{\psi}_z(p)|^2 = \int_{-\infty}^{\infty} \rho_{\mathrm{w}}(x,p)\,\mathrm{d}x = \frac{\beta}{\sqrt{\pi}\hbar}\,\mathrm{e}^{-(p-P_z)^2\beta^2/\hbar^2}\,.$$

Since $|z|$ does not depend on time and $\Im(z) = |z|\sin\omega(t-t_0)$ for an appropriately chosen t_0, we have

$$\langle z|\hat{p}^2|z\rangle = \frac{\hbar^2}{2\beta^2} + \frac{2\hbar^2}{\beta^2}|z|^2\sin^2\omega(t-t_0)\,.$$

Time averaging the $\sin^2$ term gives a factor 1/2 so

$$\langle z|\hat{p}^2|z\rangle = \frac{\hbar^2}{\beta^2}\left(\frac{1}{2}+|z|^2\right)\,. \tag{4}$$

Now $\hbar^2/\beta^2 = \hbar\omega$ and $|z|^2$ is the expectation value of $\hat{b}^\dagger\hat{b}$ according to the result of part a). Equation (4) merely expresses the fact that the energy of the field is given by the harmonic oscillator Hamiltonian $\hbar\omega\left(\hat{b}^\dagger\hat{b}+1/2\right)$.

The harmonic oscillator ground state is just the coherent state $|z=0\rangle$, and, according to (5.68), its Wigner function is

$$\rho_{\mathrm{w}}(X,P) = \frac{1}{\pi\hbar}\,\mathrm{e}^{-X^2/\beta^2}\,\mathrm{e}^{-P^2\beta^2/\hbar^2}\,.$$

The density matrix for the first excited state is

$$\rho(x,x') = \frac{2xx'}{\sqrt{\pi}\,\beta^3}\,\mathrm{e}^{-[x^2+(x')^2]/(2\beta^2)}\,.$$

Introducing sum and difference coordinates, $X = (x+x')/2$, $s = x-x'$, this amounts to

$$\rho\left(X+\frac{s}{2}, X-\frac{s}{2}\right) = \frac{1}{\sqrt{\pi}\,\beta^3}\left(2X^2-\frac{s^2}{2}\right)\mathrm{e}^{-X^2/\beta^2}\,\mathrm{e}^{-s^2/(4\beta^2)}\,.$$

The Wigner function is

$$\begin{aligned}\rho_{\mathrm{w}}(X,P) &= \frac{1}{2\pi\hbar}\int_{-\infty}^{\infty}\mathrm{e}^{-\mathrm{i}Ps/\hbar}\,\rho\left(X+\frac{s}{2}, X-\frac{s}{2}\right)\mathrm{d}s \\ &= \frac{1}{\pi\hbar}\left(\frac{2}{\beta^2}X^2+\frac{2\beta^2}{\hbar^2}P^2-1\right)\mathrm{e}^{-X^2/\beta^2}\,\mathrm{e}^{-P^2\beta^2/\hbar^2}\,.\end{aligned}$$

5.4. Following the hint we calculate

$$\frac{\mathrm{d}\hat{f}}{\mathrm{d}\lambda} = \hat{A}\,\mathrm{e}^{\lambda\hat{A}}\,\mathrm{e}^{\lambda\hat{B}} + \mathrm{e}^{\lambda\hat{A}}\,\hat{B}\,\mathrm{e}^{\lambda\hat{B}} = (\hat{A}+\hat{B})\,\mathrm{e}^{\lambda\hat{A}}\,\mathrm{e}^{\lambda\hat{B}} + [\mathrm{e}^{\lambda\hat{A}},\hat{B}]\,\mathrm{e}^{\lambda\hat{B}}\,. \tag{1}$$

Now

$$[\mathrm{e}^{\lambda\hat{A}},\hat{B}] = \sum_{n=0}^{\infty}\frac{\lambda^n}{n!}[\hat{A}^n,\hat{B}]\,, \tag{2}$$

and it is easy to show by induction that

$$[\hat{A}^n,\hat{B}] = n\hat{A}^{n-1}[\hat{A},\hat{B}]\,, \tag{3}$$

(remember that $[\hat{A},\hat{B}]$ commutes with both $\hat{A}$ and $\hat{B}$). Inserting (3) into (2) gives

$$[\mathrm{e}^{\lambda\hat{A}},\hat{B}] = [\hat{A},\hat{B}]\sum_{n=1}^{\infty}\frac{\lambda^n}{(n-1)!}\hat{A}^{n-1} = \lambda[\hat{A},\hat{B}]\,\mathrm{e}^{\lambda\hat{A}}\,.$$

Thus (1) becomes

$$\frac{\mathrm{d}\hat{f}}{\mathrm{d}\lambda} = \{\hat{A}+\hat{B}+\lambda[\hat{A},\hat{B}]\}\hat{f}(\lambda)\,. \tag{4}$$

The differential equation (4) is obviously also fulfilled by the operator function

$$\hat{f}_1 = \mathrm{e}^{\lambda\hat{A}+\lambda\hat{B}+(\lambda^2/2)[\hat{A},\hat{B}]}\,.$$

Since $\hat{f}(\lambda)$ and $\hat{f}_1(\lambda)$ go through the same point, namely unity (i.e. unit operator) at $\lambda = 0$, they must be identical solutions of the differential equation (4). Equating the values of $\hat{f}(\lambda)$ and $\hat{f}_1(\lambda)$ at $\lambda = 1$ yields the required special form of the Baker-Campbell-Hausdorff relation.

5.5. Let E_i be the energy and p_i the momentum of a free electron. The relativistic energy momentum relation is

$$E_\mathrm{i} = c\sqrt{m_0^2c^2 + p_\mathrm{i}^2}\,.$$

After absorbing a photon of energy $\hbar\omega$ and momentum $\hbar\omega/c$, the final energy E_f and momentum p_f of the electron obey

$$E_\mathrm{f} = c\sqrt{m_0^2c^2 + p_\mathrm{f}^2}\,.$$

Obviously the energy difference is

$$E_\mathrm{f} - E_\mathrm{i} = \hbar\omega = c\sqrt{m_0^2c^2 + p_\mathrm{f}^2} - c\sqrt{m_0^2c^2 + p_\mathrm{i}^2}\,.$$

Since the maximum final momentum of the electron is $p_\mathrm{i} + \hbar\omega/c$,

$$\begin{aligned} E_\mathrm{f} - E_\mathrm{i} &\le c\sqrt{m_0^2c^2 + (p_\mathrm{i} + \hbar\omega/c)^2} - c\sqrt{m_0^2c^2 + p_\mathrm{i}^2} \\ &= c\sqrt{m_0^2c^2 + p_\mathrm{i}^2}\left(\sqrt{1 + \frac{2p_\mathrm{i}\hbar\omega/c + (\hbar\omega/c)^2}{m_0^2c^2 + p_\mathrm{i}^2}} - 1\right) \\ &< c\,\frac{p_\mathrm{i}\hbar\omega/c}{\sqrt{m_0^2c^2 + p_\mathrm{i}^2}}\,. \end{aligned}$$

The right-hand side of the last inequality is always smaller than $\hbar\omega$ showing that even a maximal transfer of momentum is insufficient to produce the required energy gain for the electron.

The corresponding calculation swapping the roles of initial and final states shows that a free electron cannot emit a single photon. Note however, that the inelastic scattering of photons, which can be pictured as simultaneous absorption and emission of a photon, is kinematically allowed (Compton effect).

5.6. Assume $t_0 \le t_1 \le t_2$ and consider the propagation of an infinitesimal deviation $\Delta x(t_0)$ from a given trajectory. According to (5.75) the corresponding deviations $\Delta x(t_1)$ at time t_1 and $\Delta x(t_2)$ at time t_2 are

$$\begin{aligned} \Delta x(t_1) &= \mathbf{M}(t_1, t_0)\Delta x(t_0)\,, \\ \Delta x(t_2) &= \mathbf{M}(t_2, t_1)\Delta x(t_1) = \mathbf{M}(t_2, t_1)\mathbf{M}(t_1, t_0)\Delta x(t_0)\,. \end{aligned} \tag{1}$$

On the other hand, the defining equation for $\mathbf{M}(t_2, t_0)$ is

$$\Delta x(t_2) = \mathbf{M}(t_2, t_0)\Delta x(t_0)\,.$$

Since (1) holds for all infinitesimal $\Delta x(t_0)$, the matrix $\mathbf{M}(t_2, t_0)$ must be equal to $\mathbf{M}(t_2, t_1)\mathbf{M}(t_1, t_0)$.

If t_0 and t_1 define two different starting points on a given trajectory, then the Liapunov exponent defined by (5.77) is

$$\lambda_0 = \lim_{t\to\infty} \frac{\ln \|\mathbf{M}(t, t_0)\|}{t - t_0}\,, \quad \lambda_1 = \lim_{t\to\infty} \frac{\ln \|\mathbf{M}(t, t_1)\|}{t - t_1}\,,$$

depending on which starting point we choose. According to the chain rule however,

$$\begin{aligned} \lambda_0 &= \lim_{t\to\infty} \frac{\ln \|\mathbf{M}(t, t_1)\mathbf{M}(t_1, t_0)\|}{t - t_0} \\ &\le \lim_{t\to\infty} \frac{\ln \|\mathbf{M}(t, t_1)\|}{t - t_0} + \lim_{t\to\infty} \frac{\ln \|\mathbf{M}(t_1, t_0)\|}{t - t_0}\,, \end{aligned} \tag{2}$$

the lower line following from the inequality in the hint. The second term on the right-hand side in the lower line in (2) vanishes. The first term can be rewritten as

$$\begin{aligned} \lim_{t\to\infty} \frac{\ln \|\mathbf{M}(t, t_1)\|}{t - t_0} &= \lim_{t\to\infty} \ln \|\mathbf{M}(t, t_1)\| \left[\frac{1}{t - t_1} - \frac{t_1 - t_0}{(t - t_1)(t - t_0)}\right] \\ &= \lim_{t\to\infty} \frac{\ln \|\mathbf{M}(t, t_1)\|}{t - t_1}\,, \end{aligned}$$

which is just the definition of λ_1. We have thus shown: $\lambda_0 \le \lambda_1$.

From $\mathbf{M}(t, t_1) = \mathbf{M}(t, t_0)[\mathbf{M}(t_1, t_0)]^{-1}$ we have,

$$\begin{aligned} \lambda_1 &= \lim_{t\to\infty} \frac{\ln \|\mathbf{M}(t, t_0)[\mathbf{M}(t_1, t_0)]^{-1}\|}{t - t_1} \\ &\le \lim_{t\to\infty} \frac{\ln \|\mathbf{M}(t, t_0)\|}{t - t_1} + \lim_{t\to\infty} \frac{\ln \|[\mathbf{M}(t_1, t_0)]^{-1}\|}{t - t_1}\,, \end{aligned} \tag{3}$$

The second term on the right-hand side in the lower line in (3) vanishes, and the first term is equal to λ_0 by reasoning analogous to that following (2). Thus we have also shown: $\lambda_1 \le \lambda_0$.

Hence we conclude that the Liapunov exponent is the same for all phase space points along a given trajectory.

5.7. Let y_n be the vertical distance above the centre of the disc and x_n the horizontal distance from the centre of the disc to the point where the particle hits the disc the

nth time. Since all points (x_n, y_n) lie on the circle of radius a we have

$$x_n^2 + y_n^2 = a^2 . \tag{1}$$

Let Y_n be the vertical height above the middle at which the particle leaves the side of the square before the nth collision, and let T_n be the tangent of the angle to the horizontal at which it leaves the side of the square.

Initially we have $T_1 = \tan\alpha$, $Y_1 = 0$. The coordinates of the first collision can be determined from (1) together with

$$T_1 = \frac{y_1 - Y_1}{l - x_1} , \tag{2}$$

yielding

$$x_1 = \frac{T_1 Y_1 + T_1^2 l + \sqrt{(a^2 - Y_1^2)(1 + T_1^2) + T_1^2(Y_1^2 - l^2) - 2lT_1 Y_1}}{1 + T_1^2} ,$$

$$y_1 = Y_1 + T_1(l - x_1) , \tag{3}$$

where we have written a small l for $L/2$. After hitting the disc the particle is reflected at an angle to the horizontal given by

$$\alpha_2 = \alpha + 2\beta , \quad \tan\beta = \frac{y_1}{x_1} , \tag{4}$$

and it returns to the side of the square at $Y_2 = y_1 + (l - x_1)\tan\alpha_2$. Subsequently it travels to the disc (at an angle α_2) which it hits at (x_2, y_2). (See figure.)

A general recurrence formula for the coordinates of the nth collision with the disc can be derived from (1) together with the generalization

$$T_n = \tan\alpha_n = \frac{y_n - Y_n}{l - x_n} \tag{5}$$

of (2). The result is

$$x_n = \frac{T_n Y_n + T_n^2 l + \sqrt{(a^2 - Y_n^2)(1 + T_n^2) + T_n^2(Y_n^2 - l^2) - 2lT_n Y_n}}{1 + T_n^2} ,$$

$$y_n = Y_n + T_n(l - x_n) . \tag{6}$$

For the next iteration

$$\alpha_{n+1} = \alpha_n + 2\arctan\left(\frac{y_n}{x_n}\right) \quad \text{and} \quad Y_{n+1} = y_n + (l - x_n)\tan\alpha_{n+1} .$$

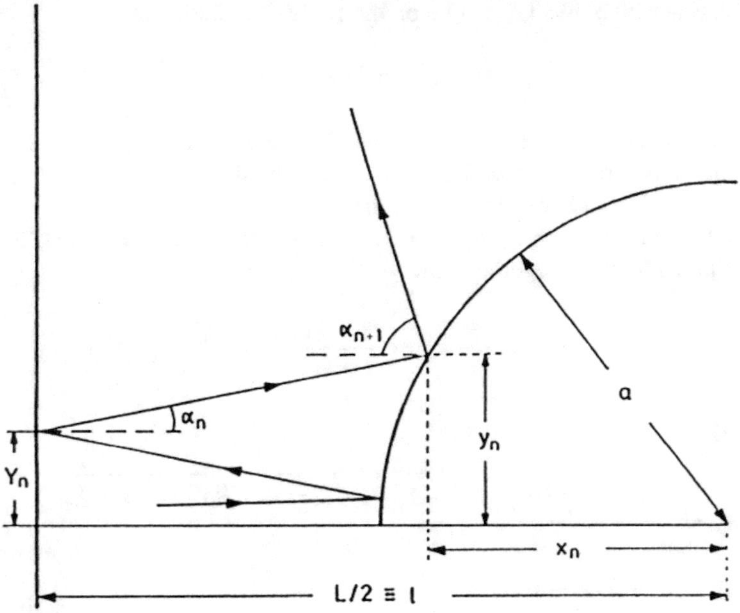

Inserting the lengths $l = 100$ cm ($L = 2$m), $a = 5$ cm given in the text, we obtain the following coordinates (x_n, y_n) (in cm) for successive collisions with the disc:

$\alpha = 0.3°$	
x_n	4.975
y_n	0.4975

$\alpha = 0.0003°$			
x_n	5.0	5.0	3.983
y_n	4.974×10^{-4}	3.880×10^{-2}	3.023

$\alpha = 0.0000003°$				
x_n	5.0	5.0	5.0	4.995
y_n	4.974×10^{-7}	3.880×10^{-5}	3.007×10^{-3}	2.330×10^{-1}

$\alpha = 0.0000000003°$					
x_n	5.0	5.0	5.0	5.0	5.0
y_n	4.974×10^{-10}	3.880×10^{-8}	3.007×10^{-6}	2.330×10^{-4}	1.806×10^{-2}

The vertical deviations y_n at collision with the disc provide a suitable measure for the deviation of a trajectory from the periodic straight-line trajectory $\alpha = 0$. Plotting these deviations on a logarithmic scale reveals the following dependence of y_n on the collision number n:

$$y_n = \text{const.} \times 10^{1.9n} = \text{const.} \times e^{4.4n} .$$

Thus the Liapunov exponent of the trajectory defined by $\alpha = 0$ is 4.4 in dimensionless units where the collision number defines the time scale. The period of the orbit at (constant) velocity v of the particle is $2(l-a)/v$ and the Liapunov exponent in physical units (s^{-1}) is $\lambda = 4.4 \times v/[2(l-a)]$. Note that the initial angle has to be accurate to roughly one ten-millionth of a degree if the particle is to hit the disc at least five times.

5.8. The probability $W(s)$ that none of the N numbers y lie in the interval $(x, x+s)$ is $[(L-s)/L]^N$. For $N \to \infty$ we have

$$W(s) = \lim_{N\to\infty}\left(1-\frac{s}{L}\right)^N = \lim_{N\to\infty}\left(1-\frac{ds}{N}\right)^N = \mathrm{e}^{-ds}\,. \tag{1}$$

At the same time, the probability defined in (1) gives us the probability for the spacing to the next number being *at least* the distance s, i.e. W is the sum (integral) over all probabilities (probability densities) for nearest neighbour spacings $P(s')$ larger or equal to s:

$$W(s) = \int_s^\infty P(s')\,\mathrm{d}s'\,, \quad P(s) = -\frac{\mathrm{d}W}{\mathrm{d}s} = d\,\mathrm{e}^{-ds}\,.$$

References

[Cha44] S. Chandrasekhar, Astrophys. J. **100**, 176 (1944)

[BS77] H.A. Bethe, E. Salpeter, *Quantum Mechanics of One- and Two-Electron Atoms* (Plenum, New York, 1977)

[FW85] H. Friedrich, D. Wintgen, Phys. Rev. A **31**, 1920 (1985)

[FJ02] H. Friedrich, G. Jacoby, C.G. Meister, Phys. Rev. A **65**, 032902 (2002)

[ME00] P. Meerwald, C. Eltschka, H. Friedrich, Few Body Syst. **29**, 157 (2000)

Appendix
Special Mathematical Functions

For completeness this appendix briefly lists without further discussion the definitions and some important properties of the special functions occurring in the book. More detailed treatments can be found in the relevant literature. The "Handbook of Mathematical Functions" [AS70], the "Tables" by Grad-shteyn and Rhyzik [GR80] and the compilation by Magnus, Oberhettinger and Soni [MO66] are particularly useful. Apart from these comprehensive works it is worth mentioning Appendix B in "Quantum Mechanics I" by Messiah [Mes70], which describes a selection of especially frequently used functions.

A.1 Legendre Polynomials, Spherical Harmonics

The lth Legendre Polynomial $P_l(x)$ is a polynomial of degree l in x,

$$P_l(x) = \frac{1}{2^l l!} \frac{\mathrm{d}^l}{\mathrm{d}x^l}(x^2 - 1)^l, \quad l = 0, 1, \dots . \tag{A.1}$$

It has l zeros in the interval between -1 and $+1$; for even (odd) l, $P_l(x)$ is an even (odd) function of x. The Legendre polynomials fulfill the orthogonality relation

$$\int_{-1}^{1} P_l(x)\, P_{l'}(x)\, \mathrm{d}x = \frac{2}{2l+1}\, \delta_{l,l'} . \tag{A.2}$$

The *associated Legendre functions* $P_{l,m}(x)$, $|x| \leq 1$, are products of $(1 - x^2)^{m/2}$ with polynomials of degree $l - m (m = 0, \dots, l)$,

$$P_{l,m}(x) = (1 - x^2)^{m/2} \frac{\mathrm{d}^m}{\mathrm{d}x^m} P_l(x) . \tag{A.3}$$

H. Friedrich, *Theoretical Atomic Physics*, Graduate Texts in Physics,
DOI 10.1007/978-3-319-47769-5

The spherical harmonics $Y_{l,m}(\theta,\phi)$ are products of exp $(im\phi)$ with polynomials of degree m in $\sin\theta$ and of degree $l-m$ in $\cos\theta$, where the θ-dependence is given by the associated Legendre functions (A.3) as functions of $x = \cos\theta$. For $m \geq 0$, $0 \leq \theta \leq \pi$ we have

$$Y_{l,m}(\theta,\phi) = (-1)^m\left[\frac{(2l+1)}{4\pi}\frac{(l-m)!}{(l+m)!}\right]^{1/2} P_{l,m}(\cos\theta)\,\mathrm{e}^{im\phi} \tag{A.4}$$

$$= (-1)^m\left[\frac{(2l+1)}{4\pi}\frac{(l-m)!}{(l+m)!}\right]^{1/2} \sin^m\theta\,\frac{\mathrm{d}^m}{\mathrm{d}(\cos\theta)^m}P_l(\cos\theta)\,\mathrm{e}^{im\phi}\,.$$

The spherical harmonics for negative azimuthal quantum numbers are obtained via

$$Y_{l,-m}(\theta,\phi) = (-1)^m(Y_{l,m}(\theta,\phi))^*. \tag{A.5}$$

A reflection of the displacement vector

$$x = r\sin\theta\cos\phi\,,\quad y = r\sin\theta\sin\phi\,,\quad z = r\cos\theta$$

at the origin (cf. (1.68)) is achieved by replacing the polar angle θ by $\pi-\theta$ and the azimuthal angle ϕ by $\pi+\phi$. This does not affect $\sin\theta$, but $\cos\theta$ changes to $-\cos\theta$. In the expression (A.4) for $Y_{l,m}$ spatial reflection introduces a factor $(-1)^{l-m}$ from the polynomial in $\cos\theta$ and a factor $(-1)^m$ from the exponential function in ϕ. Altogether we obtain

$$Y_{l,m}(\pi-\theta,\pi+\phi) = (-1)^l\,Y_{l,m}(\theta,\phi)\,. \tag{A.6}$$

The integral over a product of two spherical harmonics is given by the orthonormality relation (1.59),

$$\int Y^*_{l,m}(\Omega)Y_{l',m'}(\Omega)\,\mathrm{d}\Omega = \delta_{l,l'}\,\delta_{m,m'}\,. \tag{A.7}$$

The completeness relation is

$$\sum_{l=0}^{\infty}\sum_{m=-l}^{l} Y_{l,m}(\Omega)Y^*_{l,m}(\Omega') = \delta(\Omega-\Omega') = \delta(\cos\theta-\cos\theta')\,\delta(\phi-\phi')\,. \tag{A.8}$$

For a given l-value we have,

$$\sum_{m=-l}^{l} Y_{l,m}(\Omega)Y^*_{l,m}(\Omega') = \frac{2l+1}{4\pi}P_l(\cos\theta)\,, \tag{A.9}$$

where θ is the angle between the two directions defined by Ω and Ω'. For two vectors r and r' with $|r'| < |r|$ we have

$$\frac{1}{|r - r'|} = \sum_{l=0}^{\infty} \frac{|r'|^l}{|r|^{l+1}} P_l(\cos\theta) , \tag{A.10}$$

where θ is the angle between r and r'.

The integral over three spherical harmonics is a prototype example for the *Wigner-Eckart theorem*, which says that the dependence of the matrix elements of (spherical) tensor operators in angular momentum eigenstates on the component index of the operator and the azimuthal quantum numbers of bra and ket is given by appropriate Clebsch-Gordan coefficients (see Sect. 1.7.1). For the spherical harmonics $Y_{L,M}$ as an example for a spherical tensor of rank L we have

$$\int Y^*_{l,m}(\Omega) Y_{L,M}(\Omega) Y_{l',m'}(\Omega)\, \mathrm{d}\Omega$$
$$= \left\langle l, m | L, M, l', m' \right\rangle \left[\frac{(2l'+1)(2L+1)}{4\pi(2l+1)} \right]^{1/2} \left\langle l, 0 | L, 0, l', 0 \right\rangle . \tag{A.11}$$

The special Clebsch-Gordan coefficient $\langle l, 0|L, 0, l', 0\rangle$ is given by [Edm60]

$$\left\langle l, 0 | L, 0, l', 0 \right\rangle = \sqrt{2l+1}(-1)^{(l-L-l')/2}$$
$$\times \left[\frac{(J-2l)!(J-2L)!(J-2l')!}{(J+1)!} \right]^{1/2}$$
$$\times \frac{(J/2)!}{(J/2-l)!(J/2-L)!(J/2-l')!} . \tag{A.12}$$

The sum $J = l + L + l'$ of the three angular momentum quantum numbers must be even. The Clebsch-Gordan coefficient (A.12) vanishes for odd J.

Explicit expressions for the spherical harmonics up to $l = 3$ are given in Sect. 1.2.1 in Table 1.1. For further details see books on angular momentum in quantum mechanics, e.g. [Edm60, Lin84].

A.2 Laguerre Polynomials

The generalized Laguerre polynomials $L_\nu^\alpha(x)$, $\nu = 0, 1, \ldots$ are polynomials of degree ν in x. They are given by

$$L_\nu^\alpha(x) = \frac{\mathrm{e}^x}{\nu! x^\alpha} \frac{\mathrm{d}^\nu}{\mathrm{d}x^\nu} (\mathrm{e}^{-x} x^{\nu+\alpha}) = \sum_{\mu=0}^{\nu} (-1)^\mu \binom{\nu+\alpha}{\nu-\mu} \frac{x^\mu}{\mu!} \tag{A.13}$$

and have ν zeros in the range $0 < x < \infty$. The ordinary Laguerre polynomials $L_\nu(x)$ correspond to the special case $\alpha = 0$. In general α is an arbitrary real number greater than -1. The binomial coefficient in (A.13) is defined as follows for non-integral arguments:

$$\binom{z}{y} = \frac{\Gamma(z+1)}{\Gamma(y+1)\Gamma(z-y+1)} . \tag{A.14}$$

Here Γ is the gamma function, see Sect. A.3.

The orthogonality relation for the generalized Laguerre polynomials reads

$$\int_0^\infty e^{-x} x^\alpha L_\mu^\alpha(x) L_\nu^\alpha(x)\, dx = \frac{\Gamma(\nu+\alpha+1)}{\nu!}\,\delta_{\mu,\nu} . \tag{A.15}$$

The following recursion relation is very useful, because it enables the numerically efficient evaluation of the Laguerre polynomials for a given index α:

$$(\nu+1)L_{\nu+1}^\alpha(x) - (2\nu+\alpha+1-x)L_\nu^\alpha(x) + (\nu+\alpha)L_{\nu-1}^\alpha(x) = 0 ,$$
$$\nu = 1, 2, \ldots \tag{A.16}$$

Note: The Laguerre polynomials defined by (A.13) correspond to the definitions in [AS70, GR80, MO66]. The Laguerre polynomials in [Mes70] contain an additional factor $\Gamma(\nu + \alpha + 1)$.

A.3 Gamma Function

The gamma function $\Gamma(z)$ is defined by

$$\Gamma(z+1) = \int_0^\infty t^z e^{-t}\, dt \tag{A.17}$$

and has the property

$$\Gamma(z+1) = z\Gamma(z) . \tag{A.18}$$

For positive integers $z = n$ we have $\Gamma(n+1) = n!$. For half-integral z we can derive $\Gamma(z)$ recursively from the value $\Gamma(1/2) = \sqrt{\pi}$ via (A.18).

For small z we have

$$\frac{1}{\Gamma(z)} = \frac{z}{\Gamma(z+1)} = z + \gamma_E z^2 + O(z^3) , \tag{A.19}$$

where $\gamma_E = 0.5772156649\ldots$ is Euler's constant,

$$\gamma_E = -\left.\frac{d\Gamma}{dz}\right|_{z=1} \approx 0.577256649\ldots \tag{A.20}$$

The argument z may be complex, and

$$\Gamma(z^*) = [\Gamma(z)]^* . \tag{A.21}$$

Useful product formulae are,

$$\Gamma(iy)\Gamma(-iy) = |\Gamma(iy)|^2 = \frac{\pi}{y\sinh(\pi y)} , \tag{A.22}$$

$$\Gamma(1+iy)\Gamma(1-iy) = |\Gamma(1+iy)|^2 = \frac{\pi y}{\sinh(\pi y)} , \tag{A.23}$$

$$\Gamma\left(\frac{1}{2}+iy\right)\Gamma\left(\frac{1}{2}-iy\right) = \left|\Gamma\left(\frac{1}{2}+iy\right)\right|^2 = \frac{\pi}{\cosh(\pi y)} , \tag{A.24}$$

$$\Gamma\left(\frac{1}{4}+iy\right)\Gamma\left(\frac{3}{4}-iy\right) = \frac{\pi\sqrt{2}}{\cosh(\pi y)+i\sinh(\pi y)} . \tag{A.25}$$

From (A.23) it follows that $|\Gamma(1+iy| = \sqrt{\pi y/\sinh(\pi y)}$. By induction we can conclude,

$$|\Gamma(1+l+iy)| = \left(\prod_{n=1}^{l} |n+iy|\right)\sqrt{\frac{\pi y}{\sinh(\pi y)}} \overset{|y|\to\infty}{\sim} \sqrt{2\pi}\, e^{-\frac{\pi}{2}|y|}|y|^{l+1/2} . \tag{A.26}$$

The right-hand sides of the formulae (A.22)–(A.25) also apply if y is not real, e.g. for $y = ix$,

$$\Gamma(x)\Gamma(-x) = -\frac{\pi}{x\sin(\pi x)} , \tag{A.27}$$

$$\Gamma(1+x)\Gamma(1-x) = \frac{\pi x}{\sin(\pi x)} , \tag{A.28}$$

$$\Gamma\left(\frac{1}{2}+x\right)\Gamma\left(\frac{1}{2}-x\right) = \frac{\pi}{\cos(\pi x)} , \tag{A.29}$$

$$\Gamma\left(\frac{1}{4}-x\right)\Gamma\left(\frac{3}{4}+x\right) = \frac{\pi\sqrt{2}}{\cos(\pi x)-\sin(\pi x)} . \tag{A.30}$$

For large arguments we have *Stirling's formula*,

$$\Gamma(z) \overset{z\to\infty}{\sim} \mathrm{e}^{-z} z^{z-1/2} \sqrt{2\pi} \left[1 + \frac{1}{12z} + \frac{1}{288z^2} + O\left(\frac{1}{z^3}\right) \right] . \tag{A.31}$$

A.4 Bessel Functions

In many special cases describing realistic situations, the radial Schrödinger equation has analytical solutions in the form of Bessel functions, which makes these special functions particularly important. An excellent review of the definitions and properties of Bessel functions is contained in Olver's chapter [Olv70] in the "Handbook of Mathemetical Functions". Although the title of that chapter is "Bessel Functions of Integer Order", most results apply also for noninteger and even for complex orders.

The defining differential equation for (ordinary) Bessel functions of order ν is:

$$z^2 \frac{\mathrm{d}^2 \mathcal{C}_\nu}{\mathrm{d}z^2} + z \frac{\mathrm{d}\mathcal{C}_\nu}{\mathrm{d}z} - (\nu^2 - z^2)\mathcal{C}_\nu = 0 . \tag{A.32}$$

The connection to the radial Schrödinger equation is achieved via the transformation $u(z) = \sqrt{z}\,\mathcal{C}_\nu(z)$, which leads to the following differential equation for $u(z)$,

$$-\frac{\mathrm{d}^2 u}{\mathrm{d}z^2} + \frac{\nu^2 - \frac{1}{4}}{z^2} u = u . \tag{A.33}$$

Multiplying (A.33) by $\hbar^2/2\mu$ and writing kr for z yields the radial Schrödinger equations (4.22) and (4.192) in the free-particle case $V(r) \equiv 0$, with

$$\nu^2 - \frac{1}{4} = l(l+1) \Rightarrow \nu^2 = \left(l + \frac{1}{2}\right)^2 \text{ in 3D (4.22)} , \quad \nu^2 = m^2 \text{ in 2D (4.192)} . \tag{A.34}$$

Equation (A.33) has two linearly independent solutions, which can be defined by their boundary conditions for $z \to 0$ or for $z \to \infty$. The (ordinary) "Bessel function of the first kind" $J_\nu(z)$ has a series expansion

$$J_\nu(z) = \left(\frac{z}{2}\right)^\nu \sum_{k=0}^{\infty} \frac{\left(-\frac{1}{4}z^2\right)^k}{k!\,\Gamma(\nu + k + 1)} \tag{A.35}$$

and obeys the following boundary conditions,

$$J_\nu(z) \stackrel{z\to 0}{\sim} \frac{(z/2)^\nu}{\Gamma(\nu+1)}\left[1-\frac{(z/2)^2}{\nu+1}+O\left(\left(\frac{z}{2}\right)^4\right)\right], \tag{A.36}$$

$$\sqrt{\frac{\pi}{2}}\, z J_\nu(z) \stackrel{|z|\to\infty}{\sim} \sin\left(z-\frac{\left(\nu-\frac{1}{2}\right)\pi}{2}\right)+O\left(\frac{1}{|z|}\right). \tag{A.37}$$

When the order ν is an integer, $\nu = n$,

$$J_{-n}(z) = (-1)^n J_n(z)\,. \tag{A.38}$$

When ν is not an integer, $J_\nu(z)$ and $J_{-\nu}(z)$ are linearly independent.

The ordinary Bessel function with maximal phase difference to $J_\nu(z)$ for large z is the "Bessel function of the second kind" $Y_\nu(z)$, which is defined for noninteger order ν by

$$Y_\nu(z) = \frac{J_\nu(z)\cos(\nu\pi) - J_{-\nu}(z)}{\sin(\nu\pi)}, \tag{A.39}$$

and for integer order n by $Y_n(z) \stackrel{\text{def}}{=} \lim_{\nu\to n} Y_\nu(z)$. The large-$z$ behaviour of $Y_\nu(z)$ is

$$\sqrt{\frac{\pi}{2}}\, z Y_\nu(z) \stackrel{|z|\to\infty}{\sim} -\cos\left(z-\frac{\left(\nu-\frac{1}{2}\right)\pi}{2}\right)+O\left(\frac{1}{|z|}\right). \tag{A.40}$$

The low-argument behaviour of $Y_\nu(z)$ can be derived for noninteger order ν from (A.36) and (A.39):

$$\begin{aligned} Y_\nu(z) \stackrel{z\to 0}{\sim} & -\left(\frac{z}{2}\right)^{-\nu}\frac{\Gamma(1+\nu)}{\nu\pi}\left[1-\frac{(z/2)^2}{1-\nu}+O\left(\left(\frac{z}{2}\right)^4\right)\right] \\ & + \cot(\nu\pi)\frac{(z/2)^\nu}{\Gamma(1+\nu)}\left[1-\frac{(z/2)^2}{1+\nu}+O\left(\left(\frac{z}{2}\right)^4\right)\right]. \end{aligned} \tag{A.41}$$

For integer order, $\nu = n$, the expansion of $Y_n(z)$ in z involves logarithmic terms. For $\nu = 0$ we have

$$Y_0(z) \stackrel{z\to 0}{\sim} \frac{2}{\pi}\left[\ln\left(\frac{z}{2}\right)+\gamma_{\rm E}\right]J_0(z)+\frac{2}{\pi}\left(\frac{z}{2}\right)^2+O\left(\left(\frac{z}{2}\right)^4\right). \tag{A.42}$$

The square bracket in (A.42) contains Euler's constant $\gamma_{\rm E}$ as defined in (A.20) above. For $\nu = n \geq 1$, the leading term in the expansion of $Y_n(z)$ is $-\frac{1}{\pi}(n-1)!(z/2)^{-n}$, in agreement with the leading term in (A.41); each further term contains an additional factor $(z/2)^2$, as long as the combined exponent of $z/2$ remains smaller than n. At order $(z/2)^n$ the expansion contains a logarithmic contribution $\frac{2}{\pi}\ln\left(\frac{1}{2}z\right) J_n(z)$ (see (9.1.11) in [Olv70]). Similar to (A.38) for the J_n, we have

$$Y_{-n}(z) = (-1)^n\, Y_n(z) \tag{A.43}$$

for integer order n.

The Bessel functions of the first and second kind, which are real-valued for real argument z, can be combined with complex coefficients to define the Bessel functions of the third kind or "Hankel functions":

$$H_\nu^{(1)}(z) = J_\nu(z) + \mathrm{i}\, Y_\nu(z)\,, \quad H_\nu^{(2)}(z) = J_\nu(z) - \mathrm{i}\, Y_\nu(z)\,. \tag{A.44}$$

Their large-z behaviour follows from (A.37), (A.40),

$$\sqrt{\frac{\pi}{2}}\, z\, H_\nu^{(1)}(z) \overset{|z|\to\infty}{\sim} \mathrm{e}^{\mathrm{i}\left(z-\frac{1}{2}\nu-\frac{\pi}{4}\right)}\,, \quad \sqrt{\frac{\pi}{2}}\, z\, H_\nu^{(2)}(z) \overset{|z|\to\infty}{\sim} \mathrm{e}^{-\mathrm{i}\left(z-\frac{1}{2}\nu-\frac{\pi}{4}\right)}\,. \tag{A.45}$$

Their small-z behaviour follows, for noninteger ν from (A.36) and (A.41),

$$H_\nu^{(1)}(z) = -H_\nu^{(2)}(z) = -\frac{\mathrm{i}}{\pi}\frac{\Gamma(\nu)}{\left(\frac{1}{2}z\right)^\nu}\,, \quad z \to 0\,, \quad \Re\nu > 0\,, \tag{A.46}$$

and this leading term is also valid for integer $\nu > 0$.

For free-particle motion in 3D, the order of the Bessel functions solving the radial Schrödinger equation is half integer, $\nu = l + \frac{1}{2}$. The corresponding *spherical Bessel functions* are denoted by lower-case letters and are defined as,

$$j_l(z) = \sqrt{\frac{\pi}{2z}}\, J_{l+\frac{1}{2}}(z)\,, \quad y_l(z) = \sqrt{\frac{\pi}{2z}}\, Y_{l+\frac{1}{2}}(z)\,, \tag{A.47}$$

$$h_l^{(1)}(z) = \sqrt{\frac{\pi}{2z}}\, H_{l+\frac{1}{2}}^{(1)}(z)\,, \quad h_l^{(2)}(z) = \sqrt{\frac{\pi}{2z}}\, H_{l+\frac{1}{2}}^{(2)}(z)\,. \tag{A.48}$$

For small z we have, according to (A.36),

$$j_l(z) \overset{z\to 0}{=} \frac{z^l}{(2l+1)!!}, \tag{A.49}$$

and asymptotically according to (A.37)

$$z j_l(z) \overset{|z|\to\infty}{=} \sin\left(z - l\frac{\pi}{2}\right). \tag{A.50}$$

From (A.40) we have,

$$z y_l(z) \overset{|z|\to\infty}{=} -\cos\left(z - l\frac{\pi}{2}\right). \tag{A.51}$$

In place of the functions $y_l(z)$, some authors (e.g. [Mes70]) work with the *spherical Neumann functions*, $n_l(z) \overset{\text{def}}{=} -y_l(z)$, which tend to $+\cos(z - l\frac{\pi}{2})$ for large z.

For the derivatives of the spherical Bessel functions we have the simple formula

$$\frac{\mathrm{d}}{\mathrm{d}z} j_l(z) = j_{l-1}(z) - \frac{l+1}{z} j_l(z), \quad l \geq 1. \tag{A.52}$$

The real regular and irregular radial free-particle wave functions are the solutions of (A.33) with $z = kr$, i.e. $kr\, j_l(kr)$ and $-kr\, y_l(kr)$, as given in (4.25) in Sec. 4.1.3. The corresponding "spherical Hankel functions" give the linear combinations corresponding to incoming or outgoing spherical waves,

$$kr\, h_l^{(1)}(kr) \overset{kr\to\infty}{\sim} -\mathrm{i}\, \mathrm{e}^{\mathrm{i}\left(kr - l\frac{\pi}{2}\right)}, \quad kr\, h_l^{(2)}(kr) \overset{kr\to\infty}{\sim} \mathrm{i}\, \mathrm{e}^{-\mathrm{i}\left(kr - l\frac{\pi}{2}\right)}. \tag{A.53}$$

According to (9.1.53) in [Olv70], the differential equation

$$-\frac{\mathrm{d}^2 u}{\mathrm{d}z^2} + \left(\frac{l(l+1)}{z^2} - \frac{1}{z^\alpha}\right) u(z) = 0 \tag{A.54}$$

is solved by functions of the form

$$u(z) = \sqrt{z}\, \mathcal{C}_{\nu(l)}\left(\frac{2}{\alpha - 2} z^{1-\alpha/2}\right), \quad \text{with} \quad \nu(l) = \frac{2l+1}{\alpha - 2}. \tag{A.55}$$

If we interpret the dimensionless argument z as r/β, (A.54) is just the radial wave equation (4.65) at threshold in the partial wave l for the single-power potential $V_\alpha(r)$, as defined in (4.63), with $C_\alpha < 0$. The solutions (A.55) are of the form given in (4.66), (4.67) in Sec. 4.1.7.

The *modified Bessel functions* are solutions of the differential equation

$$z^2 \frac{\mathrm{d}^2 \mathcal{Z}_\nu}{\mathrm{d}z^2} + z \frac{\mathrm{d}\mathcal{Z}_\nu}{\mathrm{d}z} - (\nu^2 + z^2)\mathcal{Z}_\nu = 0. \tag{A.56}$$

As for the ordinary Bessel functions, the connection to the radial Schrödinger equation is achieved via the transformation $u(z) = \sqrt{z}\, \mathcal{Z}_\nu(z)$, which leads to the following differential equation for $u(z)$,

$$-\frac{\mathrm{d}^2 u}{\mathrm{d}z^2} + \frac{\nu^2 - \frac{1}{4}}{z^2} u = -u. \tag{A.57}$$

Multiplying (A.57) by $\hbar^2/(2\mu)$ and writing κr for z again yields the radial Schrödinger equations (4.22) and (4.192), in the free-particle case $V(r) \equiv 0$, but now for negative energy $E = -\hbar^2\kappa^2/(2\mu) < 0$. The order parameter ν is again related the angular momentum quantum numbers in 3D and in 2D by (A.34).

The modified Bessel function $I_\nu(z)$ of order ν solves (A.56) and is related to the ordinary Bessel function of the first kind by,

$$\mathrm{i}^\nu I_\nu(z) = J_\nu(\mathrm{i}z) \,, \quad (-\pi < \arg z \le \pi/2) \,. \tag{A.58}$$

Its behaviour for small $|z|$ is, as for J_ν,

$$I_\nu(z) \stackrel{z\to 0}{=} \frac{(\frac{1}{2}z)^\nu}{\Gamma(\nu+1)} \,, \quad (\nu \neq -1, -2, -3, \ldots) \,. \tag{A.59}$$

For $|z| \to \infty$ the asymptotic form of I_ν is

$$I_\nu(z) \stackrel{|z|\to\infty}{=} \frac{\mathrm{e}^z}{\sqrt{2\pi z}} \,, \quad (|\arg(z)| < \pi/2) \,. \tag{A.60}$$

For non-integral values of ν the modified Bessel functions $I_\nu(z)$ and $I_{-\nu}(z)$ defined by (A.58), (A.35) are linearly independent, and there is a linear combination

$$K_\nu(z) = \frac{\pi}{2} \frac{I_{-\nu}(z) - I_\nu(z)}{\sin(\nu\pi)} \,, \tag{A.61}$$

which vanishes asymptotically,

$$K_\nu(z) \stackrel{|z|\to\infty}{=} \sqrt{\frac{\pi}{2z}}\, \mathrm{e}^{-z} \,, \quad (|\arg z| < 3\pi/2) \,. \tag{A.62}$$

For integer order n, $K_n(z) \stackrel{\text{def}}{=} \lim_{\nu\to n} K_\nu(z)$.

For the modified Bessel function $K_{l+1/2}$ of half-integral order $l + 1/2$ there is a series expansion

$$K_{l+1/2}(z) = \sqrt{\frac{\pi}{2z}}\, \mathrm{e}^{-z} \sum_{k=0}^{l} \frac{(l+k)!}{k!(l-k)!} (2z)^{-k} \,. \tag{A.63}$$

The derivative of $K_{l+1/2}$ can be expressed in terms of $K_{l+1/2}$ and $K_{l-1/2}$,

$$\frac{\mathrm{d}}{\mathrm{d}z} K_{l+1/2}(z) = -\frac{l+\frac{1}{2}}{z} K_{l+1/2}(z) - K_{l-1/2}(z) \,. \tag{A.64}$$

The *Airy functions* are essentially Bessel functions of order 1/3,

$$\mathrm{Ai}(z) = \frac{1}{3}\sqrt{z}\left[I_{-1/3}(\zeta) - I_{1/3}(\zeta)\right] = \frac{1}{\pi}\sqrt{\frac{z}{3}}\,K_{1/3}(\zeta)\,,$$

$$\mathrm{Bi}(z) = \sqrt{\frac{z}{3}}\left[I_{-1/3}(\zeta) + I_{1/3}(\zeta)\right]\,, \quad \text{where} \quad \zeta = \frac{2}{3}z^{3/2}\,. \tag{A.65}$$

For large $|z|$,

$$2\sqrt{\pi}\,\mathrm{Ai}(z) \overset{|z|\to\infty}{\sim} z^{-1/4}\,\mathrm{e}^{-\zeta}\,, \quad (|\arg(z)| < \pi)\,,$$

$$\sqrt{\pi}\,\mathrm{Ai}(-z) \overset{|z|\to\infty}{\sim} z^{-1/4}\cos\left(\zeta - \frac{\pi}{4}\right)\,, \quad \left(|\arg(z)| < \frac{2\pi}{3}\right)\,. \tag{A.66}$$

The Airy functions are solutions of the differential equation

$$\frac{\mathrm{d}^2 w}{\mathrm{d}z^2} - z\,w(z) = 0\,. \tag{A.67}$$

For a linear potential $V(x)$ with a negative gradient,

$$V(x) = (x - x_{\mathrm{ctp}})\,V'\,, \quad -V' = \frac{\hbar^2}{2\mu}\,\xi > 0\,, \tag{A.68}$$

the wave function $\psi(x) = \mathrm{Ai}\left(\xi^{1/3}(x_{\mathrm{ctp}} - x)\right)$ is a solution of the Schrödinger equation (1.284).

A.5 Confluent Hypergeometric Functions, Coulomb Functions, Whittaker's Function

The confluent hypergeometric function, also called "degenerate hypergeometric function", is defined according to Chapter 13 in [AS70] and Section 9.2 in [GR80] as

$$F(a,b;z) = \sum_{n=0}^{\infty} \frac{\Gamma(a+n)}{\Gamma(a)}\,\frac{\Gamma(b)}{\Gamma(b+n)}\,\frac{z^n}{n!}\,. \tag{A.69}$$

It is a solution of the equation

$$z\frac{\mathrm{d}^2 C}{\mathrm{d}z^2} + (b - z)\frac{\mathrm{d}C}{\mathrm{d}z} = aC(z)\,. \tag{A.70}$$

Alternative notations for $F(a,b;z)$ are: ${}_1F_1(a,b;z)$, $M(a,b;z)$, $\Phi(a,b;z)$.

A linearly independent solution of (A.70), sometimes also called confluent hypergeometric function, is

$$U(a,b;z) = \frac{\Gamma(1-b)}{\Gamma(a-b+1)}\,F(a,b;z) + \frac{\Gamma(b-1)}{\Gamma(a)}\,z^{1-b}\,F(a-b+1,2-b;z)\;. \tag{A.71}$$

An alternative notation for $U(a,b;z)$ is $\Psi(a,b;z)$.

The Gaussian hypergeometric series, also called the hypergeometric function, is defined by

$$_2F_1(a,b;c,z) = \sum_{n=0}^{\infty} \frac{\Gamma(a+n)}{\Gamma(a)}\,\frac{\Gamma(b+n)}{\Gamma(b)}\,\frac{\Gamma(c)}{\Gamma(c+n)}\,\frac{z^n}{n!}\;. \tag{A.72}$$

The confluent hypergeometric functions (A.69), (A.71) are important in the context of Coulomb potentials, because they occur as components in solutions of relevant Schrödinger equations, see, e.g. (4.141) in Sect. 4.1.12 and (4.227) in Sec. 4.2.6. An important special case is the radial Schrödinger equation for motion in a pure Coulomb potential at energy $E = \hbar^2k^2/(2\mu)$, characterized by the Sommerfeld parameter η [(1.119) in Sect. 1.3.2],

$$\left[-\frac{\mathrm{d}^2}{\mathrm{d}\rho^2} + \frac{l(l+1)}{\rho^2} + \frac{2\,\eta}{\rho}\right]u_l(\rho) = u_l(\rho)\;. \tag{A.73}$$

Two linearly independent solutions are the regular Coulomb function $F_l(\rho,\eta)$,

$$F_l(\eta,\rho) = 2^l\,\mathrm{e}^{-\frac{\pi}{2}\eta}\,\frac{|\Gamma(l+1+\mathrm{i}\eta)|}{(2l+1)!}\,\mathrm{e}^{-\mathrm{i}\rho}\rho^{l+1}\,F(l+1-\mathrm{i}\eta,2l+2;2\mathrm{i}\rho)\;, \tag{A.74}$$

and the irregular Coulomb function $G_l(\rho,\eta)$,

$$G_l(\eta,\rho) = \mathrm{i}F_l(\eta,\rho) + \mathrm{e}^{\frac{\pi}{2}\eta}\frac{|\Gamma(l+1+\mathrm{i}\eta)|}{\Gamma(l+1+\mathrm{i}\eta)}\mathrm{e}^{-\mathrm{i}(\rho-l\frac{\pi}{2})}(2\mathrm{i}\rho)^{l+1}U(l+1-\mathrm{i}\eta,2l+2;2\mathrm{i}\rho)\;. \tag{A.75}$$

The small-ρ behaviour of the regular Coulomb function is, for fixed η [AS70],

$$F_l(\eta,\rho) \overset{\rho\to 0}{\sim} 2^l\,\mathrm{e}^{-\frac{\pi}{2}\eta}\,\frac{|\Gamma(l+1+\mathrm{i}\eta)|}{(2l+1)!}\,\rho^{l+1}\;. \tag{A.76}$$

For $|\eta| \to \infty$, which corresponds to approaching the threshold according to (1.119), we have via (A.25)

$$|\Gamma(l+1+\mathrm{i}\eta)| \overset{|\eta|\to\infty}{=} \sqrt{2\pi}\mathrm{e}^{-\frac{1}{2}\pi|\eta|}|\eta|^{l+1/2}. \tag{A.77}$$

In order to obtain a formula for the regular Coulomb function of small argument $\rho = kr$ close to threshold we combine (A.76) and (A.77) to

$$F_l(\eta, kr) \overset{k\to 0,\ r\to 0}{=} \sqrt{\frac{\pi}{2|\eta|}} \frac{(2kr|\eta|)^{l+1}}{(2l+1)!} \mathrm{e}^{-\frac{1}{2}\pi(\eta+|\eta|)} . \tag{A.78}$$

Expoiting the ρ-independence of the Wronskian [AS70],

$$\frac{\partial F_l}{\partial \rho} G_l(\eta, \rho) - \frac{\partial G_l}{\partial \rho} F_l(\eta, \rho) = 1 , \tag{A.79}$$

we can derive the small-ρ behaviour of the irregular Coulomb function for fixed η,

$$G_l(\eta, \rho) \overset{\rho\to 0}{\sim} \frac{\mathrm{e}^{\frac{\pi}{2}\eta}\,(2l)!}{2^l\,|\Gamma(l+1+\mathrm{i}\eta)|}\,\rho^{-l} . \tag{A.80}$$

The large-ρ behaviour of the Coulomb functions is,

$$\begin{aligned} F_l(\eta, z) &\overset{z\to\infty}{=} \sin\left(z - \eta \ln 2z - l\frac{\pi}{2} + \sigma_l\right) , \\ G_l(\eta, z) &\overset{z\to\infty}{=} \cos\left(z - \eta \ln 2z - l\frac{\pi}{2} + \sigma_l\right) . \end{aligned} \tag{A.81}$$

The constants σ_l are the Coulomb phases,

$$\sigma_l = \arg \Gamma(l + 1 + \mathrm{i}\eta). \tag{A.82}$$

Whittaker's equation,

$$\frac{\mathrm{d}^2\mathcal{C}}{\mathrm{d}z^2} - \left[\frac{m^2 - \frac{1}{4}}{z^2} - \frac{\lambda}{z}\right]\mathcal{C}(z) = \frac{1}{4}\mathcal{C}(z) , \tag{A.83}$$

acquires the form of the radial Schrödinger equation for an attractive pure Coulomb potential at negative energy $E = -\hbar^2\kappa^2/(2\mu)$ if we write $\left(l + \frac{1}{2}\right)^2$ for m^2 and replace z by $2\rho \equiv 2\kappa r$ and λ by $-\eta = |\eta| \equiv 1/(a_C\kappa)$, a_C being the Bohr radius:

$$\left[-\frac{\mathrm{d}^2}{\mathrm{d}\rho^2} + \frac{l(l+1)}{\rho^2} - \frac{2\,|\eta|}{\rho}\right] u_l(\rho) = -u_l(\rho) . \tag{A.84}$$

Two solutions of (A.84) are

$$M_{|\eta|,\,l+\frac{1}{2}}(2\rho) = (2\rho)^{l+1}\,\mathrm{e}^{-\rho}F\,(l + 1 - |\eta|, 2l + 2; 2\rho) , \tag{A.85}$$

$$M_{|\eta|,-l-\frac{1}{2}}(2\rho) = (2\rho)^{-l}\,\mathrm{e}^{-\rho}F\,(-l - |\eta|, -2l; 2\rho) . \tag{A.86}$$

The linear combination of (A.85) and (A.86) which vanishes for large ρ is Whittaker's function,

$$W_{|\eta|,\,l+\frac{1}{2}}(2\rho) = \frac{\Gamma(-2l-1)}{\Gamma(-l-|\eta|)} M_{|\eta|,\,l+\frac{1}{2}}(2\rho) + \frac{\Gamma(2l+1)}{\Gamma(l+1-|\eta|)} M_{|\eta|,-l-\frac{1}{2}}(2\rho)\,, \tag{A.87}$$

$$W_{|\eta|,\,l+\frac{1}{2}}(2\rho) \overset{\rho\to\infty}{\sim} \mathrm{e}^{-\rho}(2\rho)^{|\eta|}\left[1 + O\left(\frac{1}{\rho}\right)\right]. \tag{A.88}$$

At least one gamma function in (A.87) is ill-defined for integer l, but the expression for $W_{|\eta|,\,l+\frac{1}{2}}(2\rho)$ is well defined when taking the limit as l approaches its integer value. For integer nonnegative l, $W_{|\eta|,\,l+\frac{1}{2}}(2\rho)$ vanishes as $\rho \to 0$ when $|\eta|$ is an integer larger than l; in this case $W_{|\eta|,\,l+\frac{1}{2}}(2\rho)$ is a regular normalizable solution of (A.84).

References

[AS70] M. Abramowitz, I.A. Stegun (eds.), *Handbook of Mathematical Functions* (Dover Publications, New York, 1970)

[Edm60] A.R. Edmonds, *Angular Momentum in Quantum Mechanics* (Princeton University Press, Princeton, 1960)

[Lin84] A. Lindner, *Drehimpulse in der Quantenmechanik* (Teubner, Stuttgart, 1984)

[GR80] I.S. Gradshteyn, I.M. Ryzhik, *Tables of Integrals, Series and Products* (Academic Press, New York, 1980)

[Mes70] A. Messiah, *Quantum Mechanics*, vol. 1 (North Holland, Amsterdam, 1970)

[MO66] W. Magnus, F. Oberhettinger, R.P. Soni, *Formulas and Theorems for Special Functions of Mathematical Physics* (Springer, Berlin, Heidelberg, 1966)

[Olv70] F.W.J. Olver, Chap. 9 in [AS70]

Index

H. Friedrich, *Theoretical Atomic Physics*, Graduate Texts in Physics,
DOI 10.1007/978-3-319-47769-5

D

Q

R

T

U

Printed in the United States
By Bookmasters